Springer Proceedings in Physics 13

Managing Editor: H. K. V. Lotsch

[illegible]

Springer Proceedings in Physics

Managing Editor: H. K. V. Lotsch

Springer Proceedings in Physics is a new series dedicated to the publication of conference proceedings. Each volume is produced on the basis of camera-ready manuscripts prepared by conference contributors. In this way, publication can be achieved very soon after the conference and costs are kept low; the quality of visual presentation is, nevertheless, very high. We believe that such a series is preferable to the method of publishing conference proceedings in journals, where the typesetting requires time and considerable expense, and results in a longer publication period. Springer Proceedings in Physics can be considered as a journal in every other way: it should be cited in publications of research papers as *Springer Proc. Phys.*, followed by the respective volume number, page number and year.

The Physics and Fabrication of Microstructures and Microdevices

Proceedings of the Winter School
Les Houches, France, March 25–April 5, 1986

Editors: M. J. Kelly and C. Weisbuch

With 328 Figures

Springer-Verlag Berlin Heidelberg New York
London Paris Tokyo

Dr. Michael J. Kelly
GEC Research Ltd., East Lane, Wembley, Middlesex HA9 7PP, England

Dr. Claude Weisbuch
Laboratoire Central de Recherches, Thomson-CSF,
F-91401 Orsay, France

We acknowledge generous support from:
The European Community Through ESPRIT;
GEC; THOMSON;
Ministere de la Recherche et de la Technologie; Ministere de l'Education Nationale; Centre National de la Recherche Scientifique; Direction des Recherches, Etudes et Techniques; Direction des Industries Electroniques et de l'Informatique; US Army, US Army Research, Development and Standardisation Group (UK); US Navy, Office of Naval Research; US Air Force, European Office of Aerospace Research and Development; BULL; MATRA

ISBN-13: 978-3-642-71448-1 e-ISBN-13: 978-3-642-71446-7
DOI: 10.1007/978-3-642-71446-7

Library of Congress Cataloging-in-Publication Data. The Physics and fabrication of microstructures and microdevices. (Springer proceedings in physics; v. 13). Proceedings of the Winter School on "the Physics and Fabrication of Microstructures." Includes index. 1. Semiconductors – Congresses. 2. Integrated circuits – Very large scale integration – Design and construction – Congresses. 3. Semiconductors – Design and construction – Congresses. I. Weisbuch, C. (Claude), 1945–. II. Kelly, M. J. (Michael J.), 1949–. III. Winter School on "the Physics and Fabrication of Microstructures" (1986: Les Houches, Haute-Savoie, France) IV. Series. TK7871.85.P47 1986 621.3815'2 86-17731

Preface

Les Houches

This Winter School on "The Physics and Fabrication of Microstructures" originated with a European industrial decision to investigate in some detail the potential of custom-designed microstructures for new devices. Beginning in 1985, GEC and THOMSON started a collaboration on these subjects, supported by an ESPRIT grant from the Commission of the European Community.

To the outside observer of the whole field, it appears clear that the world effort is very largely based in the United States and Japan. It also appears that cooperation and dissemination of results are very well organised outside Europe and act as a major influence on the development of new concepts and devices. In Japan, a main research programme of the Research and Development for Basic Technology for Future Industries is focused on "Future Electron Devices". In Japan and in the United States, many workshops are organised annually in order to bring together the major specialists in industry and academia, allowing fast dissemination of advances and contacts for setting up cooperative efforts.

From the beginning GEC and THOMSON have included in the ESPRIT programme an annual event which should help to disseminate results and bring together specialists and students in order to broaden the European skills and to develop cross-collaborations. This Winter School is the first of such annual events. Next year's School should be on "Physics, Fabrication and Applications of Low-Dimensional Structures", with the emphasis on growth and applications of multilayered semiconductor and metal structures. The 1988 School is forecast to be focused more on new devices.

While new device concepts have been appearing regularly at a very fast pace, solid-state physics has known new and fascinating developments, exemplified by the discovery of the quantum Hall effect, a major breakthrough in low-dimensional physics. The enormous range of phenomena which can now be reached through the marriage of well-controlled growth techniques such as molecular beam epitaxy and fine line fabrication techniques is shown in Figure 1 (courtesy of M. Pepper, University of Cambridge and GEC).

We are therefore witnessing a rare event in which technologists, physicists and device engineers are creating their own microscopic objects of study with

Physics from the Marriage of MBE and Fine Line Lithography

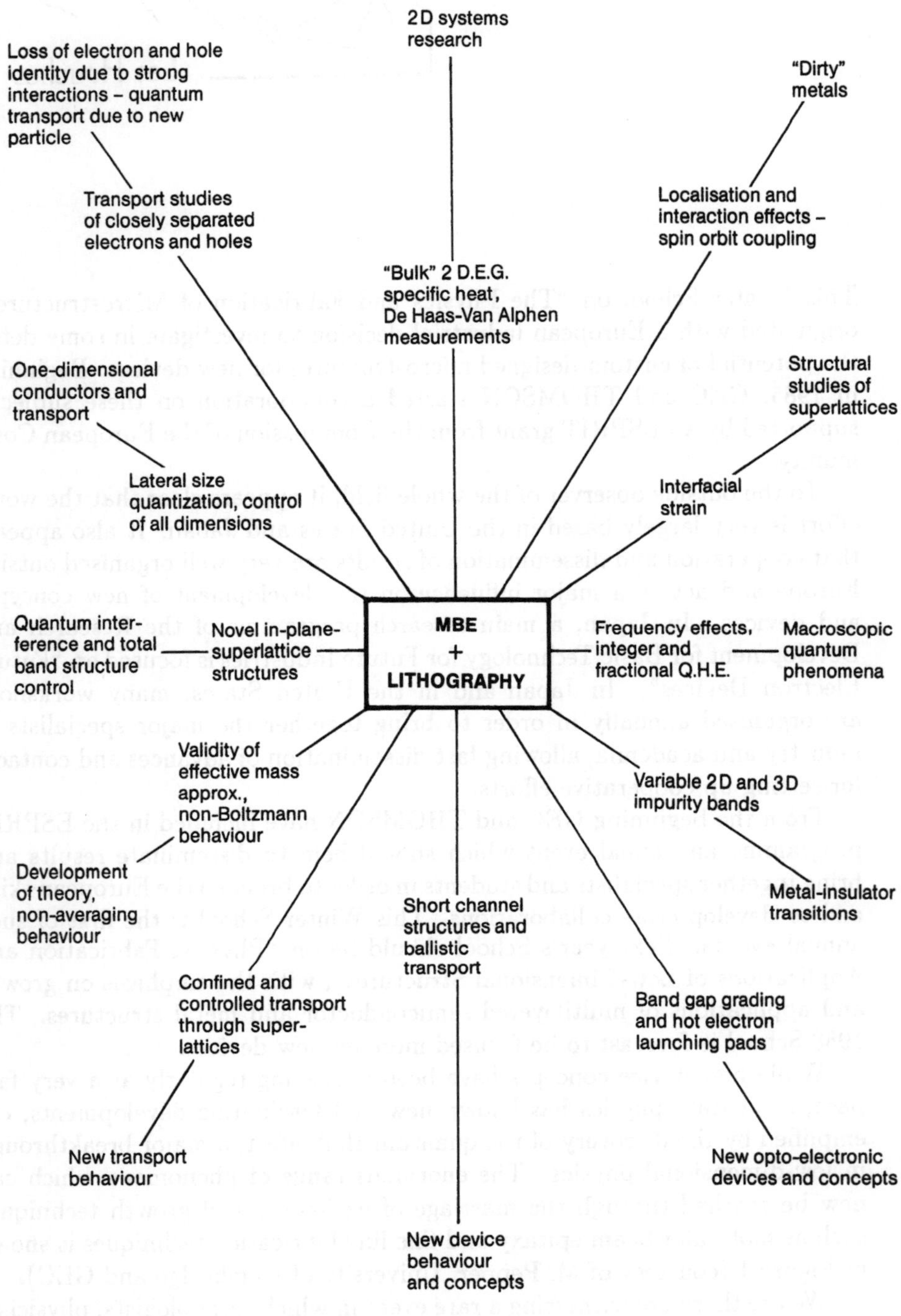

Figure 1

unforeseen (and often unpredicted) properties (for the physicist) or figures of merit (for the engineers). The aim of this School was to focus on various aspects of these fields, with the hope of getting some cross-fertilisation.

The organisation of this book reflects the structure of the School with major sections on (i) fabrication technology, (ii) physics of small structures and (iii) perspectives for applications. We hope that the book will convey some of the excitement felt by those taking part, i.e. lecturers and students who came from many countries and continents.

The editors wish to thank Mrs C. Pendl of Spinger-Verlag for her assistance with the preparation of the manuscripts.

Les Houches, April 1986 *M.J. Kelly C. Weisbuch*

... has been adopted [illegible] played [illegible] role [illegible] of [illegible] the [illegible]. The aim of this [illegible] Scott on various aspects of these fields with [illegible] radiation.

[illegible] this book [illegible] the School with [illegible] of shock structures and [illegible] for appropriate [illegible] this book will convey [illegible] and students who came from many countries [illegible].

The editor would like to thank Mrs. [illegible] for her [illegible] of the manuscript.

[illegible] 1985 [illegible] C. Wehrmann

Contents

Part II Physics of Microstructures

Part III Perspectives in Microfabrication Applications

Part IV **Poster Session, Abstracts**

Part IV Poster Session, Abstracts

Part I

Physics and Engineering of Microfabrication

High Resolution Lithography (Some Comments on Limits and Future Possibilities)

A.N. Broers

Cambridge University Engineering Department, Trumpington Street, Cambridge, CB2 1PZ, England

1. INTRODUCTION

1.1 Article Content

This paper contains comments on present day limits to microlithography. The methods discussed are those used to fabricate semiconductor integrated circuits, but the same methods are also used to fabricate integrated optical devices, thin film displays and a variety of other devices of commercial and scientific interest.

It is assumed that the reader is familiar with the basic methods. For general information about microlithography see [1], and for current developments, see [2].

2. GENERAL COMMENTS

2.1 Ultimate Resolution

Provided that the highest resolution resists are used, resolution in thin film lithography is set by the resolution of the exposing tool, or by scattering or de-localization of the energy of the exposing radiation in the resist or sample, and not by fundamental resist limitations such as the grain size limit encountered with silver halide emulsions. Resist contrast, and the resist development process, are nonetheless of vital importance in determining the dimensional accuracy of developed resist patterns. The etching, plating or lift-off processes that are used in conjunction with resist patterns are of similar importance. Eventually, at dimensions below a few tens of nanometers, non-uniformities in polycrystalline films become unacceptable and epitaxial films must be used. Lateral damage induced by dry etching also becomes important and may limit the size of useful structures.

2.2 Edge Defined Structures

Very small structures can be made by evaporating at an oblique angle onto samples with sharply defined surface steps. The evaporated material deposits more thickly on the side of the steps, and when the sample is subsequently etched from a direction normal to the surface, fillets are left adjacent to the steps. These fillets can be used themselves as devices features, or to form a mask for subsequent etching. A similar effect is accomplished by sputter deposition, chemical vapour deposition, or film growth, provided the thickness of the deposited film is greatest next to surface steps.

The minimum size of edge-defined structures is set by the degree to which the thickness of the deposited film can be controlled. It is easy, therefore, to produce dimensions <0.1 micron. The pattern shapes that can be formed are obviously limited.

2.3 Statistical Control of Dimensions and Overlay Accuracy

Lithography tools and processes for manufacturing integrated circuits must produce statistically good control of dimensions. Minimum linewidth is set by the ability to control linewidth and not simply by the ability to reproduce a given linewidth. The device designer chooses the minimum linewidth (L) that gives him an acceptable ratio (∿10:1) of linewidth to the experimentally measured standard deviation ($\sigma(L)$) of linewidth. Devices are typically made tolerant to $\pm 3\sigma(L)$ variations in linewidth. To obtain the data base necessary to make these decisions, patterns containing a range of linewidths are exposed on many wafers, and hundreds of measurements made on all the surfaces encountered in the process.

Overlay errors, which are the errors in the position of one pattern with respect to another, must also be treated statistically. Any systematic offset error (δ) due to faulty adjustment of the alignment system is determined and subtracted to leave the random errors. The error distribution is again fitted to a Gaussian and the standard deviation ($\sigma(a)$) calculated. Devices are designed to be tolerant to alignment errors of up to $\pm(\delta+3\sigma(a))$. Two kinds of error contribute to overlay errors; those due to the exposure tool, such as alignment errors and image distortions, and those in the mask.

3. ULTRA-VIOLET LIGHT CONTACT AND PROXIMITY PRINTING

3.1 General

Contact and proximity printing with ultra-violet radiation have generally been abandoned for integrated circuit production, because of defects that arise through accidental, or purposeful, contact between mask and wafer, but they are still used for the fabrication of simpler structures that are more tolerant to defects.

Minimum linewidth in proximity printing is set by Fresnel diffraction between the mask and the bottom of the resist. A 'rule-of-thumb' approximation is that the minimum linewidth is equal to $(\lambda g)^{1/2}$ (micron), where λ (micron) is the wavelength of the exposing radiation, and g (micron) is the gap between the mask and the bottom of the resist. This is the minimum linewidth at which the intensity at the centre of the line approaches that of the background radiation. Strictly, for UV radiation, g should be reduced slightly because the wavelength of the exposing radiation is reduced in the resist. It is difficult to control linewidth in this size regime, however, because the exposure varies rapidly from below, to 70% above the background level, for linewidths immediately above and below this size.

4. X-RAY LITHOGRAPHY

4.1 General

Proximity printing with soft x-rays is known as x-ray lithography [3]. It is the only practicable method for using x-rays for microlithography, because it is not yet possible to build high resolution, large area x-ray lenses. Glancing incidence mirror lenses and zone plate lenses are being worked on, but resolution and field size are not adequate for lithography.

X-ray lithography is the most developed of the alternatives to optical projection that offer comparable cost, but higher resolution. Significant advantages of x-ray lithography are that it produces high aspect ratio resist profiles for linewidths above 0.05 micron, and that it is tolerant to mask contamination. Small, low atomic weight, particles on the mask do not produce enough contrast to be printed.

4.2 X-Ray Lithography - Resolution

Two factors set the resolution for x-ray lithography: (1) Fresnel diffraction between the mask and the surface of the sample, and (2) the range of the photoelectrons formed when the x-ray photons are absorbed in the resist.

The same diffraction criterion used above for UV radiation can be applied to x-rays. The wavelength of the x-rays is so short (0.2 nm - 4 nm) that a gap of 50 micron to 100 micron is tolerable for minimum linewidths of 0.5 micron. In most cases for integrated circuit fabrication, penumbral blurring due to the finite size of the source is more likely to limit resolution. Diffraction becomes important again for dimensions below 0.1 micron.

The second factor limiting resolution, the photoelectron range, can be estimated from the Gruen dose depth relationship for electrons [4]. This relationship has only been confirmed for energies between 5keV and 54keV, but has been shown to predict the resolution for a typical resist such as PMMA (density 1.2 gm/cm^3) to within a factor of about 2 [5]. The minimum linewidth is assumed to be equal to the Gruen range R_G, where $R_G \approx 10^{-23}\lambda^{-1.75}$ (m), and (λ) is the wavelength of the exposing radiation.

For linewidths in the nanometer range (<100 nm), diffraction becomes more of a problem, and a compromise has to be made between the need for short wavelength to alleviate the deleterious effects of diffraction, and the need to use low energy, long wavelength x-rays (e.g. carbon 4.5 nm) to restrict the energy and range of the photoelectrons. Linewidths below a few tens of nanometers will probably only be possible with masking layers that are formed directly on the sample surface, thereby minimizing the mask to sample gap. In this size range, electron beam lithography offers higher aspect ratio provided relatively high electron energies (>25 keV) are used.

4.3 X-ray Lithography - Throughput

The only x-ray source that presently provides enough output to reach throughputs comparable to those of optical cameras is the electron storage ring. Laser plasma, and plasma focus sources are possible alternatives but their output is currently ten to a hundred times lower [6].

In particular, compact, stretched, storage rings that use superconducting magnets to minimise the radius of the electron orbit, promise high enough output to be practicable. A single ring will supply radiation to several exposure stations. Criticisms of storage rings as lithography sources are that they are too expensive and difficult to maintain, and that the consequences of a failure in the ring will be too serious, because all exposure stations supplied by the ring will be disabled simultaneously. The possibility of having two rings that can service a single set of exposure stations has been considered, but the practical difficulties of achieving this are obvious.

Work is still needed to perfect x-ray masks; in particular to reduce mask distortion to acceptable levels. With point sources, such as electron bombardment sources, laser/plasma or plasma discharge sources, there are also problems with run-out errors that result from non-repeatable flatness of the wafer and/or the mask. Both of these problems are alleviated by reducing the field size and making the exposure in a step-and-repeat manner, but the resultant loss in throughput is unacceptable. Throughput will only be restored by increasing source output. The storage ring provides adequate output and the additional advantage that the emerging radiation has a small divergence angle.

5. ELECTRON AND ION BEAM PROXIMITY PRINTING

5.1 General

Electrons [7] are being used for proximity printing using stencil masks, and ions [8] are being used for proximity printing using stencil or membrane masks. Both techniques compete with x-ray lithography, but are less well developed. Neither topic will be discussed here.

6. OPTICAL PROJECTION

6.1 General

Most integrated circuit exposures are made today with optical projection cameras, the best of which are capable of printing 1 micron dimensions. Reflective and refractive lenses are used in these cameras, and both types of lenses have potential for further improvement, although mirror lenses appear to offer the greatest potential because they can be designed more easily to operate over a range of wavelengths. They also allow the use of very short wavelengths (200-250 nm) where the absorption of glasses becomes a significant problem for refractive lenses. Eventually it should be possible to build lenses with numerical apertures (N.A.'s) of 0.4-0.5, that will operate at vacuum ultra-violet wavelengths ($\sim$200 nm). The minimum linewidth for lithography can be assumed under optimum conditions to be equal to $0.6\,\lambda/\text{N.A.}$, so these lenses should print linewidths approaching 0.3 micron [9].

The major difficulty with high resolution optical systems is their very small depth of focus ($\lambda/(\text{N.A.})^2$. Present systems, such as the full-field cameras of Perkin Elmer, have a numerical aperture of 0.16 and a depth of focus of 6 micron to 10 micron. Step-and-repeat cameras have smaller depth of focus, but work over relatively small exposure areas so it is easier to adequately level the sample. Their depth of focus is still 2-4 micron.

200 nm exposure wavelength, and 0.5 numerical aperture, will produce a resolution of 0.3 micron, but a depth of focus of only 1 micron. Vertical dimensions of integrated circuit devices will probably not scale with lateral dimensions, so the depth of focus will be no larger than the surface roughness. Multi-layer resists with a planarised imaging layer will help, but it seems likely that at some point around 0.5 micron linewidth, the small depth of field will become intolerable, and optical lithography will have to give way to electron or x-ray lithography.

7. ELECTRON AND ION BEAM PROJECTION

7.1 General

Electron and ion beam projection systems offer the potential for high resolution, and high enough throughput to compete with optical projection or x-ray lithography. However, improvements still have to be made in a number of areas such as mask distortion, image distortion and alignment accuracy. Multi-beam systems incorporating electrons or ions have also been proposed, but similarly, considerable work remains to be done before they will be ready for evaluation against the competing methods. The effort on these methods is small compared with the efforts on optical projection and x-ray lithography. Progress with the electron and ion beam methods can be monitored by reading, for example, the proceedings of the Microcircuit Engineering conferences [2].

8. SCANNING ELECTRON BEAM LITHOGRAPHY

8.1 General

At first sight, scanning electron beam lithography is the ideal method for writing integrated circuit patterns. No masks are required and resolution is beyond foreseeable needs for integrated circuits. Unfortunately, it is slow and very expensive. Exposure cost is at least an order of magnitude higher than it is with optical exposure, so electron beams are restricted to environments where the high cost can be offset by the savings that result from eliminating the cost and/or the time taken to make masks. This advantage is particularly significant in research and development.

Electron beams are the best method for fabricating devices with dimensions below 0.1 micron, and so far all useful devices with these dimensions have been made with electron beam lithography.

8.2 Scanning Electron Beam Lithography - Throughput

Efforts to improve the throughput of electron beam lithography systems centre on the use of electron optical columns that produce shaped electron beams. The shape is varied in such a manner as to most efficiently fill the pattern features. This maximises the average beam current whilst retaining the ability to write a variety of patterns.

A key advantage of shaped-beam systems is that throughput does not depend on the address increment of the integrated circuit pattern. The address increment, which is the accuracy with which pattern features are placed, is frequently much smaller than the linewidth. With a system which uses a small round beam, the beam diameter must be scaled down with the address increment. The number of beam addresses therefore increases and the throughput is reduced. With shaped beams, the spot edge definition and the beam position are not related and the number of exposure positions does not increase if the address increment is reduced. This advantage is particularly significant for mask making.

Even with shaped beams, the major limit to throughput remains the time taken to expose the resist, rather than the times taken to process the pattern data, deflect the beam, or move the sample. As a result, higher throughput will only be obtained by increasing resist sensitivity and/or electron beam current density. Unfortunately, there are fundamental limits to the gains that can be made in either area.

If the current density is increased above a critical level for a given electron column, electron-electron interactions along the beam path result in an unacceptable increase in the energy spread in the beam. The beam definition then degrades due to both deflection and axial chromatic aberrations. The increase in energy spread depends on the beam aperture, the beam path length, and the beam brightness. Smaller beam aperture, higher brightness, and longer path length increase the energy spread [10].

One method with potential for increasing current density, without increasing energy spread, is to use a very short focal length lens similar to those used in transmission electron microscopes [11]. These lenses have much lower on-axis aberrations than the lenses normally used in electron beam fabrication systems, and would allow up to 20 times higher current density by permitting the use of larger apertures. The deflection field would have to be very small, but a combination of mechanical movement and electronic deflection could be used to cover large samples. The sample would have to be immersed in the magnetic field of the lens, but this has already been done without difficulty with the VAIL concept [12].

Another method that has been proposed for increasing the beam current for a given spot size is to form the beam from a series of beamlets that are separated for much of their path-length, and therefore have reduced energy spread [13]. This 'shower beam' idea is yet to be put into practice.

The resist sensitivity limit is encountered when the number of electrons required to expose the resist ceases to be large enough to produce a statistically repeatable dose. This occurs, for example, when the resist sensitivity reaches 10^{-7} C/cm^2 for a pixel size of about 0.1 micron [14].

Even through the use of short focal length lenses and/or shower beams in combination with a resist sensitivity of 10^{-7} C/cm^2, it is unlikely that the throughput of scanning electron beam systems will be increased by the factor of more than 100 needed to approach the throughput of either optical projection cameras or storage-ring x-ray lithography.

8.3. Electron Beam Lithography - Proximity Effect

Proximity effect is the effect that makes exposure at one point depend upon the exposure of adjacent points. It is a fundamental characteristic of electron beam exposure that must be taken into account when writing patterns containing linewidths below about 2 micron [15]. The effect results from the fact that areas surrounding the point of impact of the electron beam are partially exposed by electrons that are backscattered from the sample. The result is a background exposure that reduces image contrast. Proximity effect is most serious for patterns containing closely spaced shapes.

The most usual technique for compensating for proximity effect is to vary the dose for each shape so that the average exposure for every shape is the same. For a complex integrated circuit pattern, the calculation required to determine the dose for each shape is very large. Efficient algorithms reduce the computation time, but the computing expense is nonetheless a significant factor that must be taken into account in assessing the feasibility of the technology. Perfect compensation for proximity effect is not possible because negative exposure dose would be required.

Proximity effect can be minimised, at the expense of process complexity, by using multi-layer resists. To ensure high contrast and high resolution, the upper imaging layer is kept thin, and the underlayer thick. The thick underlayer backscatters fewer electrons than a typical substrate because it has lower atomic weight.

8.4 Electron Beam Lithography - Resolution and Nanolithography

The resolution of electron beam resist exposure is set by electron scattering in the resist and sample, and/or by a combination of the range of the interactions between the electrons and the resist molecules, and the range of the low-energy secondary electrons in the resist. It is always possible, provided exposure speed is not important, to keep the beam size small compared to the other effects. Ultimately, the minimum dimensions in resist are about 10 nm, whereas beam diameters can be as small as 0.5 nm.

It is difficult to measure the ultimate resolution of electron beam resist exposure because the electron microscopes needed to make the measurements distort the resist test patterns. This has led to the development of a method that determines the effective exposure distribution (the equivalent to the Airy distribution in assessing the resolution of an optical microscope) without the need to measure linewidth [16]. All that is required is a knowledge of the 'written' linewidth, which is a function of the electron beam pattern generator, and a determination of which is the smallest linewidth in which the resist developed through to the substrate. To help in this determination, resist patterns are either shadowed, or delineated by lift-off metalization.

The test pattern used to determine the exposure distribution is shown in Fig.1. The pattern in Fig.1 was written with a 1 nm diameter electron beam in a 60 nm thick layer of PMMA. The minimum linewidth is about three times smaller than the sigma (σ) for the distribution, and the largest linewidth is about 10 times greater than σ. About 10 different exposure doses are used for each experiment. The lightest dose is below that needed to open up the largest shapes. The heaviest dose is high enough to ensure that the resist developed through to the substrate in the site of the narrowest line.

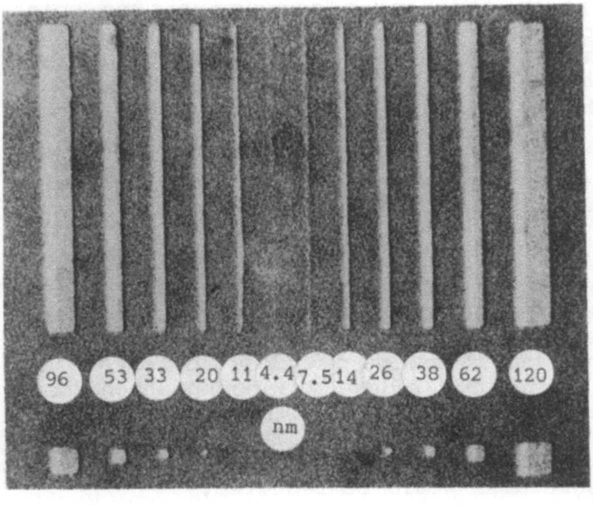

Fig.1. Test pattern used to measure electron exposure distribution in PMMA resist [16]. Measurement is made by determining the exposure level at which the centre of lines of a given width develop through to the substrate, and not by measuring linewidth. The pattern has been shadowed from the left at 45° with 3 nm of gold-palladium

If it is assumed that the exposure distribution is Gaussian, and that the lines are equivalent to infinitely long rectangles, then the exposure dose (Q_W) received at the center of a line is given by

$$Q_W = Q_O \operatorname{erf}(W/2\sigma)$$

where Q_O is the exposure dose in the center of an infinitely large shape, and is the standard deviation of the distribution. Q_W is measured for each linewidth by determining the dose at which the line first develops through to the substrate. As the exposure increases, narrower and narrower lines develop through to the substrate. The exposure distribution is calculated from the Q_W versus linewidth data.

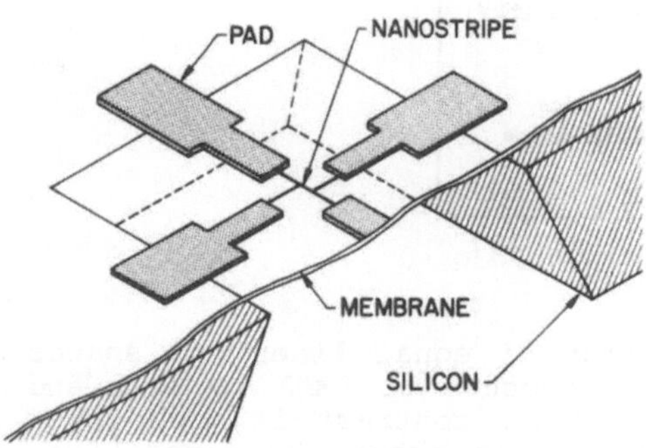

Fig. 2. Thin film substrate developed for nanolithography by Sedgwick et al [17]. The substrate allows samples to be examined with transmission electron microscopy, and yet allows electrical contact to be made to the test structures or devices

These measurements are made on the thin film substrate shown in Fig. 2 which was developed by Sedgwick et al. [17] for the fabrication of sub-tenth micron structures. The substrate not only allows samples to be examined and written upon in a transmission electron microscope (scanning or projection), as in the evaluation of resist resolution, but also allows contact to be made to structures fabricated on the fragile membrane by means of leads that extend out from the bulk area. Thin membranes of a variety of materials have been made and it is possible, for example with silicon or GaAs membranes, to fabricate devices using the membrane as part of the device.

In Fig.3, the data for the thin substrate are combined with data for the backscattering distribution from bulk substrates to come up with an overall estimate of the resolution of electron beam exposure of PMMA on bulk and thin substrates. It can be seen that the backscattering reduces contrast but does not change the ultimate resolution very much. In practice a contrast of 50% is adequate for the fabrication of high quality structures, so minimum dimensions approaching 10 nm can be produced.

Measurements on different PMMA samples, and on other resists, all yield a similar distribution width. The 10 nm limit is thought to be due to the fact that exposure is largely due to secondary electrons. Secondaries are generated up to 2 nm from the primaries, and can straggle several nanometers further into the resist before their energy is dissipated.

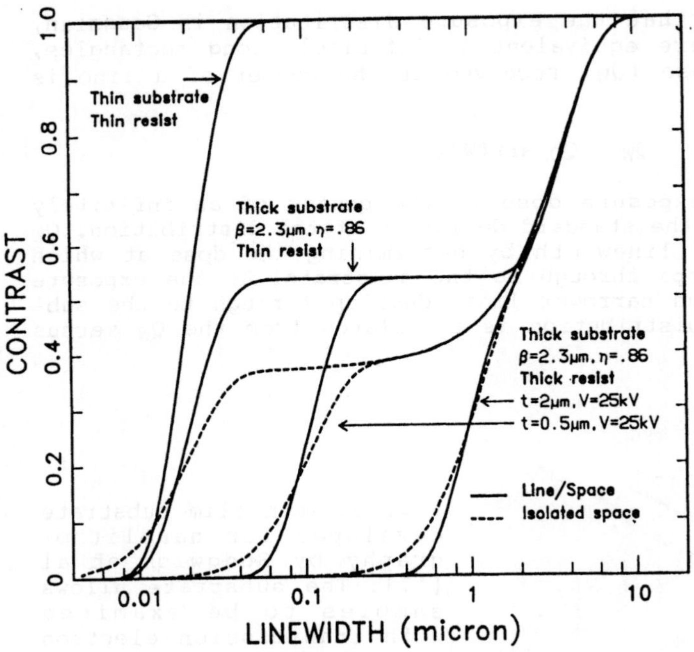

Fig. 3. Contrast for an infinite array of equal lines and spaces versus linewidth for electron beam exposure of PMMA on thin and thick substrates. For smaller linewidths, contrast is much lower for the thick substrate, but resolution is about the same. Linewidths less than 0.1 micron are readily produced on thick substrates [21-23]

8.5 Sub-10 nm Structure Fabrication

It is possible to avoid the 10 nm limit encountered with conventional electron resists such as PMMA, by using a recording medium that is sensitive only to high energy primary electrons, and not to low energy secondaries.

The most promising method is the direct sublimation under electron bombardment of a variety of materials including NaCl, MgF_2, AlF_3, LiF and Al_2O_3. The high resolution of this technique was discovered in the experiment proposed by J.W.Matthews that produced the 5 nm diameter holes shown in Fig.3. The holes were formed in a NaCl crystal by a 1nm diameter 50kV electron beam [18]. The crystal was 0.25 micron thick. The convergence half-angle of the beam was 10^{-2} rad., and the beam therefore formed a conical-shaped hole with the base of the cone being about 5nm in diameter, assuming that the beam is focused on one face of the crystal. This explains the discrepancy between the apparent size of the hole and the beam diameter, and suggested that the resolution of the process was much better than 5 nm. Isaacson et al. [19] and Mochel et al. [20] have confirmed that higher resolution is possible by writing structures down to 1.5 nm in thin NaCl and AlF_3 films.

8.6 Nanostructure Devices

The first useful devices with dimensions below 0.1 micron were made on bulk substrates in the early 1970's [21]. Since then a variety of devices, including very short gate field effect transistors, have been made using multi-layer resists [22,23].

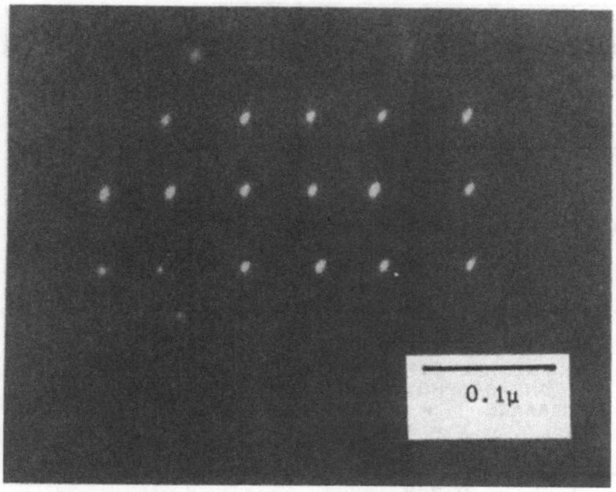

Fig. 4. 5 nm diameter holes drilled in a 0.25 micron thick NaCl crystal with a 0.5 nm diameter electron beam. These holes were the first demonstration of the high resolution available with direct electron beam induced sublimation [18]

Whilst it is possible to fabricate nanostructures with resists such as PMMA, a number of relatively simple, but scientifically valuable structures, have been made using 'contamination resist' and ion etching. Contamination resist is formed from a thin layer of hydrocarbons condensed onto the sample surface at a pressure of 10^{-4}-10^{-8} mm Hg in an oil-pumped vacuum system. The electron beam 'cracks' the hydrocarbons, and a layer that is highly resistant to ion bombardment builds up at the point of impact of the beam. The dose of about 1 C/cm^2 required to produce an adequate resist thickness for protecting metal layers as thick as 100 nm from ion etching is much higher than the doses delivered during microscopical examination, so the sample can be examined before and after resist formation with little effect. This makes it easy to accurately position the structures with respect to contact pads, or other device layers, and to examine the resist pattern after it has been formed. The build-up of resist can also be accurately monitored by observing the decay in the transmitted signal. It is also possible to remove the layers of hydrocarbons by heating the sample to about 100° C. This permits the extended examination of the samples without further build up of resist.

A variety of devices has been made with this method including microbridges [24], SQUID's [25] (Superconducting QUantum Interference Devices), and fine wires for exploring localization effects [26].

9. SCANNING ION BEAM LITHOGRAPHY

9.1 General

Ion probes offer great potential for microanalysis and mask repair, but cannot compete with electron probes for microlithography, or nanolithography. The sources and lenses available today do not provide enough final beam current (>0.1 microamp) for the range of beam sizes (0.1 micron to 0.5 micron) needed for integrated circuit lithography, or small enough beam size (<2nm) for nanolithography. The unfavourable comparison with electrons exists despite the fact that ions scatter over much smaller distance in the resist than electrons so there is no proximity effect, and that ions dissipate their energy more fully in the resist so the required exposure dose (Coulomb/cm^2) is much lower. Several papers on ion beam lithography can be found in [2].

Table 1.

LITHOGRAPHY TYPE	RESOLUTION LIMITS	
	NORMAL	ULTIMATE
ULTRA-VIOLET LIGHT CONTACT/ PROXIMITY	2.5 μ FRESNEL DIFFRACTION AT MINIMUM USABLE GAP AND WAVELENGTH λ = 250 nm t = 25 μ	0.2 μ FRESNEL DIFFRACTION CONTACT PRINT, RESIST THICKNESS = 200 nm λ = 200 nm t = 200 nm
X-RAY PROXIMITY	0.2 μ FRESNEL DIFFRACTION AT USABLE GAP FOR STEP & REPEAT OPER. λ = 0.8 nm t = 50 μ	0.01 μ FRESNEL DIFFRACTION AND/OR PHOTO-ELECTRON RANGE λ = 4.4 nm t = 25 nm
OPTICAL PROJECTION	0.5 μ FRAUNHOFER DIFFRACT. AT N.A. SET BY LENS FABRICATION/FIELD SIZE CONSTRAINTS λ = 250 nm N.A.= 0.36 DEPTH OF FOCUS = 2 μ	0.2 μ FRAUNHOFER DIFFRACT. WAVELENGTH SET BY ABSORPTION OF RESISTS AND OPTICAL COMPONENTS λ = 200 nm N.A.= 0.9 DEPTH OF FOCUS = 250nm FIELD SIZE <200 μ
ELECTRON BEAM	<0.1 μ RESOLUTION/THROUGHPUT TRADE-OFF, AND LATERAL SCATTERING IN NORMAL RESIST THICK-NESSES [>0.5 μ FOR <~30 kV ACCEL. POTL.]	0.02 μ (CONVENTIONAL RESIST) DE-LOCALISATION OF SECONDARY ELECTRON ELECTRON EXPOSURE. 0.001 μ (DIRECT SUBLIMATION) COMBINATION OF ELECTRON INTERACTION RANGE, AND DIFFRACT./ SPHERICAL ABERRATION
ION BEAM	<0.1 μ RESOLUTION/THROUGHPUT TRADE-OFF	0.02 μ (CONVENTIONAL RESIST) DE-LOCALISATION OF SECONDARY ELECTRON EXPOSURE 0.01 μ (DIRECT MILLING) ION OPTICAL LIMITS CHROMATIC ABERR. SPUTTER INTERACTION RANGE

10. SUMMARY

Table 1 summarises, in abbreviated form, the resolution limits for the different lithographies, and the factors that set these limits.

11. REFERENCES

1. A.N.Broers: In Materials for Microlithography, ed. by L.Thompson et al., (Am. Chem. Soc., Washington, D.C. 1984) p.11, and A.N.Broers and T.H.P Chang: In Microcircuit Engineering 78, ed. by H. Ahmed and W.C.Nixon, (Cambridge University Press 1978), p.1

2. Microcircuit Engineering 83, ed. by H.Ahmed, J.R.A.Cleaver and G.A.C.Jones (Academic Press, London 1983); Microcircuit Engineering 84, ed. by A.Heuberger and H.Beneking (Acadamic Press, London, 1985); Microcircuit Engineering 85, ed. by K.D.van der Mast and S.Radelaar (North Holland, Amsterdam, 1985)

3. D.L.Spears and H.I.Smith: ELectron. Lett. 8, p.102, (1972)

4. A.E.Gruen: Z. Naturforsch, 12a, p.89, (1957)

5. E.Spiller and R.Feder: In H.J.Quesser X-Ray Optics, (Springer, Berlin 1977)

6. A.Heuberger: Solid State Technology, February, (1986)

7. H.Bohlen: IBM J. Res. Dev. 26, p.568, (1982)

8. R.L.Seliger et al.: J. Vac. Sci. Technol., 16, p.1610, (1979)

9. D.Markle: Solid State Technology, September, (1984)

10. H.C.Pfeiffer: IEEE Trans. Electron. Dev. ED-26, p.663, (1979)

11. A.N.Broers and P.Coane: to be published Appl. Phys. Lett. (1986)

12. M.A.Sturans and H.C.Pfeiffer: In Microcircuit Engineering 84, ed. by A.Heuberger and H.Beneking (Academic Press, London 1984), p.107

13. K.D.Van der Mast, G.H.Jansen and J.E.Barth: In Microcircuit Engineering 85, (North Holland, Amsterdam 1985), p.43

14. T.E.Everhart: In Materials in Microlithography, ed. by L.F.Thompson et al., (Am. Chem. Soc. Washington, D.C. 1984) p.5

15. T.H.P.Chang: J.Vac. Sci. Technol. 12, p.127, (1975)

16. A.N.Broers: J. Electrochem. Soc., 128, p.166, (1980)

17. T.O.Sedgwick, A.N.Broers and B.J.Agule: J. Electrochem. Soc., 119, p.1769, (1972)

18. A.N.Broers, J.J.Cuomo, J.Harper, W.Molzen, R.Laibowitz, M.Pomerantz: Electron Microscopy 1978, Vol. 111, ed. by J.M.Sturgess, (Microscopical Soc. of Canada, Toronto 1978) p.343-354, and A.N.Broers, J.J.Cuomo and W.Krakow: IBM Technical Disclosure Bulletin, 24 , p.1534, (1981)

19. M.Isaacson and A.Muray: J. Vac. Sci. Technol., 19 , p.1117, (1981)

20. M.E.Mochel, C.J.Humphreys, J.M.Mochel and J.A.Eades: Proc. 41st Annual Meeting of the Electron Microscopy Soc. of America, (San Francisco Press 1983), p.100

21. A.F.Mayadas and R.B.Laibowitz: Phys. Rev. Lett., 28 p.156, (1972)

22. W.S.Mackie, W.Patrick, S.P.Beaumont and C.D.W.Wilkinson: Microcircuit Engineering 84, (Academic Press, London 1985), p.213

23. R.E.Howard, L.D.Jackel and W.J.Skocpol: Microcircuit Engineering 85 (North Holland, Amsterdam 1985), p.3

24. R.B.Laibowitz et al.: Appl. Phys. Lett., 35, p.891, (1979)

25. R.Voss, et al.: Appl. Phys. Lett., 37, p.656, (1980)

26. P.Chaudhari, A.N.Broers, C.C.Chi, R.B.Laibowitz, E.Spiller and J.Viggiano: Physics Review Letters, 45, p.930, (1980)

Question by Chris Wilkinson.

You mentioned that the minimum linewidth achievable by x-ray lithography is of the order of 10 nm. While Dale Flanders, in a beautiful demonstration experiment, obtained 17 nm lines with 17 nm gaps, the mask was made by shadowing an edge, and thus had metal of high aspect ratio. What do you feel is the linewidth achievable using e-beam lithography to write the mask pattern?

Answer by A.N.Broers.

If the mask is a separate entity that has to be brought into contact with a resist coated sample, then the minimum practically achievable gap between the mask and the bottom of the resist pattern will be about 0.1 micron. For a wavelength of 4.4 nm, this will result in a minimum linewidth of 20nm, according the definition given above ($(\lambda g)^{1/2}$). This assumes that the metal has adequate thickness, but this should be possible for the soft x-rays (4.4 nm) needed if the photo-electron range is not to limit the resolution. A gold layer a few tens of nanometers thick should provide adequate contrast.

The practical difficulties of reliably obtaining a gap of <0.1 micron are formidable, and it will probably be difficult to reproduce linewidths smaller than about 50nm. The value of 10 nm that I used relates to the fundamental limit that should be achievable in the absense of practical constraints.

Submicron Lithography Tools

G. Pircher

79, rue Vatonne, F-91191 Gif sur Yvette, France

J. Perrocheau

Thomson-CSF, Laboratoire Central de Recherches, Domaine de Corbeville
B.P. 10, F-91401 Orsay Cedex, France

1. INTRODUCTION

During the last 50 years, the feature size of components has been reduced from a few centimeters for valves to a few microns for transistors in integrated circuits. Four orders of magnitude have therefore been gained in the race for miniaturization.

By virtue of this miniaturization, it is possible to increase the density of integration of given functions as well as the speed of circuit operation. Thus within dimensions that are compatible for every day usage, electronic systems are able to achieve ever more impressive performances in terms of memory capacity, computational power, rate of throughput, etc.

This progress is possible as a result of two major advances: (i) the discovery of the transistor in the forties and the subsequent pervasive use of solid state technology; (ii) the development of the planar process by which a circuit is created from a series of planar images and may be fabricated using batch processing.

At the present time electronic systems are designed around "cells"; these are chips of silicon of the order of one centimeter square with a considerable number of transistors.

These chips may be characterized by a number of different electrical properties but two particular parameters indicate the advance of these components: (i) the complexity or number of transistors per chip; (ii) the resolution of the minimum feature size of these transistors.

Less complex chips may have higher resolution and in the case of components such as microwave devices, a feature size, or linewidth, of a few tenths of a micrometer can be involved.

Complexity and resolution depend directly on the planar process employed and in particular the microlithography. Microlithography concerns the different means necessary to achieve the inplane geometry of the circuits. It is directly responsible for the miniaturization.

The main questions are: What progress in miniaturization can be expected from electronic circuits over the next few years and what type of microlithography will allow submicron dimensions to be achieved?

Thus although there are fundamental physical limits to the smallness these limits represent a fascinating field of research and discovery, with unpredictable applications in electronics, biology and chemistry.

2. GENERAL ASPECTS OF MICROLITHOGRAPHY

2.1 Principles of Microlithography[1, 2]

A solid state circuit is always an assembly of several layers resembling a planar jig-saw puzzle[3]. Each element of the puzzle has a size of a few microns, and a thickness of about one micron or less. The material of an element may be: a semiconductor with different doping, a dielectric insulator, or a conductor (Fig.1).

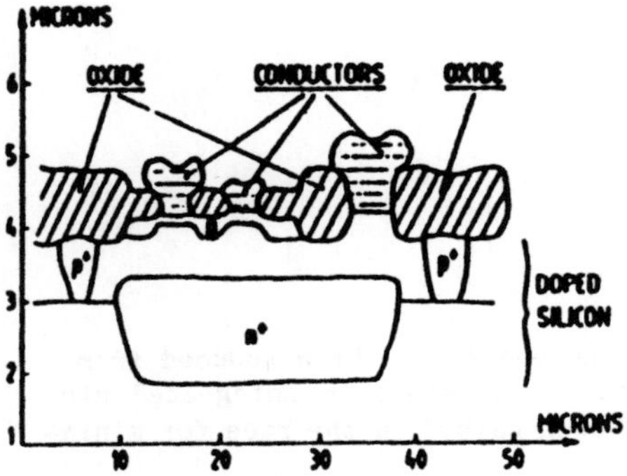

Fig.1: View of a bipolar transistor on silicon substrate

Building a circuit consists of producing the elements of the puzzles on the surface of a substrate, layer after layer, subject to tight tolerances in the definition of the dimensions of all the elements, and in the respective positions of the different layers.

Each layer is formed by a pattern obtained by a selective transfer of matter through apertures of a special material, normally silicon dioxide, deposited onto the substrate (Fig.2).

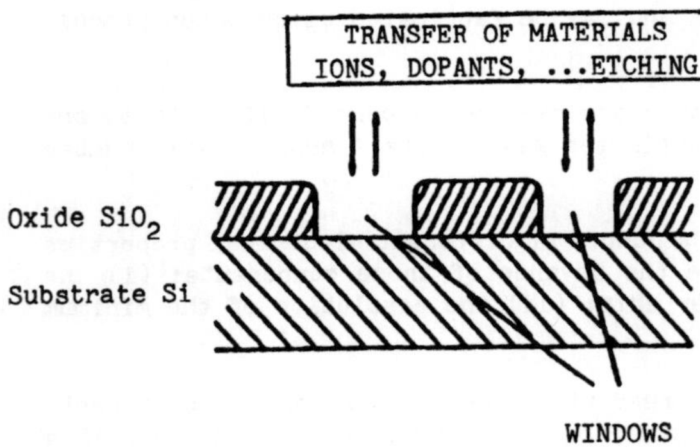

Fig.2: Principle of lithography

Apertures are obtained by etching through other windows in a resist deposited on the silicon dioxide (Fig.3).

This resist is generally a polymer sensitive to a radiation; visible light, UV light, X-rays, electrons or ions.

The desired windows are obtained by developing the resist after exposure to the radiation.

This exposure is obtained by formation of an image of the desired pattern, on the surface of the resist.

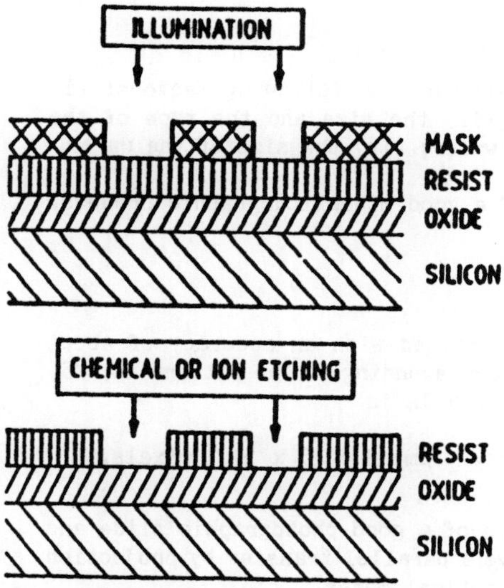

Fig.3: Principle of window transfer

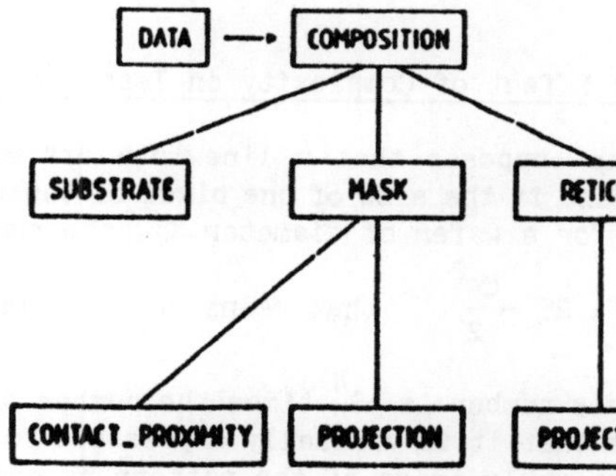

Fig.4: Means for obtaining a pattern on a substrate

Image is obtained from the pattern design data stored on discs or tapes and the subsequently realized using special apparatus, sometimes by the intermediate stage of a mask (Fig.4).

For processes not involving a physical mask, the image is formed directly by a machine acting as a composer; projecting elementary spots of fixed size (pixel), or fixed or variable shape.

If the process uses masks at actual size reproducing the entire pattern, the mask is created by the previous process need and the transfer of the mask to the wafer is obtained by contact, proximity or projection (Fig.5).

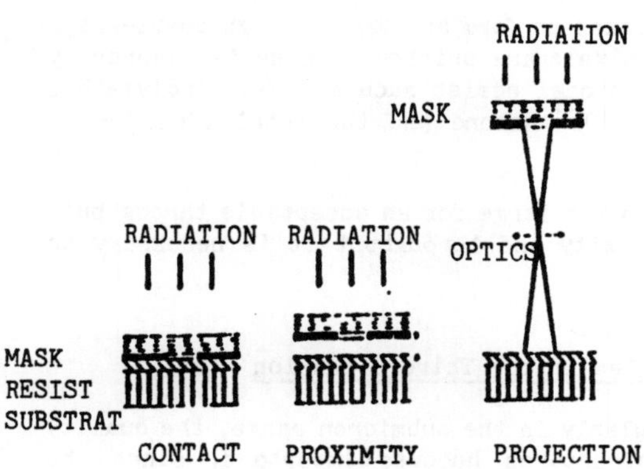

Fig.5: Transfer of a mask by contact, proximity, projection

If the process uses masks at scale X/1 it is referred to as a reticle. This reticle is very often the enlarged image of one or several circuits and the total pattern on the resist is obtained by projection with reduction, and stepping.

Besides these classical masking processes, other techniques exist for particular applications, for instance selective local material creation or erosion assisted by focused radiation[4].

2.2 Resolution, Complexity, Throughput, Profiles

The choice of a microlithographic process depends on the following factors: (i) the resolution or linewidth of the circuits; (ii) the size and the type of the circuits; (iii) the number of the circuits per wafer; (iv) the size of the wafer.

These aspects highlight the requirements of a good microlithographic process.

2.2.1 Effect of Complexity on Image Transfer

If one supposes that a linewidth "t" must be defined with an accuracy of about t/5, and if the size of one pixel of the image corresponding to the pattern is t/5, then for a wafer of diameter Φ, the number of pixels is

$$N = 25 \frac{\Phi^2}{t^2} \qquad \text{that means } \Phi = 100 \text{ mm and } t = 1\ \mu m;\ N = 25 \times 10^{10} \text{ pixels.}$$

This number is 10^4 times the number of pixels of a good photographic slide and shows that it is virtually impossible to achieve a parallel transfer by projection of the whole image of the pattern by ordinary optical means.

2.2.2 Complexity and Accuracy

The different parts of an image must be positioned to within one pixel in order to obtain accurate overlay registration. With the values given above, the relative positional accuracy is 2×10^{-6}. For instance in the case of silicon this value is of the order of the coefficient of thermal expansion. This accuracy is a function of various parameters that influence the stability of the image and that of the substrate.

This problem is one of the most difficult to solve and increases with decreasing linewidth.

2.2.3 Throughput and Complexity

Another difficulty arises from the time taken to form an image of high complexity, i.e. of the order of 10^{11} pixels. If the pixels are printed in a serial manner by 20 keV electron beam exposure of a conventional resist such as PMMA, (polymethyl metacrylate) the time per pixel is about 10^{-6} second and the total time for an image is 25×10^4 seconds or 70 hours.

This time is more than a thousand times too large for an acceptable throughput. In order to solve this conflict of complexity and throughput it is necessary to combine parallel and serial exposure.

2.2.4 Resolution, Aspect Ratio and Profiles in the Third Dimension

As the feature size decreases and particularly in the submicron range, the quality of the profile of the edge of micro-structures becomes important, since the thickness of the different materials used have minimum values dictated by the circuit requirements.

For instance in order to avoid electromigration, the current density must not exceed 10^5 amp. per cm^2, requiring an increased thickness of metallic conductors when lateral dimensions decrease. For 1 mA of current and a linewidth of 0.5 µm, the thickness must be 2 µm, giving an aspect ratio of 4.

Low resistance electrodes in field effect transistors also require high aspect ratio submicron gates.

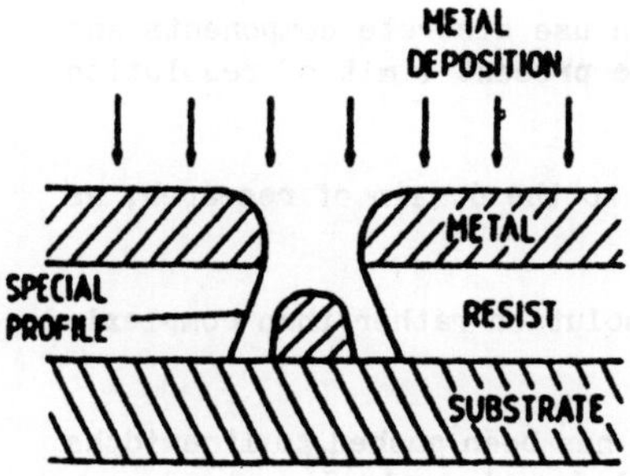

Fig.6: Example of special profile for lift-off

Special profiles are often necessary for particular techniques of lift off, self alignment, etc (Fig.6).

From another point of view an accurately defined linewidth in a material of finite thickness involves knowledge and reproducibility of profiles. Thus, microlithography in the submicron range must not only have specific properties concerning the inplane dimensions but must also be capable of achieving special profiles in the third dimension.

2.3 Characteristics of Microcircuits

The conditions described above make microlithography an especially difficult technology involving the latest refinements in engineering physics and chemistry. At the present time, microlithographic constraints are among the most important factors limiting increases in complexity, resolution and throughput.

Practical overall characteristics of commercial integrated circuits are: (i) size of the wafers: $\leqslant$ 150 mm; (ii) size of the circuit: $\leqslant$ 10 mm; (iii) linewidth: $\geqslant$ 1 µm; (iv) throughput of exposing a complete pattern: $\leqslant$ 30 per hour.

In practice most of the current commercial integrated circuits have linewidths larger than 2 µm, a few have 1 µm linewidth and only rarely is the linewidth less than 1 µm.

There is therefore a major challenge to determine a suitable process capable of submicron dimensions in large and very high speed integrated circuits (Fig.7).

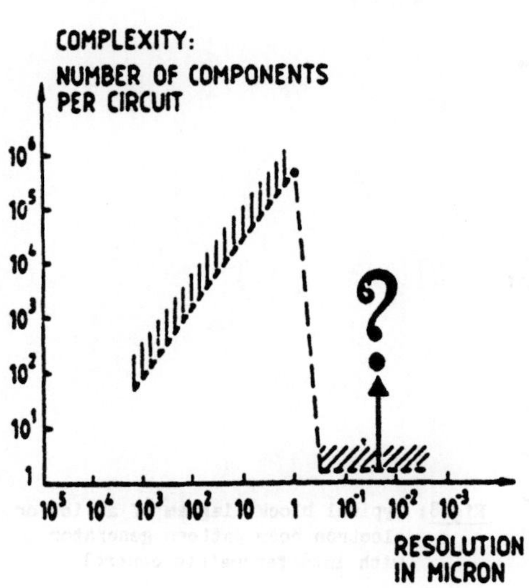

Fig.7: Complexity versus resolution for microcircuits

Commercial microcircuits with submicron linewidth use discrete components and are generally devoted to microwave applications. The present limit of resolution for these components is 0,25 µm.

All other circuits in the submicron range belong to the domain of research, as described in various technical articles.

Most of them are discrete with an emphasis on resolution rather than complexity[5].

The limit of resolution for artificial structures has been pushed to linewidths of 15 Å by erosion with electrons in a very thin halide substrate[6].

This represents, therefore, the limit of possible range for submicron structures although much work remains to be done before usable complex microstructures can be realized and their interesting theoretical properties exploited.

3. EQUIPMENT FOR SUBMICRON LITHOGRAPHY

3.1 Machines for Direct Composition of Patterns from Design Data

Although this type of machine is predominantly based on electron beam exposure, there are some examples of machines using ion beam techniques.

These types of machine have certain features in common, including (Fig.8): (i) a source of particles having a brightness of the order of 10^6 amp.cm^{-2}.sr^{-1}; (ii) particle optics for condensing and focusing; (iii) some electrodes for beam blanking and deflection; (iv) in certain cases electrodes for beam shaping; (v) a movable substrate holder, whose position is very accurately, controllable, fre-

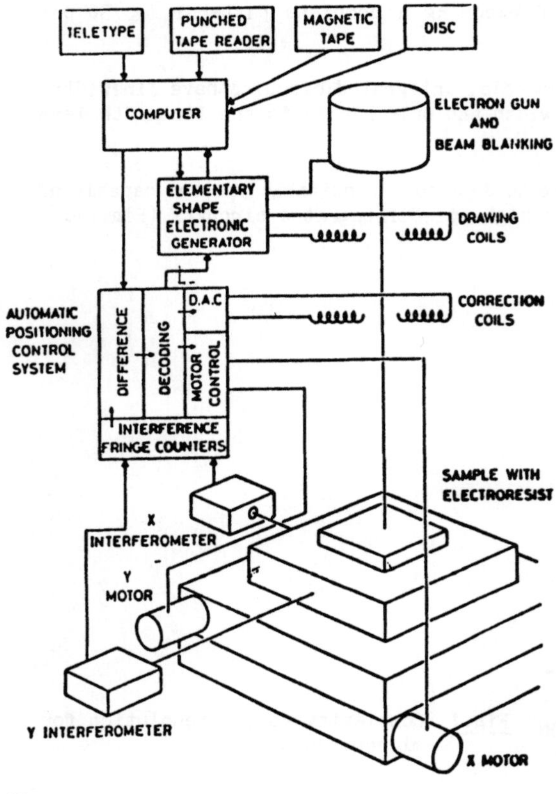

Fig.8: Typical block diagram of an ion or electron beam pattern generator with interferometric control

quently by laser interferometry; (vi) electronic circuitry for driving the beam blanking, deflection systems, and substrate holder displacement; (vii) detection of backscattering and secondary electrons in order to analyse the image formed by the particle beam at the surface of the substrate; (viii) detection of transmitted electrons in the case of use of scanning transmission electronic machines.

3.1.1 Resolution

The resolution of these machines depends on the source, the focusing properties of the optics and the interaction of the beam with the resist coated substrate. In practice the resolution is relatively unaffected by diffraction effects.

In the case of electron beam machines, beam diameter of 5 Å have been obtained, which is below the minimum size for a stable artificial microstructure[6, 7].

Thus from the point of view of beam diameter, electron beam pattern generators are capable of any resolution within the range of microstructure.

Very often the limitation arises from interaction between the electron beam and the coated substrate, which creates a significant proportion of backscattered and secondary electrons[8]. These electrons exhibit a lateral spreading beyond the diameter of the incident beam (Fig.9).

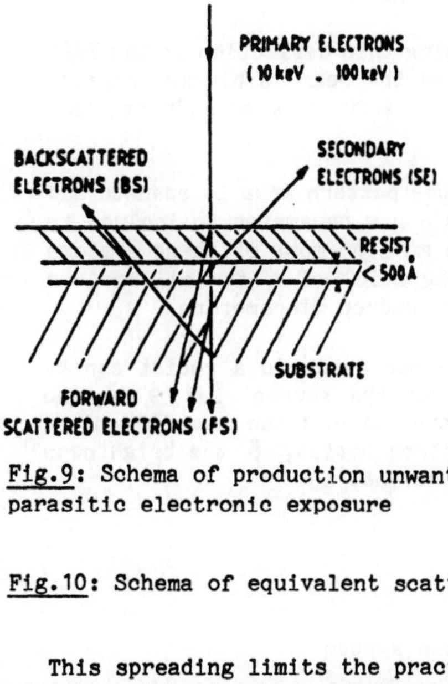

Fig.9: Schema of production unwanted parasitic electronic exposure

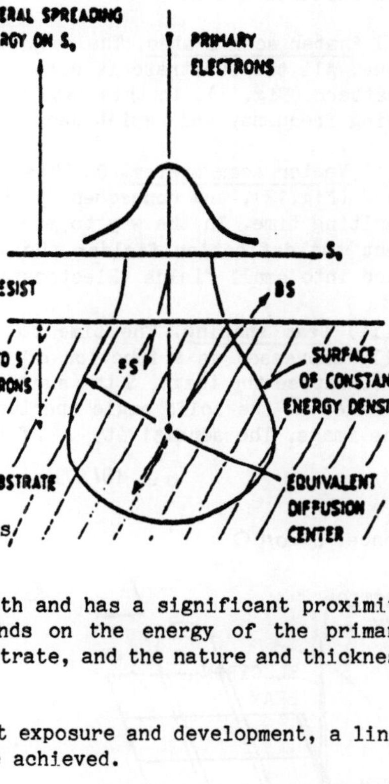

Fig.10: Schema of equivalent scattering process

This spreading limits the practical linewidth and has a significant proximity effect on the image contrast. Spreading depends on the energy of the primary electrons, the nature and thickness of the substrate, and the nature and thickness of the resist (Fig.10).

In practice for optimum conditions of resist exposure and development, a linewidth of 0.2 µm in 0.6 µm PMMA resist could be achieved.

For smaller linewidths it is necessary to decrease the influence of backscattering (BS) and the forward scattering (FS) in the resist[9]. Many solutions have been tried:

Increasing the energy of the electrons to 50 keV or more[10], in order to decrease the exposure due to BS by displacing the equivalent center of diffusion for BS away of the surface.

Separating the sensitive resist from the surface of the substrate by an intermediate layer [11].

Using a very thin resist, with thickness less than 0.2 μm in order to decrease FS.

For example, by particularly careful control of exposure conditions, it has been possible [12] to achieve 80 Å wide lines in 1000 Å thick PMMA resist deposited on GaAs.

Using a very thin substrate of thickness less than the distance between the surface and the equivalent center of diffusion in a thick substrate.

Even when the effects of BS and FS are negligible, there is still a residual effect due to straggling of secondary electrons in the resist, with a range corresponding to σ = 125 Å [13].

Smaller linewidths are obtainable by avoiding the effect of secondary electron straggling in the resist; in this case, the density of power given by the electron beam must be sufficient to vaporise thin layers deposited on a thin substrate. In this case the electron beam machine is a scanning transmission microscope [6, 14].

Electron beam machines are characterised by their writing strategy depending on the deflection system and the beam characterisation.

(i) Raster scan system. The most known system using this deflection is the EBES machine. All the substrate is completely scanned and the beam is blanked outside the pattern (Fig.11). In this case the writing time is a function of address grid, scanning frequency and resist sensitivity.

(ii) Vector scan system. On these systems only the pattern area is scanned and exposed (Fig.12), and consequently the coverage is a new parameter to include in the writing time. In the way to minimize the stage motion, most of these systems present two deflection fields: the main field (Magnetic) of a few millimeters divided into small fields (Electrostatic) of a few hundred micrometers.

(iii) Beam shaping. The time to expose an electron image in a resist can be readily expressed as a function of the brightness of the source. Let θ be the time to expose the image, S the surface exposed in the image, i the beam current on the image, Ω the solid angle aperture of the electron optics, β the brightness at the image. The sensitivity σ of the resist is defined as

$$\sigma = i\theta/S \quad \text{where} \quad \beta = i/S\Omega.$$

Hence, $\theta = \sigma/\beta\Omega$.

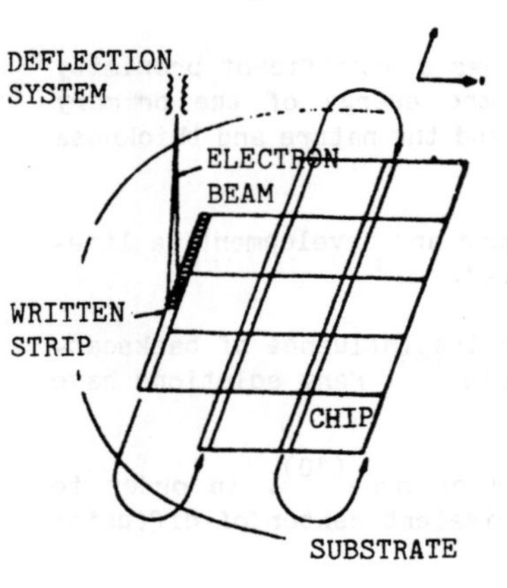

Fig.11: Raster scan system

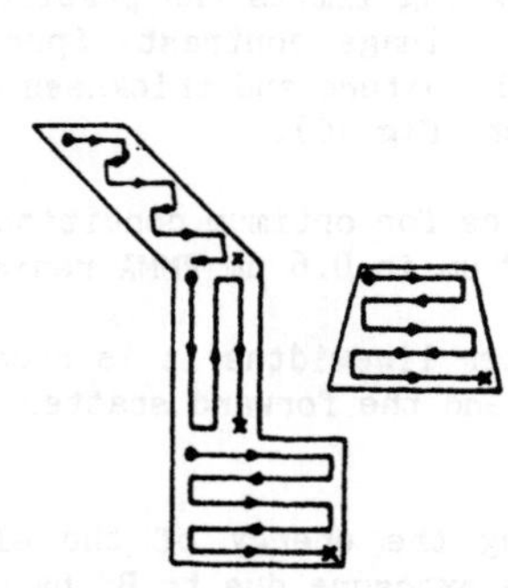

Fig.12: Vector scan system

The minimum time to expose an image is only a function of the resist sensitivity, the brightness of the source and the aperture of the optics. We see here the interest of shaped beam systems compared to Gaussian beam systems; there is more information in a shaped beam than in a Gaussian beam, but the time to draw is the same if other functions are kept constant. Shaped beam systems which have been proposed in the beginning were fixed square beams so the dimension of the beam was limited by the minimum dimension of the pattern to be drawn. The situation now is different with the variable shaped beam. Figure 13 gives an example of a simple pattern and the way it is described in three approaches.

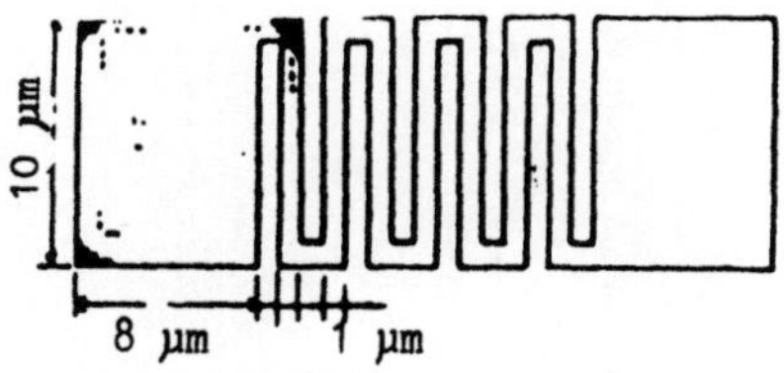

(a) 0,25 µm Gaussian beam description

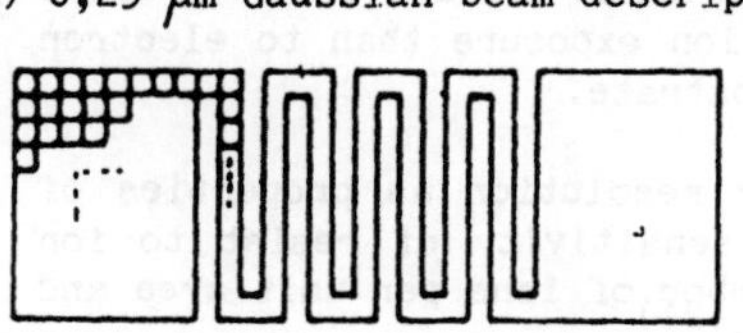

(b) 1 µm x 1 µm square beam description

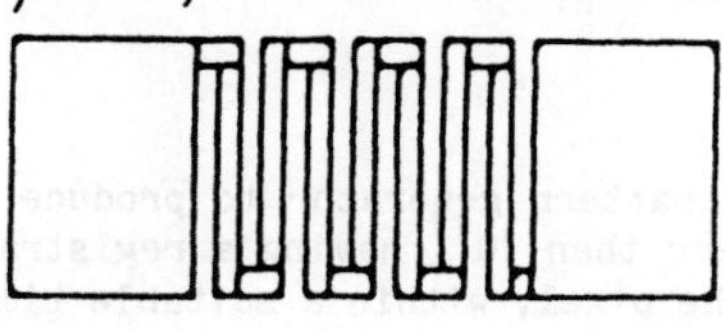

(c) variable rectangular beam description

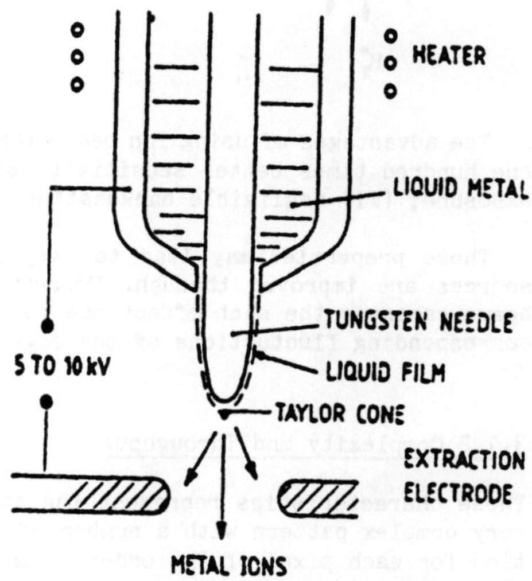

Fig.14: Schema of liquid metal ion source

Fig.13: (a) the pattern is described with a 0.25 µm-diameter Gaussian beam requiring 3806 exposures; (b) the pattern described with a fixed 1 x 1 µm-square beam requiring 238 exposures; (c) the pattern is described with a rectangular-variable beam requiring 17 exposures.

The time to expose the pattern is proportional to be the number of exposures so the time for the Gaussian beam description is 224 times the time for the variable shaped beam description.

The ion beam machines have beam diameters larger than those of electron beam machines as a result of larger chromatic aberrations due to energy spreading of the ion in the source which is related to Boersch effect.

To achieve sufficient ion probe current, bright sources (frequently using liquid metal field effect ion sources) are necessary but tend to have an associated energy spread $>$ 10 eV and corresponding beam diameters of the order of a few 100 Å[15, 16] (Fig.14).

Numerous metals and alloys such as: gallium, tin, gold, indium, gold, silicon eutectic, etc. are studied[17, 18]. These sources are used either for resist exposure or for directly focused ion implantation.

Gas field ionization sources have lower energy spread (of the order of a few eV[19]) and this allows ion probe diameters of less than 100 Å (Fig.15).

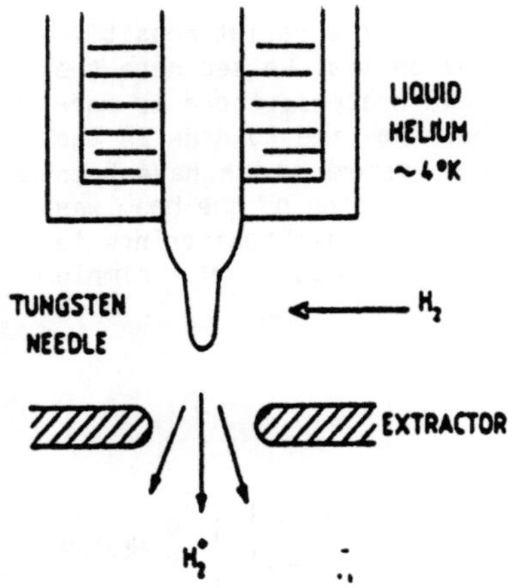

Fig.15: Schema of hydrogen field emission source

The advantages of using ion beams for pattern generation include[20]: (i) about one hundred times better sensitivity of resist to ion exposure than to electron exposure; (ii) negligible backscattering in the substrate.

These properties may lead to very good writing resolution as properties of sources are improved through. Unfortunately good sensitivity of resist to ion beams enhances the shot effect due to the small number of ions per unit area and corresponding fluctuations of the edge of the lines[21].

3.1.2 Complexity and Throughput

These characteristics represent the ability of a pattern generator to produce a very complex pattern with a number of pixels larger than 10^{11} having a registration for each pixel of the order of the size of the pixel, within a suitable time corresponding to a total exposure time of a few minutes.

In the case of electron beam machines, advances in the following areas have enabled the problem of throughput to be resolved for micron dimensions: (i) high brightness sources; (ii) variable shaped beams; (iii) fast deflection systems with small dwell time; (iv) decomposition of a pattern into individual zones mechanically stitched together with an accuracy controlled by a laser interferometric system; (v) correction of the inplane distortion of the wafer and overlay registration using reference masks located on the wafer; (vi) sophisticated deflection systems having a large field (more than $10^4 \times 10^4$ pixels) of deflection with low aberration and automatic corrections.

In the submicron range, the problem becomes more and more difficult as the linewidth decreases even if the number of pixels remains the same.

Stitching becomes very difficult due to mechanical and electronic instabilities; but main difficulties are related to the increased charge density.

In order to avoid fluctuations due to the shot effect the number of electrons in a pixel must be about n = 100.

If t is the linewidth, t/5 is the size of a pixel and q the electron charge, the minimum density of charge is

$$\sigma = 25 \frac{nq}{t^2} \text{ in Coulomb per unit area.}$$

It is apparent that if the linewidth is decreasing, the density of charge increases as the square of the inverse of the linewidth.

An increased charge density has various consequences such as an increase in the time T_o necessary to write a pixel, with the electron beam. This time is

$$T_o = \frac{\sigma}{B\Omega} = 25\,\frac{nq}{t^2 B\Omega} \quad ,$$

where B is the brightness of the source and Ω is the solid angle of acceptance of electrons at the surface of the wafer.

T_o increases as the square of the inverse linewidth.

For instance if $B = 5 \times 10^6\,A.cm^{-2}.sr^{-1}$, which is a typical value for a LaB_6 cathode, $\Omega = 4 \times 10^{-6}$ and t = 100 Å; then values of σ and T_o are: (i) $\sigma = 4 \times 10^{-4}$ cm^{-2} which is close to the value indicated by K. Lee and H. Ahmed, for a linewidth of 150 Å[15]; (ii) $T_o = 2 \times 10^{-5}$ second which is twenty times the time necessary for writing a pixel on PMMA in micron range.

In this way the time necessary to write a complex pattern in the submicron range will become very long and practically impossible.

Another effect of increased charge density is the reduced need for high sensitivity resist but rather resists with high resolution.

Power density increases when the linewidth decreases and for linewidths in the range of 10 Å this density is capable of direct vaporization of halide materials[6].

If complex patterns are required it becomes necessary to transfer the pattern onto the substrate by means of masks or reticle.

3.2 Transfer Systems

These systems use UV light, X-rays, electrons or ions for transferring a pattern from a mask to a substrate.

3.2.1 UV Transfer Systems

3.2.1a Contact or Proximity Systems

Most of the circuits in the micron range are made using contact printing to transfer patterns from mask to wafer. The process is very simple, the mask creates a shadow when interposed between a wafer and a beam of collimated UV light. In principle duplication by contact has no limitations in resolution and complexity[22], but practically it has many drawbacks: (i) the wear of the mask surface; (ii) the difficulty of accurate registration since relative motion between mask and wafer is impossible; (iii) a limitation in resolution because neither mask nor wafer are perfectly flat (Fig.16); (iv) a limitation in complexity due to imperfect registration, and deformation of a non perfectly flat surface during contact.

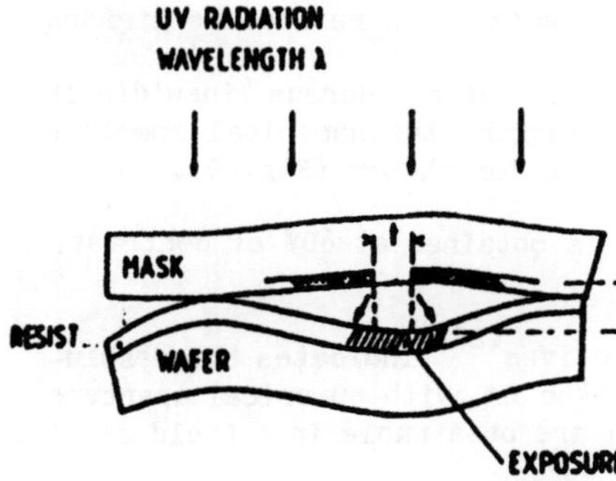

Fig.16: Bow of resolution by diffraction through the gap between mask and wafer

If a residual spacing S exists between mask and wafer for a wavelength λ the lower limit for the linewidth t is given by a practical formula[9] where all dimensions are in meters

$$t = 15\sqrt{\frac{\lambda\ S}{2000}}$$

For w = 1 µm and λ = 0.4 µm, the spacing must be less than 3 µm, which is frequently difficult in view of the non-flatness of the wafer. This has many causes, in particular the polishing defects, the thermochemical process and the lithography steps on the surface (dielectric, conductor).

Some progress has recently been made by using smaller wavelengths of light between 222 nm and 308 nm produced by excimer lasers with the advantage of a very short exposure time corresponding to a few pulses of 10 ns each[23]. At a wavelength of 220 nm, a linewidth of 0.3 µm on 1.1 µm thick PMMA had been presented.

3.2.1b Projection Systems

To avoid drawbacks of the contact process it is possible to use UV light in projection systems. These systems use optics to transfer the image of a mask at scale 1 or a reticle at scale 5 or 10, on the surface of the wafer. The optics may be refractive or reflective (Figs.17, 18).

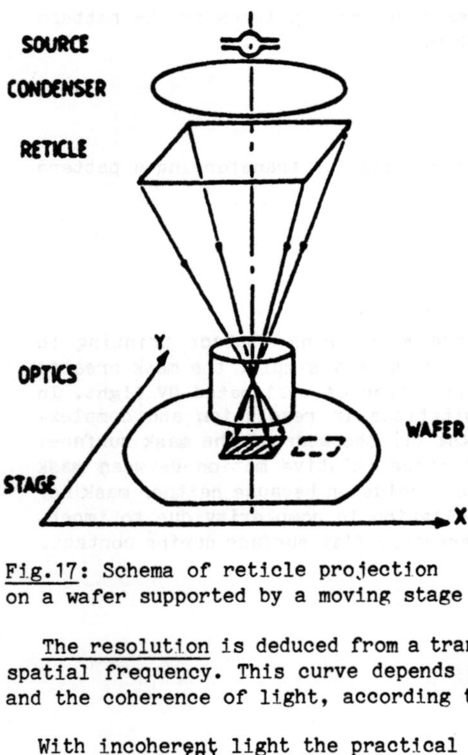

Fig.17: Schema of reticle projection on a wafer supported by a moving stage

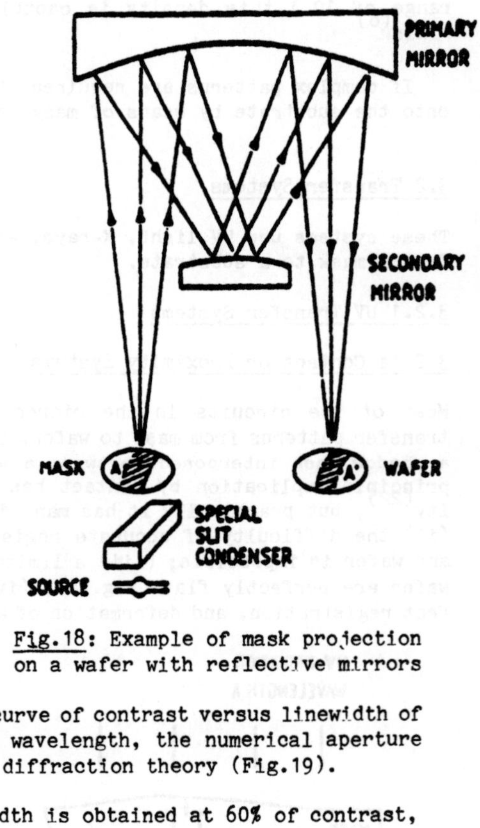

Fig.18: Example of mask projection on a wafer with reflective mirrors

The resolution is deduced from a transfer curve of contrast versus linewidth of spatial frequency. This curve depends on the wavelength, the numerical aperture and the coherence of light, according to the diffraction theory (Fig.19).

With incoherent light the practical linewidth is obtained at 60% of contrast, by the relation[9]: $t = \lambda/1.28\ NA$

A review of some of the modern refraction objectives[24] indicates the resolution and complexity obtainable. At a wavelength of 436 nm, with numerical aperture of 0.35 and a 5x reduction, linewidths of 0.75 µm are obtainable in a field of 21 millimeter diameter.

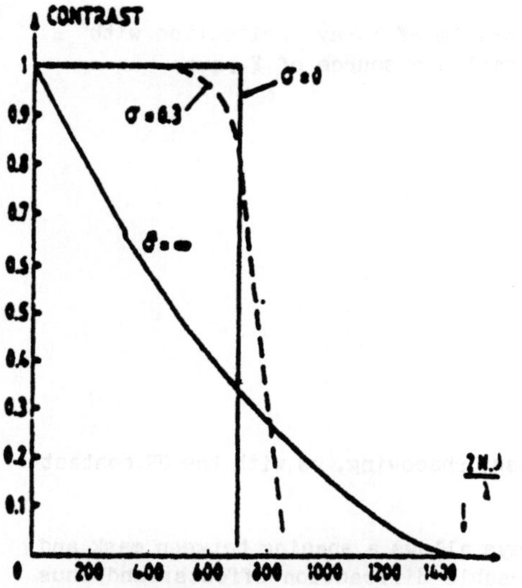

SPATIAL FREQUENCY (line pair per millimeter)

Fig.19: Example of curve of contrast versus spatial frequency for lens with numerical aperture (N.A.) = 0.3, $\lambda = 0.436\ \mu m$, σ : coherence parameter

At wavelength of 365 nm with a numerical aperture of 0.35 and a 10x reduction, linewidths of 0.65 µm are obtainable in a field of 14 millimeter diameter.

Research is underway aimed at decreasing the wavelength 220 nm for use with excimer laser sources.

Linewidths of the order of 0.5 µm across a field of one centimeter square are predicted.

Large numerical apertures are obtainable with microscope objectives but the image field is very small, i.e. of the order of 100 µm.

Decreasing the wavelength below 220 nm seems particularly difficult, since the index of materials becomes very small with increasing absorption.

It is possible that reflective optics could be used but the number of adjustable design parameters is small, as a result of which it is necessary to tolerate aberrations and small fields.

Reflective objectives are used at 313 nm for transferring a mask at actual size onto a wafer with practical resolution in the micron range [25]. Some research is underway for using shorter wavelengths in order to reach the submicron range with reflective objectives.

At the present time submicron linewidths have only been obtained with refractive objectives [26].

The throughput depends on the projection time, the time of stage motion, the alignment time, resist sensitivity, etc.

Bright sources of UV light and efficient optics are necessary and among these the most promising sources are excimer laser sources [27].

With commercial resist, the throughput of wafer steppers, in the micron range, is of the order of a few tens of wafer levels per hour which is about ten times higher than the fastest electron beam pattern generators.

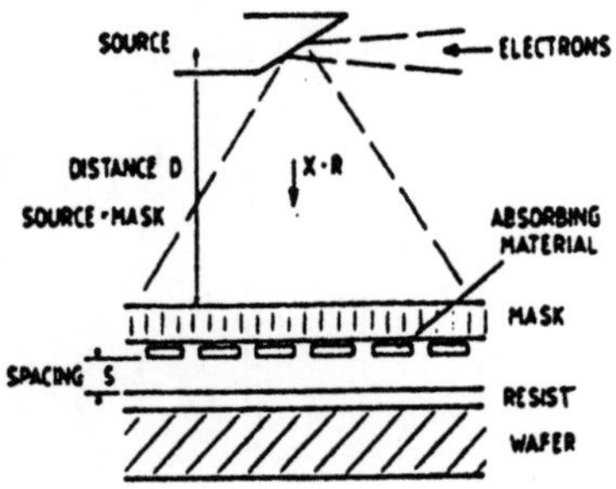

Fig.20: Principle of X-ray duplication with thermoionic source of X-rays

3.2.2 X-Rays Transfer Systems

These systems operate by contact or near contact shadowing, as with the UV contact system (Fig.20).

However the shorter wavelength of the X-rays allows a spacing between mask and wafer of a few tens of microns with no noticeable diffraction effects; and thus permits a relative displacement between mask and wafer for alignment.

An X-ray duplicating system includes: (i) a source with the highest possible brightness emitting a narrow band, in the range 5 Å to 30 Å of wavelength; (ii) a special mask suited to X-rays having good contrast, resolution, dimensional stability, and sufficiently transparent; (iii) a wafer holder with accurate displacement; (iv) an automatic alignment system for registration; (v) a wafer coated with special resist.

These systems are very simple in principle and represent a development of the UV contact systems although the area to be duplicated is limited for various reasons[28]: (i) stability in the range of 10^{-6} (50 nm on 5 cm field) is required and on a laboratory scale 5.10^{-6} has been reached; (ii) conical distortion due to finite distance of the source; (iii) problems of realizing masks of large diameter; (iv) waviness of the wafer surface producing spacing variations and nonreproducible distortion; (v) in practice, in the submicron range, step-and-repeat systems are necessary with steps of a few centimeters.

3.2.2a Resolution

It depends on many parameters: (i) the diffusion of photoelectrons created by the X-rays in the metallized mask and the coated substrate; (ii) the diffraction of X-rays the importance of which increases with spacing between mask and wafer[29]; (iii) the penumbral effect due to the finite size of the source and the finite thickness of the absorbing material of the mask; (iv) the properties of resist.

D.C. Flanders[30] has shown the possibility of duplicating lines and spaces of 175 Å in PMMA, in special conditions, although linewidths of microcircuits are limited to a few tenths of a micron[28] for practical conditions of duplication by X-rays.

Complexity is limited by the aforementioned limitation of surface replication, the accuracy of alignment, and the number of defects per mask: (i) for 0.15 µm per centimeter, of mask distortion and a linewidth of 0.2 µm: field size is limited to about 7.5 mm; (ii) accuracy of alignment of 0.01 µm has been measured with a prototype[31]. This accuracy, if confirmed on a practical system, gives a possibility of overlay and stitching subfields, compatible with linewidths lower than 0.1 µm; (iii) the number of defects per mask is due to the complex fabrication process, and is less than ten per cm^2 at the present time. This tends to limit the possibility of stitching subfields with a good yield but there is a lack of experimental results in this area.

3.2.2b Throughput

The time of exposure, for a resist, depends on the sensitivity in energy per unit area and the density of power received by the source of X-rays.

With PMMA resist, linewidth of a few tenths of a micron, spacing of about 40 µm, and conventional thermionic sources having an electronic power of a few kilowatts, exposure is around 1 hour[31].

This time is too long for practical industrial use and research is in progress towards more sensitive resists and higher brightness sources.

3.2.2c High Brightness Sources

Two kinds of high brightness sources are studied: plasma sources and synchrotron sources[32].

Plasma sources of soft X-rays use electrical discharges in an inert gas such as krypton or argon[33].

The sources have an efficiency about hundred times that of a conventional source.

The X-ray emission is pulsed with a pulse length of a few tens of a nanosecond and a pulse repetition frequency of the order of 1 Hz.

The X-ray energy per pulse is a few kilojoules and the emitted spectrum is very large and needs filtering.

The radiation is not directional.

A factor 10 in exposure time has been achieved[33] experimentally.

Merits of these sources are the brightness, the efficiency, the absence of rotating parts and a relatively compact geometry.

However, there are numerous drawbacks: (i) the non-reproducibility of the intensity and the spectrum; (ii) the wide spectrum and consequent need for filtering; (iii) the damage of the filters due to long wavelength radiation; (iv) the wear of the electrodes and the debris, which damages masks and filters.

In fact conventional thermoionic sources and plasma sources suffer from a large spatial spectrum of radiation giving a small spatial efficiency of the order of 10^{-4}, because of the small solid angle of radiation used.

It is necessary to have directional sources of X-rays in order to improve considerably the exposure time.

Some possibilities are the use of soft X-ray mirrors[34] for concentrating the radiation, the use of synchrotron radiation from storage rings or wigglers[28, 32] and perhaps, in the future, X-ray lasers (Fig.21).

The use of a synchrotron has been simulated; and the computed exposure time is between a few seconds and tens of minutes[28, 35].

Some experimental data give a time between minutes and tens of minutes[35] depending on the experimental conditions, for duplicating structures with linewidths larger than 0.2 µm.

One possibility is to use many ports on the perimeter of the storage ring but the complete system becomes cumbersome and very expensive (Fig.22).

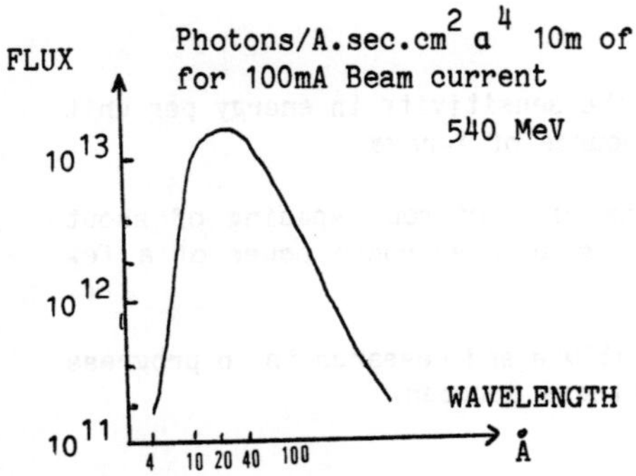

Fig.21: Schema of typical synchrotron radiation

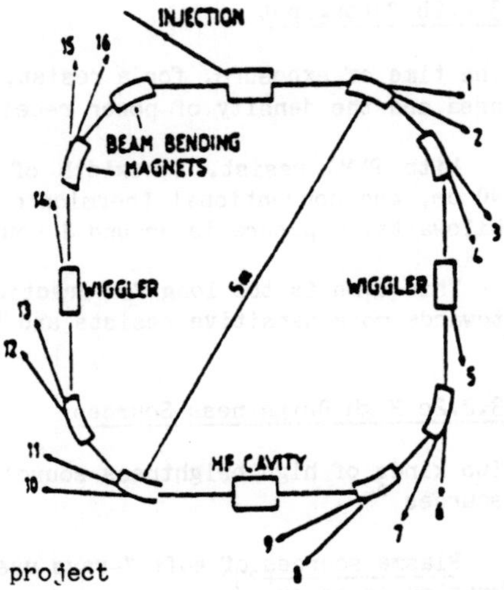

Fig.22: Scheme of a multiport (16) storage ring project

Radiation transition[36] and wigglers give in principle a directional radiation with a very high brightness but there is a lack of experimental results with application to microlithography.

However, even if the power density could be drastically increased on the wafer, some technological problems will limit the exposure time, for instance the energy absorbed in the mask and the corresponding distortion.

3.2.3 Electronic Transfer Systems

Proximity[37], projection at actual size[38], and reduction projection[39] have been studied.

Special masks are necessary to transmit or to emit electrons.

For transmission, masks have holes for self supporting, or some complementary masks are necessary for one pattern.

The masks must be very thin and tend to suffer distortion by electronic heating, and contamination, the corresponding systems often are at a research stage with difficult application in submicron range.

Emitting masks form photoelectric cathodes excited by UV light. Projection could be made by reduction either by electron optics[40] or with no reduction using a special electron optics with magnetic field and electric and a very large image area[41] (Fig.23).

Systems with emitting masks and imaging at actual size have been developed and have shown 0.2 µm linewidth capabilities[39].

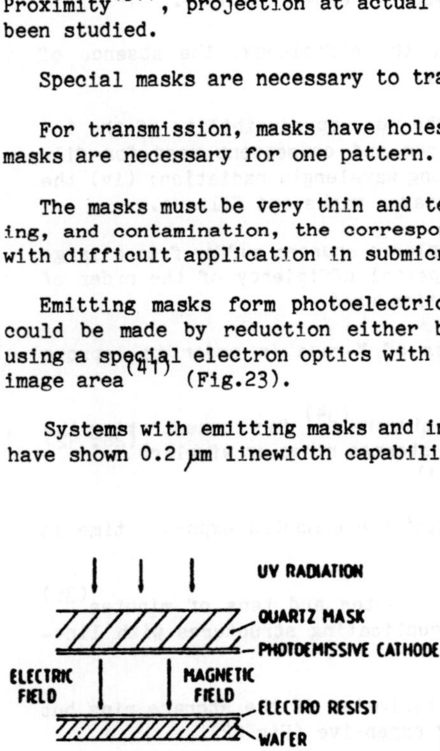

Fig.23: Scheme of electronic planar projector

Advantages of these systems are: (i) a good throughput because the exposure time of a full wafer is of the order of a few tens of seconds[42] giving a capability of more than 25 wafers per hour; (ii) no limitation (in principle) to the complexity from the point of view of image field: (iii) a good resolution, in principle, limited only by electron diffusion in the resist and backscattered electrons in the substrate.

There are also certain drawbacks: (i) distortion of the image due to the waviness of the wafer surface[43], the stray fields, the charge in surface, the non homogeneity of the magnetic and electric field; (ii) non-corrigible sensitivity to random in-plane distortion of the wafer; (iii) limited life and evolution of the cathode; (iv) in-plane contrast variation due to backscattered electrons from the wafer reintroduced in the electron optics; (v) special expensive masks.

It seems that electron projection systems are an interesting solution for special purposes from the point of view of patterns and substrates, but unlikely to find general applications.

3.2.4 Ion Transfer Systems

These resemble electron transfer systems with the same varieties using contact[44] proximity[45, 46, 47, 48] and projection[49].

Advantages of using ion transfer systems are: (i) compared to using X-rays, the possibility of a large spacing because particles are collimated; and the resist sensitivity which is higher with a factor of about 100 in energy doses[50]; (ii) compared to using electrons the absence of backscattering and proximity effects and a high resist sensitivity with a factor of the order of 100 in charge doses[50]. Compared to using visible photons a better resolution for a given spacing due to the small diffraction effect and a larger depth of focus in projection systems due to the smaller acceptance angle of the particles at the surface of resist; (iii) the possibility of using different ions, according to different applications: H, D, He, N, Ne, Ar; (iv) the possibility of using inorganic resist (SiO_2, Si_3N_4) developed by etching, or other inorganic resist as $GeSe_2$ covered with thin layers of Ag_2Se; (v) the masks must be transparent to ions; in some cases a mask is supported by a grid and uses reduction projection[49], in others thin silicon oriented membranes or fine grids are used in proximity printing to support the mask[47, 45] (Fig.24).

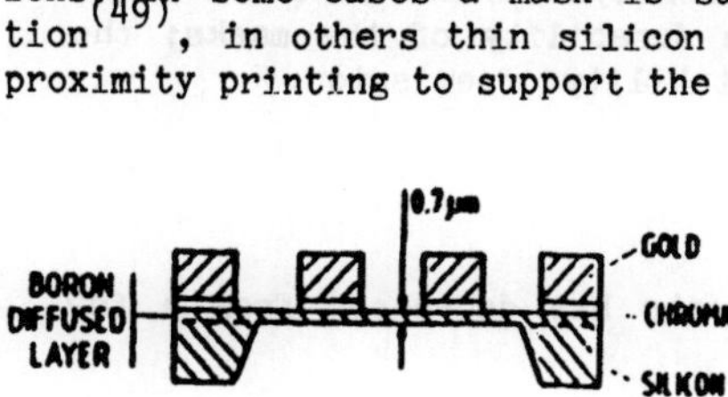

Fig.24 : Channeling mask for ion proximity printing

Resolution in proximity depends on the scattering of ions within the membrane and the quality of optics in projection systems.

Linewidths of a few 1000 Å have been obtained in practice[45, 47].

Contact systems eliminate effects of scattering; a linewidth of more than 4000 Å has been obtained in PMMA by proton bombardment through a gold conformal mask[44] and a polyimide mask[51].

A complete ion wafer stepper has been described[49] with a 10X ion reticle, reducing optics and a laser controlled stage; a linewidth of 0.2 µm is forecast (Fig.25).

Fig.25: Principle of an ion projection stepper (IPS)

Because of the sensitivity of organic resists to ions, shot effects may limit the edge definition at very fine linewidths(21), the use of inorganic resists may resolve this limitation.

The throughput of ion beam proximity systems is high as a consequence of the high sensitivity of the resist to ions.

Exposure times with proximity systems are of the order of one second.

In the projection system, exposure times of 0.01 second for organic resists and one second for inorganic resists have been reported(49).

Complexity is limited by the distortion of the thin masks due to heating by ion bombardment - 0.1 µm of in-plane distortion, for masks of 1 cm, has been measured(17) in case of proximity printing. This distortion limits the practical linewidth to about 0.5 µm for surfaces of the order of 1 cm^2.

All the aforementioned properties of ion transfer systems are attractive but the main problems remain: the stability and the durability of the masks; the quality of source from the point of view of the stability of emission.

4. CONCLUSIONS

Over the last fifty years, the size of components has decreased from a few centimeters to a few microns.

At the present time, most advanced commercial circuits use a resolution of about one micron and complexity of around 1 000 000 transistors.

This race for smallness and complexity is justified by the increasing performance in speed; density of functions and cheapness of the functions.

Four decades separate the actual circuit resolution from atomic dimensions, and the problems are how far will the race go and with what kind of lithography tools.

It seems that the use of UV light is limited to linewidths larger than 0.5 µm for transferring mask patterns of reasonable complexity and throughput.

The X-ray, electron or ion systems give the possibilities of transferring masks with linewidths larger than 0.2 µm and circuit sizes of a few millimeters, and with an interesting throughput.

Below 0.2 µm the use of masks seems very difficult and direct writing by focused particles is a necessity.

The quality of ion sources and electrostatic optics limits the use of focused ions for linewidths larger than 0.1 µm.

The focused electron systems may have spots of few Angströms and are able to write artificial structures at the limit of stability, near 10 Å.

Although some tools exist for closing the resolution gap between 1 µm and 10 Å, the problems of complexity and throughput become more and more difficult as the linewidth decreases because of the granularity of radiation.

Even if there is some hope of the microcircuit industry developing into the range between 1 µm and 0.2 µm, it seems unlikely that artificial complex circuits with linewidths below 0.1 µm could be made with acceptable throughput using the existing tools and conventional materials.

REFERENCES

1. Special Issue on Pattern Generation and Microlithography, IEEE E-D, ED-22, N°7 (75)
2. G. PIRCHER: Proc. of 9th Int. Vac. Congress and 5th Int. Conf. on Solid Surfaces, Invited and Speakers' Volume p.427 (1985)
3. Integrated Circuit Engineering, Basic Technology, Boston Technical Publishers Inc. (1966)
4. MC WILLIAMS, J.P. HERMAN, F. MITLISKY, R.A. HYDE, L.L. WOOD: Appl. Phys. Lett. 43, 10 (1983)
5. J.M. BALLANTYNE: Some Examples of Microfabrication Techniques for Submicron Devices., Microcircuit Engineering - Grenoble (1982)
6. M. ISAACSON, A. MURRAY: J. Vac. Sci. Technol. 19 (4), (Nov/Dec 1981)
7. A.N. BROERS, J.M.E. HARPER, W.W. MOLZEN: Appl. Phys. Lett. 33, 5 (1978)
8. H. NIEDRIG: J. Appl. Phys. 53 (4), (1982)
9. A.N. BROERS: IEEE Trans. ED-28, N°1 (1981)
10. K. LEE, H. AHMED: J. Vac. Sci. Technol. 19 (4), p946 (1981)
11. L.D. JACKEL, R.E. HOWARD, E.L. HU, D.M. TENNANT, P. GRABBE: Appl. Phys. Lett. 39 (3), p368 (1981)
12. F. EMOTO, K. GAMO, S. NAMBA, N. SAMOTO, R. SHINIZU, N. TAMURA: Microcircuit Engineering 85 (Rotterdam)
13. A.N. BROERS: J. Electrochem. Soc. Solid State Science and Technol. 128 N°1, p166 (Janv. 1981)
14. M.E. MOCHEL, C.J. HUMPHREYS, J.A. EADES, J.M. MOCHEL, A.M. PETFORD: Appl. Phys. Lett. 42 (4), p392 (1983)
15. G. GARRY, D. DIEUMEGARD, M. CROSET, P. SUDRAUD, J. VANDE VALLE: Microcircuit Engineering, p154 - Grenoble (1982)
16. R.L. SELIGER, R.L. KUBENA, R.D. OLNEY, J.W. WARD, V. WANG: J. Vac. Sci. Technol. 16 (6), p1610 (1979)

17. L. SWANSON: Microcircuit Engineering, p267 - Amsterdam (1980)

18. K. GAMO, T. UKEGAWA, Y. INOMOTO, Y. OCHIAI, S. NAMBA: J. Vac. Sci. Technol. 19 (4), p1182 (1981)

19. G.R. HANSON, B.M. SIEGEL: J. Vac. Sci. Technol. 19 (4), p1176 (1981)

20. G.M. MLADENOV, B. EMMOTH: Appl. Phys. Lett. 38 (12), p1000 (1981)

21. H. RYSSEL, K. HABERGER: Microcircuit Engineering - Lausanne (1981)

22. U. CH. FISCHER, H.P. ZINGSHEIM: Appl. Phys. Lett. 40 (3), p195 (1982)

23. K. JAIN, C.G. WILSON, B.J. LIN: SPIE, 334, p259 (1982)

24. J.H. BRUNINGS: Microcircuit Engineering 85 (Rotterdam)

25. J.P. ROBIC, S. KNIGHT, W. STRAUB: Microcircuit Engineering, p79 - Grenoble (1982)

27. G.M. DUBROEUCQ, D. ZAHORSKI: Microcircuit Engineering, p73 - Grenoble (1982)

28. A. HEUBERGER: Microcicuit Engineering 85 (Rotterdam)

29. J. TROTEL, B. FAY: Colloque International sur la Microlithographie, p201 Paris (1977)

30. D.C. FLANDERS: Appl. Phys. Lett. 36 (1), p93 (1980)

31. B. FAY, A. CORNETTE, J.P. NIVOLIERS: Microcircuit Engineering, p116 - Grenoble (1982)

32. N.P. ECONOMOU, D.C. FLANDERS: J. Vac. Sci. Technol. 19 (4), p.868 (1981)

33. J.S. PEARLMAN, J.C. RIORDAN: J. Vac. Sci. Technol. 19 (4), p1190 (1981)

34. R.P. HAELBICH, A. SEGMULLER, E. SPILLER: Appl. Phys. Lett. 34, p184 (1979)

35. W.D. GROBMAN: Handbook on Synchroton Radiation, I, Ed. by E.E. Kock, D.E. Eastman and Y. Farge, North Holland, Amsterdam

36. M.A. PIESTRUP, P.F. FINMAN, A.N. CHU, T.W. BARBEE, J.R. RICHARD, H. PANTELL, R.A. GEARHART, F.R. BUSKIRK: IEEE QE 19 N°12

37. P. NEHMIZ, H. BOHLEN, J. GRESCHNER: J. Vac. Sci. Technol. 19 (4), p971 (1981)

38. R. WARD, A.R. FRANKLIN, P. GOULD, M.J. PLUMMER: J. Vac. Sci. Technol. 19 (4), p966 (1981)

39. J. FROSIEN, B. LISCHKE, K. ANGER: J. Vac. Sci. Technol. 16 (6), p1827 (1979)

40. R. SPEIDEL, M. MAYER: Microcircuit Engineering, p43 - Aachen (1979)

41. T.W. O'KEEFE, J. VINE, R.M. HANDY: Solid State Electron. 12, 841 (1969)

42. I. MORI, K. SUGIHARA, C. ITOH, MITABATA, T. SHINOZAKI: Microcircuit Engineering 85 (Rotterdam)

43. G.A. WARDLY: J. Vac. Sci. Technol. 12 (6), p1313 (1975)

44. L. KARAPIPERIS, C.A. LEE: Appl. Phys. Lett. 35 (5), p457 (1983)

45. J.N. RANDALL, D.C. FLANDERS, N.P. ECONOMOU, J.P. DONNELLY, E.I. BROMLEY: Appl. Phys. Lett. 42 (5), p457 (1983)

46. L. CSEPREGI, P. EICHINGER, A. HEUBERGER: Microcircuit Engineering, p159 Grenoble (1982)

47. J.L. BARTELT, C.W. SLAYMAN, J.E. WOOD, J.Y. CHEN, C.M. MC KENNA, C.P. MINNING, J.F. COAKLEY, R.E. HOLMAN, C.M. PERRYGO: J. Vac. Sci. Technol. 19 (4), p1166 (1981)

48. U. BEHRINGER, R. SPEIDEL: Microcircuit Engineering, p369 - Lausanne (1981)

49. G. STENGL, R. KAITNA, H. LOSCHNER, R. REIDER, P. WOLF, R. SACHER: Microcircuit Engineering, p345 - Lausanne (1981)

50. H. RYSSEL, H. KRANZ, K. HABERGER, J. BOSCH: Microcircuit Engineering, p293 - Amsterdam (1980)

51. N.P. ECONOMOU, D.C. FLANDERS, J.P. DONNELLY: J. Vac. Sci. Technol. 19 (4), p1172 (1981)

Electron Beam Nanolithography

C.D.W. Wilkinson and S.P. Beaumont

The University, Glasgow, G12 8QQ, United Kingdom

INTRODUCTION

Lithography plays a central role in the fabrication of electronic devices, and is essential in the preparation of samples for studies of transport in 1-D. The minimum feature size required in the device is important in the choice of lithographic method. Generally for linewidths above 1μm linewidth optical lithography is used whilst for smaller dimensions it is necessary to use electron, ion beam or X-ray lithography. The linewidths in current production VLSI circuits range between 1 and 3μm; there are programs in the U.S and the U.K to reduce the linewidth to 0.5μm over the next few years [1]. Single conventional semiconducting electronic devices have been made with gates as small as 0.1μm, and some experimental superconducting devices employ features with sizes in the 10-30nm range.

The performance of field effect transistors depends on the width of the gate and marked deviation from conventional behaviour can be expected in silicon for gate widths of the order of 0.1 to 0.3μm and in GaAs for gate widths less than 0.1μm.

Very small structures with dimensions of the order of tens of atoms exhibit effects not seen in larger scale structures.

Overall, the challenge of making very fast electronic devices and of exploring the behaviour of material at sizes where quantum mechanical behaviour is directly observable, has led to wide international interest in very high resolution lithography.

1. BASIC LITHOGRAPHIC PROCESS

Lithography is used to delineate a pattern in metal, or in a dielectric or in doping concentration on a semiconducting substrate. A radiation sensitive layer (often polymeric) is deposited upon the substrate, and exposed to the radiation. If in the subsequent development, the areas exposed to the radiation are removed, the resist is said to have positive tone (see Fig.1a), whereas if the exposed areas are hardened by exposure and the unexposed region removed in the developer, the resist has negative tone (see Fig.1b). The patterned resist is then used in the subsequent processing in one of two ways (etching).

In subtractive processing (etching) (see Fig.1d) the resist is used to protect the material underneath it, while any unprotected material is removed by chemical action (acids or alkalis), or by etching in a reactive gas or by inert gas ion bombardment.

In additive processing (lift-off or stencil processing) (see Fig.1c) the material to be patterned is deposited onto the patterned resist and underlying substrate, and the resist is dissolved away, taking with it the material deposited on the resist but leaving behind the metal deposited on the substrate. For lift-off to be successful, the resist solvent must be able to penetrate into the resist, and thus, ideally, the metal coating on the resist must not be continuous with the metal coating on the substrate (see Fig.2). Lift-off is most readily performed when the resist profile is

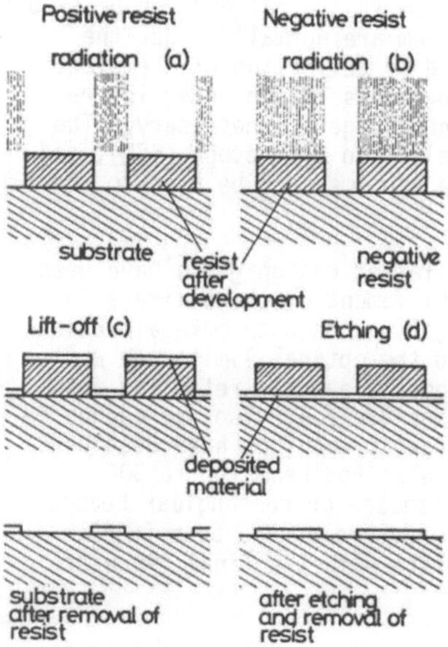

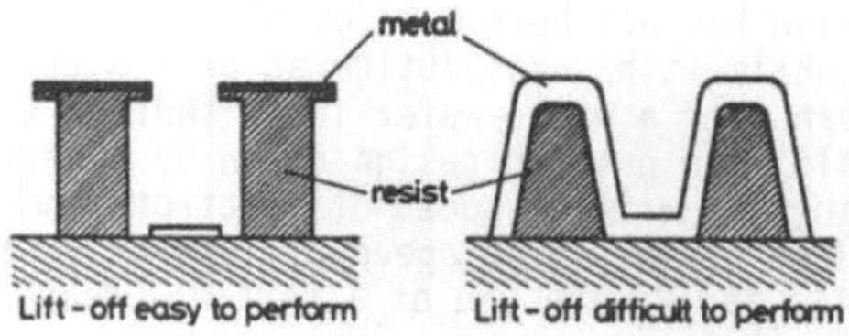

Fig.2. Lift-off ideally requires an undercut resist profile

Fig.1a-d. (a) In positive tone resists exposed material is removed by developer. (b) In negative tone resist exposed material is hardened by exposure and unexposed material is removed by developer. (c) Lift-off or additive processing. (d) Subtractive or etching processing

slightly undercut and when the material is deposited in a collimated beam. Plating through a resist pattern on an underlying plating base deposit is an interesting variation of additive processing which does not require an undercut resist profile.

Lift-off processing allows the definition of a high resolution pattern in materials for which no high resolution etching technique is available. Lift-off processing in electron beam lithography on thin substrates [2,31,32] has been demonstrated to have very high resolution (16nm lines). For lift-off processing to be successful the undercut profile established in the resist during exposure and development must be preserved during the deposition of metal. Using polymeric resists this implies that the resist temperature during metal deposition must be maintained well below the glass transition temperature of the polymer.

2. EXPOSURE METHODS

2.1 Optical Exposure

Photolithography is the workhorse of the silicon microelectronics industry. However the resolution of optical lithography is limited by the wavelength of light and the practical limit at the moment appears to be around 0.5μm lines with 0.5μm gaps printed in positive photoresist.

2.2 Electron Beam Exposure

An electron beam can be focused to a spot of diameter less than a nanometre, and so provides a potentially very high resolution means of exposing an electron resist. An electron beam lithography machine consists of a source of electrons, a series of condensor lenses to collect the electrons and an objective lens to focus the electron beam in a narrow probe onto the resist [3]. Deflection coils are also provided so that the electron probe can be scanned over the specimen and some means of turning the electron beam on and off (blanking) is essential so that a pattern can be exposed in the resist. The deflection and blanking of the beam are normally under the control of a computer. The area which can be exposed by deflection of the electron beam is limited; to expose larger areas accurate movements (often under interferometric control using a laser) of the specimen on a stage are necessary. The machine is similar in overall design to a scanning electron microscope (SEM), and as in an SEM, secondary electrons scattered from the specimen can be used to form an image on a CRT.

Many electron beam lithographic machines with different design goals have been built in the past ten years. Until recently the achievement of the ultimate in resolution has not been a major aim, rather the target has been to make a machine which, while it has resolution at or a little beyond the optical limit, can write a pattern over a whole wafer in as short a time as possible while retaining the flexibility in pattern design given by direct computer control. Two approaches in improving the writing speed of electron beam lithographic machines have been used 1) the use of very high speed deflection of a round electron beam (up to 300 Mb/s [4]) and 2) the use of a variably shaped beam (square or rectangluar beam at will) so that a larger area can be written at each exposure of the beam [5,6]. The beam size employed in such machines is commensurate with the final linewidth desired and so is rarely less than 50nm.

For very high resolution (linewidths below 0.1μm), modified SEM's and Scanning Transmission Electron Microscopes (STEM) have been employed [7,8,9]. Speed of writing is not of the essence but a spot size smaller than the narrowest line to be written is clearly vital. In a SEM usually the specimen is placed at the focus of the electron beam which is some tens of millimetres from the objective lens pole piece and in many SEM's large objects (such as 2" diameter wafers) can be accommodated on the stage. With such an objective lens the minimum probe size which can be obtained is of the order of 2-10nm (depending particularly on the choice of electron source-tungsten filament, LaB_6, etc.).

The STEM is a modification of a conventional transmission electron microscope (TEM) and as with a TEM it is customary to use specimens which are transparent to an electron beam. In a STEM the beam after passing through the specimen is often refocused onto an electron detector. In order to form the smallest possible probe, an immersion objective lens in which the specimen is placed in the middle of two coils is commonly employed. In such an objective lens a probe of diameter 0.5-2nm can be formed; however, space limitations often restrict the specimen size to 10mm or less.

In high resolution machines the distribution of electrons in the focused probe is of great importance. Ideally this should be Gaussian and this Gaussian profile should be followed accurately, far away from the centre of beam. Any additional background current leads to a reduction of contrast in the final image in the electron resist.

In order to obtain an adequate writing speed, the brightness (current per unit area per steradian) of the source should be as high as possible. The highest brightness is obtained with field emission and LaB_6 sources, and tungsten filament sources are less bright; on the other hand, a perfect Gaussian profile is more achieved with a tungsten filament than with other sources.

The shape of the electron beam can be measured by passing the beam over a sharp vertical edge. Such an edge must process very high contrast to an electron beam; an edge formed by anisotropic etching of Silicon has been found to be atomically smooth and very suitable for this purpose, allowing measurement of probe sizes down to 4nm [10].

3. RESISTS

3.1 Electron Beam Resists (a) Positive

Electron irradiation of positive polymeric resists causes degradation of the resist and the formation of fragments of lower molecular weight. The development step consists of the removal of the degraded material by dissolution in a solvent-non-solvent solution which does not dissolve unexposed resist. Under a fixed set of conditions of beam energy, resist thickness, substrate material and developer conditions, positive resists are characterised by a plot of the remaining resist thickness versus the log of the dosage. The contract of a resist is a concept borrowed from photographic practice, a low photographic film reproduces a wide range of greys while a high film is used to give sharp white-black transitions. The contrast γ is often defined [11] as the slope of the linear portion of the thickness versus log(dosage) curve where Q_0 and Q_i are found as shown in Fig.3. For lithographic purposes a high γ is desirable, so that a sharply defined resist pattern can be obtained from a less sharply defined electron dosage distribution.

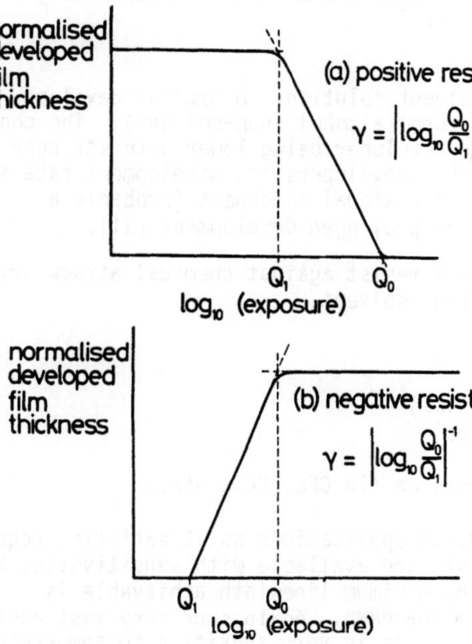

Fig.3. Definition of contrast for positive and negative resist

A widely used polymeric electron resist is poly(methylmethacrylate) - PMMA [12]. It is a glassy solid below 120°C (the glass transition temperature, T_g).

PMMA is normally spin-coated from solution onto the substrate. The uniformity of the coating depends on the rheology of the polymer-solvent system and on the parameters of the spinner (speed, acceleration rate, etc.) [13]. After spinning the PMMA is baked above the glass transition temperature for a number of hours to promote adhesion to the substrate.

Upon exposure to an electron beam the polymer is broken up into fragments of lower molecular weight, the secondary electrons caused by the incident electron beam, breaking bonds within the polymer. The molecular weight distribution of PMMA (Elvacite 2041) as determined by gel permeation chromatography is shown in Fig.4 [15], before and after exposure to a 20kV electron dosage of 50×10^{-6}C/cm^2. The effect of the electron irradiation is to reduce the average molecular weight by a factor of five, permitting development of a relief pattern by dissolution of the

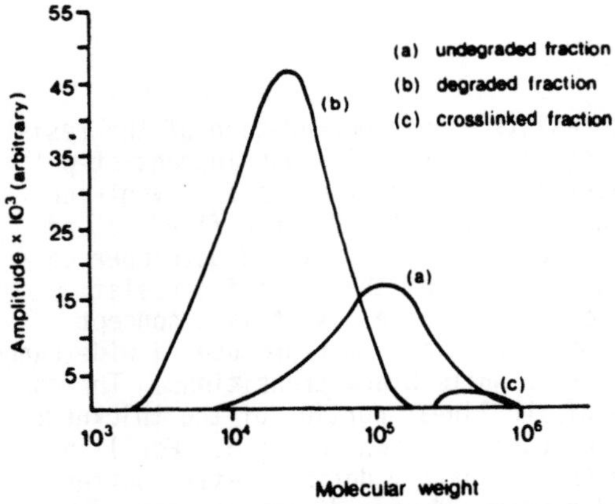

Fig.4. Molecular weight of PMMA after exposure to 50 C/cm^2 of 20kV electrons. Note that while the major effect of exposure is to reduce the average molecular weight a small fraction (2%) of the resist is crosslinked.

degraded polymer in a suitable solvent-non-solvent solution. A typical developer [16] is 1:3 Methyl isobutone ketone (solvent):Isopropol alcohol (non-solvent). The contrast of PMMA varies with the strength of the developer being lower with stronger developers (1:1 MIBK:IPA). However, with weaker developers the development rate is not linear with time for low exposures,- indeed residual thickness (probably a higher molecular weight fraction) remains after prolonged development [31].

After development PMMA is used directly as a resist against chemical attack or, for lift-off processing is dissolved in a strong solvent.

The requirements for a resist are:

1) high sensitivity (to reduce exposure time)
2) good resolution (for desired linewidths)
3) high gamma (γ)
4) good dry etch resistance substrate adhesion (in CF_4, CCl_4 etc.)

PMMA is widely used as a resist for sub 0.2µm applications as it satisfies requirements 2 and 3. More sensitive positive resists are available with sensitivities as low as 1µC/cm ; however with such resists the minimum linewidth achievable is greater than 0.1µm compared to less than 20nm for PMMA. Again such very fast resists use very dilute developers and the development rate is very sensitive to temperature (which is not surprising - high sensitivity implies low activation energy). High dry etch resistance can be obtained using aromatic compounds but it has been observed that higher sensitivity implies lower dry etch resistance in positive resists [17].

Very high resolution positive resist behaviour has been observed in inorganic materials under high current electron beam irradiation. 1.5nm wide lines have been etched into a thin (<50nm) film of NaCl using a 0.5nm electron probe [9], the salt being vapourised by the beam. The dosages used were in the range 5 - 50 C/cm^2. Holes 2nm in diameter and lines 2nm wide have been drilled into metal beta - aluminas [18]. The 2nm hole can penetrate a 200nm thick film albeit at a dosage of 5×10^6C/cm^2.

(b) Negative Resists

Electron beam irradiation of negative resists causes toughening of the exposed area so during development the unhardened material is removed. In polymeric negative resists, crosslinking of a low molecular weight starting polymer occurs. As chemical groups with high radiation sensitivity such as vinyls and epoxies can be introduced into the resist, negative resists are usually more sensitive than positive. However, in general, they exhibit lower contrast (γ) and also lower resolution.

Many negative resists are susceptible to swelling during development; this can lead to pattern distortion. High resolution can be obtained with negative resists if very low molecular weight material is used but at the expense of sensitivity [19]. An interesting limiting case of high resolution, low sensitivity and low molecular weight is the use of hydro-carbon contamination in a scanning transmission electron microscope as a negative resist. 8nm wide lines [8] were etched in W-Re using this resist.

Non-polymeric negative resist systems have been investigated and found capable of high resolution. The chalcogenide glasses As-S and Ge-Se can be deposited by evaporation or sputtering, sensitized by a silver layer (of Ag_2S or Ag_2Se) formed by immersion in silver nitride or silver cyanide, and upon exposure to an electron beam, regions of silver rich glass are formed which are insoluble in a mild alkali developer. These systems can also be dry developed in CF_4 on SF_6 by reactive ion etching. The contrast is very high and 30nm wide lines [52] can be obtained in the As_2S_3/Ag_2S system. The sensitivity is low ($5 \times 10^{-3} C/cm^2$) but acceptable for very high resolution applications.

3.2 Statistical Considerations in Image Formation

Suppose that in order to expose a given area of resist, N primary electrons, X-ray photons or ions are required, then the statistical nature of the emission of the particles implies that the fluctuation in the number of particles will be of the order of $\sqrt{N}$. Clearly, for adequate definition of the final image N must be large, certainly not less than a few hundred. For 16nm wide lines, made by e-beam exposure at 50kV [2], an exposure dose of $3 \times 10^{-4} C/cm^2$ was used implying that 2000 electrons were used to expose a 10 by 10nm area. Clearly, this resist would not be suitable for the definition of a 1 by 1nm pattern even if all other constraints on resolution were removed.

In Fig.5, the number of particles per picture element is plotted against the linewidth for resists of various sensitivity. It can be seen that statistical considerations play a much greater role in ion beam lithography at sub 50nm linewidths.

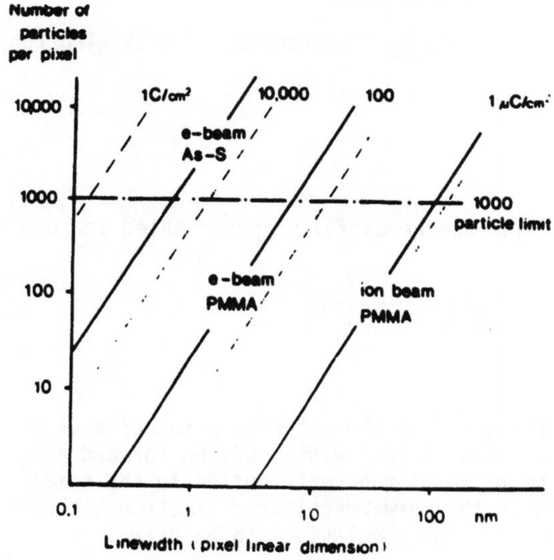

Fig.5. Number of particles per picture element (pixel) against linewidth for resists of different sensitivty. One pixel is taken as the area (linewidth)^2

4. ELECTRON BEAM WRITING

4.1 Interaction of Electron Beam with Resist

When a tightly focused beam of electrons hits a solid material, it is scattered both elastically and inelastically. The elastic collisions change only the direction of the electrons while in the inelastic collisions, energy is exchanged with the material and secondary electrons are formed. The secondary electrons are vital as it is they that expose resist. The incident electrons will spread out as they penetrate the solid until all their energy is lost or they leave the surface after back scattering interactions. A Monte Carlo simulation of the path of 100 electrons initially incident as a point beam of 50kV is shown in a 0.3μm thick resist film on a silicon substrate (see Fig.6). Some electrons penetrate 10 microns into the material while others are scattered back re-emerging in the resist up to 10 microns away from the entrance point of the beam. For comparison the trajectories of a 10kV electron beam is shown in Fig.7. Here more of the original energy is lost in the resist film and the initial beam spreads more widely in the resist.

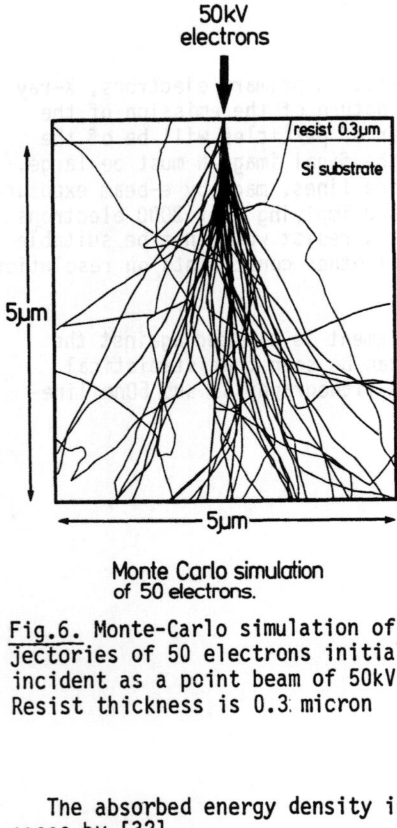

Fig.6. Monte-Carlo simulation of trajectories of 50 electrons initially incident as a point beam of 50kV. Resist thickness is 0.3 micron

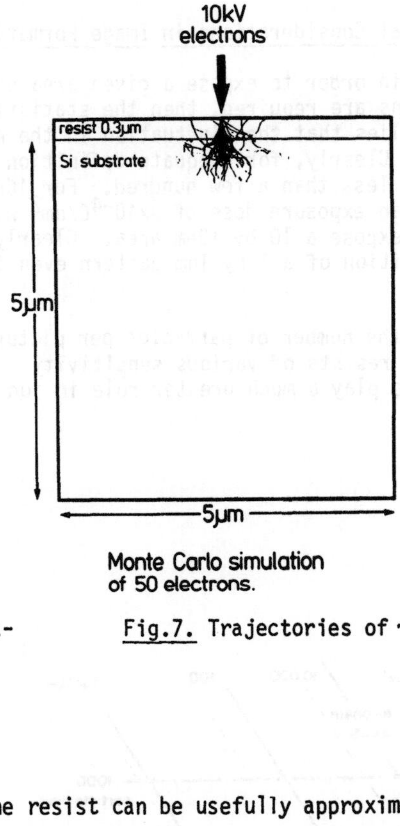

Fig.7. Trajectories of 10kV electrons

The absorbed energy density in the resist can be usefully approximated in some cases by [32]

$$\varepsilon(r) = k \exp\left(\frac{-r^2}{\beta_f^2}\right) + \eta_e \exp\left(\frac{-r^2}{\beta_e^2}\right)$$

where $\varepsilon(r)$ is the energy density at distance r from the entrance point of a very narrow beam, and β_f and β_e are the characteristic half widths of the forward and backscattered distributions and η_e is the ratio of the contribution to the total energy dissipation of the back scattered to the forwardscattered electrons. This model is by no means exact but does enable useful conclusions to be drawn.

At 20kV for a 0.5 m film of PMMA on Si,

$\beta_f \sim 0.08\mu m$ and $\beta_e \sim 2\mu m$ while at 50kV

$\beta_f \sim 0.04\mu m$, $\beta_e \sim 9\mu m$.

For thinner resist β_f is smaller, still, while β_e is essentially unaffected.

The effect of the backscattered electrons is to reduce the overall contrast in the latent image formed by electron exposure in the resist. BROERS [26] has used this model to predict the contrast in the latent image for electron exposure of equal lines and spaces in PMMA on a silicon substrate. For a 1μm thick resist layer on silicon exposed at 20kV, the effects of the backscattered electrons wipe out the contrast in the image for equal spaced lines and gaps less than 0.15μm. Using thin resist on a thick substrate, the backscattered electrons are widely spread and a contrast of 0.5 can be obtained down to a linewidth of 50nm. If the effects of the backscattered electrons are removed entirely (and this can be done by employing a membrane as a substrate), excellent contrast should be obtained for linewidths of 40nm or more.

The double Gaussian model also predicts that very narrow lines should be more easily obtained on a solid substrate at higher voltages so both the width of the forward scattering is reduced and the width of the backscattered contribution is increased.

4.2 Very High Resolution Lithography on Membrane Substrates

It is clear that elimination of the backscattered electrons by using a very thin substrate opens a route to the fabrication of 20nm linewidth structures. BEAUMONT et al. [2] exposed PMMA on a thin membrane (50nm) of carbon using a 50kV probe 8 nm in diameter. After development, the sample was overcoated with 10nm of platinum/palladium and lift-off processing was used to produce metal lines on a carbon substrate which were examined in a transmission electron microscope, (one advantage of using thin substrates is that the samples are immediately ready for examination by TEM at nanometre resolution). The narrowest centre to centre spacing achieved was 55nm, and the linewidth was 16nm.

BROERS [3], using PMMA on an Si_3N_4 membrane, fabricated well-defined 12nm wide isolated lines using a 3nm diameter probe.

For both these experiments, performed on thin substrates, electron backscattering has been eliminated. However, in both cases the linewidth obtained was some 8-9nm wider than the diameter of the electron probe. A major cause of this broadening is the effects of the low energy secondary electrons produced by the 50kV beam. The range of low energy electrons has been measured directly [27]. A low-voltage electron beam was used to expose PMMA through a wire mesh as shown in Fig.8. The depth after development of the exposed regions can be taken as a measure of the range of these low-energy electrons (see Fig.9). The exposure range above 400eV follows closely calculated values for the range of electrons in polyethylene. At lower voltages the range falls to approximately 4+2nm at an energy of 50eV. In order to calculate the exposure dosage across the width of the line it is necessary not only to know the range of a low energy secondary electron but also the distribution of secondary electron energy. It is possible to measure the distribution of energy of the primary beam after passing through a membrane of PMMA - a technique known as energy loss spectrometry. Convolution of the energy loss spectra with the range of the secondary electrons leads to the conclusion that the minimum linewidth obtainable with PMMA should be less than the 10nm reported by different experimenters [26,35].

While the range of the secondary electrons is a major factor determining the ultimate resolution of electron beam lithography, probably resist-dependent factors also play a role. The minimum centre to centre spacing of lines which can be

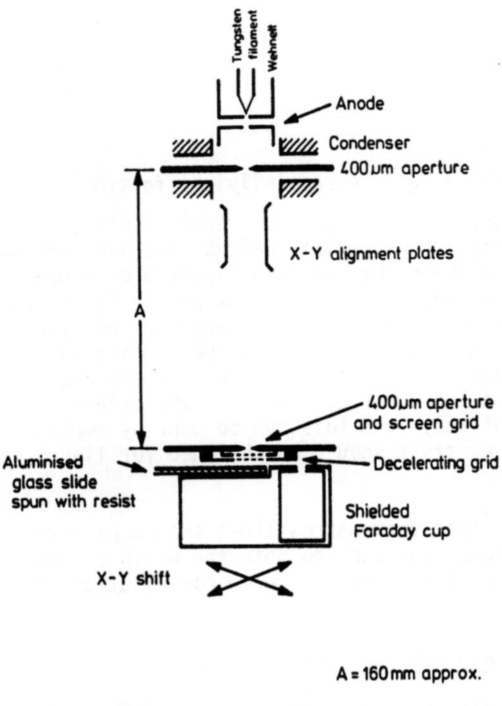

Fig.8. Exposure of PMMA by low-energy electron beam

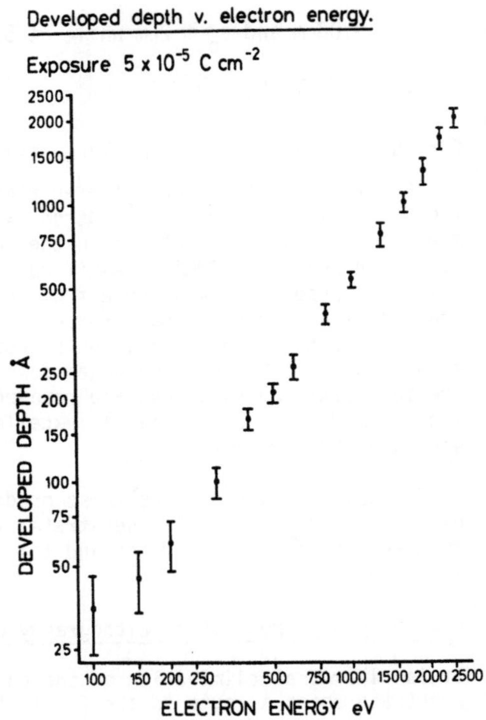

Fig.9. Depth of pit in resist as measured by Talystep after exposure to low-energy electrons

achieved on thin substrates is 45-55nm [2]; this is somewhat wider than twice the minimum linewidth, the expected value if the secondary range were the only factor. It is believed that when PMMA is spin-coated onto a substrate from solution, the polymer chains are not left randomly interwined but they tend to coil so in any small volume (1cm cube) only a limited number of polymer chains will be found. When the exposure of a small volume has reduced the molecular weight of this group of chains sufficiently, all the affected polymers will develop. The present evidence is that such a molecular size limitation does exist in positive polymeric resists - but further study is required to elucidate the details.

4.3 Very High Resolution Electron Beam Lithography on Solid Substrates

On solid substrates coated with a thin layer of resist, back scattered electrons will reduce the contrast in the latent image (see Section 4.1). It has proved possible to compare directly the linewidths of lines written on a solid and a thin substrate [29]. A 60nm thick membrane of Si_3N_4 is prepared by selective etching of silicon and after coating with PMMA, the exposing beam crosses from solid silicon to thin substrate. After metallisation and lift-off, the resulting pattern is protected by resist while a second etch of the silicon leaves the whole pattern on a thin membrane of Si_3N_4 and thus ready for examination in a TEM. Figure 10 shows the linewidths as a function of exposure dose (normalised to the minimum dose required to form a pattern) for thick and thin substrates. It should be noted that on a thick substrate, and for closely spaced lines, the linewidth is a strong function of exposure dosage and lift-off fails completely at 1.3 times the minimum dose while for widely spaced lines the linewidth is the same over the exposure range explored for both thin and thick substrates.

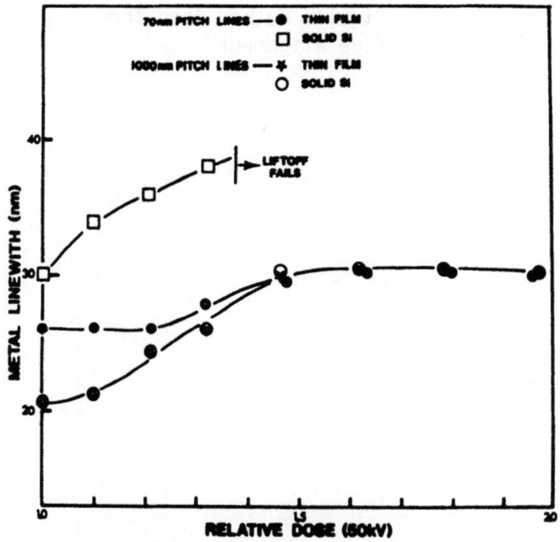

Fig.10. Linewidth as a function of exposure dose for lines of 1 micron and 65nm spacings on thin and thick substrates

HOWARD et al. [7] have produced 20nm lines on 70nm centre to centre spacing at 20kV and 15nm wide lines on 50nm centres at 120kV in gold-palladium using lift-off processing of PMMA on a GaAs substrate. STEM with a 2nm electron probe was used. It should be noted that improved resolution was obtained at the higher electron beam voltages.

Semiconducting materials can be formed into thin membranes which allow semiconducting devices to be made using very high resolution electron beam lithography and for immediate TEM examination of these devices. Membranes of active GaAs 150μm in diameter and 50nm thick have been made by selective etching of multi-layered wafers of GaAs/GaAlAs on a GaAs substrate [29].

4.4 Alignment

As a part of the production of semiconductor devices it is often necessary to align one feature made during one lithographic step with another produced by a subsequent lithographic step. The process of alignment is performed by examining special alignment features created on the surface at earlier lithographic steps. The precise way in which this task is done varies from machine to machine; the common aim is to produce an easily reproduced feature which has high contrast on examination. If the area which can be written under the electronic control of deflection of the beam (the field) is less than the area of the desired device, the stage has to be moved between exposure of fields. In machines designed for feature sizes of 0.1μm or more, a laser interferometer is normally used to provide local dimensions on one level of exposure to a typical accuracy of 0.1μm at the 3σ confidence level. In high resolution machines, it is necessary to produce a series of alignment marks for each frame on each level.

The issues involved in designing an alignment system can be illustrated by consideration of an alignment technique used to obtain re-alignment to better than 4nm on thin substrates [29] using an 8nm probe. Initially a T-bar pattern was fabricated on the substrate using lift-off of gold-palladium. As the substrate was a 50nm membrane of Si_3N_4, the 20nm thick, 20nm lines of gold-palladium produce a high contrast picture viewed using a transmission detector. When re-aligning, it is important to minimise the area scanned so that as much of the surface as possible is available for use. In this case a particular choice of scanning

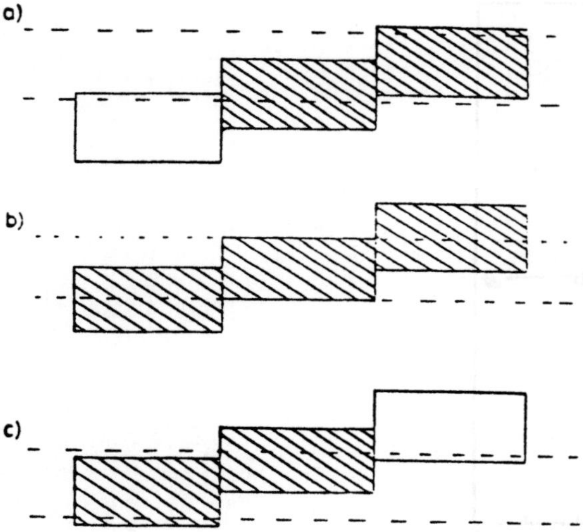

Fig.11. Method used for fine alignment

pattern of the electron beam during re-alignment was used so that alignment with an accuracy of less than the beam width can be achieved. By dead reckoning the stage was moved to approximately the correct position. Fine positioning was achieved using a dog-leg scan as shown in Fig.11. The off-set in scan was one pixel - 4nm in this case. Thus, when perfectly aligned, the top third of the alignment mark appeared light, the middle dark and the bottom third dark.

4.5 Multiple Level Resists

When resist is exposed, developed and then used as a mask in a subsequent etching, the resist is playing two distinct roles; firstly, as a radiation sensitive pattern transfer layer and secondly, as a chemically resistant layer. In high resolution processing these roles may be incompatible - often thin resist is required to achieve the minimum linewidth with a given resist, while thick resist is required to avoid total erosion of the resist during the etching process.

Multiple level lithographic techniques [27-31] have been developed to overcome this difficulty. Here the upper layer is radiation sensitive and is used to create the pattern. The pattern is then transferred into an intermediate layer which itself is used as a mask to allow etching into the thick bottom layer. This final thick layer is used in the subsequent processing. A common choice for this bottom layer is polyimide, an organic polymeric coating material widely used in the electronic industry as it withstands high temperatures (350°C) and attack by many inorganic chemicals. A property of polyimide (which it shares with other organic polymers) is that under reactive ion etching in oxygen, vertical walls are formed and ridges of very high aspect ratio can be formed.

A tri-level resist system [31] was used by JACKEL et al. to make 50nm wide structures in silicon. The silicon wafer was overcoated with a co-polymer of methylmethacrylate and methacrylic acid 280nm thick, then a 25nm layer of Ge was deposited and finally a 100nm thick of PMMA electron beam resist. The resist was exposed and developed to give a mask for the dry etching of Ge in a CF_4 plasma. The thin germanium layer then withstood an oxygen reactive ion etch which cut through the polyimide. The resulting polyimide mask was used in the final step of the fabrication - the formation of 15nm high Ni-Cr lines which were used as the mask in the reactive ion etching of silicon in CF_3Br.

Multilayer resists are also used to create a resist profile which is suitable for lift-off (see Fig.2). A simple example is the use of two layers of PMMA of different molecular weight [29]. In order to obtain a 100nm long aluminium gate for a silicon MOSFET [30], a faster resist (PMMA of molecular weight 85K) was coated to a thickness of 200nm and a 100nm of PMMA of molecular weight of 361K deposited on top. After exposure both resists were developed simultaneously in 1:3 MIBK: IPA, and after lift-off a gate 120nm long, 160nm high was formed.

5. ETCHING

One dimensional structure can be made using additive or subtractive techniques. Additive techniques are very effective in the formation of narrow metallic structures but for the formation of wires in semiconducting material, etching must be used.

In general wet chemical etches remove material isotropically. Thus when etching takes place through a window in an etch resistant layer on top of the layer to be etched, undercutting of the window occurs. If the aspect ratio (that is the ratio of the height to the width) of the desired final feature is small then the undercutting can be tolerated; however, for the high aspect ratio features needed for 1-D experiments and for the gates of sub 0.1 micron FET devices, anisotropic etching is required.

In certain cases anisotropic etching of crystalline materials can be achieved using chemical etches which attach different crystallographic faces at different rates (NaOH attacks the (100) plane of silicon but hardly erodes a (111) plane [4]; however, a more generally applicable method is to use dry etching in a gaseous atmosphere.

In dry etching, material is removed as a volatile product in a vacuum system. The material may be removed by physical means by the bombardment of ions or atoms (sputtering) or by the formation of gaseous product by the chemical action of ions, atom or radicals. The active species can be formed in a variety of ways. In ion beam etching a gas is ionised by a hot filament and the ions are extracted as a beam through a series of charged grids. Alternatively an r.f. discharge is used to ionise the gas. It is helpful to differentiate two regimes of etching using an r.f. discharge. In plasma etching a relatively high pressure gas is ionised by electrodes (which can be outside the vacuum chamber); as the ions suffer many collisions before colliding with the substrate they do so with relatively low energy (10-30eV say) and in a random direction. The substrate is immersed in the plasma and the etching is mainly chemical in nature and is often isotropic. On the other hand in reactive ion etching a relatively low pressure is ionised by an r.f. field between two electrodes. The substrate to be etched is placed on the driven electrode fed capacitively from the r.f. source while the other electrode is earthed and a considerable average a.c. voltage (50-300V say) is built up between the electrodes. This d.c. voltage accelerates the ions so that they have considerable energy on impact with the substrate. The etching is both chemical and physical and the etching is often anisotropic. The distinction between the two regimes is not sharp but is convenient.

The desired pattern is transferred into the substrate using a mask of resistant material; photo resists dielectrics (silicon oxide and nitride) and metal all being used for this purpose.

It is necessary not only that the desired material be removed but that this be done without damage to the remaining substrate and without the addition of residual material (e.g. redeposited substrate material or polymer formed from the etch gases).

The extent of damage due to dry etching can be evaluated by a variety of techniques: photoluminescence [45,46], deep level trap spectroscopy [48], Rutherford backscattering [46], Raman light scattering [44], X-ray photoelectron spectroscopy (XPS) and the measurement of the ideality factor of metal-semiconductor junctions have all been used on semiconductor samples and XPS on polymeric materials [47]. Direct transmission electron microscope examination on suitably thinned specimens is also helpful. The minimum area which can be examined varies from method to method and the examination of the damage on the walls of the etched specimen presents particular problems.

The field of dry etching is large and developing rapidly. In order to give an impression of the factors involved in high resolution etching, the formation of sub 100nm GaAs wires will be considered.

A variety of dry etching techniques are available for GaAs. Purely physical etching in argon is suitable for the formation of 2μm wide 1μm high ridges for optical waveguides [39]. The wall profile depends on the angle of the ion beam to the substrate; at normal incidence a trench is formed (local overetching at the bottom of the wall), at 30°, there is re-deposition close to the walls, while at 22.5^o vertical walls are formed. This suggests that the formation of 0.1μm ridges in GaAs could be difficult by dry etching.

Dry etching in a variety of chlorine containing gases has been reported. With gases such as Freon 12(CCl_2F_2) which contains carbon there is a tendency for a polymeric coating to be formed on the sample, and while this can be removed by the addition of oxygen to the reactive plasma, a single reactive gas must be preferred for its simplicity. $SiCl_4$ is a suitable choice. However, the erosion rate of PMMA in a chlorine containing plasma is high and so the resist itself cannot be used as a mask for $SiCl_4$ etching of GaAs.

The GaAs conducting wires on a semi-insulating substrate shown in Fig.12 were formed by exposing and developing a double layer of PMMA, the lift-off of NiCr to form an ion-resistant mask reactive ion etching of the GaAs in an $SiCl_4$ discharge followed by removal of the NiCr (which is not affected by the $SiCl_4$ etch) in a wet etch.

Fig.12. 150nm wide, 150nm high wire of GaAs on an undoped substrate

The resulting wires have the desired topographic shape - but that is not to say that the GaAs is unchanged by the etching process. Evaluation of the electrically significant damage in this specimen is currently being investigated.

CONCLUSION

Very exciting progress has been made in the past decade in the fabrication of devices with dimensions less than 100nm. Ultimately the atomic nature of matter itself will present the final limitation; a pattern can only be defined to within one atomic spacing.

ACKNOWLEDGEMENTS

Our thanks go to our colleagues at Glasgow University; in the Department of Electronics and Electrical Engineering, Professor John Lamb, Dr. Stuart Mackie, Dr. Steve Rishton, Dr. Calum MacGregor, Dr. Steve Thoms, Dr. Andy Heraud, Dr. Gordon Doughty, Dr. Will Patrick, Mr. Kim Lee, Mr. Clive Rees and Mr. Jim Adams, and in the Department of Natural Philosphy, Professor Bob Ferrier and Dr. John Chapman. The technical assistance of Mr. David Gourlay, Mr. Andy Stark and Mr. Victor Law was important to the work.

REFERENCES

1. J.H. Hains: "The Government Role in VLSI", in VLSI Electronics Microstructure Sciene, Vol.1, ed. by N.G. Einspruch (Academic, New York 1981) Chap.7
2. S.P. Beaumont, P.G. Bower, T. Tamamura, C.D.W. Wilkinson: Sub 20nm wide metal lines by electron beam exposure of thin PMMA films and lift-off. Appl. Phys. Lett. 38, 436 (1981)
3. G.R. Brewer: Electron Beam Technology in Microelectronic Fabrication (Academic, New York 1980)
4. J. Kelly, T. Groves, H.P. Kuo: A high-current high speed electron beam lithography column. J. Vac. Sci. Technol. 19, 936-940 (1981)
5. R.D. Moore, G.A. Caccoma, H.C. Pfeiffer, E.V. Weber, O.C. Woodward: EL-3 a high throughput high resolution lithography tool. J. Vac. Sci. Technol. 19, 950-952 (1981)
6. M. Fujinami, T. Matsuda, K. Takamoto, H. Yoda, T. Ishiga, N. Saitu, T. Komoda: Variable shaped electron beam lithography system Eb-55. J. Vac. Sci. Technol. 19, 941-945 (1981)
7. H.G. Craighead, R.E. Howard, L.D. Jackel, P.M. Mankiewich: 10nm linewidth electron beam lithography on GaAs. Appl. Phys. Lett. 42, 38-40 (1983)
8. A.N. Broers, W.W. Molzen, J.J. Cuomo, N.D. Wittels: Electron beam fabrication of 80 metal structures. Appl. Phys. Lett. 29, 596-598 (1976)
9. M. Isaacson, A. Murray: In-situ vaporization of very low molecular weight resists using 1/2nm diameter electron beams. J. Vac. Sci. Technol. 19, 1117-1120 (1981)
10. S.A. Rishton, S.P. Beaumont, C.D.W. Wilkinson: Measurement of the profile of finely focused electron beams in a scanning electron microscope. J. Phys. E: Sci. Instrum. 17, 296-303 (1984)
11. M.J. Bowden: Electron irradiation of polymers and its application to resists for electron beam lithography. CRC Critical Reviews, Solid State Sci. 8, 223-264 (1979)
12. M. Hatzakis: Electron resists for microcircuit and mask production. J. Electrochem. Soc. 116, 1033-1037 (1969);
R.A. Harris: Polymethyl methacrylate as an electron sensitive resist. J. Electrochem. Soc. 120, 270-274 (1974)
13. W.J. Daughton, F.L. Givens: An investigation of the thickness variation of spun-on thin films. J. Electrochem. Soc. 129, 173-179 (1982)
14. J.M. Shaw, M. Hatzakis: Developer temperature effects on e-beam and optically exposed positive photoresist. J. Electrochem. Soc. 1266, 2026-2031 (1979)
15. V.K. Sharma, R.A. Pethrick, S. Affrossman: Poly(methyl methacrylate): influence of tacticity on its use as an electron resist. Polymer 23, 1732-1736 (1982)

16. J.S. Greeneich: Developer characteristics of poly(methyl methacrylate) electron resist. J. Electrochem. Soc. 122, 970-976 (1975)
17. K. Harada, O. Kogure, K. Murase: Poly(phenylmethacrylate-co-methacrylic acid) as a dry-etching durable positive electron resist. IEEE Trans. Electr. Dev. ED-29, 518-524 (1982)
18. M.E. Mochel, C.J. Humphreys, J.A. Eades, J.M. Mochel, A.M. Petford: Electron beam writing on a 20nm scale in metal-aluminas. Appl. Phys. Lett. 42, 392-394 (1983)
19. T. Tamamura, K. Sukegawa, S. Sugaward: Resolution limit of negative electron resist exposed on a thin film substrate. J. Electrochem. Soc. 129, 1831-1835 (1982)
20. J.M. Moran: High resolution resist patterning using reactive ion etching techniques. Solid State Technol. 24, 195-198 (1981)
21. M. Hatzakis, J. Paraszczak, J. Shaw: "Double Layer Resist Systems for High Resolution Lithography", in Proceedings of Microcircuit Engineering 81 (Lausanne, Switzerland) p.386-396
22. S.P. Beaumont, T. Tamamura, C.D.W. Wilkinson: "A Two-Layer Resist System for Efficient Lift-Off in Very High Resolution Electron Beam Lithography", in Proceedings of Microcircuit Engineering (Delft University Press, Amsterdam 1981) p.381-388
23. C.E. Binnie, S.P. Beaumont, C.D.W. Wilkinson, J.C. White: "The Fabrication of Very Short Gate Length n-Channel MOSFET's by Direction Electron Beam Exposure", in Proceedings of Microcircuit Engineering 82 (Grenoble, France)
24. L.D. Jackel, R.E. Howard, E.L. Hu, D.M. Tennand, P. Grabbe: 50nm silicon structures fabricated with trilevel electron beam resist and reactive-ion etching. Appl. Phys. Lett. 39, 268-270 (1981)
25. T.H.P. Chang: J. Vac. Sci. Technol. 12, 1271 (1975)
26. A. Broers: Resolution limits of PMMA resist for exposure with 50kV electrons. J. Electrochem. Soc. 128, 166-170 (1981)
27. S.A. Rishton, S.P. Beaumont, C.D.W. Wilkinson: "Measurement of the Effect of Secondary Electrons on the Resolution Limit of PMMA", Proceedings of Microcircuit Engineering 82 (Grenoble, Switzerland)
28. S.P. Beaumont, B. Sing, C.D.W. Wilkinson: "Very High Resolution Electron Beam Lithography - Thin Films on Solid Substrates", in Proceedings of the 10th Electron and Ion Beam Sci. Technol. Conference (Montreal, Canada 1982)
29. W.S. Mackie, S.P. Beaumont, C.D.W. Wilkinson, J.S. Robert: "High Resolution Lithography on Thin Active Semiconductor Membranes", in Proceedings of the 10th Electron and Ion Beam Sci. Technol. Conference (Montreal, Canada 1982)
30. M.P. Lepsetter, W.T. Lynch: "Resolution Limitation for Submicron Technology", in VLSI Electronics, ed. by N.G. Einspruch (Academic, New York 1981)
31. S.P. Beaumont, P.G. Bower, T. Tanamura, C.D.W. Wilkinson: "Replication of Very High Resolution e-Beam Written Masks by Carbon k X-Ray Contact Printing," in Proceedings of Microcircuit Engineering 81 (Lausanne, Switzerland)
32. H.G. Craighead: 10nm resolution electron beam lithography. J. Appl. Phys. 55, 4430-4435 (1984)
33. S. Mackie, S.P. Beaumont: High resolution lithography. Solid State Technol. (August 1985)
34. D.B. Lee. Anisotropic etching of silicon. J. Appl. Phys. 46, 4569-4574 (1969)
35. S. Semura, H. Saitch, K. Asakawa: Reactive ion etching of GaAs in CCl_4/O_2. J. Appl. Phys. 55, 3131 (1984)
36. H. Yamada, H. Ito, H. Inaba: Anisotropic reactive ion etching of GaAs and AlGaAs materials for integrated optical device fabrication. J. Vac. Sci. Techol. B3, 884-888 (1985)
37. M. Kawabe, N. Kanzaki, K. Musuda, S. Namba: Effects of ion etching on the properties of GaAs. Appl. Opt. 17, 2556-2561 (1978)
38. G.A. Lincon, M.W. Geis, S. Pang, N.N. Efrenow: Large area ion beam assisted etching of GaAs with high etch rates and controlled anisotropy. J. Vac. Sci. Technol. B1, 1043-1046 (1983)
39. A.P. Webb, C.D.W. Wilkinson: Ion beam etching GaAs for integrated optical applications. Vacuum 34, 159-162 (1984)

The Spin Coating Process Mechanism

A. Weill

Centre National d'Etudes des Télécommunications, BP 98,
F-38243 Meylan Cedex, France

1. THE PROCESS

The spin coating process is extensively used, particularly in microelectronics, for coating planar or pseudo-planar surfaces. Spin coating has long been accepted as the best coating method for obtaining a thin (0.5-2 µm), uniform, adherent, homogeneous film over the wafer. It is accomplished by floading the substrate with a polymer solution and rotating it at a constant speed (1000-8000 rpm) until the solvent is evaporated.

2. INFLUENCE OF PARAMETERS ON THE THICKNESS

2.1 Rotating Speed

For a given polymer solution and a given equipment, the thickness of the film depends on the rotating speed. Figure 1 gives the example of a spinning curve showing the final polymer layer thickness, t, as a function of rotating speed, ω.
Several papers have already noted that Log(t) versus Log(ω) is always linear

$$t = k\omega^{a}$$

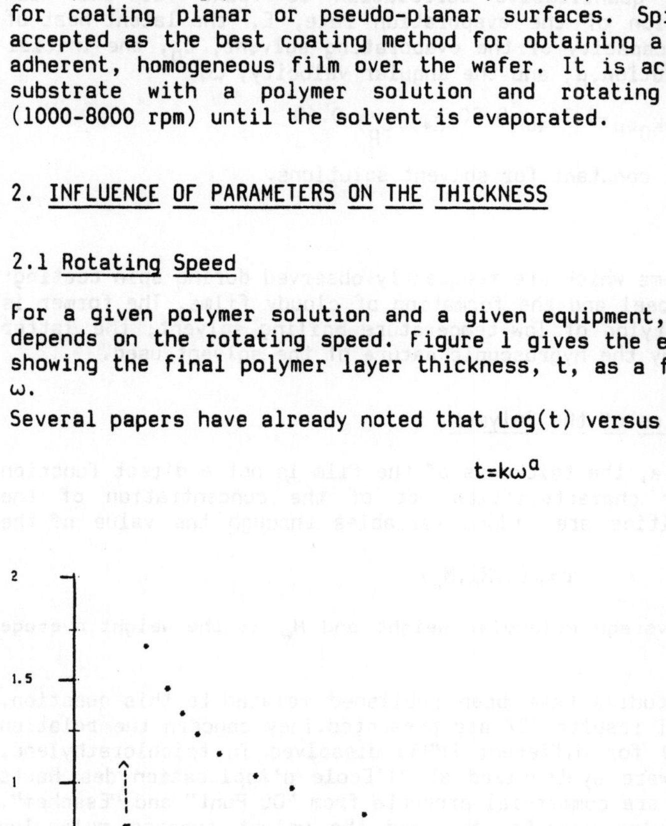

Fig.1 Spinning curve showing the polymer layer thickness, t, as a function of rotating speed, ω. PMMA 2041, 9% in chlorobenzene, bake 10 min at 160 °C in a belt oven

and that α is found to vary from 0.4 to 0.7. The theoretical value of 0.67 is foreseen by the work of WASHO/1/, which is based on the following assumptions:
The solution is a newtionian fluid.
The volumetric flow rate is constant.
There is no evaporation of solvent during spinning.

But it is reasonable to assume with CHEN /2/, MAYERHOFER /3/ and LAI /4/ that some evaporation of the solvent occurs during the spinning process, simultaneously with the flow of the solution on and off the wafer, although current solvents (cellosove acetate, butyl acetate, chlorobenzene ,etc...) are not highly volatile fluids.

Let us consider the investigation of the solvent-evaporation effect presented by CHEN. The film layer begins to be formed during the rotation. Therefore, it is necessary to include in the hydrodynamic equations the effect of heat transfer in addition to momentum transfer (fluid flow) and to mass transfert (solvent evaporation). A quantitative correlation is found for the film thickness, t, as a function of the evaporation rate, E, the latent heat of evaporation, λ, the heat capacity of the evaporating solvent, C_p, the initial shear viscosity of the solution, μ, and the angular velocity, ω,

$$t=K_0*\mu^{-0.36}*\omega^{-0.50}(E*\lambda/C_p)^{-0.60}$$

where K_0 is a proportional constant for solvent solutions.

2.2 Nature of the Solvent

LAI /4/ studied two problems which are frequently observed during spin coating: the appearance of orange peel and the formation of cloudy films. The former is attributed to the rapid drying of low-temperature boiling solvent; the latter is believed to be caused by the hygroscopic nature of the solvent used.

2.3 Molecular Charactristics of the Polymer

According to CHEN's formula, the thickness of the film is not a direct function neither of the molecular characteristics nor of the concentration of the polymer. These two quantities are hidden variables through the value of the viscosity of the solution

$$\mu=\mu(C,M_n,M_w)$$

where M_n is the number average molecular weight and M_w is the weight average molecular weight

As far as we know, no studies have been published related to this question. Here after CNET's original results /5/ are presented.They concern the relation thickness=f(concentration) for different PMMAs dissolved in trichlorethylene, C_2HCl_3. Several of them were synthesized at "l'Ecole d'Application des Hauts Polymères" /5/ and others are commercial products from "Du Pont" and "Esschem". The number average molecular weights, M_n, and the weight average molecular weigths, M_w, were measured by Size Exclusion Chromatography /7/.

For two polymers of completely different origin, but having quite the same molecular characteristics, the two curves thickness =f(concentration) fit very well together and it is concluded that the control of the molecular weight seems to be enough for characterising the polymer, regardless of the polymerisation process.

It is now advisable to determine the more pertinent parameter beTween M_n and M_w. For this purpose, two PMMAs of the same M_w but of very different M_n (ratio 1:2) were synthesized. The results, reported in Fig /2/, show that M_w is the pertinent parameter.

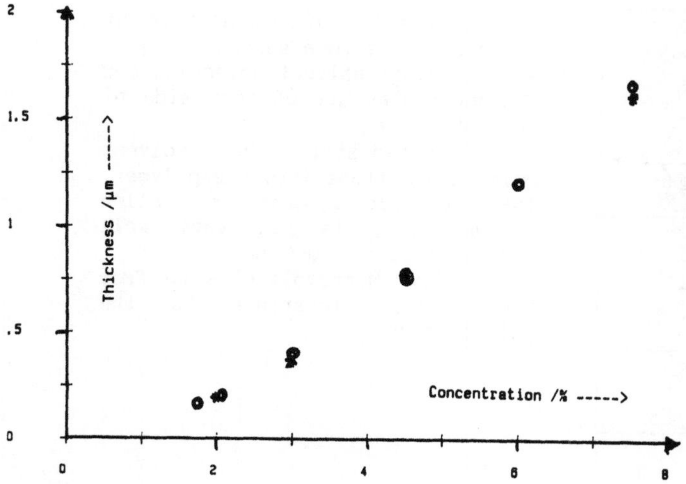

Fig.2 Thickness layer, t, versus the concentration, C, of the polymer. Solvent: trichloroethylene; rotating speed: 2000 rpm.
O: PMMA 613 ; M_w= 288,000 ; M_n= 150,000
*: PMMA 577 ; M_w= 260,000 ; M_n= 66,000

The shear viscosity of the solution, µ, is a function of the concentration, C, and of the weight average molecular weight, M_w. The problem to be answered is: Will different solutions of different M_w, but exhibiting the same shear viscosity, µ,yield identical thickness? Four solutions of PMMA in trichlorethylene *of* the *same viscosity,* µ=1.8 mPa s , have been prepared and spin on at the same rotating speed, 2000 rpm, on the same equipment. Thickness measurements are given in table 1.

Table 1 : Thickness of PMMA film as a function of the molecular weight of four solutions of the same shear viscosity. Thickness directly varies with M_w.

REF	M_w g/mol	concentration %	thickness µm
570	223,000	2.12	0.198
613	288,000	1.74	0.165
2041	500,000	1.23	0.121
562	2000,000	0.75	0.89

It leads to the conclusion that the mechanism presented by Chen is not complete. In our opinion it fails in the sense that it does not take account of the specific conformation and mechanical behavior of macromolecules in organic solvents. In order to complete the mechanism, the basic knowledges on polymer solutions are hereafter described in terms of solubility, crosslinking, entanglements and viscoelasticity.

3 POLYMER SOLUTIONS

3.1 Solubility

The cohesive properties of a polymer find their expression in their solubility in the organic solvents. The concept of solubility is based on the equilibrium thermodynamic of polymer solutions and has been developed by Hildebrand /8/ and others /9,10/ in the 1950's. The chemical nature of the polymer is the most important determinant of the problem.

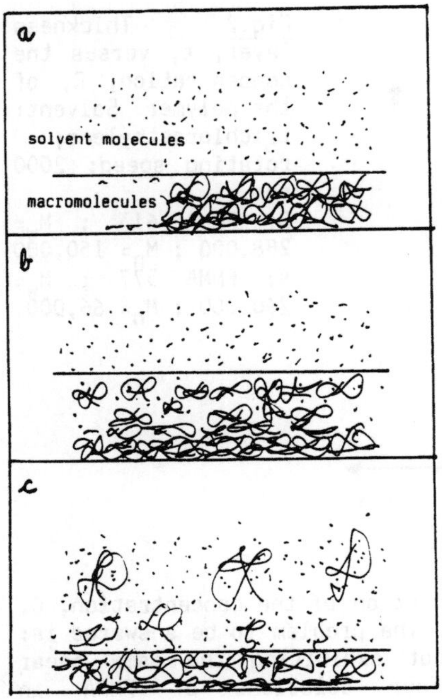

Fig.3 Mechanism of dissolution of macromolecules in a solvent
a) beginning: solvent molecules and macromolecules are on each side of the interface
b) intermediate: The solvent molecules diffuse into the polymer. The interface , made of swollen polymer gel, is not very well defined (10^{-2} - 10^{3} µm)
c) finally: Macromolecules go from the swollen interface to the solvent phase.

Dissolving a polymer is a slow process that occurs in two steps (Fig.3). First, the solvent slowly diffuses into the polymer to produce a swollen gel. If polymer-polymer intramolecular forces can be overcome by the polymer-solvent interactions, the second step of the process, the dissolution itself, can take place.

But the solubility of polymer systems is more complex than that of small molecular compounds because of the high value of the viscosity of the polymer solution which controls the kinetics of the dissolution. As will be mentionned later the viscosity is an increasing function of the molecular weight of the polymer. The solution process can be speeded by agitation or by heating. Nevertheless it can take several hours, days or weeks for materials of very high molecular weight.

3.2 Test of Degree of Crosslinking

For crosslinked materials, the molecular weight is infinite and the dissolution rate is zero. They do not dissolve but only swell if indeed they interact thermodynamically with the solvent. A lightly crosslinked material (like insolated negative resist) swells in solvents which dissolve the same uncrosslinked material (non insolated part of the resist). A highly crosslinked resist, like a deep UV cured positive photoresist, may not swell appreciably in contact with any solvent.

The behavior of a polymer layer, of thickness t_0, in contact with a good solvent is a very good test to appreciate the degree of crosslinking of the molecules. Let us put the wafer in the solvent and observe four cases (Fig.4):
a) There is no polymer left on the substrate: non-crosslinked
b) After the film has been dried, the thickness t is lower than t_0: lightly crosslinked. Some molecules were not linked to the others and have been dissolved.
c) The thickness t, before drying, is higher than t_0, but after drying it is equal to t_0: highly crosslinked, but solvent molecules can penetrate the bulk.

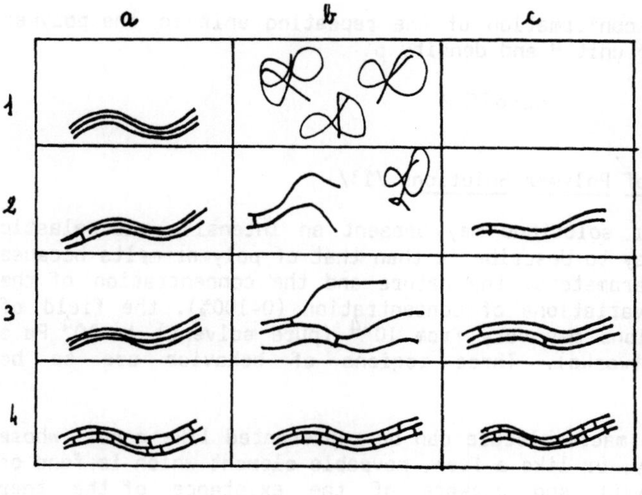

Fig.4 Schematical representation of the behavior of a polymer thin layer in contact with a good solvent. a) before the interaction b) during the contact of the layer and the solvent c) after drying
1- non crosslinked
2- lightly crosslinked
3- highly crosslinked
4- completely crosslinked

d) The thickness t is equal to t_0, even in contact with the solvent, the layer has the same color in the solvent or in the air: completely crosslinked, the solvent molecules cannot penetrate the layer.

3.3 Solubility Parameter /11,12/

Solubility occurs when the free energy of mixing

$$\Delta S = \Delta H - T\Delta S$$

is negative. The entropy of mixing ΔS is determined by the sign and magnitude of the heat of mixing ΔH. In absence of strong polar or hydrogen bonds ΔH is assumed to be the same as for the mixing of small molecules. According to HILDEBRAND /8/, the heat of mixing is

$$\Delta H = V_s * V_p \ (\delta_s - \delta_p)^2$$

where V is the volume fraction and δ^2 is the cohesion energy. The solubility parameter, δ, can be calculated for both the polymer and the solvent. As a first approximation, solubility can be expected if $\delta_s - \delta_p$ is less then 1.7-2.0, but not if it is much larger. A few typical values of δ ($J^{1/2}$ $cm^{3/2}$) are listed
first approximation, solubility can be expected if $\delta_s - \delta_p$ is less then 1.7-2.0, but not if it is much larger. A few typical values of δ ($J^{1/2}$ cm $^{3/2}$) are
listed in table 2.

The simplest way to estimate δ_p of a given structure is by the use of the molar-attraction constants E which are listed in tables /11,12/. Values of E are

Table 2: Typical values of the solubility parameter of solvents and polymers.

n-HEXANE	14.5	POLYETHYLENE	16
CCL_4	17	POLYPROPYLENE	17
2-BUTANONE	18	POLYSTYRENE	18
BENZENE	18.5	PMMA	19
ACETONE	19	PVC	21
METHANOL	29	PVA	28

summed over the structural conformation of the repeating unit in the polymer chain, with repeat molecular unit M and density ρ.

$$\delta_p=\rho\Sigma E/M$$

3.4 Rheological Properties of Polymer Solutions /13/

Like polymer melts, polymer solutions may present an intensive viscoelastic behavior. But it is less easy to describe it than that of polymer melts because there are two additional parameters: the nature and the concentration of the solvent.According to the variations of concentration (0-100%), the field of viscosity of polymer solutions may vary from 10^{-4} (pure solvent) to 10^{4} Pa s (viscosity of swollen networks). Three regions of behavior are to be distinguished:

In very dilute solutions, a macromolecule can be represented like a coil whose conformation changes rapidly, or like a free, moveable element which is fond of the molecule of the solvent and unaware of the existence of the other macromolecules. The solution is a pure viscous fluid whose viscosity is a linear function of the concentration

$$\mu_{sp}=(\mu-\mu_0)/\mu_0=\langle\mu\rangle*C$$

with μ being the solution viscosity
μ_0 being the solvent viscosity
and $\langle\mu\rangle$ being the proportionality constant, called "intrinsic viscosity"

In dilute solutions, the macromolecule is sometimes, for a finite time, encroaching upon the volume of another. Due to the size of the swollen molecules and due to the density of their population, there are numerous problems of vicinity. Although each macromolecule keeps going its own way, the internal mobility of the system suffers from this situation. The solution is a viscoelastic fluid and the linear behavior no longer holds

$$\mu_{sp}=\langle\mu\rangle*C+k_h\langle\mu\rangle^2*C^2$$

with k_h Huggins constant /13/.

In concentrated solutions, like in polymer melts, the phenomenon of critical conditions appears/13/. This means that in a $\log(\mu)$-$\log(M)$ or $\log(\mu)$-$\log(C)$ plot, the slope changes drastically if M or C exceed a critical value M_{cr} or C_{cr}. This is attributed to entanglements between molecules. Different authors (14-16) have proposed a power law dependence

$$\mu=\kappa*C^5*M^{3.4} \quad .$$

Beyond the critical conditions, the concentration of the macromolecules in the solvent is so high in regard to the length and the flexibility of the chain that a sort of "crisis of housing" is reached. It is impossible to distinguish two seperate volumes for two neighbouring molecules. They totally tangle up together. The macromolecules are tied together through sorts of knots, called *entanglements.*

The theory of entanglements has been developed by Greassly/17/. The existence of entanglements is now universely accepted and is thought of as the basic mechanism for explaining the viscoelastic mechanical behavior of polymers.

Here appears the notion of *mechanical relaxation time* which is related to the time needed for creating or undoing these knots. Without going into all the details of the theory of viscoelasticity, for example the existence of a spectrum of relaxation times, it can be easily understood that when the characteristic time of the investigated system is

a)smaller than the mean relaxation time, the inter-molecular bondings through the entanglements prevent any flow. The main mechanical behavior of the solution is that of a solid elastic body;

b)greater than the mean relaxation time, in spite of the entanglements, the major mechanism involved then is the viscous flow.

4. PROPOSED MECHANISM

Like others, we consider two simultaneous transport phenomena: the radial flow of the solution along the x axis and the evaporation of the solvent perpendicular to the surface.

Due to the radial flow rate, Q, the height , h, of the solution is decreasing with time, $dh/dt<0$. Solvent molecules diffuse through the solution layer, and a concentration gradient along the z axis is created. The concentration of polymers, C, increases from $z=0$ to $z=h$. When $C(h,t)$ reaches the critical value C^*, for which the solution behaves like an elastic solid, the layer is, from a mechanical point of view, separated into two phases (fig.5).

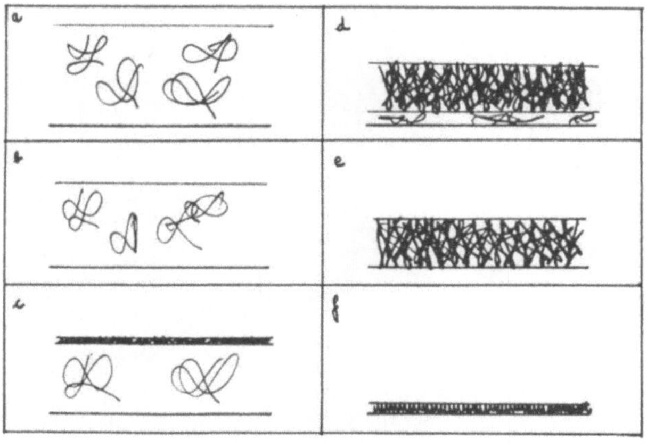

Fig.5 Schematical mechanism for spin coating
a) and b): Due to the radial flow , the height, h, of the solution layer is decreasing. Due to evaporation of the solvent perpendicular to the surface, there is a concentration gradient, $dC/dz >0$. c) The concentration at the top of the layer, z=h, reaches the critical value C^*, for which the solution is a gel and behaves like an elastic "crust". Two phases, solid and fluid, superimpose.
d) The thickness of the crust is increasing and that of the dilute solution is decreasing. e) The radial flow stops because there is no more dilute solution on the wafer. The gel layer is going to dry because the wafer keeps on rotating. f) The gel layer has been completely dried. A "soft bake" is needed for removing the few solvent molecules which remain in the layer.

From $z=0$ to $z=h_v$, $C(h,t)$ is $< C^*$; the solution is a viscous phase which flows radially and whose thickness, t_v, decreases continually with time.From $z=h_v$ to $z=h$, $C(h,t)$ is $> C^*$; the solution is an elastic "crust" made of polymer gel whose thickness, $tg_{el}=h-h_v$, increases with time.

The radial flow of the solution stops when $t_v=0$, that is when $t_{gel}=h$. At that time the final mass of the polymer on the wafer is obtained, and the only process which is going on is solvent evaporation. The height, h, decreases because the gel dries, giving rise to the well-known procession of the interference colors of the film. When the solvent is evaporated, the color becomes stable and the process ends up although the films enclose some residual molecules which have to be removed by heating the substrate.

REFERENCES

1. B.D.Washo: IBM J. Develop.190 (1977)
2. B.T.Chen: Pol. Eng. Sc. 23, 399 (1986)
3. D.Mayerhofer: J. Appl. Phys. 49, 3993 (1978)
4. J.H.Lai: Pol. Eng. Sc. 19, 1117 (1979)
5. P.Paniez, A.Schiltz, A.Weill: In CNET Technical Report NT/CNS/RPT/08 (1983)
6. C.Goett, J.Journé: Caoutch. Plast. 53, 47 (1976)
7. A.Weill, P.Paniez: Microelectron. Eng. accepted for publication
8. J.H.Hildebrand, R.L.Scott: In The Solubility of non-Electrolytes (Rheinhold, New York 1949)
9. P.A.Small: J. Appl. Chem. 3, 71 (1953)
10. C.W.Bunn: J. Polym. Sci. 16, 323 (1955)
11. D.W.Van Krevelen: In Properties of Polymers (Elsevier, Amsterdam (1976) p.129
12. F.W.Billmeyer: In Textbook of Polymer Science (Wiley, New york 1972) p.21
13. D.W.Van Krevelen: In Properties of Polymers (Elsevier, Amsterdam 1976) p.376
14. S.Onogi: J. Polym. Sci. C15, 381 (1966)
15. M.F.Johnson, W.W.Evans , T.Jordan: J. Coll. Sci. 7, 498 (1952)
16. G.Pezzin, N.Giglio: J. Appl. Polym. Sci. 10,1 (1966)
17. W.W. Graessley: In the Etanglement Concepts in Polymer Rheology (Springer , Berlin, Heidelberg 1974)

Resists Patterning

A. Weill

Centre National d'Etudes des Télécommunications, BP 98,
F-38240 Meylan Cedex, France

1. IMAGING

Physical images in resists are produced when radiations (photons, electrons, ions or neutral particules) interact with the macromolecules and initiate physical or chemical changes which are then *developed* in order to create geometrical *patterns*. The *mask* defines the material areas to be exposed and therefore the geometry of the patterns to be produced .Resist materials can be classified /1/ as positive or negative resists, according to their behavior when submitted to radiations (Fig.1). But resist can also be subdivided into two categories depending upon the number of the basic chemical components.

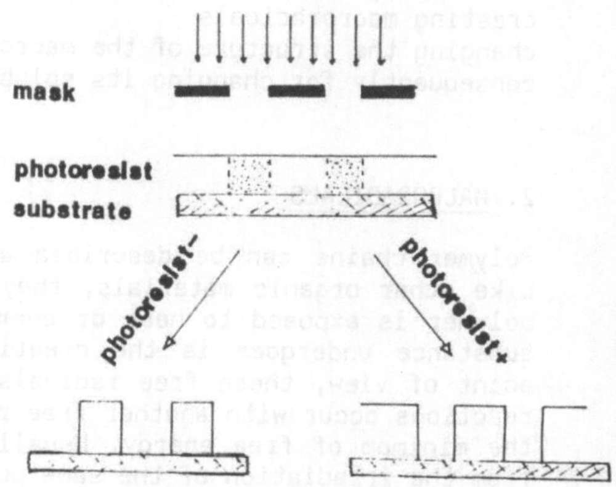

Fig.1 : Schematic patterns obtained from the same mask using a negative or a positive photoresist.

1.1 TWO-COMPONENT SYSTEM

If the radiation energy (400-450 nm) is not high enough for breaking the C-C polymer bonds, the resist must consist of a solid solution of a polymer and a sensitizer. The polymer acts as a matrix (film forming and chemical resistance), whereas the photosensitizer molecule is only involved in the photomodification.

The near UV *positive* photoresists formulation consists of diazonaphtoquinone sensitizer (a) and cresylic novolac polymer (b) dissolved in an organic solvent (Fig.2). The sensitizer acts as a dissolution inhibitor for aqueous base development of the novolac polymer. Upon exposure the diazonaphtoquinone molecules react with light and water. A succession of reactions takes place and the final product is a carboxylic acid which accelerates the development by basic developers of the exposed areas.

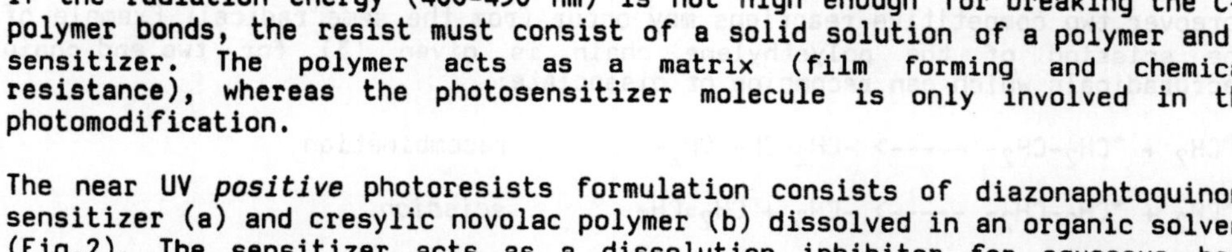

Fig.2 : Chemical representation of (a) the diazonaphtoquinone sensitizer and (b) of the cresylic novolac polymer.

Typical two component *negative* resists consist of a cyclised polyisoprene (low glass transition temperature and high solubility in non-polar solvents) and a bis-arylazide sensitizer whose photoreaction produces a nitrene

$$R\text{-}N_3 \xrightarrow{h\nu} R\text{-}N: + N_2$$

which then undergoes different reactions which result in covalent polymer-polymer linkage.

1.2 ONE-COMPONENT SYSTEM

One-component organic resists are polymers that combine reactivity toward radiations and other properties like film forming and etch resistance properties. The radiation particles which are to be considered (deep uv photons, electrons, ions, X rays) have sufficient energy for
interacting with covalent bonds
creating macroradicals
changing the structure of the macromolecule
consequently for changing its solubility in organic solvents.

2. MACRORADICALS

Polymer chains can be described as a succession of covalent chemical bonds. Like other organic materials, they are more or less easily degradable. When a polymer is exposed to heat or energetic radiations, the main effect which the substance undergoes is the creation of macroradicals.From a thermodynamical point of view, these free radicals are not stable. They chemically evolve and reactions occur with another free radical or a molecule according to the law of the minimum of free energy. Usually several types of radicals may be produced from the irradiation of the same polymer sample:

$$\text{-}CH_2\text{-}CH_2\text{-}CH_2\text{-} \text{ -----> } \text{-}{}^*CH_2 + {}^*CH_2\text{-}CH_2\text{-}$$

$$\text{-}CH_2\text{-}CH_2\text{-}CH_2\text{-} \text{ -----> } \text{-}CH_2\text{-}{}^*CH\text{-}CH_2\text{-} \quad .$$

Moreover,two competitive reactions may occur from the same radical. Example of the scission of the polyethylene chain is given (3) for two end chain macroradicals which can recombine or dissociate:

$\text{-}{}^*CH_2 + {}^*CH_2\text{-}CH_2\text{-} \text{ -----> } \text{-}CH_2\text{-}CH_2\text{-}CH_2\text{-}$ recombination

$\text{-}{}^*CH_2 + {}^*CH_2\text{-}CH_2\text{-} \text{ -----> } \text{-}CH_3 + CH_2\text{=}CH\text{-}$ *scission.*

From the point of view of microlithography, recombination is an insensitive event. But when scission within the main chain occurs, the polymer molecule is divided into two smaller molecules so that the molecular size distribution becomes lower. Consequently physico-chemical properties, such as solubility, refractive index, density, mechanical characteristics, are modified.

Crosslinking is an intermolecular reaction between two free radicals, whose consequence is the creation of a covalent bond between two macromolecules, giving rise to a larger macromolecule. The molecular size distribution increases and
physico-chemical properties are also modified. The chemical equation is

$$\text{-}CH_2\text{-}{}^*CH\text{-}CH_2\text{-} + \text{-}CH_2\text{-}{}^*CH\text{-}CH_2\text{-} \text{ -----> } \begin{array}{l}\text{-}CH_2\text{-}CH\text{-}CH_2\text{-}\\ \text{-}CH_2\text{-}CH\text{-}CH_2\end{array} .$$

3. THERMODYNAMIC

As a polymer degradation reaction evolves by different radical mechanisms, it is very useful to know the thermodynamical data of free-radical formations in comparison with that of bonded molecules in order to predict which reaction, scission or crosslinking, predominates. A very good compilation of such data has been published by Van Krevelen /3/ yielding to

*free enthalpy of formation of small molecules

*related group contribution to the free enthalpy of formation of large molecule

*free enthalpy of formation of radicals

As an illustration, the thermodynamical prediction of the polyethylene behavior is proposed: will molecular weight increase or decrease under irradiation?

	Group contribution	$\Delta G(T)/J.mol^{-1}$
	$-{}^*CH_2$	142,000 - 4 T
	$-CH_2-$	-22,000 + 102 T
free radicals	------------------	
$-{}^*CH_2 + {}^*CH_2-CH_2-$		262,000 + 96 T
	$=CH-$	38,000 + 38 T
	$=CH_2$	23,000 + 30 T
	$-CH_3$	-46,000 + 95 T
scission	---------------	
	$-CH_3 + CH_2=CH-$	15,000 + 163 T
	$-CH_2-$	-22,000 + 102 T
recombination	----------------	
	$-CH_2-CH_2-CH_2-$	-66,000 +306 T
	ΔGr is estimated as	-328,000 + 210 T
	ΔGs is estimated as	-247,000 + 67 T
	$\Delta=\Delta Gs-\Delta Gr$	+81,000 - 143 T

$\Delta < 0$ only if $T > 567$ K.

It is concluded that scission in polyethylene occurs only at high temperature (end chain macro-radicals). The same estimation based on radicals located in the middle of the chain yields the following equations respectively related to scission and crosslinking

$$-CH_2-{}^*CH-CH_2- + -CH_2-{}^*CH-CH_2- \text{ ---> } -CH=CH-CH_2 + -CH_2-CH_2-CH_2$$

$$-CH_2-{}^*CH-CH_2- + -CH_2-{}^*CH-CH_2- \text{ ---> } \begin{array}{l} -CH_2-CH-CH_2- \\ \quad\quad\;\; | \\ -CH_2-CH-CH_2- \end{array} .$$

Calculations show that $\Delta Gx-\Delta Gs$ is <0 for T< 400 K and that crosslinking is predominant at room temperature. These simple estimations are in good agreement with experiments.

4. "G" VALUE

Although scission and crosslinking are competitive reactions, it is usually assumed that they are independent of each other and randomly produced. The efficiency of the two reactions is generally accepted to be quantitavely estimated as the G value /4,5/. G(scission), or G(s), indicates the number of main chain sission per 100 eV of absorbed energy. On the contrary G(crosslinking), or G(X), measures the number of crosslinking per 100 eV of absorbed energy.

Mathematically if G(s) is greater than 4G(X), complete crosslinking (gelation) of the material cannot be produced, and the polymer is considered as a *positive* resist. Of course G values are related to the variations of the molecular characteristics, M_n and M_w, of the irradiated materials. Two relations have been established /2,3,6/ between G_s,G_x and M_n,M_w

$$1/M_n = 1/M_{n0} + (G_s - G_x)\ Q\ /\ 100\ N_A$$

$$1/M_w = 1/M_{w0} + (G_s - 4G_x)\ Q\ /\ 200\ N_A\ .$$

M_{n0} and M_{w0} are the initial number and weight average molecular weights, Q the dose of irradiation, and N_A the Avogadro number.

Schiltz (7), defines a macroscopic sensitivity, Q_{sm}, which is the minimal dose for which a change in the solubility of the polymer layer can be detected. If this macroscopic sensitivity is considered as a threshold in the polymer response, Charlesby's law has to be modified by replacing the dose Q with the difference $(Q-Q_{sm})$ which is responsible for the actual dose irradiation effects

$$1/M_n = 1/M_{n0} + (G_s - G_x)\ (Q-Q_{sm})\ /\ 100\ N_A\ .$$

He applies this equation to analyze polymer degradation profiles. From development curves, thickness versus development time, he reaches the dissolution rate profile, R_d = f (thickness). Using the Greeneich equation /8/ that relates dissolution rate to the number average molecular weight,

$$R_d = A + B\ (Mn)^{-a}\ ,$$

he obtains the molecular weight profile through the irradiated polymer layer.

5. MOLECULAR WEIGHT DISTRIBUTION AND MEASUREMENTS

Although they could be very helpful (quality control or new resist development), polymer molecular weight measurements are not familiar in the microelectronic laboratory. They are shortly described hereafter.

Polymeric products always contain molecules having many different chain lengths and may be characterized by their molecular weight distribution. If there are N_i molecules of mass M_i, the order j moment is defined as

$$Xj = N_i*(M_i)^j\ .$$

The distribution is usually limited to

*M_n, the *number average molecular weight*: It is the total weight of all the solute species divided by the number of moles present and is very sensitive to the presence of small number fraction of low molecular weight macromolecules

$$M_n = \Sigma N_i M_i / N_i\ .$$

M_n can be obtained from membrane osmometry, ebulliometry, cryoscopy and end group analysis, which are all dependent on the number of analyzed molecules t.

*M_w, the *weight-average molecular weight* defined as

$$M_w = \Sigma N_i (M_i)^2 / \Sigma N_i M_i\ .$$

It is sensitive to the presence of small amounts weightwise of high molecular weight molecules and can be measured by laser light scattering or by equilibrium centrifugation.

*The *width of the distribution* is generally taken as the ratio M_w/M_n.

A commonly used technique for measuring the polymer weight distribution /9/ is Size Exclusion Chromatography (SEC). This technique separates products by molecular size using porous media (silica gels) having different sizes of pores ($10-10^4$nm). The fractionation mechanism is that the largest molecules cannot penetrate the pores, and thus elute first. Smaller molecules are retained in the interstices of the pores and therefore require more time to elute .In order to convert elution profiles in molecular weight distribution informations, proper calibration methods are required. The use of standards, which are samples of narrow molecular weight distribution, is generally required.As an example three elution curves are given in Fig.3

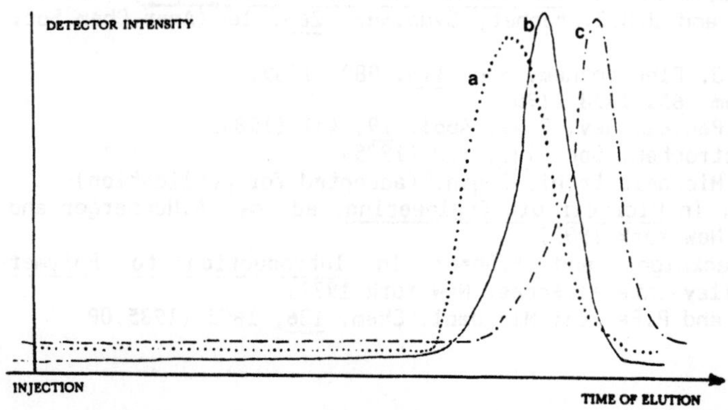

Fig.3 The molecular degradation of PMMA when exposed to deep UV radiations (200-250 nm; FUSION ILLUMINATOR 150) is shown by Size Exclusion Chromatography. The three curves are chromatograms of three PMMA. a)reference b)exposed 10 seconds c)exposed 40 seconds. The greater the time of elution, the smaller the molecular weight.

6. BAKING

During the lithographic process, the polymer layer is heated twice. After spin coating it is annealed (prebaked) in order to remove residual solvent, to relaxe mechanical stresses and to reduce the free volume created by the rapid loss of the solvent. After development a second annealing (postbaking) is required because the developer molecules are to be removed and the polymer-wafer adhesion to be improved. For that reason the knowledge of thermal properties is important.

Due to the necessary optical homogeneity of the sensitive layer all over the wafer, resists always consist of non-cristalline polymers. As there is no fusion point, the thermal properties of amorphous polymers are governed by the value of their glass transition temperature, T_g, /10/. This transition is not due to thermodynamic but to kinetic effects. It may be approximately described as the limit between rigidity and flexibility of the macromolecular chain, or as the transition between the glassy state and the elastic state.

The coefficient of diffusion of small molecules /11/ that diffuse through the resist layer rapidly increases when temperature reaches the T_g value (125°C for PMMA, 60°C for novolak resist).When temperature is again increased, material is no longer elastic but becomes a viscous fluid. This is the well-known thermal flow effect: the pattern blocks are submitted to irreversible creep deformations, creating dramatic losses in the geometrical definition of the image. On the other hand, this viscous flow phenomenon may be used for obtaining a better planarization of the wafer surface.

REFERENCES

1. C.G.Wilson: In Introduction to Microlithography, ed. by L.F.Thompson, C.G.Wilson, M.J.Bowden, Symp. Ser. 219, 3 (Amer. Chem. Soc., Washington DC 1983)
2. M.Dole: In The Radiation Chemistry of Macromolecules, Vol.1 (Academic , New York 1972) Chap.4
3. Van Krevelen: In Properties of Polymers, (Elsevier, Amsterdam, 1976) Chap.20
4. K.U.Pohl and F. Rodrigez: In Materials for Microlithography, ed. by L.F.Thompson, C.G.Wilson and J.M.J.Frechet, Symp.Ser. 266, 16 (Amer.Chem.Soc. ,Washington 1984)
5. H.Y.Ku and L.L.Scala: J. Electrochem. Soc. 116, 980 (1969)
6. R.W.Kilb: J. Phys. Chem. 63, 1838 (1959)
7. A.Schiltz, A.Weill, P.Paniez: Rev. Phys. Appl. 19, 431 (1984)
8. J.S.Greeneich: J. Electrochem. Soc. 122, 970 (1975)
9. P.Paniez and A.Weill: Microelectronic Engin. (accepted for publication)
10. P.Paniez and A.Weill: In Microcircuit Engineering, ed. by A.Heuberger and H.Beneking, (Academic, New York 1985)
11. J.J.Aklonis, W.J.Macknight and M.Shen: In Introduction to Polymer Viscoelasticity,Vol 4 (Wiley-Intersciences, New York 1972)
12. F.Sebillote, A.Weill and P.Paniez: Macromol. Chem. 186, 1695 (1985)OP

Dry Etching: Concepts, Methods and Applications

H.W. Lehmann

Laboratories RCA Ltd., Badenerstr. 569, CH-8048 Zürich, Switzerland

1. INTRODUCTION

The ever decreasing dimensions of modern ICs and the demand for increasing packing density of devices on Si chips is providing continued motivation to improve manufacturing techniques in VLSI fabrication and in particular to refine the lithographic processes. The requirement to conserve "real estate" on the chip demands that features must be placed very closely together which again requires excellent control over any pattern transfer process. Pattern transfer from the mask into the substrate can in principle be done by an additive process like lift-off or a subtractive process like etching. However since lift-off processes are rather difficult to be reproducibly executed in a manufacturing environment, most pattern transfer processes are performed today by etching. Over the past ten years we have seen a steady transition from traditional wet chemical etching to anisotropic dry etching.

At the same time as the device density has increased there has been an ever increasing demand to manufacture these chips on larger and larger wafers. While the standard wafer size was 3" only a few years ago, there are today manufacturing lines where memory chips are produced on 6" wafers and experimentally even on 8" wafers. Close dimensional control over patterns with micron size coupled with the handling of very large substrates requires very sophisticated and expensive equipment. This is particularly true for dry etching. Whereas the cost of etching equipment was below 10% of the total capital cost of a VLSI-manufacturing line a few years ago, it amounts to over 30% today.

Since dry etching is becoming such an important technique in the field of VLSI manufacturing, there is a steadily increasing number of scientific and technical papers published in the open literature. Many of these papers simply describe etch rates and profile control of one particular material in a given reactor. Since the parameter space in dry etching is relatively large and since many of these parameters are interdependent on one another and furthermore since there is also a large dependence on geometrical factors of the particular reactor, it is impossible to present something close to a "unifying theory" of dry etching

It is the goal of this paper to present an introduction into the field of dry etching and only present as much of the basics as is required to understand the various dry etching processes as they are used today in the IC industry. Frequent references to this field which has been the major driving force behind much of the research and development work will be made. It is impossible to present much more than a brief overview of the basics within the relatively small amount of space available. There are already a number of monographs on the subject of dry etching available [1-4] to which the interested reader is referred.

After an outline of the general concepts and the terminology of dry etching (which sometimes can be rather confusing!) we will discuss some physical and chemical aspects of gas discharges. This is followed by the description of the actual etching process namely the plasma-surface interaction. Then the monitoring of plasma processes will be discussed. In order to illustrate the points made in previous paragraphs three typical examples of etching of materials related to Si-technology will be described. Finally some limitations and also a few novel concepts in dry etching will be discussed.

2. GENERAL CONCEPTS AND TERMINOLOGY IN DRY ETCHING

Although in principle any etching process in which a solid surface is etched in a gaseous environment is a dry-etching process, we are using the term "dry etching" in this paper as a synonym for plasma-assisted etching. Dry etching has evolved over the past 20 years from a novel method to ash organic materials in an O_2-plasma [5] to a method capable of transferring patterns with nanometer dimensions with extremely good control over size and etch profile [6,7].

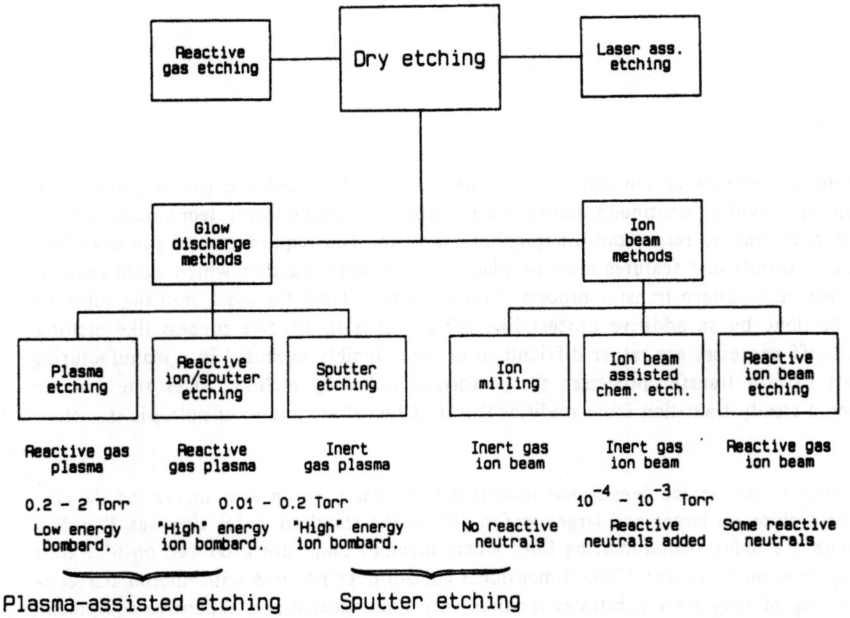

Fig. 1. The relationship between the various dry etching methods (from [8])

Figure 1 shows the relation between the various dry etching methods. The plasma-assisted dry etching methods can be broadly categorized into glow discharge methods and ion beam methods. In the *glow discharge* methods the plasma is generated in the vacuum chamber where the substrate is located whereas in *ion beam* methods the plasma is generated in a separate chamber from which ions are extracted and directed towards the substrate by a number of grids. In *sputter etching* and *ion beam milling* etching is only due to a physical effect, namely the momentum transfer between energetic Ar^+-ions (in the range between 0.1 and 1 keV) and the substrate surface. In all the other methods indicated in Fig. 1 there is also some chemical reaction taking place. *Plasma etching* and *reactive ion/sputter etching* are the techniques most commonly used today in VLSI fabrication. In plasma etching (at gas pressures in the 1 Torr range) the substrate sits on the grounded electrode. At this relatively high gas pressure chemical effects dominate which means that etch rates are usually higher and the selectivity is better than in reactive ion/sputter etching which is performed in the 0.01-0.1 Torr range. In this lower pressure range it is usually easier to obtain anisotropic etching but physical effects like faceting and trenching can have unwanted effects.

Figure 2 shows in a schematic way all the essential components of a dry etching system which consists of a vacuum chamber, a gas handling system to meter and control the flow of the reactant gases and a pumping package to maintain the pressure required for a glow discharge (in the 1 - 1000 mTorr range). In addition, the figure also lists the important system parameters. The parameters labelled in italic are the theoretically interesting parameters and are determined by the other system parameters like pressure, flow rate etc. The rf-generator supplies the voltage for starting and maintaining the glow discharge. In the so-called "barrel" reactor the rf-energy is coupled into the discharge by means of a coil wound around the quartz reactor vessel. But most modern plasma reactors are of the parallel-plate type as illustrated in Fig. 2. Excitation of the plasma is ac-

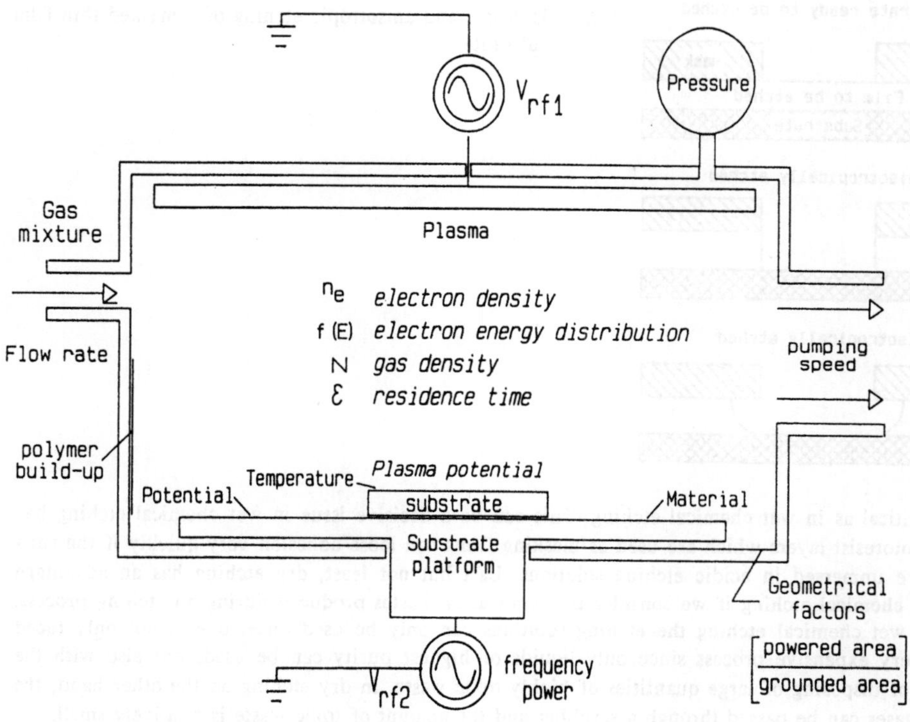

Fig. 2: Schematic diagram of a parallel plate dry etching system. (Important basic parameters are labelled in italics)

complished through parallel electrodes inside the vacuum chamber. In the *plasma etching* mode the substrate sits on a grounded platform ($V_{rf1}>0$, $V_{rf2}=0$ in Fig.2) opposite to the powered electrode, whereas in the so-called *reactive ion etching* mode the substrate sits directly on the powered electrode ($V_{rf1}=0$, $V_{rf2}>0$ in Fig.2). The expression reactive ion etching is actually somewhat misleading since it implies that the etching is done by reactive ions which is not always the case. As will be discussed in the appropriate section, the role of ions in the etching process can be very different depending on the material being etched. This method is sometimes also called reactive sputter etching in order to point out that it is similar to inert sputter etching but done in a reactive atmosphere.

Figure 2 furthermore gives some indication as to the large parameter space in dry etching. Many of these parameters (e.g. flow rate, pressure, power etc.) can independently be chosen for a particular experiment whereas others (e.g. geometrical factors) are determined by the reactor used. While these two types of parameters can relatively well be controlled and monitored, there is a number of parameters which either slowly vary during one particular run (e.g. temperature of substrate surface) or change from run to run (e.g. polymer build-up on reactor walls and fixtures). Since these latter parameters cannot be easily controlled, it is very difficult to obtain good run-to-run reproducibility in almost any dry etching system. It is important to realize this fact before the advantages of dry vs. wet chemical etching are discussed.

Wet chemical etching of a solid surface, which has long been the standard method of pattern delineation in Si-technology [9] has the big advantage of being very selective and fast but it usually leads to an isotropic etch profile (see Fig. 3) because the etch front moves at the same rate in the vertical as well as in the horizontal direction. Anisotropic liquid etching only occurs in crystalline materials when etch rates along different crystallographic orientations can vary by a large extent (e.g. in Si, see [10]). Dry etching on the other hand can result in anisotropic etch profiles in crystalline as well as in polycrystalline or amorphous materials. Furthermore dry etching has the advantage over wet chemical etching that the adherence of the masking film to the substrate surface is

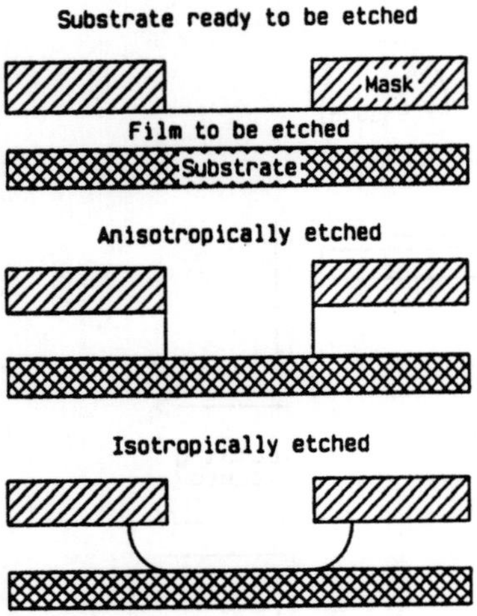

Fig. 3: Isotropic and anisotropic etching of a masked thin film substrate

not as critical as in wet chemical etching. This can be a decisive issue in wet chemical etching because photoresist layers which are used as masking films can loose adhesion very quickly if the substrates are immersed in acidic etching solutions. Last but not least, dry etching has an advantage over wet chemical etching if we consider the removal of wastes produced during an etching process. Since in wet chemical etching the etching solutions can only be used once, one is not only faced with a very expensive process since only liquids of highest purity can be used, but also with the problem of disposing of large quantities of highly toxic waste. In dry etching on the other hand, the exhaust gases can be passed through a scrubber and the amount of toxic waste is relatively small.

In order to improve dry etching processes some understanding of the physical and chemical effects in glow discharges is required. This will be treated in the next two paragraphs.

3. PHYSICS OF A GLOW DISCHARGE

The energy of ions and electrons bombarding the surfaces inside and around a glow discharge is mainly determined by three different potentials which are established as soon as the plasma is turned on, namely the potential of the powered electrode, the floating potential which solid surfaces assume when they are not grounded or externally biased and the potential of the glow region which is usually called the plasma potential. A plasma always assumes a positive potential even with respect to a grounded surface. The bombarding energy is proportional to the potential difference between the plasma potential and the surface which the ions strike. In other words, all the surfaces surrounding the glow discharge receive a certain amount of positive ion bombardment, including walls and fixtures in a vacuum chamber.

The plasma potential is essentially uniform throughout the volume of the glow discharge. Between the visible glow and the electrode is a narrow region in which the potential changes from the plasma potential to the potential of the surface. This region is called the dark space or sheath (see Fig. 4). The plasma potential is mainly determined by geometrical factors by the following expression [11]:

$$V_2 = V_1/(A_2/A_1)^4$$

where V_1 is the potential difference between the powered electrode and the plasma and V_2 the potential difference between the grounded electrode and the plasma and A_2/A_1 is the ratio of the respective electrode areas. From this expression it is obvious that unwanted effects of sputtering of the walls and fixtures, which occur due to large plasma potentials, can be avoided by properly choosing the geometry of the reactor. The plasma potential inside a reactor can be estimated by visually comparing the sheath thicknesses at the powered and the grounded electrode. If they are roughly

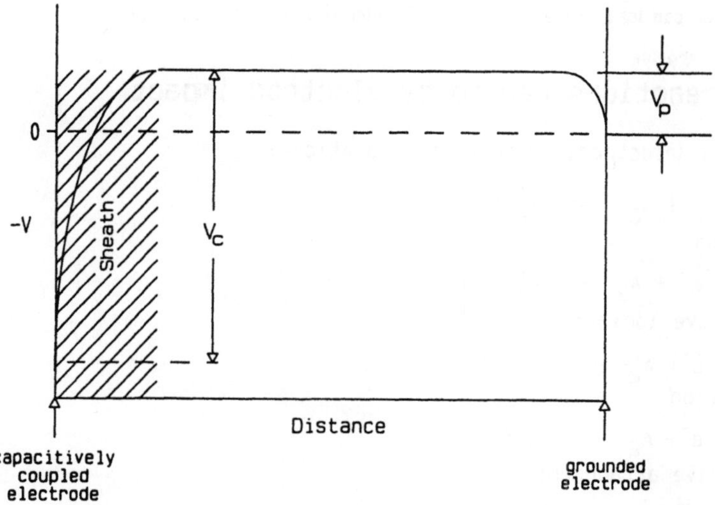

Fig. 4: Potential distribution between electrodes in a parallel plate etcher (-V: rf-induced dc-bias, V_2: plasma potential, $V_1=|V+V_2|$ determines energy of ions striking the surface)

equal then the system is symmetric and the plasma potential relatively high. As will be discussed later the plasma potential which is one of the most important parameters in dry etching can be determined by a Langmuir probe measurement.

In the above discussion on potentials of glow discharges we have tacitly assumed that we are dealing with dc-potentials. But most modern plasma equipment is powered by an rf-generator usually working at the standard industrial frequency of 13.56 MHz. Some generators also work in the 100 kHz range. The main reason for using rf excitation is that also insulating substrates and substrate platforms can be used. As we will see later in this paper, the choice of the proper electrode material is very important. Furthermore electrons can also acquire energy from the rf fields in rf discharges. Therefore plasmas can be operated at lower pressures than dc discharges.

The rf electrode which is powered through a blocking capacitor to prevent the net flow of charge assumes a negative potential slightly smaller than the rms value of the applied rf voltage. This negative potential is caused by the fact that electrons have a much larger mobility than ions in a glow discharge. The potential distribution in a parallel-plate etcher is shown in Fig. 4.

4. PLASMA CHEMISTRY IN A GLOW DISCHARGE

There are basically 4 different processes which have to be considered in an analysis of the plasma chemistry in a glow discharge. They are *production* of reactive species in the plasma, *adsorption* of the reactant on the surface, *reaction* with the surface material and *desorption* of the reaction products. All these processes have to occur at a finite rate otherwise the whole etching process stops. In this chapter we will restrict ourselves to a discusiion of the plasma chemical processes as they occur in a glow discharge.

The glow discharge generates a plasma which is a collection of electrically charged and neutral particles. Since the plasma has to be electrically neutral, the number of positive ions and negatively charged particles (electrons and negative ions) must be the same. Homogeneous gas-phase collisions of electrons with etch gas molecules generate the species which are responsible for plasma etching, namely reactive ions and radicals (neutral atoms or molecular fragments with an unpaired electron, eg. $F^{\cdot}$ or $CF_3^{\cdot}$). The collision of an energetic electron with a gas molecule can lead to a number of different excitations. Some of the possible reactions which can occur between an electron and a molecule A_2 are listed in Table I [12].

Table I: Reaction paths, which can lead to reactant species by electron-molecule collisions

Excitations and reactions caused by electron impact

1) Excitation (electronic, vibrational, rotational)

$$e^- + A_2 \;-->\; A_2^* + e^-$$

2) Ionization

$$e^- + A_2 \;-->\; A_2^+ + 2e^-$$

3) Dissociative ionization

$$e^- + A_2 \;-->\; A^+ + A + 2e^-$$

4) Dissociation

$$e^- + A_2 \;-->\; 2A + e^-$$

5) Dissociative attachment

$$e^- + A_2 \;-->\; A^- + A$$

Here are some examples to illustrate the reactions listed in Table I: In some of our own work we have found that excited O_2-molecules are most probably responsible for etching of organic polymers [13]. Direct electron impact ionization is the main process for generating positive ions in a discharge. Whereas one finds mainly O_2^+-ions in an O_2-plasma, a Cl_2-plasma can contain Cl^+ as well as Cl_2^+-ions depending on the excitation conditions [14]. Under certain conditions a molecule only dissociates into atoms or radicals. Since many etch gases (e.g. O_2, CF_4, CCl_4, BCl_3) show relatively strong electronegative character, electron attachment occurs and therefore negative as well as positive ions are produced. The negative ions have very little effect on the etching process since they are repelled by the negatively biased surface. Thus the main etchant species are either radicals or positive ions.

The generation rate of reactive species R is given by the following expression

$$R = kn_eN$$

where n_e is the electron concentration and N the concentration of the reactant. Here k is the rate coefficient which is a function of the reaction cross section and the electron distribution function, the latter of which is very difficult to determine quantitatively. Accurate expressions for these parameters are usually not available and therefore R can only be determined qualitatively. Figure 5

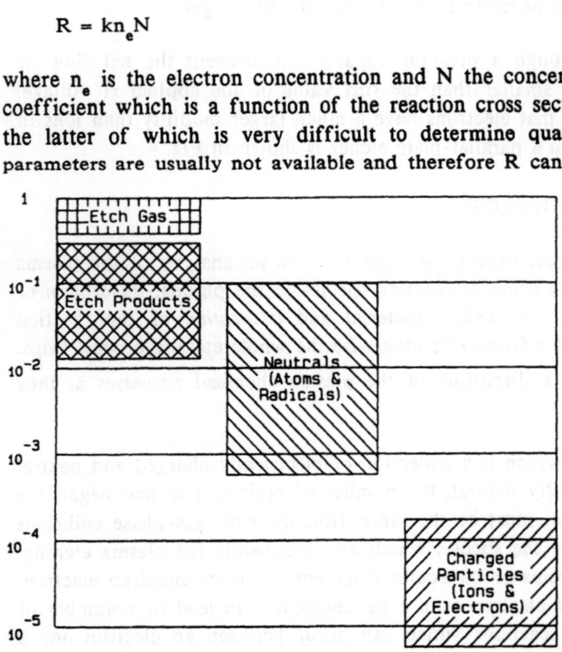

Fig. 5: Relative concentration of various gaseous species in a plasma (from [8])

compares the relative density of the various species in a glow discharge. It is interesting to note that the density of charged particles (electrons and ions) is generally much lower than that of neutral particles.-

Besides electron-molecule collisions one also has to take into account secondary processes, namely the collision of heavy species (ions and radicals) generated by electron impact with neutral gas molecules. This type of collision can again lead to a whole spectrum of different processes, but in addition to the processes listed in Table I, one also has to consider the different possibilities for chemical interactions like recombination or molecular rearrangement.

The type of reaction taking place upon electron-molecule collision not only depends on the particular gas but to a large extent also on the actual electron energy distribution. This function again is determined to a large extent by the various potentials inside a reactor. Therefore the relative concentration of etchant species in a given gas mixture might be different in different reactors. This could explain why reported etch results (rate and etch profile) for a given material/gas system sometimes widely differ.

From this very brief discussion of the generation of reactive species in a glow discharge it is obvious that it is definitely impossible to make any meaningful predictions about a particular etch process based only on first principles. In addition to this multitude of homogeneous gas phase reactions, we now also have to consider the heterogeneous processes occurring at the substrate surface which eventually lead to the etching of solid material.

5. PLASMA-SURFACE INTERACTIONS

In an glow discharge ignited in an inert gas atmosphere the etching reaction takes place due to momentum transfer between the inert gas ions (e.g. Ar^+) and the surface. This process is called sputter etching (see Fig. 1). If the discharge occurs in a gas where reactive species are formed as described in the previous paragraph, the etching reaction is much more complex. Primarily there is an interaction between the reactive species created in the glow discharge and the substrate surface. These reactive species can either be atoms, radicals or ions. The neutral particles arrive at the surface through diffusion whereas ions are accelerated towards the surface by the self-induced dc-bias. The impinging ions can either enhance or suppress the chemical reaction. In addition to the chemically active species we also have to take into account the effect of electron bombardment and irradiation by visible and UV-radiation emanating from the plasma. The reaction between the plasma-generated chemically active species and the surface which can either lead to etching <u>or</u> to deposition can be influenced by any one of these energetic particles, but it has been demonstrated that ions have the largest influence.

Before there can be any reaction on the solid surface a radical or an atom has to be adsorbed at a surface site. Ions strike the surface with an energy determined essentially by the applied voltage. In most cases there is not only one species responsible for the etching, but usually a whole variety (e.g. in the etching of Si with CF_4 one of the reactive species is certainly the F-atom, but the CF_x radicals play an equally important role, especially when polymer deposition is considered which is essential for a good selectivity with respect to SiO2). After the reaction has taken place the reaction product has to be desorbed from the surface again. The higher the vapor pressure of the product, the faster it will be desorbed.

There are essentially four different ways how ions can participate in enhancing an etching reaction:

1) By chemically enhanced physical sputtering. The role of the ions is to activate the desorption process and help remove reaction products held on the substrate surface [15].

2) By creating surface damage. Ion bombardment creates reactive sites on the substrate surface [16].

3) By chemical sputtering. Ion bombardment supplies energy to drive surface reactions [17].

4) By direct reactive ion etching. Energetic reactive ions dissociate into atoms upon impact on the substrate surface and thus react directly with it [18].

Most of the etching reactions can be categorized empirically by one of these types. But there remains the question to be answered: "What kind of reaction is actually taking place at the surface and how is it influenced by ion bombardment?" Since all chemical reactions taking place inside a plasma chamber are influenced by a large number of parameters, it is impossible to find a good answer to this question by simply monitoring the reaction in one of the commercial reactors. If the power is changed in a parallel plate system, the concentration of active species changes but also at the same time, the energy of the ions and their density. Basic experiments can best be done by using separate beams of active neutrals and independently accelerated inert ions. This kind of experiment has been pioneered and perfected by the group of COBURN and WINTERS at IBM in San Jose [19].

Coburn and Winters propose that etching reactions involving halogen gases are likely to be similar to reactions which involve oxygen, i.e. etching is similar to oxidation. If a clean surface is exposed to oxygen or halogen a thin chemisorbed layer is formed which changes to a compound whenever the number of oxygen or halogen atoms increases beyond a critical minimum value. Etching will not occur if only a chemisorbed layer is formed. It will take place as soon as there is three-dimensional compound formation occurring and if the product formed at the surface has a large enough vapor pressure. The reaction products found in most cases are those expected from simple vapor pressure arguments (e.g. Al_2Cl_6 at low temperature and $AlCl_3$ at high temperature). There are a few notable exceptions, namely the simultaneous formation of SiF_2 and SiF_4 if Si is exposed to XeF_2. Furthermore the dominant product from Si with Cl_2 under ion bombardment is not $SiCl_4$ as expected but rather $SiCl_2$ [20].

Simultaneous impingement of inert ions (e.g. Ar^+-ions) onto a surface which is exposed to a chemically active neutral can have very different effects: It can result in either a *suppression* of the etching (e.g. Cl_2 on Al) or to a large *enhancement* in the rate of a reaction which already proceeds slowly without ion bombardment (e.g. XeF_2 on Si). Furthermore SiO_2 can be etched in XeF_2 only if it is simultaneously bombarded by ions (e.g. Ar^+-ions).

In all cases where etching is enhanced by simultaneous ion bombardment one also observes directional or anisotropic etching, whereas in systems where the ion beam either has no influence or even suppresses the reaction to a certain extent, it is difficult to obtain directional etching.

In order to understand why under certain conditions it is also possible to obtain anisotropic etching e.g. for Al one also has to consider condensable species which are also formed in a plasma (e.g. CF_2, CCl_2 etc.). These species are removed from surfaces bombarded by ions but condense on sidewalls which are not subjected to ion bombardment (e.g. sidewalls of the masking resist profiles). This so-called sidewall blocking, which sometimes is intentionally enhanced by carefully "blending" the processing gas with certain additives, is responsible for anisotropic etching. But it is often difficult to decide if directional etching is due to ion-assisted etching of the substrate or ion-assisted etching of the condensable species. Figure 6 illustrates the two different mechanisms by which anisotropic etching can occur.

Most recently KOLFSCHOTEN et al. [20] have reported on plasma-surface interactions in the Si(Cl_2,Ar^+)-system. In addition to conventional mass spectrometry the kinetic energy distribution of the ejected particles was also determined by a time-of-flight method. The main conclusion from this study was that bombardment of the Si-surface by energetic Ar^+-ions has essentially *two* effects: creation of defects on the Si- surface *and* at the same time induced mixing of adsorbed Cl into the Si-lattice. The implanted Cl forms bonds with the Si-lattice but also reacts with interstitials forming SiCl and $SiCl_2$-molecules. Finally the Ar^+-ions sputter off these reaction products. Etching is therefore a consequence of the mixing of a relatively large amount of Cl (approx. 4 atoms) into Si by a single Ar^+-ion and the low binding energy of the products SiCl and $SiCl_2$ to the Si-lattice. Ion

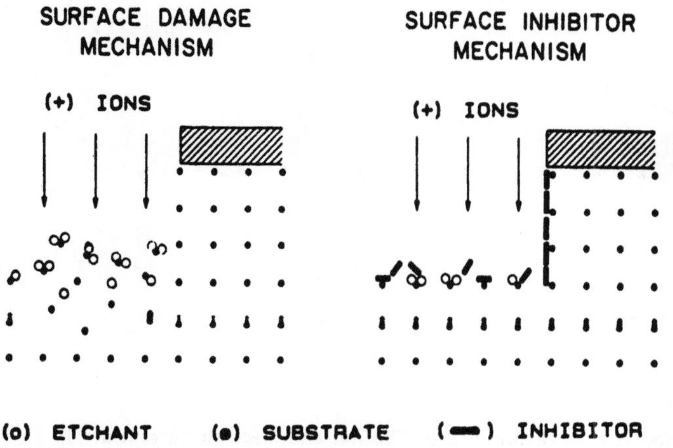

Fig. 6: Two suggested mechanisms for ion-induced anisotropic etching (from [21])

beam created amorphization of the surface is not sufficient to explain the large yields. Lattice damage *and* subsequent mixing - mainly by enhanced diffusion - are both required.

6. MONITORING OF DRY ETCHING

Because of the complexity of the physical and chemical environment in a processing plasma, a large number of monitoring tools is required to fully monitor and characterize the many aspects of a glow discharge. The four most frequently used techniques to monitor plasmas will be described, namely 1) the *Langmuir probe* which can be used to measure the plasma potential, ion density and electron temperature, 2) *optical emission* to characterize metastable species in the plasma, 3) *mass spectrometry* to analyze products of a discharge and 4) *interferometry* to directly determine and compare etch rates during the etching process. In addition to these "standard" methods there are other techniques which are used less frequently due to their various limitations: quartz crystal monitoring which is the standard method in evaporation can really only be used in conjunction with beam experiments [22]. Laser induced fluoresecence [23] has been developed into a very powerful diagnostic tool in plasma fusion, but is only rarely used in dry etching because the equipment which is required is rather expensive. This is unfortunate since the technique is very sensitive and can be used to *quantitatively* determine concentrations of reactive species.

6.1 Langmuir Probe

As the name implies this technique dates back to the work of LANGMUIR [24] in the 1920's. Probe measurements yield information on plasma potential, ion density and electron temperature. All these parameters of the discharge are of critical importance, in particular the electron temperature which is a measure for the energy of the electrons which collide with neutral gas molecules to create the reactive species.

The method is actually very simple and only requires standard laboratory equipment [25]. The electrostatic probe consists of a wire that is immersed into the plasma to measure the electron and ion current drawn by the probe as a voltage is applied. The resultant current-voltage plot is shown schematically in Fig. 7. Qualitatively it is straightforward to explain the characteristic curve of a Langmuir probe. At large negative potential a sheath of positive ions forms around the probe such that only ions are collected and electrons are repelled. LAFRAMBOISE [26] has shown that under certain conditions (which are usually fulfilled in etching processes at low pressures, but do not work properly at pressures above 100 mT) the *ion density* can be calculated from a plot of the square of the ion current vs. applied voltage. At large positive potential the probe current consists only of electrons. Due to the large difference in mobility of electrons and ions the electron saturation current is larger than the ion saturation current.

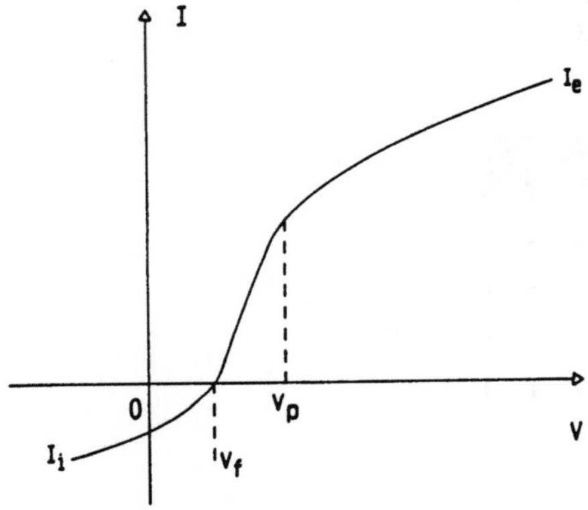

Fig. 7: Current-voltage characteristics of a Langmuir probe. The ion and electron current are I_i and I_e and the floating and plasma potentials are V_f and V_p respectively

The other important plasma parameters can be extracted from the transition region located between these two extreme ranges where ions and electrons contribute in comparable numbers to the total probe current. At small negative potential where the slope of the characteristic changes rapidly, the probe acts as an energy selector for electrons. It is in this region where the *electron temperature* can be computed from the slope of the characteristic. The potential at which the current goes through zero is called the *floating potential.* As the potential is increased the electron current starts to rise exponentially. The knee in the current-voltage characteristic indicates the *plasma potential.*

Although it is qualitatively easy to interpret the data, it is rather difficult to extract precise values for the above-mentioned parameters due to intrepretational difficulties. In addition the situation is made more difficult whenever reactive gases are used which either attack the probe or cause deposition of foreign material on it. Nevertheless we have used Langmuir probe measurements extensively in our laboratory to help to interpret etching results of SiO_2 in CHF_3 etc. [18] and polymers in O_2-plasma [13]. These measurements gave ion densities of the order of 10^{10} - $10^{11}/cm^3$ for a 13.56 MHz discharge in the pressure range of 20 -100 mTorr (for a comparison of the relative concentrations see Fig. 5).

6.2 Optical Emission

The characteristic glow of a discharge had attracted the interest of spectroscopists already in very early times. A major advantage of spectroscopic analysis is that the visible or UV-light from a glow discharge can be observed through a window in the process chamber and no probes interfere in any way with the physics and chemistry of the plasma. Furthermore optical emission analysis is well established and a number of spectral catalogues exist [27,28] which help to interpret emission spectra from plasmas. The equipment (a quartz lens to collect the light, a monochromator with a resolution equal to or better than 0.5 nm and a strip chart recorder) is probably already available in most laboratories. It can be used to record spectra over a certain wavelength region or record the time dependence of a particular emission line. This latter mode is frequently used for endpoint detection. An application of this technique will be given later (see Fig. 10).

If photodetectors with rapid response are used it is even possible to measure the phase lag of the optical emission of excited species with respect to the applied rf voltage. From these measurements it is possible to extract information on the kinetics of the plasma chemical reactions [29].

Although optical emission spectroscopy is easy to analyze and can give much useful information about the species and mechanisms involved, the technique has several limitations. The emission spectrum consists only of lines from species that emit strongly in the spectral region being analyzed. It has been found that the emission from O^+-ions is two orders of magnitude lower than that of neutral O-atoms. It is usually not possible to observe optical emission lines from all species involved in a plasma chemical reaction. Furthermore it is extremely difficult to make quantitative measurements because the emission of species in the plasma depends on the optical oscillator strength of the electronic transition as well as on the concentration of the particular species. COBURN and CHEN [30] have developed a so-called actinometer technique which consists of adding a small amount of an inert gas such as Ar to the reactive gas mixture. All emission intensities are normalized with respect to the emission intensity of a suitable spectral line of the inert gas. These normalized emission intensities are then approximately proportional to the concentrations of the corresponding species in their ground states if the excited states lie in the same energy range and if the excitation of the inert gas and of the species of interest occur via the same mechanism. There is also a limit to this method: Only lines due to transitions from the ground state can be considered for actinometric measurements.

6.3 Mass Spectrometry

Mass spectrometry is one of the most commonly used analytical techniques for characterizing plasma systems. The wealth of information contained in a complete mass spectrum is very large. It not only contains information about the etching process under investigation but also about the state of the vacuum system. If mass spectra are taken routinely and the spectra are interpreted carefully one might notice general vacuum problems like small leaks or malfunctioning of a pump before it will have a deleterious effect on the etching process. Figure 8 shows such an example. It was taken on a system in which Al was etched in BCl_3-plasma for about two months [31]. It was noticed that the pumpdown times increased from run to run and that the base pressure which could be attained got worse with every run. Analyzing the mass spectrum and the time dependence of some of its pertinent peaks, it became immediately obvious that the polyfluorinated ether which was used in the diffusion pump had started to decompose.

Mass spectrometry can also be used to determine the endpoint of a given etching reaction. In this case the time dependence of a mass peak due to a reactant gas or a product is followed as a

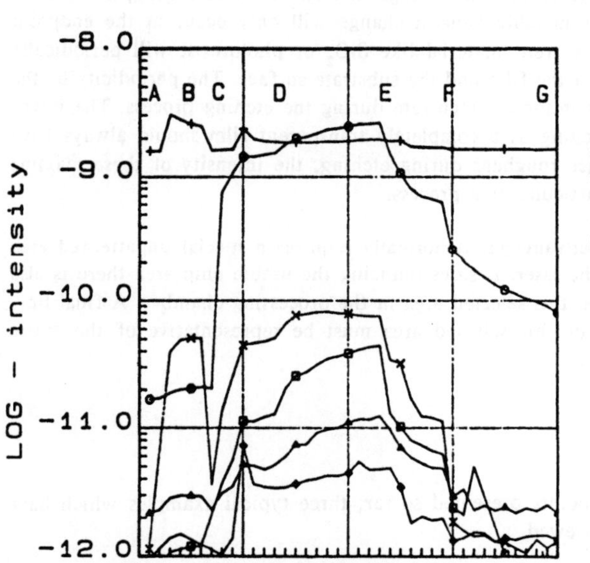

Fig. 8: Time dependence of a number of pertinent mass peaks before and during etching of Al in BCl_3-plasma in a diffusion pumped vacuum system. Region A: Background in vacuum chamber (10^{-4} Torr); B: Plasma ignited without external gas feed; C: Plasma off, BCl_3 admitted; D: BCl_3-plasma (notice the large rise of C-containing peaks!)(from [32])

function of time. Most instruments on the market today have the capability to either record a full mass spectrum or the time dependence of individual peaks.

Since a mass spectrometer only operates properly at a pressure < 10^{-4} Torr, it is usually housed in a separate vacuum chamber with its own pumping system and connected to the main chamber via a small aperture of 10 - 20 µm. Due to this geometrical arrangement most of the gaseous species arriving in the ionizer of the mass spectrometer have already undergone one or several collisions with walls. Therefore one is only able to observe relatively stable species of high vapor pressure. When we investigated the etching of Al [32] we were never able to observe any signal due to Al-containing species. The etch product Al_2Cl_6 was apparently condensing on the walls and never reached the mass analyzer.

By suitable experimental arrangement it is also possible to directly observe ionized species with a mass spectrometer. But the experimental set-up is much more complicated since an energy analyzer has to be included [33] if one wants to maintain the mass resolution. However this additional information about the composition of the ionized species does not yield much more information which is not already available from an analysis of the effluent gas.

Whereas conventional mass spectrometry is a very useful technique for analyzing plasma chemical reactions, its analytical power can greatly be enhanced by performing simultaneous time-of-flight measurements [34] which can yield direct information on the composition *and* the energy of products emitted from etched surfaces. An example of this technique was already given earlier [20].

6.4 Laser Interferometry

Whereas the three techniques described so far monitor an average state of the discharge, the technique to be described now monitors only a relatively small spot on a selected surface. A laser beam is directed onto the surface to be investigated and the change in reflectivity as etching proceeds is recorded as a function of time [35]. For metallic films a change will only occur at the endpoint whereas the signal reflected from a transparent material like SiO_2 or photoresist will periodically change due to interference effects between the film and the substrate surface. The periodicity of the reflected signal can directly be used to monitor the etch rate during the etching process. The interference maxima appearing during the etching of a completely transparent film should always have the same intensity. If the substrate surface roughens during etching, the intensity of these maxima decreases indicating problems with the particular etch process.

Laser interferometry also has its disadvantages: It normally requires a special unpatterned area to be positioned directly in the path of the laser. Besides reducing the usable chip area there is also the requirement of accurate positioning on this selected area in the processing chamber. A final limitation of the technique is that etching of this selected area must be representative of the entire wafer or even of a whole batch of wafers.

7. APPLICATIONS

In order to illustrate some of the basic points presented so far, three typical examples which have been studied in our laboratory will be reviewed.

7.1 Etching of Polymers in O_2-Plasma

The removal of organic materials by an O_2-plasma was the first application of dry etching [5]. The high rate removal of organic material in a high pressure (0.1 - 1 Torr) O_2-plasma is still used today in photoresist strippers etc. But more recently the need for anisotropic etching of polymers has arisen in connection with multilevel resist and multilevel metal structures. Patterning of organic polymers for theses applications is usually performed in a low-pressure (~0.01 Torr) O_2-discharge.

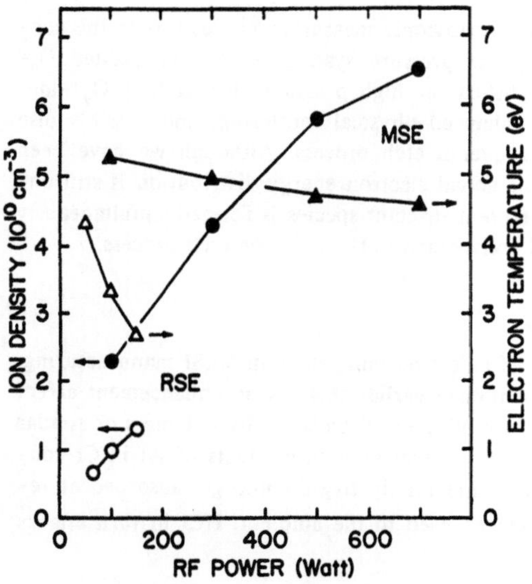

Fig. 9: Ion density and electron temperature in an O_2-plasma in a reactive sputter etching system (RSE) and magnetron etcher (MSE) as determined by Langmuir probe measurements (from [18])

For our studies [13] positive photoresist HPR 204 was used as a model material for the organic polymer. The etching of this material was investigated in two different types of reactors, namely in a conventional parallel plate etcher in an RSE (reactive ion/sputter etching) mode and also in a magnetron etcher (MSE) [36]. Besides etch rates and SEMs of etched profiles, the following diagnostic tools were used: Langmuir probe, optical emission actinometry and mass spectrometry.

Fig. 9 shows a plot of ion density and electron temperature as determined by Langmuir probe measurements as a function of rf power. In both reactors the ion density increases and the electron temperature decreases with increasing rf power. But the absolute value of the ion density is higher by a factor of 2 in the magnetron system due to the higher ionization efficiency caused by the application of a permanent magnetic field. Because the wafer cooling in the magnetron is much more efficient than in the RSE reactor, it was also possible to use higher power levels than in RSE. Table II summarizes some results found for the two reactors. Despite the fact that the ion flux to the substrate is about three times higher in the magnetron system, one still observes roughly the same etch rate at a power level of 100 Watts. This is because the ion energy is much lower in the magnetron system than in the RSE system (100 eV vs. 800 eV). But since substrate cooling is very efficient in the magnetron system, the power can be increased to 1 kW without damaging the resist. At this power level the etch rate can be as high as 1.5 µm/min at a substrate bias of only 300 Volts (compared to about 0.5 µm/min for an RSE system at the same voltage).

Table II: Typical plasma parameters at p = 10 mTorr (MSE: Magnetron etcher; RSE: Reactive sputter etcher)

	MSE	MSE	RSE
rf power (Watts)	100	500	100
Ion density (cm^{-3})	$2x10^{10}$	$5x10^{10}$	1x1010
Electron temperature (eV)	5	5	3
Ion flux to substrate (cm^{-2}sec^{-1})	$1x10^{16}$	$2x10^{16}$	$3x10^{15}$
Ion energy at cathode (eV)	100	200	800
Etch rate of HPR 204 (nm/min)	160	700	160

Optical emission actinometry coupled with mass spectroscopic measurements lead us to the conclusion that the etching of polymers in these two low pressure systems is due to excited O_2-molecules and not due to O-atoms as has been established for high pressure etching [21]. O_2^+-ions enhance the etch rate probably through chemically enhanced physical sputtering, and possibly also surface damage promoted etching contributes to the overall etch process. Although we have been able to determine an average electron temperature, the actual electron energy distribution is still unknown. If the distribution is wide, a multitude of different reactant species is formed simultaneously and it is very difficult to sort out the individual and collaborative effects in the etch process.

7.2 Etching of Al in BCl_3-Plasma

Anisotropic etching of Al films is one of the most difficult processing steps in VLSI manufacturing. There are two main reasons for this: As has been mentioned earlier there is no enhancement effect due to ion bombardment. Therefore one has to make use of sidewall blocking by polymers or similar reactions. In addition one also has to deal with the highly corrosive etch products of Al in Cl-containing plasma namely Al_2Cl_6. These compounds which are highly hygroscopic get adsorbed at remaining photoresist surfaces and form HCl if they are exposed to the ambient. HCl in turn attacks the remaining Al.

First we have studied the etching of Al in BCl_3-plasma in a cryopumped system [32]. The directionality of the etching is critically dependent on pressure: Below 20 mTorr the etching is anisotropic, whereas increasing undercutting is observed at higher pressure. Optical emission, SEM and mass spectrometry were the tools by which Al etching was investigated.

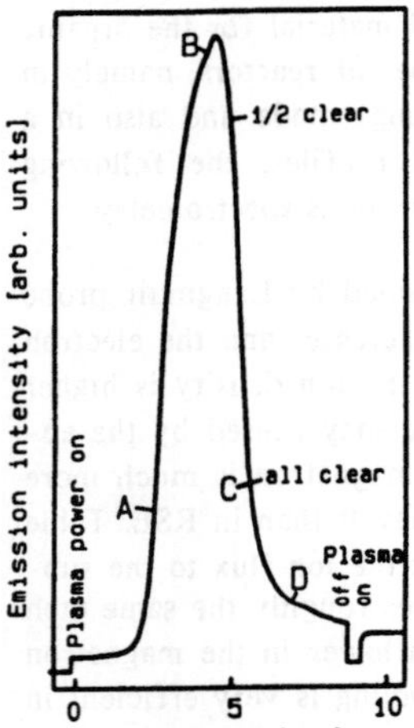

Fig. 10: Time dependence of optical emission of AlCl (at 261.4nm) during etching of Al in BCl_3-plasma at 20 mTorr. The capital letters indicate etch times for the SEM micrographs shown in Fig. 11 (from [34])

Figure 10 shows the time dependence of the optical emission of AlCl (at 261.4 nm). One can clearly distinguish three separate regions: The emission does not rise immediately after the plasma has been turned on; there is an "induction period" of about 1 min during which the always present thin layer of Al_2O_3 is etched. After this period the emission rises to a peak value and decays again as more and more Al is removed. Etching is complete after about 8 min. The last part of the curve clearly shows that even after all of the Al has been etched there is still emission due to AlCl which only decays very slowly. This is due to the interaction of the plasma with the low vapor pressure etch product Al_2Cl_6 which has been deposited on walls and fixtures in the vacuum chamber. These highly hygroscopic residues will react with water vapor in the ambient whenever the chamber is

Fig. 11: SEM micrographs of crossed square wave grating structures etched for various amounts of time at 20 mTorr in BCl_3. a: 2.5 min, b: 4.5 min, c: 5.5 min and d: 7 min (from [36])

vented. In parallel experiments Al-films were etched for short times corresponding to the times indicated with letters AD in Fig. 10. The SEM corresponding micrographs are shown in Fig. 11. They demonstrate very clearly that etching in BCl_3 can be highly anisotropic (see e.g. the "stringers" in Fig. 11c!) but that etching proceeds at rather uneven rates. The slowest etching areas are the grain boundaries of the individual Al grains. Overall etch rates in a BCl_3-plasma in an RSE system are between 0.1 and 0.2 µm/min.

Etching of Al was also studied in a magnetron system [36] in an $SiCl_4/Cl_2$-plasma. As Fig. 12 shows it was again possible to etch anisotropic profiles this time at a rate of 1 µm/min. It turned out that a parameter which is usually not very well controlled, namely the substrate temperature, had a crucial influence on the outcome of the experiment. Whereas etching was slow and badly reproducible when the substrates were resting on a water-cooled substrate platform, the process became much more controllable whenever the chilled water was replaced by hot tap water (temperature approx. 60° C). It is apparently very important that the surface of an Al-film to be etched has a

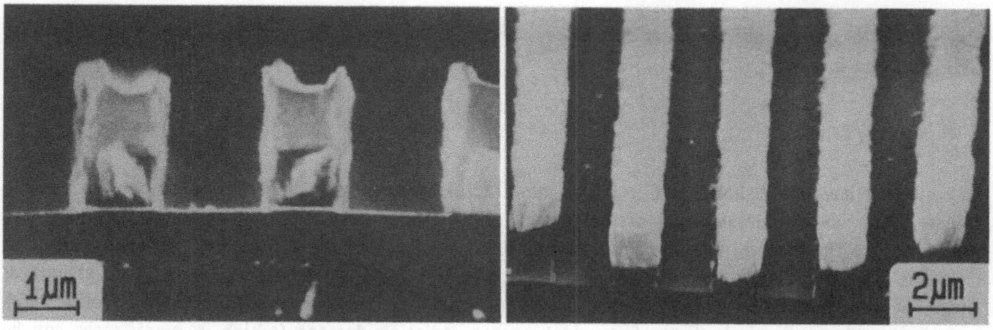

Fig. 12: SEM micrographs of Al-grating structures etched for 1 min. in $Cl_2/SiCl_4$ mixture in magnetron etcher. Left: Cross section after etching with resist still in place. Right: After removal of resist. Notice slight etching of underlying Si

certain minimum temperature otherwise the reaction products do not become volatile enough. But at the same time sidewall blocking (by some, as yet unknown cause) also has to occur otherwise it is not possible to obtain anisotropic profiles.

7.3 Fabrication of Crossed 0.25 µm Periodicity Gratings in SiO_2

In order to test the limits of our in-house capabilities to produce submicron structures by optical contact printing and dry etching, we have shown that it is possible to fabricate crossed square wave grating structures with a periodicity of 0.25 µm over areas as large as 1 cm^2 by these relatively simple and inexpensive methods [37]. The primary image forming device is a Cr-grating on glass which has been made by holographic methods [38]. The grating structure in SiO_2 was made as follows: The Cr mask was placed in hard contact with a thin layer of AZ 1350 photoresist and was exposed by filtered radiation from a mercury lamp. After the first exposure the mask was rotated by 90° and a second exposure was made. The pattern was first transferred into a layer of amorphous Si deposited by LPCVD on the actual substrate (quartz glass). In a first etching step in CCl_2F_2-plasma the am. Si was defined and subsequently this Si served as an etch mask to define the SiO_2 in a CHF_3- plasma. Figure shows an SEM micrograph of a crossed grating structure etched into SiO_2. The individual posts are slightly conical. This is due to faceting effects of the etch mask occurring during etching with CHF_3.

Fig. 13: SEM micrograph of crossed square wave grating (periodicity = 0.25 µm) etched in CHF_3-plasma into SiO_2 (from [37])

8. SUMMARY

Dry etching has established itself in VLSI manufacturing as a very powerful technique for pattern transfer. The basic understanding of the physics and the chemistry in the glow discharge and of the interaction of the plasma with the solid surface has made good progress over the past few years. It is now possible to design processes for specific applications and to define extremely fine structures which would not have been possible with wet-chemical or isotropic and non-selective etch techniques. The main limitations of dry etching are radiation damage (which in many cases can be annealed out at relatively low temperatures), the heterogeneous particle generation in the plasma which leads to contamination of the substrate surface and the run-to-run reproducibility in a production environment.

The designers of present day equipment are trying very hard to come up with new systems in which these problems are minimized. Two designs for future reactors which look rather promising are the magnetron and the microwave etcher. In the magnetron etcher one makes use of a permanent magnetic field to achieve large ionization efficiencies and one is able to work at much lower voltages (and thus reduce the radiation damage) and still achieve the same etch rates. In a microwave plasma etcher the plasma is generated outside the chamber where the substrates are located. In a recent listing [39] of dry etching equipment 29 manufacturers with a total of 92 types of equipment with a variety of different options are listed. It is unlikely that this diversity of systems can continue to survive for very much longer. Only the manufacturer who is able to apply the basics of plasma physics and chemistry to his systems will survive in the market place.

References

[1] C.M. Melliar-Smith, C.J. Mogab: In Thin Film Processes, ed. by J.L. Vossen and W. Kern (Academic, New York 1978) p. 497

[2] B. Chapman: Glow Discharge Processes (Wiley, New York 1980)

[3] J.A. Mucha, D.W. Hess: In Introduction to Microlithography, ed. by L.F. Thompson, C.G. Willson and M.J. Bowden (ACSSymposium Series 219, Washington D.C. 1983) p. 215

[4] N.G. Einspruch, D.M. Brown (eds): Plasma Processing for VLSI, VLSI Electronics Microstructure Science, Vol. 8 (Academic New York 1984)

[5] S.M. Irving: Kodak Photoresist Seminar Proc. 2 (1968)

[6] D.C. Flanders, N.N.Efremov: J. Vac.Sci.Technol. B1, 1105 (1983)

[7] R.E. Howard: In Plasma Synthesis and Etching of Electronic Materials, Mat. Res. Soc. Symp. Proc. Vol 38, ed. by R.P.H. Chang and B. Abeles (Materials Research Society, Pittsburgh PA. 1985) p. 247

[8] J.W. Coburn: Lecture notes for AVS courses (unpublished)

[9] W. Kern, C.A. Deckert: In Thin Film Processes, ed. by J.L. Vossen and W. Kern (Academic New York 1978) p. 401

[10] K.E. Bean: IEEE Trans. Electron Dev. ED-25, 1185 (1978)

[11] H.R. Koenig, L.I. Maissel: IBM J. Res.Dev. 14, 168 (1970)

[12] A.T. Bell: Solid State Technol. 21 (4), 89 (1978)

[13] C. Steinbrüchel, B.J. Curtis, H.W. Lehmann, R. Widmer: IEEE Trans. Plasma Sci. (April 1986) (accepted for publication)

[14] R.H. Bruce, D.L. Flamm, V.M. Donnelly, B.S. Duncan: Electrochem.Soc. Ext.Abstr. 81-2, No. 260 (The Electrochemical Society 1981)

[15] J.L Mauer, J.S. Logan, L.B. Zielinski, G.S.Schwartz: J.Vac.Sci.Technol. 15, 1734 (1978)

[16] J.W. Coburn, H.F. Winters: J.Appl.Phys 50, 3189 (1979)

[17] Y.Y. Tu, T.J. Chuang, H. Winters: Phys. Rev. B23, 823 (1981)

[18] C. Steinbrüchel, H.W. Lehmann, K. Frick: J.Electrochem.Soc. 132, 180 (1985)

[19] H.F. Winters, J.W. Coburn: In Plasma Synthesis and Etching of Electronic Materials, Mat. Res. Soc. Symp. Proc., Vol. 38, ed. by R.P.H. Chang and B. Abeles (Materials Research Society Pittsburgh PA. 1985) p. 189

[20] A.W. Kolfschoten, R.A. Haring, A. Haring A.E. de Vries: J.Appl.Phys. 55, 3813 (1984)

[21] D.L. Flamm, V.M. Donnelly: Plasma Chem.Plasma Proc 1, 317 (1981)

[22] J.W. Coburn, H.F. Winters, T.J. Chuang: J.Appl.Phys. 48, 3532 (1977)

[23] T.A. Miller: Plasma Chem. Plasma Proc. 1, 3 (1981)

[24] I. Langmuir, M. Mott-Smith: Phys.Rev 28, 727 (1926)

[25] C. Steinbrüchel: J.Electrochem.Soc 130, 648 (1983)

[26] J.G. Laframboise: Univ. of Toronto Inst.Aerospace Studies Rept. No. 100 (1966) (see also Adv.Appl.Mech.Suppl. 3, Vol. II (1966))

[27] R.W.B. Pearse, A.G. Gaydon: The Identification of Molecular Spectra(Chapman & Hall, London 1963.

[28] G.R. Harrison MITWavelength Tables (MIT Press, Cambridge, MA 1969)

[29] D.E. Ibbotson, V.M. Donnelly, D.L. Flamm, B.S. Duncan: Electrochem.Soc. Ext.Abstr. 81-2, No. 267 (The Electrochemical Society 1981)

[30] J.W. Coburn, M. Chen: J.Vac.Sci.Technol. 18, 353 (1981)

[31] H.W. Lehmann, E. Heeb, K. Frick: Solid State Technol. 24 (10), 69 (1981)

[32] H.W. Lehmann, R.Widmer: Microelectron. Eng. 1, 1 (1983)

[33] J.W. Coburn: Rev.Sci.Instrum. 41, 1219 (1970)

[34] F.H.M. Sanders, A.W. Kolfschoten, J. Dieleman, R. Haring, A. Haring, A.E. deVries: J.Vac.Sci.Technol. A2, 487 (1984)

[35] M. Sternheim, W. van Gelder, A.W. Hartman: J.Electrochem.Soc. 130, 655 (1983)

[36] Model MIE-710, made by Materials Research Corporation, Orangeburg N.Y.

[37] H.W. Lehmann, R. Widmer, M. Ebnöther, A. Wokaun, M. Meier, S.K. Miller: J.Vac.Sci.Technol. B1, 1207 (1983)

[38] W.E. Ng, C.S. Hong, A. Yariv: IEEE Trans Electron. Devices ED25, 1193 (1982)

[39] In 1986 Master Buying Guide of Semiconductor International 8, No. 13, 54 (1986)

Overlayers

S. Furukawa

Tokyo Institute of Technology, Graduate School of Science and Engineering,
4257 Nagatsuda, Midoriku, Yokohama 227, Japan

1. INTRODUCTION

It can be said that formation of overlayers is one of the most important techniques in semiconductor devices and IC (Integrated Circuit) fabrication, since planar technology is usually employed in the fabrication of such devices.

Recent semiconductor technology has the following features related to the formation of overlayers:

1) Layer thicknesses are decreasing,
2) multi-layers are very often used,
3) overlayers with different kinds of materials such as metal, insulator and semiconductor are used,
4) low-temperature fabrication is strongly preferred,
5) organic materials are sometimes used as well as inorganic ones.

In particular item 4) will become more important in the near future, since impurity redistribution and defect introduction in wafers during high-temperature processes cannot be neglected for obtaining a good VLSI and ULSI performance. For these reasons, low-temperature fabrication techniques will be mainly considered in the following.

2. AMORPHOUS MATERIAL OVERLAYERS

2.1 Thermal Oxidation

As a typical amorphous material, silicon dioxide will be considered. Silicon dioxide layers can be formed by various methods: 1) plasma CVD, 2) anodization, 3) thermal oxidation. The last one is most frequently used in industrial practice. We shall begin the study by reviewing a classical theory (Deal-Grove Model) of thermal oxidation. This is because of the fact that any overlayer formation by chemical process is usually controlled by diffusion and/or chemical reaction, as treated in the Deal-Grove Model.

2.2 Kinetics of Oxide Growth in Thermal Oxidation

There are three consecutive steps for the oxidizing species to reach the silicon surface in thermal oxidation, as follows [1]:

1) Species must be transported from the bulk of the gas to the oxide-gas interface,
2) species must diffuse across the oxide layer already present, then
3) the species react with the silicon surface.

Considering that the fluxes corresponding to the three steps should be equal in steady-state oxidation, a differential equation can be obtained.

The solution of the differential equation leads to the general relationship, the so-called parabolic relation, for the oxidation of silicon:

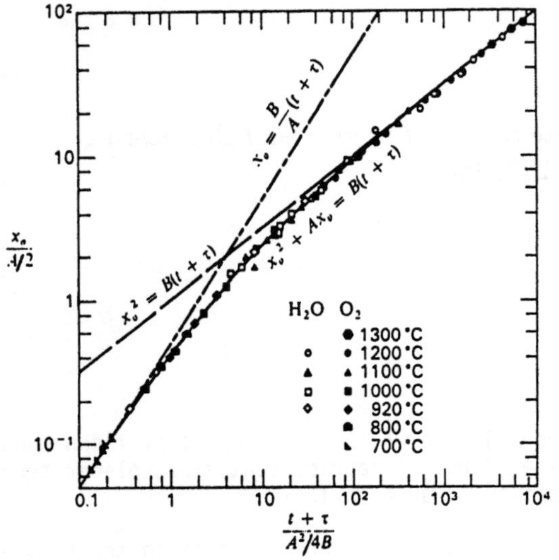

Fig.1. The general relationship for silicon oxidation and its two limiting forms

$$X_0^2 + AX_0 = B(t + \tau) \quad . \tag{1}$$

Figure 1 shows that the theoretical prediction given by (1) deduced by means of the simple model of oxidation can be used to explain the experimental results over a wide range of conditions.

There are two limiting cases of this general relationship. For large times, i.e $t \gg A^2/4B$, we obtain the so-called parabolic relationship which shows the diffusion-controlled case. For small times, i.e. $t \ll A^2/4B$, we obtain the linear law which shows the reaction-controlled case. A more detailed experimental study shows the following. The constant B is proportional to the partial pressure of the oxidant in the gas, indicating that the assumption of Henry's law, i.e. the concentration of species within a solid is proportional to the partial pressure of that species in the surrounding gas, holds. It follows that, for both dry and wet oxidations, the oxidizing species moving through the oxide layer are apparently molecular.

2.3 High Pressure Thermal Oxidation

In the case of diffusion-controlled thermal oxidation, the oxide growth rate is determined by the parabolic rate constant B in (1) which is proportional to the partial pressure in the bulk of the gas. Therefore, increasing the partial pressure of oxidants (H_2O or O_2) is effective in getting the high growth rate at relatively low temperatures. Figure 2 shows a cross-sectional view of a typical high pressure thermal oxidation furnace [2]. The experimental data in Fig.3 show that the oxide thickness is proportional to the pressure-time product, as theoretically predicted.

We can expect that the Deal-Grove model is also applicable for the case of the high-pressure oxidation. A conventional pressure vessel cannot withstand a pressure as high as 15 atm. Therefore, the high-pressure oxidation technique can enhance the oxide growth rate by only one order at the same furnace temperature. In other words, this technique does not have all the advantages of the low-temperature process.

2.4 ECR Plasma Oxidation

Plasma oxidation, utilizing a highly activated oxygen plasma, seems to be a candidate for a low-temperature technique. There are two techniques concerning the plasma

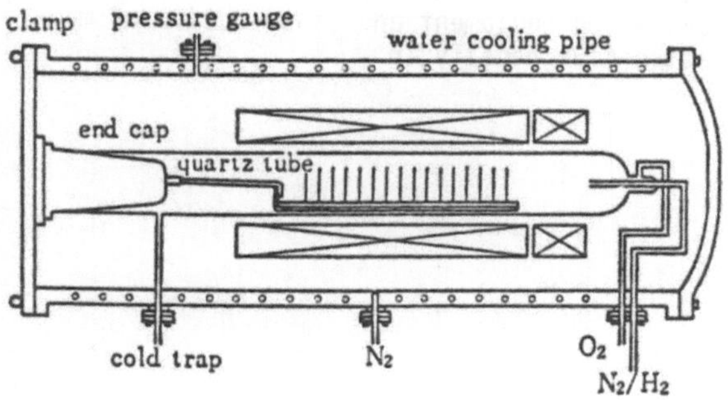

Fig.2. Cross-sectional view of a high-pressure thermal oxidation furnace

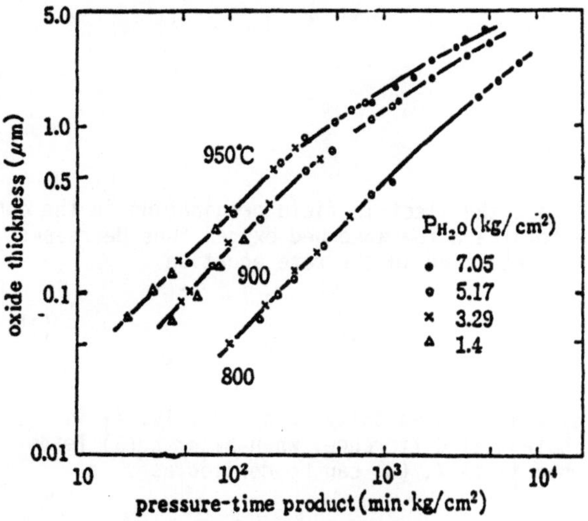

Fig.3. Relation between oxide thickness and pressure-time product in high pressure oxidation

oxidation, i.e. 1) plasma anodization in which the specimen to be oxidized is positively biased to enhance the diffusion of oxygen, and 2) plasma oxidation without external bias. By the plasma anodization, a considerably higher oxidation rate can be obtained. For example, an oxidation rate of about 1 μm/h has been reported [3] at 600°C for Si oxidation. This rate is equivalent to that in wet oxidation at 1200°C. Contrary to such merits, there are such demerits in the plasma anodization as contaminations caused by electrode sputtering and heavy damage by irradiation of charged particles and UV light.

On the other hand, plasma oxidation has not been investigated in comparison with the case of plasma anodization, since high oxidation rates are not expected. However, as the gate oxide thickness becomes increasingly thinner in accordance with the decrease in device geometry, plasma oxidation might be used in place of plasma anodization, because the above-mentioned demerits can be avoided. Figure 4 shows a schematic diagram of plasma oxidation equipment [4]. A 2.45 GHz microwave is guided through the waveguide to the quartz discharge tube filled with oxygen gas. The discharge tube is surrounded by magnetic coils, the field intensity of which is tuned to meet the Electron Cyclotron Resonance (ECR) condition.

According to FULLER et al. [5], we can express the relation between the oxide thickness (X) and oxidation time (t) by the power law relation as

$$X^n = kt \quad . \tag{2}$$

In thermal oxidation (Deal-Grove Model), n varies from 1 to 2, as mentioned. In contrast to the thermal oxidation, plasma oxidation shows different n values which exceed 2. In order to explain such a difference, the Cabrera-Mott Model seems to

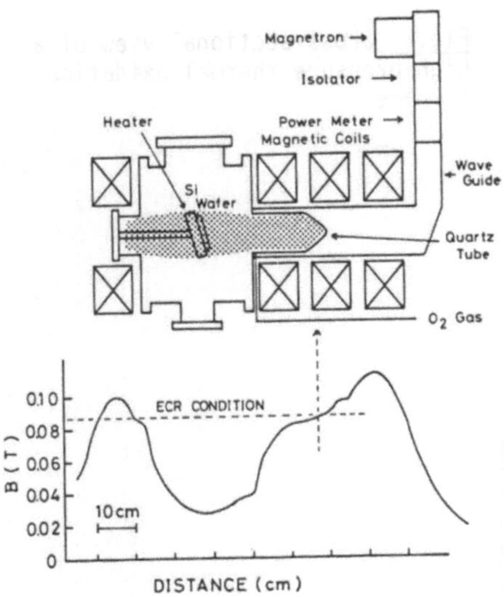

Fig.4. Schematic diagram of microwave plasma oxidation equipment and distribution of magnetic flux density, B

be valid. In the model, it is assumed that the electric field originating in the contact potential difference between the metal and the absorbed oxygen ions decreases the activation energy for ion movement, resulting in the rate equation

$$\frac{dX}{dt} = 2U\sin\left(\frac{X_1}{X}\right) \quad , \tag{3}$$

where X_1 and U are a characteristic distance and velocity, respectively. X_1 is linearly proportional to the contact potential difference. When $X_1 \ll X$, (3) leads to the parabolic rate law. However, when $X_1 \gg X$, (3) can be derived as

$$\frac{dX}{dt} = U\exp\left(\frac{X_1}{X}\right) \quad . \tag{4}$$

The results of the analysis are indicated in Fig.5, in which U and X_1 are determined by applying the repetition method. It is obvious that the measured oxidation rate

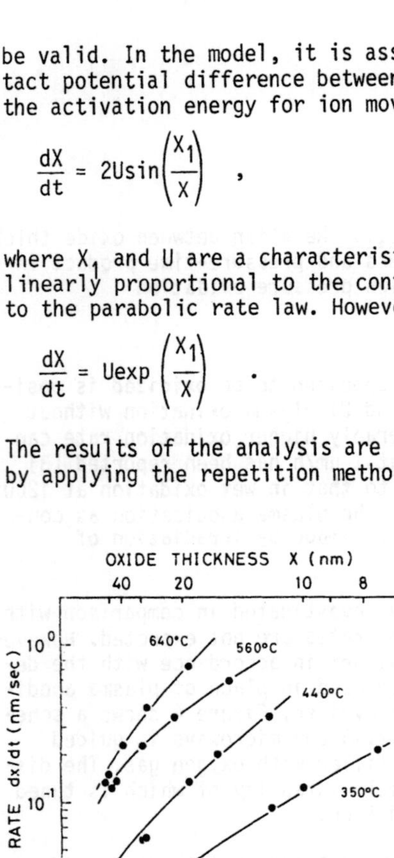

Fig.5. Plot of reciprocal oxide thickness (1/x) vs oxidation rate (dx/dt) using the Cabrera-Mott equation. Solid lines are calculated curves

(closed circles) coincides well with the calculated curve, so as to show the Cabrera-Mott Model be valid for the case of plasma oxidation. The Hitachi group succeeded in obtaining good MOSFET in which the oxide films were formed by this technique at 600°C [6].

2.5 Formation of Amorphous Semiconductor Layers

Conducting or semiconducting materials are also used in IC and semiconductor devices. The most important one is polycrystalline silicon. However, this material is usually used as an electrode and has no serious problem from the viewpoint of industrial practice. Recently, systems composed of an amorphous semiconductor overlayer on crystalline silicon have attracted attention in device applications. An example is the amorphous silicon carbide (a-SiC) film, which is chemically and mechanically stronger and more durable than other materials used as protecting coatings for thermal heads. Another example is seen in a wide-band gap emitter HBT (Hetero-junction Bipolar Transistor), which uses an a-SiC overlayer on crystalline Si [7].

PCVD (Plasma Chemical Vapor Deposition) is a useful technique for forming a-SiC overlayers on crystalline Si. A PCVD apparatus employing the L-couple method is shown schematically in Fig.6. A glow-discharged plasma is excited in the flowing gas mixture ($CH_4 + SiH_4 + He$, in this case) by an RF (Radio Frequency) coil. This coil is driven by a 13.56 MHz power supply. Figure 7 shows the film deposition rate as a function of the pressure of gas. By changing the RF power gas mixing ratio, the growth rate and composition ratio of SiC can be controlled. Since this apparatus uses activated species excited by RF power, the substrate temperature needed during deposition is only 300 - 400°C, a really low-temperature process. Study of the kinetics of PCVD in such systems is still continuing.

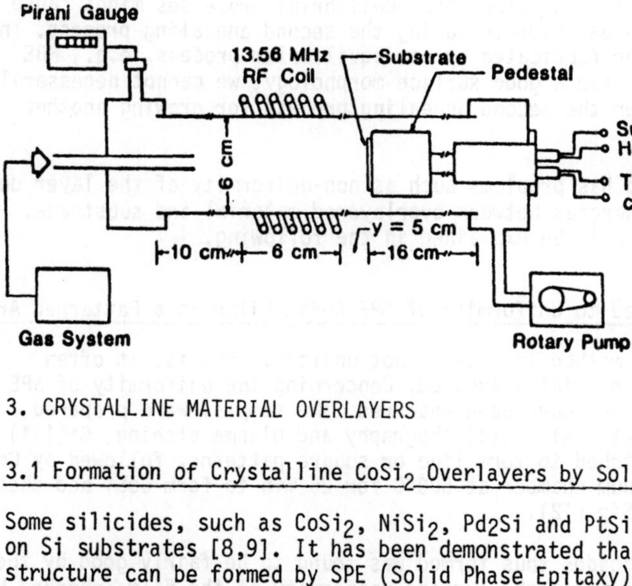

Fig.6. Schematic diagram of the PCVD apparatus

3. CRYSTALLINE MATERIAL OVERLAYERS

3.1 Formation of Crystalline $CoSi_2$ Overlayers by Solid Phase Epitaxy

Some silicides, such as $CoSi_2$, $NiSi_2$, Pd_2Si and PtSi, are known to grow epitaxially on Si substrates [8,9]. It has been demonstrated that an Si/$CoSi_2$/Si double heterostructure can be formed by SPE (Solid Phase Epitaxy). Such heterostructures can be used to fabricate novel Si devices such as a metal-base transistor [10] and a PBT (Permeable-Base Transistor). From a practical point of view, SPE is considered to be a suitable technique for growing epitaxial silicide films on top of crystalline substrates, since it has the following features:

1) good productivity,
2) possible low-temperature processing,
3) fine patterning capability,
4) thermal equilibrium processing.

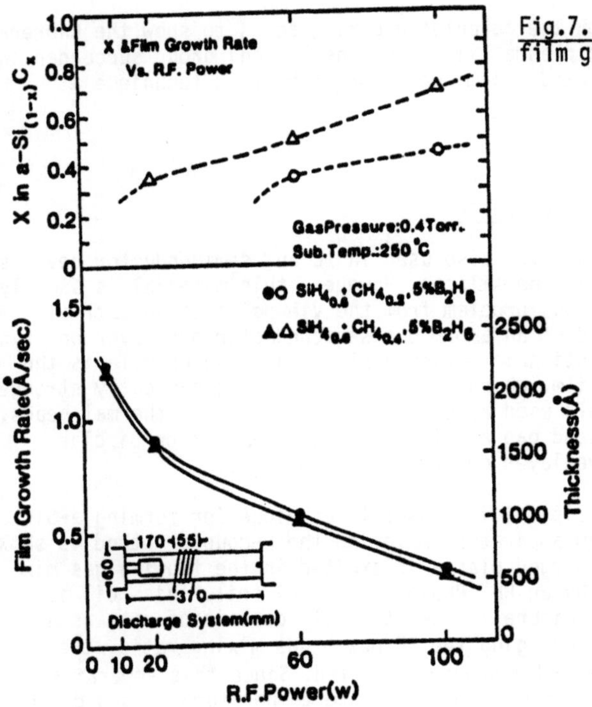

Fig.7. RF power dependence of the film growth rate and x

Concerning item 4), it is noted that thermal equilibrium processes are preferable for forming epitaxial silicide layers, since non-equilibrium processes might cause the morphology of the layers to deteriorate during the second annealing process. In other words, even though a layer fabricated by non-equilibrium process, e.g., MBE (Molecular-Beam Epitaxy) might have a good surface morphology, we cannot necessarily expect the same morphology after the second annealing process for growing another layer.

However, the SPE method also has problems such as non-uniformity of the layer due to the difference of surface energies between overlayered material and substrate. This problem can be solved as will be explained in the following.

3.2 Experimental Results Related to Uniformity of SPE $CoSi_2$ Films in a Patterned Area

A $CoSi_2$ film formed by the SPE method is usually not uniform. That is, it often breaks and the Si substrate is partially exposed. Concerning the uniformity of SPE $CoSi_2$ films, interesting phenomena have been observed. The samples were prepared as follows. By employing conventional photolithography and plasma etching, Si(111) wafers were selectively mesa-etched to form line or square patterns, followed by Co deposition, annealing in a vacuum chamber at 500°C for 25 min to form CoSi and then at 900°C for 30 min to form $CoSi_2$ [12].

The crystallinity of the silicides thus formed was found to be fairly good by the Rutherford backscattering method. But the surface uniformity of the films checked by SEM (Scanning Electron Microscopy) and Nomarski optical microscopy was not necessarily good. The film uniformity was evaluated by the quantity R_{Si} defined as the area of the exposed Si region normalized by the total area. Figure 8 shows the R_{Si} values as a function of the pattern size for the silicide films formed on stripe patterns and square patterns. The parameter is the thickness of the evaporated Co film before the silicide formation, t_{Co}. The estimated thickness of the $CoSi_2$ is about 3.5 times t_{Co}. It is obvious from this figure that there is some evidence for the presence of a principal mechanism controlling the film uniformity.

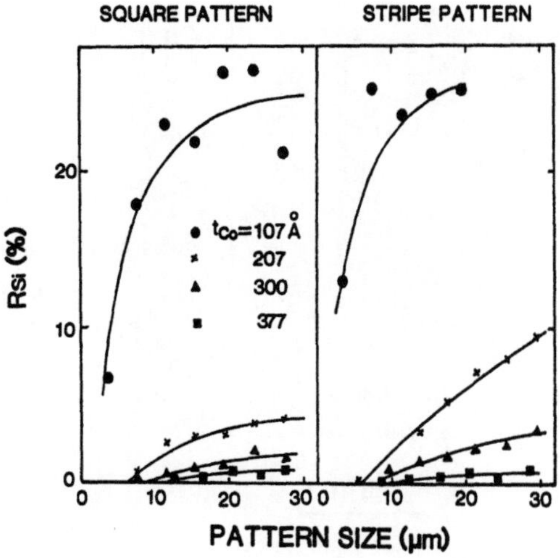

Fig.8. R_{Si} of silicide on striped patterns and square patterns as a function of pattern width. The parameter is the thickness of the evaporated Co before the silicide formation

3.3 Discussion on the Mechanism of Surface Morphology by Surface Energy

3.3.1 Surface Energies of Si and $CoSi_2$

The surface of a solid is known to have a surface energy which corresponds to the amount of work needed to cleave the solid at the boundary to form new surfaces. First of all, the surface energies $H_s(hkl)$ of the (hkl) planes in $CoSi_2$ and Si are calculated assuming that bonding energies of only the nearest-neighbor atoms are considered (the first-order assumption):

$$H_s(hkl) = \frac{n(hkl)}{Z} \cdot N(hkl) \cdot \frac{\Delta H}{A_0} \quad , \tag{5}$$

where Z is the number of nearest-neighbor atoms, n(hkl) is the number of nearest-neighbor atoms which would be separated when the solid is cleaved in the (hkl) plane, N(hkl) is the atomic density of the solid in the (hkl) plane, A_0 is the Avogadro constant and ΔH is the latent sublimation heat. The value of Z, n(hkl), ΔH and the calculated values of $H_s(hkl)$ using (5) for Si and $CoSi_2$ are summarized in Table 1. The experimental value of $H_s(111)$ for Si has been reported [12] as 1240 erg/cm^2, which is almost the same as the theoretical value derived from (5). Even though the calculated data are obtained under the first-order assumption, it is noted that the surface energy for Si is lower than that for $CoSi_2$ in the same plane, and that the energies of the (111) planes are lower than those of the (100) planes for both Si and $CoSi_2$.

3.3.2 Difference of Surface Morphology of Silicide Between Si(111) and Si(100) Substrates

Figure 9 shows the surface morphologies of $CoSi_2$ films on Si(100) and Si(111) substrates. The $CoSi_2$ film on Si(100) has changed its shape to agglomerate. On the other hand, though the $CoSi_2$ film on Si(111) contains holes, it still forms a con-

Table I. Summarized data of number of nearest neighbor atoms (Z), number of nearest neighbor atoms which would be separated when the solid is cleaved in (*hkl*) plane (*n*(*hkl*)), atomic density of the solid in (*hkl*) plane (*N*(*hkl*)), latent sublimation heat (ΔH) and the calculated surface energies for $CoSi_2$ and Si(Hs(*hkl*)).

	plane	Z	*n*(*hkl*)	*N*(*hkl*) (cm^{-2})	ΔH (kcal/mol)	ΔH (erg/mol)	*Hs*(*hkl*) (erg/cm^2)
Si	(111)	4	1/2	1.57×10^{15}	107.7	4.5×10^{12}	1.47×10^3
	(110)		1	9.59×10^{14}			1.80×10^3
	(100)		2	6.78×10^{14}			2.55×10^3
$CoSi_2$	(111)	16	2	8.01×10^{14}	341.1	1.42×10^{13}	2.38×10^3
	(110)	(8(Co)+	4	4.90×10^{14}			2.92×10^3
	(100)	2×4(Si))	4	6.94×10^{14}			4.13×10^3

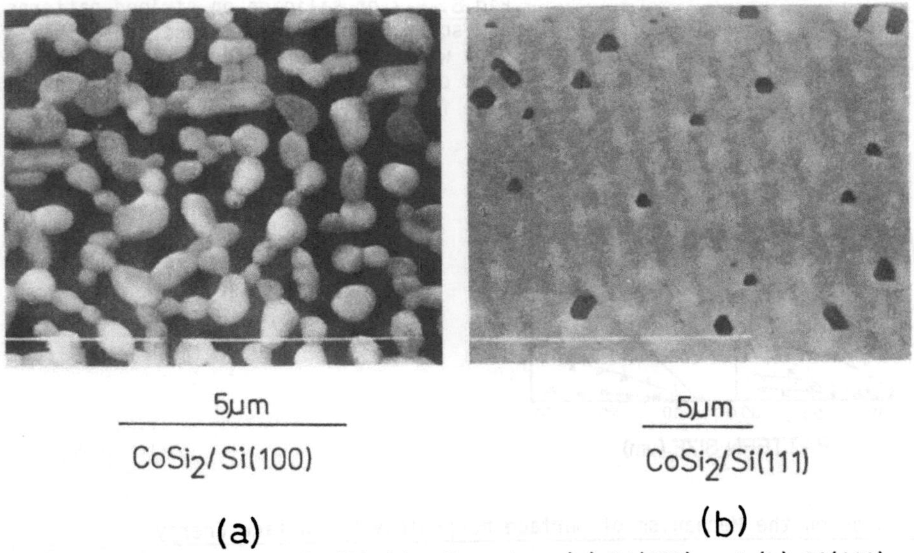

Fig.9. SEM photographs of silicides formed on (a) Si(100) and (b) Si(111) substrates. The bright regions correspond to the silicide regions and the dark regions to the Si substrate regions

tinuous film. However, the common feature shared by these films is that the substrate Si tends to become exposed during SPE. This can be interpreted qualitatively from the viewpoint of the surface energy, since, if there are holes in the $CoSi_2$ film exposing the Si substrate, the system is more stable because the surface energy of Si is lower than that of $CoSi_2$, as shown in Table 1. Further, since the surface energy of $CoSi_2(100)$ is higher than that of $CoSi_2(111)$, it is natural for the $CoSi_2(100)$ film on the Si(100) substrate to agglomerate, decreasing the area of $CoSi_2(100)$ and increasing the area of $CoSi_2(111)$.

3.4 Considerations on the Patterning Effect by the Surface Energy

The phenomena related to the patterning method can also be interpreted by introducing the concept of surface energy as described below. The model shown in Fig.10 can be assumed to be a reasonable one from the experimental results. Since there is no $CoSi_2$ just outside the pattern, the area of the $CoSi_2$ is defined by the pattern in the shape of squares. The total surface energy of a square patterned area of size "b" is estimated. A regular hexagonal hole of size "a" is assumed to exist in the center of the pattern. The total surface energy of the area, composed of the $CoSi_2$ (hatched area) and the exposed Si area is calculated numerically with the following additional assumptions:

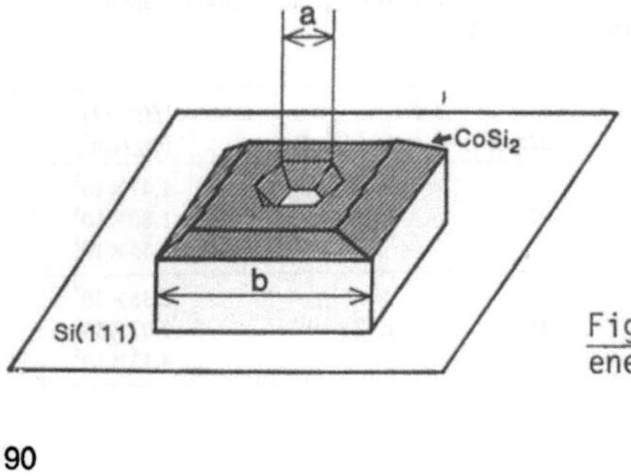

Fig.10. Model used for calculation of surface energy of silicide on square pattern

1) There is no change in the volume of the $CoSi_2$.
2) The sidewalls of the hole are faceted in (110) planes. This assumption is made because the angle between the wall and the silicide-silicon boundary was found by SEM to be about 35°.
3) The temperature dependence of the surface energy can be neglected.
4) The boundary energy between the $CoSi_2$ and Si boundaries can be neglected. This assumption is made because the boundary energy between solid-solid interfaces is usually much lower than the surface energy.

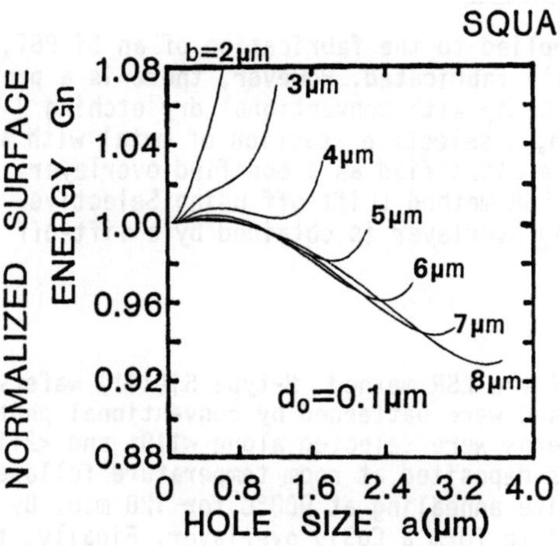

Fig.11

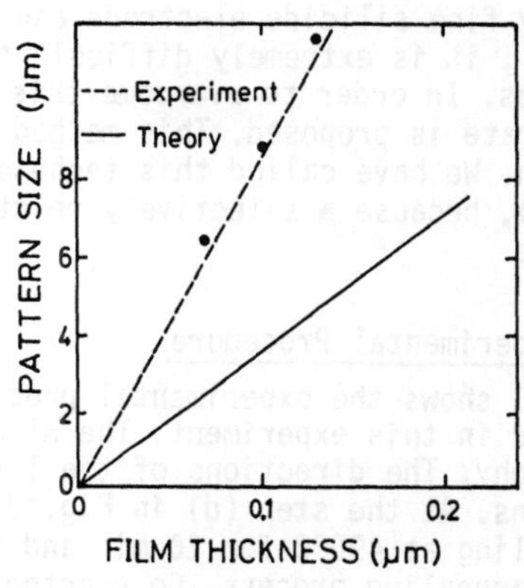

Fig.12

Fig.11. Surface energies as a function of hole size "a". The energies are normalized by the energy in the case of a completely uniform film. The parameter is the pattern size "b"

Fig.12. Critical length of pattern size at less than which the silicide should be uniform, as a function of the silicide thickness

Figure 11 shows the surface energies normalized by the energy for the uniform film as a function of the hole size "a", where the original $CoSi_2$ thickness is 100 nm. When the pattern size "b" is less than 3 μm, the total surface energy increases monotonically as a function of "a". This means that even if there is a hole at the first step of the process, there is a tendency for the hole to shrink and the film to become uniform in the subsequent annealing process, which reduces its surface energy. On the other hand, if "a" is larger than 4 μm, the surface energy increases and reaches a maximum value before decreasing. This means that the film will not become uniform if there is a hole in the film at the first step of the process. The critical value for "a", at less than which the film can be expected to be uniform after the subsequent annealing process, is calculated to be 3.5 μm.

Figure 12 shows the critical values of the pattern size "b" as a function of the initial $CoSi_2$ thickness. The calculated values, shown by the solid line, are a linear function of the initial $CoSi_2$ thickness. The broken line shows the experimental data obtained from the intercepts of the horizontal axes in Fig.8. The experimental data agree qualitatively with the theoretical values. Since experimental and theoretical values are linearly related to the thickness of the $CoSi_2$ films, the effect of patterning is explained in principle by the surface energy model. Two reasons can be considered to explain the quantitative discrepancy between the theoretical and experimental values. The first is the first-order assumption made in deriving the values of the surface energies of $CoSi_2$ and Si in the theoretical calculation. The second is that there is some ambiguity in the film thickness, because the thickness at the edge of the pattern is greater than the average $CoSi_2$ thickness owing to the swelling at the edge.

As discussed, the behavior of the epitaxial silicide/silicon system is regulated by the surface-energy minimum principle, and the silicides tend to change their shapes during annealing. The surface energy must therefore be carefully considered when uniform epitaxial silicides are required.

3.5 Patterned Overlayer by Using Selective Reaction

$Si/CoSi_2/Si$ double heterostructures can be applied to the fabrication of an Si PBT, if a very fine silicide electrode can be easily fabricated. However, there is a problem, i.e., it is extremely difficult to etch $CoSi_2$ with conventional dry etching techniques. In order to overcome this problem, a selective reaction of metal with an Si substrate is proposed. This method might be classified as a modified overlayer formation. We have called this technique the LSR method (Lift-off using Selective Reaction), because a selectively reacted $CoSi_2$ overlayer is obtained by a lift-off technique.

3.5.1 Experimental Procedure

Figure 13 shows the experimental procedure of the LSR method. N-type Si(111) wafers were used in this experiment. The Al masks used were patterned by conventional photolithography. The directions of the line patterns were selected along <110> and <211> directions. At the step (d) in Fig.13, Co was deposited at room temperature followed by annealing at 490°C for 20 min and successive annealing at 900°C for 120 min. By such an annealing process, Co reacted with Si to form a $CoSi_2$ overlayer. Finally, the masks were removed by hot H_3PO_4 and a buffered HF solution leaving $CoSi_2$ lines.

3.5.2 Experimental Results and Discussions

SEM observations showed that fine $CoSi_2$ lines had been fabricated by the LSR method along the <110> direction. Furthermore, a $CoSi_2$ line as narrow as 0.6 µm has been fabricated along the <110> direction using a mask having the space of 3 µm. This suggests that the phenomenon of shrinkage of $CoSi_2$ lines had occurred. $CoSi_2$ lines along the <211> direction have also shrunk, but the line edges have become zigzag in such

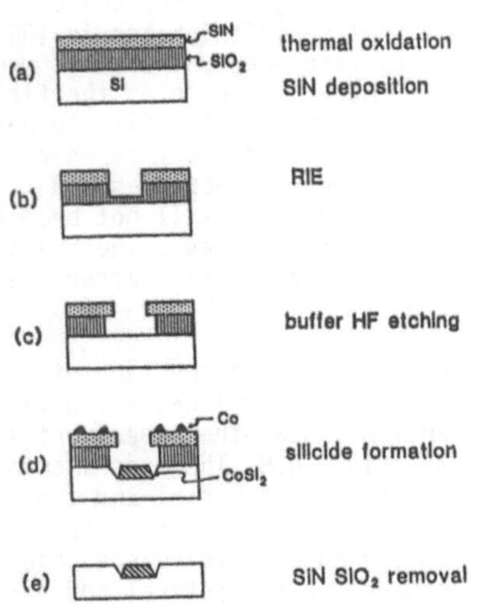

Fig.13a-e. Experimental procedure of the LSR method

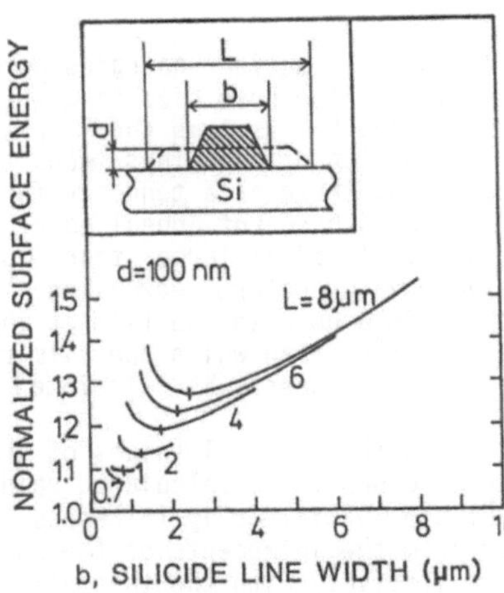

Fig.14. Surface energy as a function of the silicide line width. The inset shows a model, which indicates the cross section of the sample used in the calculation. The energy is normalized by a surface energy of a 10 µm width Si(111) surface

a manner that the line edges have been milled along the <110> direction. To give an interpretation for these observations, we have calculated the total surface energy of the system, as shown in Fig.14. This figure predicts that shrinkage of the line width can occur and that there are thermal-equilibrium line widths at which the total surface energy becomes a minimum. Figure 15 shows the relation between theoretical and experimental thermo-equilibrium line widths. It can be said from the good agreement between theoretical and experimental data that the physical model assumed is fairly reasonable.

By using the LSR method and molecular-beam epitaxy, we have successfully fabricated a PBT using a metal base to observe SIT (Static Induction Transistor) operation [14].

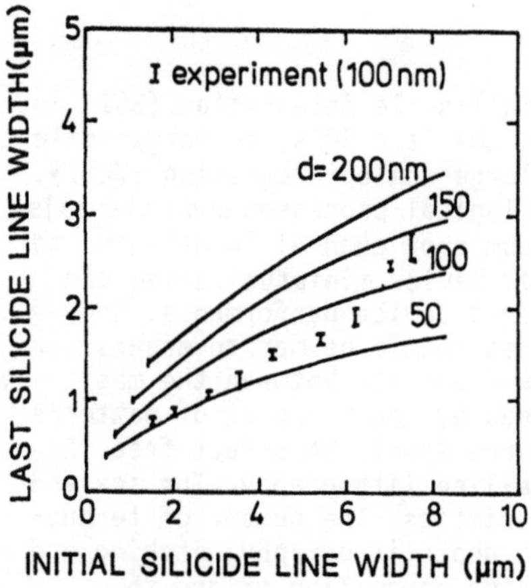

Fig.15. The silicide line width corresponding to the thermal equilibrium condition as a function of the initial line width

4. SUMMARY

There are many combinations of overlayers, that is, materials and crystalline phases respectively are divided into three categories. Even though we considered only a single overlayer, there would be cases of as many as $(3\times3)^2$. In this paper, only two typical cases have been discussed, but the author believes that the above discussions would be useful for considering and discussing other cases of overlayers.

References

1. B.E. Deal, A.S. Grove: J. Appl. Phys. 36, 3770 (1965)
2. N. Tsubouchi et al.: Jpn. J. Appl. Phys. 16, 1055 (1977)
3. V.Q. Ho, T. Sugano: Thin Solid Films 95, 315 (1982)
4. S. Kimura et al.: J. Electrochem. Soc. 132, 1460 (1985)
5. C.R. Fuller, F.J. Strieter: The Electrochm. Soc. Ext. Abstr. 64, 1 (1964)
6. T. Warabisako et al.: Tech. Dig. Intern. Electron Device Meeting '85, No.8.5, 216 (1985)
7. M.M. Rahman, S. Furukawa: Jpn. J. Appl. Phys. 23, 515 (1984)
8. S. Furukawa, H. Ishiwara: Jpn. J. Appl. Phys. Suppl. 22-1, 21 (1982)
9. S. Saitoh et al.: Appl. Phys. Lett. 37, 203 (1980)
10. E. Rosencher et al.: Electr. Lett. 20, 762 (1984)
11. K. Ishibashi et al.: Ext. Abstr. 15th Conf. Solid State Devices and Materials, No.A-1-2, 111 (1983)
12. K. Ishibashi, S. Furukawa: Jpn. J. Appl. Phys. 24, 912 (1985)
13. K. Ishibashi, S. Furukawa: Trans. IEEE, ED-33, No.3 (1986) (to be published)
14. K. Ishibashi, S. Furukawa: Tech. Dig. Intern. Electron Device Metting '84, No.34.8, 868 (1984)

Self-Aligned Growth of Microstructures

Y. Pauleau

Centre National d'Etudes des Télécommunications, B.P. 98,
F-38243 Meylan, France

1. INTRODUCTION

Integrated circuit complexity has progressed from small-scale integration (SSI) in the early 60's, to medium-scale integration (MSI) in the late 60's, to large-scale integration (LSI) in the 70's, and finally to very large-scale integration (VLSI). For ultra large-scale integration (ULSI), new technological processes and materials are needed to reduce device dimensions (typically submicron channel length) and to increase the number of components per chip (10^6-10^8). Device miniaturization can result in reduced unit cost per function and in improved device performance. Some basic requirements for microminiaturization of devices should be met to ensure performance and reliability of integrated circuits. The disparity between the mask layout dimensions and the final device geometry caused by the transfer of patterns should be minimized. The materials in device structures should be defect free. A planar surface topology is highly desirable for fine-line lithography. The scaled-down structures should maintain good device characteristics. The number of technological steps such as oxidation, diffusion, masking, photolithography, etching and deposition should be reduced. Consequently, methods for deposition or growth of materials in selected areas on thin film structures defined by optical lithography or by electron beam delineation techniques have become increasingly attractive in the production of micron-size devices.

In this paper, an overview of recent techniques used for self-aligned growth of microstructures is presented. The most important results in the selective epitaxial growth of silicon are reported in the first section. In the second section, the selective chemical vapor deposition of metals (Ag, Mo, Re and W) is discussed and the major applications of selective W deposition are given. Finally, in the third section of this article, the principle of the self-aligned silicide technology as well as the electrical characteristics of silicided devices are described showing the interest of this technique in submicron technology.

2. SELECTIVE EPITAXIAL GROWTH OF SILICON

Various technologies for lateral isolation of submicron devices are under intensive investigations in order to overcome the problems associated with the local oxidation of silicon (LOCOS) [1]. The well-known drawbacks of the LOCOS technique include an uncontrolled reduction of usable chips and a line width variation caused by the bird's beak encroachment, a lower yield attributed to the white-ribbon effect, and a degeneration of the electrical characteristics due to the outdiffusion of the channel stopper implantation [2]. Moreover, numerous process steps are needed to perform a planar surface topology. The selective epitaxial growth (SEG) of silicon can provide an answer as a lateral isolation technology. The SEG technique consists of the deposition of an epitaxial layer on selected areas of single crystal Si wafers which are exposed via windows etched through protective insulating films (SiO_2, Si_3N_4, Al_2O_3,...). The selective epitaxy offers such unique advantages as the growth of silicon in controlled geometries on a substrate surface (self-aligned growth), the formation of independent n- and p-type wells required in complementary metal-oxide-semiconductor (CMOS) technology, and the deep isolation which is highly effective in the prevention of latch-up phenomena. Furthermore, a significant reduction in the number of technological steps (highly desirable in VLSI technology) can be accomplished through the use of SEG instead of LOCOS techniques.

2.1 Selective Epitaxy at Atmospheric Pressure

The selective epitaxy of silicon on oxide masked silicon substrates was first achieved by JOYCE and BALDREY [3] from $SiCl_4$-H_2 gas mixtures at atmospheric pressure. These experiments were carried out at 1200°C with a molar ratio, $SiCl_4/H_2$, of 0.02; at total flow rates below 1 l/min, the epitaxial growth of Si layers up to 25 µm thick occurred without any polycrystalline silicon (poly-Si) deposition on the oxide masks. This promising pioneer work has spurred on the research in this field.

The feasibility of SEG could easily be demonstrated with thick oxide masks, thin preferential epitaxial layers, and highly symmetrical patterns providing a large percentage of unmasked area. Nevertheless, a lot of problems remain to be solved in order to meet the requirements for fabrication of improved devices. SCHNABLE et al. [4] achieved acceptable SEG of silicon by the H_2 reduction of $SiCl_4$ at 1200°C employing growth rates of 0.3-1 µm/min. However, the Si layer was thicker at the edges immediately adjacent to the oxide mask than in the center of the unmasked regions, and the density of crystallites grown on the oxide mask was lower in regions close to the unmasked area than in regions further away from the unmasked areas (ribbon geometry). These results were confirmed by OLDHAM and HOLMSTROM [5] who proposed a model from which the concentrations and fluxes of the interesting gaseous species in the neighborhood of the window can be derived. This model assumes an equilibrium at the Si surface and a purely diffusive transport through the gas phase. The concave growth surfaces are explained by the constriction of the flow near the edges of the windows. Furthermore, the derived concentration profiles explain the appearance and relative size of the band of oxide free from Si overgrowth which surrounds each window in the growth experiments.

The two major problems in the SEG of silicon are the prevention of etching of the oxide mask, and the avoidance of the nucleation and growth of poly-Si on the surface of the masking material. These problems are a consequence of the following reactions:

$$Si\ (s) + SiO_2\ (s) \rightarrow 2\ SiO\ (g) \tag{1}$$

$$SiCl_4\ (g) + 2\ H_2\ (g) \rightarrow Si\ (s) + 4\ HCl\ (g) \quad . \tag{2}$$

The interaction between the thermally-grown oxide mask and the underlying silicon [reaction (1)] can produce undesirable effects such as oxide lifting. Elementary Si can be deposited on the mask regions resulting from the reaction (2) over 1000°C and this Si deposit would disappear partly by reacting with the underlying SiO_2 according to the reaction (1).

Silicon deposition on SiO_2 was inhibited when HCl or HBr gas was injected into the gas stream simultaneously with $SiCl_4$ [6]. Moreover, the flatness of the epitaxial Si layer was closely controlled by varying the amount of halide injected into the epitaxial reactor. If the supersaturation of the vapor phase with silicon in the reaction zone is kept low, epitaxial growth on exposed Si surfaces was performed while nucleation on the amorphous masking layer was still inhibited [7]. These investigations have shown that selective growth can be successfully achieved independently of the ratio of the surface area of silicon to the surface area of the oxide mask. The best growth conditions correspond to a substrate temperature of 1200°C, and a growth rate of 0.5-1 µm/min. However, ridges were observed at the boundary of the oxide mask. This effect could be reduced either by lowering the deposition temperature or the growth rate, or by using Si_3N_4 as a masking material. The formation of ridges results from the reaction of the SiO_2 mask with the Si substrate [7, 8].

A good selectivity was obtained by controlling the amount of HCl injected with $SiCl_4$-H_2 mixtures. 5 µm thick Si layers were grown at 1120°C under $SiCl_4$ and HCl partial pressures of 5.6 Torr and 4.2 Torr, respectively [9]. The growth rate of Si islands was about 1 μm/min whereas the growth rate of Si films on the accompanying unmasked Si wafer was only 0.4 μm/min. A similar effect of the deposition surface area on the growth rate was also observed using SiH_4-H_2 mixtures as a starting material [10].

The SEG of Si (or Ge) could also be achieved by pyrolysis of SiH_4 (or GeH_4) in hydrogen at atmospheric pressure [10]. Using the SiH_4-HCl-H_2 gas system, the growth rate, selectivity, and Si layer profile was found to be directly dependent on the HCl concentration in the gas phase [8]. Furthermore, the formation of ridges was completely eliminated with Si_3N_4 masked Si wafers. Without HCl in the gas phase, the epitaxial growth from SiH_4-H_2 mixtures was found to be selective for temperatures greater than 1200°C [9]. The prevention of growth on the SiO_2 mask was probably the result of the reaction (1); above 1270°C, this reaction leads to a lifting of the oxide mask.

Using the $SiHCl_3$-HCl-H_2 gas system, the refilling of deep grooves with vertical walls etched in (110)-oriented Si wafers was demonstrated without using a mask on top of the surface [11]. The growth rate on the groove walls increased with an increase in the depth into the grooves so that a planar surface was created after refilling the grooves with variables depths. The adjustment of the Si/Cl ratio accomplished by a convenient addition of HCl in the gas mixture was found to be the key factor in this SEG technique.

2.2 Selective Epitaxy at Reduced Total Pressure

More recently, the SEG of Si layers was carried out under reduced total pressure in epitaxial reactors with radiant heating from either $SiCl_4$ [12] or SiH_2Cl_2 [1, 2, 13-16] as a starting material.

2.2.1 Selective Epitaxial Growth from the $SiCl_4$-HCl-H_2 Gas System

2.2.1.1 Experimental Procedure

U-grooves were etched anisotropically in (100)-oriented Si wafers using a chlorine plasma with SiO_2 as a masking material. In order to coat the sidewalls with an SiO_2 layer, the wafers were conformally covered with an oxide layer produced by chemical vapor deposition (CVD) and then the oxide was removed from the bottom of the grooves by reactive ion etching. The ratio, bare silicon to total surface area is only 2%. The selective epitaxy was carried out at 1015°C with a total flow rate of 140 l/mn and a total pressure of 50 Torr.

2.2.1.2 Results and Discussion

The epitaxy is selective only within a relatively narrow range of HCl concentration in the gas phase (Fig.1). For a non-selective epitaxy, the growth rate within the grooves is greater than the growth rate of silicon films on unmasked wafers. For a selective epitaxy, the reverse situation can be observed. Moreover, the growth rate within the grooves is very strongly dependent on the HCl partial pressure (Fig.1).

The HCl concentration effect on the growth rate of silicon on unmasked wafers is very slight, probably due to the fact that the growth is mass-transport limited. For deposition on SiO_2-masked Si wafers, smaller amounts of reactants and products are consumed and evolved; the deposition process may be limited by the surface reaction and the growth rate is strongly dependent on the HCl partial pressure.

The major difficulty in this process arises from the effect of the pattern geometry on selectivity and uniformity. The probability of poly-Si growth increases with an increase in the masked surface area, i.e., local supersaturation.

The decrease in the window size leads to an enhanced sensitivity of the growth rate to the HCl concentration and may cause reproducibility problems. The best pat-

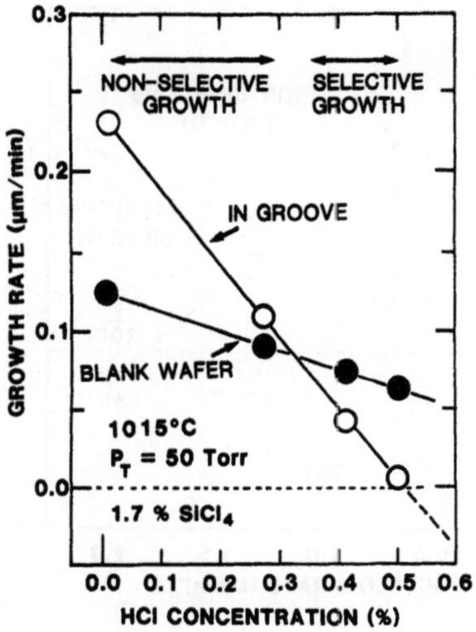

Fig.1. Growth rate of Si layers in grooves and on blank wafers as a function of HCl concentration (from [12])

tern uniformity is obtained by reducing the total pressure, lowering the deposition temperature, and with HCl in the gas phase. Under these conditions, the Sherwood number [17] can be reduced and the growth kinetics is limited by the surface reaction.

2.2.2 Selective Epitaxial Growth from the SiH_2Cl_2-HCl-H_2 Gas System

The fabrication of electronic devices with a lateral isolation technology based on the SEG process has only been investigated for silicon epitaxial layers grown from dichlorosilane at reduced pressure. Using SiH_2Cl_2-HCl-H_2 gas mixtures under a total pressure of 1 atm, Si epitaxial layers are grown at 1060°C, and the growth kinetics is controlled by the mass-transport process. By reducing the total pressure, the process passes from a transport-controlled regime towards a surface reaction controlled regime [2]. In addition, the growth temperature is lowered by 100°-150°C, epitaxial layers with a nearly planar surface are obtained and a perfect selectivity can be reached more easily. The degree of selectivity also depends on the pretreatment of the Si surface and the ratio of unmasked to masked area [16].

As the deposition temperature decreases, the thickness uniformity of the layers is improved and the amount of HCl required for good selectivity decreases (Fig.2). The selectivity decreases when the total flow rate is increased. The growth rate of Si layers depends on the crystallographic orientation of Si substrates. In the temperature range 900°-1000°C, the epitaxial growth is rate-limited by the surface reaction and the activation energy is 24 kcal mol^{-1}, regardless of the orientation of the substrates [13].

To ensure the best surface morphology of Si layers, the windows through the SiO_2 mask should be opened by a partial (90%) reactive ion etching followed by a wet etching to remove the last 200 nm of oxide, and the pre-cleaning treatment by HCl should be carried out at reduced total pressure [1]. The oxide lifting is avoided by coating the oxide sidewalls with either nitride or poly-Si layers. With an Si_3N_4 coating, four symmetrical facets corresponding to (311) planes are formed at the edge of the epitaxial layer. Unfaceted epitaxial layers with fairly flat surfaces are grown in poly-Si coated windows [13,14]. The defects induced at the sidewall interface can be suppressed by re-orienting the oxide mask so that the sidewall is along a (100) direction [1].

A schematic diagram of the isolation fabrication process using SEG of Si layers is represented in Fig.3.(100)-oriented Si wafers are covered with a thermally-grown oxide layer 1-2 μm thick. Windows with steep sidewall profiles are opened through the oxide by anisotropic reactive ion etching. A 0.1 μm thick Si_3N_4 or poly-Si film is deposited by a low pressure chemical vapor deposition (LPCVD) process followed

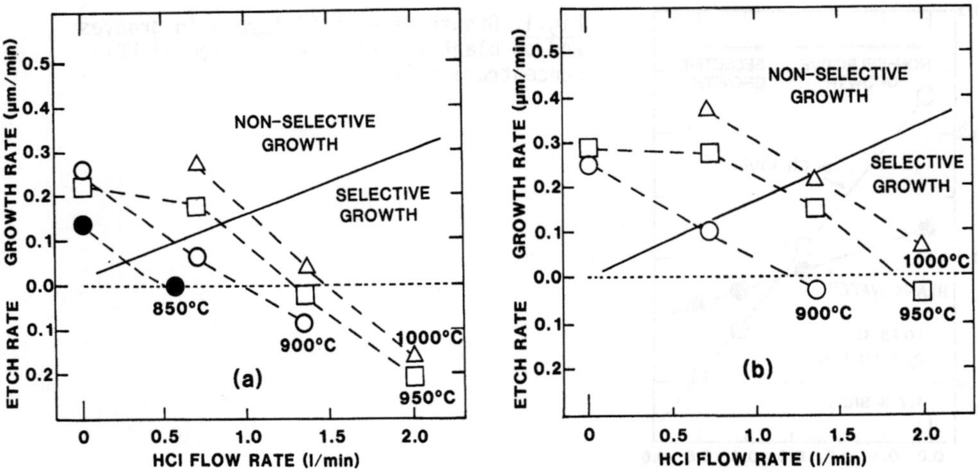

Fig.2. Growth rate and etch rate of Si layers. SiH_2Cl_2 flow rate = 0.30 l/min (a) total pressure = 25 Torr; total flow rate = 77 l/min. (b) total pressure = 80 Torr; total flow rate = 133 l/min (from [1])

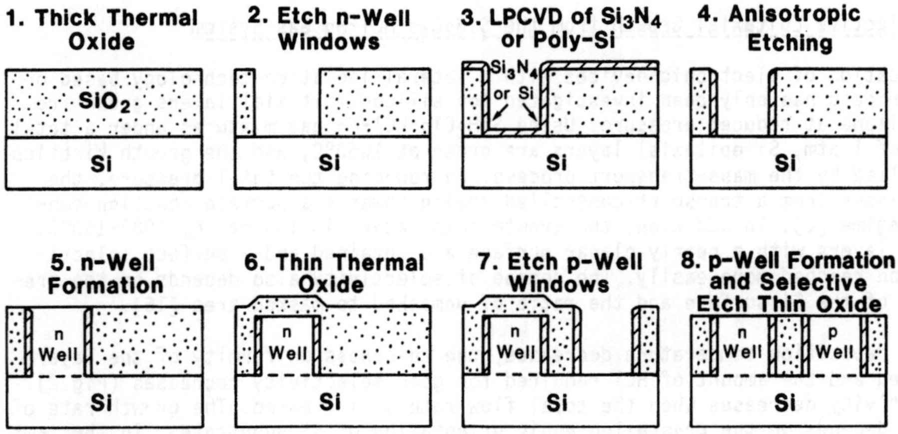

Fig.3. Schematic diagram of the isolation fabrication process using SEG of Si layers

by an anisotropic etching. Then, the n-type well formation is performed by SEG using SiH_2Cl_2-HCl-H_2 gas mixtures and AsH_3 as dopant. After thermal oxidation, the windows for p-type wells are etched through the SiO_2 layer. The sidewalls are coated with nitride or poly-Si, and the formation of p-type wells occurs by SEG using B_2H_6 as dopant. Then, the oxide layer on the n-type well surface is selectively etched. SEG processing allows the formation of independent p- and n-type wells for CMOS technology.

CMOS devices with different channel widths have been fabricated on these Si wafers [15]. The gate oxide and the poly-Si for gate metallization were 20 and 400 nm thick, respectively. By extrapolating the proportional relationships between the channel width and the channel conductance for n-channel devices, the bias values in the channel width with SEG and LOCOS isolation techniques are 0.2 and 1.2 µm, respectively. For p-channel devices, similar differences can be observed. Consequently, the narrow channel effect is less for SEG than for LOCOS isolation structures.

3. SELECTIVE CHEMICAL VAPOR DEPOSITION OF METALS

New materials and processes capable of matching the increased packing density and performance requirements are investigated for metallization of VLSI circuits. For example, diffusion barrier layers with low contact resistance are needed to protect shallow junctions from aluminum spiking, and low resistance materials are required to shunt the poly-Si gate metallization. Refractory metals present desirable properties such as high conductivity, ability to withstand high temperature, good diffusion barrier capability and are promising candidates as contact and diffusion barriers as well as poly-Si gate shunts and interconnect lines.

The selective deposition process of metals on Si wafers can be based on the Si reduction of metal halides. A necessary condition is that the following reaction

$$4/m\ MX_m\ (g) + Si\ (s)\ \rightarrow\ 4/m\ M\ (s) + SiX_4\ (g) \qquad (3)$$

proceeds until the completion, i.e., without any formation of a stable metal subhalide. In the reaction (3), M is a metal of valency, m, and X is a halogen. Numerous metal halides can be thermodynamically reduced by Si [18]. Nevertheless, the selective deposition was demonstrated only for Ag, Re, Mo and W using metal fluorides as a starting material (Table I). Since the halide reduction can only occur on Si regions to the exclusion of SiO_2, this process is strictly selective. However, these metallic layers are limited in thickness because of the passivation effect of metal film on the underlying silicon.

Table I. Selective metal deposition

Metal	Source	Deposition temperature [°C]	Reaction	Reference
Ag	AgF	80-600	Si reduction	[18]
Mo	MoF_6	300	Si reduction	[19]
Re	ReF_6	300	Si reduction	[20]
W	WF_6	280-320	Si or H_2 reduction	[26]

To produce thicker metal films, the selective deposition can proceed by the H_2 reduction of metal fluoride:

$$MF_m\ (g) + m/2\ H_2\ (g)\ \rightarrow\ M\ (s) + m\ HF\ (g)\ . \qquad (4)$$

The capability of this reaction in selective growth has been explored only for W deposition.

3.1 Selective Growth by the Si Reduction of Metal Fluorides

The process may be carried out in a flow system for metal fluoride with high volatility such as WF_6, MoF_6, and ReF_6 or in a vacuum evaporator for metal fluoride with low vapor pressure such as AgF. Owing to problems of supplying and the high cost of pure metal halides, only few experimental attempts deal with selective growth of Mo and Re [19, 20].

The selective growth of silver on oxide-patterned Si wafers with (111) orientation was accomplished through the reaction of Si with AgF vapor [18]. Silver fluoride was evaporated from a graphite crucible in a vacuum chamber. The Si reduction of AgF can occur only after removing the native oxide layer on the Si surface by etching in a diluted HF solution just prior to loading of samples in the reaction chamber. Smooth and adherent Ag films were produced at a substrate temperature within a range of 80°-600°C. 1 µm thick Ag layers were grown at 400°C for 2 to 5 min. The thickness of consumed Si is lower than that of deposited Ag films by a factor of 3.4. Between 100° et 200°C, the thickness limit of Ag films was about 100-200 nm. At lower temperatures, AgF condensed on Si wafers, this temperature being less than the evaporator temperature. After the selective Ag growth and unloading of samples, AgF deposited on the wafer surface was removed immediately by rinsing in deionized water because unreacted fluoride can rapidly deteriorate silver and oxide surfaces exposed to air. Polycrystalline silver is preferentially oriented with the (111) planes parallel to the substrate surface. The possibility of fluorine incorporation in the films and electrical characteristics of the material have not been reported. Consequently, this study demonstrating the feasibility of the process is largely exploratory.

Since tungsten hexafluoride is a gaseous compound at room temperature, the Si reduction of WF_6 for selective deposition can be carried out in a conventional CVD reactor. Tungsten films are grown at temperatures within a range of 250°-800°C using WF_6-N_2 or Ar gas mixtures at atmospheric or reduced total pressure. Prior to deposition, the Si surface must be etched in a diluted HF solution to remove the passivative native oxide layer. This etching step leads to a slight widening of the windows etched through the oxide mask. The W films attain a thickness limit of about 5-100 nm; the value of this thickness as well as the corresponding amount of Si consumed are dependent on the substrate temperature and deposition time [21]. The ratio, thickness of Si consumed to thickness of W deposited is about 1.9, according to the stoichiometry in reaction (1). Under given deposition conditions, a conventional wet chemical cleaning of phosphorus-doped poly-Si samples in diluted HF solution provides a maximum obtainable thickness of W films of about 40 nm whereas, with a low power argon plasma treatment, approximatively 90 nm thick W films can be deposited reproducibly [22]. Under atmospheric pressure, the growth rate of the films is dependent on the flow rate of carrier gas and WF_6 partial pressure. The activation energy of the Si reduction of WF_6 is only 3 kcal mol^{-1} over the temperature range of 400°-700°C. This very little temperature dependence of the growth rate suggests that the diffusion of WF_6 towards the Si surface is the rate-limiting step [23].

An encroachment of tungsten along the SiO_2-Si interface can occur when the metal layer is grown on oxide-patterned Si wafers [21, 24]. The depth of the encroachment is dependent on the substrate temperature and WF_6 partial pressure [21].

3.2 Selective CVD of Tungsten Films via the H_2 Reduction of WF_6

The selective feature of tungsten deposition via the H_2 reduction of WF_6 was first mentioned by CUOMO [25] who used borosilicate glass or soda-lime glass covered with metal dots as substrates and a cold wall CVD reactor. The feasibility of selective W deposition on oxide-masked Si wafers using hot wall reactors (offering higher wafer throughput than cold wall reactors) was reported in 1978-1980 [26]. Both these promising results and the requirements for VLSI technology have spurred on research and development efforts in this selective deposition process because of its unique advantages such as self-aligned growth, inherent etching of the Si surface leading to well-adherent films and reduction in photolithographic and etching steps.

The selective deposition of tungsten from the WF_6-H_2 gas system is performed at a low temperature (less than 300°C) in a low pressure CVD reactor. Argon can be used as a buffer gas. The native oxide layer on the Si surface is etched in a diluted HF solution by dipping the sample only for a few seconds to avoid an undesirable widening of windows. During the first few minutes of the deposition process, tungsten films are grown by the Si reduction of WF_6 which is considerably faster than the H_2 reduction. Then, once the tungsten film has masked the underlying Si substrate, the W deposition proceeds by the H_2 reduction of WF_6, and the film thickness increase linearly with increasing deposition time [21, 27]. The activation energy of the H_2 reduction of WF_6 is 16.5 kcal mol^{-1} over the temperature range 280°-750°C. For relatively long deposition times, nodules of tungsten appear on the oxide surface near the pattern edges. Similar failures of selectivity can be observed for shorter deposition times using higher substrate temperatures.

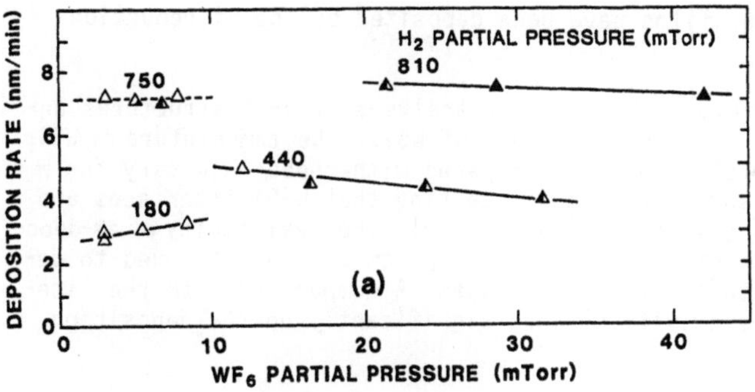

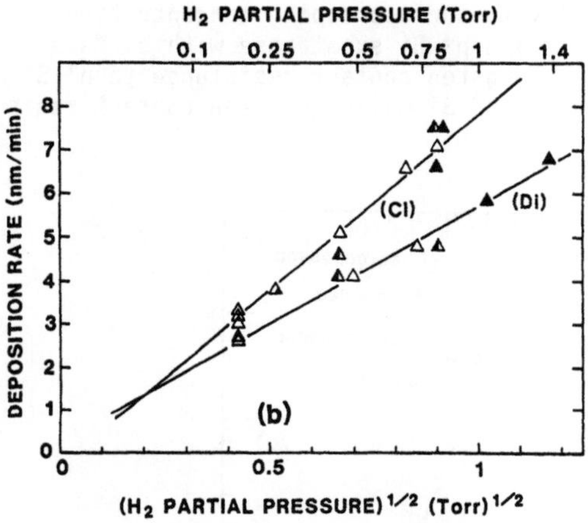

Fig.4. Deposition rate of W films versus (a) WF_6 partial pressure for films produced from WF_6-H_2Ar (solid line) or WF_6-H_2 (dashed line) gas systems; (b) square root of H_2 particle pressure (from [28])

The deposition rate of W films produced from WF_6-H_2-Ar or WF_6-H_2 gas systems at 285°C was found to be essentially independent of the WF_6 partial pressure (Fig.4a). The H_2 partial pressure varied between 180 and 810 mTorr, and the Ar partial pressure was adjusted to maintain a total pressure of 1.4 Torr. The linear dependence of the growth rate of these films on the square root of the H_2 partial pressure is shown in Fig.4b. During these experiments, the WF_6 partial pressure could be varied over a wide range (5-50 mTorr) without showing any effect on the deposition rate. The curve (Cl) results from experiments carried out with the quartz tube, quartz boat, and dummy wafers free from any W deposit. After several runs, the quartz ware is covered progressively with a W deposit, and the growth rate versus the square root of the H_2 partial pressure is represented by curve (Di). In other words, the growth rate is dependent on the deposition surface area, i.e., on the cleanliness of the quartz ware. This decrease in deposition rate was attributed to the effect of the HF partial pressure on the surface reaction. A reaction mechanism based on the Langmuir-Hinshelwood model was proposed [28]. In Figs. 4a and b, open, half-closed and closed symbols mean perfect selectivity, selectivity failures and non-selective deposition processes, respectively. The selectivity deteriorates with increasing WF_6 or H_2 partial pressures.

These W films display a slight (100) preferential orientation, as already noticed, for metallurgical W coatings produced at higher temperatures. Micrographs in scanning electron microscopy of either the surface of the samples after metal stripping or the cross-section exhibit a cavity in the Si substrate at the border line of the oxide pattern [29], i.e., a larger amount of silicon is etched around the periphery of the windows than in the center. A similar result has also been reported for selective growth of silver via the Si reduction of AgF [18]. Micrographs obtained by transmission electron microscopy show a tunnel-type defect in the Si substrate [24, 30]. The tunnels are observed only at the rim of the windows suggesting that the interfacial stress could aid in their formation. These defects are not visible in

Si substrates onto which tungsten films have been deposited by the Si reduction of WF_6.

X-ray diffraction and Rutherford backscattering analyses of W-Si structures annealed at 650°C for 1 h exhibit the tetragonal form of WSi_2. The temperature of WSi_2 formation by interdiffusion is relatively low compared with those necessary for W films deposited by evaporation and sputtering suggesting that W-Si interfaces and W layers are free of impurities (oxygen, in particular). The resistivity of as-deposited W films was found to be within the range 12-17 μΩ cm and was observed to decrease when the thickness of metallic films increases. A comparison with the literature data shows that the W resistivity depends significantly on the deposition temperature [29].

The contact resistance of Al/W/Si (substrate) metallization was measured [31]. The sensitivity of specific contact resistivity to surface doping concentration was investigated for selective W contact to n^+ and p^+ Si substrates with surface concentrations from 10^{18} to 10^{20} cm^{-3} (Fig.5). Tungsten contact resistance to n^+ Si is about a factor of 20 lower than that of Al to n^+ Si while tungsten contact resistance to p^+ Si is comparable to that of Al.

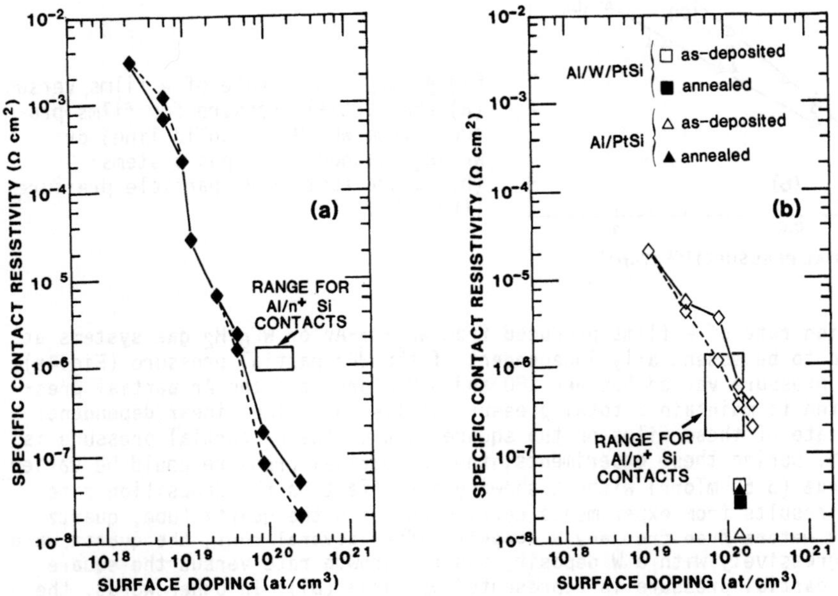

Fig.5. Specific contact resistivity for as-deposited (solid line) and annealed (dashed line) Al/W/mono-Si structures as a function of (a) phosphorus surface concentration; (b) boron surface concentration. Annealing conditions: 450°C, 30 min in forming gas (from [31])

The main applications of selective W deposition expected in VLSI technology consist of (i) the formation of contact and diffusion barriers between Si and Al or other metallization (100 to 150 nm thick W films would be convenient for this purpose), (ii) the refilling of contact windows or vias through dielectric layers to alleviate planarization and step coverage problems (considerable additional work remains to be done in order to selectively deposit the 500-800 nm thick W film needed in this application).

4. SELF-ALIGNED GROWTH OF METAL SILICIDES

The speed gains resulting from short channel transistors can be seriously affected by the RC time delays caused by the high resistivity of fine lines in poly-Si. In the scaling-down of MOS devices, silicides associated with an underlying layer of poly-Si (polycide structures) are increasingly used as gate metallization in order to reduce the sheet resistance to about 1 Ω/□. However, the sheet resistance of shallow n^+ and p^+ regions remains relatively high (5-50 Ω/□ and 50-200 Ω/□, respectively). Even with a polycide technology, the conductivity of source/drain diffused regions can become a limiting factor. The reduction in sheet resistance (to about 10 Ω/□ or less) of source/drain (S/D) zones is highly desirable for improving the device and circuit performance when devices are scaled down to submicron dimensions. The silicidation of S/D regions can also lead to acceptable contact resistances for contact areas below 1 μm². For this purpose, a localized growth process of metal silicide appears to be convenient so that to take full advantage of submicron technology.

Noble metal silicides (PtSi and Pd_2Si) formed by interdiffusion have already been used as contact barriers to diffused zones. However, this silicidation technology presents certain drawbacks including the need for an extra mask level and the lack of self-alignment resulting in an adverse effect on the current density [32]. Furthermore, PtSi agglomerates at high temperature ($\geqslant$ 850°C) and reacts with aluminum at temperatures as low as 350°C. This low stability of layered structures has an adverse impact on the device reliability and process flexibility [33]. Refractory metal silicides are preferable owing to their thermal and chemical stability.

4.1 Fabrication of Self-Aligned Silicide Structures

The most common process sequence for producing self-aligned metal silicide is shown in Fig.6. After ion implantation of S/D regions, an oxide layer is chemically deposited at low temperature and directionally etched down to the diffused zones to leave sidewall spacers along the poly-Si gate. A metal blanket about 40 nm thick is deposited on the sample and the metal silicide forms on the silicon regions by interdiffusion of elements during heat treatment at temperatures within the range of 600°-800°C. Then a selective etch is required to remove the unreacted metal over the oxide areas.

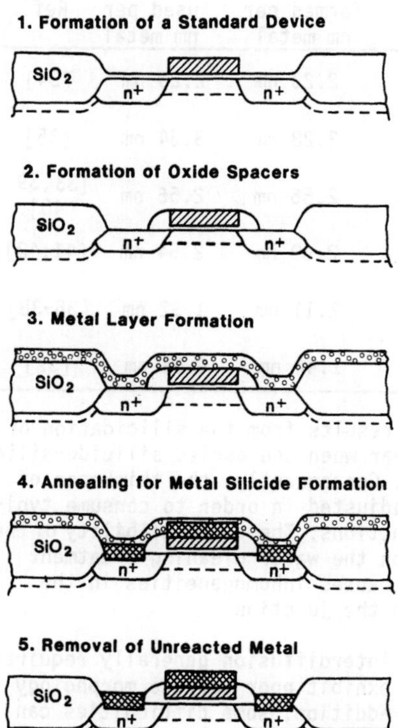

Fig.6. Process sequence of the self-aligned metal silicide (salicide) technology

To ensure the formation of metal silicide and a satisfactory result from the process, various requirements should be met. An effective etching of the native oxide layer on Si surface should be made to prevent any inhibition. The metal should react similarly with both poly-Si and single crystal Si. The salicidation should occur without any bridging between gate and source/drain zones as a result of Si diffusion in the metallic layer. The metal silicide with low resistivity should have good thermal stability during subsequent annealing treatments and device passivation in order to preserve gate and junction integrity. The contact metallurgy system should provide low contact resistance to S/D diffused regions and protect shallow junctions from degradation during contact annealing [32, 34].

Several metals have already been investigated for the self-aligned silicide (salicide) technique (Table II). Titanium disilicide appears to be an attractive candidate owing to its low resistivity and the Schottky barrier height of 0.55 eV to n-type silicon which is advantageous in the formation of ohmic contact [34]. Furthermore, $TiSi_2$ is chemically stable with Al up to 550°C. If a diffusion barrier is proved to be necessary to preserve the contact integrity, Ti or TiN can easily be deposited on $TiSi_2$ by sputtering. The native oxide layer on Si surface can be reduced by titanium which makes the silicide formation easier. This desirable property becomes a drawback when titanium reacts with the oxide mask. The use of $CoSi_2$ as a shunt of poly-Si is prohibited at temperatures above 700°C since Co atoms diffuse through the underlying unreacted poly-Si to the oxide below [35]. Tungsten disilicide, WSi_2, presents many advantages over PtSi, $TiSi_2$ and $MoSi_2$ [41]. Its good thermal stability makes high temperature (> 900°C) processing possible without detriment to the gate oxide and p-n junctions. A passivative SiO_2 film can be conveniently produced by thermal oxidation of the polycide structure. Moreover, WSi_2 is a good diffusion barrier against Al-Si interaction and the reaction of W with SiO_2 is thermodynamically unfavorable.

Table II. Properties of metal silicides used in salicide technology

Metal Silicide	Reaction Temperature [°C]	Selective etchant	Resistivity [μΩ cm]	Silicide formed per nm metal	Silicon used per nm metal	Ref.
$TiSi_2$	550-600	NH_4OH/H_2O_2	13-16	2.23 nm	2.26 nm	[34]
$CoSi_2$	600	HCl/H_2O_2	22-28	3.28 nm	3.64 nm	[35]
$MoSi_2$	600	NH_4OH/H_2O_2	20	2.56 nm	2.56 nm	[33,39 40]
WSi_2	850	$K_3Fe(CN)_6/KOH$	24	2.69 nm	2.54 nm	[41,42]
PtSi	500	aqua regia	28-35	2.11 nm	1.32 nm	[36-38]
Pd_2Si	400	KI/I_2	30-35	1.44 nm	0.68 nm	[32]

A significant consumption of monocrystalline Si results from the silicidation of S/D regions. Detrimental leakage currents can appear when the buried silicide-silicon interface is closer than 20-50 nm to the junction. Consequently, the thickness of the metal layer deposited on Si wafers should be adjusted in order to consume typically only 0.1 μm of silicon above the shallow junctions. The reproducibility of the whole salicide technique is critically dependent on the wafer cleaning treatment [43]. A partial removal of the native oxide layer causes inhomogeneities in the silicide formation and a local penetration through the junction.

The formation of refractory metal silicides by interdiffusion generally requires high temperature annealing; these silicide layers exhibit poor surface morphology and irreproducible electrical characteristics. In addition, some difficulties can arise from heat treatments at high temperatures such as bridging of gate and S/D regions, lateral formation of silicide, interaction between metal and the SiO_2 masks in the case of titanium. The reduction in temperature for silicidation by interdiffusion is highly recommended. For this purpose, various experimental approaches can be utilized. With the help of ion bombardment, refractory metal silicides ($MoSi_2$, WSi_2) are readily formed at a low temperature (about 350°C) [44]. This ion beam induced (IBI) mixing technique consists, for example, of an As^+ implantation through the metal blanket deposited on the Si wafer. Owing to this implantation, metal and Si atoms are intermixed in the vicinity of the interface, and the impurity atoms responsible for inhibition in silicide formation are dispersed [42]. Furthermore, when group III or V ions are used as implants, both silicide-Si ohmic contacts and shallow p-n junctions located directly below the contact can be formed simultaneously [40, 45, 46]. Since very high implanted energies are required to pass through a layer of metal with high atomic number, this technique could not be easily extended to small dimension devices (the depth of shallow junctions would be difficult to control).

To reduce the lateral diffusion of Si atoms in the metal layer, the interdiffusion is carried out via a rapid thermal annealing (RTA) of samples. The RTA process allows unwanted dopant redistribution and outdiffusion of dopants from the poly-Si and S/D regions to be minimized. Another technique for avoiding lateral silicide formation (and metal-SiO_2 interaction) consists in a partially converting metal into sub-silicide at a low temperature, then the free metal is selectively removed and finally both the complete silicide formation and the reduction in resistivity of the material are achieved during a thermal annealing at a high temperature [32, 35]. For example, an Si wafer covered with a Ti layer is annealed at 550°-600°C for a pre-reaction, the unreacted metal is stripped in an H_2O_2-NH_4OH-H_2O (1:1:5) solution, and finally the sample is annealed at 800°C for 30 min to form stoichiometric $TiSi_2$ with a low resistivity [34]. In the case of titanium, the lateral silicide formation can also be blocked by introducing nitrogen in the reaction chamber during thermal annealing of the samples. Nitrogen atoms diffuse readily through Ti and saturate all grain boundaries leading to a probable reduction in Si atoms moving through the silicide layer into the titanium interface region [43].

A diffusion barrier between the Al(Si) metallization and the silicided S/D regions is needed because for many applications a complete removal of radiation damage by annealing the samples is required [32]. Many technological steps and difficulties in the salicide technology would be eliminated by using the selective deposition process to produce metal films on poly-Si and S/D regions. The selective deposition of W films about 40 nm thick appears to be feasible via the Si or H_2 reduction of WF_6. The formation of tungsten disilicide occurs at 650°C by annealing the W_{CVD}/Si structures [29]. This relatively low temperature makes the selective W deposition in salicide technique very attractive. However, a considerable amount of silicon is consumed by both the reduction of WF_6 and the silicide formation; this point remains a major concern questioning the viability of the process.

4.2 Electrical Characteristics of Salicided Devices

The integrity of 0.25 µm deep junctions with self-aligned $TiSi_2$ has been investigated [34]. Excellent electrical characteristics and very low leakage currents have been obtained from Al-Cu-Si or Al-Cu-Si/Ti metallurgy systems associated with $TiSi_2$ used as a contact barrier.

The removal of the native oxide layer on the Si surface by etching in a diluted HF solution followed by a sputter-etching of the surface ensures a reproducible and uniform Ti-Si interface and a lower sheet resistance [43]. The sheet resistance measured on p^+ diffused zones is reduced by a factor of 10 using $TiSi_2$ (Table III), and the leakage current of n^+/p diodes of 7×10^{-9} A/cm² is comparable to leakage currents of devices produced without silicide (9×10^{-9} A/cm²). A minimum sheet resistance of 1.6 Ω/□ is reached when $CoSi_2$ is produced by annealing at 600°C for 20 min from 45 nm thick cobalt films deposited on n^+ poly-Si [35]. Tungsten disilicide obtained by RTA of 35 nm thick W films deposited on As^+ implanted monocrystalline Si

Table III. Comparison of sheet resistances of gate and S/D regions, and series resistances of S/D regions measured on (a) silicided devices; (b) unsilicided devices (from [47])

	Sheet resistance of gate [Ω/□]	Sheet resistance of S/D [Ω/□]	Series resistance of S/D [Ω]
n-channel			
(a)	4.9	4.4	25
(b)	27	24	89
p-channel			
(a)	4.5	2.3	48
(b)	42	83	868

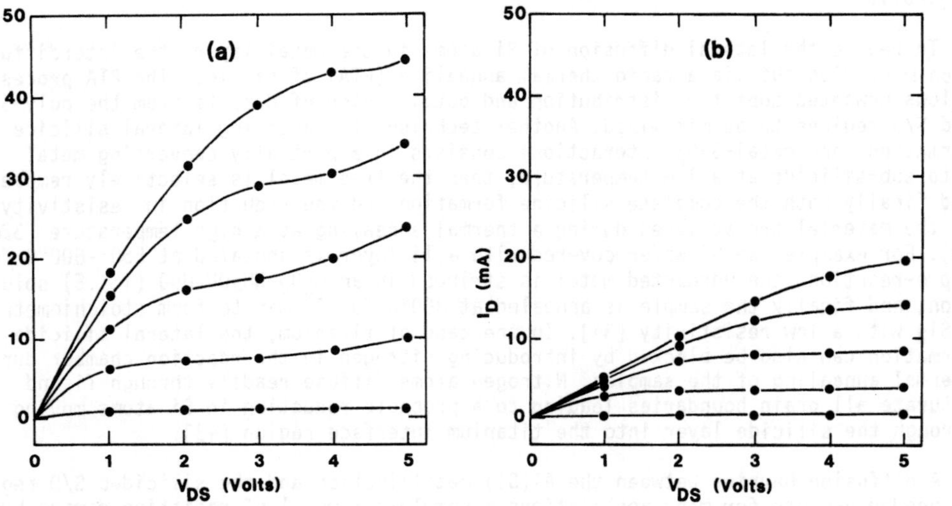

Fig.7. I-V characteristics of (a) $TiSi_2$ salicide and (b) unsilicided p-channel MOS transistors. Channel length = 1.3 μm; channel width = 730 μm; gate oxide thickness = 25 nm (from [47])

exhibits a sheet resistance value of about 7 Ω/□. Consequently, the silicidation of S/D regions allows a reduction in the sheet resistance by a factor of 10 which is very favorable for submicron technology.

Owing to the reduction in S/D resistance, the current capabilities of $TiSi_2$ silicided p-channel transistors are higher by a factor of 2 (Fig.7) compared with p-channel transistors without silicide. This result is obviously obtained without increasing the device size.

The feasibility of the self-aligned refractory metal silicide process, and the interest in reducing the sheet resistance of diffused zones by silicidation have been demonstrated. However, additional improvements still remain to be made; complementary works are necessary to determine the best silicide and to alleviate certain technological difficulties consisting of a decrease in annealing temperatures for silicide formation, in an improvement of silicon surface cleaning and a reduction in the lateral formation of silicide.

REFERENCES

1. R.E. Logar, J.O. Borland: Solid State Technol. 28, 133 (1985)
2. H. Kürten, H.-J. Voss, W. Kim, W.L. Engl: IEEE Trans. Electr. Dev. ED-30, 1511 (1983)

3. B.D. Joyce, J.A. Baldrey: Nature 195 485 (1962)
4. G.L. Schnable, W.J. Hillegas, Jr., C.G. Thornton: Electrochem. Technol. 4, 485 (1966)
5. W.G. Oldham, R. Holmstrom: J. Electrochem. Soc. 114, 381 (1967)
6. D.M. Jackson: Trans. Met. Soc. AIME 233, 596 (1965)
7. E. Sirtl, H. Seiter: In Semiconductor Silicon, Proc. of the Intern. Symp. on Silicon Mater. Sci. and Technol., ed. by R.R. Haberecht, E.K. Kern (The Electrochemical Society Softbound Proceedings Seris, Pennington, NJ 1969) p.189
8. M. Druminski, R. Gessner: J. Cryst. Growth 31, 312 (1975)
9. P. Rai-Choudhury, D.K. Schroder: J. Electrochem. Soc. 118, 107 (1971)
10. D.J. Dumin: J. Cryst. Growth 8, 33 (1971)
11. R.K. Smeltzer: J. Electrochem. Soc. 122, 1666 (1975)
12. M.W.M. Graef, B.J.H. Leunissen, J. Soeteman: Proceedings of the 5th European Conference on C.V.D., Uppsala, Swede 1985, ed. by J.-O. Carlsson, J. Lindström (University of Uppsala, Sweden 1985) p.465
13. K. Tanno, N. Endo, H. Kitajima, Y. Kurogi, H. Tsuya: Jpn. J. Appl. Phys. 21, L564 (1982)
14. N. Endo, K. Tanno, A. Ishitani, Y. Kurogi, H. Tsuya: Techn. Digest of the Intern. Electr. Dev. Meeting, San Francisco, CA 1982 (IEEE New York, 1982) p.241
15. N. Endo, N. Kasai, A. Ishitani, Y. Kurogi: Techn. Digest of the Inern. Electr. Dev. Meeting, Washington, D.C. 1983 (IEEE New York 1983) p.31
16. H.-J. Voss, H. Kürten: Techn. Digest of the Intern. Electr. Dev. Meeting, Washington, D.C. 1983 (IEEE New York 1983) p.35
17. C.H.J. Van den Breke: Philips Res. Rep. 32, 118 (1977)
18. R.J.H. Voorhoeve, J.W. Merewether: J. Electrochem. Soc. 119, 364 (1972)
19. G.G. Pinneo: Proceedings of the 3rd Intern. Conf on C.V.D., Salt Lake City, UT 1972, ed. by F.A. Glaski (Am. Nucl. Soc., Hindsdale, IL 1972) p.462
20. M.L. Green, R.A. Levy: J. Metals 6, 63 (1985)
21. T. Moriya, K. Yamada, Y. Tsunashima, S. Nakata, M. Kashiwagi: Extended Abstracts of the 15th Conf. on Solid-State Devices and Materials, Tokyo 1983 (Jpn. Soc. of Appl. Phys., Tokyo 1983) p.225
22. K.Y. Tsao, H.H. Busta: J. Electrochem. Soc. 131, 2702 (1984)
23. C.E. Morosanu, V. Soltuz: Thin Solid Films 52, 181 (1978)
24. M.L. Green, R.A. Levy: J. Electrochem. Soc. 132, 1243 (1985)
25. J.J. Cuomo: Proceedings of the 3rd Intern. Conf. on C.V.D., Salt Lake City, UT 1972, ed. by F.A. Glaski (Am. Nucl. Soc., Hindsdale, IL 1972) p.270
26. Y. Pauleau: Thin Solid Films 122, 243 (1984)
27. E.K. Broadbent, C.L. Ramiller: J. Electrochem. Soc. 131, 1427 (1984)
28. Y. Pauleau, Ph. Lami: J. Electrochem. Soc. 132, 2779 (1985)
29. Y. Pauleau, Ph. Lami. A. Tissier, R. Pantel, J.C. Oberlin: Thin Solid Films (to be published)
30. W.T. Stacy, E.K. Broadbent, M.H. Norcott: J. Electrochem. Soc. 132, 444 (1985)
31. S. Swirhun, K.C. Saraswat, R.M. Swanson: IEEE Electr. Dev. Lett. EDL-5, 209 (1984)
32. C.M. Osburn, M.Y. Tasi, S. Roberts, C.J. Lucchese, C.Y. Ting: In VLSI Science and Technology, Vol.82-7, ed. by C.J. Dell'Oca, W.M. Bullis (The Electrochem. Soc. Softbound Proceedings Series, Pennington, NJ 1982) p.213
33. H. Obakayashi, E. Nagasawa, M. Morimoto: Techn. Digest of the Intern. Electr. Dev. Meeting, San Francisco, CA 1982 (IEEE New York 1982) p.556
34. C.Y. Ting, S.S. Iyer, C.M. Osburn, G.J. Yu, A.M. Schweighart: In VLSI Science and Technology 1982, Proceedings of the First Intern. Symposium on VLSI Science and Technology, Vol.82-7, ed. by C.J. Dell'Oca, W.M. Bullis (The Electrochemical Society Softbound Proceedings Series, Pennington, NJ 1982) p.224
35. C.J. Lucchese: In VLSI Science and Technology 1982, Proceedings of the First Intern. Symposium on VLSI Science and Technology, Vol.82-7, ed. by C.J. Dell'Oca, W.M. Bullis (Electrochemical Society Softbound Proceedings Series, Pennington, NJ 1982) p.232
36. M. Morimoto, M. Sugimoto, K. Terada, K. Takahashi, T. Ishijima, Muta, S.-I. Suzuki: Proceedings of the 12th Conference on Solid-State Devices and Materials, Tokyo 1980, Jpn. J. Appl. Phys. 20 (Suppl. 20-1), 123 (1981)
37. D.B. Scott, Y.C. See, C.K. Lau, R.D. Davies: Techn. Digest of the Intern. Electr. Dev. Meeting, Washington, D.C. 1981 (IEEE New York 1981) p.538

38. T. Shibata, K. Hieda, M. Sato, M. Konaka, R.L.M. Dang, H. Iizuka: Techn. Digest of the Intern. Electr. Dev. Meeting, Washington, D.C. 1981 (IEEE New York 1981) p.647
39. M. Morimoto, E. Nagasawa, H. Okabayashi, M. Kondo: Techn. Digest of the Intern. Electr. Dev. Meeting, Washington, D.C. 1981 (IEEE New York 1981) p.655
40. D.L. Kwong, D.C. Meyers, N.S. Alvi, L.W. Li, E. Norbeck: Appl. Phys. Lett. 47, 688 (1985)
41. B.-Y. Tsaur, C.H. Anderson, Jr.: Appl. Phys. Lett. 47, 527 (1985)
42. B.-Y. Tsaur, C.K. Chen, C.H. Anderson, Jr., D.L. Kwong: J. Appl. Phys. 57, 1890 (1985)
43. C.K. Lau: Abstract 371, The Electrochemical Society Extended Abstracts, San Francisco, CA 1983 (The Electrochemical Society, Pennington, NJ 1983) p.569
44. M.Y. Tsai, C.S. Petersson, M.J. d'Heurle, V. Maniscalco: Appl. Phys. Lett. 37, 295 (1980)
45. D.L. Kwong, D.C. Meyers, N.S. Alvi: IEEE Electr. Dev. Lett. EDL-6, 244 (1985)
46. T.P. Chow, W. Katz, G. Smith: Appl. Phys. Lett. 46, 41 (1985)
47. C.K. Lau, Y.C. See, D. B. Scott, J.M. Bridges, S.M. Perna, R.D. Davies: Techn. Digest of the Intern. Electr. Dev. Meeting, San Francisco, CA 1982 (IEEE New York 1982) p.714

Laser-Induced Growth of Microstructures

Y. Pauleau and G. Auvert

Centre National d'Etudes des Télécommunications, B.P. 98,
F-38243 Meylan, France

1. INTRODUCTION

In the field of microelectronic devices, laser-induced processes are currently under investigation for various applications such as wafer marking, substrate cleaning, doping, oxidation, etching, deposition of thin films, exposure or removal of photoresists and recrystallization of silicon on insulators [1]. Laser irradiation of reactant gases can generate activated species on the substrate surface and localized growth of microstructures can be achieved by focusing a laser beam on absorbing substrates. Owing to small laser beam diameters and high energy densities (fluences), very high reaction rates are obtained. The laser spot can be positioned and automatically moved over the substrate surface making one step fabrication of microstructures possible. This "direct writing" technique is very promising to establish new interconnection networks in very-large scale integrated (VLSI) circuits through laser-induced chemical vapor deposition (LCVD) and etching. With such processes, semiconductor devices have been fabricated without using masks [2]. Laser processing is a way of obtaining the precision required in the microelectronic industry for integrated circuits with micron and submicron sizes. Review papers have been published on laser-generated microstructures [3] as well as laser-induced CVD [4-7] and laser-microchemical processing for VLSI [8-12].

The intent of the present article is to give a short introduction to the area of laser-induced microchemical reactions which can be used for laser-direct writing of microstructures. The reaction mechanisms and principles involved in LCVD processes, namely, the interactions of photons with reactive materials and substrates, growth kinetics of thin films and spatial resolution are discussed. The major characteristics of photolytic and pyrolytic processes as well as typical experimental arrangements are described. Finally, some experimental results are reported to illustrate various aspects introduced in the first section of this paper.

2. MECHANISMS OF THE LASER-INDUCED CHEMICAL VAPOR DEPOSITION

Laser irradiation of a substrate surface placed in a reaction chamber containing reactant gases can induce gas phase or surface reactions leading to the localized growth of microstructures. The laser beam penetrates into the gas cell equipped with transparent windows, passes through the gas phase and is focused on the substrate surface which is covered with a physically or chemically adsorbed layer (Fig.1). A thin film of solid material can be deposited according to a given pattern geometry by moving either the substrate or the laser spot (with a diameter of $2w_0$) in the direction of x or y.

2.1 Interactions of Photons with Gas Molecules, Adsorbed Species and Substrates

2.1.1 Interactions of Photons with Gas Molecules

Interactions between photons and gas molecules occur when the laser light is absorbed by the gas phase. Depending on the optical absorption of gases, the laser wavelength is chosen either within the infrared (IR) or ultraviolet (UV) range and very seldom

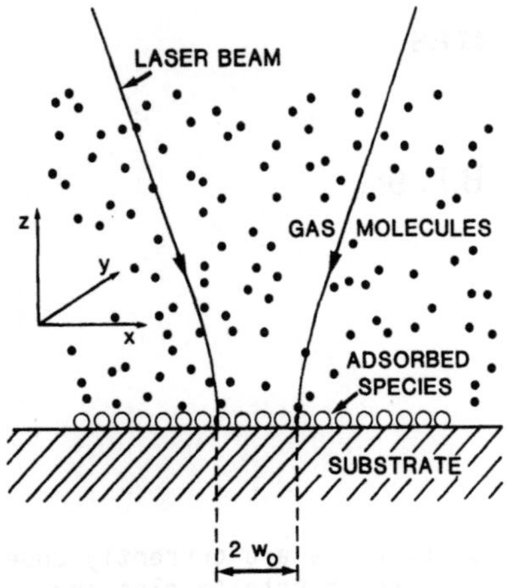

Fig.1. Schematic arrangement for laser-induced chemical vapor deposition

within the visible range. The photodissociation of gas molecules consists of the excitation of one of the molecular bonds resulting from the absorption of one or several photons. This process involves transitions between vibrational and/or electronic energy levels.

Gas molecules can be vibrationally excited by the absorption of IR Photons and can be dissociated by a multiple IR photon absorption or by a collisional process. The absorption of UV photons through a single photon process can lead directly to the dissociation of molecules. In both cases, the deposition of material resulting from the dissociaton of molecules occurs via the "photolysis" of gas. Pyrolytic reactions induced by the heating effect of the laser beam on the substrate can be excluded at the onset of deposition using transparent substrates. However, in most cases, the deposited material absorbs the laser radiation. The resulting increase in temperature remains very low since the laser power or laser energy density necessary for inducing photolytic processes is very low. Any heating effect on the substrate can be definitively eliminated by positioning the laser beam parallel to the substrate surface. Nevertheless, this arrangement is not convenient for direct writing but can be used rather for deposition on large surface areas.

2.1.2 Interactions of Photons with Adsorbed Species

In the case of physisorption processes, the weak interaction between the adsorbed molecules and the substrate surface results in a broadening of the absorption bands in the optical absorption spectrum of gas. In other words, gas molecules and physisorbed species absorb at the same wavelengths. On the other hand, the absorption spectrum of chemisorbed species is quite different from the spectrum of gas molecules. A laser wavelength can be chosen to be selectively absorbed by the chemisorbed phase. This photon absorption induces an excitation of adsorbed species followed by their dissociation and the formation of a solid material on the substrate surface. This laser-induced dissociation is also a photolytic process since the heating effect of the laser beam is negligible considering the very low laser power or energy density used in these experiments. Besides, the rapid increase in temperature resulting from a pulsed laser irradiation of a substrate surface can lead to the desorption of adsorbed species.

2.1.3 Interactions of Photons with Substrates

By a suitable combination of reactant gas, substrate and laser, the laser wavelength can be absorbed only by the substrate. Using an appropriate power or energy density, the absorption of the laser beam allows the substrate to be heated through various processes such as energy transfer from hot electrons, non-radiative recombining process and vibrational lattice excitation. This heating effect on the substrate surface leads to a thermal dissociation of molecules impinging the heated zone, i.e., to a "pyrolysis" of gas via a surface reaction process. Owing to this localized selective heating, gas phase nucleation of particles is avoided. Furthermore, the surface reaction can proceed even at high gas pressures enabling high deposition rates to be obtained.

2.2 Growth Kinetics of Microstructures

The growth rate of microstructures is a function of numerous parameters dependent on the type of laser employed, the material deposited, the gas phase and substrate used. Recently, some theoretical considerations have been proposed to establish a relationship between the growth rate and the laser fluence or the laser power [3].

The deposition velocity, v, is defined as the velocity of the gas-deposited material interface in the direction of z. The value of the deposition velocity is a function of time as well as of the position within the laser spot on the substrate surface and can be expressed in the following form:

$$v(x,y,t) = \frac{dz}{dt} \quad . \tag{1}$$

The coordinates x, y and z are represented in Fig.1.

Assuming that the deposited material has a density, ρ, the mass of microstructures, r, can be derived from the deposition velocity and is given by the following equation:

$$r(t) = \rho \int_0^\infty \int_0^\infty v(x,y,t)dx\,dy \quad . \tag{2}$$

Therefore, the value of the deposition velocity, v(x,y,t), is obtained by interrupting the process at different deposition times and measuring on the point of coordinates x and y, the thickness of the film on a cross-section of the processed zone. On the other hand, the deposition rate, r(t), is determined from weight measurements of the deposit at different deposition times using a crystal quartz monitor. A detailed knowledge of v(x,y,t) allows deposition profiles to be given by appropriately shaping the laser beam and enables the maximum translation speed of the laser spot or writing speed to be determined. Furthermore, the deposition velocity depends on various properties of the deposited material such as reflectivity and thermal conductivity. For a given deposited material, the velocity of the gas-deposit interface is proportional to the net flux of reactants, ϕ_p, towards the interface. This flux of active species contributing to the growth of microstructures is linked to the excitation and deactivation processes, the sticking coefficient of these active species on the surface, the local temperature and the chemical reaction paths. The flux is obviously dependent on the laser fluence, ϕ_L; however, a simple relationship between these quantities cannot easily be established.

For a photolytic process, ϕ_p can be considered as the flux of activated species contributing to the growth of microstructures. This flux may be expressed as a function of the laser fluence, ϕ_L, by the following equation [3]:

$$\phi_p = \frac{\phi_L}{h\nu} (\eta_g F_g + \eta_a F_a) \tag{3}$$

where $h\nu$ is the photon energy, F is a probability factor for activated species to reach the reaction zone and η is the quantum yield, i.e., the ratio between the number of transitions and the number of photons necessary to induce the transition; the subscripts g and a refer to the gas phase and to the adsorbed layer, respectively. Generally, the deposition velocity is not linearly dependent on the laser fluence and gas pressure, especially for experiments in which the adsorbed layer and the gas phase contributes simultaneously to the growth of a microstructure.

The contribution of the adsorbed layer can be neglected using either a laser beam which is not absorbed in the adsorbed layer or a laser beam parallel to the substrate surface. The probability factor corresponding to the gas phase, F_g, depends on collisional parameters in the gas phase, geometry, sticking coefficient of active species and deactivation processes on the surface. These parameters are assumed to be constant for uniform distribution of substrate temperatures. Under these conditions, the deposition velocity, v, can be found proportional to the laser fluence. However, various phenomena can limit the validity of the linear law within a specific range of laser fluence. The nucleation rate of material on the substrate surface at low laser fluences and the depletion of the gas phase at high laser fluences or the saturation under high reactant partial pressures can lead to deviation with regard to the linear law.

A linear relationship between deposition velocity or deposition rate and laser fluence is not a sufficient condition to conclude that the deposition process from the gas phase is the dominant step. Indeed, a linear dependence can be expected when the terms, $\eta_g F_g$ and $\eta_a F_a$, are constant. To distinguish whether the growth of microstructures originates from the gas phase or from the adsorbed layer, the deposition velocity has to be investigated as a function of the radius, w_0, of the laser spot. According to the model proposed by WOOD et al. [13, 14], the growth rate is proportional to w_0^{-n}. The exponent n equals unity for a pure contribution of the adsorbed layer. However, in general, both adsorbed and gas phase molecules participate in the growth of microstructures by direct writing and the individual contribution of these two processes depends on the photon energy.

The relationship between the deposition velocity and laser fluence (3) is essentially valid for photolytic processes. The growth rate of microstructures formed by a pyrolytic process will mainly depend on the temperature in the processed zone. Various models have been proposed to evaluate the temperature of the substrate surface within the laser spot. Assuming a Gaussian laser beam with a radius w_0 at $1/e^2$ intensity, LAX's calculation [15] gives the maximum temperature rise, ΔT, at the center of the laser spot,

$$\Delta T = P \frac{1 - R_s}{2(\pi)^{\frac{1}{2}} w_0 \kappa} \tag{4}$$

where P is the laser power impinging a semi-infinite substrate with a thermal conductivity, κ, and R_s is the surface reflectivity of the deposited material. Equation (4) gives only an approximate temperature value since heat losses due to radiation or transport via the gas phase are neglected. This equation remains valid as long as the thermal conductivity and reflectivity do not change significantly with the change in temperature.

The deposition velocity is given as a function of substrate temperature, T, by the Arrhenius law

$$v = v_0 \exp(-\frac{E_a}{kT}) \tag{5}$$

where k is the Boltzmann constant and E_a the activation energy; the term v_0 is a function of partial pressures of gases according to the reaction orders. Since the gas phase remains approximately at room temperature, very high partial pressures of reactant gases can be used without formation of powdery material by homogeneous reactions. Consequently, very high deposition rates by surface reactions can be achieved. A possible limitation in growth rate can originate from the depletion of reactant gases in the vicinity of the deposition zone.

2.3 Spatial Resolution

The spatial resolution for a laser writing process is defined as the minimum diameter, δ, of the deposited zone. For a photolytic process, this diameter can be approximated by the following expression:

$$\delta = \delta_0 + 2 L \tag{6}$$

where δ_0 is the optical resolution and L the mean free path of the active species contributing in the growth process. For laser writing purposes, the laser beam is generally focused with a microscope objective. Assuming a Gaussian laser beam with a radius w_0 (value at $1/e^2$) measured at the entrance of the focusing optics in air, the optical resolution is given by [3]

$$\delta_0 = \frac{2\lambda f}{\pi w_0} \qquad \text{with } w_0 < D \tag{7}$$

where f is the local length, D the diameter of the entrance pupil and λ the laser light wavelength. Consequently, the best resolution can be expected from an UV laser irradiation of the reactive system.

The mean free path depends on the cross-section, σ, and the concentration, N, of the active species, i.e.,

$$L = \frac{1}{(2)^{\frac{1}{2}} N\sigma} \quad . \tag{8}$$

The resolution can be improved by decreasing L. A decrease in the mean free path is achieved by increasing the pressure of the buffer gas or using larger molecules with an increased collision cross-section.

For a pyrolytic process, the spatial resolution depends essentially on the laser-induced temperature profile, i.e., on the coupling of the laser beam to the substrate, the irradiation time (or laser scan speed) and the thermal conductivity of the substrate. A significant reduction of the spatial resolution on a low heat conducting substrate is reached when the heat diffusion length in the substrate is lower than the laser spot diameter. The exponential temperature dependence of the deposition velocity limits the deposition process whithin the hottest zone of the temperature profile. Furthermore, the resolution can be improved because the reaction only occurs above a "threshold" temperature.

3. MAJOR CHARACTERISTICS OF LCVD PROCESSES

A typical experimental arrangement is schematically represented in Fig.2. The laser direct-writing equipment is essentially composed of 5 parts: a visible, UV or IR cw laser, a high-magnification optical microscope, translation stages for substrate positioning, a gas cell containing the test substrate and a computer controller. The microscope is used to focus the laser beam onto the substrate as well as to view the surface for accurate positioning and process monitoring. The translation stages are driven by the computer to generate specific patterns. The gas cell can be evacuated and filled with the reactant gas. The deposition occurs in a static atmosphere. Various in situ monitoring tools can be incorporated into the system such as quartz-crystal microbalance.

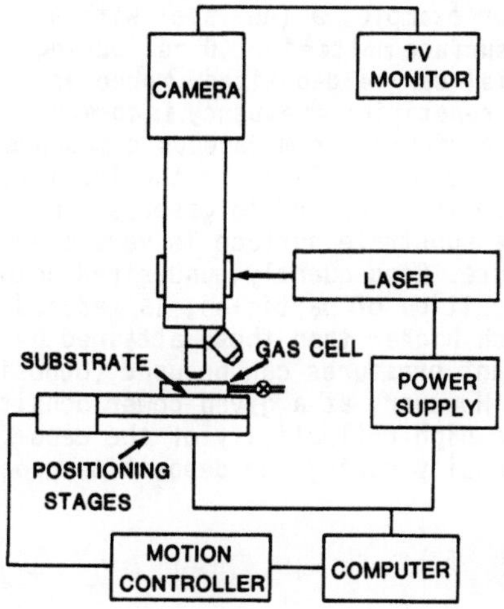

Fig.2. Experimental arrangement used for laser writing

3.1 Photolytic Process

The most critical parameters involved in a laser-induced photolysis of gas molecules are the laser wavelength, the adsorbed layer formation, the partial pressures of reactant gas and buffer gas. In order to induce only a "pure" photolytic process, the laser heating effect must be excluded by using a substrate transparent to the laser beam. Generally, photolytic processes are generated with low energy densities. Consequently, thin-film materials can be deposited on substrates with poor thermal stability. Moreover, the process is insensitive to the nature of substrates and the growth of microstructures can be performed over several materials (oxides, semiconductors or metals) during the same experiments.

Photolytic decomposition of molecules is usually stimulated by a cw or pulsed laser beam with a wavelength in the near or deep UV range, i.e., frequency doubled Ar^+ or excimer lasers. Using CO_2 lasers with IR wavelengths associated with a transparent substrate, the photolysis of molecules (SiH_4 for example) can occur via a multiphoton process. The microstructures are currently grown at room temperature and the deposited material is amorphous or polycrystalline. Under experimental conditions corresponding to large mean free paths of active species, the growth of whiskers can be observed far from the processed zone leading to a decrease in spatial resolution. Metal alkyls, MR_n, metal carbonyls, $M(CO)_n$, metal halides, MX_n and hydrides are mostly used as gaseous starting materials to deposit metal, semiconductor and insulating films. The photodeposition process is characterized by a relatively low deposition rate (up to 1 µm/s, typically) compared with the pyrolytic process. Particles are produced by homogeneous decomposition of gas and are deposited both on the wall reactor and on the front window of the gas cell. This deposit can absorb the laser beam rapidly excluding any photon-molecule interaction. Consequently, the thickness of films deposited on the substrate can be limited. This effect can be alleviated by introducing a buffer gas in the vicinity of the front window.

3.2 Pyrolytic Process

The pyrolytic decomposition of molecules within the irradiated zone depends considerably on various parameters such as thermal conductivity, optical absorption coefficient, reflectivity of the substrate and the deposit as well as reactant partial pressures. Substrates absorbing the laser wavelength and high power density are required to reach the temperature which is necessary to induce the pyrolytic reaction. The deposited material is generally polycrystalline; however, the growth of single crystals has been demonstrated.

Continuous working (cw) laser beams with visible or IR wavelengths are preferable for this type of process. Using a pulsed laser, the deposition zone remains at a high temperature for only a very short duration. For example, a YAG laser with a pulse duration of 7 ns is able to leave a silicon surface molten for 40 ns. During such a time interval, a negligible amount of material can be deposited. Moreover, the fractional duty cycle is quite small since the repetition frequency is commonly less than 100 Hz. The growth of microstructures is performed from gaseous compounds similar to those employed in a photolytic deposition process. Owing to the localized heating zone and the high thermal conductivity of solids compared to gaseous compounds, the thermal gradient in the vicinity of the substrate surface is very deep and gas molecules are essentially at room temperature. Consequently, undesired homogeneous pyrolysis, which leads to uncontrolled deposition of particles, is reduced or totally eliminated. The deposition rates are much higher than those attained by a photolytic decomposition since much higher reactant pressures can be used (deposition rates up to 100 µm/s have been demonstrated). However, at a given power density, a rapid decrease in temperature can result from the high reflectivity of the deposited material, and without an increase in power density during the deposition process, the thickness of films may be limited.

4. EXPERIMENTAL RESULTS AND DISCUSSION

Numerous metals, oxides and semiconductors have been deposited via laser irradiation of reactant gases. A review paper recently published [3] gives an exhaustive list of the materials deposited by a photolytic and pyrolytic process as well as the substrates and gases used, the resolution, thickness, growth velocity and growth mechanisms. Consequently, only few experimental results are reported in the present paper in order to illustrate the major theoretical aspects of the LCVD process discussed in the first part.

4.1 Photolytic Deposition Process

4.1.1 Deposition from Photodissociation of Gas Molecules

Silane molecules are vibrationally excited by absorption of the P(20) CO_2 laser line at 10.59 µm. Under appropriate silane pressure-laser fluence combinations, the radicals SiH_2 and SiH_3 can be generated in the gas phase. Operating with a pulsed CO_2 laser beam parallel and tangent to the substrate surface, amorphous Si films were deposited at temperatures lower than 400°C [16, 17]. The pulse duration and the repetition frequency of the laser were 100 ns and 1 Hz, respectively. Under 15 and 20 Torr of silane, the deposition rate of amorphous Si films is linearly dependent on the laser fluence (Fig.3a). This result is in good agreement with the dependence deducted from (3) in which the term $\eta_a F_a$ is zero since, in these experiments, interactions of photons with the adsorbed phase are excluded. At a silane pressure of 20 Torr, the deposition rate falls sharply for laser fluences higher than 1.1 J/cm² since a large depletion of reactant gas results from the homogeneous formation of powdery Si at vicinity of the vertical substrate surface.

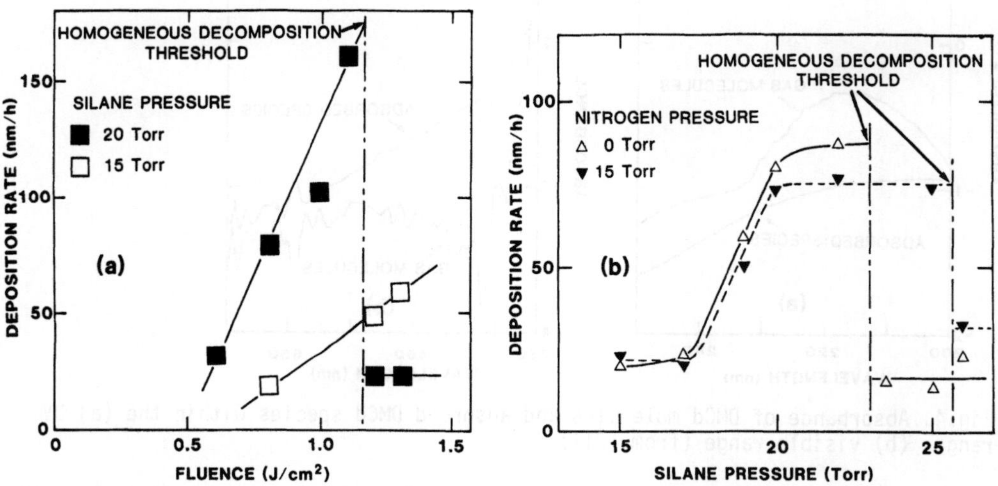

Fig.3. Deposition rate of amorphous Si films as a function of (a) pulsed CO_2 laser fluence; (b) silane pressure at 350°C (from [17])

The silane pressure effect on the deposition rate of amorphous Si films obtained at 350°C with a laser fluence of 0.8 J/cm² is exhibited in Fig.3b. The deposition rate decreases sharply for silane pressures higher than those corresponding to the homogeneous decomposition threshold of gas. With silane-nitrogen mixtures, higher silane pressures are necessary in order to reach the homogeneous decomposition threshold. However, the deposition rate remains essentially within the same order of magnitude. Nitrogen molecules act as a buffer gas screening the vibrationally excited species and volume reactions are prevented.

The net growth rate of films achieved with a pulsed CO_2 laser is very low because of the very short duty cycle of this type of laser. Considerably higher deposition rates (up to 40 nm/min) can be reached using a cw CO_2 laser parallel to the substrate surface [18-20]. Furthermore, the deposition of Si films could not be performed using the pulsed CO_2 laser perpendicular to a quartz surface [16]. The very high flux of photons impinging the surface would probably lead to the desorption of active species.

4.1.2 Deposition from Photodissociation of Adsorbed Molecules

The optical absorption spectra of gaseous cadmium dimethyl (DMCd) and adsorbed DMCd molecules on quartz surface are shown in Figs.4a and b. The absorption coefficient for DMCd molecules reaches a maximum of 10^3 l $mol^{-1}cm^{-1}$ at 216 nm (Fig.4a), whereas above 300 nm, the absorption coefficient is less than 2.5 l $mol^{-1}cm^{-1}$. Consequently, a direct photodissociation of DMCd molecules in the gas phase cannot be achieved by irradiating the gas molecules with a visible laser light. The maximum absorption of adsorbed DMCd molecules appears at a shorter wavelength than that of gas molecules and the absorption band is much broader. The decreased slope in the long wavelength tail leads to a measurable absorption at visible wavelength (Fig.4b). The formation of the adsorbed layer is critically dependent on the surface preparation of the quartz substrates. This adsorbed DMCd layer may result from dipole interactions with residual H_2O molecules on the quartz surface.

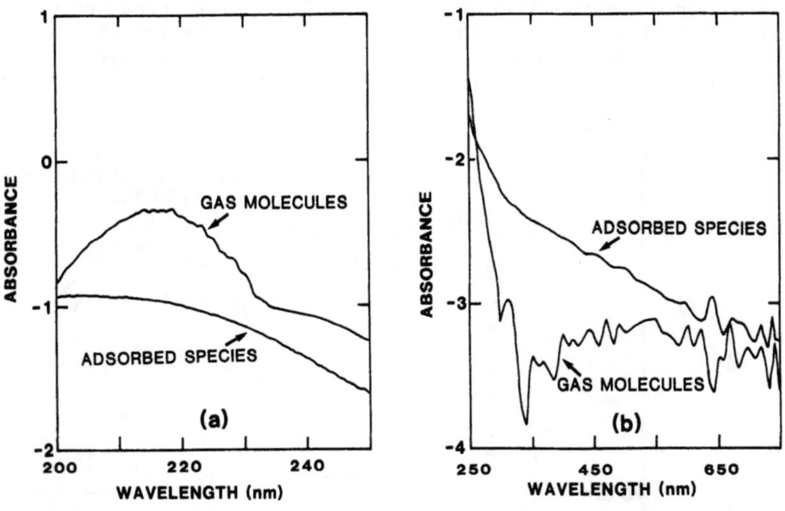

Fig.4. Absorbance of DMCd molecules and adsorbed DMCd species within the (a) UV range; (b) visible range (from [21])

The experimental arrangement for deposition of Cd films consists of a gas cell equipped with two quartz windows and the beam of a cw Kr laser tuned at 530.9 nm is focused on the inner side of the entrance window of the gas cell. The spot diameter of the Kr laser is 12 µm. An He-Ne laser with a beam diameter of 8 µm is used for transmission measurements through the gas cell and for monitoring the deposition process. The deposition of cadmium dots was performed using DMCd-H_2 mixtures at atmospheric pressure. The partial pressure of DMCd was 10 Torr.

The transmission measurements and the growth rate of Cd films are given as a function of time in Fig.5. Typically, three different steps can be distinguished. First, the transmission of the He-Ne laser beam remains essentially constant for 6 s. The duration of this first period is strongly dependent on laser power density and photon energy. Induction times up to 15 min can be observed. At the beginning of the second step, particles of submicron size are formed on quartz surface. These

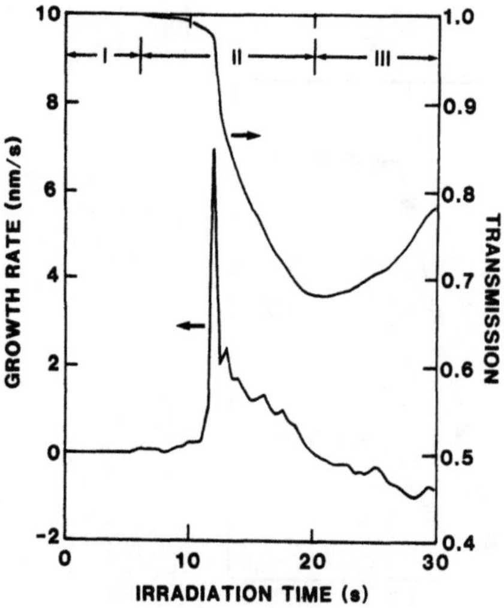

Fig.5. Growth rate of Cd films and He-Ne laser beam transmission as a function of irradiation time; wavelength = 530.9 nm; power density = 130 kW/cm² (from [21])

particles appear as scattering centers by observing the sample in reflection. Then, this speckle pattern moves, indicating controlled growth of the Cd film. The growth rate of metal films is deduced from the slope of the transmission curve. The deposition of metal film proceeds during this second step via a thermal dissociation of DMCd molecules on the substrate surface. The heating effect of the laser beam leads to a rapid increase in the growth rate. The thermal resistance of the processed zone decreases with increasing film thickness, i.e., the growth rate reaches a maximum value and then decreases with increasing deposition time. During the third step, the transmission of the He-Ne laser beam through the metallic deposit increases surprisingly. At high laser power densities, the temperature in the center of the laser spot is higher than 320°C. Consequently, melting and evaporation of metal can occur leading to the observed increase in the optical transmission and to decrease in deposition rate.

The induction time period may correspond to the formation of Cd nuclei on the quartz surface. At a photon energy of 2.41 eV (λ = 514.5 nm), the nucleation time was found to be inversely proportional to the square of the laser power [21]. This power law suggests that two photons are involved with the dissociation of the adsorbed DMCd species. The dissociation of adsorbed species may be initiated by a two-step optical absorption process. This dissociation process is plausible considering the values of the bond dissociation energies of gaseous DMCd molecules and the photon energy. In addition this mechanism is supported by the experimental relationship established from nucleation time measurements as a function of laser power density and photon energy [21].

4.2 Pyrolytic Deposition Process

Localized deposition of polycrystalline silicon can be achieved by pyrolytic dissociation of silane on quartz and silicon substrates irradiated by a cw CO_2 laser [22, 23] and a cw Ar laser [24, 25], respectively. A cw CO_2 laser operating at 10.59 μm is focused on silicon coated quartz substrates placed in a gas cell containing pure silane at reduced pressure [23]. The laser spot diameter on the quartz substrate is between 250 and 300 μm. With an Si coating on quartz substrates, the changes in reflectivity in the irradiated zone could be reduced or eliminated during LCVD experiments. Since silicon is transparent to the laser light, only a heat transfer occurs through the Si coating during laser irradiation. The laser beam is maintained at a fixed position for a given irradiation time so that Si dots are deposited and grown both laterally and vertically. The dot profile obtained from profilometer measurements is observed to be Gaussian for low laser powers. The height of the Si dots was found to be linearly dependent on the irradiation time which varied

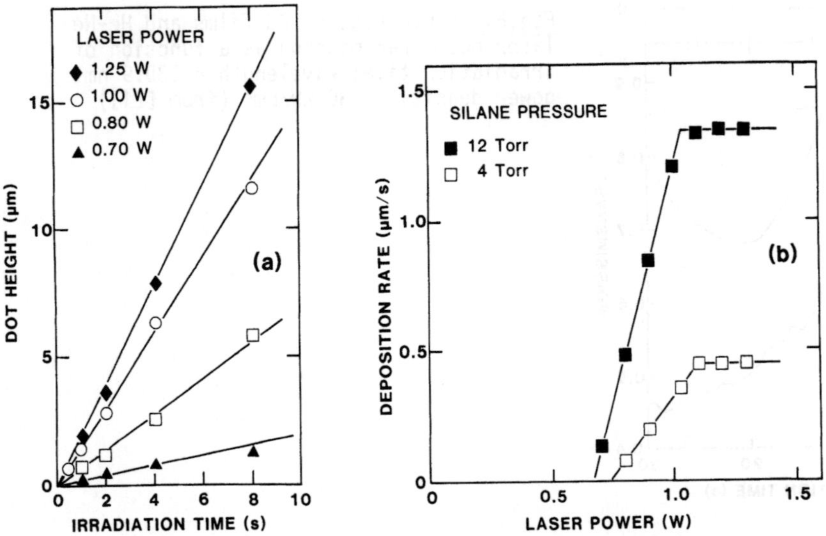

Fig.6. (a) Height of Si dots versus irradiation time for various cw CO_2 laser powers; silane pressure = 15 Torr. (b) Deposition rate of Si dots versus cw CO_2 laser power (from [23])

between 1 and 10 s (Fig.6a). In other words, the deposition rate is not time dependent, i.e., the irradiated quartz zone probably reaches a constant temperature within an irradiation time shorter than 1 s. This result is concordant with the value of the characteristic time which can be evaluated for a heat transfer through a 1 mm thick quartz plate.

The laser power is measured by a thermopile with an accuracy of 0.05 W. One part of the laser power is absorbed by silane molecules. The values of laser power on quartz substrates are given after correction due to silane absorption. The laser power dependence of the deposition rate of Si dots is exhibited in Fig.6b. The onset of Si deposition occurs for a laser power corresponding to a substrate temperature of about 600°C. The value of this laser power threshold is not constant because of laser instability. For laser powers higher than 1.1 W (Fig.6b), volcano-shaped Si dots are grown and the maximum height which can be measured is independent of the laser power. This dot profile may result from the melting and evaporation of silicon in the center of the laser spot. After cooling down of samples, very large grains of about 20 µm can be observed by Raman scattering in this central zone.

Assuming that the thermal conduction parallel to the substrate surface is negligible with respect to the perpendicular thermal conduction, the surface temperature of substrates can be estimated for a given laser power [26]. The temperature dependence of the deposition rate of silicon obtained by pyrolysis of silane under laser irradiation as well as in a furnace heated CVD reactor is given in Fig. 7. An activation energy of (40 ± 10) kcal mol^{-1} is deduced within the low temperature range from the slope of the straight line corresponding to LCVD at a silane pressure of 10 Torr. This value is in good agreement with that of activation energy obtained for the pyrolytic deposition of silicon in a conventional CVD reactor [27]. Consequently, the LCVD process is thermally activated within the range of low temperatures. Above 850°C, the decrease in slope may indicate that either the deposition process is mass-transport limited, as is the case in a conventional cold wall CVD reactor, or there is a depletion in reactant gas resulting from the onset of particle formation by volume reactions. The upper curve in Fig.7 shows the temperature dependence of the deposition rate of silicon which was prepared via thermal decomposition of silane on Si substrates heated by a visible laser beam [5, 25]. A com-

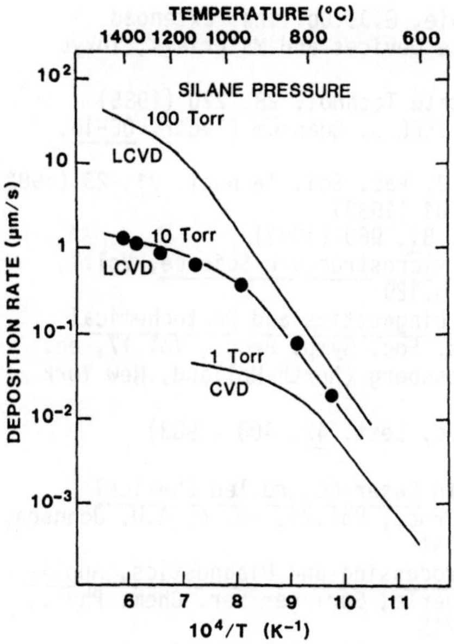

Fig.7. Arrhenius plot for deposition rate of silicon in LCVD and conventional CVD processes

parison of LCVD and CVD curves shows that the deposition rate is proportional to silane pressure within the range of high temperatures. Moreover, the deposition rate obtained by an LCVD process is much higher than in a conventional CVD system since much higher partial pressures of silane can be used without any formation of powdery Si by volume reactions. A more accurate determination of the substrate temperature resulting from the laser irradiation is necessary for an effective comparison of LCVD and CVD processes within the range of low temperatures.

5. CONCLUSION

The major theoretical aspects and characteristics of photolytic and pyrolytic processes involved in laser-induced growth of microstructures have been described in this paper. The deposition process can result from interactions of photons with gas molecules, with adsorbed species and with substrate. Experimental difficulties in the control of the LCVD processes can originate from this very large diversity of interactions resulting in the activation of localized reactions. Owing to specific features of laser-induced deposition processes including high deposition rate and spatial resolution, various applications are possible in microelectronic technology such as custom metallization of integrated circuits. However, applications of laser-direct writing are at an early stage of development and additional work has to be done to improve the capabilities of this technique as well as the understanding of laser-induced processing.

REFERENCES

1. T.Mc Grath: Solid State Technol. 26, 165 (1983)
2. B.M. Mc Williams, I.P. Herman, F. Mitlitsky, R.A. Hyda, L.L. Wood: Appl. Phys. Lett. 43, 946 (1983)
3. Y. Rytz-Froidevaux, R.P. Salathé, H.H. Gilgen: Appl. Phys. A37, 121 (1985)
4. D. Bäuerle: In Laser Diagnostics and Photochemical Processing for Semiconductor Devices, Mat. Res. Soc. Symp. Proc., Vol.17, ed. by R.M. Osgood, Jr., S.R.J. Brueck, H.R. Schlossberg (North-Holland, New York 1983) p.19
5. D. Bäuerle: In Surface Studies with Lasers, ed. by F.R. Aussenegg, A. Leitner, M.E. Lippitsch, Springer Ser. Chem. Phys., Vol.33 (Springer, Berlin, Heidelberg 1983) p.178

6. P.J. Boyer, C.A. Morre, R. Solanki, W.K. Ritchie, G.J. Collins: Extended Abstracts of the 15th Conference on Solid State Devices and Materials, Tokyo 1983, p.109
7. R. Solanki, C.A. Moore, G.J. Collins: Solid State Technol. 28, 220 (1985)
8. D.J. Ehrlich, R.M. Osgood, Jr., T.F. Deutsch: IEEE J. Quantum Electr. QE-16, 1233 (1980)
9. D.J. Ehrlich, R.M. Osgood, Jr., T.F. Deutsch: J. Vac. Sci. Technol. 21, 23 (1982)
10. R.M. Osgood, Jr.: Ann. Rev. Phys. Chem. 34, 7701 (1983)
11. D.J. Ehrlich, J.Y. Tsao: J. Vac. Sci. Technol. B1, 969 (1983)
12. D.J. Ehrlich, J.Y. Tsao: In VLSI Electronics: Microstructure Science, Vol.7, ed. by N. Einspruch (Academic, New York 1983) p.129
13. T.H. Wood, J.C. White, B.A. Thacker: In Laser Diagnostics and Photochemical Processing for Semiconductor Devices, Mat. Res. Soc. Symp. Proc., Vol.17, ed. by R.M. Osgood, Jr., S.R.J. Brueck, H.R. Schlossberg (North-Holland, New York 1983) p.193
14. T.H. Wood, J.C. White, B.A. Thacker. Appl. Phys. Lett. 42, 408 (1983)
15. M. Lax: J. Appl. Phys. 48, 3919 (1977)
16. Y. Pauleau, R. Stawski, Ph. Lami, G. Auvert: In Laser-Controlled Chemical Processing of Surfaces, Mat. Res. Soc. Symp. Proc., Vol.29, ed. by A.W. Johnson, D.J. Ehrlich (North-Holland, New York 1984) p.41
17. Y. Pauleau, D. Tonneau, G. Auvert: In Laser Processing and Diagnostics, Applications in Electronic Materials, ed. by D. Bäuerle, Springer Ser. Chem. Phys., Vol.39 (Springer, Berlin, Heidelberg 1984) p.215
18. R. Bilenchi, I. Gianinoni, M. Musci: J. Appl. Phys. 53, 6479 (1982)
19. R. Bilenchi, I. Gianinoni, M. Musci, R. Murri: In Laser Diagnostics and Photochemical Processing for Semiconductor Devices, Mat. Res. Soc. Symp. Proc., Vol.17, ed. by R.M. Osgood, Jr., S.R.J. Brueck, H.R. Schlossberg (North-Holland, New York 1983) p.215
20. R. Bilenchi, M. Musci, R. Murri: In Laser-Assisted Deposition, Etching and Doping, Proc. SPIE, Los Angeles, January 1984, Vol.459 (Int. Soc. Opt. Eng. 1984) p.61
21. Y. Rytz-Froidevaux, R.P. Salathé, H.H. Gilgen, H.P. Weber: Appl. Phys. A27, 133 (1982)
22. C.P. Christensen, K.M. Lakin; Appl. Phys. Lett. 32, 254 (1978)
23. D. Tonneau, G. Auvert, Y. Pauleau: Proceedings of the International Congress on the Applications of Lasers and Electro-Optics, Symposium on Materials Processing, November 1985, San Francisco, CA (Laser Institute of America) (to be published)
24. D.J. Ehrlich, R.M. Osgood, Jr., T.F. Deutsch: Appl. Phys. Lett. 39, 957 (1981)
25. D. Bäuerle, P. Irsigler, G. Leyendecker, H. Noll, D. Wagner: Appl. Phys. Lett. 40, 819 (1982)
26. D. Tonneau, G. Auvert, Y. Pauleau: Thin Solid Films (to be published)
27. W.A.P. Claassen, J. Bloem: In Semiconductor Silicon 1981, Proceedings of the 4th Intern. Symposium on Silicon Material Science and Technology, ed. by H.R. Huff, R.J. Kriegler, Y. Takeishi, Vol.81-5 (The Electrochemical Society Softbound Proceedings Series, Pennington, NJ 1981) p.365

Structural Characterization of Superlattices by X-Ray Diffraction

M. Quillec

Centre National d'Etudes des Télécommunications, Laboratoire de Bagneux, 196 Avenue Henri Ravera, F-92220 Bagneux, France

Abstract : We give an overview on the characterization of superlattices by X-rays diffraction. We show that this technique provides excellent determination of structural parameters as well as departure from ideality. We will limit ourselves to practical considerations for the user.

1. The Structural Parameters of Superlattices (SL)

The SL can be described by a one dimension modulation. This modulation is generally a mixture of chemical and geometrical modulations. An example of purely chemical superlattice would be obtained by periodic alternating amorphous blocks of materials A and B. On the other hand, a good example of structures showing a dominant geometrical contribution are the Strain Layer Superlattices as will be shown later. Among the SL structures, we will concentrate on the coherent ones, obtained by epitaxial growth; we will consider cubic materials (typically, III-V compounds).

1.1 The Ideal Superlattice

1.1.1 Strains

A coherent SL always shows alternate strains in the sublayers due to the differences in atomic size. There is a single lattice parameter in the directions parallel to the interfaces : $a_{/\!/}$. The non-vanishing strains are :

$\Sigma_{xx}^{1,2}$, $\Sigma_{yy}^{1,2}$, $\Sigma_{zz}^{1,2}$ (indexes ^{1,2} refer to material 1 and 2).

with $\Sigma_{xx}^{1,2} = \Sigma_{yy}^{1,2} = (a_{/\!/} - a_0^{1,2})/a_0^{1,2}$ (1)

$$\Sigma_{zz}^{1,2} = (a^{1,2} - a_0^{1,2})/a_0^{1,2}. \qquad (2)$$

$a_0^{1,2}$ are the lattice parameters of the bulk materials and $a^{1,2}$ are the lattice parameters of materials 1 and 2 in the direction of the modulation.
These parameters are related to the stresses :

$$\sigma_{ii}^{1,2} = C_{ij}^{1,2} \; \Sigma_{jj}^{1,2}$$

through the elastic constants C_{ij}. As no external stress is applied,

$$\sigma_{zz}^{1,2} = 2\, C_{12}^{1,2} \; \Sigma_{xx}^{1,2} + C_{11}^{1,2} \; \Sigma_{zz}^{1,2} = 0$$

$$\Sigma_{zz}^{1,2} = -\, 2\, C_{12}^{1,2} / C_{11}^{1,2} \; \Sigma_{xx}^{1,2} \qquad (3).$$

From (1), (2) and (3), the strain free lattice parameters of materials 1 and 2 can be calculated from the measured actual parameters in the SL, $a_{/\!/}$ and $a^{1,2}$. In the case of III-V ternary compounds, the composition can then be deduced from these strain free lattice parameters (Vegard's law).

1.1.2 Parameters

The ideal coherent SL is composed of alternating layers of materials 1 and 2. The useful parameters are the compositions and thicknesses of the sublayers. In the case of ternary III-V alloys, the composition can be obtained from the strain free lattice parameter. That lattice is related, as seen above, to the measured strained one, through elastic constants. It has to be recalled that a relation between the structural parameters is often provided by the growth conditions; for instance, in the case of a SL built by MBE from III-V ternary layers: $A_{X1}B_{1-X1}C(n1)$ and $A_{X2}B_{1-X2}C(n2)$, it is easy to show (QUILLEC et al. /1/) that $(n_1t_2)/(n_2t_1) = (1-x_2)/(1-x_1)$; t_1 and t_2, growth times of the sublayers, are measured with high precision.

1.2 Departure from Perfection

1.2.1 Curvature

If the SL equilibrium lattice parameter $a_{/\!/}^e$ in the directions parallel to the interfaces is different from that of the substrate, a_s, the sample shows a curvature ; the radius was be estimated by REINHART et al. /2/, ROZGONYI et al. /3/

$$R = (L_S^2/L_{SL})\ a_s/6\ \ a_{/\!/}^e - a_s$$

where L_S ans L_{SL} are the thicknesses of the substrate and the SL.

This curvature leads to a broadening of the diffracted profile, which can be easily detected (and reduced) by reducing the size of the diffracting area. In practice, the radius is generally more than 10 m and the broadening, less than a few seconds of arc, can be neglected.

1.2.2 Composition

Composition variations in the growth plane occur due to imperfections in the growth techniques. However, these variations generally do not exceed 10^{-3}/cm and thus like the curvature are responsible only for a few seconds of arcs broadening in the diffracted peaks. Both curvature and composition variations can of course be measured with high accuracy through X-rays diffraction by moving the sample in its plane.

1.2.3 Thickness

Departure from periodicity can be due to variations of the growth rate during epitaxy. X-ray diffraction is very sensitive to such imperfections.

1.2.4 Interfaces

Real interfaces are not flat and abrupt. Some diffusion (imperfection in the Z direction) and roughness (affecting directions x and y) may occur. The second ones can be analysed from X-rays diffracted by planes not parallel to the interfaces. The first ones can be identified and followed versus annealing time or temperature (FLEMING et al. /4/, JONCOUR et al. /5/). We will come back to this in a next section.

1.2.5 Strain Relaxation

It is well known that beyond some critical thickness, a mismatch epilayer generates misfit dislocations. This holds for each sublayer in a multilayer and also for a coherent structure mismatched as a whole with the substrate ($a_{/\!/}^e \neq a_s$). Such strain relaxation affects the X-ray profiles through displacement and broadening of the peaks. The presence of misfit dislocation (even a single one), is detected by X-ray topography. Although this technic is one aspect of X-rays diffraction, we will not develop it here.

2. X-ray Diffractometers

2.1 Simple Diffraction

Fig.1 shows a typical set-up. The beam diverges from the X-ray tube and is diffracted to form a convergent beam detected by a counter. The angle of the sample, θ and of the counter, 2θ are coupled. The tuning of this system is easy ; it provides a maximum intensity on the sample and the whole angular range can be scanned, so that reflexions near GaAs (000), (002), (004), (006) for instance, can be measured in a row. However, the problems of beam divergence and dispersion cannot be well solved, and the resulting angular dispersion is at least a few tens of seconds of arc. An accurate determination of a SL period can be obtained, but a better resolution is necessary to get full information from the profiles.

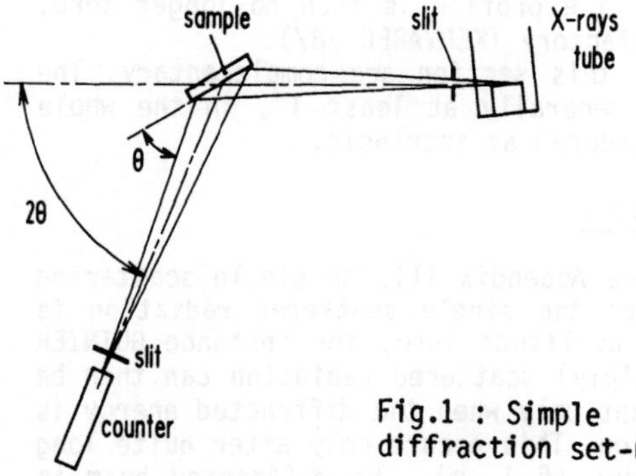

Fig.1 : Simple diffraction set-up

2.2 Double Diffraction (DD)

It has been shown (R.W. JAMES /6/) that in the parallel setting shown in fig.2, the diffracted profile does not depend on divergence nor dispersion of the incident beam. The "second" crystal is the studied sample, provided that the first crystal has a similar structure, it is possible to get extremely high resolution on (00L) diffractions. It is also possible to study asymetric reflections, although with lower resolution. It has to be pointed out that the first crystal, in the parallel setting, provides more than a plain monochromator.

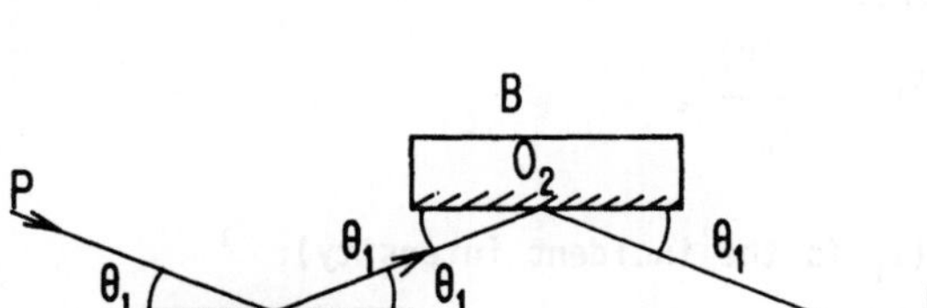

Fig.2 : Double diffraction in the parallel setting (principle : JAMES /6/)

This setting is useful to measure accurately small peak separation, for instance between the zero order of a GaAs/GaAlAs SL and the next GaAs related peak, leading to the average composition $\bar{x}$, the relation between $\bar{x}$ and the angle between GaAs 004 and zero order of SL θ is $\theta''=375.x$. As the resolution is better than 1", the accuracy on $\bar{x}$ is excellent (KERVAREC /8/).

The draw back is the difficulty of tuning two crystals, some loss in incident power, and the limited angular scan for which the resolution advantage remains. For instance, it is necessary to move and tune the first crystal to shift from GaAs 004 to GaAs 002.

For both techniques, the studied area, which is commonly a few mm^2, can be reduced to a fraction of mm^2 by appropriate diaphragms. In order to keep the high resolution advantage of DD in the whole angular range, it is necessary to go to three crystal diffractometers (using a crystal monochromator) or even to a five crystal setting, with a four crystal monochromator, thus keeping the incident beam's direction constant (BARTELS /7/). The improvement obtained with these systems is however at the expense of incident beam intensity, so that the noise and counting time are the final practical limitations. In most published works, the authors use a DD system, in conditions close to the above parallel setting. It is often more convenient however to choose, as a first crystal, a material different from that of the studied sample. For instance, in order to get high incident intensity near the 002 GaAs order, InP as a first crystal will be preferred to GaAs, since the structural factors F(002) are respectively 118.2 and 6.9 for the two materials. (See Annexe I). A much higher power is thus obtained with InP as a first crystal. Of course, the broadening of the profile is then no longer zero, but 0.73" in that case, which is still satisfactory (KERVAREC /8/).

The two types of settings described in this section are complementary. The second is used for high resolution; it is generally at least 1", in the whole angular range. Thus the profiles can be considered as intrinsic.

3. Diffraction of X-Rays by a Real Superlattice

The use of the kinematical approximation (see Appendix II), or single scattering approximation, implies that the amplitude of the single scattered radiation is very small compared with the incident beam amplitude (see, for instance,GUINIER /9/). The amplitude of the doubly (and multiply) scattered radiation can then be negligible. Multiple scattering is significant only when the diffracted energy is concentrated in some sharply defined direction. This occurs only after quite long path lengths in case of X-rays (of the order of 1 μm). The diffracted beam is then diffracted again and the result refers to dynamical theory (see Appendix II). However, this takes a very high degree of coherency on a long range, which often does not occur in real structures.

The SL for instance, are generally only a few 1000 Å thick. So the multiple scattering is (i) very weak (ii) not well phase correlated and can be neglected. This assumption is very well verified in practical cases. Neglecting the dynamical effects may however be incorrect when thick encapsulating layers are present.

The diffracted intensity I is maximum near Bragg angles (see Appendix I) ; it covers a small angular amplitude :(θ_B - Σ, θ_B + Σ). The function I(θ) depends on the incident wave shape. The relevant experimental parameters is the reflecting power, defined as (GUINIER /9/):

$$P_{hkl} = \int_{\theta_B-\Sigma}^{\theta_B+\Sigma} \frac{I(\theta)}{I_0} ,$$

which is independant of the wave shape (I_0 is the incident intensity).

In the kinematical approach, see Appendix 1

$P_{hkl} = C \; |F_{hkl}|^2$

We will come back to the term C, which is detailed by GUINIER /2/. The important feature is its slow angular variation.

3.1 Reciprocal Lattice of the Superlattice

Let us consider an SL composed of III-V compounds like GaAs-GaAlAs or GaAs-GaInAs (QUILLEC /1/). In direct space, the unit cell is composed of one period in the z direction and a unit GaAs cell in the xy plane.

Clearly, the points hk0 in the reciprocal space are not affected by the superlattice effect occuring in the z direction in direct space. We will focuse on the 00ℓ pattern, which is affected by the new periodicity and is

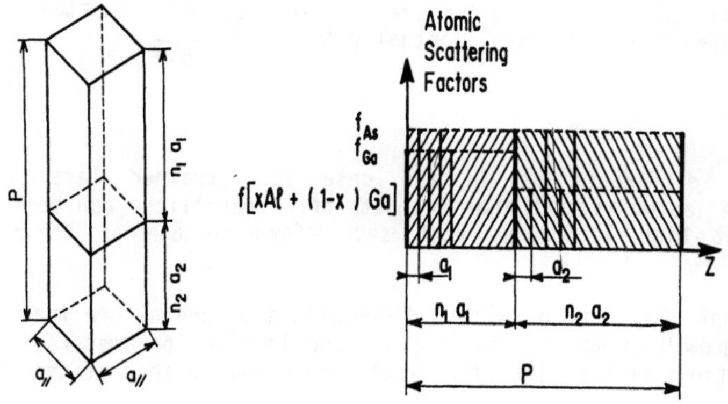

Fig.3 : Unit cell of the SL.

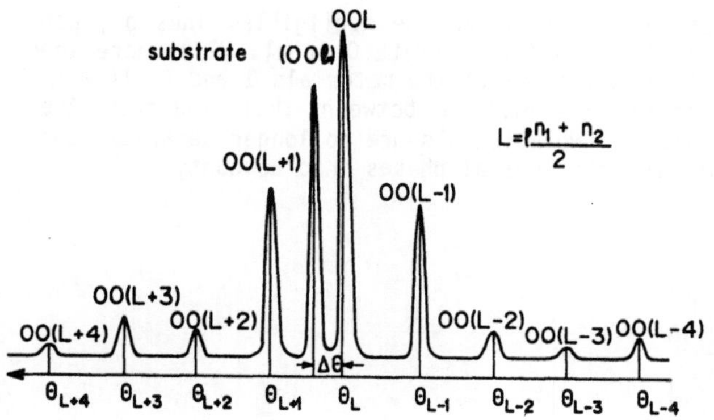

Fig.4 : X-rays diffraction profile of a SL (schematic)

a one dimensional problem. In direct space, the one variable function is represented by fig.3. The reciprocal lattice indicates at what angle diffraction occurs ($2\sin\theta/\lambda=L/P$). The intensities of the peaks is given by the structural factor, i.e. the Fourier transform of one period, where the atoms are represented by their atomic scattering factors.

This function has significant values only near the diffraction orders of the bulk materials they are made of. For instance, a GaAlAs/GaAs SL grown on a GaAs substrate gives the set of satellites presented in Fig.4 near the 001 order of GaAs. We call ℓ the diffracting orders of the basic materials and L those of the SL. A virtual crystal of parameter $\bar{a} = (n_1a_1 + n_2a_2)/(n_1 + n_2)$ in the z direction would have diffracting orders located between those of materials 1 and 2. Since $\bar{a}$ and P are commensurate, these orders are particular SL orders OOL, with $L=\ell(n_1+n_2)/2$. (The factor 1/2 is due to the definition of n_1 and n_2 : number of molecular layers (m.l.), whereas ℓ refers to a two m.l. unit cell). The ℓ orders related to $\bar{a}$ are called the SL "zero" orders, and the SL orders are often called satellites, with numbers -2,-1,+1,+2 (see Fig.5). It has to be noted that as the

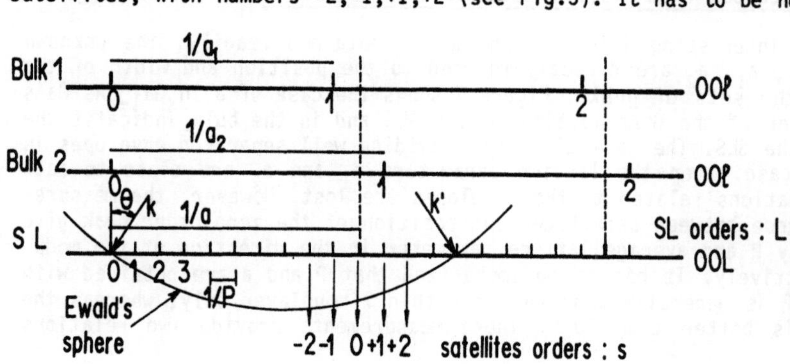

Fig.5 : Diffraction order along OOL; a_1 a_2 are taken as the thickness of one monolayer of materials 1 and 2.

scanning angle around a given bulk order is quite small (a few degrees), scattering angles and orders are almost proportionnal (actually $\Delta\Theta = \Delta(\frac{L}{P})\ \frac{\lambda}{2\cos\Theta}$).

3.2 The Diffracted Profile

If a_1 and a_2 are quite different, as in the case for strained layer superlattices, the structure factor is such that two sets of SL satellites can be separated, especially about high l orders. Each set refers to one of the constituant materials.

In the case of a coherent SLS, free of misfit dislocations a_1 and a_2 are the lattice parameters in the growth direction (see Fig.3). The in-plane parameters, needed to calculate diffractions orders other than (ool) are equal to that of the substrate, as shown above.

The overlap of the F.T. from each material may be negligible. Thus a plain decomposition can be performed as shown in Fig.6 (QUILLEC et al. /1/) where the final profile is obtained near the 0 0 l order of the materials 1 and 2. If a_1 and a_2 are too close, the spatial correlation between them modifies the description; the contributions from the two materials are no longer separated and the previous decomposition should take the spatial phases into account.

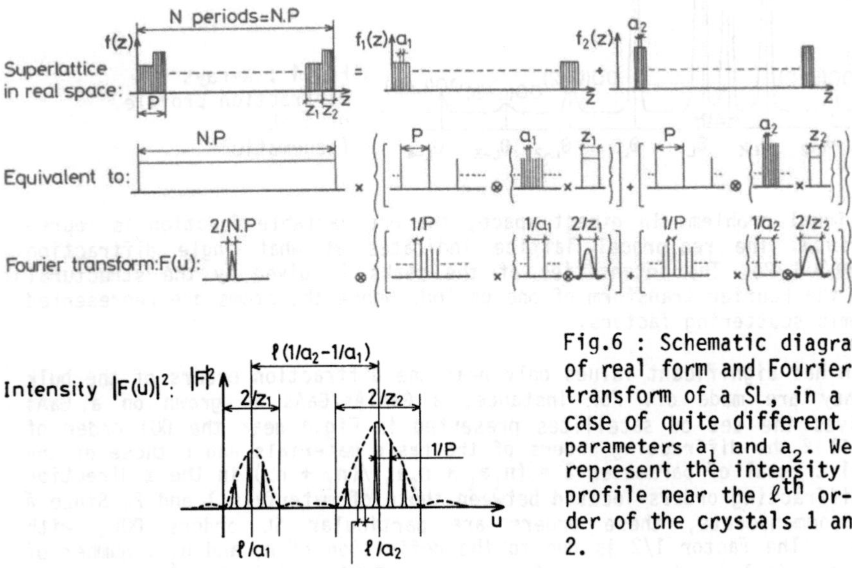

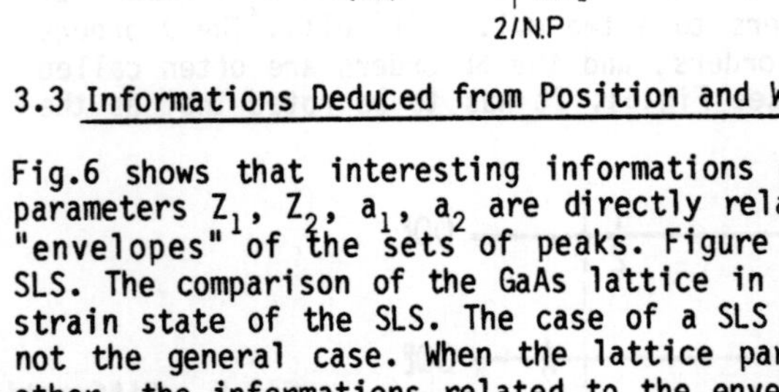

Fig.6 : Schematic diagram of real form and Fourier transform of a SL, in a case of quite different parameters a_1 and a_2. We represent the intensity profile near the ℓth order of the crystals 1 and 2.

3.3 Informations Deduced from Position and Width of the Peaks

Fig.6 shows that interesting informations can be obtained readily: the unknown parameters Z_1, Z_2, a_1, a_2 are directly related to the position and width of the "envelopes" of the sets of peaks. Figure 7 shows the case of a $In_xGa_{1-x}As/GaAs$ SLS. The comparison of the GaAs lattice in the SLS and in the bulk indicates the strain state of the SLS. The case of a SLS providing well separated envelopes is not the general case. When the lattice parameters a_1 and a_2 are close to each other, the informations related to the envelopes are lost. However, the measurement of the distance between satellites and position of the zero order peak give the SL periodicity P and average lattice parameter in the direction of the modulation, $\bar{a}$, respectively. It has to be emphasized that P and a are obtained with high accuracy : P is generally measured to within 1 monolayer only, whereas the precision on $\bar{a}$ is better than 10^{-4}. These measurements provide two relations

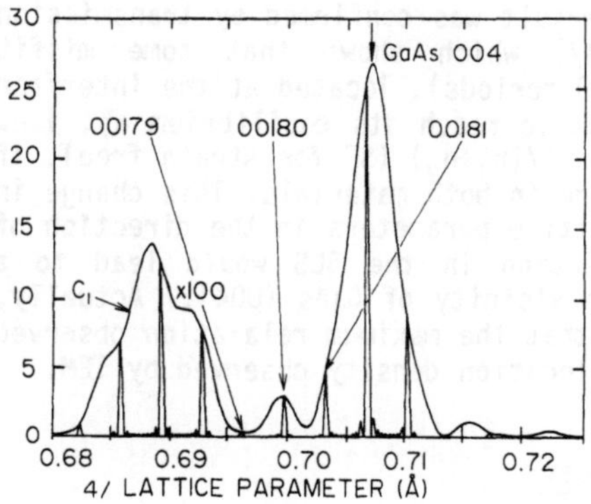

Fig.7 : 00L structure factor for a $(In_{0.3}Ga_{0.7}As)30/(GaAs)60$ Strained Layer Superlattice.

between the unknown parameters with high precision and without calculation. Besides, the width of the satellites are directly related to the total SL thickness, if the whole structure is spatially coherent. Otherwise, its width gives the "bloc size", in the Z direction, i.e. the maximum correlation length in the direction of the modulation.

3.3.1 Correlation Lengths

The broadening mentioned above is constant along the 00L direction, in units of the scattering vector. In terms of scattering angle, it is $\Delta(2\theta)=0.9\lambda/L \cos\theta$ at half maximum (Scherrer formula) (Mc WHAN /10/). Departure from perfect periodicity also leads to a broadening of the diffracted satellites. From Fig.5 it is clear that this broadening should increase linearly with SL diffraction order. Such variation is reported by Mc WHAN for Cu/Ni or GaAs/AlAs SL.

Up to now, we have not considered departure from perfection in the in-plane directions. However, it is well known that the interfaces show steps and other in-plane imperfections even in the best samples. Correlation lengths Lx and Ly can thus be defined in the x and y directions, like in the Z direction, so that in the reciprocal space, the points have finite dimensions Lx^{-1}, Ly^{-1}, Lz^{-1} in the principal directions. FLEMING et al. /4/ measured the in-plane correlation lengths CL of (GaAs)m (AlAs)n SL. They used the second harmonic around the 002 GaAs order. These CLs are found to be anisotropic. The maximum and minimum CL are obtained in the $[1\bar{1}0]$ and $[110]$ directions respectively. This result, as well as the dimensions of the CL, can be related to the surface observations made by VAN HOVE et al. /11/ by in situ Reflection High Energy Electron Diffraction during Molecular Beam Epitaxy of GaAs.

3.3.2 Strains

Another departure from perfection of SL is the relaxation of strains through mismatch dislocations. The assumption of a uniform parameter in the directions parallel to the interfaces is then wrong. Direct measurement of this parameter can be obtained from diffraction performed in a direction parallel to the interfaces. This was done by GYORGY et al. /12/ on Cu/Ni SL. These structures, grown along [111] direction, were analysed along the [hho] direction around [220] reflection of the average in-plane parameter ($a_{/\!/}$). The diffracted peak in that direction is directly related to $a_{/\!/}$. They found average value expected from a coherent structure. However, for increasing period, the peak was broader, suggesting a loss of coherency and an evolution towards a splitting into the two peaks related to pure Cu and Ni. Such measurement of the $a_{/\!/}$ variation in an SLS can be obtained from asymetric reflections, which involve a combination of $a_{/\!/}$ and $a_{\perp}$. FANTNER /13/ obtained a very precise determination of the angle between the (8 2 14) reflection of PbTe and PbSnTe sublayers of an SLS. It was obtained from a series of θ-2θ scans obtained by varying the offset angle. In the case of $In_xGa_{1-x}As/GaAs$ SLS, grown on a GaAs substrate, some loss of coherency was found between the whole SL and the GaAs buffer layer, whereas the structure itself was found coherent (QUILLEC et al. /1/). This result was confirmed by transmission electron microscopy (GOLDSTEIN et al. /14/) which shows that some misfit dislocation occur for periods beyond 200 Å (10 periods), located at the interface with the buffer layer. The coherent SLS tends to reach its equilibrium $a^e_{/\!/}$, i.e. the average lattice parameter $(a_1^{SF} n_1 + a_2^{SF} n_2)/(n_1+n_2)$ (SF for strain free), if the elastic coefficients are taken equal change in both materials. This change in $a_{/\!/}$ leads to an equal change in a_1 and a_2, lattice parameters in the direction of the modulation. So, a partial strain relaxation in the SLS would lead to a general shift of the diffracted peaks in the vicinity of GaAs (004). Actually, the absolute position of the satellites show that the maximum relaxation observed was a few per cent, in agreement with the dislocation density observed by TEM.

3.4 Measurements Related to the Refractive Index

A simple calculation (JAMES /6/) shows that the composition of the remaining wave with the diffracted part in the forward direction after crossing a plane of atoms (density n) leads to a phase shift :

$$- q_0 = (n\lambda/\sin\theta)\ f(o)\ e^2/mc^2$$

where f(o) is the atomic scattering factor in the forward direction. Thus, crossing p planes separated by a distance a leads to a phase lag $2\pi x/\lambda + pq_0$ (with $x=pa/\sin\theta$) instead of $2\pi x/\lambda$ in vacuum. In other words, the refractive index is slightly less than one : $1-\delta$ with $\delta=\lambda^2 e^2 N f(o)/2\pi mc^2$ (N=density of atoms). One consequence of this is the existence of a critical angle for total reflection : $\theta_c \simeq \sqrt{2\delta}$. Another is that the Bragg's law should be slightly modified : the actual Bragg angle is θ, with $\sin^2\theta = \sin^2\theta_B + 2\delta$.

This change is very small, since δ is of the order 10^{-6}. However, MICELI et al /15/ have shown recently that it can be measured with good precision near 000 diffraction, where the (00L) SL satellites are all shifted : $\sin^2\theta = (\lambda/2P)^2 L^2 + 2\delta$. A good precision on δ is obtained from the plot of $\sin^2\theta$ as a function of L^2. δ provides a determination of the sum of the atomic scattering factors, and thus a relation between X and n_1/P, for instance, in the case of GaAs $(n_1)/Al_xGa_{1-x}As$ (n_2) SL. This needs somes simplifying assumptions, like an averaging of the index in the whole structure and neglecting dynamical effects. The calculation of diffraction near θ=0 was done earlier by CHANG et al /16/, by a classical optical description, each sublayer being given its own index. The fit with experiments was remarkable showing for instance that the sublayer thicknesses are defined to withing about one monolayer. The particular feature at small angle scattering is that the cristal periodicity has not to be taken into account (see Fig.5). The SL behaves as if it were made of amorphous sublayers, so that an average parameter like the refractive index can describe quantitatively the diffracted profile.

4. Calculated X-Ray Diffraction Profiles

The relation between the measured diffraction intensity (by double diffraction) and the structural factor F(hkl) is given by FLEMING et al. /4/ :

$$I_{measured} = C\ F(hkl)^2\ A(\theta) \times [1/\sin\theta]\ P_{hkl}\ D$$

with $A(\theta)$ = absorption term = $(1/2\ \mu_1) \exp(-2\mu_c t_c/\sin\theta) \times [1 - \exp(-2\mu_1 t_1/\sin\theta)]$

P_{hkl} = polarization term = $(1+\cos^2 2\theta_m \cos^2 2\theta)/(1+\cos^2 2\theta_m)$

D = temperature term = $\exp(-B\sin^2\theta/\lambda)$

C = θ independent adjusting variable.

Here θ and θ_m are the diffraction angles from the sample and the first crystal respectively, B is the Debye Waller factor (0.6 for GaAs), $\mu_{c,1}$ and $t_{c,1}$ are the linear absorption coefficients and thickness of the cap layer and SL respectively. It has to be noted that the whole coefficient varies slowly with θ compared to F(hkl).

Two approaches are considered in the literature :

4.1 Direct Calculation of the Real Space SL Structure

Considering a one dimension modulation, SEGMULLER et al /17/, PALATNIK et al. /18/ and FLEMING et al./4/ developed a calculation of the amplitude of the first harmonics in the Fourier sum in real space of both the concentration function and the position of the atomic planes. This calculation was performed first by DE FONTAINE /19/ in a general case. FLEMING et al./4/ derived a simple expression of these harmonic amplitudes as a function of the ratio between integrated intensities related to satellites (00L), and zero order. From this expression, these authors were able to calculate directly the composition modulation of a GaAs/AlAs SL for various annealing times. This approach is well suited to that case, since the profiles were not abrupt after interdiffusion, so that a limited Fourier sum was satisfactory. It has to be noted that a problem arises as to the sign to be affected to the calculated Fourier terms since only their absolute value is obtained. This is solved by comparing with possible profiles. As is noted by FLEMING et al./4/, the intensity of the SL satellites near (000) is less sensitive to the lattice spacing modulation, and should thus be preferred when the composition modulation in needed ; this can be deduced from Figs.5 and 6.

4.2 Fitting the X-Ray Profiles

The second approach is to assume a given profile of the modulation, (generally a step one), to derive the structural factor and compare the calculated diffraction profile to the measured one : the structural parameters can be obtained from this fitting process. In the case of a step model, the relevant parameters (compositions and thicknesses of the sublayers) can sometimes be calculated only from the position of the satellites (period and zero order, as seen above) and from growth considerations, as seen in section I.

The profile can then be calculated without any fitting parameter. QUILLEC et Marzin /20/ obtained, for instance, very good fits in the case of $In_{x1}Ga_{1-x1}As/In_{x2}Ga_{1-x2}As$ SLS grown on an InP substrate in such conditions. The structural factor near InP 004 is

$$F = \sum_{m=0}^{n-1} \exp(i4m\pi.P \;.\; \sin\theta/\lambda).C$$

$$\text{with } C = \sum_{n=0}^{n_1-1} e^{2i\pi a.\frac{\sin\theta}{\lambda}} (F_A^{(1)}+F_B^{(1)} e^{i\pi a_1 \frac{\sin\theta}{\lambda}})$$

$$+ \sum_{n=0}^{n_2-1} e^{2i\pi(na_2+n_1a_1)\frac{\sin\theta}{\lambda}} (F_A^{(2)}+F_B^{(2)} e^{i\pi a_2 \frac{\sin\theta}{\lambda}})$$

where $F_{A,B}^{1\;2}$ are the atom scattering factors of the constituent atoms in the two materials.

The good fit shows that the step model fits and also that the kinematical approach is suitable.

Figure 8 shows the experimental and calculated profiles obtained near InP 004 on an $In_{x1}Ga_{1-x1}As/In_{x2}Ga_{1-x2}As$ SLS grown on InP, with 20 periods, P=1150 Å, $L_1 \approx L_3$, x_1 and x_2 such that the equilibrium parameter $a_{\overline{F}}$ was nearly lattice matched to InP (i.e. $x_{1,2}$ = 0.53 ± 0.07). This SLS provides a good example of well separated envelopes as illustrated schematically in Fig.6. The "zero order"

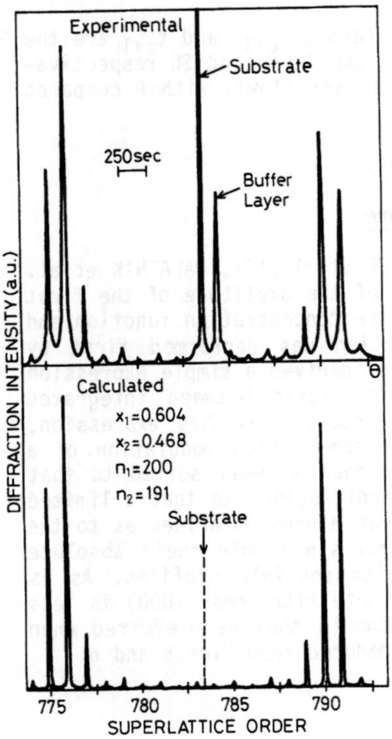

Fig.8 : Experimental and calculated X-ray profiles of a $In_{X1}Ga_{1-X1}As/In_{X2}Ga_{1-X2}As$ SLS grown on InP by MBE. This is a good illustration of the case considered schematically in fig.6

of the SL, in that case the 0 0 782 diffraction order is very weak and the envelopes, referring to each material, are roughly symmetrical on both sides, as expected.

Acknowledgements : Some experimental work exposed in this paper result from a collaboration with L. Goldstein, M.C. Joncour, G. Le Roux, J. Primot and J. Burgeat, whom I want to thank for their cooperation and stimulating discussions.

APPENDIX 1

Diffraction

We recall the principles of diffraction and how the Fourier transform appears in its calculation. Diffraction by an electron gives, in the classical Thomson scattering theory, an intensity I in M (if E_0 is the incident wave (Fig.9))

$$E = E_0 \left(\frac{e^2}{mc^2}\right) \frac{1}{r} \quad [(1+\cos^2\theta)/2]^{\frac{1}{2}} \tag{1}$$

The second bracket is the polarization factor. It is usually ignored in the calculation until the last step. The amplitude e^2/mc^2r is the electron "unit of scattering". It is usually taken as unit. Due to the term m^{-1} in the diffracted intensity, only electrons are to be considered. Scattering matter is thus represented by an electron cloud, with density $n(\vec{\rho})$.

So, let the incident wave be represented by

$$E(\vec{\rho}) = E_0 \exp(i[\vec{k}\,\vec{\rho} - \omega t]).$$

We recall here the kinematical calculation of diffraction, which holds strictly only for very small diffracting crystals.

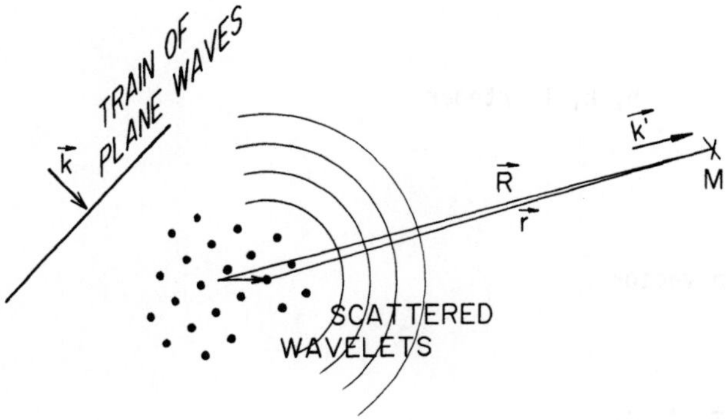

Fig.9 : Calculation of the diffracted amplitude in the kinematical approach

The classical treatment [see for instance KITTEL/21/] is as follows : A scattering center lies in $\vec{\rho}$; at a distance r, the diffracted wave is [see expression (1)]

$$E_d = C\,E\,(\vec{\rho})\,\frac{e^{ikr}}{r} \qquad \text{(see Fig.9)}$$

$$\left(C = \frac{e^2}{mc^2}\sqrt{\frac{1+\cos^2\theta}{2}}\right).$$

(Solution of the propagation equation $\nabla^2 E=(1/C^2)\ \ ^2E+\partial t^2)$.

As $r \approx R - \rho\cos(\vec{\rho},\vec{R})$

$$\vec{k}\,\vec{\rho} + k\,r = \vec{k}\,\vec{\rho} - k\,\rho\cos(\vec{\rho},\vec{R}) + k\,R = -\,\vec{\rho}\,\Delta\vec{k} + k\,R$$

$$(k' = k\cos(\vec{\rho},\vec{R}))$$

$$\Delta\vec{k} = \vec{k}' - \vec{k}.$$

The diffracted amplitude is proportionnal to

$A \sim \sum_\rho \exp i\,\vec{\rho}\,\Delta\vec{k}$ (the coefficient is $C\,\frac{Eoe^{ikR}e^{-i\omega t}}{R}$).

So the diffraction by an element dV : n (ρ) dV is

$$A \sim \int dV\,n(\rho)\,\exp(-i\vec{\rho}\,\Delta\,\vec{k}) \qquad (2)$$

Expression (2) shows that the diffracted amplitude in that approximation is nothing but the Fourier transform of real space, represented by the electron density.

If this function shows periodicity (crystal) :

$$n(\vec{\rho}) = \sum_{\vec{K}} n_{\vec{K}}\,\exp(i\,\vec{K}\,\vec{\rho})$$

$$n(\vec{\rho} + \vec{\rho}_{mnp}) = n(\vec{\rho})\ [\vec{\rho}_{mnp} = m\vec{a} + n\vec{b} + p\vec{c}]\ m,n,p \text{ integer} \qquad (3)$$

$$\rightarrow \vec{K}\,\vec{\rho}_{mnp} = 2\,\Pi.\text{Integer}$$

we define the reciprocal lattice as the family of vectors $\vec{G}$ such that $\vec{G}\,\vec{\rho}_{mnp} = 2\Pi.\text{Integer}$

It can easily be shown that

$$\vec{G} = h\,\vec{A} + k\,\vec{B} + l\,\vec{C} \qquad h,\ k,\ l \text{ integer}$$

with $A = 2\pi \dfrac{\vec{b} \wedge \vec{c}}{(\vec{a},\ \vec{b},\ \vec{c})}$, etc.

(3) shows that $\vec{K}$ is one particular vector $\vec{G}$

$$n(\vec{\rho}) = \sum_{G} n_G \exp\,(i\,\vec{G}.\vec{\rho})$$

and, from (1) $A \sim \sum_{G} n_G \int dV \exp i\vec{\rho}\,(\vec{G}-\Delta\vec{k})$

this is significant only if $\Delta\vec{k} = \vec{G}$ (then, $A \sim n_G$). (4)

The geometrical representation of (4) is the Ewald's construction (see fig.10)

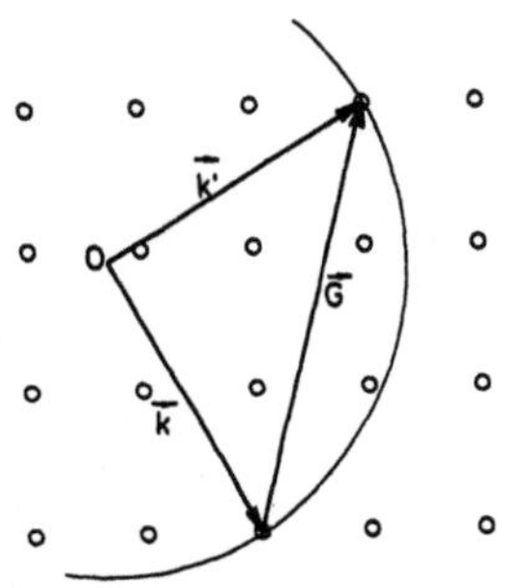

Fig.10 EWALD's CONSTRUCTION

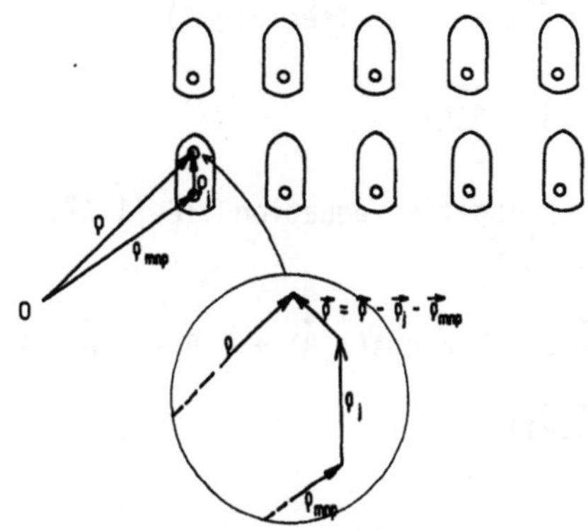

Fig.11:Details of the diffracting element.

The details on the diffracting element can be represented by

$$\vec{\rho}_j = x_j\vec{a} + y_j\vec{b} + z_j\vec{c} \quad \text{(see Fig.11)}$$

with j atoms, with an electronic distribution

$$C_j(\vec{\rho}\,') \qquad \vec{\rho}\,' = \vec{\rho} - \vec{\rho}_j - \vec{\rho}_{mnp}$$

so $n(\rho) = \sum_{mnp} \sum_{j} C_j\,(\vec{\rho} - \vec{\rho}_j - \vec{\rho}_{mnp})$

(1) becomes

$$A \sim \sum_{mnp} \sum_{j} \int dV\ C_j\,(\vec{\rho}\,')\ \exp(-i\vec{\rho}\Delta\vec{k})$$

$$\sim \Sigma \quad \Sigma\ f_j \exp[-i\Delta\vec{k}\ (\vec{\rho}_j + \vec{\rho}_{mnp})]$$

with $f_j = \int dV\ C_j\ (\vec{\rho}\,')\ \exp\ (-i\vec{\rho}\,'\Delta\vec{k})$ (5)

$A(\Delta k) \sim [\sum_{mnp} \exp(-i\vec{\rho}_{mnp}\ \Delta\vec{k})]\ [\sum_{j} f_i\ \exp(-i\vec{\rho}_j\ \Delta\vec{k})]$; (6)

where f_j is the scattering factor of atom j.

Due to the first term of (6), $A(\Delta k)$ is significant only if $\vec{\Delta k}=\vec{G}$; then, $A(\vec{G})$ is proportional to

$$A(\vec{G}) \sim \sum_j f_j \exp\,[-i\vec{\rho}_j\vec{G}]. \tag{7}$$

$\vec{G}$ is represented by the coordinates h,k,l. This expression depends only on the unit cell. Each atom of this unit cell is represented by its location $\rho_j(u,v,w)$ and its atomic scattering factor f_j. The atomic scattering factors for atoms and ions can be found in the International Tables. They refer to the scattering amplitude from a single electron. They are a (slow) function of λ and θ. A small complex quantity is added to expression (5) to take into account the proximity of an X-ray absorption edge.

The diffracted intensity is thus, in the kinematical approximation

$$I \sim \sum_j f_j \exp[-i(uh + vk + wl)]^2 \quad \text{(summation over the unit cell).}$$

In expression (6), the second term F, called structure factor, depends only on the elementary cell, whereas the first term G, called the network factor, reflects the periodical arrangement of these elementary cells.

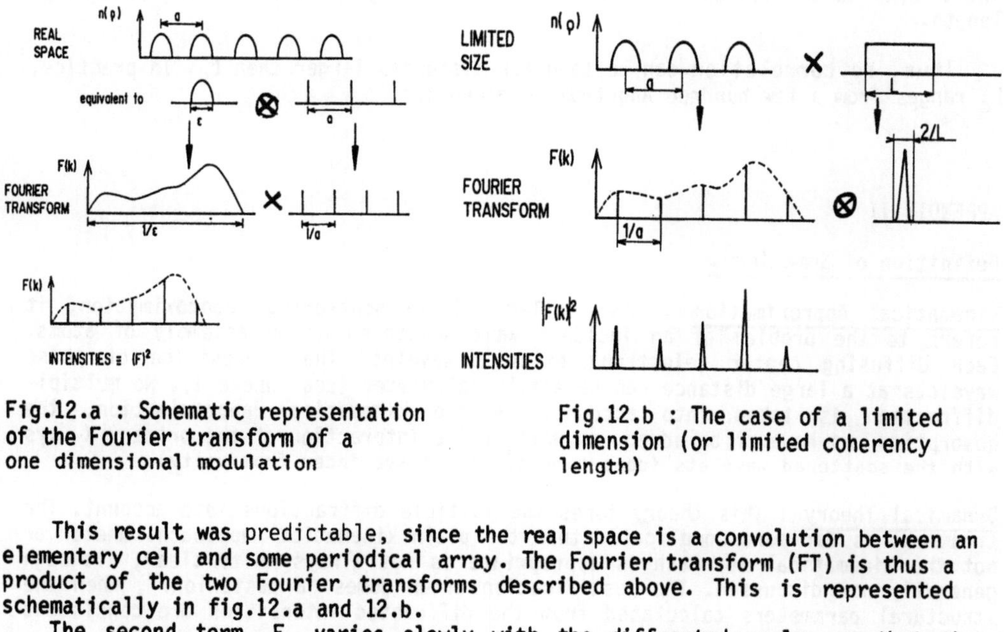

Fig.12.a : Schematic representation of the Fourier transform of a one dimensional modulation

Fig.12.b : The case of a limited dimension (or limited coherency length)

This result was predictable, since the real space is a convolution between an elementary cell and some periodical array. The Fourier transform (FT) is thus a product of the two Fourier transforms described above. This is represented schematically in fig.12.a and 12.b.

The second term, F, varies slowly with the diffracted angle, so that the existence and shape of the diffracted peaks will only depend on G, whereas the intensity will be a function of F.

For a "large" perfect arrangement, G has a significant value only for $\vec{\Delta k} = \vec{G}$, so that

$$A\,(\vec{\Delta k}) \sim \delta\,(\vec{\Delta k} - \vec{G}) \times F.$$

The measured quantity is the intensity, i.e. the square of the Fourier transform FT of real space :

$$I\,(\vec{\Delta k} = G) \sim |FT|^2.$$

The inverse of the FT of I thus does not give back the real electron arrangement (unfortunately), but a function of it :
$P(\vec{\rho})$, which is the autocorrelation function (called the Patterson function).

$P(\vec{\rho})$ is the number of pairs separated by $\vec{\rho}$. The phase information has been lost and lacks to reconstruct the object in real space.

So, in general, the user assumes a model describing the real object, calculates FT^2 and compares it to the diffracted pattern or profile (trial and error).

<u>Departure from Ideality</u>

In experiments, the incident wave is slightly different from a plane wave.

$$I_{exp}(\vec{k}) = I_{th}(\vec{k}) * T(\vec{K})$$

where $T(\vec{K})$ is the apparatus function.

The F.T. is thus $P(\vec{x}) \times t(\vec{x})$,

where $t(\vec{x})$ is a Gaussian function negligible for $x > L_c$, the limiting coherency length.

Thus, no correlation can be seen for distances larger than L_c. In practice, L_c ranges from a few hundred Angström to microns.

APPENDIX II

<u>Definition of Some Terms</u>

<u>Kinematical Approximation</u> : Also called "single scattering" approximation, it refers to the problem of an incident wave scattered by an assembly of atoms. Each diffusing center (electron) emits a wavelet. The composition of these wavelets at a large distance can be easily calculated (see Annexe I). No multiple diffraction is taken into account. In the kinematical approximation, the absorption can however be added, as well as the interaction of the primary X-rays with the scattered wavelets (departure of refractive index from unity).

<u>Dynamical Theory</u> : This theory takes the multiple diffractions into account. The treatment is far more complicated than the plain kinematical approach. Wheter or not the kinematical approach is appropriate is not always quite clear, although generally not discussed. The justification often comes "a posteriori", when the structural parameters calculated from the diffracted intensities are consistent with other measurements.

<u>Absorption</u> : Exponential decay of the incident beam due to photoelectric absorption (independent of the Bragg condition). The absorption depth is the thickness of a layer which reduces the reflectivity by a factor of e. For III-V compounds, it ranges typically from 1 to 10 μm. This effect is taken into account in the kinematical approximation.

<u>Extinction</u> : Power loss of the incident beam in conditions where Bragg's law is fulfilled. This loss is larger than that due to absorption. The associated depth, t_e, is such that 93% of the power is reflected by $2t_e$. It ranges typically from 1 to 5 μm. Extinction is due to multireflections and must be treated through the dynamical theory. The correction in diffracted intensity due to extinction is given by GUINIER /9/ p.138.

References

1. M. Quillec, L. Goldstein, G. Le Roux, J. Burgeat and J. Primot: J. of Appl. Phys. 55, 2904 (1984)
2. F.K. Reinhart and L.A. Logan, J. Appl. Phys. 44, 3171 (1973)
3. G.A. Rozgonyi and T.J. Ciesielka, Rev. Sci. Instrum. 44, 1053 (1973)
4. R.M. Fleming, D.B. Mc Whan, A.C. Gossard, W. Wiegman and R.A. Logan, J. Appl. Phys. 51, 357 (1980)
5. M.C. Joncour, M.N. Charasse and J. Burgeat, J. Appl. Phys. 58, 3373 (1985)
6. R.W. James, In the Optical Principles of the Diffraction of X-rays, Ed by G. Bell and Sons, London 1962
7. W.J. Bartels, J. Vac Sci. Technol. B 1, 2 (1983)
8. J. Kervarec, thesis (1983)
9. A. Guinier, "Théorie et technique de la radiocristallographie", (Dunod, Paris 1964)
10. D.B. Mc Whan, In Synthetic Modulated Structures, Ed. by L.L. Chang and B.C. Giessen (Academic London 1985) p. 43
11. J.M. Van Hove, P.I. Cohen and C.S. Lent, J. Vac. Sci. Technol. A1, (2), 546 (1983)
12. E.M. Gyorgy, D.B. Mc Wham, J.R. Dillon, L.R. Walker and J.V. Waszczak, Phys. Rev. B 25, 6739 (1982)
13. E.J. Fantner, In Semiconductor Quantum Well Structures and Superlattices, Ed by K. Ploog and N.T. Linh, Les Editions de Physique, Les Ulis, France
14. L. Goldstein, M. Quillec, K. Rao, P. Henoc, J.M. Masson and J.Y. Marzin, J. Phys. Paris 12, C5, 201 (1982)
15. P.F. Miceli, D.A. Neumann and H. Zabel, Appl. Phys. Lett. 48, 24 (1986)
16. L.L. Chang, A. Segmüller and L. Esaki, Appl. Phys. Lett. 28, 39 (1976)
17. A. Segmüller and A.E. Blakeslee, J. Appl. Cryst.6, 19 (1973)
18. L.S. Palatnik, A.A. Koz'ma, I.F. Mikhailov and V.N. Maslov, Sov. Phys. Crystallogr. 23, 316 (1978)
19. D. de Fontaine, In Local Atomic Arrangements Studied by X-ray Diffraction. Metallurgical Society Conferences 36, Ed by J.B. Cohen and J.E. Hilliard, pp.51-94, N.Y., London, Paris
20. M. Quillec and J.Y. Marzin, Interm. Symp. on GaAs and Related Compounds, Karnizawa 1985
21. C. Kittel, Introduction to Solid State Physics, (Wiley, New York 1971)

Recently Developed TEM Approaches for the Characterisation of Semiconductor Heterostructures and Interfaces

W.M. Stobbs

Department of Metallurgy and Materials Science, Pembroke Street, Cambridge, United Kingdom

The concepts needed to understand the use of the transmission electron microscope for the microstructural characterisation of semiconductor systems are first exemplified by applications of the basic methods, as are the various approaches required for the preparation of suitable thin foils. Developments in the techniques which can now be applied to the atomic level structural and chemical characterisation of interfaces in such systems are then described mainly by considering specific problems for which the new methods have been developed. The applications considered range from the study of semiconductor/metal and semiconductor/insulator interfaces to the problems in analysing the quality of current III-V heterostructures. In conclusion the future trends in technique which will be required are discussed.

1. INTRODUCTION

The exciting new developments currently being mastered in the engineering and physics of semiconductor device systems are mirrored in a parallel change in approach in the detail to which such systems need to be characterised. This is, in turn, requiring rapid improvement in the quantification of data which can be obtained by the use of a very broad range of analytic techniques, many of which are as expensive and frustratingly complex as the modern methods needed to fabricate the systems examined. Nearly every imaginable method for the structural and chemical analysis of a system has its special rôle to play, but in many ways Transmission Electron Microscopy (TEM) is of central importance partly because of the rich variety of individually different ways in which the instrument can be used and partly because, in most of its modes of use, it is ideally suited to the characterisation of localised change in structures. It is the way an engineering approach to the scaling down of a multicomponent system very often neither makes best use of the available physics, nor faces up to the changes in materials science which then need to be dealt with, which thus makes the use of TEM so important. Equally, however, the level to which modern device systems need to be characterised is requiring very rapid development of the basic traditional TEM methods which have been used for many years for the examination of structural irregularities.

Without straying too far into the problems involved, my aim here is to provide an understanding for the non-specialist of some of the more exciting TEM methods which are currently being developed for the quantitative structural characterisation of semiconductor systems. I have not attempted to review the subject and the references cited are neither exhaustive nor necessarily fully representative, equally several areas are not covered at all. I do not for example consider the important ways in which Scanning Electron Microscopy (SEM) is now being used diagnostically and non-destructively nor have I attempted to discuss either the specialised applications of STEM in chemical analysis using Electron Energy Loss Spectroscopy (EELS) or the increasing usefulness of cathodoluminescence methods, at least for bulk specimens, in SEM. The reader wishing to gain an appreciation of such techniques is referred to the variety of international EM conference series publications now readily available.

2. THE STANDARD TECHNIQUES IN TEM

2.1 The Basic Concepts

A high energy (1 keV to 1 MeV) electron can interact with matter in a great many ways with elastic scattering cross-sections increasing with atomic number (Z) rather faster that those for inelastic processes, the respective cross-sections being similar for Z~10. This is often forgotten when the microscope is used as an imaging instrument mainly because of the way the periodic Bloch potentials of a crystalline material result in the inherent addition of amplitude for elastic scattering yielding bright spots in the diffraction pattern despite the summed diffuse background intensity frequently being predominant. The electron microscope itself is disarmingly simple. Electrons accelerated from a normally thermionically excited source are focused electromagnetically into either a fairly parallel ($\alpha \lesssim 0.001$ rad) or convergent ($\alpha \sim 0.01$ rad) probe onto the specimen depending on the technique to be used. The probe size can be as small as 1 nm or so even in a non-specialised STEM, when chemical analysis is required using EELS or EDS methods, and up to a micron or more in diameter when imaging conventionally. Equally specimens only a few nanometres in thickness are required when obtaining quantitatively useful data at very high resolution whether chemically by EELS or structurally by the interference of the elastically scattered beams.

The dynamical nature of the elastic scattering process has to be considered for very small thicknesses, arguably down to the unit cell size unless, as in the weak beam method, the elastic scattering is weakened by orienting the region examined well away from a Bragg condition. The weak phase object approximation, for which the scattered beams are considered to be $\pi/2$ out of phase with the straight through beam, would thus seem at first sight to be of limited application, and yet it is central both to qualitative and to quantitative interpretation of high resolution imaging. Quantitatively the multi-slice method for the computation of image simulations to be compared with those obtained experimentally relies on this limit to the scattering process, amplitude then being added from slice to slice up to the full thickness of the model structure simulated. It is the phase changes caused by the objective lens of the microscope which then have to be incorporated into the calculation, as a function of the defocus, to determine how the amplitude emerging at different angles from the specimen is added differently to form the final intensity distribution in the image plane. It should by now be clear that a high resolution image of even a uniform structure can have a multitude of different forms as the thickness and defocus are changed. It is virtually never directly comparable with the projected potential. This, coupled with the general difficulty in evaluating these parameters accurately and independently, necessitates the comparison of image series with computed simulations if a unique characterisation of the structure is to be obtained. Even so it is very rare that the real space intensity distribution is analysed fully and in this context we will see why future development of the theory is required to take account of inelastic scattering with sufficient accuracy.

Returning to more qualitative approaches to high resolution image interpretation, the Weak Phase Object (WPO) approximation is assumed. Specific Fourier components of the scattered amplitude can now be considered to predominate in the interference image as a function of how the defocus is arranged to make the phase change at the appropriate angle an odd multiple of $\pi/2$. The characteristics of the microscope can in fact be assessed easily by imaging an amorphous specimen and obtaining the power spectrum in the image, as a function of defocus, either by optical diffraction or by image digitisation and Fourier transforming the data. An important parameter for the microscope is then the angle to which the power spectrum exhibits no zeros, this being maximised at "extended Scherzer defocus" when the maximum negative phase change produced by the lens lies between $\pi/2$ and π. Even in microscopes for which this angle can correspond to spacings less than about 0.16 nm, the power spectrum is then usually weakened by chromatic aberration effects to such an extent that the above figure approximates to current realistic machine resolution limits. However, given full computation there is no reason why images obtained at higher defoci, with enhanced transfer at slightly higher angles, should not be used for structure assessment. The antiphase contributions of different Fourier components does then complicate image interpretation, but provided the WPO approximation is valid then the effects of the transfer function can be deconvoluted from an image series to provide an approximation to the potential distribution in the specimen without recourse to model calculations. At the same time, it should be noted that information about spacings very much less than the resolution of the microscope can often be extracted from the image using surprisingly simple methods.

One way of thinking about the microscope is to consider it to be a rather efficient fast Fourier transform machine. The image plane is Fourier related to the diffraction pattern in the back focal plane of the objective just as this is Fourier related to the object plane, the beauty of the process being that aperture functions (determining the "transfer function" and thus modifying the relationship between object and image) are introduced either by physically placing apertures around different beams and/or by changing the defocus. On this type of approach, and with the realisation that the lenses below the objective conveniently allow either the object plane or diffraction pattern to be imaged, we are now in a position to summarise the various basic modes of operation of the microscope.

2.2 Convergent Beam Electron Diffraction

The examination of the diffraction pattern allows the symmetry of the crystal to be determined and in the "Convergent Beam Electron Diffraction" method (CBED) the use of a small probe of high convergence allows the space group of a laterally small region to be determined, information about "High Order Laue Zones" then being retained in the HOLZ pattern. The greater the thickness of the foil examined (up to limits determined by inelastic scattering) the higher the accuracy to which relative changes in lattice parameter can also be measured for neighbouring regions. This is done by examining the relative shifts of HOLZ lines, these becoming progressively thinner as the specimen thickness is increased up to and beyond the relevant extinction distance (inversely proportional to the related Bloch wave potential). This technique has been applied to the measurement of, for example, the In content of layered GaInAs/GaAs systems but is, as the basic approach described, of insufficient accuracy to be useful for GaAlAs/GaAs heterostructures for which the lattice parameter change is small (e.g. BRITTON et al. [1]).

2.3 Bright and Dark Field Imaging

The most basic imaging method is to orient the material examined in a specific Bragg condition and then to examine the image plane with either the central (bright field) or diffracted (dark field) beam apertured in the back focal plane and on the optic axis of the objective lens. Basic though the approach is, it remains powerful in allowing the characterisation of defects in a material as can be useful in for example assessing growth characteristics in MOCVD or MBE techniques across a chemical interface as a function of the imposed misfit (e.g. NEWCOMB et al. [2]). While the method can also obviously be used to characterise the structure of second-phase inclusions, the approach can also be applied to the assessment of both short-range order in a crystalline phase (e.g. STOBBS et al. [3]) and the differentiation of microcrystallinity from an amorphous state down to size limits below which such a differentiation would be meaningless (eg. STOBBS [4]). The method, as applied to an amorphous structure has also been shown to be sensitive to localised changes in density (DONOVAN et al [5]) and thus, potentially, chemistry and so might usefully be applied to the characterisation of amorphous barriers as can form in the solid state by interdiffusion at a metallic contact (e.g. BOOTHROYD et al.[6]).

2.4 Weak Beam Technique

The simplicity of the above imaging method masks the considerable complexity of the contrast that is generally associated with a localised displacement field and while a Bloch wave approach to the contrast interpretation is generally highly successful, as described in basic texts such as that by HIRSCH et al.[7] and THOMAS et al.[8], it is less applicable in the presence of near vertical boundaries, as recently discussed by HOWIE et al. [9] in connection with the anomalous contrast behaviour of high angle faults. The particular charm of the Weak Beam technique (COCKAYNE et al. [10]), in which the dark field reflection used to form an image is weakened by tilting it away from the Bragg condition, is thus that the contrast is to first order simplified to the extent that it can often be treated kinetically. The method, the general uses of which are described elsewhere (eg STOBBS [11]), is particularly beneficial in the characterisation of the closely spaced defects which can be associated with an interphase boundary (DONOVAN et al. [12]) and is often usefully applied in conjunction with high resolution methods in the assessment of the different types of silicide/-silicon boundary (as has been discussed by CHERNS [13]).

2.5 High Resolution Method

The principles behind the high resolution approach have already been discussed sufficiently to allow an appreciation of the basic dangers of misinterpretation of the very beautiful images which can be produced. It is noteworthy in this context that since the discovery of quasicrystals exhibiting icosahedral ($m\bar{3}\bar{5}$) symmetry (SCHECHTMAN et al. [14]) a plethora of papers has presented high resolution images which have generally been misinterpreted as revealing the atomic structure involved. While such images do exhibit the symmetry of the Penrose tiling they are much less readily used to solve the fascinating problem of the atomic occupancy of the tiling framework. It has, for example, been demonstrated that most of the atomic occupancy models so far suggested, when modelled quantitatively, do not even yield appropriate intensity distributions in the diffraction pattern (KNOWLES et al. [15]). Current work in Cambridge is demonstrating that the solution of this problem, admittedly in the end by matching HREM images, requires the input of information obtained from both EXELF techniques (to provide a measure of the coordination), as well as CBED. In a discussion of the choice of methods for the solution of different types of structural problems (STOBBS [16]) it has been argued that the overexposure of the high resolution technique in the study of silicide/silicon interfaces has biased such work too heavily towards the planar interface silicides, for which the method is applicable, when it is remembered that the real problem is to interpret the electrical contact behaviour in terms of the structure of generally rather irregular boundary. HREM has, however, seen outstanding success in this field, as reviewed by CHERNS [13], though in the interpretation of images of interfaces problems on the effect of ionicity remain difficult to assess, as discussed by SPENCE [17] when examining the use of HREM in relation to other new methods. In very general terms HREM has progressively less application as the dimensionality of the defect studied is reduced from planar to line to point. Nevertheless outstanding results on the structural changes at line as well as planar defects, as associated with oxygen segregation in silicon, have been reported by BOURRET et al. [18] and most recently by BERGHOLZ et al. [19].

For the microscopist perhaps the most interesting problem in assessing the application of the high resolution method is to balance the ease of qualitative interpretation against the frequently months required for the full quantitative analysis of an image series. It is largely for this reason that other less direct methods for the extraction of quantitative high resolution data are sometimes preferable or usually required, in conjunction with the high resolution imaging (SMITH et al. [20], STOBBS [21]). This simplifies the testing of the uniqueness of a structural model.

3. SPECIMEN PREPARATION METHODS FOR INTERFACES

There are two related problems in deciding the most appropriate geometry for a specimen to be used for the examination of an interface. Firstly the interface of interest need not be planar on a scale ranging from the atomic to several microns and secondly the structures separated by the interface are rarely sufficiently similar for the successful application of conventional chemical or electrochemical thinning methods. The main advantages of viewing the interface edge-on is that at least the spatial geometry of the structures present can then be determined and equally, if a reaction has taken plane, the temporary sequence of the reactions involved can be ascertained given care in the interpretation. Furthermore if the interface of interest is planar, then the edge-on orientation is ideal for the application of high resolution methods. A disadvantage for this geometry is however that if a phase is formed over only a limited depth range relative to the interface then conventional selected area diffraction techniques are rarely usable for the assessment of the structure of the phase formed. Consequently most interfaces are usually best examined in both plan-view orientation as well as edge-on. If the plan-view specimen is fabricated so that a slightly inclined section is taken through the interface reaction zone then several of the different phases which might have been formed can be viewed separately in different parts of the thin foil; the edge-on foil then provides information on the depth inter-relationships between these phases. It is worth emphasising that even when an interface is apparently planar, the projection problem can lead to alarming averaging effects, particularly in high resolution imaging. Consequently, interface steps associated with low strain, which could otherwise be examined in tilted edge-on sections by weak beam methods, are often best examined using disallowed reflections in thin plan view specimens, though then care is required to differentiate the steps on the upper and lower free surfaces from those on the interface.

Most methods for making an edge-on specimen are perforce based on ion beam thinning techniques for the production of the final thin area across the interface. The methods involved were pioneered in the semiconductor industry at least fifteen years ago but have seen considerable development more recently, mainly in increasing the speed of fabrication and the rehability of the process given the frequently fragile nature of the interfaces. In the latter context the idea of encapsulating several interface specimens in a slotted rod, which is itself held in a tube, prior to sectioning and grinding has greatly simplified the foil preparation procedure. This approach as well as several others, are discussed by NEWCOMB et al. [22] and the basic geometries and procedures involved for both edge-on and plan-view specimens may be understood with reference to this paper.

It should perhaps be emphasised that ion-beam thinning is a last resort. A wide range of artefacts can result from the damaging effect of the ions and these have to be taken into account in the characterisation of the interface structures unless the damaged surfaces can be removed by subsequent chemical dipping. Furthermore, while various tricks have been developed in attempts to make disparate materials across an interface thin at similar rates, several materials of past and current interest in the semiconductor industry are particularly difficult to deal with. Niobium, for example, generally thins particularly slowly because of its tendency to oxidise as does sapphire for different reasons. Amusingly the thinning rate of niobium is in principle increased by purifying the oxygen supply to the ion guns while the thinning rate of sapphire would be expected to be increased by bleeding oxygen into the system. Some effort has been expended on the use of chemically assisted ion-beam thinning and iodine ions can be usefully used in thinning GaAs/InP structures (CHEW et al. [23]). The current challenge is to find a reliable method of obtaining thin edge-on sections of specifically chosen elements of a device which has been electronically characterised on a semiconductor chip. It would appear that industrial expertise in reactive ion-beam etching steep steps, coupled with masking procedures, should be beneficially applicable, starting in plan and completing fabrication edge-on.

4. CURRENT DEVELOPMENTS IN METHODS

Current trends in the development of methods for the structural and chemical characterisation of an interface tend towards an increased requirement for quantification to provide, in particular, structural information which can be correlated with electrical and, for example, photoluminescence data. the classic problem of the relationship between a surface state and any associated localised inhomogeneity in structure and chemistry remains. TEM techniques do not compete with SIMS or RBS methods for the analysis of the average chemistry as a function of depth but both these latter methods assume regular erosion during ion beam thinning (while effectively assuming lateral uniformity anyway) and as the TEM specimen preparer knows only too well ion-beam erosion is a far from regular process! Rather than list the specific current advances being made in the TEM approach to the problem I emphasise the way the development work has been led by the needs of the problems by examining a range of different structural interface studies each of which is requiring a different new approach.

4.1 Irregular Silicon/Silicide Interfaces by Improvement in Plan View Specimen Usage

It has already been suggested that an over emphasis on the use of high resolution techniques has led the study of silicide/silicon interfaces away from many of the more interesting reactive boundaries simply because those best suited to high resolution study should be planar. Here we examine a "development" in approach which might appear at first sight to be retrogressive in that conventional bright field, dark field and selected area diffraction methods are often the best techniques to apply when studying an irregular interface. The innovation involved is to examine the inclined plan-view section as well as the classic edge-on sample. Following the work of OTTAVANI et al. [24] on the reactive behaviour of a noble metal-rare earth pair, Pd-Er, on Si using RBS, Auger and X-ray data to correlate with Schottky barrier values, K.N. TU [25] suggested that the interface reaction should be examined by electron microscopy. The PdEr was laid down in a microcrystalline form and though some Er_2O_3 formed occasionally on the upper surface no development of this outer oxide was seen on subsequent annealing treatments in a purified d He f furnace. A typical edge-on micrograph of a reacted contact is shown in Fig.1a: the layer above A was readily demonstrated to be predominantly Er_2O_3 with PdEr in various forms below this down to the flat silicon interface beneath which localised Pd_2Si island ingrowths can be seen. Little good structural work can be done on such a sample however and we see the power of the correlated plan view examination of such an interface system in Figs. 1b,c and d showing, in this section, the successive Er_2O_3 and crystalline PdEr layers (Fig. 1b), the lower amorphous PdErSi immediately above the silicon interface (Fig.1c), and finally the epitaxially oriented Pd_2Si islands growing into the silicon (Fig.1d). In particular it is doubtful whether the interesting amorphous interlayer would ever have been noticed if edge-on samples alone had been examined while it would also then have been extremely tedious to characterise the silicide crystallography at all, let alone to the detail allowed by the moire fringes seen in Fig.1d. The further interesting effect on ion-beam promoted room temperature Pd/Si reactivity, as demonstrated in Fig.1e, was observed in preparing plan view specimens of unannealed interfaces.

4.2 High Resolution Method Improvements : Metal/Insulator Structures and Rigid Body Displacements

Arguably the most important improvement in approach in the high resolution field recently has been to stop applying the method where it can be misleading. The image shown in Fig.2a is of a non-planar Pd_2Si island/Si interface (as in Fig.1) which though of good quality is quantitatively unanalysable given interface curvature through the foil depth. Figure 2b shows both a planar and stepped region of a Nb/Al_2O_3 interface and the planar region, as part of a through focal series is currently being analysed quantitatively. Work on interfaces of this

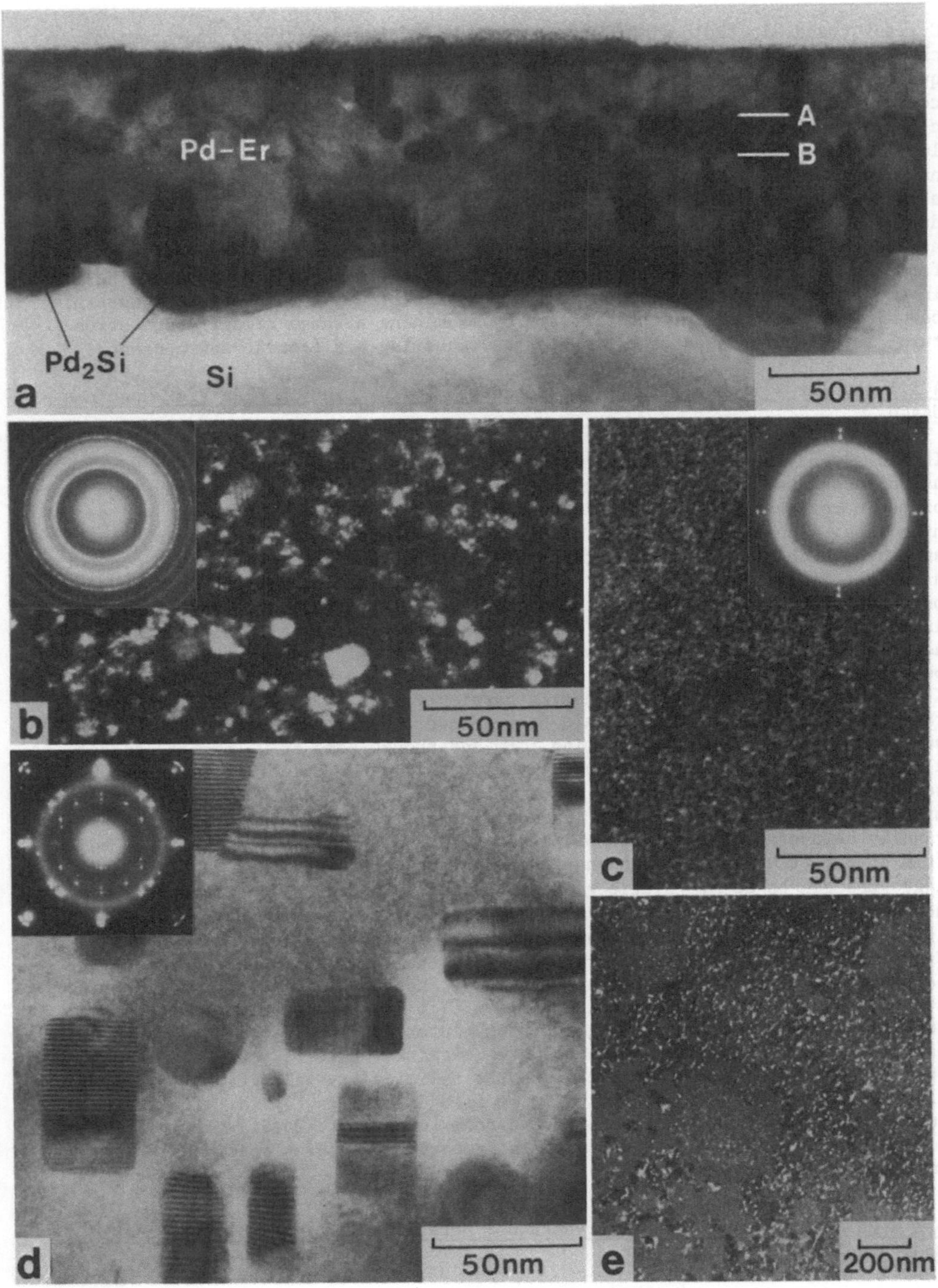

Fig.1 a-e. An edge-on image of annealed PdEr on Si with plan view sections at sequential depths while, e) shows localised interface reactivity in unannealed material by ion-beam thinning (courtesy of C.B. Boothroyd and I.B.M. Yorktown Heights for specimen preparation)

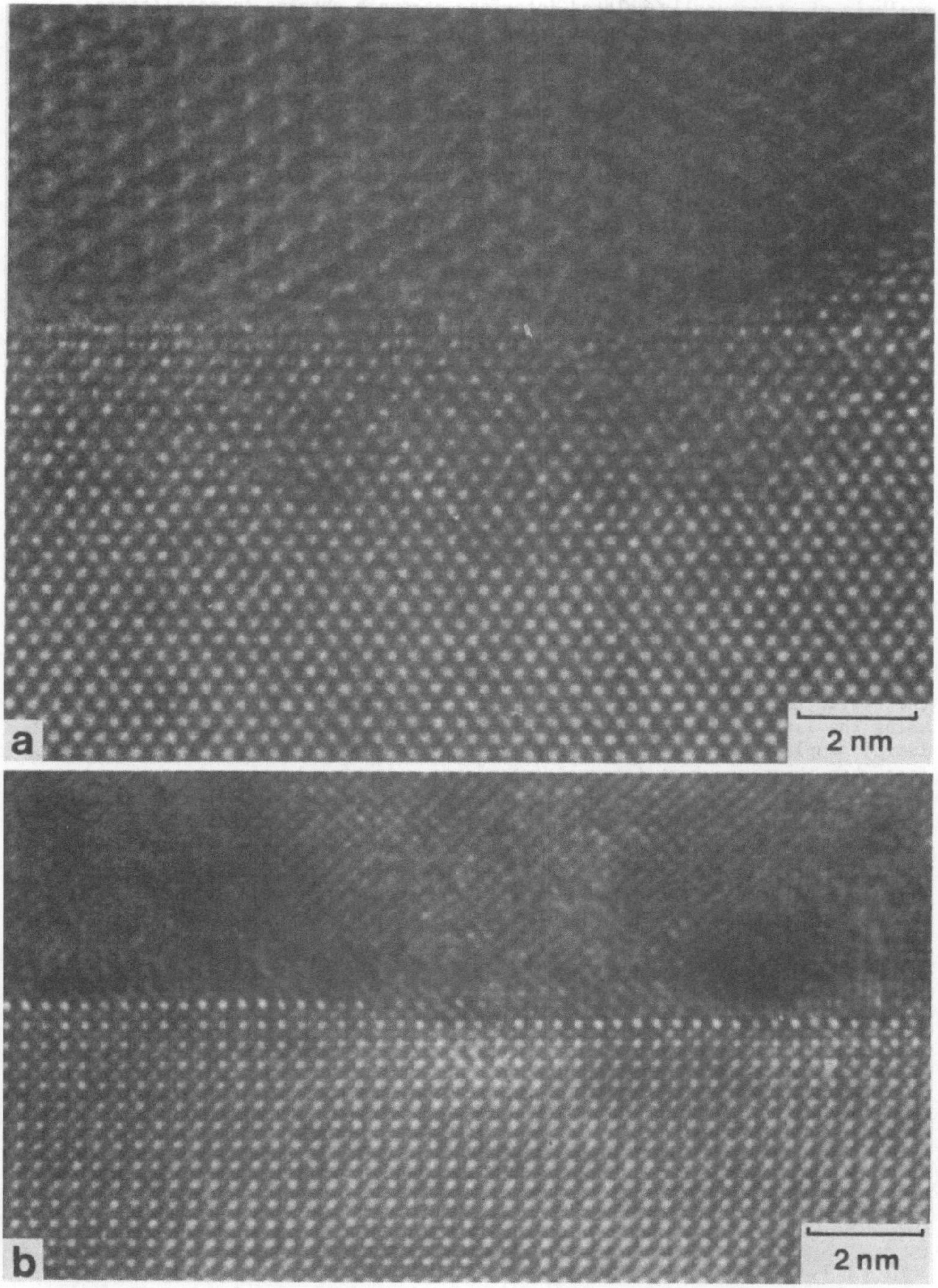

Fig.2 a) shows an irregular Pd_2Si/Si interface while b) shows a flat and a stepped region of Nb/Al_2O_3 (a) by courtesy of C.B. Boothroyd and b) by courtesy of R. Somekh, K.B. Alexander and K.M. Knowles)

or a similar type is now of interest in several fields, these ranging from the analysis of structurally related interface strength changes in metal/fibre composites to the examination of a range of metal/ceramic interfaces as a function of thermal fatigue for potential packaging applications. In the HREM field generally, rather more significant advances have been made in the accuracy to which rigid body displacements may be measured between the components on either side of an interface. This work has centred to date on the Investigations of periodic strains as of relevance in the correlation of superconducting layer behaviour and Josephson junction reliability is, we are finding, largely better addressed using weak beam methods. In the HREM field generally, rather more significant advances have been made in the accuracy to which rigid body displacement may be measured between the components on either side of an interface. The work has centred to date on the characterisation of metal/metal interfaces at boundaries of low Σ (WOOD et al. [26] and STOBBS et al. [27]) and a Σ3 interface displacement can be measured to yield values considerably less than the vibration amplitude of the atoms in the structure and ~1/100 the microscope resolution. While there remain problems in the application of the methods involved to, for example, planar silicide/silicon interfaces, published images show indications of these displacements and the problem is worthy of reexamination.

4.3 HREM for Surface Reconstructions

While scanning tunneling microscopy is by far the best way of mapping the patterns of a free surface reconstruction, the pioneering work on the free surface of gold by MARKS et al. [28] has demonstrated that such reconstructions can be viewed edge-on by HREM and the planar spacing change at the upper layer measured and compared with image simulations for appropriate models. Unfortunately it would appear that unless the surface is truly clean then there is considerable atomic mobility during examination in the TEM. It is thus possible that while this is one of the prime potential problems suggested for examination using the latest high vacuum, high resolution, TEMs the approach will have only limited application. This might well not be the case generally for reconstructions when these occur at internal interfaces. An example of such a reconstruction is shown in Fig.3 for a Si/SiO_x interface. Current work is showing that both the surface roughness and the degree of reconstruction (note that the outermost silicon layer appears to be at an increased spacing from its neighbouring plane) is the greater the thinner the oxide of certain specific types. It should be noted that the spacing irregularity visible in the image

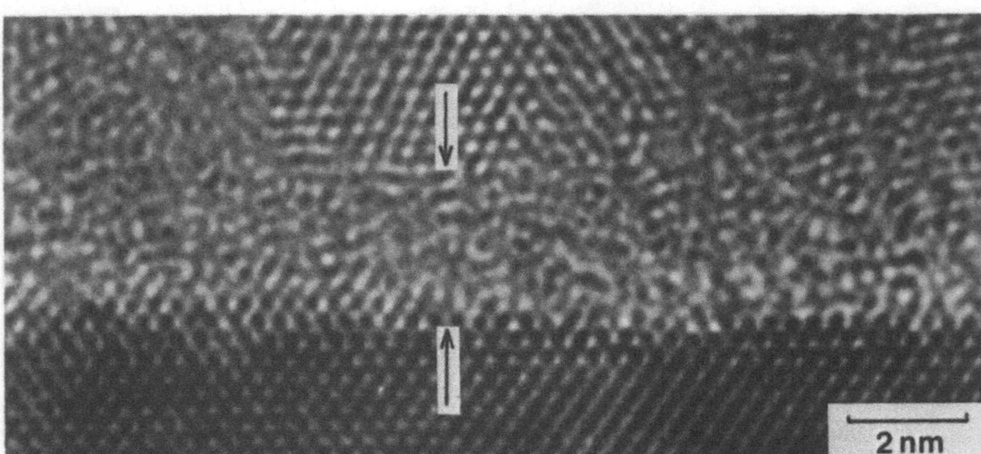

Fig.3 An image demonstrating surface silicon spacing irregularities adjacent to a thin oxide layer itself covered by polysilicon (courtesy of K.M. Knowles)

exaggerates the real spacing change and is also possibly a function of the local ionicity [17].

4.4 A Fresnel Approach to Local Chemistry Changes

Traditionally the best way of obtaining chemical data about, for example, segregation to an interface when using a TEM is to apply EELS techniques with a small probe. Using plasmon shifts the method is very old, but now the more usual approach is to examine and to attempt to quantify the respective characteristic edges (including their shifts). While there are problems in the quantification (e.g. BOURDILLON et al. [29]) the method has seen successful application in the semiconductor field for the characterisation of equilibrium As grain boundary segregation in polysilicon using a specialised STEM (BATSON et al [30]). In some circumstances site occupancy in the unit cell can be measured using the crystallographic orientation dependence of EDS (eg BOURDILLON et al [31]) but this method also is rather difficult to apply with any accuracy. Recently however, partially as a result of the examination of the way Fresnel behaviour at interfaces can be altered as a function of the local displacements [26, 32], a new method has been developed relating the intensity of the elastic contrast changes at an interface, as a function of defocus, to the local change in scattering potential. Given qualitative EELS data, which can then be correlated with the change in potential, this new approach has been demonstrated to yield quantitative information on the chemistry at sub 1 nm amorphous grain boundary layers in ceramics (NESS et al. [33]). Furthermore the thinner the layer the more sensitive the method so that the technique is ideally complementary to EELS, whose proper role is perhaps in obtaining EXELF data. It remains to be seen whether or not useful information can be obtained for semiconductor interface segregation but monolayer interdiffusion has been demonstrated by a variant of the approach for thin metal multilayers (BAXTER et al. [34]).

4.5 Multilayer Heterostructure Characterisation

There have been paralleled and exciting advances in the understanding of the anomalous property behaviour, as a function of wave length, of both metallic and semiconductor heterostructures. The quantitative characterisation of such structures with wavelengths less than about 2 nm is currently one of the more difficult challenges in transmission electron microscopy. The problem is all the more difficult if the layers exhibit a degree of interdiffusion as well as periodic strains so that it is at first sight surprising that the first steps in the solution of this problem have already been made for the more difficult metal/metal systems in that BAXTER et al [35] have demonstrated that at low wavelengths for (100) epitaxial Cu/NiPd systems a tetragonal "phase" transformation occurs. In fact the problem is equally difficult when the layers have low strains and similar scattering potentials, as is the case for AlAs/GaAs heterostructures. A high resolution image at (110) shows very little interlayer contrast because of 111 double diffraction. Provided, however, that edge-on specimens can be obtained at (100) satisfactorily interpretable images (Fig.4a) are attainable given also that the layer normal is accurately crystallographically in the cube direction; unfortunately, but probably significantly, this is not generally the case so that dealing with the projection problem is added to the microscopists' misery. Vicinality is evident in the graded heterostructures shown in Fig. 4b (as imaged in the Al content and order sensitive 200 reflection) by the reduction in contrast for the thinner layers which arises simply because they are leaning over.

A variety of approaches is currently being developed to improve the degree to which Laue condition high resolution images can be forced to be more chemistry-sensitive at high resolution. The most interesting of these involves preventing the central beam from contributing to the image (LONG et al. [36]) but interestingly there are problems in contributions to the image due to the inelastic scattering (SAXTON et al. [37]). Perhaps, however, the most important experi-

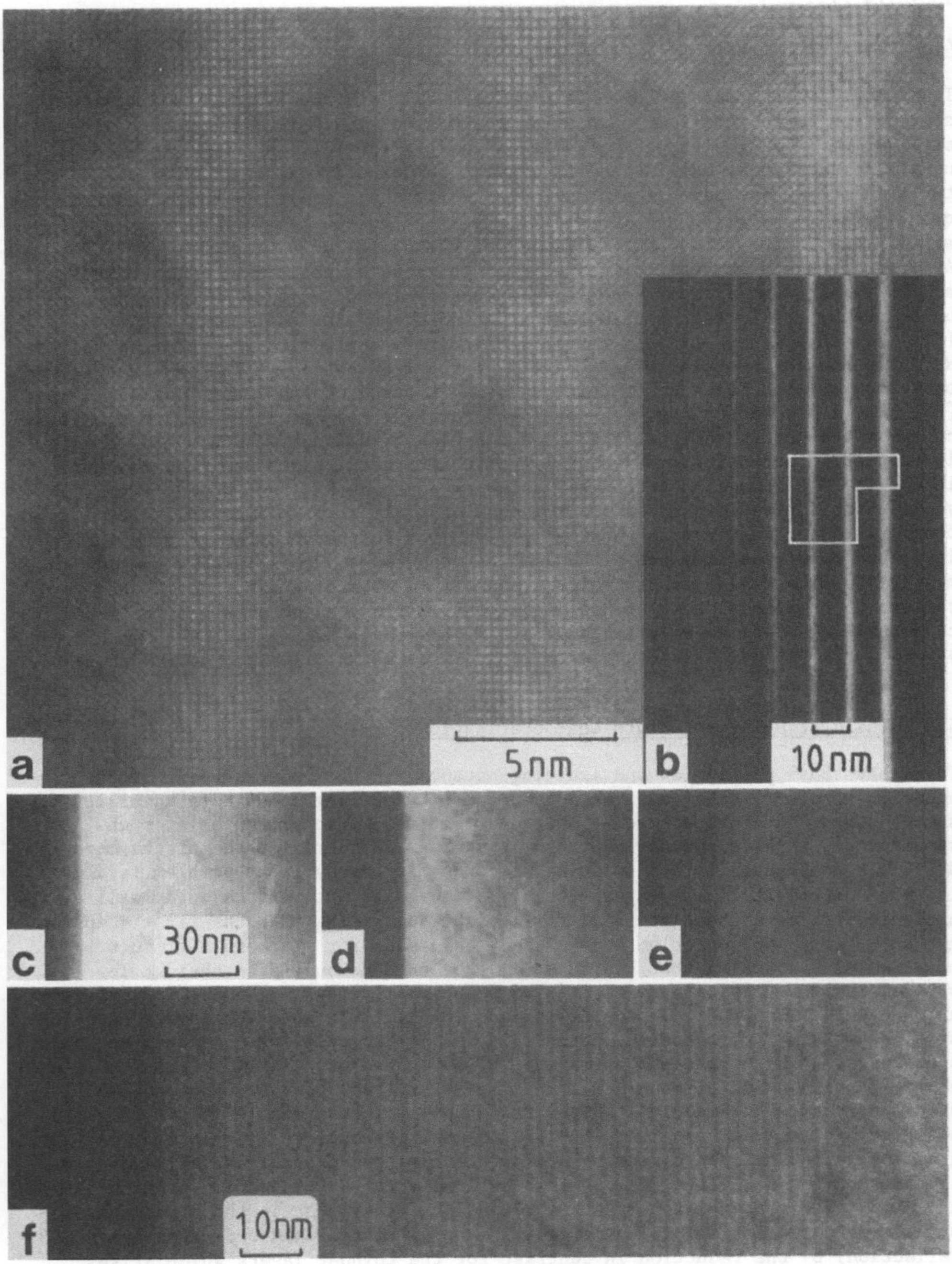

Fig.4a-f. a) Conventional high resolution image of an AlAs/GaAs heterostructure with b) showing in the a a 200 d.f image c) d) and e) are respectively unfiltered, filtered and plasmon loss images, printed for the same exposure, of an area of accidental superlattice such as that shown unfiltered in 200 in f) (courtesy of C.B. Boothroyd and E.G. Britton as well as Zeiss for use of their energy loss filtering TEM)

mental advance now required is a reliable set of chemical etches to remove ion-beam thinning damage evenly for the variety of superstructure layers which now need to be characterised. It is worth mentioning that a supposedly accurate method of using measurements of the scattered intensity in a 200 reflection to give the Al content [e.g. 36] is apparently as unreliable as other methods (e.g. [1]) are insufficiently sensitive. While this is not because of vicinality problems, inelastic contributions at 200 by inelastic/elastic scattering can be significant at surprisingly low thicknesses (Fig. 4c,d,e). However this too would not appear to have sufficient effect, at least for GaAs/AlGaAs, so that anti-site ordering effects are probably significant as is highlighted by Fig 4f which shows a supposedly uniform AlGaAs layer grown by MBE by A.N. OTHER in a direction from the (black) GaAs substrate. While such ordering effects are well known for (110) growth it is alarming that the instability of the growth process, at certain concentrations and growth temperatures, allows a near perfect "accidental" superlattice to suddenly change into a uniform alloy (the grey region). It is effects such as these which demonstrate the need to correlate physical property measurements with high quality structural analysis.

5. FUTURE DEVELOPMENTS

Potential trends in microscope design have been recently reviewed (STOBBS [38]) but none the less there are some current advances which are very exciting. The use of parallel recording in EELS (e.g. BOURDILLON et al. [39]), given the way this facilitates EXELF data collection, will soon become common place. However the way a Wien filter can be incorporated beneath the objective in a high quality TEM (Zeiss) will revolutionise image interpretation not only in the biological field but also in materials science. Arguably, and for the reasons discussed above, this particular advance will have considerably more useful applications than the current trend towards the highly expensive high vacuum, high resolution TEM, though high vacuum transfer of material from a growth environment to a TEM remains an attractive possibility.

Currently the specific reason why efficient energy loss (even if only plasmon) and energy filtered imaging are such an important necessity is that while it has been demonstrated that these low losses can contribute high resolution detail in an unfiltered image [37] no current image simulation program takes account of this properly and the theory is only poorly developed. It is actually fairly surprising that high resolution imaging has been so successful in the quantitative interpretation of structures using images matched to simulations by pattern rather than by intensity correlations.

6. ACKNOWLEDGEMENTS

I thank Prof. D. Hull for Laboratory facilities, the SERC, Philips, GEC, STL, IBM (Yorktown) and IBM (Zürich) for their support in various aspects of the work described and am equally grateful for the provision of some of the materials examined from some of the above firms. I also thank my group for permission to report some of their more recent unpublished data, in particular, K.B. Alexander, C.S. Baxter, C.B. Boothroyd, E.G. Britton, K.M. Knowles and F. Ross.

REFERENCES

1. E.G. Britton, W.M. Stobbs: In Electron Microscopy and Analysis, 1985, ed. by G.J. Tatlock IOPCS Vol 78 (Hilger, Bristol, 1985) p.45

2. S.B. Newcomb, W.M. Stobbs: J. Cryst Growth (in press)

3. W.B. Stobbs, S.H. Stobbs: Philos. Mag. (in press)

4. W.M. Stobbs: In The Structure of Non Crystalline Materials, ed. by P.H. Gaskell (Taylor and Francis, London, 1977) p.253

5. P.E. Donovan, W.M. Stobbs: Acta Met. 29, 1419 (1986)

6. C.B. Boothroyd, W.M. Stobbs: In Proc. EMSA Vol 44 ed. by G.W. Bailey (San Francisco, U.S.A., 1986)

7. P.B. Hirsch, A. Howie, R.B. Nicholson, D.W. Pashley, M.J. Whelan: Electron Microscopy of Thin Crystals (Butterworths, London 1965)

8. G. Thomas, M.J. Goringe: Transmission Electron Microscopy of Materials (Wiley, New York 1979)

9. A. Howie, J.L. Hutchison: J. Microsc. 142, 131 (1986)

10. D.J.H. Cockayne, I.L.F. Ray, M.J. Whelan: Philos. Mag. 20, 1265 (1969)

11. W.M. Stobbs: In Electron Microscopy in Material Science ed. by E. Ruedl and U. Valdre (Commission European Communities 1975)

12. P.E. Donovan, W.M. Stobbs: J. Microsc. 130, 361 (1985)

13. D. Cherns: In Proc EMSA, Vol 42, ed. by G.W. Bailey (San Francisco, U.S.A., 1984) p. 376

14. D. Schechtman, I. Blech, D. Gratius, J.W. Cahn: Phys. Rev. Lett. 53, 1951 (1984)

15. K.M. Knowles, W.M. Stobbs: In Proc EMSA, Vol 44, ed. by G.W. Bailey (San Francisco, U.S.A., 1986)

16. W.M. Stobbs: J. Microsc. 136, 137 (1984)

17. J.C.H. Spence: In Proc EMSA, Vol 42, ed. by G.W. Bailey (San Francisco, U.S.A., 1984) p. 364

18. A. Bourret, J. Thibault-Desseaux, B.N. Seidmann: J. Appl. Phys. 55, 825 (1984)

19. W. Bergholz, J.L. Hutchison, P. Pirouz: J. Microsc., 141, 143 (1986)

20. D.J. Smith, V.E. Cosslett, W.M. Stobbs: Interdiscip. Sci. Rev. 6, 155 (1981)

21. W.M. Stobbs: J. Ultramicrosc., 9, 221 (1982)

22. S.B. Newcomb, C.B. Boothroyd, W.M. Stobbs: J. Microsc. 140, 195 (1985)

23. N.G. Chew, A.G. Cullis: Appl. Phys. Lett. 44, 142 (1984)

24. G. Ottavani, K.N. Tu, R.D. Thompson, J.W. Mayer, S.S. Lau: J. Appl. Phys. 54, 4614 (1983)

25. K.N. Tu: private communication

26. G.J. Wood, W.M. Stobbs, D.J. Smith: Philos. Mag. A50, 375 (1984)

27. W.M. Stobbs, G.J. Wood, D.J. Smith: J. Ultramicrosc. 14, 145 (1984)

28. L.D. Marks, V. Heine, D.J. Smith: Phys. Rev. Lett. 52, 656 (1984)

29. A.J. Bourdillon, W.M. Stobbs: Ultramicrosc. 17, 147 (1985)

30. P.E. Batson, C.R.M. Grovenor, D.A. Smith, C. Wong: In Proc EMSA, Vol 41, ed. by G.W. Bailey (San Francisco, U.S.A., 1983) p. 154

31. A.J. Bourdillon, P.G. Self, W.M. Stobbs: Philos. Mag. A44, 1335 (1981)

32. C.B. Boothroyd, W.M. Stobbs: Philos. Mag. (in press)

33. J.N. Ness, W.M. Stobbs, T.F. Page: Philos. Mag. (in press)

34. C.S. Baxter, W.M. Stobbs: In Electron Microscopy and Analysis 1985 ed. by G.J. Tatlock IOPCS Vol 78 (Hilger, Bristol, 1985) p. 387

35. C.S. Baxter, W.M. Stobbs: In Kyoto International E.M. Conference (1986)

36. A.P. Long, M.J. Kelly, T.M. Kerr, K.M. Knowles, E.G. Britton, W.M. Stobbs: Phys, Rev. Letts. (in press)

37. W.O. Saxton, K.M. Knowles, W.M. Stobbs: in Electron Microscropy and Analysis 1985 ed. by G.J. Tatlock IOPCS Vol 78 (Hilger, Bristol, 1985) p. 75

38. W.M. Stobbs: J. Microsc. 131, 47, (1983)

39. A.J. Bourdillon, W.M. Stobbs, K. Page, R. Home, C. Wilson, B. Ambrose, L.J. Turner and G.P. Tebby: in Electron Microscopy and Analysis 1985, ed. by G.J Tatlock, IOPCS Vol 78 (Hilger, Bristol, 1985) p. 161

Fabrication of Small Structures of Semiconductors and Metals

H.G. Craighead

Bell Communications Research, Room 4E-208, Holmdel, NJ 07733, USA

1. Introduction

The drive for higher speed and higher density electronic circuits has been pushing fabrication technology to increasingly smaller, dimensions. It is clear that past advances of electronic circuit development cannot be extrapolated much further into the future, because in straightforward scaling-down existing device concepts will fail. As the dimensions of structures approach and shrink below characteristic length scales in solids, new physical phenomena come into play. This may be viewed as an impediment to extending current device technology. It may also be viewed as strong motivation to the physicist and engineer to consider the basic assumptions of device technology and stimulate consideration of physical phenomena of small structures.

Advances in fabrication technology have enabled new directions in research and spurred ideas for devices that take advantage of geometrical confinement in more than one dimension. Ballistic and quantum based devices have been suggested and explored. Single electron trapping phenomena have been seen in small volumes. The relevant dimensions for realization of such concepts are of the order of tens of nanometers [1-3].

The control of the structure of materials, obtained by epitaxial growth techniques, has yielded an unprecedented ability to tailor the electronic properties of semiconductors [4]. Compositions can be controlled at the level of lattice spacings in perfect crystals, on the order of less than one nm. This has enabled the construction of new electronic and optoelectronic devices such as high electron mobility transistors and heterostructure lasers, and has revealed new phenomena such as the quantized Hall effect. Growth of atomic clusters and "zero dimensional crystals", with optical properties dominated by quantum confinement in all three dimensions, has been demonstrated [5]. These levels of structural control have not yet been obtained in lateral patterning. However, the obtainable dimensions are shrinking and the advances are extending to a broader range of materials and geometrics [6,7].

In this article, we will consider the methods available in a research environment, for lithographically fabricating ultra small objects. We hope to develop a sense of the potential for realizing zero and one-dimensional structures. The emphasis will be on the fabrication options available for patterning in general geometries and the limits and results obtainable. The III-V materials, in particular, show promise as being useful and interesting at small dimensions. We will consider not only the fabrication process but some of the problems associated with imaging and observing fabricated structures. Examples of fabrication methods and structures will be given, with emphasis on the most relevant structures and discussion of future possibilities. By lithography we specifically mean a method of transferring a desired pattern onto a substrate. This is normally done by exposing an image recording "resist" layer.

2. Lithographic Techniques

In lithographic patterning, a range of particles and radiation is available. Photolithography has been, and will continue in the immediate future to be, the most widely used manufacturing technique. The drive to smaller dimensions requires shorter wavelengths in order to limit diffraction effects. The choice is either projection, with a diffraction limit of approximately the wavelength divided by the numerical aperture of the optics, or proximity printing, where the size limit can become vanishingly small for true contact printing. However, the need for finite resist thickness and the impracticality of damaging contact, restrict the printing possibility. As is true for all printing techniques, the pattern can be no smaller or of no better quality than the original mask, so the resolution problem is transferred to that of mask fabrication. The diffraction limit can be effectively reduced by using X-rays. In this case, the fidelity of mask replication is limited by penumbral blurring, the range of photoelectrons, and diffraction. The smallest structures made by X-ray lithography are about 20 nm wide. The use of ions eliminates diffraction effects and high energy particles scatter minimally in the thickness of typical resist thicknesses. The fabrication of durable ion masks with adequate stopping power is somewhat difficult. The ion sources and ion optics presently available do not allow focused scanned ion beams as small in size as the smallest electron beams. Because of advantages in particle scattering, exposure efficiency and chemical possibilities, ion beams are likely to be of future importance. Electron sources are highly advanced and readily available with electron optics of high quality. The easily-scanned, finely-focused electron beam can define very small patterns. It is electron beam lithography that is used to make most of the fine line masks for replication by any of the above mentioned techniques. Electron beams have been used to fabricate the smallest structures to date of about 2 nm [8]. Table 1 shows, for comparison, relevant wavelengths, particle energies, and resolution limiting effects for these various lithographic methods and an estimate of the smallest structures yet produced by each technique.

Table 1. Wavelengths and Particle Energies

TYPE	(nm)		EFFECT	STRUCTURE
UV	400-150	3-8eV	diff.	100nm
X-ray	0.4-5	0.2-3eV	diff./scatt.	30nm
Electron	0.01-0.003	10-120keV	scatt.	2nm
Ion	<.001	.1-1000keV	scatt.	50nm

3. High Resolution Electron Beam Lithography

Electron beam lithography is readily adapted to the fabrication of small structures. Figure 1 schematically shows the conventional case of a focused beam impinging on a resist-coated substrate. Typically, the electron beam energy is 20 - 100 keV focused to a diameter of 0.2 - 125 nm and exposes an organic polymer resist. The beam renders the resist either soluble or insoluble in a developer, and is said to be a positive or negative resist respectively. The electron beam exposure apparatus can range from a simple scanning electron microscope to a highly sophisticated high-speed beam-writing machine. The difference among these instruments is their speed, ease of handling complex patterns, and alignment possibilities. We have used STEM/TEM type instruments because of their flexibility, range of electron energies, and small beam diameter. Common to all types of exposure and resist systems are the limits of electron scattering and beam geometry.

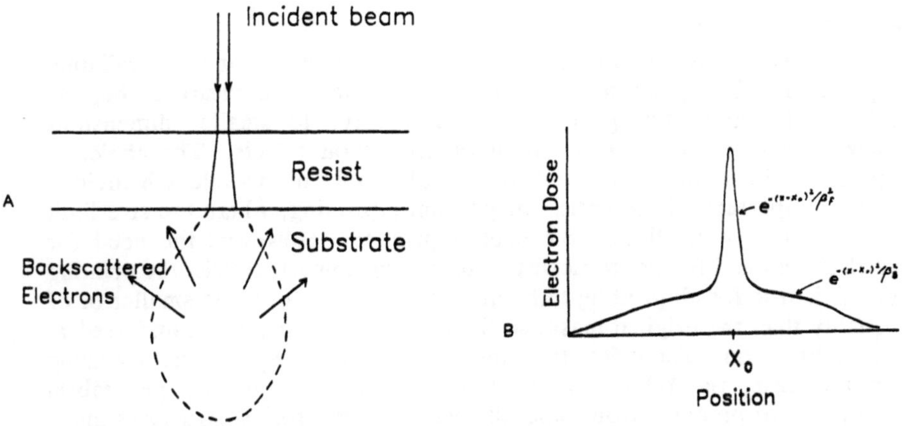

Fig. 1 (a) Schematic view of electron beam exposure of resist on a bulk substrate.
(b) Two Gaussain model of electron beam exposure.

Because incident electrons interact strongly with the electrons of the solid, they scatter and deposit energy over extended volumes. Careful consideration of this scattering must be taken to obtain the desired patterns. The scattering and beam geometry lead to "proximity effects" and limit the resolution. Resist exposure convolution of the beam geometry, small angle forward scattering, and lateral secondary electrons. The broader Gaussian results from the trajectories of electrons that have backscattered, that is, undergone large angle scattering, from the bulk of the resist and substrate. The relevant electron ranges and scattering angles depend on the incident electron energy and the materials through which they pass, and we have some control over these variables.

The ultimate resolution limit of electron beam lithography is limited by the beam size and lateral secondary electrons [9]. Estimates of this have been done by Monte Carlo simulation that agree with experimental data for polymer resists. With the smallest beams available and the thinnest polymer resists this limit is about 10 nm. Processes that use denser resists with correspondingly shorter low-energy secondary electron ranges or exposure processes that involve only high-energy electron interactions can have finer resolution.

The effect of backscattered electrons, the β_F component of the two Gaussian models, has more to do with process latitude and proximity effects [10]. The qualitative features can be understood by considering the energy dependent Grün range, R_G, of electrons in a solid which behaves as

$$\beta_F \sim R_G \, \alpha E^{1.7} \tag{1}$$

where E is the electron energy. For low energies, where the range is comparable to the pattern feature size the contribution of backscattered electron dose to the local exposure can be great [11]. The total exposure from the backscattered electrons is nearly independent of incident electron energy [12]. Therefore, for higher energies the range of the backscatter contribution can be large (tens of microns at 150 keV) and add a nearly uniform background dose of

$$D_B = A \, \eta \, D_0 \, , \tag{2}$$

where D_0 is the incident dose, A the areal density and η the effective backscatter coefficient. The peak to background exposure therefore is

$$\approx 1 + 1/A\eta \sim 1 + 1/A \,. \tag{3}$$

Since η is of the order of one, low density patterns will have high contrast exposure.

The scattering from the substrate can be eliminated by using thin film substrates or using higher energy election beams on thick substrates. Both have been done effectively, but we have done primarily the latter to make experimental structures using modified electron microscopes with beam energies up to 150 keV. For normal low-density patterns, therefore, proximity effects are nonexistent, and the exposure contrast can be excellent on bulk substrates. It is the additional processing that remains critical, and advances in resists, development, and pattern transfer are important. Let us consider a few examples of processing steps used to fabricate small structures in several different materials.

4. Examples of Lithography and Pattern Transfer

One illustrative example of electron beam writing was a resolution test on Si [13]. The electron-beam resist was a 50-nm-thick layer of 950,000 molecular weight PMMA spun onto 3 × 8 mm crystals. In a series of tests, fields 15 μm square were filled with 70-nm spaced lines and exposed with a line dose of 3×10^{-9} C/cm. The exposed resist was developed in a cellosolve-methanol solution of 5 sec. Following this, a 15-nm-layer $Au_{0.6}Pd_{0.4}$ alloy was deposited on the substrate at normal incidence from a thermal source. The polymer and the metal it supported were removed in acetone in a so called lift-off step to leave the series of parallel lines, shown in Fig. 2.

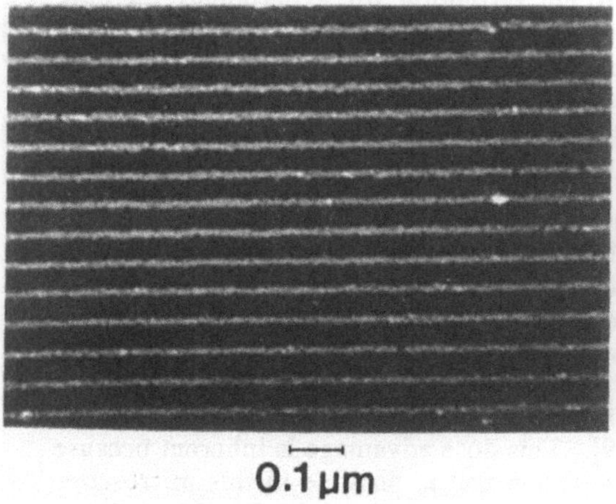

Fig. 2 Scanning electron micrograph of a pattern of metal wires formed on a silicon crystal substrate.

Fig. 3 Au particles of sapphire

Arrays of small metal particles were fabricated for optical and electromagnetic studies. In one example, a total of 3.6×10^5 gold particles were produced on a sapphire substrate. Problems of sample charging during the electron-beam writing required the deposition of a conducting film on the resist layer. The electron-beam exposure time was on the order of one msec per point. Figure 3 is a scanning electron micrograph of a section of this 600 × 600 array of particles demonstrating the regularity in shape and position [14]. Heating the system after formation, yielded rounder appearing particles while maintaining the spatial periodicity.

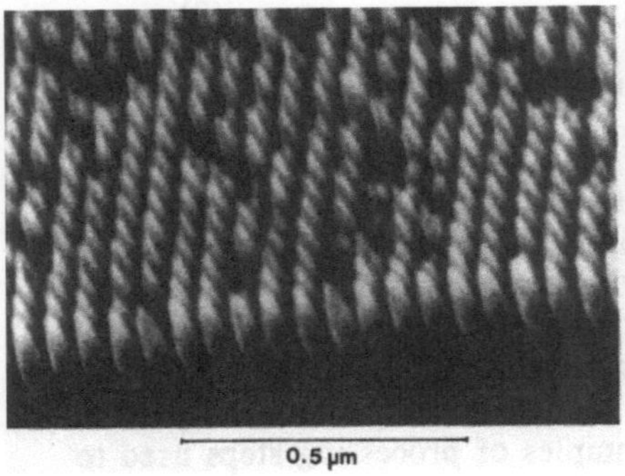

Fig. 4 Columns etched in GaAs

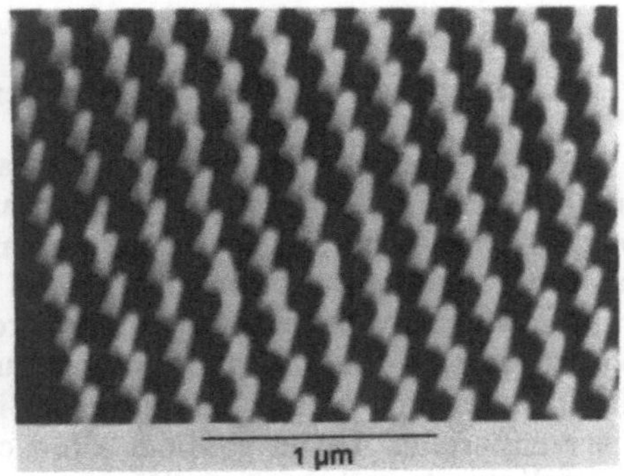

Fig. 5 Columns etched in GaAs/A1As quantum wells

The lithographically defined patterns can also be used as masks for etching of patterns into bulk materials. By using anisotropic reactive ion etching, the metal pattern on the surface of a GaAs crystal can be transferred into the GaAs as in Fig. 4, where a pattern of columns was defined by NiCr metal dots and etched into the GaAs in the $SiCl_4$ plasma [15]. Similar structures can be formed in GaAs/AlAs quantum well materials (Fig. 5) [16]. These particle structures are being studied for size-limited optical properties in semiconductors.

These examples suggest the versatility of the electron microscope in fabricating a range of structures with 10 nm resolution for experiments on bulk substrates. The indication is that we have neared the ultimate limit of conventional electron-beam lithography. Novel studies on thin films and unconventional techniques have also been done in the same microscope system. These include the fabrication of arrays of 7 nm diameter Ag particles formed on carbon support films by contamination lithography and ion milling [17].

A novel inorganic resist scheme has been demonstrated with electron beam radiolysis of fine grain polycrystalline CaF_2, SrF_2 and BaF_2 films and reaction to oxide followed by development in H_2O. The exposure processes were monitored in real time by both electron diffraction and electron energy loss spectroscopy. Figure 6 shows a TEM image of lines 30 nm wide on 200-nm centers in CaF_2 produced by irradiation with a 120-keV electron beam at a dose of $\sim 10 C/cm^2$ [18]. This exposure dose is about one order of magnitude less than that required to vaporize fine grain alkali-halide films [19]. This dose advantage is inherent because of the efficiency of radiolysis over vaporization and is possible in this microscope because of the oxygen in the background gas. On radiolysis of the fluoride to metal and F_2 the metal reacts with the available oxygen to form oxide. Because CaO is much more soluble in H_2O, this inorganic resist can be simply developed in water. A similar process forms a more sensitive negative resist with BaF_2/SrF_2 mixtures [20]. All of the fluorides are useful as highly resistant masks for halogen gas etching.

These examples illustrate several forms of pattern transfer of the electron beam exposed pattern. The lift-off process can produce fine lines but it is not possible to make thick films with a single layer of resist. The lift-off can be used for a reversal of the pattern to be used as a mask for an etch step. Ion milling, by physical sputtering, works for all materials. This is because the sputter yields do not vary dramatically for different materials, but this also means it is generally difficult

Fig. 6 Lines formed in a CaF_2 film

to find good etch masks. There can also be deposition of sputtered material on vertical walls making etching narrow grooves difficult. Reactive ion etching (RIE) can be highly anisotropic and the chemical selectivity can be exploited to enable the use of thin masking layers. The ultimate resolution of reactive ion etching is not known because even the smallest masks generated have been reproduced by RIE. Reactive ion beam etching may be even more directional, enabling higher aspect ratio structures. With focused ion beams, direct ion beam milling, etching, and doping may be possible. A method of ion compositional disordering of III-V materials may be an alternative to lithography and etching.

5. Structure Imaging

The study of new physical phenomena arising from ultra small dimensions and the development of nanometer scale fabrication techniques taxes the capabilities of standard imaging methods. Transmission electron microscopy (TEM) has been the primary technique for imaging and structural determination of thin films, but this requires thinning the specimens to obtain spatial resolution of a few angstroms. This is generally inappropriate for the study of small lateral lithographically defined structures and impossible for small features formed on thick substrates. For this technologically and scientifically important class of structures, scanning electron microscopy has been the available imaging tool. The limited resolution and lack of contrast mechanisms hampers the study of structures that can be made by nanolithography on thick substrates. This has led us to employ the old and recently revitalized method of reflection electron microscopy (REM) [21-23]. By using a conventional transmission electron microscope and imaging the surface of a specimen using diffracted beams, surface topographical information as well as crystal structure can be obtained with the spatial resolution of the TEM. Figure 7 shows a schematic of this imaging geometry.

A reflection electron micrograph of a sample similar to Fig. 5 is shown in Fig. 8. One immediately observes the dramatic foreshortening, in this case a factor of about 20. Also, because of the limited depth of focus and high degree of tilt, the image is in focus only in the center of the field. The contrast, due to the shadowing of the surface features, is very strong and dominates the appearance of the image. The bright appearance of the top of the surface is due to strong transmission diffraction from the peaks. This indicates the utility of this method in examining side wall roughness resulting from the etching process. The individual conical structures are barely

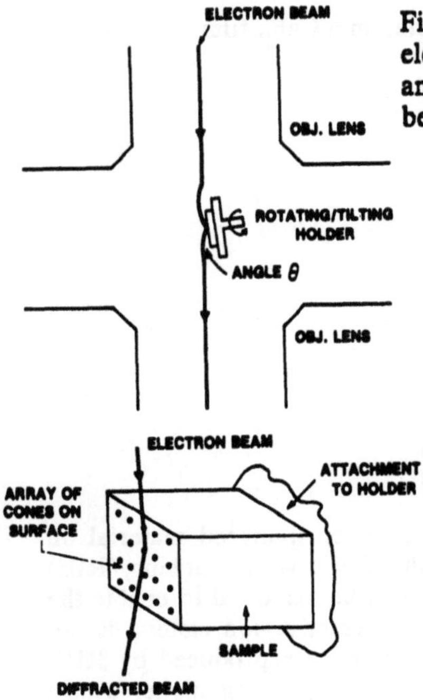

Fig. 7 Schematic diagram of the sample and electron beam orientation for REM. The sample and beam are tilted and an appropriate diffracted beam is selected for dark field imaging.

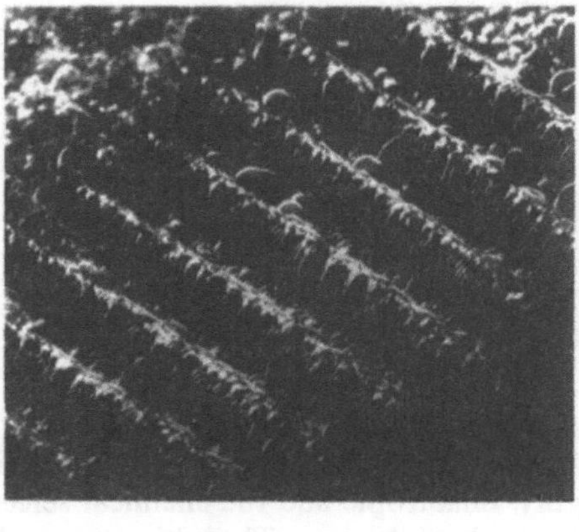

Fig. 8 Reflection electron micrograph of an array of etched columns in GaAs/A1As

resolvable in this view, because of the overlap produced by the forshortening. One can observe flattened tops on the cones that reflect the original mask dimension. The general roughness in the background, because of the bright appearance, is established to be etched structures in the III-V material probably due to inadvertent etch mask formation by surface contaminants rather than material redeposited during the etching. Deposited material, such as a gold film, elsewhere on the sample appears dark because the image is formed only from the semiconductor substrate diffracted beam. This determination of crystal structure cannot be done by SEM nor does one obtain the high contrast.

6. Conclusion

By electron-beam lithography and high resolution pattern transfer processes like RIE, it is possible to construct, in a range of materials, essentially arbitrary shaped structures at dimensions less than 100 nm. This has enabled a new class of physics experiments. Specialized techniques for practical high resolution lithography, pattern transfer, and imaging continue to evolve.

I acknowledge the collaboration of many people listed in the references and A Scherer, the help of L. Schiavone and E. Beebe and my former Bell Laboratories associates where some of this work was done.

References

[1] H. Sakaki: Jpn. J. Appl. Phys. **19**, L735 (1980)

[2] C. T. Rogers and R. A. Buhrman: Phys. Rev. Lett. **53**, 1272 (1984): K. S. Ralls et.al., Phys. Rev. Lett. **52**, 228 (1984)

[3] R. E. Howard, P. F. Liao, W. J. Skocpol, L. D. Jackel and H. G. Craighead: Science **221**, 117 (1983)

[4] See, for example: A. Cho: Thin Solid Films **100** , 291 (1983)

[5] C. Sandroff, D. Hwang and Chung (to appear in Phys. Rev. B)

[6] A. N. Broers: J. Electrochem. Soc. **128**, 166 (1981)

[7] H. G. Craighead: J. Appl. Phys. **55**, 4430 (1984)

[8] M. Mochel, C. Humphreys, J. Eades, J. Mochel and A. Petturd: Appl. Phys. Lett. **42**, 38 (1983)

[9] D. F. Kyser: J. Vac. Sci. Technol. B. **1**, 1391 (1983)

[10] N. Jamoto and R. Shimizu: J. Appl. Phys. **54**, 3855 (1983)

[11] R. E. Howard, H. G. Craighead, L. D. Jackel, and P. M. Mankiewich: J. Vac. Technol. **B 1** 1101 (1983)

[12] L. D. Jackel, R. E. Howard, P. M. Mankiewich, H. G. Craighead, and R. W. Epworth: Appl. Phys. Lett. **45**, 698 (1984)

[13] H. G. Craighead, R. E. Howard, L. D. Jackel, and P. M. Mankiewich: Appl. Phys. Lett. **42**, 38 (1983)

[14] H. G. Craighead and G. A. Niklasson: Appl. Phys. Lett. **44**, 1134 (1984)

[15] M. B. Stern, H. G. Craighead, P. F. Liao and P. M. Mankiewich: Appl. Phys. Lett. **45**, 410 (1984)

[16] H. G. Craighead and A. Scherer (to be published).

[17] H. G. Craighead and P. M. Mankiewich: J. Appl. Phys. **53**, 7168 (1982)

[18] P. M. Mankiewich, H. G. Craighead, T. R. Harrison, and A. Dayen: Appl. Phys. Lett. **44**, 468 (1984)

[19] M. Isaacson and A. Murray: J. Vac. Sci. Technol. **19**, 166 (1981)

[20] A. Scherer and H. G. Craighead: Proc. Electron Ion and Photon Beam Meeting, Boston (1986)

[21] M. E. Haine and W. Hirst: Brit. J. Appl. Phys. **4**, 239 (1953)

[22] P. E. Hojlund Nielsen and J. M. Cowley: Surf. Sci. **54**, 340 (1976)

[23] T. Hsu and J. M. Cowley: Ultramicroscopy **11**, 239 (1983)

Atomically Controlled Growth in 2, 1 and OD, and Applications

H. Watanabe

NEC Corporation, Fundamental Research Laboratories, Kawasaki, Japan

1. INTRODUCTION

Today's most advanced semiconductor memory, a 4 Mbit DRAM (Dynamic Random Access Memory) [1] uses about 10^{13} silicon atoms for a 1 bit cell (which includes the silicon in the passive region). As shown in Fig.1, by extrapolating the number with the current slope, 1 Gbit DRAM in 1998 will need 10^{11} - 10^{12} atoms/bit. It is frequently said that 1 Gbit memory is the last device based on the conventional IC architecture from the viewpoint of microfabrication technology and transistor operation principle. On the other hand, according to the recently proposed concept of molecular devices, molecular memory or sometimes bio-memory, the number of atoms (mainly C, H, O, N) per 1 bit is of the order of 10^3 atoms/bit [2]. However, I believe there is no reason to obey the assumption that, at the beginning of the 21st century, we must give up semiconductor research, leaving seven orders of magnitude difference of atoms/bit. 50 or 100-year research with innovated concepts is definitely necessary to reach a 10^3 atoms/bit memory, either extrapolating the trend so far or with a somewhat slower trend. It means that the 21st century will still be a semiconductor research age.

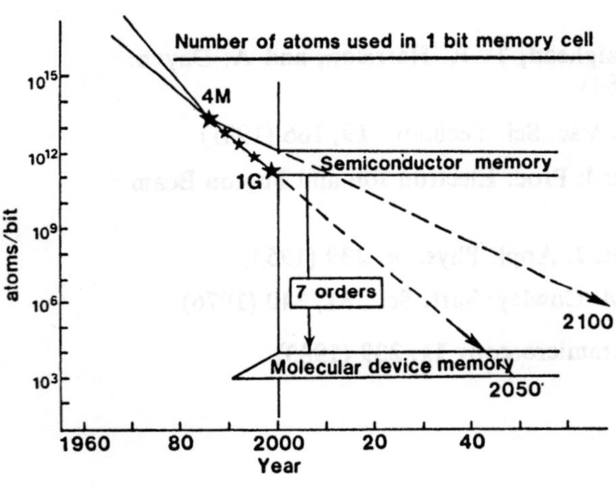

Fig.1. Trend of number of atoms in a 1 bit memory cell

2. DIGITAL EPITAXY

2.1 Concept

Even if an extremely high speed device is developed, it has only a small significance if it is discrete. All the semiconductor functional devices should eventually be integrated. Many ideas have been proposed for high-speed or optoelectronic devices, using atomically sharp interface multilayer structures. MBE and MOCVD are very powerful tools in obtaining such atomically controlled multilayers. MBE can in situ control monolayer thickness by monitoring RHEED oscillation. Advanced sophisticated MOCVD also enables us to make monolayer superlattice. However, for future LSIs with a monolayer-controlled multilayer structure, we need a more advanced epitaxial technique, which has atomic-scale high uniformities of thickness and composition, perfect selective epitaxy free from the thickness non-uniformity caused by the local loading effect, and perfect mirror grown surface free from any spitting, mounds or hillocks.

SUNTOLA [3] proposed the concept of Atomic Layer Epitaxy (ALE). The sequence of monolayer (molecular layer) growth is schematically shown in Fig.2. Type I is for compounds such as II-VI semiconductors which consist of two volatile components. Type II is applicable to not only II-VI but also III-V compounds. Type III is for elemental semiconductors such as Si or Ge. For simplicity, the idea is explained for the type I ZnS epitaxial growth. Consider the (111)S surface of a ZnS substrate. Zinc vapor is supplied and adsorbed on the (111)S surface (Step 1). After the zinc coverage has become unity, zinc vapor is purged (Step 2). Sulfur vapor is introduced and chemi-adsorbed on the zinc-adsorbed surface (Step 3), then purged (Step 4). At this stage, precisely one molecular monolayer of ZnS is formed. The next step is the same as Step 1, and by repeating this procedure ZnS growth proceeds stepwise with an atomic layer unit.

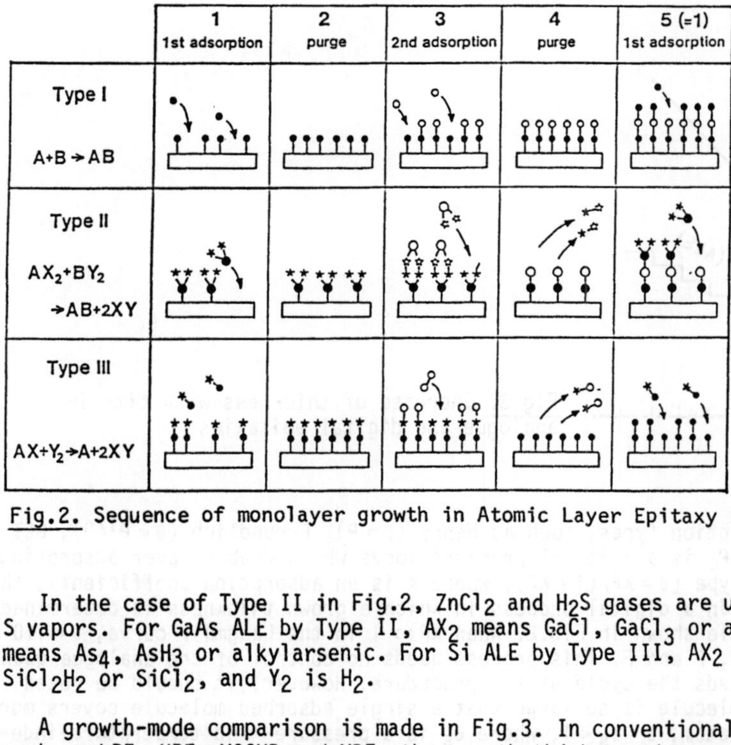

Fig.2. Sequence of monolayer growth in Atomic Layer Epitaxy (ALE)

In the case of Type II in Fig.2, $ZnCl_2$ and H_2S gases are used instead of Zn and S vapors. For GaAs ALE by Type II, AX_2 means GaCl, $GaCl_3$ or alkylgallium and BY_2 means As_4, AsH_3 or alkylarsenic. For Si ALE by Type III, AX_2 is, for example, $SiCl_4$ $SiCl_2H_2$ or $SiCl_2$, and Y_2 is H_2.

A growth-mode comparison is made in Fig.3. In conventional epitaxial methods, such as LPE, VPE, MOCVD and MBE, the growth thickness increases linearly with time, where the gradient is a function of analogue quantities such as temperatures of source and substrate, pressures and flow rates of source gases and a growth time. On the other hand, the thickness in ALE increases stepwise. Here it should be noted that ALE is divided into two categories, analogue ALE and digital ALE. The comparison of them, together with conventional epitaxies, is shown in Table 1. If the surface coverage θ, which is Na/Ns (Na and Ns being adsorbant and surface atom densities, respectively), is a function of the substrate temperature T and adsorbant vapor phase pressure P and flow rate F, the gradient of the stepwise increase in Fig.3 is dependent on these analogue quantities. So, this ALE growth mode is regarded as an analogue process even if the thickness increases stepwise.

Table 1. Concept of digital epitaxy

	Analogue Epitaxy		Digital Epitaxy	
Method	LPE, VPE, MOCVD, MBE	ALE		
Growth mechanism	step/2D nucl.	adsorption layer reaction		
		$\theta = f(T,P,F)$	$\theta \neq 1$	$\theta = 1$
Control parameters	temperature pressure flow rate		none	
Thickness control	time		number of cycles	
Atomic abruptness	possible	possible at $\theta=1$	impossible	easy
Thickness uniformity	good		perfect	
Selective epitaxy	non-uniform		uniform	
Surface quality	wavy,hillock,spitting,mound		no local enhancement	
Ordered alloy	possible	possible at $\theta=1$	impossible	easy

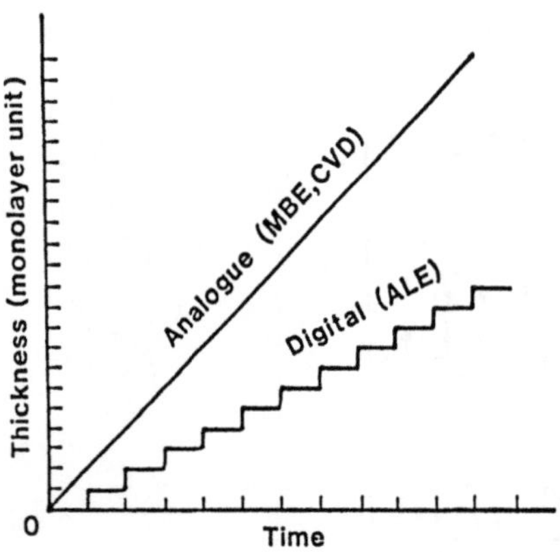

Fig.3. Increase of thickness with time in analogue and digital epitaxies

Among several adsorption types, such as Henry ($\theta \propto P$), Freundlich ($\theta \propto P^{1/n}$), BET ($\theta > 1$ at $P > P_c$, where P_c is a critical pressure above which double-layer adsorption occurs) and Langmuir type [$\theta = KP/(1+KP)$, where K is an adsorption coefficient], the Langmuir type results in a digital process in which a grown thickness is determined only by the growth cycle shown in Fig.2. When $KP \gg 1$ in the Langmuir curve, $\theta = 1.0$ being independent of T, P and F. This process needs no control of the analogue quantities, but it only needs the cycle of the procedure. However, it should be noted that if an adsorbed molecule is so large that a single adsorbed molecule covers more than a single surface atom, $\theta = \theta_s < 1$, where θ_s is a pressure- and temperature-independent saturation coverage in the Langmuir-type adsorption. This growth process is still digital, but the single-cycle growth thickness is thinner than the crystallographic monolayer thickness.

As can be easily understood, the ALE process with $\theta \neq 1$ has neither monolayer controllability nor atomically sharp interfaces. We should pay attention to a possibility of a small deviation of θ from unity, for example, $\Delta = (1-\theta) = 10^{-6}$. The deviation means an atomically local degradation of a monolayer abruptness. But after two ALE cycles, the probability of the existence of the nonabruptness ranging two monolayers becomes negligibly small (Δ^2). Effects of Δ for defect generation should be studied.

2.2 Example of Digital Epitaxy

To date, several ALE experiments have been reported. Among them, NISHIZAWA et al.'s [4] and USUI and SUNAKAWA's [5] GaAs ALE results seem to have the digitality. The former was carried out by the alternative supply of trimethyl gallium and AsH_3. Figures 4 and 5 show the results of Usui and Sunakawa's method by the GaCl and AsH_3 alternative supply. The growth thickness per single alternate cycle is independent of GaCl pressure and growth temperature and is in good agreement with calculated monolayer thickness in different substrate surface orientations, except (110). These results support the process digitality with $\theta = 1$. They obtained highly uniform-grown layers with mirror surfaces. Furthermore, no local loading effect with only small growth enhancement was observed in the selective epitaxial growth experiment. These are features expected from the adsorbed layer reaction model for ALE.

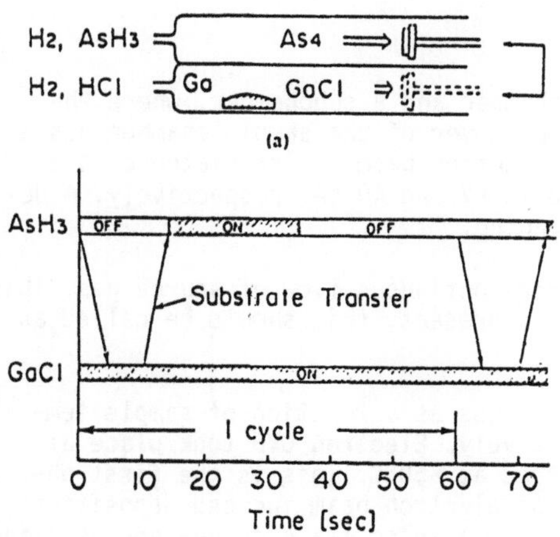

Fig.4. Growth apparatus and schedule in GaAs ALE [5]

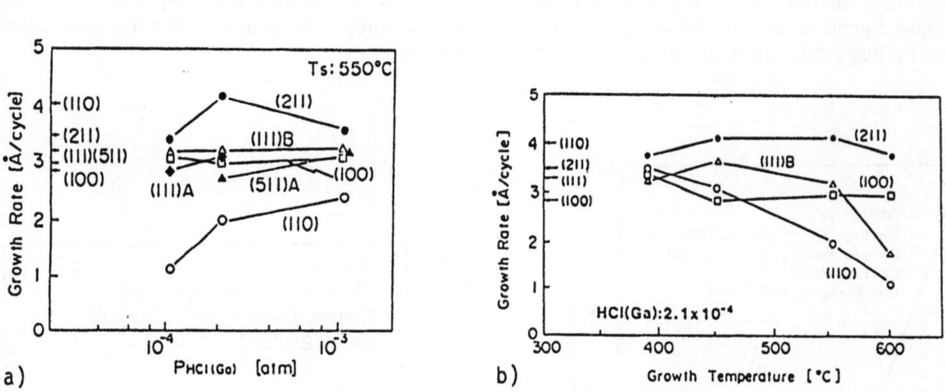

Fig.5a,b. Dependence of growth rate per sinlge ALE cycle on HCl gas pressure (a) and growth temperature (b)

3. ELECTRON BEAM INDUCED SELECTIVE ETCHING AND DEPOSITION (EBISED) TECHNOLOGY

BROERS et al. [6] proposed an electron beam induced localized deposition technique. They deposited localized (grid-like) 80 Å line-width carbon film by irradiating a hydrocarbon adsorption layer with an extremely narrow electron beam (5 Å diameter) and used it as an etching mask for the metal film underneath. MATSUI and MORI [7] recently succeeded in a direct deposition of localized (line) tungsten and chromium film by irradiating WF_6, WCl_6 and $Cr(C_6H_6)_2$ adsorbed layer with a focused electron beam. They used two vacuum systems, one is for fundamental studies on reaction process and the other, shown in Fig.6, is for the localized deposition and is a modified

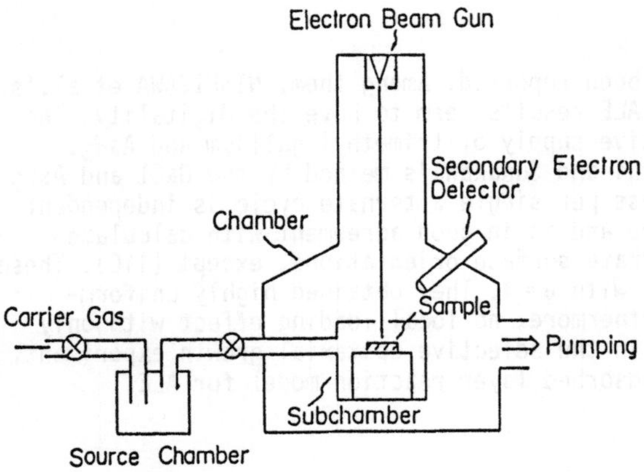

Fig.6. Experimental apparatus for Electron Beam Induced Selective Etching and Deposition (EBISED) technology [7]

SEM (Scanning Electron Microscope). A source chamber and a subchamber, where the sample is placed, were added to the SEM. The top cover of the sample chamber has a pinhole with 1 mm diameter, through which the electron beam is irradiated on the sample. Electron beam voltage and current were 15 kV and 40 pA, respectively. A deposited pattern was observed in situ by SEM imaging.

Deposition experiments were carried out under continuous flow of source gas. This means that the deposition process is not ALE. At present, this should be called an analogue process.

Figures 7 and 8 show deposited (etched) thickness as a function of sample temperature and source gas (WF_6) pressure, respectively. Electron CVD took place at < 50°C and electron beam induced etching occurred at >50°C. This is the first observation of temperature-controlled switching of electron beam induced deposition and etching, which is a very attractive feature as a selective etching-and-refilling technique because of the high position control accuracy. Monolayer etching and deposition is possible in the Langmuir-type adsorption.

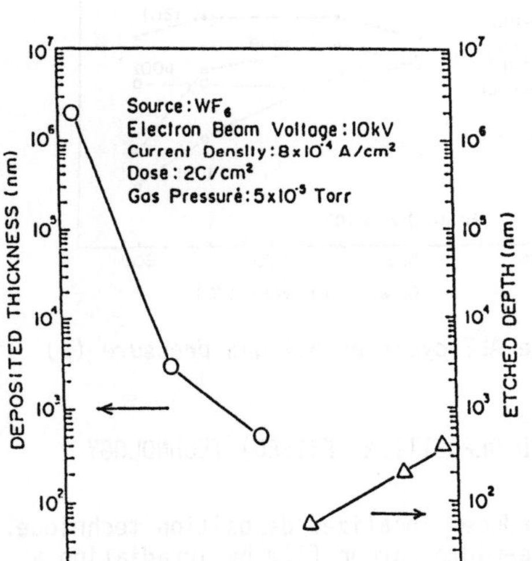

Fig.7. Temperature dependence of deposited thickness and etched depth in tungsten EBISED [7]

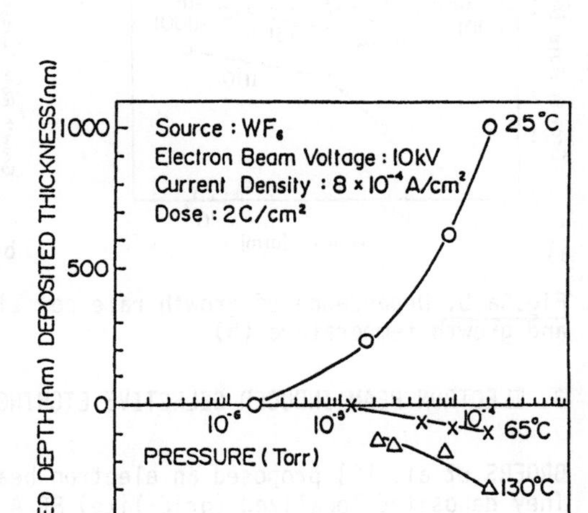

Fig.8. Source gas pressure dependence of deposited thickness and etched depth [7]

They obtained a 0.15 μm linewidth Cr deposited pattern on a silicon substrate by irradiating $Cr(C_6H_6)_2$ adsorbed layer with a focused electron beam in about 0.1 μm diameter which was not measured but estimated for their modified SEM.

Recently, KURODA et al. [8] developed an ultra-high resolution SEM which enables us to observe atomical step arrangements of the crystal surface. They observed a tungsten <110> field emitter tip and succeeded in revealing an atomical surface structure by using a 5 Å diameter electron beam with 30 kV acceleration.

In the experiment of BROERS et al. [6], a substrate was thinned in order to minimize the pattern broadening due to backscattered electrons. Matsui and Mori's EBISED experiments show a remarkably high efficiency. In some cases, three electrons are enough to cause a single chromium atom deposition. The high efficiency also minimizes not only the scanning spatial fluctuation but also the backscattered electron broadening. A nanometer resolution etching and deposition will be realized by the monolayer EBISED technology.

4. APPLICATIONS

4.1 Quasi-Dimensional Structures

The new technologies described above lead us to imagine formation of not only 2, 1 and 0D structures such as wells, wires and boxes, respectively, but also more complicated structures such as pipes, snakes and balls shown in Fig.9, where electrons are confined in dotted regions with quasi-dimensionalities between 3D and 0D. A quasi-dimensional example such as a quantum coil which is a buried helical quantum wire can be constructed only by the new method with selective deposition during growth, not by any lithographic method.

The fabrication of essentially arbitrary geometries could lead to dramatic control of the electrical and optical properties of solids.

4.2 Quantum Ball, Superatom and Super-Superlattice

One of the most interesting applications is a quantum ball. (A quantum ball is not a spherical quantum box which confines electrons in a box.) Figure 10 shows two types of the quantum ball. One is a spherical quantum well, where electrons and/or holes are spherically confined in a thin skin.

The skin region has a larger electron affinity than those of the core and the surrounding materials, where the core and the surrounding materials are not necessarily identical. Impurity doping into one or all of the three regions affects the locali-

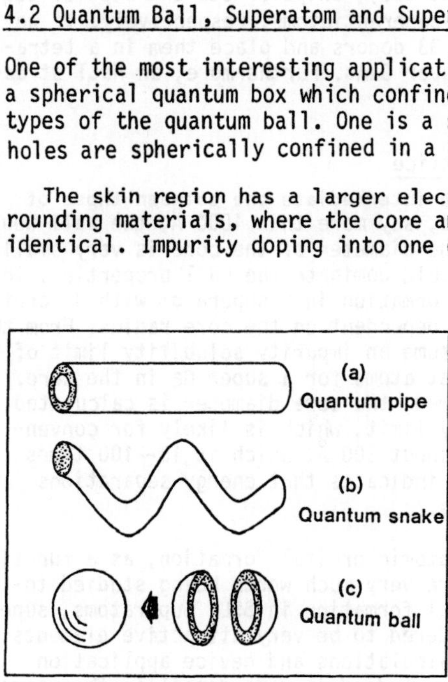

Fig.9. Quasi-dimension structures

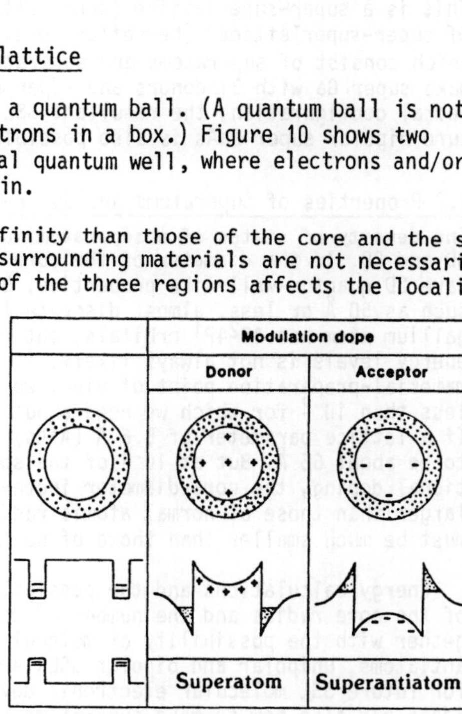

Fig.10. Concept of quantum ball

zation properties. The size of the core through which confined electrons can tunnel is the most important parameter.

The other one is a modulation-doped quantum ball. The concept of the modulation doping was originally proposed by DINGLE et al. in order to increase electron mobility by separating a free electron region from the parent donor region [9]. In the present case, the idea of the spatial separation of parent donors (acceptors) and electrons (holes) is utilized. If n donors are doped in a spherical core which has a smaller electron affinity than that of the surrounding material, n electrons are confined in a spherical skin outside the core. The charge neutrality is established as a quantum ball. As shown in Fig.11, if n = 31, the physical situation looks like a gallium atom, which is an atom of the atomic number 31 having 31 protons in the core and 31 electrons in the spherical cloud. This similarity of the charge balance in the spherical geometry leads us to call the modulation-doped quantum ball a superatom. If acceptors are doped it must be called a superantiatom.

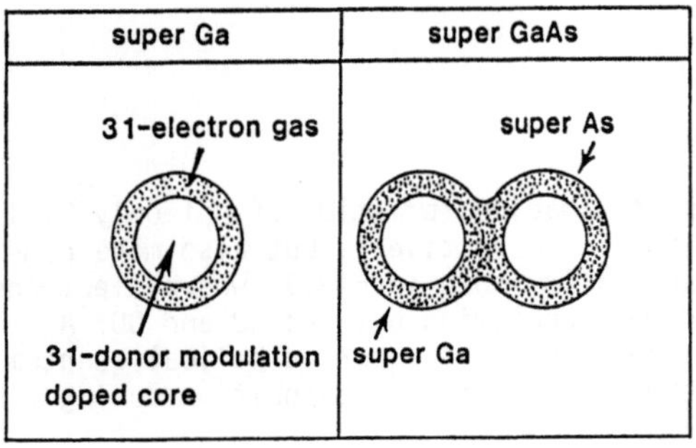

Fig.11. Superatom and super-superlattice

By using superatoms, we can construct 1, 2 and 3D lattices with superatom arrays. This is a super-superlattice (superlattice made of superatoms). There are two kinds of super-superlattices (hereafter referred to as SSL), unipolar SSL and bipolar SSL, which consist of superatoms and superatoms-and-superantiatoms, respectively. If we make super Ga with 31 donors and super As with 33 donors and place them in a tetrahedral configuration, the resultant SSL is a super GaAs. Furthermore, an NaCl structure bipolar super GaAs is also possible.

4.3 Properties of Superatoms and Super-Superlattice

The density of states of a superatom must be an intermediate one between those of 3D and 0D. If the diameter of the core is large, say more than 1000 Å, the ball would have 2D quantum-well-like properties. But if the diameter of the core is very small, such as 50 Å or less, almost discrete levels could dominate the ball properties. The gallium atom has $4S^2 4P^1$ orbitals, but orbital formation in a superatom with discrete energy levels is not always likely. It must be dependent on the core radius. From the material-preparation point of view, we must assume an impurity solubility limit of less than 10^{-2} for which we need about 3100 host atoms for a super Ga in the core. If a lattice parameter of 5.6 Å (AlAs) is assumed, the core diameter is calculated to be about 65 Å. But at 10^{-4} of the solubility limit, which is likely for conventional doping, the core diameter increases to about 300 Å, which is 10 ~ 100 times larger than those of normal atomic radii. This indicates that energy separations must be much smaller than those of normal atoms.

Energy calculations and the possibility of atomic orbital formation, as a function of the core radius and the number of donors, are very much worth being studied together with the possibility of molecular orbital formation in SSL. Superatoms, superantiatoms, unipolar and bipolar SSLs are considered to be very attractive elements for future SSL molecular electronic devices. Calculations and device application ideas and also how to make superatoms should be studied in more detail in the years to come.

ACKNOWLEDGEMENTS

The author very much thanks Mr. A. Usui, Drs. S. Matsui and K. Mori for supplying and discussing their very recent works on GaAs ALE and EBISED technologies. He also thanks Drs. T. Inoshita, S. Ohnishi and S. Oshiyama for discussing quantum ball properties. He is greatly indebted to Drs. C. Weisbuch and M.J. Kelly for inviting him to talk about his and his co-workers' recent experimental results and new concepts of superatoms and super-superlattices. He also wants to express his great appreciation to the Organizing Committee by giving the superatoms a new name, "Les Houcheons", to commemorate this Winter School at Les Houches on The Physics and Fabrication of Microstructures and Microdevices. He is much indebted to Mrs. C. Pendl for helping with the preparation of this improved paper. Finally, the author deeply thanks Drs. D. Shinoda and S. Asanabe for their continuous encouragement.

REFERENCES

1. M. Takada, T. Takeshima, M. Sakamoto, T. Shimizu, H. Abiko, K. Katoh, M. Kikuchi, S. Takahashi, Y. Sato, Y. Inoue: ISSCC Digest of Technical Papers, p.270 (February 1986)
2. T. Kamiya (ed.): Investigation Reports on Molecular Devices (Research and Development Association for Future Electron Devices, Tokyo 1985) [in Japanese]
3. T. Suntola: Extended Abstract of the 16th (1984 International) Conference on Solid State Devices and Materials, Kobe (1984) p.647
4. J. Nishizawa, H. Abe, T. Kurabayashi: J. Electrochem. Soc. 132, 1197 (1985)
5. A. Usui, H. Sunakawa: Jpn. J. Appl. Phys. 25, 28 (1986)
6. A.N. Broers, W.W. Molzen, J.J. Cuomo, N.D. Wittels: Appl. Phys. Lett. 29, 596 (1976)
7. S. Matsui, K. Mori: J. Vac. Sci. Technol. B4(1), 299 (1986)
8. K. Kuroda, S. Hosoki, T. Komoda. J. Electron. Microsc. 34, 179 (1985)
9. R. Dingle, H. Störmer, A.C. Gossard, W. Wiegmann: Appl. Phys. Lett. 33, 665 (1978)

Microstructure of Organic Mono- and Multilayers

H. Möhwald

TU Munich, Physics-Dept., D-8046 Garching, Fed. Rep. of Germany

1. INTRODUCTION

Biological membranes are less than 100 A thick planar structures formed by self-assembly of lipids and proteins. They separate two electrolyte solutions and enable selective material and information transfer. Their features would be highly desirable in many technical systems, but are presently impossible to mimic due to a lack of understanding the complex biological system.

In recent years investigation of different types of phospholipid model membranes offered much insight into the physics of biomembranes /1/. Among these model systems monolayers at the air/water interface have proven to be extremely interesting because they allow for studies of intermolecular interactions while varying many independent parameters like surface pressure, phase state, molecular density and subphase conditions. They also represent those structures that after purposeful manipulation and incorporation of functional molecules like dyes or proteins may be transferred onto solid supports /2/. Thus multilayers, so-called Langmuir-Blodgett (LB) films, can be formed that offer a variety of application perspectives /3/.

This report addresses the controlled manipulation of the microstructure of these films and discusses the underlying forces in view of classical as well as modern theories, taking into account specific two-dimensional and interface properties. This is done predominantly for films at the air/water interface before technologically more relevant features of films on solid supports will be considered.

2. MONOLAYERS AT THE AIR/WATER INTERFACE

Classically surfactant films at interfaces are studied by measuring the surface pressure as a function of surface area per molecule. In analogy to the behaviour of pressure/volume diagrams of real gases discontinuities in these two-dimensional relationships are indicative of phase transitions. This also holds for films of many lipids where by increasing the lateral pressure a transition from a fluid to a solid state can be induced /4/. Special characteristics of this first-order phase transition and of the solid phase will be discussed below. As an extremely useful technique fluorescence microscopy at the air/water interface has been established /5-7/. Basically one observes the distribution of fluorescent dye probes incorporated in a concentration below 1% into the monolayer. As generally these probes are only weakly soluble in the solid phase crystalline domains appear dark under the microscope. Hence crystal formation, shapes and interactions can be studied with lateral resolution of about 1 µm and with time resolution of 0.1 sec.

On increasing the surface pressure Fig. 1 shows a series of fluorescence micrographs that starts at the fluid/solid phase transition point. One observes the onset of nucleation and the growth of crystallites (dark areas), but this crystallization is obviously very peculiar. There is no critical domain size above which one domain would grow on the expense of all the others thus minimizing line tension. Instead, all domains are of uniform size and repel one

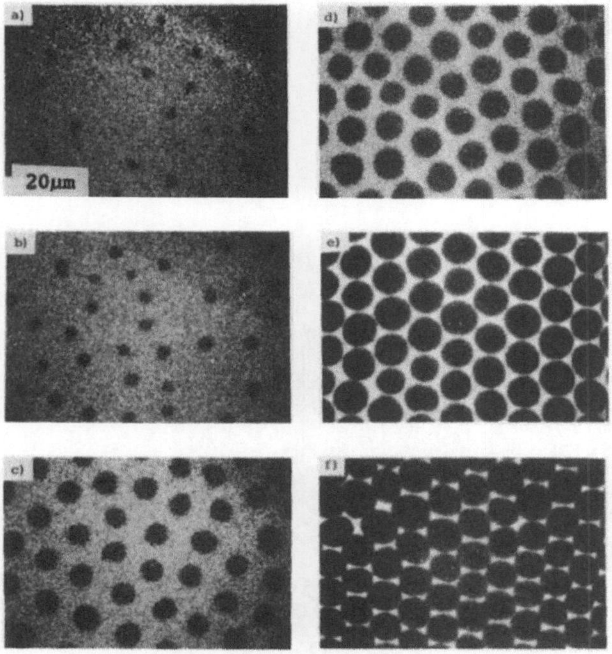

Fig. 1. Fluorescence micrograph of a phospholipid monolayer on increasing the surface pressure from top left to right bottom

another thus forming a hexagonal lattice. This can be understood as due to electrostatic forces arising at the interface /8,9/.

The condensing molecules are charged or/and exhibit a dipole moment normal to the surface which is not compensated. Molecular densities in solid and fluid phases differ by up to 50%. Hence solid domains repel one another due to excess charge or dipole density, respectively. To explain uniform domain size we need a force favouring small domains on the expense of large ones thereby counteracting line tension. Again the long range nature of electrostatic forces has to be accounted for because the contribution of Coulombic repulsion to the heat of melting is larger for a large compared to a small domain.

The above forces are not yet quantitatively, but qualitatively understood. This provides the feasibility to control domain structure via electrostatic forces changed in a predictible way via ionic conditions of the water subphase /10/.

The theoretical treatment of solid domains as two-dimensional colloids is also supported in inhomogeneous electric field experiments /11/.

Figure 2 shows the result of such an experiment where a tip electrode is positioned some 10 µm above the electrolyte solution at zero potential. Solid domains obviously are collected or repelled by the inhomogeneous fields of opposite respective polarities. This technique of manipulating crystallites is also interesting as it is possible to fix crystallites or to measure interdomain forces by means of oscillating electric fields.

Impurities may influence crystal growth basically in two different ways: (i) They affect the crystallization process by increasing both the number of critical nuclei and the melting pressure. (ii) They may change the line tension of different crystal faces differently thus causing peculiar crystal shapes.

As an example on the latter action figure 3 presents the pressure induced growth of spiral crystallites in monolayers of phospholipids containing 1 mol % of cholesterol. Here the lipid head group charge is increased via subphase pH

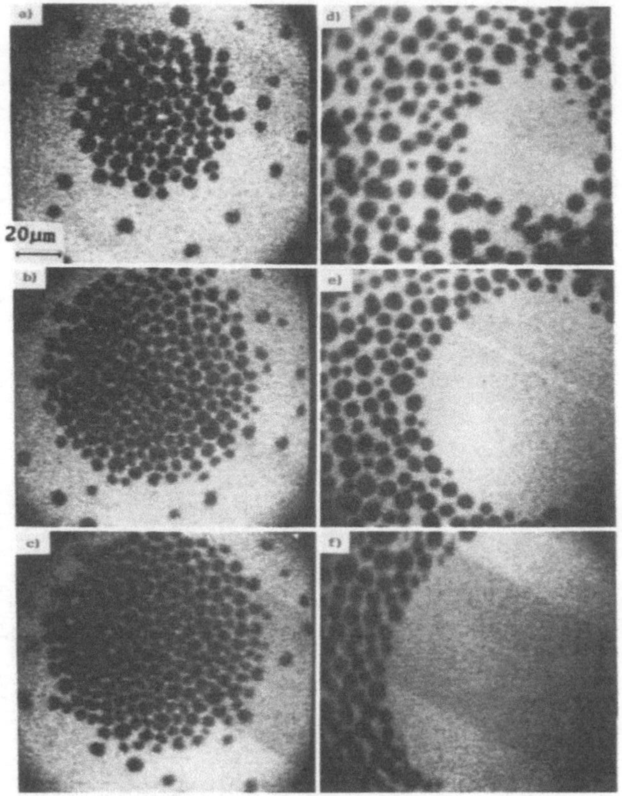

Fig. 2. Fluorescence micrograph of particle movement under tip electrode. Left column: time sequence with negative tip potential (−100 V). Right column: positive tip potential + 100 V)

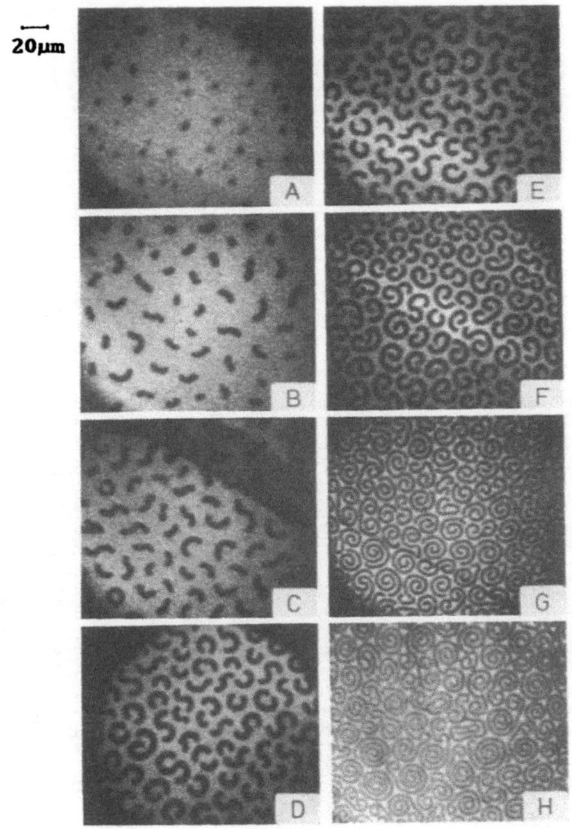

Fig. 3. Pressure induced growth of spiral crystals of a monolayer of L-α-dimyristoyl-phosphatidic acid containing 1 mol % cholesterol (pH 11.4, T = 10°C)

thus causing a tilt of the hydrocarbon chains with respect to the surface normal and a reduction of crystal symmetry to less than threefold /12/. Cholesterol presumably stabilizes one or two of the crystal faces thus elongating the crystallite. Due to molecular chirality the macroscopic axes are bent and an additional repulsive force from the center turns other- wise forming circles into spirals. The widths of these domains depend on temperature, pH, pressure and cholesterol content, and spiral formation is reversible and independent of growth kinetics /13/.

This kinetics can be studied by quickly compressing the film thus increasing the pressure from below to above its transition value and then observing solidification patterns as demonstrated in Fig. 4. We found formation of dendritic structures exhibiting similar features as observed in calculations on diffusion-limited aggregation (DLA). The corresponding analysis of circumference and area, however, yields a fractal exponent of 1.5 /14/ well below the value of 5/3 determined for DLA /15/. This indicates that the DLA model is not sufficient to describe the process.

On the other hand impurity (dye) enrichment in the fluid phase near the solidification interface clearly shows that diffusion limited hindrance of the growth process is involved. These impurities, rejected from the solid phase, are enriched near the solid/liquid boundary thus causing a local melting point reduction /14/. For the typical surface active impurity dye this can be visualized and quantified analyzing images as given in figure 4.

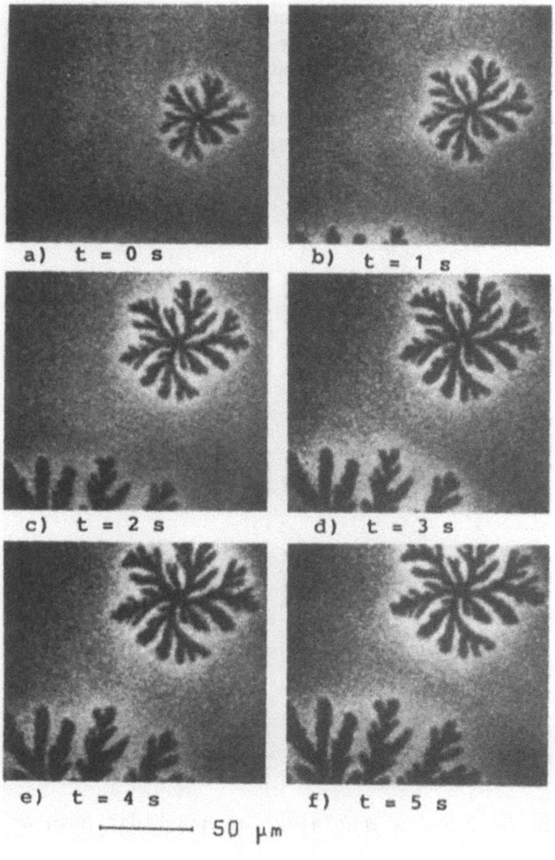

Fig. 4. Crystallite growth following a surface pressure jump by about 0.5 mN/m. Under the micrographs the time delay from the first observation is given

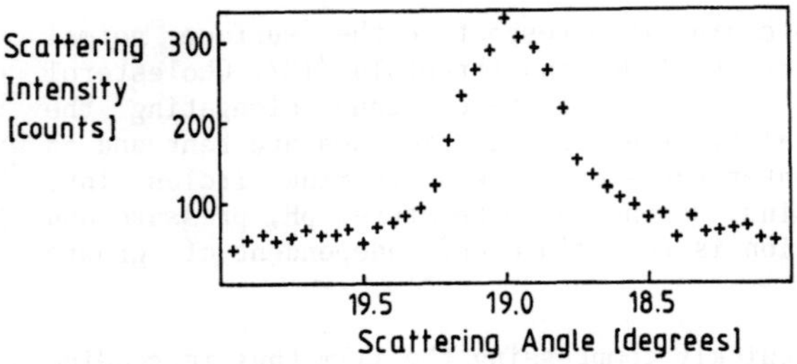

Fig. 5. X-ray intensity as a function of scattering angle for a monolayer of dimyristoyl-phosphatidylethanolamine on the water surface (T ≈ 20°C, pressure 30 mN/m, angular resolution 0.3°)

An important, but also challenging experiment is the determination of lattice structures of crystalline films on water surfaces. Figure 5 gives the first record of an X-ray diffraction peak of a monolayer on fluid substrate /16/. The lattice constant derived from the center of gravity of the peak (4.19 Å) is as expected for densely packed hydrocarbon chains, but the coherence lengths, determined from the widths (270 Å) are much smaller than expected for conventional single crystals. This may be a peculiar feature of two-dimensional solids, as often speculated.

3. MONOLAYERS AND LB FILMS ON SOLID SUPPORT

Films manipulated at the water surface can be transferred onto solid supports without drastic change of domain shapes /17/. This can be visualized again by fluorescence and by electron microscopy. Transferability includes optical properties and lattice dimensions, the latter within an uncertainty of 0.1 Å. This is derived from electron diffraction measurements which show that crystallographic axes within one domain are parallel demonstrating the existence of long range orientational order /18/. Going along one spiral of figure 3, however, the local axes are then shown to rotate.

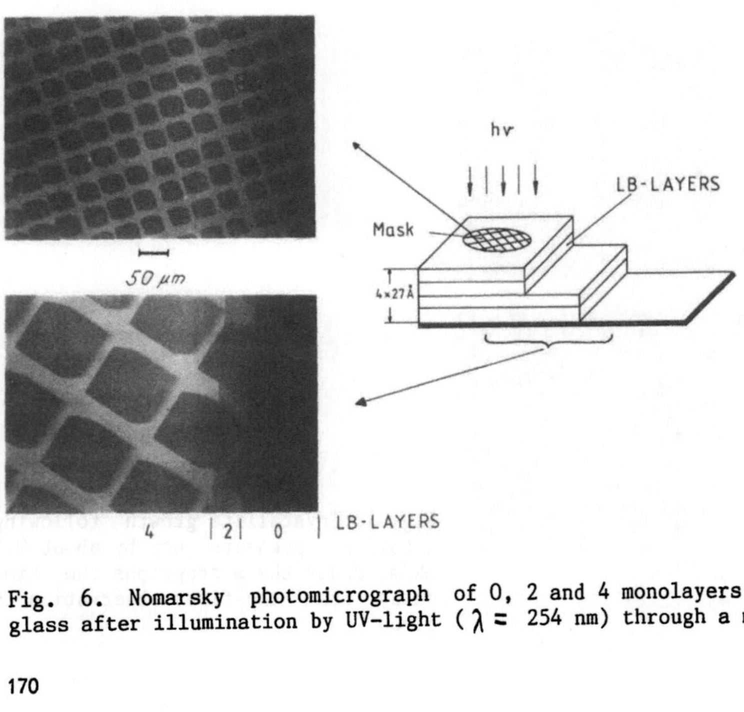

Fig. 6. Nomarsky photomicrograph of 0, 2 and 4 monolayers of arachidic acid on glass after illumination by UV-light (λ = 254 nm) through a mask

Thermal stability of mono- and multilayers was studied measuring thermally programmed desorption /19/. Binding energies can be increased via proper subphase conditions and also by photopolymerization of e.g. diacetylene groups introduced into the hydrocarbon chain region /20/. The latter process can be studied kinetically using the simple and fast technique of interference enhanced reflection measurements developed during the progress of this work. With this technique one also investigates photodesorption which may be used as a dry lithographic process. This is demonstrated in figure 6 where the film of two and four monolayers of Cd arachidate is illuminated by UV light across an electron microscope grid. Subsequently patterns due to photodesorption can be visualized by Nomarsky microscopy.

As lipids are natural matrices of membrane proteins one may also envision far future devices where a protein is coupled to an inorganic surface, e.g. that of a wave guide or of a FET. In order to achieve this, however, many technological steps will have to be studied and developed before designing a reliable system. As a first step, the incorporation of proteins into the film was assessed by means of fluorescence and electron microscopy using proteins of the photosynthetic apparatus. The latter are structurally well known, easy to handle and can be studied functionally by optical techniques. The functional reconstitution of these proteins into monolayers and transfer onto solid supports could be demonstrated /21/.

ACKNOWLEDGEMENT

I reported on parts of the PhD thesis works of my enthusiastic and skillful co-workers M. Lösche, L.A. Laxhuber, A. Miller, C.A. Helm and W.M. Heckl. We profited from discussions and collaborations with the groups of E. Sackmann, W. Knoll and J. Als-Nielsen. Our work was supported by the Deutsche Forschungsgemeinschaft, Stiftung Volkswagenwerk and the German Bundesministerium für Forschung und Technologie (BMFT).

REFERENCES

1) E. Sackmann: Ber. Bunsenges. Phys. Chem. 82, 891 (1978)
2) H. Kuhn and D. Möbius, Angew. Chem. 83, 672 (1971)
3) G.G. Roberts: Adv. Phys. 4, 475 (1985)
4) O. Albrecht, H. Gruler and E. Sackmann: J. de Physique 39, 301 (1978)
5) M. Lösche, E. Sackmann and H. Möhwald: Ber. Bunsenges. Phys. Chem. 87, 848 (1983)
6) R. Peters and R. Beck: Proc. Natl. Acad. Sci. USA 80, 7183 (1983)
7) R.M. Weiss and H.M. McConnell: Nature 310, 5972 (1984)
8) A. Fischer, M. Lösche, H. Möhwald and E. Sackmann: J. de Phys. Lett. 45, L785 (1984)
9) D. Andelmann, F. Brochard, P.G. de Gennes and J.F. Joanny: Comptes Rendus 301, 675 (1985)
10) C.A. Helm, L.A. Laxhuber, M. Lösche and H. Möhwald: Coll. Poly. Sci. 264, 46 (1986); M. Lösche, C.A. Helm, H.D. Mattes and H. Möhwald: in press in Thin Solid Films
11) A. Miller and H. Möhwald: in press in Europhys. Lett.
12) W.M. Heckl, M. Lösche, D.A. Cadenhead and H. Möhwald: in press in Europ. Biophys. J.
13) H.E. Gaub, V.T. Moy and H.M. McConnell: in press in J. Phys. Chem.
H.M. McConnell, D. Keller and H.E. Gaub: in press in J. Phys. Chem.
R.M. Weiss and H.M. McConnell: J. Phys. Chem. 9, 4453 (1985)
14) A. Miller, W. Knoll and H. Möhwald: Biophys. J. 49, 317a (1986), subm. to Phys. Rev. Lett.
15) B. Witten and L.M. Sander: Phys. Rev. B 27, 5686 (1983)
16) K. Kjaer, J. Als-Nielsen, C.A. Helm, L.A. Laxhuber and H. Möhwald subm. to Nature

17) M. Lösche, J. Rabe, A. Fischer, B.U. Rucha, W. Knoll and H. Möhwald: Thin Solid Films 117, 269 (1984)
18) A. Fischer and E. Sackmann: J. de Physique 45, 517 (1984)
19) L.A. Laxhuber, B. Rothenhäusler G. Schneider and H. Möhwald: Appl. Phys. A39, 173 (1986)
20) L.A. Laxhuber, U. Scheunemann and H. Möhwald: Chem. Phys. Lett. 24, 561 (1986)
21) W.M. Heckl, M. Lösche, H. Scheer and H. Möhwald: Biochim. Biohys. Acta 810, 73 (1985); W.M. Heckl, M. Lösche and H. Möhwald: in press in Thin Solid Films

Part II

Physics of Microstructures

Fundamentals of Low Dimensional Physics

M.J. Kelly

GEC Research Limited, Long Range Research Laboratory, East Lane, Wembley, Middlesex HA9 7PP, United Kingdom

1 INTRODUCTION TO THE PHYSICS OF MICROSTRUCTURES

In the last decade the technological drive towards smaller microelectronic devices has given rise to the concept of low-dimensional physics in two distinct contexts. The earlier referred to low meaning less than three, and the properties of two-dimensional electron systems (as in the inversion layer in Si MOSFETs or at the modulation doped GaAs/$Al_xGa_{1-x}As$ heterojunction) is now a mature, even aging, subject. More recently, low has come to small, as in the physics of microstructures, where the feature size of the sample is small compared with intrinsic length scales of the physics under investigation (e.g. mean-free path of electrons or even the inter-electron separation). In this introductory set of lectures, some of the basic physics pertaining to both meanings will be described, with a tutorial emphasis. Many of the following speakers will take up the contemporary points of interest, and expand where appropriate on the ideas expressed here. The introductory overview of fabrication technologies covers the preparation of the samples that we investigate.

The purpose of this first lecture is to introduce the systems that are available, or might soon be so. In addition some of the key aspects of physics are introduced. In Table I (courtesy M PEPPER and H AHMED, and see the Chapter by H AHMED for further details) we present a logarithmic length scale from 0.1 nm to 10 μm (5 orders of magnitude) which encompasses some of the phenomena under investigation. One sees the wavelength of visible light, the coherence length of superconductors and the interatomic spacing in solids presented along with the feature size of devices in today's chips, the size of research devices, and of microstructures that will be discussed more fully during this school. Some of the important derived lengths are also plotted - the typical spacing between electrons in silicon inversion layers or at heterojunctions, the cyclotron orbit of electrons in modestly high magnetic fields, and the mean-free-path of electrons between inelastic scattering collisions at low temperatures. One knows from elementary studies of light that the condition for strong diffraction or ray optics depends on whether the wavelength of light is rather greater or much smaller than the feature size of the aperture or stop. Similar considerations apply in our present studies: the inelastic mean-free-path can be greater than the relevant size of the test structure, with electron transport being ballistic, or much smaller, with the drift motion as used to describe carriers in larger scale semiconductor devices.

The basis of solid state physics since the advent of quantum theory is Bloch's theorem. The nature of wavefunctions and the energy-momentum relation (i.e. the bandstructure) in infinite crystalline solids are closely constrained by the periodicity of the crystal. The early 1970's saw the first extensive theoretical and experimental investigations of non-crystalline solids, and in particular amorphous semiconductors. The new concept of localised electron states was introduced, and the ideas of energy bands were considerably generalised to describe the qualitative similarities in many aspects of the optical properties of crystalline and amorphous semiconductors. At the same time, the availability of ultra-high vacua meant that surfaces could be cleaned and retain their integrity for several hours (before the absorption of a monolayer of residual

TABLE I

Length scales in solid state physics

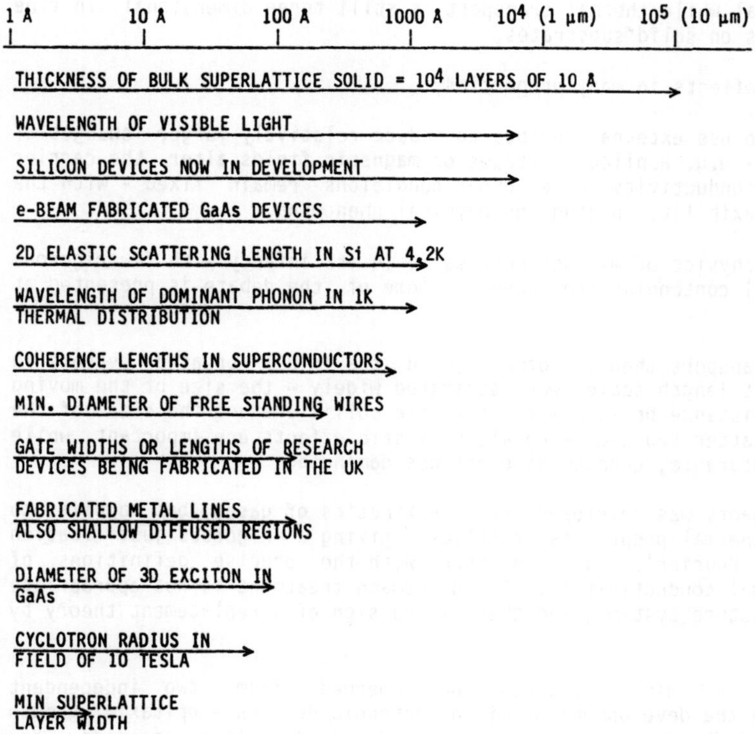

gases). The solids under investigation were then semi-infinite, and the physics determined by the way bulk Bloch functions terminate at the surface, and the nature of two-dimensional surface electron states trapped by energy gaps in the bulk and the vacuum level outside the crystal.

The physics of fabricated microstructures represents the current frontier of condensed matter physics where, although the periodicity of the crystal may, or may not, be retained, the infinity (in practice the irrelevance of the boundaries) is certainly not. Indeed, many of the basic ideas, built upon the assumption of the long-range homogeneity of the system, used over the last six decades to describe the electronic, optical, magnetic, elastic and other properties of solids are of limited validity at best, and quite irrelevant at worst.

The new conditions in microstructures include

(i) the low inertia (mass, magnetisation, specific heat, ...) which mean that external probes (light, electric fields, etc.) drive the system into non-linear regimes of response,

(ii) the large surface to volume ratio, with properties often dominated by the surface condition,

(iii) the fluctuations in feature size, becoming increasingly large as structures get smaller (and can exceed 10% mass fluctuations along the length of a 50 nm diameter wire), come to dominate the physical properties of very fine structures,

(iv) the intrinsically small number of carriers in fine-scale microstructures leading to enhanced noise characteristics,

(v) mixed dimensionality of phenomena, as when electrical transport is one dimensional while thermal transport is still three dimensional in fine metallic wires on solid substrates,

(vi) quantum size effects in many properties,

(vii) the ability to use external probes to induce relatively larger changes of one variable - e.g. applied voltages or magnetic fields alter the carrier density or conductivity while other conditions remain fixed - with the consequent flexibility in studying physical phenomena.

The theoretical physics of microstructures is in its infancy with few rigorous results, and several contending approaches. Some of the debate is presented at the school.

The Boltzmann transport theory formulated in the 1870's treated systems in which three important length scales were separated widely - the size of the moving particle, the mean distance between inter-particle collisions and the size of the sample. When the latter two become equal, ballistic effects are important, while if all three are comparable, quantum size effects dominate.

The Boltzmann theory was developed for the kinetics of gases, but applies to the electrical and thermal properties of solids - giving a rigorous justification for Ohm's law and Fourier's law, together with the precise definitions of electrical and thermal conductivities. The Boltzmann treatment is not appropriate in the new microstructure systems, and there is no sign of a replacement theory by 1986.

The fabrication of microstructures has emerged from two independent technologies used in the development of microelectronic devices - epitaxial growth and lithography. Early devices in silicon, and indeed gallium arsenide, were based on bulk substrate material implanted with doping of various species and implantation energies to get specified doping profiles. At an early stage the advantages of epitaxial overgrowth (at first by liquid phase epitaxy) were sought so that doping profiles could be controlled more closely and junctions made more abrupt. The more recently developed techniques that are in wide use today are Molecular-Beam-Epitaxy (MBE) and Metal-Organic-Chemical-Vapour-Deposition (MOCVD). They are essentially very highly controlled versions of evaporation and gas-phase cracking, respectively. In MBE atomic and/or molecular beams of Si (or Ga and As_4 for GaAs) impinge in a heated substrate in an ultra-high-vacuum environment. In MOCVD a closely controlled flow of $(CH_3)_3$ Ga and AsH_4 in a H_2 carrier gas passes over a heated GaAs substrate. Dopant species can be incorporated as an atomic beam or an ingredient of the gas flow. The net result is now independent spatial control over the doping profile, and the semiconductor band-gap (in the III-V systems where Al atomic beams or $(CH_3)_3$ Al replaces a fraction of the Ga equivalent) that reaches to the level of single atomic layers. Nanometre-scale layers are available to those studying two-dimensional systems. The materials research with epitaxial growth has now been extended to cover multilayers with not only other semiconductors, but metals and dielectrics as components. Systems include the transition metal silicide-silicon multilayers, the metal-metal multilayers (although most of the work to date on these systems involves sputter growth), Ca(Ba)F_2/GaAs, and others.

The miniaturisation of electronic device components has meant that the linear feature size in commercially available Si integrated circuits has halved every two years since the late 1960's and is at ~ 1 μm today. The research devices are over 10 times smaller, but the technologies for fabricating fine geometry features in the x-y plane are capable of fabricating test structures that are a further 10 times smaller again. In practice the main technique is electron-beam-lithography, although there are other contenders - focussed ion beams being one.

Examples of the successful fabrication of microstructures include

(i) the processing of a III-V semiconductor heterojunction that contains a two-dimensional electron gas into an array of 3 μm diameter discs in a 5 μm diameter lattice - the observed coupling of the electron gas discs being the first example of synthetic two-dimensional crystalline behaviour in microstructures /1/

(ii) the processing of multilayers of GaAs/AlGaAs into 0.25 μm square pedestals, 0.5 μm apart and into 0.25 μm wide lines for the study of the optical properties where the electronic excitations are confined to 0D, 1D or 2D /2/

(iii) arrays of hemispherical metal dots of diameter 30 nm and on a lattice of 50 nm /3/

(iv) free-standing wires of Au-Pd of dimensions 0.03x0.04x15 μm and of α-Si_3N_4 of dimensions 0.1x0.14x125 μm /4,5/

(v) supported wires of 0.01 μm diameter and less, by e-beam lithography, and in the finest cases using inorganic 'resists' where the metal is effectively the debris associated with e-beam drilled holes in β-alumina for examples /6/

(vi) metal gratings with sub 0.5 μm pitch /7/

(vii) quantum well wires of submicron diameter /8/

Pictures are not included here, as they appear elsewhere, but original references are cited.

The highlights of the physics derived from these systems include the observation of

(i) quantum electronic and optoelectronic effects revealing the one-, two-and three-dimensional properties of electron gases

(ii) non-Ohmic electrical behaviour due to the effects of disorder (strong localisation) and reduced dimensionality (weak localisation or 'quantum corrections to Ohm's law')

(iii) finite size effects where a single electron trap produces ~0.2% fractional changes in the resistance of a device

(iv) no evidence for one-dimensional phonon effects in free-standing wires at low temperatures except under strong electric field conditions where a phonon bottleneck is formed

(v) small scale fluctuations in local stress leading to optical variations in the exciton features in quantum well wires

(vi) enhanced optical non-linearities due to quantum confinement.

The fabrication and the physics of microstructures have been largely separated activities until recent years, but engineers and physicists are now joining forces, both because the methods of fabrication have a larger element of physics (c.f. the discussions on electron and ion beams and their interaction with solids), and because the structures are entering the length regime where the interesting physical phenomena are large rather than minor perturbations. Once two or more of the length dimensions of a structure are ~0.1 μm or smaller, the mode of operation of any device becomes qualitatively different from that of the larger devices in current use. Miniaturisation, as such, seems to be reaching an end, and while optical and other radical alternatives to information processing may emerge, there is a possibility of total crystal engineering by the end of this century. The ability to tailor three-dimensional nanometre scale microstructures in a wide range of materials may lead to synthetic solids with more desirable device properties than those provided by nature - modulation doped superlattices are a precursor of such solids. The combination of MBE crystal growth with the methods of fine-line lithography is the current research frontier en route to the fully synthetic materials.

This first section ends with Table II which provides some of the interrelationships of the nanometre length scale, the millielectron volt energy scale and the picosecond time scale, which are relevant to the physics of microstructures and which will be used throughout the school.

TABLE II

Energy, length and time scales

(A) $1\ \text{meV} = 11.6\ \text{K} = 0.24 \times 10^{12}\ \text{Hz}$

In Si $Ry^* = 19.55$ meV
In GaAs $Ry^* = 5.25$ meV

In a typical Si MOSFET

$1\ \text{meV} = \hbar\omega_C$ at 1.64 T ($m^* = 0.19\ m_e$)
$= g\mu_B B$ at 8.64 T ($g = 2$)
$\approx 100\ \text{N mm}^{-2}$
$\approx$ 1V on GATE

(B) In Si $a^* = 2.145$ nm $\ell_C = 25.66\ (B)^{-\frac{1}{2}}$ nm
GaAs $a^* = 10$ nm

(C) 1 ps = 0.66 meV

The present school is, to my knowledge, the first of its kind. A conference on 'Microstructures, Microdevices and Superlattices' held in Champaign-Urbana, Illinois, in 1984 was heavily weighted in favour of semiconductor multilayers. The proceedings appear as the first four issues of a journal 'Superlattices and Microstructures' (Pergamon Press). In Table III, a list of recent review-type publications is given that is certainly useful background material for this meeting.

2 SOME BASIC ELECTRONIC CONCEPTS IN LOW-DIMENSIONAL PHYSICS

In this lecture we consider the range of electron states that are encountered in various microstructure systems and the implications for their electronic properties. The ideal systems we consider first come straight from the quantum mechanical textbooks - potential wells of different shapes bounded by infinitely high barriers:

(i) a purely one-dimensional line bounded by z=0 and L

(ii) a purely two-dimensional plane bounded by x,y=0 and L

(iii) a three-dimensional volume bounded by x,y,z=0, L

TABLE III

Reviews on quantum well topics

Ando, T., Fowler, A.B. and Stern, F.: Electronic properties of two-dimensional systems. Rev. Mod. Phys. 54 437-625 (1982)

The definitive review of the work in the 1970s using the Si MOSFET device, with some mention of III-V superlattices. The review concludes with an exhaustive bibliography of 2000 relevant papers.

Ploog, K. and Dohler, G.: Compositional and doping superlattices in III-V semiconductors. Adv. Phys. 33 285-360 (1963)

A review of multilayer systems with emphasis on the optical properties as a probe of electronic structure.

Wood, C.E.C.: Progress, problems and applications of molecular beam epitaxy. Physics of Thin Films 11 35-103 (1980)

A review of growth of different III-V alloy system, the incorporation of dopants and device applications.

Chang, L.L.: Molecular beam epitaxy. Handbook of Semiconductors. Volume 3, ed S.P. Keller Amsterdam: North-Holland, 563 (1980)

A general review of the principles and development of MBE growth.

Mullin, J.B., Irvine, S.J.C., Moss, R.H., Robson, P.N. and Wright, D.R. (ed): Metalorganic vapour phase epitaxy. J. Cryst. Growth 68 (1984)

The proceedings of the Second International Conference on MOVPE or (MOCVD), which contains recent advances, and also reviews of the different aspects of this growth technique.

Bauer, G., Kuchar, F. and Heinrich, H. (ed): Two-dimensional systems, heterostructures and superlattices. (Berlin: Springer) (1984)

A collection of invited review papers given at a winter school, which gives a good coverage of the different topics.

Kelly, M.J. and Nicholas, R.J.: The Physics of Quantum Well Structures. Reports of Progress in Physics 48, 1699-1741 (1985)

A review concentrating on the physics.

Weisbuch, C.: Fundamental Properties of III-V Semiconductor 2-D quantized Structures: The Basis for Optical and Electronic Device Applications. To be published in Semiconductors and Semimetals, R. Willardson and A.C. Beer (ed): Academic Press (1986)

A review of physics with a discussion of the device implications.

(iv) a quasi two-dimensional system bounded by $x,y=0$, L and $z=0$, λ, where λ is very small compared with L and

(v) a quasi one-dimensional system bounded by $x,y=0$, λ and $z=0$, L.

The one-electron Schrödinger equation, relating the kinetic and potential energy to the total energy E of that electron, is given by

$$\frac{\hbar^2}{2m^*}\left(\nabla^2 + V\right)\psi = E\psi$$

where ∇ and V is 1,2,3,3,3 dimensional in the five cases above, with V being zero or infinity inside or outside the well. The eigenfunctions and eigenenergies are given respectively by

$$\psi_n = \left(\frac{2}{L}\right)^{\frac{1}{2}} \sin\frac{n\pi z}{L}, \quad E_n = \frac{\hbar^2}{2m^*}\left(\frac{n\pi}{L}\right)^2 \qquad (1)$$

$$\psi_{lm} = \left(\frac{2}{L}\right) \sin\frac{l\pi x}{L} \sin\frac{m\pi y}{L}, \quad E_{l,m} = \frac{\hbar^2}{2m^*}\left(\left(\frac{l\pi}{L}\right)^2 + \left(\frac{m\pi}{L}\right)^2\right) \qquad (2)$$

$$\psi_{lmn} = \left(\frac{2}{L}\right)^{3/2} \sin\frac{l\pi x}{L} \sin\frac{m\pi y}{L} \sin\frac{n\pi z}{L} \qquad (3)$$

$$E_{l,m,n} = \frac{\hbar^2}{2m^*}\left(\left(\frac{l\pi}{L}\right)^2 + \left(\frac{m\pi}{L}\right)^2 + \left(\frac{n\pi}{L}\right)^2\right)$$

and similar results for (iv) and (v) where appropriate alterations from L to λ are made. Note that with $\lambda \ll L$, the kinetic energy of confinement in the λ dimensions are very high, and in systems containing many electrons the usual quantum number to vary is in the L dimension. If we define $k = (l\pi/L, m\pi/L, n\pi/L)$, the density of modes in k-space is uniform at L/π in each spatial dimension. If we place many electrons into the wells satisfying Fermi Dirac statistics, the distribution of states in energy is an important quantity, obtained by equating the number of states between E and E+dE to the length/area/volume of k-space bounded by the two energies and the constant density of states in k-space. The results are

$$N(E) = \frac{L}{2\pi}\left(\frac{2m^*}{\hbar^2}\right)^{\frac{1}{2}} \frac{1}{\sqrt{E}} \qquad (4)$$

$$N(E) = \frac{L^2}{\pi}\,\frac{2m^*}{\hbar^2} \qquad (5)$$

$$N(E) = \frac{2L^3}{\pi^2}\left(\frac{2m^*}{\hbar^2}\right)^{3/2} \sqrt{E} \qquad (6)$$

$$N(E) = \sum_n \frac{L^2}{\pi}\,\frac{2m^*}{\hbar^2}\, H(E - E_n) \qquad (7)$$

$$N(E) = \sum_{\ell m} \frac{L}{2\pi}\left(\frac{2m^*}{\hbar^2}\right)^{\frac{1}{2}} (E-E_{lm})^{\frac{1}{2}}\, H(E-E_{lm}) \qquad (8)$$

where $H(\theta)$ is the Heaviside function ($H(\theta)=1$ $(\theta>0)=0$ $(\theta<0)$), E_n and E_{lm} are as defined in (1) and (2) above. These functions are shown in Fig. 1. In the limit $L\gg\lambda$, the density of states in energy is high for the L dimensions, and we use the contimumn limit, speaking of energy bands with dispersion relations $E = \hbar^2k^2/(2m^*)$.

A whole range of approximations is implicit or explicit before applying the simple picture of idealised quantum wells to the micro-structures of this school. First, the idea that the simple one-electron model applies at all in solid

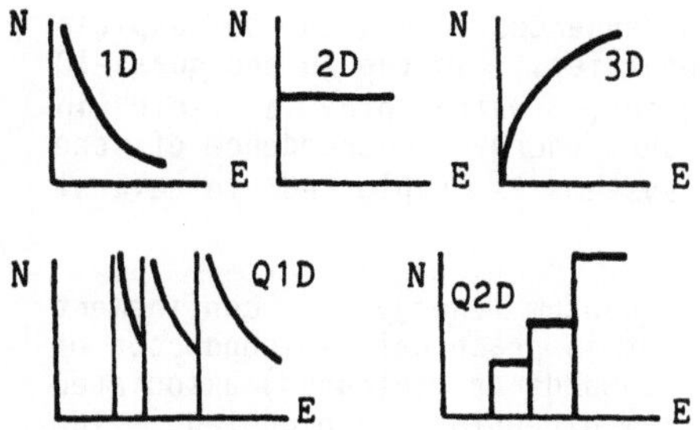

Fig. 1 : The density of states for the 1D, 2D, 3D, quasi-1D and quasi-2D systems

systems is backed by much detailed research into the nature of many-electron systems /9/. In practice the potential V(r) is a subtle simplification of the many-electron-many-proton interparticle potential. Indeed the fact that the simple ideas above are a useful guide in 1D, 2D or 3D systems turns the question around to 'why does the potential work so well?' and is considered in other lectures here. The breakdown of the simple picture and the role of many-electron effects is also considered. Here we assume that we can fill the lowest energy electron states (to a degeneracy two to account for spin) subject to Fermi statistics: a Fermi level separates the occupied and empty electron states. If we apply the model to actual microstructures, the precise form of V needs to be considered. In simple crystalline metals, the free-electron model works well, but the potential outside is a few eV above the Fermi energy (the work-function) rather than infinite. Until recently the finiteness of the potential has only been encountered in specific tunnelling structures (metal-oxide-metal in most cases). With the close proximity of adjacent microstructures, tunnelling through finite potential barriers is an important mechanism for their interaction, and the finiteness will become increasingly important. In semiconductor heterostructures the simple model applies but the justification is even more subtle. For n-doped systems, the valence bands are full and excess electrons occupy conduction band states. With two different semiconductors with different band-gaps, V is simply a measure of the offset in the two conduction bands, although the precise value (0.81x eV for $Al_xGa_{1-x}As$ and GaAs) is a matter of active research and debate.

In other semiconducting systems with large variations in the doping profile, V may refer to the electrostatic potential set up by local imbalances in the density of free-carriers and of donor and/or acceptor ions. In turn this V may depend on applied voltages, as in the case of Si MOSFETs /10/.

When the potential barriers are finite the appropriate exponential wavefunctions (i.e. complex momentum solutions) are used to describe electrons in the classically forbidden region. For an isolated well, the fact that an electron spends time in the confining layers results in a generally small decrease in the kinetic energy of confinement compared with that of the same well with infinite barriers. Whereas in metallic systems the mass m^* is essentially the free-electron mass, it is the conduction or valence band effective masses in semiconductors, obtained from the curvature of the energy bands adjacent to the gap. The precise derivation of this result for bulk semiconductors, and the treatment of donor ions leading to the simple modified Bohr theory (electron mass → effective mass, and the free space dielectric constant being multiplied by the relative dielectric constant of the semiconductor) is a beautiful result dating from the 1950's /11/.

Why start with the density of states? In calculating many response functions, the changes to the individual electron states or to their occupation - particularly those near the Fermi energy - dominate /9/. The specific heat, the magnetisation, the thermoelectric power and the conductivity all depend on the density of states at the Fermi energy or its energy derivative there. A number of points can be made about the density of states in Fig. 1:

(i) Dimensionality: the differing energy dependences are used to explain several of the optical and electrical properties. In the 1D and quasi-1D cases, there are square-root divergences, which play a role in destabilising the simple structures. The energy independence of the density of states in 2D and quasi 2D systems is exploited in several optical devices.

(ii) In the quasi 1D and 2D systems, the quantum energy of confinement associated with the thickness λ is large but in practical semiconductor or metal systems, several of the replicas (sub-bands or minibands) associated with the higher energy states in the narrow direction are occupied. The scattering of electrons between minibands is an important contribution to the resistance. In systems where the Fermi energy can be moved strong fluctuations are associated with the onset of the occupation of further minibands and are measured in magnetoresistance and magnetisation etc.

(iii) In semiconductors, the offset in the valence band edges leads to a set of hole levels that mirror the electron levels. The optical properties reflect the size quantisation effects [the E_n, and E_{lm} in (7) and (8) above] in that optical transitions couple the electron and hole states across the gap and in cases where the narrow gap material forms the wells; transitions corresponding to the bulk of the material are not available.

(iv) Perturbations to the simple picture of the electron states came from several sources: (a) Fluctuations in the dimensions (particular in the λ direction) can result in relatively large fluctuations in the E_{lm} and E_n terms, and the Schrödinger equation no longer decouples easily (in the quasi 1D system these fluctuations may smear out the sharp structure associated with the onset of the new minibands, particularly if the correlations of the fluctuations have long length components of appreciable magnitude). (b) Applied fields in the lateral dimensions alter the Schrödinger equation in the λ directions, which are used to induce large relative changes in both the energy of optical transitions and their associated optical non-linearities, and in special 3D systems to induce on dimensional transition from an L to a λ length scale. (c) Applied fields in the L direction when one or both of the other dimensions are of size λ with dimensionality-dependent hot-electron transport or quenching of optical non-linearities.

(v) Many thermodynamic quantities are related to the electronic density of states, and the differing energy dependences as a function of dimensionality are reflected in differing temperature dependences. The self-consistent analysis of $U = \int^{E_f} f(\varepsilon)\ n(\varepsilon)\ d\varepsilon$, [the one electron contribution to the total energy where the Fermi energy is E_f given by $N = \int f(E)\ dE$, and $f(E)$ is the Fermi occupation factor $[\exp((E-\mu)/kT)+1]^{-1}$ for μ the zero-temperature limit of E_f] gives a specific heat $U \propto N(E_f)T$. The low-temperature phonon contribution varies as T^3 for three-dimensional systems or low-dimensional electronic systems embedded in 3D most material, but for free-standing membranes (wires) the variation is T^2,(T) as the lateral modes of vibration are quenched. (As a guide, the dominant phonon wavelength varies as 100 nm/T(K) and for the λ dimension of 0.1 μm the dimensionality change over is at ~1K). This is to be added to the electronic contribution in extracting or predicting thermodynamic properties.

The magnetoresistance, even for low electric and magnetic fields, is a strong function of dimensionality reflecting the density of states and the nature of available final states for scattering. An extreme example is in the Quantum Hall regime where the magnetoresistance and magnetoconductivity of a two-dimensional system vanish [note in this case that $\rho_{xx} = \sigma_{xx}/(\sigma_{xx}^2 + \sigma_{xy}^2)$ and while σ_{xx} vanishes σ_{xy} is quantised to a value e^2/hn where n is an integer]. In systems where the dimensionality can be varied the <u>sign</u> of the magnetoresistance is a signature of the dimensionality (see next section).

The shape of the optical absorption edge also reflects dimensionality, as the total absorbing power of a system expressed to light of energy $\hbar\omega$ contains a factor

$$\int |M|^2 J(E,\hbar\omega) f(E)(1-f(E+\hbar\omega))dE$$

where f are the occupation factors guaranteeing filled initial states and empty final states, $J(E,\hbar\omega)$ is the joint density of states, a convolution of initial and final states, and M is a dipole matrix element coupling the initial and final states. (Actually the form above is a considerable simplification as the full expression is a summation over all filled initial and all available final states with the dipole matrix element). The difference between the $E^{1/2}$ and E^0 band edge shows up in the joint density of states and the shape of the optical absorption edge. The sharper features associated with the $E^{-1/2}$ edge in 1D have not been sought in microstructures, but the effects that have been seen in some molecular solids, where very weak intermolecular interactions simulate the 1D condition, are now attributed to excitons.

With sub-band gap light it is possible to create bound electron-hole pairs, excitons, in semiconductors. The lowest energy exciton corresponds to a 1s hydrogen level, with an effective Bohr radius a^* of ε/m^* times the hydrogen Bohr radius (i.e. 1.9 nm in Si and ~10 nm in GaAs). The constraint that the electron wavefunction should be confined in linear dimensions of $\lambda \lesssim a^*$, and closer to the hole, implies a modification of the binding energy - indeed the 2D exciton is four times more tightly bound than the 3D exciton, and the quasi 2D exciton binding is intermediate. The equivalent effects in quasi 1D systems should be more dramatic, but experimental results have not been reported in semiconductors.

So far we have considered the electrons as independent particles, but the full Coulomb interaction is divided into a screened interaction capable of an independent particle approximation and a strong unscreened interaction that gives rise to the plasma oscillations. There are several ways of obtaining this division, and one, due to Stern, shows that in 2D, the electron gas oscillations have an ω_p^2 dispersion, in contrast to the $\omega_p^2 = 4\pi ne^2/m^*$ in 3D. In 3D metals $\hbar\omega_p \sim 10$ eV while in semiconductors $\hbar\omega_p \sim 10$ meV. The regime of $\omega_p^2 \propto q$ does not extend all the way to $q \to 0$ but eventually gives way to a q^2 dependence. The one-dimensional case is often both analytically soluble and physically pathological. The quasi 1D systems microstructures have the λ degrees of freedom which relieve the pathologies, but often in very subtle ways which are still under active investigation. In semiconductors the plasma and phonon frequencies are often in the same range, and coupled modes play a role both in hot-electron phenomena and in spectroscopic studies in 2D and 3D systems. In metals the plasmon frequency is much higher and generally enters only the ultraviolet optical properties.

We have described the basic one-electron states in LDS with emphasis on dimensionality in altering their analytic nature which in turn has implications for thermodynamic and transport properties. We turn to the latter now.

3 FUNDAMENTALS OF TRANSPORT THEORY IN L.D.S.

The Boltzmann transport theory starts with the introduction of a distribution function $f(\underline{r},\underline{k},t)$ which describes the number of carriers in $d\underline{r}$ at $\underline{r}$, $d\underline{k}$ at $\underline{k}$ at a time t /9,12/. It must be assumed that the solid is infinite and homogeneous for $\underline{r}$ and $\underline{k}$ to be well defined. It is then assumed that externally applied fields (electric, magnetic or thermal) induce only small changes in f from its equilibrium value, and that the changes are linear in the field. In analogy with the continuity equation in gas dynamics, the Boltzmann equation is

$$\frac{df}{dt} = \frac{\partial f}{\partial t} + \dot{\underset{\sim}{k}}.\nabla_{\underset{\sim}{k}} f + \dot{\underset{\sim}{r}}.\nabla_{\underset{\sim}{r}} f = \frac{\partial f}{\partial t}\Big|_{coll} \qquad (9)$$

where the right-hand side refers to the collision suffered by the carriers in their interaction with each other, or with impurities or lattice vibrations; these collisions provide the mechanism for returning a disturbed system to equilibrium. In crystalline solids we use a semi-classical transport theory for for the individual carriers and write $\underline{v} = \underline{\dot{r}} = (1/\hbar)\, \nabla_k E(k)$, relating the carrier velocity to the bandstructure, and the Lorentz force is just $\underline{\dot{k}} = (e/\hbar)(\underline{E} + \underline{v} \times \underline{B})$. This semiclassical description is valid when the electron wavepacket extends over several unit cells, the mean distance between scattering is very much larger than this, and in turn is much smaller than the sample size. One can write a Boltzmann equation for a distribution $g(\underline{r},\underline{q},t)$ of phonon modes and a coupled electron-phonon Boltzmann equation. Solutions to the general equation have proved intractible, and further approximations are required. The simplest form for the collision term is to assume some dominant relaxation mechanism is operating, characterised by a time scale τ, so that

$$\frac{\partial f}{\partial t}\Big|_{coll} = \frac{-f-f_0}{\tau}$$

where f_0 is the equilibrium value. One assumes that collisions destroy all previous phase-space memory of the carriers and that the scattering is quasi-elastic and isotropic. These allow one to obtain a solution of the form, for a system under uniform fields,

$$f(\underline{k},t) = f^{\circ}(\underline{k}) + \int_{-\infty}^{0} dt'\, e^{-(t-t')/\tau(E(\underline{k}))} \left(\frac{-\partial f}{\partial E}\right) \underline{v}(\underline{k}(t')) \times \left[-e\,\underline{E}(t') - \nabla\eta(t') - (E(\underline{k})-\eta)\nabla T(t')/T\right]. \tag{10}$$

(The distinction between electric field $\underline{E}$ and bandstructure energy $E(\underline{k})$ is obvious, η is the chemical potential, and other symbols have the usual meaning). From this expression, one can calculate the steady state current as

$$j = \frac{-e}{4\pi^3} \int f(\underline{k})\, \underline{v}(\underline{k}) d\underline{k} = \underline{\underline{\sigma}}\, \underset{\sim}{E}$$

$$\text{where } \underline{\underline{\sigma}} = \frac{e^2}{4\pi^3} \int (-\partial f(\underline{k})/\partial E)\, \underline{v}(\underline{k})\, \underline{v}(\underline{k})\, \tau(E(\underline{k})) \tag{11}$$

and eventually $\sigma = ne^2 \tau/m$, Ohm's law for free electrons. The other transport coefficients (thermopower etc.) also follow, representing the end point of many approximations to the most general transport theory. Attempts to generalise this solution have not been productive. Technically we can regard $\underline{k}$ in the final forms of the expression for σ as merely a label of the eigenstates, so in the case of amorphous materials, where k has no technical meaning and some other expressions for v are used, a qualitatively similar expression can be obtained. Attempts to treat the magnetic fields beyond the simplest use of (10) have been unsuccessful, and indeed the anomalies in the sign of the Hall coefficient of amorphous materials is without a rigorous explanation.

The original arguments by THOULESS /13/ on the maximum metallic resistance of a wire are based on assumptions quite removed from those above, but the conditions under which Ohm's law is recovered are a special case. We consider a metallic sample of length L and area A, in which the density of states per unit volume is dn/dE. One might estimate how sensitive the individual energy levels are to changes at the boundary as $\Delta E \sim \hbar/\tau$ where τ is the time taken for the electron to diffuse to the end of the wire and sample the boundary, i.e. L^2/D where D, the diffusion constant, is related through the Einstein relation to the conductivity and density of states: $\sigma = \frac{1}{2}e^2 D\, dn/dE$. One arives at an equality

$$\frac{\hbar}{\tau} = \frac{\hbar}{e^2}\,\frac{2}{r}\,\frac{dE}{dN}$$

where r is the conventional resistance of the wire and dE/dN is the average separation of energy levels in the sample. If we now join two similar, but not identical wires, we can ask about the resistance of the combined system. If $\hbar/\tau \gg dE/dN$, all the energy levels in one system interact with those in the other, and the resulting calculation scales up for a sample of length 2L, and so reproduces the conditions where Ohm's law applies in macroscopic systems. If $\hbar/\tau \ll dE/dN$, the electron levels only weakly interact, and motion of electron will involve some hopping motion from one unit to the other and this leads to a multiplicative resistance, i.e. one which increases exponentially rather than linearly with length. The condition $r > 2\hbar/e^2 \sim 50.4$ kΩ is met in very fine wires ($\sqrt{A} \sim 10$-30 nm, $L \sim 10$-100 μm). This derivation borrows the concept of hopping between localised electron states as derived to explain activated conductivity in disordered systems. Indeed a more detailed analysis of the conductivity charge in a quasi one-dimensional system based on these concepts leads to an expression of the form

$$\frac{\delta R}{R_0} \approx \frac{\rho\,(4D\tau_i)^{\frac{1}{2}}}{A\;R_T}$$

where R_0 is the residual resistance at T=0, A the cross-section area, D the electron diffusivity, τ_i the inelastic scattering time and $R_T = 50.4$ kΩ. The inverse area dependence is well verified, and the temperature dependence is associated with the $\tau_i(T)$. Independently of this work, a Russian group (AL'TSHULER et al.) examined the many-body Coulomb interaction effects in quasi one-dimensional disordered systems and deduced

$$\frac{\delta R}{R} = \rho\frac{(2-F)}{AR_T}\left(\frac{4D\hbar}{kT}\right)^{\frac{1}{2}}$$

where F $(0<F<1)$ is a screening parameter. For plausible choices of $\tau_i(T)$ the functional form of the two expressions is identical, and an unambiguous separation of the two effects is not possible. Much research, theoretical and experimental, has been devoted to this topic, and the equivalent effects in two-dimensional systems (see below) have been differentiated in magnetotransport studies.

The contrast between the starting assumptions, the methods of analysis and the approximations of the Boltzmann and the modern theories of transport is wide. One common strand is the role of different length scales. In the Boltzmann transport it was the carrier size (de Broglie or Debye wavelength), the mean-free-path and the sample size. The mean-free-path in Boltzmann transport is a statistical average. In the modern theories, as applied to low dimensional structures, the different nature of elastic and inelastic scattering, the partial or complete retention of quantum phase memory at collisions, the radius of electron states localised by disorder in the electron potential and intrinsic lengths (e.g. the cyclotron radius in the case of magnetic fields) all interact in a wide variety of ways. The central role of disorder, first studied in imperfect crystalline and then amorphous or glassy three-dimensional structures, is becoming better appreciated in reduced dimensional structures, whether the disorder is intrinsic to the material, or, as is becoming more prevalent, is a function of the inability to grow or fabricate ideal specimens. A scaling theory is emerging to tie the various strands of transport in non-Boltzmann systems.

The basic concept of localisation follows from the effects of disorder on extended wavefunctions in solids. In a weak random potential the Bloch electron would lose phase coherence over a distance l, the mean-free-path, but still the wavefuntion is extended throughout a three-dimensional crystal. Stronger disorder will not allow the electron to propagate, but rather it becomes localised with an envelope to the Bloch function of $|\psi(\underline{r})| \sim \exp(-|\underline{r}-\underline{r}_0|/\xi)$. The nearly degenerate states are too far apart on average, and the nearby states have too small matrix elements because of a difference in energy. At zero-temperature there is postulated a mobility edge separating in energy states which have zero or finite conductivity. Mott's (1973) conjective was that the transition would be sharp, based on the idea that the mean-free-path could not be less than k_f^{-1} or an atomic spacing, i.e. $k_f\, l>1$ and so there would be a minimum metallic conductivity $\sigma_{min}(3D) = (1/3\pi^2)(e^2/(\hbar a))$, a the lattice constant, (cf $\sigma=ne^2\tau/m$, $\tau=v_f l$, $v_f=\hbar k_f/m^*$). A similar argument in 2D gives $\sigma_{min}(2D) \sim 0.1\, e^2/\hbar$. Some recent experimental data suggest that the transition may not be sharp.

The Thouless argument above is a scaling argument, and more general progress follows from noting that the conductance G of a d-dimensional sample is dimensionless in units of $e^2/\hbar$, and the dimensionless conductance $g=G/(e^2/\hbar)$ must scale with the sample size. In macroscopic samples where Ohm's law applies $g(L) = \sigma L^{d-2}$, whereas in highly disordered systems $g(L) \propto \exp(-L/\xi)$ when $L>>\xi$. The next step is to consider that the logarithmic derivative of g, $d(\ln g)/d(\ln L)$, is a function of g alone, $\beta(g)$. In the Ohm's law regime $\beta(g) = d-2$, while in the localised regime $\beta(g) = \ln(g/g_c)$, and in the case of weak disorder ($k_f l>>1$) perturbative corrections go as $-a/g$ where $a=g_c=\pi^{-2}$ so that conduction is always less than Ohmic. This allows the construction of $\beta(g)$ as shown in Fig. 2. In 3D, $\beta(g^*)=0$ for $g^*>1$ and, if the microscopic disorder is such that the conductance g_0 at the microscopic cutoff-length exceeds g^*, then scaling $L\to\infty$ takes $\beta(g)\to 1$ and Ohm's law while, if $g_0<g^*$, the scaling takes $\beta(g)\to-\infty$ and an insulator. Studies around g^* now suggest a phase transition with finite exponents and no minimum metallic conductivity, with the localisation length diverging on one side and the correlation length on the after. In $2+\varepsilon$ dimensions we can use a perturbation theory to get a similar story as in 3D, while in 2D all states as localised, even in weak disorder with

$$g(L) = g_0 - \frac{e^2}{h\pi^2} \ln\left(\frac{L}{l}\right)$$

where g_0 is the conductance at length scale l, the lower cutoff (scattering length in large samples, sample size in smaller systems). In conventional transport $g_0 = (e^2/2\pi\hbar)\, k_f l$ and the two expressions coincide for a localisation length $\zeta^{(2)} = l \exp(\pi\, k_f l/2)$ which can be quite large making direct and detailed experiments on localisation more difficult. In 1D $\beta(g)<-1$ and all states are localised. The more recent ideas focus on the small probability of almost unity conductance that can arise in 1D resonant tunnelling systems and which will be discussed in detail in some of the later lectures.

This scaling approach has been complemented by diagrammatic perturbation theories of electron-electron scattering in disordered systems. With non-interacting electrons weakly scattered by modern impurities one can calculate the static conductivity via the Kubo formalism

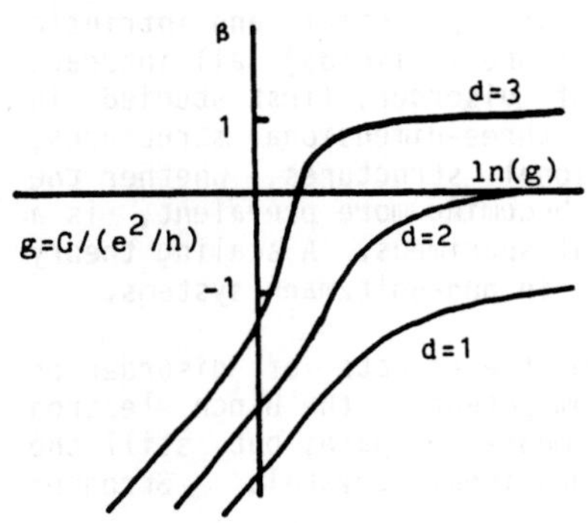

Fig. 2. The scaling function β(g) versus a dimensionless conductance

$$\sigma_{\gamma\gamma'}(\omega=0) = \frac{e^2 h}{2\pi L^d} \sum_{pp'} \left[\frac{p_\gamma}{m}\right] G^{\pi+-}_{pp'} \left[\frac{p'_{\gamma'}}{m}\right]$$

where G is the two-particle Green function describing the propagation of an electron-hole pair from momentum state $\underline{p}$ to $\underline{p}'$, and obtain the familiar

$$\sigma_{\gamma\gamma}(0) = \frac{ne^2\tau}{m} \delta_{\gamma\gamma'}$$

where $\hbar/\tau = 2\pi |V|^2 \rho(\varepsilon_f) n_i$, ($V$ = scattering potential, $\rho(\varepsilon_f)$ density of electron states and n_i = density of scattering centres), an expression that is of the lowest order in the scattering. To higher orders a number of complications arise, the most important being of the kind shown in Fig. 3 involving maximally crossed diagrams. The amplitude of this process is which

$$W(\underline{p},\underline{p}') = \frac{d(n_i V^2)^2}{(\underline{p}+\underline{p}')^2 v_f^2 \tau^2}$$

indicates that the backscattering process ($\underline{p}+\underline{p}' \approx \underline{0}$) is singular. The interference effect is important to the localisation process, and can be related to density fluctuation processes. The singular backscattering reduces the conductivity to

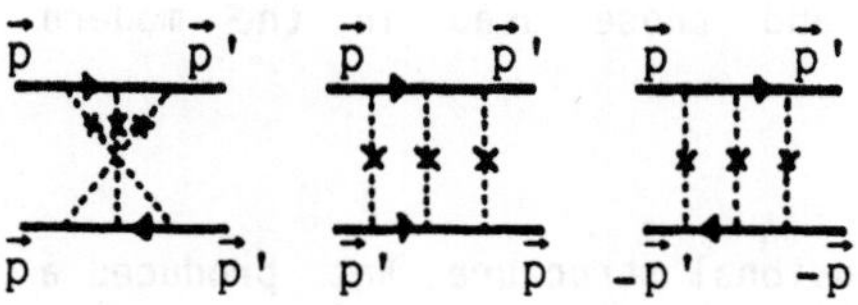

Fig. 3. The maximally crossed diagrams in a particle-hole propagator, with dotted lines denoting scattering by impurities.

$$\sigma = \frac{ne^2\tau}{m} - \frac{2e^2}{\hbar\pi} \frac{1}{L^d} \sum_Q \left[\frac{1}{Q^2}\right] .$$

The lower Q cutoff is ~ 1/L for a system of size L and the upper is ~ 1/l giving

$$\sigma_{3D}(L) = \sigma_0 - \frac{2e^2}{\hbar\pi^3} \frac{1}{l} - \frac{1}{L}$$

$$\sigma_{2D}(L) = \sigma_0 - \frac{e^2}{\hbar\pi^2} \ln\left(\frac{L}{l}\right)$$

$$\sigma_{1D}(L) = \sigma_0 - \frac{e^2}{\hbar\pi} (L - l) .$$

At finite temperatures and for microscopic samples in which experiments are performed, inelastic scattering introduces random fluctuations which limit the efficiency of the quantum interference. If the life-time of particular eigenstate is τ_{in}, and $\tau_{in} >> \tau$, the elastic scattering time, an electron diffuses $L_{Th} = \surd(D\,\tau_{in})$ between dephasing collisions ($D = V_f^2\tau/d$). If $\tau_i \sim T^{-p}$, $L_{Th} = aT^{-p/2}$ and the resulting conductivity corrections are then

$$\Delta\sigma_{3D} = + \frac{e^2}{\hbar\pi^3} \frac{1}{a} T^{p/2}, \quad \Delta\sigma_{2D} = \frac{p}{2} \frac{e^2}{\hbar\pi^2} \ln\left(\frac{T}{T_0}\right) \quad \text{and} \quad \Delta\sigma_{1D} = - \frac{ae^2}{\hbar\pi} T^{-p/2},$$

with the conductivity decreasing with decreasing temperature, a sign of localisation. This style of treatment has been extended to cover magnetic field, spin-orbit and finite frequency effects, which also introduce length scales of their area which can interfere with the transport, or in the case of microstructures, the dimensionality scales. The last few paragraphs are a summary of the introduction to the major review in this field by LEE and RAMAKRISHNAN /14/, where the theoretical work is expanded much further and an overview of the experimental work is presented.

A whole different field of study in microstructures relates to the Aharonov-Bohm /15/ effect shown schematically with reference to Fig. 4. If a loop is small enough that electrons can preserve quantum phase memory when they take one side or the other, one might expect quantum interference if a magnetc field threads the loop, as an extra phase factor $\exp(\pm i\pi HA/\Phi_0)$ is introduced according to the sense of transit round the loop (H the magnetic field, A the loop area and $\Phi_0 = \hbar c/2e$ the quantum of magnetic flux. This field has been gathering momentum over recent years and is represented here in several lectures.

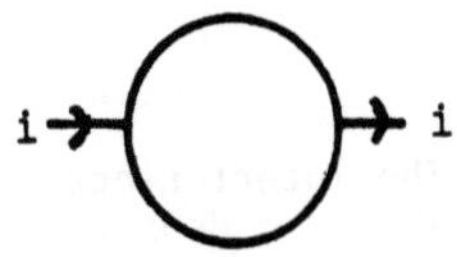

Fig. 4. The ring structure for the Aharonov-Bohm effect

The purpose of this section has been to contrast the concepts used in a conventional Boltzmann theory of transport and those used in the modern explanations of transport in microsystems.

4 TUNNELLING PHENOMENA

The transport in one-dimensional and two-dimensional structures has produced a wealth of new physics, particularly elucidating the important length scales involved in carrier transport. In multilayered semiconductors, after a slow start, recent results are starting to raise questions about the nature of tunnelling electrons. The tunnel barriers and superlattices are strictly three-dimensional structures, but until now most of the analysis is based on one-dimensional models. Some of the basic structures and ideas on tunnelling are discussed here: other lectures will treat the most recent developments.

The concept of tunnelling - the propagation of damped electron wavefunctions in regions that would preclude classical motion - is as old as quantum mechanics. One of the earliest references to its exploitation in solids is in Zener's theory of dielectric breakdown in solids (1933), based on electrons tunnelling across band-gap when a strong electric field is applied. Although avalanching mechanisms are now considered more important in many cases of breakdown, tunnelling remains an important ingredient of several electronic devices - the Esaki diode being a prominent example. If one considers a free electron of energy E moving (as a wave $\sim e^{i\underset{\sim}{k}\cdot\underset{\sim}{r}}$) in a potential V(z), the momentum is given by

$$\hbar\, k(z) = \hbar \sqrt{\frac{2m^*}{\hbar^2}\,(E-V(z))} \quad ,$$

provided V is smooth on the scale of $2\pi/k$. For $V>E$, k is imaginary and the wavefunction becomes exponential. In half-spaces only the damped modes (evanescent waves) are allowed, but normalisation of the wavefunctions is possible in layers of finite thickness allowing increasing wavefunctions as well. The phase integral $\int\ k(z)\,dz$ between the turning points $(z:E=V(z))$ determines exponentially the scale of transmission.

The simplest non-trivial semiconductor structure for the experimental study of tunnelling consists of two identical tunnel barriers separated by a narrow well, the reason being that a bound-state in the central well, if available to tunnelling electrons, can resonantly enhance the tunnelling coefficient. In the

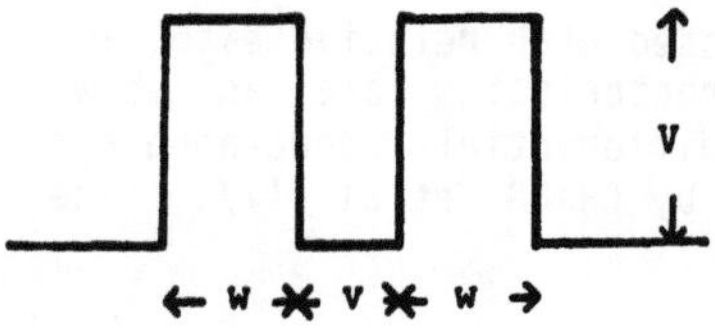

Fig. 5. A double barrier structure for resonant tunnelling

configuration shown in Fig. 5, the transmission amplitude t for an incident wavefunction of unit amplitude, is given by

$$t = \frac{\theta}{8i} \Big/ \left[\frac{1}{4\theta}(1+\theta^2)^2 \operatorname{Sin} kv + D1 + D2\right]$$

$$D1 = \frac{1}{16}\, e^{-2kw}\left[e^{ikv}\left(-\frac{1}{i\theta}+2i\theta+i\theta^3\right) + e^{-ikv}\left(\frac{1}{i\theta}+4+6i\theta-4\theta^2-i\theta^3\right)\right]$$

$$D2 = \frac{1}{16}\, e^{2kw}\left[e^{ikv}\left(\frac{1}{i\theta}-2i\theta-i\theta^3\right) + e^{-ikv}\left(-\frac{1}{i\theta}+4-6i\theta-4\theta^2+i\theta^3\right)\right] ,$$

where w is the thickness of each barrier, v the thickness of the intervening well, k the incoming electron momentum = $\sqrt{(2m^*E/\hbar^2)}$, K the damping constant in the barrier = $\sqrt{(2m^*(V-E)/\hbar^2)}$ and θ the ratio k/K. If the barriers are large, the e^{+2kw} term in D2 above suppresses the wavefunction amplitude, reflecting the total effective barrier thickness of 2w. The coefficient of e^{2kw} can be made to vanish under the condition

$$e^{2ikv} = \frac{-\frac{1}{i\theta}+4-6i\theta-4\theta^2+i\theta^3}{\left(-\frac{1}{i\theta}+2i\theta+i\theta^3\right)}$$

If the barrier is high, V and K are large and so θ is small. In this limit $e^{2ikv} \sim 1$, i.e. when $k = \pi/(2v)$ times an integer, D2 vanishes, and the tunnel amplitude is of order unity, i.e. as though the barriers were not there at all. A more general expression for different barrier parameters is given by RICCO and AZBEL /16/. Whereas off-resonance the current is proportional to the product of the two transmission coefficients, on-resonance it is proportional to the quotient of the lesser to the greater tunnelling coefficient of the two individual barriers. This resonant increase in tunnel probability is reflected in an enhanced tunnel current associated with electrons incident on the barrier with the appropriate energy. The tunnel resonance is narrow in energy, determined by Kw ~ 1, and the condition for D2 = 0 is essentially that for the energy of a 'bound' state within the well. This most simple picture has been used to explain the qualitative behaviour of resonant tunnel diodes manufactured by several groups. There are many complications in real systems, and these include (i) the need to place a bias across the structure to generate current which implies an alteration of the potential profile seen by electrons and (ii) the possible redistribution of the potential due to space charge build-up within the system in addition to that present due to doping in the contacts. Among interesting physics question are (i) what is the time-dependence of the tunnel current distribution, including transients, and the frequency performance of the total structure, (ii) what are the effects of magnetic fields or finite temperature and (iii) how do tunnelling electrons react to applied stress? Some of these questions are addressed below and in other lectures. In many early systems, the tunnel barriers were oxide layers, and excess current could be obtained by exploiting defect levels to provide a resonance condition. The advances in MBE growth have allowed the fabrication of structures that approximate the simple model - the layers being typically 3 - 10 nm thick, barriers ~< 5 nm, the voltage offset ~0.2 V (set by the alloy composition, in the $Al_xGa_{1-x}As/GaAs$ system) and with sufficient doping in the external layers to allow metallic conduction, i.e. to a Fermi energy <50 meV.

Under these conditions the resonant level can be accessed with relative ease, and the width is of order 50 meV and the I-V characteristics are as shown schematically in Fig. 6. The phenomenon of negative differential conductance has been seen by several groups since the first report by CHANG et al /17/. The simple analysis of the tunnelling current

$$J(V) = \frac{1}{4\pi^3} \int \frac{e k_z}{\hbar} \left(1 - |R|^2\right) \left(f(E) - f(E + V)\right)$$

(f the Fermi function and R the reflection coefficient of the double barrier structure) suggests that strong negative differential conductance is expected in the J(V) characteristics which have never been seen. Maximum peak-to-valley ratios of 7:1 at 4K have been reported /18/, but despite many variations of the barriers to take count of space charge etc., no greater ratios seem to be possible.

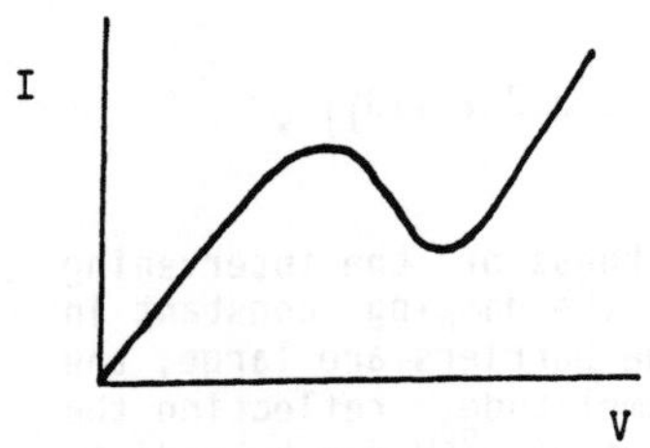

Fig. 6. Schematic I-V characteristics of a resonant tunnelling diode

The following arguments, due to RICCO and OLIVO /19/ indicate the reasons for less than ideal performance, based on space charge considerations. If one returns to the double barrier structure referred to above and solves the time-independent Schrödinger equation, the charge density is very great inside the central well, and a solution of Poisson's equation would imply a large redistribution of the potential drop and take the barrier well out of resonance. Instead of a theoretical charge Q_{TH}, a much smaller experimental charge Q_{EXP} can be placed in the well without destroying the resonance condition, and so the transmitted tunnelling current I_{EXP} is much smaller than theoretically expected. We can relate Q_{EXP} via the cathode barrier capacitance C to the voltage swing ΔV_{EXP} which in turn must be at most $\Delta E_{max}/q$ where ΔE_{max} is the tolerated energy shift in the quantum well still consistent with resonance. ΔE_{max} is essentially the width of the resonance level $\Delta E_{max} = h\nu T_{max}$ where T_{max} is the greater of the tunnelling coefficients and ν is the attempt-to-escape frequency ($\sim 10^{14}$ sec^{-1}, the natural frequency of the resonance level obtained from equating $\frac{1}{2} m^* \nu^2 v^2$ to the energy level of the well). The experimental current at resonance can be expressed as

$$I_{EXP} = Q_{EXP}\, \nu\, T_R = C\, \Delta V_{EXP}\, \nu\, T_{min} \quad \propto T_{max} \cdot T_{min}$$

(if we think of the barrier on the right being the faster barrier although the arguments go through for the reverse case). When off-resonance

$$I_{OFF} = Q'\, \nu'\, T_{min}\, T_{max} \quad ,$$

q' and ν' are the charge and attempt-to-escape frequency at the cathode. The ratio R of the tunnelling current at resonance to off-resonance is then

$$R = I_{EXP}/I_{OFF} = C\Delta E_{max}\, \nu/(Q' q\, \nu'\, T_{max}) \quad .$$

With $\nu \sim \nu'$ and $Q \sim CE_{RES}/q$ we achieve $R = h\nu/E_R$, and in all practical cases this ratio is less than 10 at most. These arguments reveal some of the practical limitations that must be considered in parallel with the simple tunnelling ideas when designing experiments.

A structure with many identical but narrow tunnel barriers separated by identical quantum wells provides a new system altogether. The wavefunctions from the bound states in adjacent quantum wells are coupled by tunnelling, and a simple Kronig-Penney model can be developed to describe a band structure intrinsic to a superlattice. In Fig. 7 we show the calculated structure for an example based on the AlGaAs/GaAs system. It consists of narrow 'minibands', separated by gaps. If a system is doped it is possible that the superlattice would act as a metal, particularly at low electric fields. The question of interest, particularly at low temperatures, concerns the nature of transport when a voltage greater than the miniband width is dropped across the superlattice. CHANG and ESAKI /20/ observed weak negative differential conductance (ndc) when a voltage equal to the miniband width was dropped over each period, but this work was never pursued.

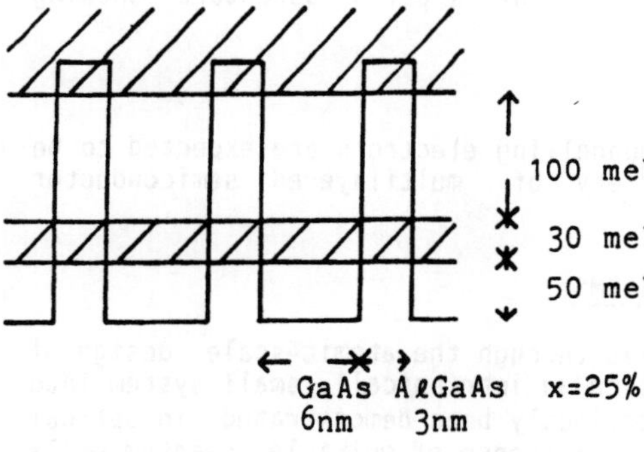

Fig. 7. The bandstructure of a simple superlattice structure

At GEC we have taken the superlattice idea one stage further by incorporating a thicker, tunnel, barrier in the middle of the superlattice /21/. Doping in the superlattice region was confined to the central 2 nm of each 6 nm GaAs well, and the Al content of each 3 nm barrier was 25%. The thicker barrier was 6 or 8 nm. The Fermi energy is in the middle of the first miniband, and to a first approximation the applied voltage is dropped across the thick barrier. The strong ndc can be explained by examining the line-up of the superlattice minibands on either side. Most recently we have observed room temperature ndc oscillations from the structure with a 6 nm central barrier. Our interpretation of the quenching of the ndc with temperature increase or at low temperatures when a magnetic field is applied perpendicular to the current, raises issues about the scattering of tunnelling electrons which have not been addressed in any of the existing theoretical formalisms.

The use of graded parameter superlattice, the CHIRP structure was first proposed by NAKAGAWA et al. /22/, and again at GEC /23/ we have examined a 5-period-graded-parameter superlattice in the i-regime of a p-i-n diode. In successive 10 nm periods, layers of 0.6, 1.2, 1.8, 2.4 and 3.0 nm thickness of AlAs are incorporated. Just inside the turn-on of a GaAs diode, we observe a strong peak due to resonant tunnelling when the local electric field just counteracts the rise in energy level in the increasingly narrow wells (Fig. 8). Preliminary results indicate that electrons use the indirect gap energy levels in the AlAs layers, so that the tunnel barrier of 0.32 not 1.2 eV is expected to reveal new aspects of the physics of tunnelling electrons.

One feature common to both of these structures, and most clearly in the recent work of CAPASSO et al /24/ is that scattering and energy loss processes are very efficient in many cases where electrons gain potential energy at heterojunctions. The sequential resonant tunnelling seen by Capasso relies on an electron relaxing from excited bound state levels in each and every well in transit through the superlattice. The idea of a Bloch oscillator involving an electron retaining quantum phase without scattering over several unit cells of a superlattice seems still to be beyond the material technology of today.

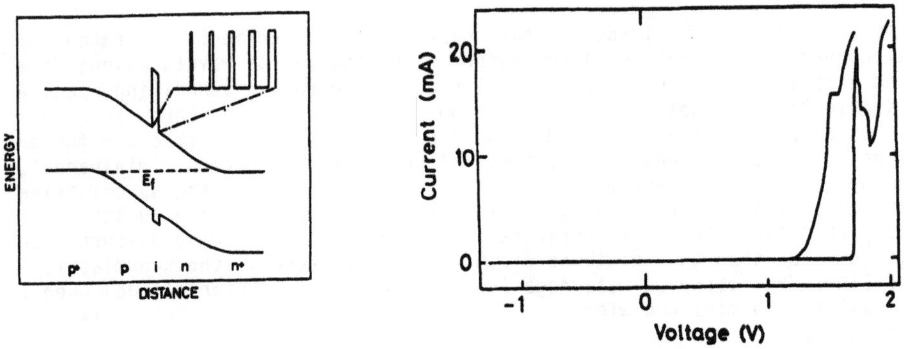

Fig. 8. The current-voltage characteristics of a p-i-n structure showing resonant tunnelling

Many further aspects of the physics of tunnelling electrons are expected to be revealed with studies of a wide variety of multilayered semiconductor structures.

5 OPTICAL PROPERTIES OF LOW DIMENSIONAL SYSTEMS

The ability (i) to tailor physical properties through the atomic-scale design of low-dimensional structures, and (ii) to excite an intrinsically small system into non-linear regimes of response, have most obviously been demonstrated in optical experiments. The highly non-linear optical response of multiple quantum wells is already showing device promise and the relevant physics will be discussed fully in other lectures. The emphasis here is on the quantum size phenomena and linear optical properties of quantum wells. For most absorbing materials, the optical response is dominated by the first micron in from the surface, where most epitaxially grown layers are to be found. With large dynamic ranges of optical absorption (as a function of wavelength), several optical features associated with quantum wells are relatively easily seen. The discussion in this lecture follows that of Weisbuch (cf. the reviews cited in Section 1).

The first point to be emphasised here in the interpretation of interband optical absorption data is that the optical matrix element

$$M = \int \phi_V(\underline{r})\, \varepsilon . \underline{r}\, \phi_C\, (\underline{r})\, d\underline{r}$$

coupling valence and conduction band states with light of polarisation ε, does not show quantum-size phenomena. The wavefunctions for electron and hole states consist of the product of envelope and Bloch functions. The integral, unit-cell by unit-cell, is essentially the same within a quantum well or bulk solid, and the envelope function preserves the normalisation in the case of a confined system. The changes in optical absorption are thus a manifestation of the energy levels and their energy distribution in quantum well systems.

A closely related point is the nature of selection rules. The dipole matrix element couples states of different parity. The symmetry of quantum wells allows us to assign a parity to the electron states in the wells. When the parity difference is taken up by the conduction and valence band Bloch wavefunctions (as in bulk material) the optical transitions are thus between electron and hole states whose envelope functions in the well have the same parity. This is a simple picture, and exceptions due to the finiteness of the confining wells, electric fields removing the well symmetry, and stress, all introduce complications that allow extra absorption features.

Among the earliest optical data on quantum wells is that of DINGLE and co-workers /25/ in 1974. The most striking feature is the strength of excitonic effects, but the general interband absorption clearly shows the two-dimensional confinement of the electron and hole levels via the flatness of the absorption spectra above the threshold, and the stepwise increase in absorption as the higher energy bound levels in the wells participate in the absorption process.

At low temperatures in bulk GaAs, one observes strong absorption with the formation of excitons, bound electron-hole pairs, many of whose properties scale from those of the hydrogen atom. The small binding energy of 4.2 meV results in a loss of this strong absorption at room temperatures. The solution of the Bohr theory of the hydrogen atom in two dimensions /25/ results in a binding energy four times greater, and a Bohr radius four times smaller, than in three dimensions. The three-dimensional exciton diameter is of order 20 nm, and the effect of exciton confinement within less thick wells is to increase the binding energy, and the thermal stability. Indeed, the room-temperature data not only show strong excitonic features, shifted in accordance with the quantum size shifts, but also a splitting corresponding to the heavy and light hole excitons. The precise energy shifts were used in the first attempts to determine one of the most important heterojunction parameters, the relative division of the band-gap discontinuity between the conduction and valence bands. Original results suggested an 85%/15% division for the GaAlAs/GaAs system but recent work from a variety of sources is now suggesting a consensus closer to 68%/32%. That a decade of research culminated over the last two years in a re-examination of this ratio, is evidence of the youth of this field of physics. The dynamics of excitons is an extensive subject which is treated separately in this School.

One important topic of research is photoluminescence /27/, especially at low temperatures, which is widely used as a non-destructive diagnostic technique of material quality, as well as for investigation impurity distributions (intentionally placed or otherwise). By recording the intensity versus frequency of light emitted from an optically excited system, or as in photoluminescence excitation spectroscopy examing the emission at a single frequency as the frequency of the optical excitation is varied, one can infer much about the structure and integrity of quantum wells. One important feature is the large quantum efficiency associated with photoluminescence that allows strong signals to be obtained from even very thin layers. There are several reasons including (i) the high efficiency of carrier collection in quantum wells at low temperature, so that even carriers in the barrier layer can be captured before recombining, (ii) the ease of exciton formation and the greater binding in confined layer, (iii) the nature of exciton luminescence, in which some of the restrictive k-selection rules in the bulk may be relaxed by the scattering from various fluctuations in the wells, and (iv) impurity gettering in the quantum wells. At present the excitation spectroscopy which probes the joint density of states often gives more useful information, and a detailed understanding of the photoluminescence line shape and shifts between luminescence and excitation spectra is not available at present. Indeed, a number of complications arises because of the ease with which non-linear optical response sets in. The gross photoluminescence spectra have been found a useful tool in the development of the materials integrity in all the semiconductor systems, with energy positions and line widths being related to well parameters and their fluctuations. In recent studies, spike-doping with donors placed in the centre or at the edge of wells have produced energy shifts in the luminescence spectra which relate to the nature of donor wavefunctions at the edge of wells /28/.

Inelastic light scattering is also providing new physics from quantum wells and heterojunctions /29/. In these experiments, one relies on a resonant enhancement of the efficiency, so that electron densities as low as 10^{11} cm^{-2} are able to be observed. Under differing conditions of the angle of incidence and the polarisation of both the incident and detected light, one can observe either a single-particle spectrum corresponding to uncorrelated particles or the collective modes of many particles. The former is seen because of spin-density excitation, the latter because of charge-density fluctuations. The polarisation dependence has been described in detail by BURSTEIN et al. /30/. Single particle excitations have been used to measure the energy levels in modulation - doped single heterojunctions, quantum wells, nipi's and superlattices, and early work on hole scattering helped start the re-examination of the band-gap offset issue. The studies of collective modes have revealed the energy of coupled LO-phonon-plasmon modes, the carrier densities in optically created plasmas and the 2D nature and 2D-3D transition in nipi structures. These studies have been extended to include magnetic fields, so that the Landau level spectra can be extracted.

We have noted that an electric field applied normally across a quantum well results in a shift in electron and hole energy levels, and therefore of optical excitation energies. In electroreflectance one monitors the principal optical features as the electric field is modulated, and this technique is providing sensitive data on well energy levels and the excitonic features. The fields can be applied either by doping profiles in the layer (as when used in a p-i-n configuration) or with a thin transparent Schottky gate at the surface /31/.

So far the pumping optical energies have been of the order of 1-2 eV, corresponding to excitations across the gap. On the milli-electron volt scale, far infrared light (~100 μm wavelength) can be used to probe intersubband energies directly. The technique was first used with MOSFET systems, where convenient wavelengths (~15 meV) would be brought into resonance with energy level separations by tuning the gate voltage. There is a lesser degree of control over energy levels within a heterojunction, and many experiments are now performed using magnetic fields to tune the energies. In the heterojunction systems it is easier to achieve a condition $\omega\tau > 1$ and to observe cyclotron resonance, with analyses of the line widths being used to give corroborative data on scattering times and mobilities. The far-infrared technique has also been used to study plasmon excitations in systems where a periodicity is imposed because the gate is replaced by a grating structue of ~ μm pitch. Modes with this wavelength are preferentially excited, and from a series of structures the excitation dispersion relations are mapped out /32/. Related techniques have also been used to extract the specific heat of the two-dimensional electron gas in MOSFETs and modulation-doped superlattices.

Within the theme of phenomena competing at the milli-electron volt scale, magnetophonon spectroscopy involves a resonance between a particular phonon mode (optic, or transverse, remote or local) and an integral number of cyclotron energies under which structure appears in the optical conductivity which in turn can be analysed to extract electron-phonon coupling strengths /33/.

Just as photoemission from cleaved surfaces has been a valuable diagnostic for the study of potential barriers, as in the case of work-functions and Schottky barriers, a form of internal photoemission, wherein the incidence of mid-infrared light excites electrons across the heterojunction potential barriers, has been studied in several laboratories to obtain band-gap offsets. While the internal photoemission works in principle, the extraction of clear data is made difficult by the corrections that must be incorporated to account for space-charge effects in different parts of the structure. Major research efforts are still not agreeing on a band-gap offset for the AlGaAs/GaAs system /34/.

REFERENCES

1 P.M. Petroff, A.C. Gossard, R.A. Logan, W. Weigmann: Appl. Phys. Letts. 41, 635 (1982)

2 M.A. Reed, R.T. Bate, K. Bradshaw, W.M. Duncan, W.R. Frensley, J. E. Lee, H.D. Smith: J. Vac. Sci. Techn. B4, 358-360 (1986)

3 H. Craighead, G.A. Niklasson: Appl. Phys. Lett. 44, 1134-6 (1984)

4 C. Smith, H. Ahmed, M.J. Kelly, M.N. Wybourne: Superlattices and Microstructures 1, 153-4 (1985)

5 K.L. Lee, H. Ahmed, M.J. Kelly, M.N. Wybourne: Elect. Letts. 20, 289-91 (1984)

6 I.G. Salisbury, R.S. Timset, S.D. Berger, C.J. Humphreys: Appl. Phys. Lett. 45, 1289-91 (1984)

7 S.P. Beaument: (1986, to be published) and S.P. Beaument, P.G. Bower, T. Tamamura, C.D.W. Wilkinson: Appl. Phys. Letts. 38, 436-8 (1981)

8 S.J. Allen, H.L. Stormer, J.C.M. Huang: Phys. Rev. B28, 4875 (1983)

9 N.W. Ashcroft, N.D. Mermin: Solid State Physics. (Holt Rinehart and Winston 1976)

10 T. Ando, A.B. Fowler, F. Stern: Rev. Mod. Phys. 54, 437-625 (1982)

11 J.M. Luttinger, W. Kohn: Phys. Rev. 97, 869-83 (1955)

12 O. Madelung: An Introduction to Solid State Theory. (Springer-Verlag 1978)

13 D. Thouless: Phys. Rev. Lett. 39, 1167 (1977)

14 P. Lee, T. Ramakrishnan: Rev. Mod. Phys. 57, 287 (1985)

15 R. Webb, S. Washborn, C. Umbach, R. Laibowitz: Phys. Rev. Lett. 54, 2696 (1985)

16 B. Ricco, M.Ya. Azbel: Phys. Rev. B29, 1970 (1984)

17 L.L. Chang, L. Esaki, R. Tsu, Appl. Phys. Lett. 24, 593 (1974)

18 T.C.L.G. Sollner, W.D. Goodhue, P.E. Tammenwald, C.D. Parker, D.D. Peck: Appl. Phys. Lett. 43, 588 (1983

19 B. Ricco, P. Olivo: (Private communication)

20 L.L. Chang, L. Esaki: Phys. Rev. Lett. 33, 495 (1974)

21 R.A. Davies, M.J. Kelly, T.M. Kerr: Phys. Rev. Lett. 55, 1114 (1985) and Elect. Letts. 22, 131 (1986)

22 T. Nakagawa, H. Imamoto, T. Sukamoto, T. Kojima, K. Ohta, N.J. Kawai: Elect. Letts. 21, 882-884 (1985)

23 N.R. Couch, M.J. Kelly, T.M. Kerr, E. Britton, W.M. Stobbs: (1986) to be published

24 F. Capasso, K. Mohammed, A.Y. Clo: Appl. Phys. Lett. 48, 478 (1986)

25 R. Dingle, W. Wiegmann, C.H. Henry: Phys. Rev. Lett. 33, 827 (1974)

26 M. Shinada, S. Sugano: J. Phys. Soc. Jap. 21, 1936 (1966)

27 R.C. Miller, D.A. Kleinmann: J. of Luminescence 30, 520 (1985)

28 B.V. Shanabrook: Proc. of EP2D5 VI, Surf. Sci. 1986 (to appear)

29 G. Abstreiter, M. Cardonna, A. Pinczuk: in Topics in Applied Physics Vol 54, Light Scattering in Solids IV, editors M. Cardona and G. Gunterhodt, Springer, (1984)

30 E. Burstein, A. Pinczuk, D.L. Mills: Surf. Sci. 98, 451 (1980)

31 C. Alibert, F. Jiahua, M. Erman, P. Frijlink, P. Jarry, J.B. Theeten: Revue Phys. Appl. 18, 709 (1983), P.C. Klipstein, P.R. Tapster, N. Apsley, D.A. Anderson, M.S. Skolnick, T.M. Kerr, K. Woodbridge: J. Phys. C 19, 857 (1986)

32 R. Hopfel, E. Gornik, Surf. Sci. 142, 412 (1984)

33 R.J. Nicholas, M.A. Brummell, J.C. Portal: in Two-Dimensional Systems, Heterostructures and Superlattices, editors G. Bauer, F. Kuchar and H. Heinrich, (Springer 1984) p69

34 c.f. G. Abstreiter et al, M. Heilbum et al: in the Proc. of the Second International Conference on Modulated Semiconductor Structures. Surface Science 1986 (to appear)

Electron States in Semiconductor Microstructures

M. Jaros

Department of Theoretical Physics, The University,
Newcastle upon Tyne, NE1 7RU, United Kingdom

1. Microscopic Theory of Confinement in GaAs-$Ga_{1-x}Al_xAs$ Superlattices

The particle-in-a-box model of confinement pictured in Fig.1 accounts for the presence and approximate positions of levels in GaAs wells confined by $Ga_{1-x}Al_xAs$ ($x < 0.35$). The only parameters required are the effective mass derived from the relevant band edge, the well width and the barrier height. In this model, it is assumed that the rapidly varying component $u_{n,\vec{k}}(\vec{r})$ of the total wave function does not change at the interfaces, i.e. that near the band edges the Bloch functions associated with a given band n and wave vector $\vec{k}$ are the same in GaAs and $Ga_{1-x}Al_xAs$. Although a number of corrections has been introduced to improve the accuracy and scope of predictions based on this simple picture, it is clear that at least further from the band edges, and in structures consisting of more dissimilar materials, the above approximation will break down.

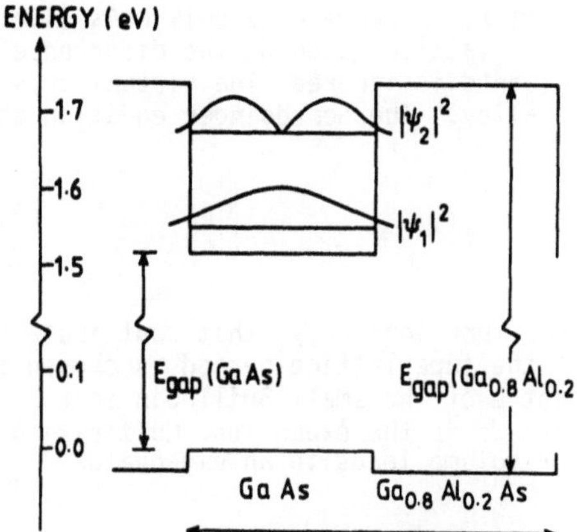

Fig.1. A sketch of the band offsets for a GaAs-$Ga_{0.8}Al_{0.2}As$ quantum well superlattice. E_g is the bulk band gap at the centre of the Brillouin zone. The envelope functions (F_1,F_2) associated with the lowest confined states derived from the bottom of the conduction band of GaAs are also included. Similar levels and envelopes are expected at the top of the valence bands for both heavy and light hole bands

Most of the sophisticated techniques employed during the last decade to model small structures (see a survey by ANDO, FOWLER and STERN [1]) exploit the abrupt nature of the potential at the interface, and the perfect regularity of well lattice matched layers, and become rather cumbersome when such conditions are not applicable. It is also apparent that, in order to provide a microscopic picture of the confinement effect, a detailed picture of the microscopic potential over the full length of the superlattice period must be retained in the theory. Unfortunately a priori supercell calculations, which do retail full flexibility as far as the choice of structure in question is concerned, must generate the 'band structure' of a superlattice on an absolute scale and fail to reproduce even some of the basic features (e.g. band gaps) correctly.

In an effort to provide accurate predictions for a wide variety of real structures, a self-consistent scheme has recently been developed [2] which combines the advantages of the accurate relative scale of energies (peculiar to the state-of-the-art envelope function effective mass calculations [3]) with the a priori self-consistent character of a supercell calculation.

Although the reader may not be impressed by the quantum mechanics leading to (3), the result of the calculations based on (3) will not be difficult to appreciate. It transpires that in addition to the familiar levels lying deep in the confining wells, the full scale calculations reveal a wealth of new states which lie above the confining barriers and whose wave functions are strongly confined either in the GaAs wells or in the barrier.

In the Newcastle method one begins by choosing a suitable bulk crystal potential for, say, GaAs such that the corresponding bulk wave functions $\phi_{n,\vec{k}}$ and band structure energies $E_{n,\vec{k}}$ can be generated in good agreement with experiment. The wave function Ψ of a superlattice system is then expanded in terms of the complete set of $\phi_{n,\vec{k}}$. The superlattice Hamiltonian is $\hat{H} = \hat{H} + \hat{V}$ where $\hat{V}$ is the difference between the potential of the atoms in the barrier material layers and that of GaAs in those layers, e.g.

$$\hat{V} = \sum_j \{\hat{V}_j(Ga_{1-x}Al_xAs) - \hat{V}_j(GaAs)\} \quad , \tag{1}$$

where j runs through all atomic sites in the barrier. $\hat{H}_0$ represents bulk GaAs (i.e. $\hat{H}_0\phi_{n,\vec{k}} = E_{n,\vec{k}}\phi_{n,\vec{k}}$). Since the two systems are well lattice matched, the difference between the lattice constants of the two constituents is ignored. The virtual crystal approximation is used in order to model the alloy. The Schrödinger equation to be solved is

$$(\hat{H}_0 + \hat{V} - E) \sum_{n,\vec{k}_i} A_{n,\vec{k}_i} \phi_{n,\vec{k}_i} = 0 \quad . \tag{2}$$

Since $\hat{V}$ in (2) is a periodic potential, the Bloch functions $\phi_{n,\vec{k}_i}$ that contribute to ψ are unambiguously determined. For example, if the superlattice period is chosen to be (70 + 70) (Å) along the <001> axis, at the centre of the small Brillouin zone, there are 51 Bloch states for each band n to include in the expansion. Multiplying (2) from the left by $\phi^*_{n',\vec{k}_j}$ and integrating over volume leads to an eigenvalue equation

$$A_{n'\vec{k}_j}(E_{n',\vec{k}_j} - E) + \sum A_{n,\vec{k}_i} \int_{\text{unit cell}} \phi^*_{n',\vec{k}_j} \hat{V} \phi_{n,\vec{k}_i} d^3\vec{r} = 0 \quad . \tag{3}$$

This problem is familiar from earlier work on impurities and dislocations [4-6], and is solved by direct diagonalisation. The energy spectrum obtained for a GaAs-$Ga_{0.8}Al_{0.2}As$ superlattice at the centre of the small Brillouin zone (i.e at Γ), is shown in Fig.2. The electron charge densities associated with the lowest conduction and valence band states shown in Fig.2, averaged over the unit cell in y and x directions, are plotted in Fig.3. We observe that the ground state (state 1') wave function is largely confined to the GaAs well and dominated by $\phi_{n,\vec{k}}$ from the Γ minimum, in agreement with previous calculations and experiment. However, state 3' has a large amplitude outside GaAs. The higher states 4'-5' are dominated by the Bloch states from the X minima. No admixture of $\phi_{n,\vec{k}}$ from the L valley is involved in the formation of any of the localised states shown in Fig.3 as can be demonstrated from the symmetry properties of $\phi_{n,\vec{k}}$.

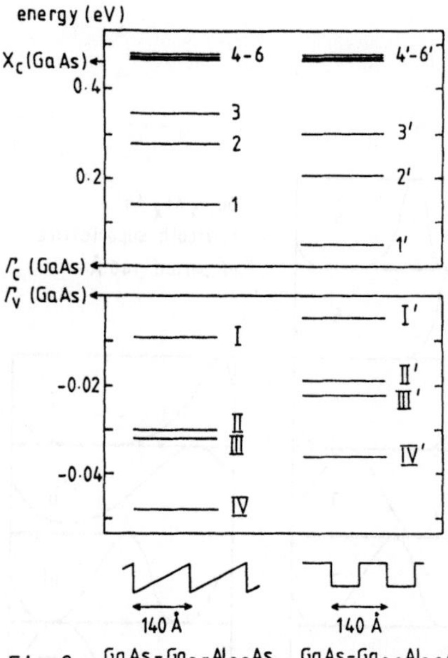

Fig.2

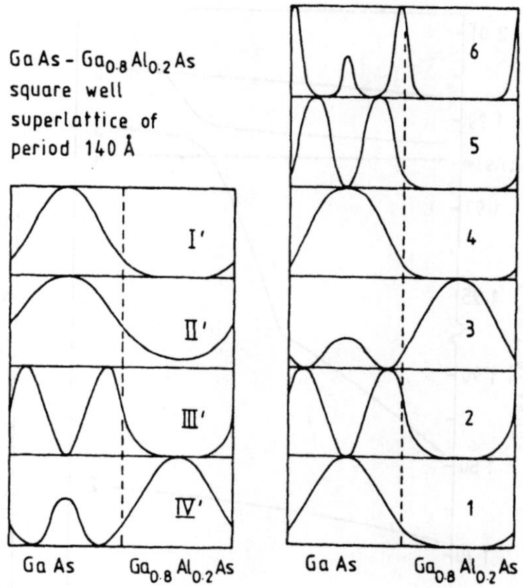

Fig.3

Fig.2. The superlattice states computed at the centre of the Brillouin zone. On the left, state 1-6 and I-IV of the sawtooth GaAs-$Ga_{1-x}Al_xAs$ structure. On the right, states 1'-6' and I'-IV' of the quantum well structure GaAs-$Ga_{0.8}Al_{0.2}As$. In both cases, the superlattice period is ~ 140 Å, along <001> direction. The conduction band states are measured in eV from the conduction band edge of GaAs at Γ_c. The valence band states are measured in eV from the valence band of GaAs at Γ_v [11]. In this calculation the ratio of the conduction-to-valence band offset was taken to be 85:15

Fig.3. The charge densities of states of the square well superlattice reported in Fig.2, averaged over the unit cell in the y and x directions, and plotted along the superlattice axis (z) [11]

The existence of state 3' has been confirmed experimentally [7]. As the aluminium fraction is increased, state 3' is pushed towards the GaAs X valley, crossing the first X-related state 4' at $x = 0.42$ (Fig.4). This behaviour can be easily understood on intuitive grounds. With the increase of the aluminium concentration, the conduction band minimum of $Ga_{1-x}Al_xAs$ rises in energy. The wave function of state 3' can be thought of as the ground state of an inverted well, since it peaks in the alloy layer. Therefore when the band edge of the alloy rises in energy, it carries state 3' with it.

The states localised above the confining barriers are not peculiar to the conduction band. It is clear from Fig.3 that similar states occur in the valence band [8,9]. Since the quantum states above the barrier are sufficiently localised, they alter the overall picture of the density of states. The existence of these states may help in understanding the real space charge transfer and hot electron transport phenomena in superlattice structures.

Very recently, CAPASSO et al. [10] drew attention to some interesting applications of graded-gap structures such as sawtooth superlattices of $Ga_{1-x}Al_xAs$. In Figs.2 and 5 the electronic structure of a (001) sawtooth superlattice of $Ga_{1-x}Al_xAs$, $0 < x < 0.3$, of period 140 Å is compared with the electronic structure of a conventional quantum well GaAs-$Ga_{0.8}Al_{0.2}As$ superlattice of the same period [11]. Although there are clearly distinguishable systematic differences, it is obvious that the signature of the square well states is retained in the sawtooth system. The same can be said about the width of the corresponding minibands which is

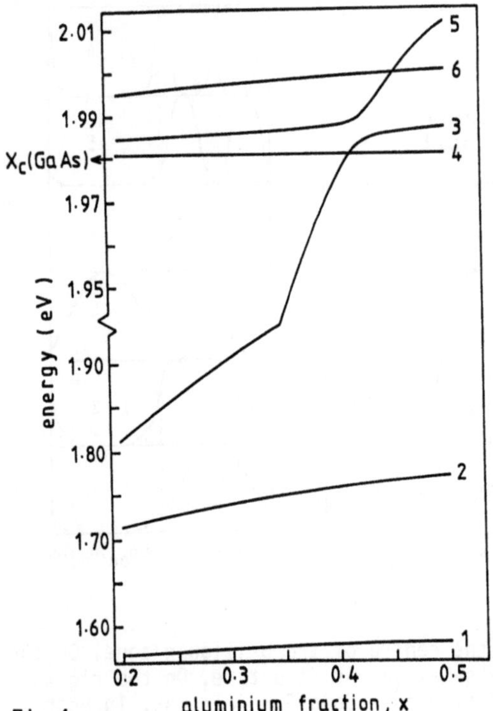

Fig.4

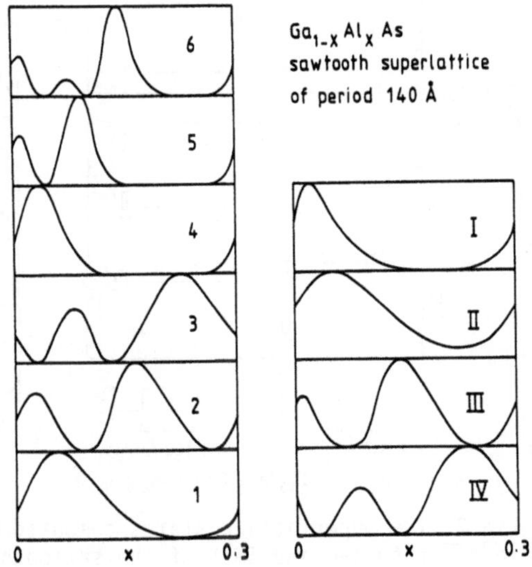

Fig.5

Fig.4. Energies calculated as a function of the aluminium fraction x. Curve 3' is the state dominated by the Γ contributions, whereas curves 4', 5' and 6' are the X-related states. X_C is the X-minimum of bulk GaAs. The zero of energy is the top of the valence band of GaAs. Note the anticrossing behaviour of states 3' and 5'. The kink at $x = 0.35$ is due to a change in the energy scale [18]

Fig.5. A summary of the electron charge densities associated with the sawtooth superlattice states reported in Fig.2. $|\psi|^2$ averaged over the interface plane is plotted along the superlattice axis in the <001> direction. The peak of $|\psi|^2$ is set to 1 to facilitate presentation

considerable for the higher states but insufficient to remove gaps. As expected on intuitive grounds the ground states of the structures described in Figs.2-5 are dispersionless.

2. Limits of Applicability of the Effective Mass Approximation

The argument given in the introductory paragraph of Sect.1 in order to justify the need for microscopic theory also spells out the most obvious limitation of the particle-in-a-box model. In this model the problem of finding the confined levels is solved by retaining only the slowly varying envelope component of the total wave function; this is the so-called effective mass approximation. Hence, by analogy with the case of chemical impurities in semiconductors, one might anticipate another limiting circumstance, namely that the model breaks down when the range of the wave function of the confined state in question becomes comparable to the host crystal (bulk) lattice constant. This means that even in those structures where the band minima of the constituent materials closely resemble each other (e.g. the situation pictured in Fig.1) the effective mass approximation is suspect for narrow wells. Such reasoning has been commonly accepted in the literature on the subject. In fact full scale calculations such as those described in Sect.1 show that although the particle-in-a-box model cannot provide an overall description of the electronic structure of narrow quantum wells, some important predictions of this model are preserved even for wells formed by a couple of monolayers! The analogy with the impurity case is misleading because in that case the increase in the wave function localisation and consequently the breakdown of the effective mass approximation is brought about by a strong (deep and highly localised) central cell potential at the impurity site. Whether one takes ten monolayers of GaAs or just one, the difference between the Ga and Al atoms that are being exchanged [one can think of $\hat{V}$ in (1) as an impurity potential obtained by replacing Ga atoms with Al to generate the barrier layers out of GaAs] is not very much affected. Indeed, a single Al impurity in GaAs is an isoelectronic defect whose electronic properties can be successfully treated within the effective mass approximation! Hence the superlattice potential $\hat{V}$ of (1) is never deep enough to mix effectively waves of very different wavelengths unless the corresponding bulk states are nearly degenerate in energy. Without this mixing the wave function remains localised in the wave vector space and the effective mass approximation survives. Of course, it is usually only the ground state at the valence band edge (the so-called heavy hole level) that remains amenable to the effective mass treatment even in ultrathin wells. This is because the Bloch functions at the top of the valence band of zinc-blende crystals run perpendicular to the superlattice axis along the (001) direction and the effect of the superlattice potential $\hat{V}$ is particularly weak; the states lying further from the bottom of the well are less well confined and are more likely to delocalise in the wave vector space as the well gets narrower. In wider wells, the validity of the effective mass approximation also deteriorates as one moves further from the ground state and it is this energy dependence of the corrections to the simple model that makes comparisons between experiment and theory so difficult. Indeed, much of the controversy generated by attempts to fit band offsets from optical spectra may well be reducible to this deficiency of the particle-in-a-box model.

3. The Effect of an External Electric Field

Previous theoretical calculations [12] of electric field effects in quantum well systems have treated the effect of the external field on the electronic structure by approximate techniques (in particular the variation method) appropriate in the weak-field limit. Such calculations correctly predict the quadratic variation of the Stark shift with field at moderate field intensities. In these calculations the confined states are viewed as bound states and their field-induced broadening is ignored. Consider the ground state of a single quantum well in an electric field.

In the envelope function approximation the problem reduces to the one-dimensional effective mass equation. For a particle of charge e and mass m*, initially localised in an isolated well of width 2a and depth V_0, the Schrödinger equation is

$$\frac{-h^2}{2m^*}\frac{d^2F}{dz^2} - (V_0 + efz)F = EF; \quad |z| < a$$

$$\frac{-h^2}{2m^*}\frac{d^2F}{dz^2} - efzF = EF; \quad |z| > a \tag{4}$$

where f is the electric field applied perpendicular to the interface plane. The exact solutions of (4) can be obtained as linear combinations of Airy functions; application of the method of phase shift analysis to the solutions enables the Stark resonance positions and widths to be calculated [13]. It was found that for moderate fields both the variational and exact calculations predict a quadratic dependence of the Stark shift on the field; this shift can be explained by perturbation theory. The quadratic Stark shift of the ground state $|0\rangle$ of a quantum system is given by

$$\Delta E = e^2f^2 \sum_{k\neq 0} \frac{|\langle k|z|0\rangle|^2}{E_0 - E_k} \tag{5}$$

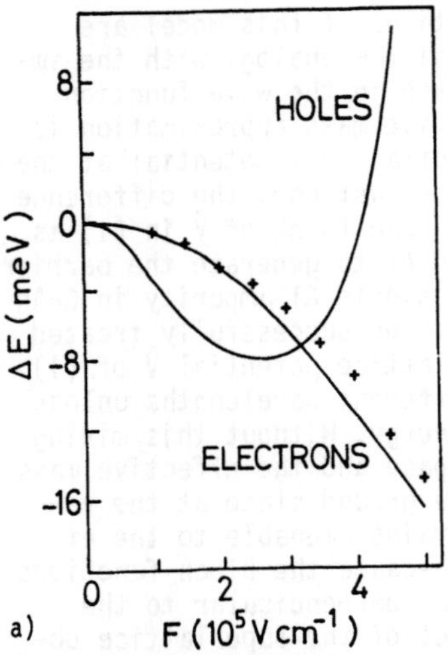

a)

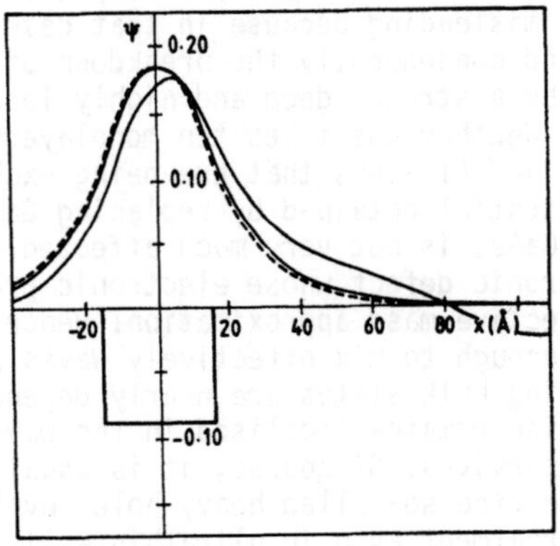

b)

Fig.6. (a) Calculated energy shifts for the electron and hole ground states. The crosses are the shifts obtained by BASTARD et al. [12], c.f. [13]. (b) Normalized hole wavefunctions for a 30 Å well with V_0 = 70 meV for $f = 2 \times 10^4$ V cm^{-1} and 8×10^4 V cm^{-1}. The zero field wavefunction is included for comparison (dotted line): the location of the well is indicated

where the states |k> are the excited states of the system. An increase in the well depth increases the binding energy E and thus decreases ΔE; an increase in the well width increases the spatial extent of the ground state wave function and thus the matrix elements <k|z|0>. At high fields the variational bound state formalism breaks down and the predictions of the two calculations differ considerably. The exact calculation predicts a decrease in the effective energy shift at high fields and the emergence of new antiresonance states from the continuum into the well.

The result is shown in Figs.6a and b for a single well. If the barrier is shallow, in fields above 10^5 V/cm the induced energy shift reaches a minimum and subsequently begins to decrease, thereby reducing the binding energy. In this treatment, the energy levels are no longer bound states. Instead, we have a resonance characterised by a density of states curve of finite width which rapidly increases with increasing external field (Fig.7).

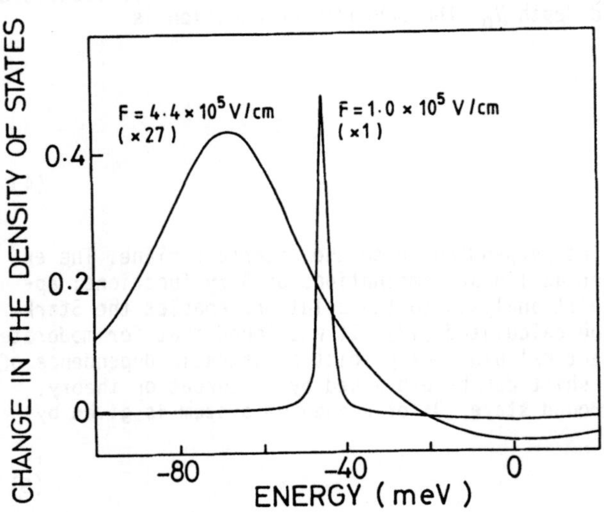

Fig.7. Plots of the change in the density of states for holes at various field intensities (in units of 10^5 V/cm). The peak height decreases with increasing field and the curves have been scaled accordingly

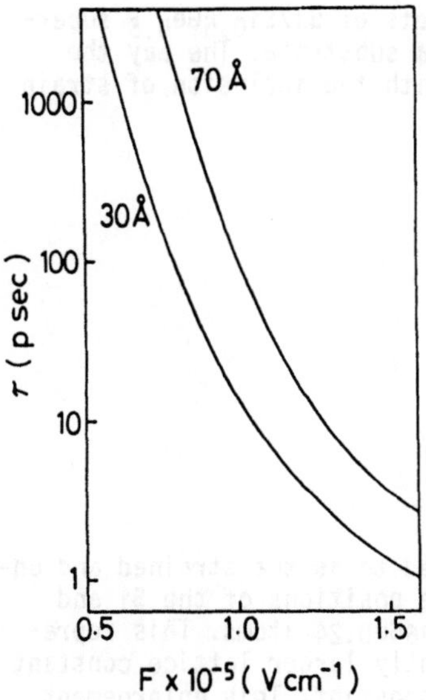

Fig.8. Calculated tunnelling lifetimes as a function of applied field f for holes in 30 and 70 Å GaAs wells. The same paramters were used as in the calculation reported above

From the resonance width it is possible to compute the corresponding lifetime which determines the rate of field-induced tunnelling. The result is shown in Fig.8 for two well widths. It can be seen that the field-induced tunnelling increases dramatically with increasing field and decreasing well width. It is possible to apply these results to experimental data concerning luminescence.

4. Strain-Induced Levels in $Si\text{-}Si_{1-x}Ge_x$ Superlattices

It has been demonstrated that structures made of lattice mismatched materials (i.e with differences of the constituent lattice constants >0.1%) can be grown without a significant number of misfit defects at interfaces provided that the layers are sufficiently thin (~100 Å). The mismatch is accomodated by uniform strain. Calculations [14] show that the strain field should be treated on an equal footing with the microscopic potential. This leads to a qualitatively different situation. One of the candidates for applications is the $Si\text{-}Si_{1-x}Ge_x$ superlattice.

Following the pioneering work of BEAN et al. [15], high-quality pseudomorphic growth of $Si\text{-}Si_xGe_{1-x}$ multilayer structures has been achieved by molecular beam epitaxy. Much of the interest in this system stems from the possibility of confining electrons in the silicon layers and exploiting the concept of modulation doping in order to increase mobility. Another interesting opportunity peculiar to this system is offered by the effect of zone folding which may enhance the optical matrix element across the fundamental superlattice gap. In both applications the key condition for successful implementation is the availability of confined electron states in the silicon layers. Unfortunately, the samples with $Si\text{-}Si_xGe_{1-x}$ superlattices grown on Si(001) substrates did not show any confinement in the silicon layers. In fact, Raman measurements implied that electrons are confined in the alloy layers [16]. However, very recently, ABSTREITER et al. [17] reported observations on an $Si\text{-}Si_{0.5}Ge_{0.5}$ superlattice grown on $Si_{0.75}Ge_{0.25}$ which were consistent with the existence of electron confinement in the silicon layers.

In this structure the substrate is chosen so as to match the lattice constant of the superlattice period and the strain is taken up by both constituent layers (compressive strain in $Si_{0.5}Ge_{0.5}$ and tensile strain in Si). If this system were not strained, it is expected that the band offsets would favour confinement of electrons in the alloy as shown schematically in Fig.9. However, when the strain is included the picture might be quite different.

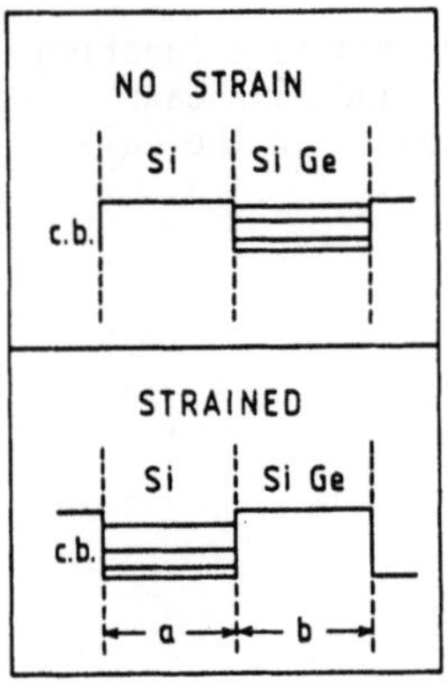

Fig.9. A sketch of the band offsets of $Si/Si_{0.5}Ge_{0.5}$ superlattices grown on an $Si_{0.76}Ge_{0.24}$ substrate. The way the offsets are expected to change with the inclusion of strain is shown

Two calculations were performed which will be referred to as the strained and unstrained calculations. In the unstrained calculation the positions of the Si and $Si_{0.5}Ge_{0.5}$ atoms are at the same sites as the bulk $Si_{0.76}Ge_{0.24}$ atoms. This represents a superlattice made from bulk silicon with a slightly larger lattice constant and bulk $Si_{0.5}Ge_{0.5}$ with a slightly compressed lattice constant. This enlargement and compression of the lattice constants is the same in all directions. In the strained calculation the positions of the silicon atoms and $Si_{0.5}Ge_{0.5}$ atoms were shifted from the bulk lattice sites of $Si_{0.76}Ge_{0.24}$ in the <001> direction to change the lattice spacing between adjacent atoms to that of the actual lattice constants of the two respective materials, the positions of the atoms in directions perpendicular to (001) being left unchanged. Thus a realistic model of a system with compressive strain in $Si_{0.5}Ge_{0.5}$ and tensile strain in Si is achieved.

The results of these calculations are summarised in Figs.10 and 11. In the unstrained case one obtains well-confined electron states in the alloy layers. The effect of strain is to shift the energies of the confined states at the conduction band edge and to localise the electron levels in the silicon layers. This confirms the intuitive view of the effect the strain is expected to have on band gaps of the constituents.

5. Optical Spectra and the Effect of Zone Folding in $GaAs-Ga_{1-x}Al_xAs$ and $Si-Si_{1-x}Ge_x$ Superlattices

Typical optical spectra of $GaAs-Ga_{1-x}Al_xAs$ superlattices obtained in luminescence experiments are shown in Fig.12. The transition energies obtained from calculations described in Sect.1 are also shown. The comparison indicates that a good agreement is obtained. It is also noticeable (Table 1) that in addition to the usual transitions involving the states lying deep in the confining wells, strong transitions are predicted for resonances at higher energies. The states contributing to these spectra are dominated by wave vectors associated with principal band edges at the centre of the Brillouin zone at Γ. The transitions involving X- or L-like conduction band states are forbidden for the same reason as in bulk materials. This is because in this structure the superlattice potential is not strong enough to mix states across the bulk Brillouin zone and the confined states are highly localised in the wave vector space at the nearest band extremum. However, as the Γ- and X-like levels get closer in energy, this mixing or zone folding which changes the nodal character of the rapidly varying Bloch component of the total wave function may become much stronger. In Fig.13, one can see a quantitative assessment of this effect for $GaAs-Ga_{1-x}Al_xAs$ superlattices whose energy levels and wave functions were discussed in Sect.1 [18].

The curves labelled 1' and 2' correspond to the Γ-like confined states lying below the top of the conventional confining barriers, curve 3' is a Γ-like resonance, and 4', 5' and 6' are X-related resonances, i.e. superlattice states lying above the

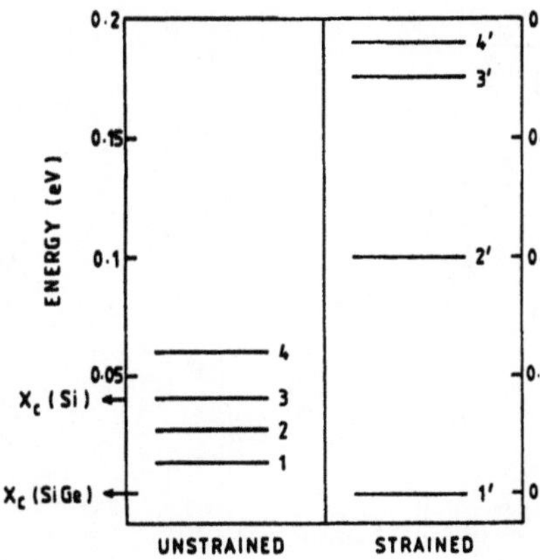

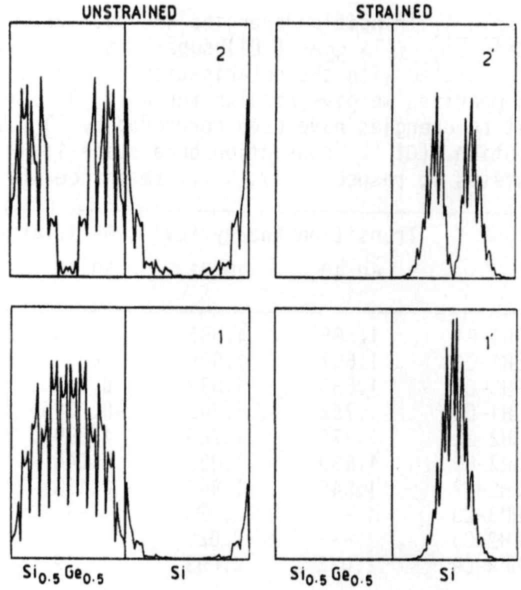

Fig.10

Fig.11

Fig.10. The energy levels of the superlattice states at the centre of the Brillouin zone obtained in the strained and unstrained calculations. All energies are measured from the bottom of the conduction band well in the unstrained case. The top of the barrier in the unstrained case is also shown

Fig.11. (a) The charge densities of the lowest confined (electron) states in the unstrained case calculated along the <001> direction through bond centres, and showing the position of the interface. (b) The charge densities of the lowest confined (electron) states in the strained case calculated along the <001> direction through bond centres, and showing the position of the interface. The actual sizes of the two systems are as follows: In the unstrained case the width of the Si and $Si_{0.5}Ge_{0.5}$ are (see Fig.9) a = 71.3 Å, b = 65.8 Å. In the strained case we have a = 70.6 Å, b = 66.5 Å. The overall period of the superlattice is the same in both cases (=a + b) and equals 137.1 Å. These parameters come from a system which comprises 52 silicon atoms and 48 $Si_{0.5}Ge_{0.5}$ atoms in the superlattice unit cell [19]

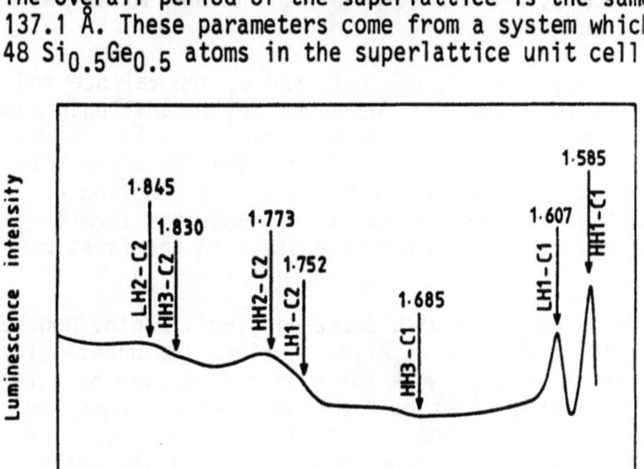

Fig.12. Photoluminescence spectra at 5 K for sample No 161 taken from DAWSON et al. [22]. The transition energies have been calculated for a 67 - 175 Å GaAs-$Ga_{0.64}Al_{0.36}As$ (001) superlattice with 60 : 40 band offset. The experimental value of the well width is 62 Å. See also Table 1 for the corresponding oscillator strengths and for a comparison with calculations based on 79 : 21 offsets

Table 1. Transition energies and oscillator strengths for a 67 + 175 Å GaAs-$Ga_{0.64}Al_{0.36}As$ (001) superlattice. The oscillator strengths have been calculated with the polarisation vector in the interface (X,Y) plane. For comparison we give results for both 60:40 and 79:21 offsets. All the transition energies have been corrected by 10 meV to allow for an exciton contribution. (C1 ... conduction band state 1; HH1, LH1 ... heavy and light hole states 1, respectively; R ... resonance state, etc.)

	Transition Energy [eV]		Oscillator Strengths	
	60:40	79:21	60:40	79:21
HH1-C1	1.585	1.587	5.428	5.512
LH1-C1	1.607	1.605	2.060	1.960
HH3-C1	1.685	1.672	0.057	0.011
LH1-C2	1.752	1.768	0.0045	0.0023
HH2-C2	1.773	1.785	4.089	4.456
HH3-C2	1.830	1.835	0.045	0.070
LH2-C2	1.845	1.840	2.138	2.343
HH3-C3	--	2.023	--	3.225
LH2-C3	--	2.028	--	0.108
HHR-CR	2.077	2.093	3.392	3.879

X-minimum of the GaAs band structure and made up of bulk X-like Bloch states. As the aluminium fraction is increased, state 3' is pushed towards the GaAs X-valley, crossing the first X-related state 4' at $x = 0.42$. The state 3', after crossing state 4', does not cross state 5', the first excited X-related state. This means that in the interval 0.4 - 0.45 the wave functions corresponding to states 3' and 5' mixed. This effect can be seen directly by inspecting the evolution of the nodes of the charge densities. In order to provide more insight into the degree of this mixing, we show in Fig.14 the modulus squared of the $A_{n,\vec{k}}$ coefficients corresponding to state 3' as a function of k along the Γ-X-line in the Brillouin zone of bulk GaAs. For simplicity, only contributions from the first GaAs conduction band are indicated. This figure shows that state 3' is dominated by the Γ-valley before the crossing ($x = 0.4$) and by the X-valley after the crossing ($x = 0.45$), whereas at points of maximum mixing with state 5' it takes contributions from both Γ and X.

The optical matrix elements $T_{i,j} = |\langle\psi_i|\hat{P}|\psi_j\rangle|^2$, where ψ_j and ψ_i are valence and conduction wave functions generated at the centre of the superlattice Brillouin zone and where $\hat{P}$ is the momentum operator whose direction has been chosen to lie in the X,Y (interface) plane, are shown in Fig.13 for transitions from the top heavy-hole state to state 3' (curve $M_{I',3'}$) and to state 5' (curve $M_{I',5'}$) as a function of composition x. We have chosen as unit of the squared matrix elements the term $T_{I',1'}$ which corresponds to the transition from the top heavy-hole state to the first conduction state (see inset in Fig.13).

Naturally, one expects the mixing to increase with decreasing well width. However, unless the width of both the GaAs wells and the $Ga_{1-x}Al_xAs$ barriers are drastically reduced, the mixing across the Brillouin zone is weak for states separated by a respectable energy interval. This relationship might be expected to change when one comes to consider strained superlattices. The effect of strain must be to enhance the mixing in momentum space to accommodate the change in the phase of the total wave function at the interfaces where there is a discontinuity in the lattice separation. In the calculation described in Sects.1 and 4 the composition of the superlattice wave function in the wave vector space is obtained automatically since the solution is given in terms of the expansion coefficients $A_{n,\vec{k}}$. In the case of Si or SiGe crystals the conduction band states at the bottom of the band are X-like. The magnitude of $A_{n,\vec{k}}$ with $k = 0$ therefore reflects the strength of the combined effect of the superlattice potential and strain which changes the character of the wave function. Since the hole state wave function is highly localised in momentum space

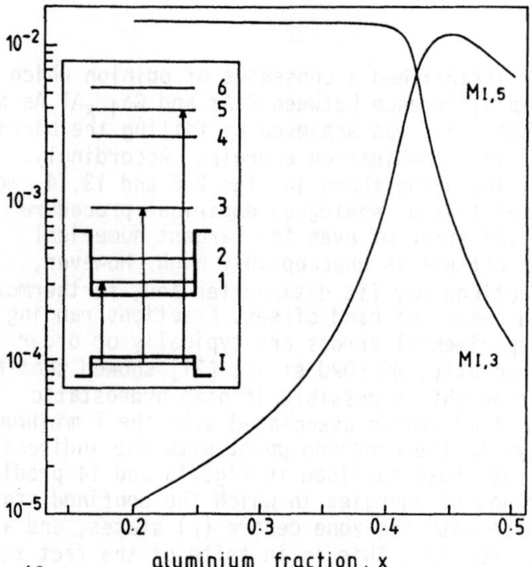

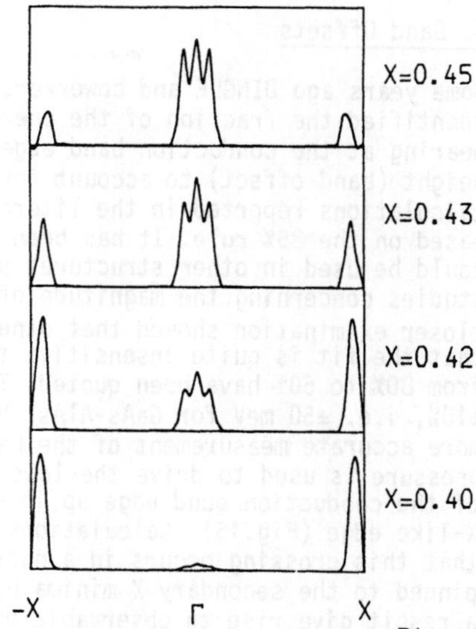

Fig.13 Fig.14

Fig.13. Squared optical matrix elements as a function of the aluminium fraction with light polarised in the interface plane. The inset contains a sketch of the relevant levels and transitions considered in this work. The conventional picture of a confining well is also used here in order to distinguish between states 1' and 2', normally discussed in the literature, and the new states 3', 4', and 5'. $M_{I',3'} = T_{I',3'}/T_{1',1'}$ and $M_{I',5'} = T_{I',5'}/T_{1',1'}$ where $T_{i,j}$ corresponds to the square of the optical matrix element between states i and j

Fig.14. A plot of $|A_{n,\vec{k}}|^2$ associated with state 3', as a function of alloy composition x (0.4 - 0.45). The wave vector $\vec{k}$ runs along the Δ line in the Brillouin zone, from the zone centre to the X point. For simplicity, only contributions from the first conduction band are shown. Because of the interaction with state 5', state 3' changes its character from Γ-like to X-like with increasing Al composition

at k = 0, the optical transition probability is given directly by $|A_{n,0}|^2$ associated with the conduction band state participating in the transition. The result of the numerical calculation is shown in Fig.15. It transpires that in the structure in question the enhancement of the optical matrix element is too small to give rise to observable transitions involving the lowest lying conduction band states of the superlattice. However, it must be stressed that the strength of the effect of strain does reflect the geometrical properties of the structure under investigation and it remains to be seen to what extent the above conclusion is generally valid [19].

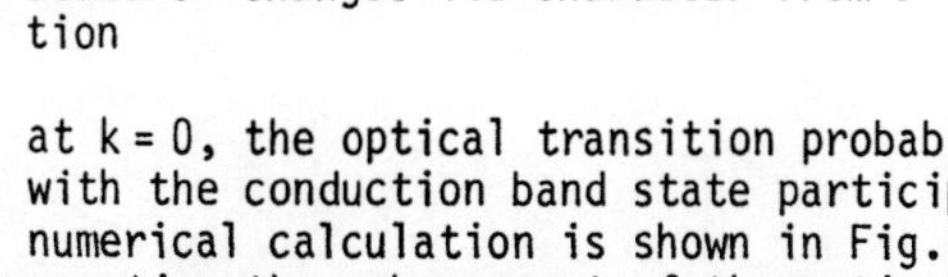

Fig.15. A plot of the $|A_{n,\vec{k}}|^2$, associated with the lowest confined electron state, in the bulk Brillouin zone of $Si_{0.76}Ge_{0.24}$ along <001>, with n corresponding to the first conduction band. The vector $\vec{k}$ runs from -X to X through Γ. The magnitude of $|A_{n,0}|^2$ is negligibly small compared to the leading coefficient

6. Band Offsets

Some years ago DINGLE and coworkers [20] established a consensus of opinion which identified the fraction of the energy gap difference between GaAs and $Ga_{1-x}Al_xAs$ appearing at the conduction band edge as 85%. This was achieved by fitting the barrier height (band offset) to account for exciton recombination energies. Accordingly, calculations reported in the literature, including those in Figs.2-5 and 13,14, were based on the 85% rule. It has been assumed that an analogous empirical procedure could be used in other structures since the error of even the largest numerical studies concerning the magnitude of band offsets is unacceptably high. However, closer examination showed that other fractions may fit data better and, furthermore, that the fit is quite insensitive to the choice of band offset. Fractions ranging from 80% to 50% have been quoted. The experimental errors are typically or order ±10%, i.e. ±50 meV for GaAs-AlAs. Very recently, WOLFORD et al. [21] showed that a more accurate measurement of the barrier height is possible if high hydrostatic pressure is used to drive the lowest confined states associated with the Γ minimum of the conduction band edge up in energy, to the crossing point with the indirect X-like edge (Fig.16). Calculations such as those outlined in Figs.13 and 14 predict that this crossing occurs in a narrow range of energies in which the confined states pinned to the secondary X minima hybridise with the zone centre (Γ) states, and as a result give rise to observable optical spectra. This is in spite of the fact that, with a lower band offset fraction (e.g. 60-70% which appears to give a better agreement with experiment) the X-like states are actually localised in the $Ga_{1-x}Al_xAs$ layers. Indeed, the theory predicts that these new X-like states are well confined and dispersionless in the superlattices in question and their spectra can be resolved with meV accuracy. This enables one to deduce the band offset as indicated in Fig.16 directly without introducing model-dependent parameters. The experimental results of Wolford et al. are in good agreement with pseudopotential calculations (Fig.17) and open the way for accurate band offset determination in other materials.

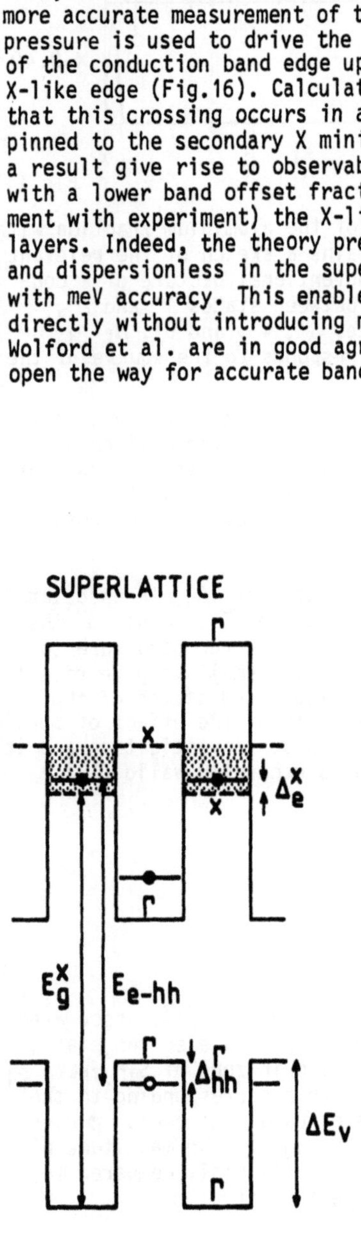

Fig.16. A rough graph showing the energies required for obtaining valence band offsets ΔE_V from spectroscopic data taken in a pressure experiment in which the lowest Γ-like state in the conduction band is driven up to cross the X-states. The position of the crossing is clearly seen in the spectra because of the abrupt change in transition probability and the appearance of a new X-related line. Although the result in Fig.13 was obtained for an incorrect band offset, the physics of the crossing phenomenon predicted there is valid and remains very much the same when the mixing involves pressure driven Γ- and X-like ground states. E_g^X is the band gap of the barrier material, E_{e-hh} is the recombination energy and Δ represents the single particle (electron or hole) energy measured from the band edge. The interrupted line shows the X point of GaAs (in the well) and that of $Ga_xAl_{1-x}As$ (in the barrier). The stipled area indicated the separation of the indirect edges

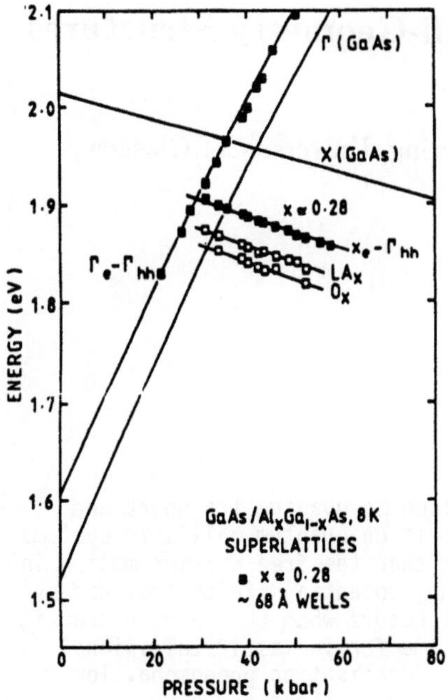

Fig.17. The experimental and theoretical (solid line) results showing the transition energy as a function of pressure. The crossing occurs at about 29 kBar for 68 Å wells and 28% Al barriers as described in Fig.16. The symbol e stands for electrons and hh for heavy holes. Al_X and O_X are phonon replicas associated with the indirect band edge of the barrier material. Γ- and X-points of bulk GaAs are also shown

References

1 T. Ando, A.B. Fowler, F. Stern: Rev. Mod. Phys. 53, 437 (1982)
2 M. Jaros: Rep. Progr. Phys. 48, 1091 (1985)
3 M. Altarelli: Phys. Rev. B28, 842 (1983)
4 M. Jaros: Deep Levels in Semiconductors (Hilger, Bristol 1982)
5 M.J. Kirton, P.W. Banks, D.I. Lu, M. Jaros: J. Phys. C17, 2487 (1984)
6 W. Andreoni, R. Carr: Phys. Rev. B21, 3334 (1980)
7 J.E. Zucker, A. Pinczuk, D.S. Chemla, A. Gossard, W. Wiegmann: Phys. Rev. B29, 7065 (1984)
8 G. Bastard, U.O. Ziemelis, C. Delalande, M. Boos, A.C. Gossard, W. Wiegmann: Solid State Commun. 49, 671 (1984)
9 Y.C. Chang, N. Schulman: Phys. Rev. B31, 2069 (1985)
10 F. Capasso, S.Luigi, W.T. Tsang, C.G. Bethea, B.F. Levine: Phys. Rev. Lett. 51, 2318 (1983)
11 M. Jaros, K.B. Wong, M.A. Gell: Phys. Rev. B31, 1205 (1985);
M. Jaros, K.B. Wong, M.A. Gell, D.J. Wolford: J. Vac. Sci. Technol. B3, 1051 (1985)
12 G. Bastard, E.E. Mendez, L.L. Chang, L. Esaki: Phys. Rev. B28, 3241 (1983)
13 E.J. Austin, M. Jaros: Phys. Rev. B31, 5569 (1985);
E.J. Austin, M. Jaros: Appl. Phys. Lett. 47, 274 (1985)
14 G.C. Osbourn: J. Appl. Phys. 53, 1586 (1982)
15 J.C. Bean, L.C. Feldman, A.T. Fiory, S. Nakshara, I.K. Robinson: J. Vac. Sci. Technol. A2(2), 436 (1984)
16 F. Cerdeira, A. Pinczuk, J.C. Bean: Phys. Rev. B31, 1202 (1985)
17 G. Abstreiter, H. Brugger, T. Wolf, H. Jorke, H.J. Herzog: Phys. Rev. Lett. 54, 2441 (1985)
18 D. Ninno, K.B. Wong, M.A. Gell, M. Jaros: Phys. Rev. B32, 1586 (1985)
19 I. Morrison, M. Jaros, K.B. Wong: J. Phys. C (in press) (1986)
20 R. Dingle: Festkörperprobleme 15, 21 (1975)
21 D.J. Wolford, T.F. Kuech, J.A. Bradley, M.A. Gell, D. Ninno, M. Jaros: J. Vac. Sci. Technol. (in press) (1986)
22 P.G. Dawson, G. Duggan, H.I. Ralph, K. Woodbridge, G.W. Hooft: Superlattices Microstruct. 1, 231 (1985)

Quantum Transport Theory for Small-Geometry Structures

J.R. Barker

Department of Electronics and Electrical Engineering, University of Glasgow, Glasgow G12 8QQ, United Kingdom

1. QUANTUM BALLISTIC TRANSPORT

1.1 Introduction

The following lectures are an elementary introduction to quantum transport and tunnelling in very small structures. The emphasis is on quantum ballistic systems for which collision processes are less significant than the free carrier motion in the strongly inhomogeneous (and possibly quantising) potential fields provided by the device structure. Quantum ballistic transport occurs when the carrier transit time is substantially shorter than the mean free time for inelastic collisions. Effects which may arise include size quantisation, localisation phenomena, low-dimensional effects, tunnelling and resonance phenomena. These are beyond the scope of classical Boltzmann-Bloch transport theory [1-3] (there is evidence that quantum effects can occur in quite "large" classical devices with channel dimensions of a few hundred nanometers). The pursuit of these effects has led to a very wide range of novel device structures in recent years [4-12].

If the conduction band edges are regarded as space-dependent in superlattice and quantum well devices we are led to contemplate effective local built-in force fields equivalent to electric fields in excess of 10^7 V cm^{-1}. The strong fields may well lead to a distortion of the scattering processes [1], but the more serious problems arise because the effective potentials vary rapidly in space and the classical picture of free drift motion becomes invalid. Tunnelling provides an obvious non-classical transport effect and it is important for the operation of a range of novel devices as well as providing the ultimate limitations to circuit isolation. However, the traditional theory of tunnelling [13] is inadequate for the description of hot electron effects and time-dependent phenomena in semiconductor microstructures [14]. Some progress has been made recently towards a unified many-body theory of transport which covers the classical and quantum regimes [15,16]. A more phenomenological approach will be taken in the lectures.

1.2 Classical Phase Space and Ballistic Transport

The classical problem of a set of point particles interacting with each other and a potential field V is described by a probability distribution function defined for all the particles over classical phase-space. Ordinary mechanics is obtained when the distribution factorises into a string of delta function on phase-space. Statistical mechanics follows otherwise. Many properties of the system are determined by the one particle distribution function $f(\underset{\sim}{r},\underset{\sim}{p},t)$, and indeed all the properties follow from a knowledge of f if the particles are independent. Thus if $A(\underset{\sim}{r},\underset{\sim}{p})$ is a dynamical variable, its expectation value is determined by

$$\langle A\rangle = \iint d^3r\,d^3p\; A(\underset{\sim}{r},\underset{\sim}{p})f(\underset{\sim}{r},\underset{\sim}{p},t) \tag{1}$$

$$\iint d^3r\,d^3p\;\; f(\underset{\sim}{r},\underset{\sim}{p},t) = 1 \quad . \tag{2}$$

Lioville's theorem exploits the continuity of matter to state df/dt=0, or

$$\partial_t f + (d\underset{\sim}{r}/dt).\partial_{\underset{\sim}{r}} f + (d\underset{\sim}{p}/dt).\partial_{\underset{\sim}{p}} f = 0 \quad . \tag{3}$$

Hamilton's equations then permit us to replace the terms in dr/dt and dp/dt:

$$\partial_t f + \underset{\sim}{v}\cdot\partial_{\underset{\sim}{r}} f + \underset{\sim}{F}\cdot\partial_{\underset{\sim}{p}} f = 0 \qquad (4)$$

where $\underset{\sim}{v}=\partial_{\underset{\sim}{p}}H$ is the velocity, $\underset{\sim}{F}=-\partial_{\underset{\sim}{r}}H$ is the force and $H=H(\underset{\sim}{r},\underset{\sim}{p})$ is the Hamiltonian. Equation (4) describes ballistic motion for which no energy dissipation takes place and the motion is reversible. Ballistic transport theory may be set up by choosing a potential field to generate an electrical field $\underset{\sim}{E}$ which will accelerate independent electrons through the geometry of a device. The problem is non-trivial if the field $\underset{\sim}{E}$ is determined self-consistently from the space-charge and boundary conditions via Poisson's equation. The current response is never ohmic under these conditions.

1.3 Boltzmann Transport

If very many additional degrees of freedom are introduced, for example by deploying a random distribution of fixed ions or introducing interactions between particles, in an open system (communication with the rest of the universe across boundaries), then apparently irreversible processes are set up and exist over very large time scales (the Poincaré period). Although the many body variant of the Liouville equation (4) is still valid, the full complexity of determining the projected one-particle behaviour may be avoided by the tempting idea of Boltzmann that the carrier motion should be split into drift-diffusion motion in the slowly varying component of the fields and a stochastic motion which is described by a gain-loss collision integral. Equation (4) then goes over into Boltzmann's equation:

$$\partial_t f + \underset{\sim}{v}\cdot\partial_{\underset{\sim}{r}} f + \underset{\sim}{F}\cdot\partial_{\underset{\sim}{p}} f = -\sum\{R(\underset{\sim}{p},\underset{\sim}{p}')f(\underset{\sim}{r},\underset{\sim}{p},t)-R(\underset{\sim}{p}',\underset{\sim}{p})f(\underset{\sim}{r},\underset{\sim}{p}',t) \qquad (5)$$

where the ordered quantities $R(\underset{\sim}{p},\underset{\sim}{p}')$ describe the transition probability per unit time for scattering from a momentum state $\underset{\sim}{p}$ at a point $\underset{\sim}{r}$ instantaneously into a new momentum state $\underset{\sim}{p}'$. This irreversible transport equation was adapted to solid state electronics by Bloch by supposing that the velocity $\underset{\sim}{v}$ was just the group velocity of a wavepacket moving with crystal momentum $\underset{\sim}{p}=\hbar\underset{\sim}{k}$ in the presence of an electric force F; the rates R were identified as golden rule transition rates for the appropriately averaged scattering processes (electron-impurity, electron-phonon etc). The Boltzmann-Bloch equation has had considerable success in explaining the transport properties of bulk semiconductors and large scale devices. It is mainly solved numerically using Monte Carlo simulation or phase-space trajectory techniques. A critique is given in [1].

1.4 Density Matrices and Wigner Functions

Attempts to derive the Boltzmann equation or to determine its limitations and generalisations are hindered by the problem that the underlying dynamics is many-body quantum field theory. The analogue of the many-body distribution f is the density matrix ρ which satisfies the quantum equivalent of Liouville's theorem in the form

$$i\hbar\partial_t\rho = [H,\rho] \qquad (6)$$

where H is the Hamiltonian operator and expectation values are given by

$$\langle A\rangle = \mathrm{TRACE}\ (A\ \rho) \quad . \qquad (7)$$

Equation (6) follows from Schrodinger's equation if we set $\rho=|\Psi\rangle\langle\Psi|$ (for pure states) or $\rho=\sum P_n|\Psi_n\rangle\langle\Psi_n|$ (for mixed states).

The outlook for a connection with classical theory appears grim and phase-space description (particularly the notion of a particle trajectory) seems to be ruled out by the Heisenberg uncertainty relations. But all is not lost. A remarkable transformation introduced by Weyl and applied by Dirac and Wigner to the density matrix allows us to construct a statistical distribution function $f(\underset{\sim}{r},\underset{\sim}{p},t)$(the Wigner function [17]) defined over a c-number phase-space and which generates the correct quantum statistical expectation values of any dynamical observable. The transformation is essentially a mixed matrix element representation of the density matrix:

$$f(\underset{\sim}{r},\underset{\sim}{p},t)=(2\pi\hbar)^{-3}\int d^3Q e^{-i\underset{\sim}{p}\cdot\underset{\sim}{Q}}\langle \underset{\sim}{r}+\underset{\sim}{Q}/2|\rho|\underset{\sim}{r}-\underset{\sim}{Q}/2\rangle \quad . \tag{8}$$

Expectation values are given by the classical-like expression (1) where $A(\underset{\sim}{r},\underset{\sim}{p})$ is derived from the matrix elements of its operator counterpart A by Wigner-Weyl transform. Unfortunately, it can be proved that the Wigner function is not positive-definite and in general does not yield a simple classical probabilistic interpretation. Only for a minimum Gaussian wavepacket does the Wigner function become truly positive definite and indeed it then coincides with a classical Gaussian distribution. Nevertheless, the contours of constant f for stationary states can be interpreted as trajectories in phase-space, and show up such quantum concepts as barrier penetration very nicely.

It has been known for a long time that the Wigner distribution for a uniform system is a function of momentum only. It follows that f(p) is positive definite, a fact which has enabled various researchers to "derive" the space-independent Boltzmann-Bloch equation under a range of restrictions. But as we noted earlier small geometry systems cannot be expected to be uniform. To study this case we ignore collision processes and look at the quantum analogue of the Lioville equation this time in the Wigner representation.

1.5 Quantum Ballistic Transport

The time-dependent Wigner function for a Hamiltonian $H=\underset{\sim}{p}^2/2m + V(\underset{\sim}{r})$ satisfies [16]

$$\partial_t f + \underset{\sim}{v}\cdot\partial_r f + (2\pi\hbar)^{-3}\iint d^3Q d^3P \cos(\underset{\sim}{Q}.\underset{\sim}{P}/\hbar)\, \underset{\sim}{F}(\underset{\sim}{r},\underset{\sim}{Q}).\partial_{\underset{\sim}{p}} f(\underset{\sim}{r},\underset{\sim}{p}+\underset{\sim}{P},t) = 0 \tag{9}$$

where F is derived from the relation $\underset{\sim}{F}(\underset{\sim}{r},\underset{\sim}{Q}).\underset{\sim}{Q}=V(\underset{\sim}{r}-\underset{\sim}{Q}/2)-V(\underset{\sim}{r}+\underset{\sim}{Q}/2)$. In the limit $\hbar \to 0$, F goes over into the classical force and (9) becomes the Liouville equation; the limit is approached non-uniformily however. Remarkably, equation (9) reduces exactly to the classical Liouville equation if V is a general quadratic form. Quantum effects can then only appear in f via the initial state and not through the dynamics. The initial state for any evolution of f may be derived from a superposition of a complete set of stationary Wigner distributions. The latter satisfy an adjoint equation to (9) which is derived from the Wigner-Weyl transform of

$$\varepsilon\rho = \{H,\rho\}/2 \tag{10}$$

where { } denotes the anti-commutator.

Equation (9) is the fundamental starting point for quantum ballistic transport and the solutions are the Green functions for the more complex task of incorporating electron-phonon and other scattering processes. Equation (9) is non-local in momentum space (and in position space if the kinetic energy is non-quadratic). Ideally we would use this equation in analogy with the collisionless Boltzmann equation (or Vlasov equation) to describe short geometry structures. The effects of crystal structure can be taken into account through effective mass theory and the band structure.

1.6 Quantum Phase Space

The quantum ballistic transport equation cannot be emulated by an ensemble of point particles in phase-space except when the non-locality is removed. However, a classical-like picture does underly the representation. It has been shown [16] that f may be represented as the linear superposition of a set of distributions $f_c(\underset{\sim}{r},\underset{\sim}{p},\underset{\sim}{Q},\underset{\sim}{P},t)$ which obey classical equations:

$$f(\underset{\sim}{r},\underset{\sim}{p},t) = (2\pi\hbar)^{-3}\iint d^3Qd^3P\cos(\underset{\sim}{Q}.\underset{\sim}{P}/\hbar)f_c(\underset{\sim}{r},\underset{\sim}{p},\underset{\sim}{Q},\underset{\sim}{P},t) \tag{11}$$

$$\partial_t f_c + \underset{\sim}{v}.\partial_{\underset{\sim}{r}} f_c + \underset{\sim}{F}(\underset{\sim}{r},\underset{\sim}{Q}).\partial_{\underset{\sim}{p}} f_c(\underset{\sim}{r},\underset{\sim}{p},\underset{\sim}{Q},\underset{\sim}{P},t) = 0 \tag{12}$$

The causal boundary conditions are $f_c(\underset{\sim}{r},\underset{\sim}{p},\underset{\sim}{Q},\underset{\sim}{P},0) = f(\underset{\sim}{r},\underset{\sim}{p}+\underset{\sim}{P},0)$.

It follows that quantum distributions can be emulated by ensembles of classical particles in a family of phase-spaces and Monte Carlo techniques can be restored, but at considerable computational cost. Phenomena like tunnelling can be readily pictured in this representation: For some values of the parameters P and Q the particle trajectories are at higher energies and the barrier potentials related to $F(\underset{\sim}{r},\underset{\sim}{Q})$ are smeared out in comparison to the classical limit Q=0, P=0; penetration to forbidden regions is thus allowed off the classical shell.

1.7 High Field Effects in Scattering and Screening

Wigner function theory has been used [18] to derive the field-dependent quantum collision integrals for a range of processes including dynamically screened interactions. Deviations from Boltzmann collision integrals are enhanced in low-dimensional geometries, especially in the screening. A strong uniform field also leads to field-dependent scattering rates but very little is known about the effect on transport in small structures. The so-called intra-collisional field effect [1] is essentially due to the acceleration of a carrier during the time it takes to complete the scattering process. The effects of a strong non-uniform field on scattering, which is inherent in small geometry devices remain unexplored.

1.8 Disorder Effects

A range of fascinating quantum transport effects are associated with disorder and localisation phenomena in low dimensional structures. These are beyond the scope of my lectures, but it is worth pointing to a few representative papers [19-21] which use concepts such as transmission matrix theory that impinge on the following topics.

2. BALLISTIC TUNNELLING

2.1 One-Dimensional Models

Part of the charm of low dimensional devices lies in the opportunity to make use of one-dimensional quantum-mechanical models of text-book simplicity. As a start let us take a semiconductor heterostructure or n-i-p-i device to be described in one dimension by a generic electron Hamiltonian

$$H_e = \varepsilon(p) + V(z) \qquad (p = -ih\,\partial/\partial z) \tag{13}$$

Here $\varepsilon(p)$ is an effective-mass conduction-band structure and $V(z)$ describes the spatial modulation of the band-edge. In keeping with simplicity we shall for the present assume $\varepsilon(p)$ is parabolic: $\varepsilon=p^2/2m^*$, so that the envelope wave-functions satisfy a basic Schrodinger equation. For later convenience we shall also let $V(z)$ include the potential of an applied electric field. Figure 1 sketches the sort of "potential" profiles which occur in recent devices.

Carriers traversing such systems may be supposed to be injected into travelling-wave plane-wave states, or into wave-packet pure states or as mixed states. Whatever the choice, the basic ballistic transport problem requires the separation of the system into a "free" component on which is superposed a "potential barrier" structure represented by $V(z)$. Without loss of generality we may take V=constant outside a "barrier domain" $0<z<a: V=V_1(z<0),\ V=V_2(z>a)$.

Analytical expressions for the reflection and transmission coefficients of the barrier potential may be obtained in only a few over-simplified cases although the

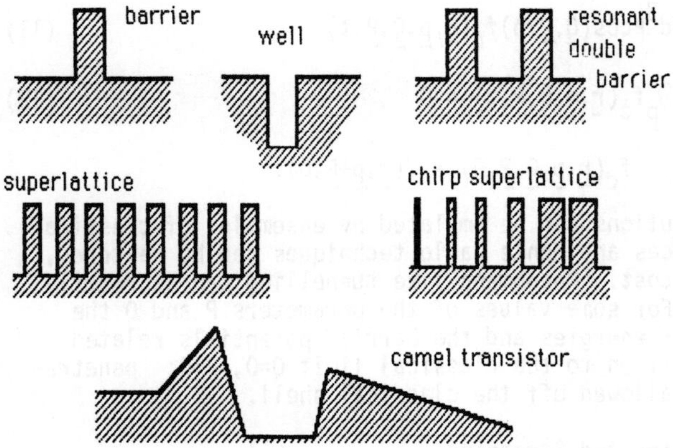

Fig.1 Device profiles

important step junction, finite superlattice, chirp structure and triangular barrier are included. It has been common practise to use WKB methods for more general potentials but although this sometimes gives a good "first pass" at the problem it is rarely valid and cannot cope with self-consistent potentials which arise from space-charge effects. A useful modular approach to complex structures exists as an analogy with transmission line theory (see section 2.5). However, one cannot escape the eventual need to numerically compute the transmission properties of V(z).

Fortunately the numerical problem may be handled very easily by discretising the Schrodinger equation:

$$H\psi(z)= E\psi \to (-\hbar^2/2\Delta m_*)[\psi(z_{n+1})+\psi(z_{n-1})-2\psi(z_n)] + V(z_n)\psi(z_n) = E\psi(z_n) \quad (14)$$

where $\Delta = z_{n+1}-z_n$ defines the grid spacing and n=0,1,2..(N+1). This equation may be solved exactly by setting up the recurrence relation:

$$S_{n-1} = b_n(\varepsilon) - 1/S_n$$

$$S_n \equiv \psi(z_n) / \psi(z_{n+1})$$

where we introduce the variables $K^2 \equiv \varepsilon \equiv (2m^*E/h^2)$, $U=(2m^*V/h^2)$ and

$$b_n(\varepsilon) \equiv 2 + \Delta^2[U(z_n)-\varepsilon]$$

Suppose a unit plane wave of energy E is incident from $z<z_0=0$, and free propagation is renewed beyond $z=z_N=a$; the complex reflection amplitude r(k) and the complex transmission amplitude t(k) are then defined by

$$\psi(z)=e^{ikz} + r(k)e^{-ikz} \quad (z<z_0) \quad (15)$$

$$\psi(z)=t(k)e^{ikz} \quad (z>z_N) \quad (16)$$

where we assume $V_1=V_2$ for the present. The total reflection (transmission) coefficient R(T) is defined as the ration of the reflected (transmitted) flux to incident flux:

$$R(k)=|r(k)|^2 \quad , \quad T=|t(k)|^2 \quad , \quad R + T = 1 .$$

VIGNERON and LAMBIN [22] have shown that R (and hence T) may be obtained as

$$R = |S_{-1} - S^{(-)}|^2/|S_{-1} - S^{(+)}|^2 \quad (17)$$

where

$$S^{(\pm)} = \beta/2 \pm i(1-\beta^2/4)^{1/2} \tag{18}$$

$$\beta = 2 + \Delta^2[U_1-\varepsilon] \tag{19}$$

and U_1 is the constant value of $U(z)$ for $z>Z_N$. S_{-1} is obtained by forward recursion and good convergence requires $\Delta^2[U-\varepsilon]<<1$. COLLINS and BARKER [23] have generalised the continued fraction method to extract the complex amplitudes r(k) and t(k) which are required to evaluate tunnelling times. The method also generalises to the case $V_1 = V_2$.

2.2 The Perfect Square Barrier

The square barrier defined by

$$V(z) = V_0(z<0);\ V(z) = 0\ (z>a)$$

is the archetypal heterostructure problem although practical structures (fig.2b) have definite distortions (as well as spatially-varying effective mass).

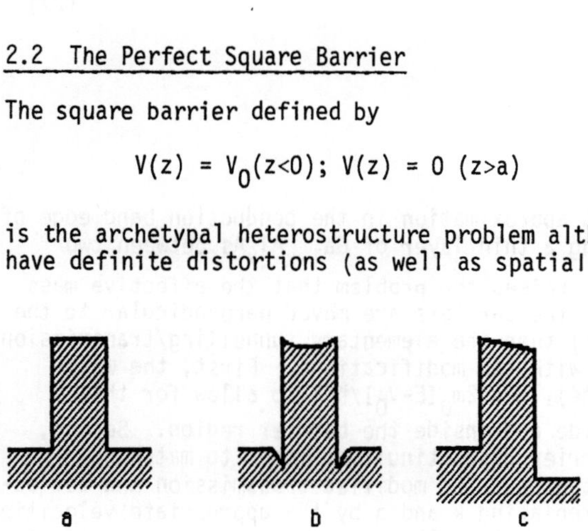

Fig.2 Variants on the square barrier

Typical parameters for GaAs=GaAlAs systems are: V_0 = 0.2-0.35eV, m^*/m_e=0.067, a = 2.5 - 100 nm. The transmission properties may be found exactly by standard wave-function matching across the two boundaries to give for $E>V_0$

$$t(k)=4kqe^{i(q-k)a}/[(k+q)^2-(k-q)^2e^{2iqa}] \tag{20}$$

$$r(k)=k^2-q^2)[1-e^{2iqa}]/[(k+q)^2-(k-q)^2e^{2iqa}] \tag{21}$$

$$T=4E(E-V_0)/[4E(E-V_0)+V_0^2\sin^2(qa)] \quad . \tag{22}$$

Here $k=\sqrt{(2m^*E/\hbar^2)}$, $q=\sqrt{(2m^*[E-V_0]/\hbar^2)}$

For $E<V_0$

$$t(k)=4kqe^{i(q-k)a}/[(k+q)^2-(k-q)^2e^{-2Ka}] \tag{23}$$

$$r(k)=(k^2-q^2)[1-e^{2iqa}]/[(k+q)^2-(k-q)^2e^{-2Ka}] \tag{24}$$

$$T=4E(V_0-E)/[4E(V_0-E)+V_0^2\sinh^2(Ka)] \quad . \tag{25}$$

Here q=iK where $K=\sqrt{(2m^*[V_0-E]/\hbar^2)}$.

This problem exhibits the phenomena of resonances and tunnelling. The resonant energies occur where the transmission passes through a maximum, which here is determined by the zeros of $\sin^2(qa)$(when $E>V_0$). The resonance condition is thus

$$a=n\lambda/2 \quad (n=1,2....) \tag{26}$$

where $\lambda = 2\pi/q$ is the de Broglie wavelength in the barrier region. These resonances are perfect resonances since $T = 1$.

For $E<V_0$ the square barrier permits tunnelling. The tunnelling current is largest when $Ka\leq 1$. For the opposite extreme, $Ka>>1$, we find the rapid exponential decrease of the tunnelling current with barrier width:

$$T \sim 16[E(V_0-E)/V_0^2]e^{-2Ka} \tag{27}$$

The WKB method gives $T \sim e^{-2Ka}$.

2.3 Modifications

The square barrier is realisable as an approximation in the conduction band edge of a heterostructure formed by sandwiching a thin layer of $Ga_{1-x}Al_xAs$ between two layers of GaAs (narrow bandgap). This raises the problem that the effective mass differs from material to material. If the carriers are moved perpendicular to the layers (no transverse energy component) then the elementary tunnelling/transmission problem follows the simple theory but with two modifications. First, the wave-vectors k,q are modified to $k=\surd(2m_1E/\hbar^2)$, $q=\surd(2m_0[E-V_0]/\hbar^2)$ to allow for the different effective masses m_1,m_0 outside and inside the barrier region. Second, the matching of current across the barrier discontinuities leads to matching $(1/m^*)(d\psi/dz)$ rather than the gradient of ψ. The modified transmission and reflection amplitudes are finally found by replacing k and q by the appropriate velocities $v(k)=\hbar k/m_1$, $v(q)=\hbar q/m_0$ except in the exponents. The resonances again are perfect and occur at $\sin(qa)=0$.

Observation of current flow across a square barrier heterostructure is complicated by the need to apply a bias field. This will distort the barrier profile (fig.2c shows the sort of distortion). Suppose for simplicity we take the entire voltage drop across the right-hand barrier edge, so that $V(z)=V_2 = V_1$ for $z>a$. If we ignore the mass variations we may obtain the corrected transmission and reflection amplitudes for carriers incident from the left by using the appropriate wave-vectors $k_1=\surd(2m(E-V_1)/\hbar^2)$, $q=\surd(2m(E-V_0)/\hbar^2)$, $k_2=\surd(2m(E-V_2)/\hbar^2)$ in the three regions. The transmission coefficient becomes

$$T = T_0\surd[(E-V_2)/(E-V_1)] \tag{28}$$

where

$$T_0=4(E-V_1)(E-V_0)/[(\surd(E-V_1)+\surd(E-V_2))^2(E-V_0)+(V_0-V_1)(V_0-V_2)\sin^2(qa)] .$$

The resonances still occur but are shifted away from the condition $\sin(qa)=0$, and they are no longer perfect resonances since T is always less than unity if $k_1=k_2$. This problem is very general and poses difficulties for devices which are based on negative resistances associated with transmission resonances or resonant tunnelling.

2.4 Transmission Matrix Methods

A finite superlattice (N barriers) or infinite superlattice may be regarded as a chain of identical square barriers. Chirp superlattice structures, which are useful for emulating continuously-graded potentials, are organised arrays of square barriers of identical height but different widths. In many other cases of interest, device configurations consist of a series of barrier or quantum-well structures separated by regions of constant potential. The question therefore arises: How do we construct the transmission/reflection properties of an array of elementary potentials? Heuristically one expects the coherence of the carrier wave-function to be an important factor. Current folk-lore suggests that the critical coherence length is limited by Λ, the mean free path or inelastic scattering. In such a case it may be argued that if two barriers are separated by substantially more than Λ their combined transmission properties are determined by the incoherent superposition of their individual transmission properties. For the opposite extreme, infinite coherence lengths, interference between the reflected and transmitted waves within the barrier system will lead to coherent superposition. A new phenomenon may then occur: resonant tunnelling, for which perfect transmission may be set up at the resonance energy, even though the individual barriers are virtually opaque. We shall see later that resonant tunnelling is intimately related to the existence of band structure in a periodic potential and is connected with the existence of minima in the energy density of states for disorder potentials.

For one-dimensional systems there exists an elegant analytical technique for handling arrays of elementary potentials. Transmission matrix theory is constructed in a precise analogy with the theory of waves on transmission lines [24]. An advanced group-theoretical treatment of the method is given by PERES [19,20,25] and applied to the determination of resistance and localisation length in 1-D disordered systems. These papers also contain a nice attempt at showing that the band structure of an almost periodic potential is chaotic (forms a Cantor set).

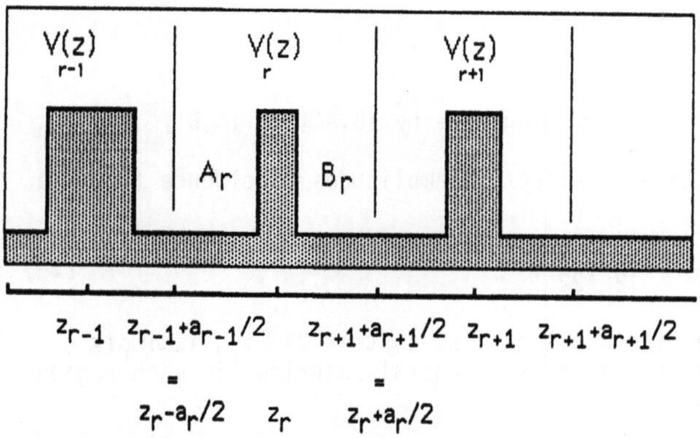

Fig.3 Construction for transmission matrix theory.

Figure 3 illustrates the idea. Consider a potential barrier defined by $V(z)=V_r(z)$ for $|z-z_r|<a_r/2$, $V=0$ elsewhere. To the left and right of the barrier there are two independent solutions:

$$z-z_r<-a_r/2:\ u_k(z)=e^{ik(z-z_r)};\ u'_k(z)=e^{-ik(z-z_r)}$$

$$z-z_r>a_r/2:\ u_k(z)=T_{11}(r)e^{ik(z-z_r)}+T_{21}(r)e^{-ik(z-z_r)}\ ;$$

$$u'_k(z)=T_{12}(r)e^{ik(z-z_r)}+T_{22}(r)e^{-ik(z-z_r)} \qquad (29)$$

The general solution is then

$$\psi_k(z)=A_r e^{ik(z-z_r)}+A'_r e^{-ik(z-z_r)} \qquad (z-z_r<-a_r/2)$$

$$\psi_k(z)=B_r e^{ik(z-z_r)}+B'_r e^{-ik(z-z_r)} \qquad (z-z_r>a_r/2) \qquad (30)$$

where $k=\sqrt{[2mE/\hbar^2]}$. In matrix terms the transmission matrix relates the coefficients A,B on either side of the barrier:

$$\underset{\sim}{B}_r = \underset{\approx}{T}_r \cdot \underset{\sim}{A}_r \tag{31}$$

where (suppressing the r-subscript for the moment) $\underset{\sim}{B}=[B,B']^t$, $\underset{\sim}{A}=[A,A']^t$ and

$$\underset{\approx}{T} = \begin{pmatrix} T_{11} & T_{12} \\ T_{21} & T_{22} \end{pmatrix} \tag{32}$$

It may be shown that $T_{22}=T_{11}{}^*$ and $T_{12}=T_{21}{}^*$ and $|T_{11}|^2-|T_{12}|^2=1$ [24]. The transmission matrix is therefore a member of the SU(1,1) group [26]. If $V(z)$ is symmetric about $z=z_r$ we find T_{21} is pure imaginary. T is related to the S-matrix of scattering theory by

$$\underset{\approx}{S} = (1/T_{22}) \begin{pmatrix} 1 & T_{12} \\ -T_{21} & 1 \end{pmatrix} \tag{33}$$

The outgoing wave is related to the incoming wave by $[B,A']^t = \underset{\approx}{S}.[A,B']^t$.

In terms of the usual transmission and reflection amplitudes (incidence from the left) we have

$$t(k)=1/T_{22} \;;\; r(k)=-T_{12}/T_{22} \,. \tag{34}$$

The general problem of superposing N potentials, so that $V=\Sigma V_r$, is simply managed by imposing matching conditions for the general solutions in each region. Continuity at each boundary gives

$$A_{r+1}=B_r e^{ik\Delta_r} \quad ; \quad A'_{r+1}=B'_r e^{-ik\Delta_r}$$

where $\Delta_r=(a_r+a_{r+1})/2=(z_r+z_{r+1})/2$. This rule may be implemented using the iteration matrix $\underset{\approx}{D}_r$ defined by

$$\underset{\approx}{D}_r = \begin{pmatrix} e^{ik\Delta_r} & 0 \\ 0 & e^{-ik\Delta_r} \end{pmatrix} . \tag{35}$$

Then

$$\underset{\sim}{A}_{r+1} = \underset{\approx}{D}_r \cdot \underset{\sim}{B}_r \tag{36}$$

from which we get the recurrence relation

$$\underset{\sim}{B}_{r+1} = \underset{\approx}{T}_{r+1} \cdot \underset{\approx}{D}_r \cdot \underset{\sim}{B}_r = \{\underset{\approx}{T}_{r+1} \cdot \underset{\approx}{D}_r \cdot \underset{\approx}{T}_r\} \cdot \underset{\sim}{A}_r \tag{37}$$

where $\underset{\approx}{T}_{(2)}=\underset{\approx}{T}_{r+1} \cdot \underset{\approx}{D}_r \cdot \underset{\approx}{T}_r$ is the effective transmission matrix for two adjacent barriers. For N barriers we evidently have

$$\underset{\approx}{T}_{(N)} = \underset{\approx}{T}_N \prod \underset{\approx}{D}_r \cdot \underset{\approx}{T}_r \qquad (38)$$

The theory is easily amended to allow for non-zero potential values at the boundaries of the piecewise potential. Mass variation may be similarly included.

2.5 Resonant Tunnelling

A resonant tunnelling barrier has perfect resonances in the transmission coefficient for tunnelling. They arise in multiple simple barrier situations and have been extensively studied [11-14,27-30]. Resonant tunnelling systems have technological importance because they have potential as fast switching devices and multi-state logic.

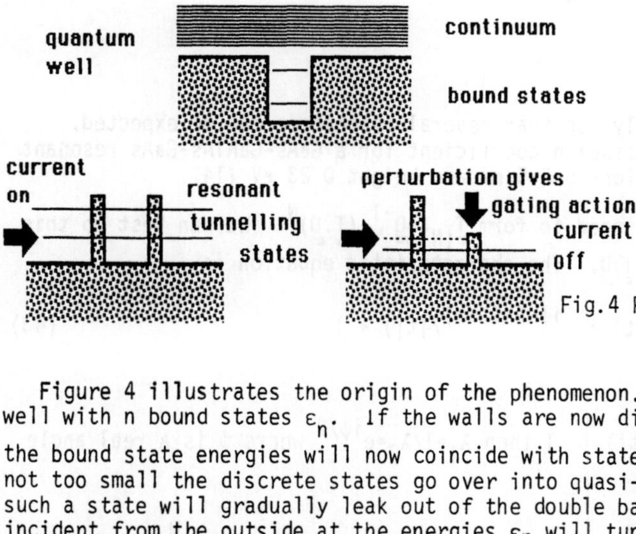

Fig.4 Resonant tunnelling as a switch

Figure 4 illustrates the origin of the phenomenon. Consider a single quantum well with n bound states ε_n. If the walls are now diminished to finite width d the bound state energies will now coincide with states in the continuum. If d is not too small the discrete states go over into quasi-bound states; an electron in such a state will gradually leak out of the double barrier system. Electrons incident from the outside at the energies ε_n will tunnel through without reflection even if the individual barriers are nearly opague. The perfect transmission energies correspond to conduction bands when a long array of barriers or wells is considered.

Resonant tunnelling is easily analysed with the aid of transmission matrix theory. For example, consider a double barrier of spacing b and individual barrier height V_0 and width a respectively. Let the transmission amplitude for the single barrier be $t_1 = |t_1| e^{i\phi}$ where ϕ is the phase angle. Using the product $\underset{\approx}{T}_{(1)} \cdot \underset{\approx}{D} \cdot \underset{\approx}{T}_{(1)}$ we obtain

$$t_2(k) = T_1 e^{-ik(a+b)} / \{R_1 + e^{i2\Theta}\} \qquad (39)$$

where $T_1 = |t_1|^2$, $R_1 = 1 - T_1$ and

$$\Theta = \phi - k[a+b] \qquad (40)$$

The transmission coefficient is

$$T_2 = T_1^{\,2} / \{ T_1^{\,2} + 2(1+\cos 2\Theta)(1-T_1)\} \qquad (41)$$

Evidently the resonance energies are determined by

$$\phi - k[a+b] = n\pi/2 \text{ where n is an odd integer} \qquad (42)$$

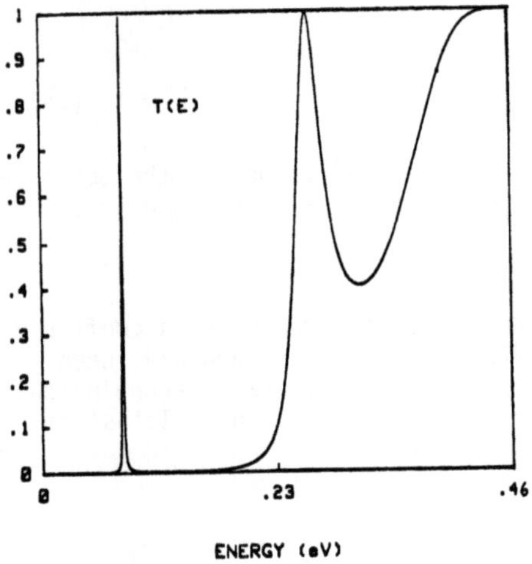

Fig.5 Transmission Coefficient for a 5-5-5nm double barrier

If b is large, kb varies rapidly, so that several resonances may be expected. Figure 5 illustrates the transmission coefficient for a GaAs-GaAlAs-GaAs resonant tunnelling structure of dimensions 5-5-5 nm and height 0.23 eV [14].

For N identical barriers we need to form $\underset{\approx}{T}_{(N)}=\underset{\approx}{D}^{-1}.(\underset{\approx}{T}.\underset{\approx}{D})^N$. We can best do this by diagonalising the matrix $\underset{\approx}{G}=\underset{\approx}{T}.\underset{\approx}{D}$. The characteristic equation is

$$\lambda^2-\lambda(e^{-i\phi+ik(a+b)}/|t| + e^{i\phi-ik(a+b)}/|t|) + 1 \qquad (43)$$

There are three cases:

Case (i) : $|\mathrm{Re}(g=e^{i\phi-ik(a+b)}/|t|) < 1$ then $\lambda_1=1/\lambda_2=e^{i\psi/2}$ where ψ is a real angle.

The eigenvalues of G^N are thus $e^{\pm iN\psi/2}$. If ψ/π is rational there will be some finite N such that $\underset{\approx}{G}^N=\underset{\approx}{I}$ where $\underset{\approx}{I}$ is the unit matrix. If ψ/π is rational there will still be some finite N such that G^N will be arbitrarily close to the unit matrix. The transmission coefficient will thus quasi-periodically peak up to unity as N increases. In the limit of infinite N we identify the resonant energies as belonging to the conduction band for the superlattice.

Case (ii): $|\mathrm{Re}\ g| > 1$. Here we may put Re g = $\pm\cosh\psi/2$ and $\lambda_1=1/\lambda_2=\pm e^{\psi/2}$. In this case the transmission coefficient decreases exponentially as $e^{-N\psi}$. This situation corresponds to the energies in the band gaps when N is infinite.

Case (iii) occurs exceptionally when Re $|g|=1$, the elements of G^N then grow linearly rather than exponentially with N.

2.6 Current-Voltage Characteristics

The most widely used elementary theories of tunnelling are the stationary-state method and the equivalent transfer-Hamiltonian method [13]. The generic tunnelling model involves a tunnelling barrier which is sandwiched between two degenerate three dimensional conducting materials, each electrode (1 and 2) being described by an equilibrium Fermi function $f_1=f_0(E)$, $f_2=f_0(E+eV)$ where V is the applied bias across the barrier. The net current density is asserted to be:

$$J = 2e/(2\pi)^3 \iiint d^2k_1 dk_x v(k_x) T(E,V)\{f_1 - f_2\}$$

or

$$J = 2e/(2\pi)^3 h \int dE_x [f_0(E) - f_0(E+eV) \int d^2k_1 T(E_x+E_1,V) \quad (44)$$

where T is the current transmission probability (ratio of incident to transmitted probability current=tt*(1→2)v_2/v_1 where t is the particle transmission probability amplitude) which depends on the bias V and energy.

For an ideal barrier of width a the barrier potential is

$$U(x,V)=U(x,0)-eV(x+a/2)/a \quad (45)$$

The bias potential distribution is reminiscent of the staircase potentials currently in vogue [31]. So the staircase is obtained if U(x,0) is zero, and a classical staircase is obtained if T(E,V)=1. For the true staircase we know that pronounced hot electron effects [4], particularly velocity and energy overshoot can occur if the drop V is large (few meV) over a short distance (few hundred Angstroms). Such a situation must be reconciled with the elementary tunnelling theory which incorporates no explicit details of scattering or hot electron effects.

The elementary theory requires modifications for heterostructures where there is an effective mass variation from material to material [29] which mixes the transverse and longitudinal wave motion. The non-parabolicity of the effective mass Hamiltonian also leads to modifications to T, but is manageable by a new technique described elsewhere [32]. It has been argued [33,34] that for sufficiently high conduction band discontinuities at GaAs-AlGaAs heterojunctions the commensurability between Γ-valley and satellite valley edges will lead to a failure of effective mass theory and deeper methods will be required to find T.

2.7 I-V Characteristics for Resonant Tunnelling

The condition for unit transmission probability for a double barrier may be shown to require that the transmission coefficients for each isolated barrier T_1 and T_2 are equal at the resonance energy. Although this symmetry holds for zero bias the application of a bias voltage may both shift and significantly reduce the transmission peak.

Most practical experiments have thus been in a regime for which the elementary theory of the I-V characteristics should hold at all but the lowest temperatures. This is borne out by calculations, although there are some subtleties. In fig.6a we show experimental data [11] compared with the elementary

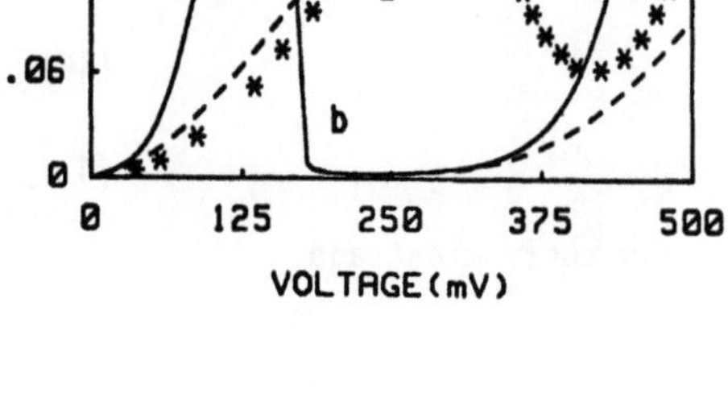

Fig.6 Current-voltage for 5-5-5nm double barrier

tunnelling theory for a GaAs-GaAlAs double barrier of dimensions 5-5-5 nm and 25% aluminium at T=100K.

The erroneous peak position cannot be accounted for by vagaries of barrier heights/widths or effective mass variation. However, if we assume that an effective resistance R is in series with the double barrier so as to alter the effective voltage drop across the barrier. The choice R=843 ohms gives agreement with the peak position and the positive resistance part of the I-V curves. The calculations give a best fit for a conduction band discontinuity of $\Delta E_c = 0.65\Delta E_g$, fig.6c. The origin of the series resistance may be traced to the distortion of the perfect barrier due to band-bending and charge accumulation. The total voltage drop thus divides between the true barrier and small regimes in the vicinity of the barrier. It is not yet understood why the large peak to valley ratios predicted by theory are not observed experimentally. There is some evidence that the quality of the interfaces is crucial and very recent Japanese data give ratios of 30:1.

Data for lower temperatures are in worse agreement with theory; carrier heating does lead to better agreement but the number of uncertainties is too large for a definitive result. This situation is discussed in section 3.

2.8 Tunnelling and Traversal Times

In recent years a considerable controversy has developed over the definition and evaluation of barrier traversal times. Early theories swung wildly between instantaneous tunnelling to very slow tunnelling and we can illustrate the confusion by quoting various estimates for the traversal time for a thick square barrier of width a:
(i) $\tau = 2m/(\hbar kK)$ [35-39]; (ii) $\tau = 0$ [40]; (iii) $\tau = ma/(\hbar K)$ [41-43]; $\tau = \hbar k/(KV_0)$ [44,45]; τ = reciprocal of tunnelling transition rate [46]. Early numerical calculations brought further confusion because different times were found for different wavepacket shapes.

On the other hand we might expect that wavepackets with sharp momentum distributions which are a good approximation to plane wave trains would have comparable traversal times at the same energies. Indeed the elementary tunnelling theory works quite well by assuming the states are plane wave outside the barrier region. To investigate these ideas we have recently completed extensive computer studies of a wide range of wave-packet v barrier tunnelling problems where the space-time trajectory of the peak in the transmitted asymptotic wavepacket is tracked to evaluate the delay τ. The results have been compared [14] with various analytical models [35-46]. The wavepackets were sized such that the transmission coefficient varies slowly across the momentum distribution. The numerical results were obtained in three independent ways: (i) solving the time-dependent Schrödinger equation; (ii) solving the Wigner equation (9); (iii) constructing a superposition of stationary waves. The conclusion is that a generalisation of the stationary phase time is an excellent fit to the numerical curves under the above conditions and for narrow packets where velocity filtering is anticipated.

The phase time is easily obtained by noting first that the Wigner phase-space distribution for the transmitted wavefunction (to right of barrier) is obtained by the convolution theorem from a Wigner-Weyl transform as

$$f(z,p,t) = \int dX\, T(z-X,p)\, g(X,p,t) \tag{46}$$

$$T(z,p) = (1/2\pi\hbar)\int dP e^{iPz/h} t(p+P/2)t^*(p-P/2) \tag{47}$$

where $t(p) = |t|\exp(i\phi)$ is the complex transmission coefficient and

$$g(z,p,t)\sim\exp[-(p-p_0)^2\sigma^2/2\hbar^2-2(z-z_0-pt/m)^2/\sigma^2] \qquad (48)$$

is the Wigner function for the unperturbed motion of an incident Gaussian wave-packet (initial position z_0,p_0). Provided t varies slowly over the width h/σ the strongest contribution to f comes from $z\sim X$, we may then expand $T(z,p)$ as

$$T(z,p) \cong |t(p)|^2\delta(z)-(i\hbar/2)[t^*(p)\partial_p t(p) - t(p)\partial_p t^*(p)](d/dz)\delta(z) \qquad (49)$$

$$\cong |t(p)|^2\{\delta(z) + \hbar\partial_p\phi(d/dz)\delta(z)\}$$

so that

$$|\Phi|^2 = \int dpf \cong \int dp|t(p)|^2\{1-\hbar\partial_p\phi d/dz\}g(z,p,t)$$

$$\cong \int dp|t(p)|^2\exp(-\hbar\partial_p\phi d/dz)g(z,p,t)$$

$$= \int dp|t(p)|^2 g(z-\hbar\partial_p\Phi,p,t) \qquad (50)$$

The traversal time follows as

$$\tau=m[a-z(p_0)]/p_0 \qquad (51)$$

where $z(p_0)\text{-}\hbar\partial_p\phi$ for a slow variation in the phase.

In general an incident wavepacket will be limited in size by either the mean free path or by the dimensions of the electrode region. The momentum spread may thus be larger than the conditions we have discussed. In that case there is substantial velocity filtering which can however be modelled by a different expansion scheme. Rather than pursue that line we turn instead to a much more rigorous treatment which establishes the phase-time on a sound footing and which holds for arbitrary wave-packets.

2.9 Rigorous Theory of Traversal Time

Let us define τ_I as the average time spent by a particle within a domain D:{0<z<a} associated with a tunnelling barrier of width less than or equal to a. For a wave-packet state $\Psi(z,t)$ we may use the Born interpretation to find

$$\tau_I = \int_0^a dz \int_{-\infty}^{\infty} dt|\Psi(z,t)|^2 \qquad (52)$$

This interaction time may be compared with that of the same wavepacket incident on a null potential. Denote this wavepacket by $\Psi_0(z,t)$. The time delay τ_d between the two is then

$$\tau_d = \int_0^a dz \int_{-\infty}^{\infty} dt\{|\Psi(z,t)|^2 - |\Psi_0(z,t)|^2\} \qquad (53)$$

If D is extended over all space the resulting delay time must vanish. We may thus expand as an integral over the range to the left and right of the domain D:

$$\tau_d = \underbrace{\left\{ \int_a^{\infty} dz \int_{-\infty}^{\infty} dt\right.}_{=\ I_2} + \underbrace{\left.\int_{-\infty}^{0} dz \int_{-\infty}^{\infty} dt \right\}}_{+\ I_1} \{|\Psi_0(z,t)|^2 - |\Psi(z,t)|^2\} \qquad (54)$$

If the wavepackets are incident from the same location deep to the left of the barrier we may expand them in terms of scattering states defined by the transmission and reflection amplitudes:

$$\Psi_0(z,t) = \int dkG(z,k,t) = \int dkg)k)e^{ikz}e^{-i\varepsilon t/\hbar} \quad (55)$$

$$\Psi(z,t) = \int dkG(z,k,t)|t(k)|e^{i\phi(+)} \qquad (z>a) \quad (56)$$

$$\Psi(z,t) = \int dkG(z,k,t)|\{1-|r(k)|e^{-i2kz}e^{i\phi(-)}\} \qquad (z<0) \quad (57)$$

where g(k) is non-zero for positive k values and $\phi(\pm)$ are the phase angles for transmission and reflection.

The evaluation of the integrals in equation (54) involves two important steps. The first uses the asymptotic result that the time integral converges to a delta function representing conservation of energy. Because the initial condition largely limits the k-integrals to positive k regimes implies that the energy conservation can be re-written in terms of momentum conservation.

$$\int_{-\infty}^{\infty}dt\ e^{-i\{\varepsilon(k)-\varepsilon(k')\}t/\hbar} = 2\pi\hbar\ \delta(\varepsilon(k)-\varepsilon(k')) \quad (58)$$

$$\rightarrow[2\pi/v(k)]\delta(\hbar k-\hbar k') \quad (59)$$

where $v(k)=\hbar k/m$ is the group velocity and the delta-function in k+k' is not required.

The second integral of importance has the form:

$$\int_0^{\infty}dz\ e^{\pm izu} = \pm i/[u\pm i0^+] = \pm i\mathcal{P}(1/u) + \pi\delta(u) \quad (60)$$

This integral occurs in every term provided one makes the change of variable $z \rightarrow z+a$ where appropriate.

If we now consider the integrals in (54)-(58) we find that I_1 contains the incident free wavepacket plus the interacting wavepacket which in turn gives rise to an incident, reflected and interference component in the probability densities. Using (59) and (60), performing the z integral before the time integral we find that the interference contribution vanishes and we are left with just the reflected component:

$$I_1 = \int dk\int dk'g(k)g(k')^*|r(k)||r(k')|e^{i\{\phi(-)-\phi'(-)\}}[2\pi/v(k)]\delta(\hbar k-\hbar k')$$
$$\times\ \{i\mathcal{P}(1/[k-k'])+\pi\delta(k-k')\} \quad (61)$$

Similarly

$$I_2 = \int dk\int dk'g(k)g(k')^*e^{ia(k-k')}\{1-|t(k)||t(k')|e^{i\{\phi(+)-\phi'(+)\}}[2\pi/v(k)]\delta(\hbar k-\hbar k')$$
$$\times\ \{i\mathcal{P}(1/[k-k'])+\pi\delta(k-k')\}\ . \quad (62)$$

The factor of 1 from the free wavepacket in (62) may be written identically as: $1=|r(k)|^2+|t(k)|^2$ and the $|r|^2$ component transferred into the integral in (61). We may then carry out the $k\rightarrow k'$ limits and perform the k' integration (noting that terms in d/dk(t or r) cancel out) to find

$$\tau_d = 2\pi\int dk\ \{|g(k)|^2/v(k)\}\{|t(k)|^2\ (d/dk)\phi(+) + |r(k)|^2\ (d/dk)\phi(-)\}\ . \quad (63)$$

Similarly, we find the free wavepacket has an interaction time with the null potential given by

$$\tau_0 = 2\pi\int dk\ |g(k)|^2a/v(k) \quad . \quad (64)$$

The interaction time for the interacting wavepacket is thus:

$$\tau_I = 2\pi\int dk \{|g(k)|^2/v(k)\}\{a+|t(k)|^2 (d/dk)\phi(+) + |r(k)|^2 (d/dk)\phi(-)\} . \quad (65)$$

The interaction time evidently is an average over all the possible tunnelling (traversing) and reflection processes, so it does not give the tunnelling time directly. However, we can re-write (65) as the weighted sum of an effective tunnel/traversal time τ_b and a reflection time τ_r

$$\tau_I = R_{eff}\, \tau_r + T_{eff}\, \tau_b \quad (66)$$

where

$$T_{eff} = 2\pi\int dk \{|g(k)|^2 \ |t(k)|^2\};\ R_{eff} = 2\pi\int dk \{|g(k)|^2 \ |r(k)|^2\} \quad (67)$$

$$\tau_r = 2\pi\int dk \{|g(k)|^2 \ |r(k)|^2/v(k)\}\{a + (d/dk)\phi(-)\}/R_{eff} \quad (68)$$

$$\tau_b = 2\pi\int dk \{|g(k)|^2 \ |t(k)|^2/v(k)\}\{a + (d/dk)\phi(+)\}/T_{eff} \quad . \quad (69)$$

It is the traversal time given by (69) which closely fits the numerical calculations. As such, (69) holds for arbitrary wavepackets provided they have very small negative k components in the initial state. If $|t(k)|^2$ varies slowly over the initial momentum distribution $|g(k)|^2$ which we presume is then peaked at $k=k_0$ we obtain the simple phase delay result

$$\tau_b = \{1/v(k_0)\}\{a + (d/dk)\phi(+)\} \quad . \quad (70)$$

The essence of the analysis is contained in the concept of the time a particle spends in interaction with a scattering centre; a similar three-dimensional analysis in terms of partial waves is given in [47].

Figure 7 illustrates the tunnelling time calculated for the resonant tunnelling barrier described earlier. We observe that the tunnelling time is very fast and is somewhat slowed down in the vicinity of the resonance. There is considerable scope for calculating tunnel times analytically by this method in combination with transmission matrix theory.

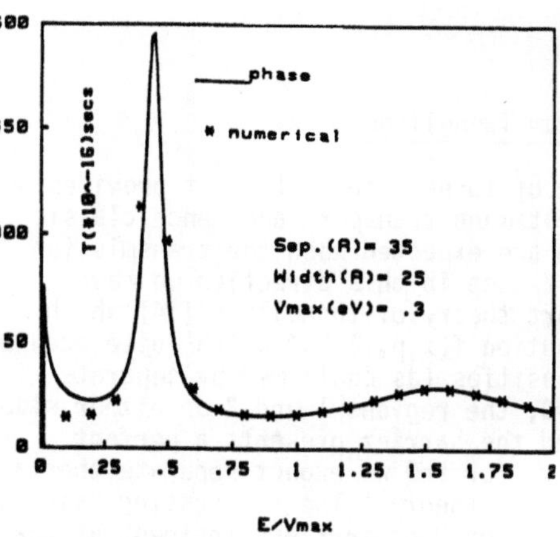

Fig.7a Tunnelling time for twin barrier

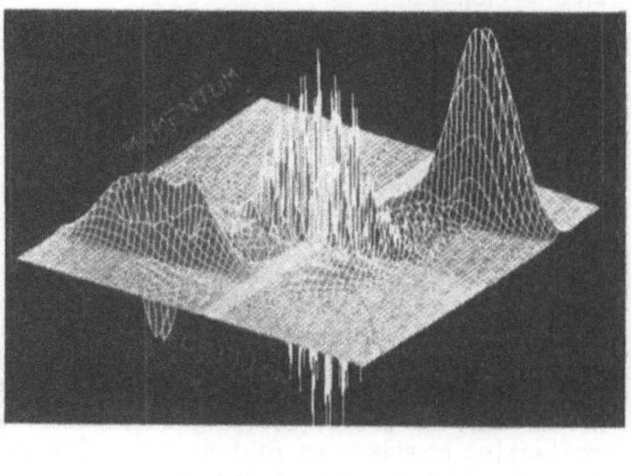

Fig.7b Wigner Distribution for a wavepacket incident on resonance

The square thick barrier case is curious [14]. The traversal time becomes independent of the barrier thickness (confirmed by direct computer simulation): τ_b=2m/(hkK). There is no violation of causality here, but a relativistic calculation would be of interest. Despite the fast effective crossing time only an exponentially small component of the probability distribution tunnels through. The effective transition rate is very small. In the limit of a step potential no carriers tunnel through at all and τ_b loses significance, although τ_r is still physical.

2.10 Escape from Quantum Wells

During resonant tunnelling there is a large build up of electron density in the well itself [32,14]. This trapped charge has a rate of escape which depends on the wavepacket momentum distribution. An interesting case occurs when a spatially very narrow (and hence broad in momentum) wavepacket is incident on resonance. The packet is almost entirely reflected when the resonance is narrow. Numerical studies indicate that a small component associated with the precise resonance does emerge quickly but most of the distribution which eventually tunnels is quickly trapped in the well and then decays very slowly. The rate of decay [14] is close to the decay rate predicted from the formula $|\Psi|^2 \sim e^{-\Gamma t}$ (see also [46]), which may be derived by assuming

$$t \sim i\Gamma/2/\{E(p)-E_0+i\Gamma/2\} \tag{71}$$

near resonance energy E_0. In such a situation the trapped carrier may make a faster transition out of the well via an inelastic process such as phonon scattering (either tunnelling or thermal excitation over the barrier). For very large resonant widths the decay may contain 1/t and $1/t^3$ dependences more characteristic of free wavepacket decay.

2.11 Inelastic Scattering

We have recently studied the tunnelling of wavepackets through simple and resonant tunnelling barriers in the presence of inelastic scattering on optical phonons associated with the barrier alone. The effect is modelled by choosing a space and time-dependent modulation of the barrier height and leads to pronounced side-bands in the momentum distribution and the appearance of several delayed transmitted and reflected wavepackets associated with emission and absorption of optical phonons in the barrier. The effect is strongest for resonant tunnelling where the delay time is long.

3. HOT ELECTRON TUNNELLING

3.1 A Semi-Classical Model for Non-Equilibrium Tunnelling

A major difficulty with the elementary theory of tunnelling is that it provides no clues as to the transition to conventional Boltzmann transport and hence classic hot electron/ballistic electron regimes which are expected when the transmission coefficient is large (of order 1). As a first step in this direction we have recently developed a phenomenological transport theory of tunnelling [14] which uses a classical electron phase-space distribution f(x,p,t) but which quite accurately predicts the correct charge and current densities (as confirmed by separate calculations for the Wigner function). If T=0, the regions 1 and 2 on either side of the barrier are effectively unconnected and the barrier presents a perfect reflecting boundary to regions 1,2. Evidently, for T=0 we expect separate thermal distributions in each region as in the elementary theory. The interesting case is of course 0<T<1, for which we may divert a fraction R of carriers incident at the left or right barrier edges into reflected streams; the transmitted (tunnelling if the incident energies are below the barrier height) carriers are assumed to keep to classical trajectories which are calculated for the appropriate barrier traversal time obtained from a separate quantum mechanical calculation.

Mathematically, we obtain a Boltzmann transport equation with addition source and sink terms g and r that describe tunnelling as a branching process:

$$\partial f/\partial t+v\partial f/\partial x+F\partial f/\partial p=g-r+\partial f/\partial t|_c \quad (72)$$

Here the force F is composed of the classical force associated with the bias potential field alone. F acts within the barrier to give a finite barrier traversal time. The advantage of this model is that we may deploy Monte Carlo particle simulation techniques without any difficulty. It has been shown elsewhere [14] that this formalism leads to the conventional tunnelling result when $T\sim 0$ and the energy relaxation is sufficiently strong.

Typically the transmitted carriers have energies in a range up to around a few hundred meV. If the energy and momentum relaxation processes are insufficiently fast the transmitted ($1\to 2$) carriers will be quasi-ballistic and fill-up the state $p'=[2m(eV + p^2/2m)]^{1/2}$; similarly the states p in region 1 will become depleted. The result will be a suppression of the tunnelling current via a reduction in the relative occupancy of states of the same total energy. The situation is analysable in a simplistic fashion by introducing f_s and f_a as the symmetric and anti-symmetric parts of the distribution function and setting

$$\partial f_s/\partial t|_{collisions} = -[f_s - f_0]/\tau \quad \text{where} \quad (73)$$

$$1/\tau = 1/\tau_{elastic} + 1/\tau_{inelastic} \quad (74)$$

Introducing the tunnelling rate $1/\tau_t = vT/2d$, where d is the length of the left region 1, we find

$$f_s(-a)-f_s(a) = [f_0(-a)-f_0(a)][1-1/\{1+\tau_t/(\tau_1+\tau_2)\}] \quad (75)$$

The elementary result evidently requires fast thermalisation and small T (slow tunnelling rate).

3.2 Delays

So far we have allowed the carriers to cross the barrier at a velocity determined by motion in the bias field alone. Similarly, the reflected carriers have been re-injected instantaneously along the classical (barrier-free) trajectory. However both reflection and transmission processes will give rise to an addition quantum delay in both the asymptotic locations of the transmitted and reflected wavepackets. This effect can be calculated and allowed for phenomenologically in the classical formalism in a variety of ways. A particularly elegant method is to distort the intra-barrier phase-space trajectory by scaling the velocity on the classical trajectory by a factor τ_c/τ_b where τ_b and τ_c are the quantum and classical traversal times (which depend on the incident momentum at the barrier). This is equivalent to adding a perturbation $H'=H_0(1-\tau_c/\tau_b)[\theta(x-x_1)-\theta(x-x_2)]$ to the original classical Hamiltonian H_0. A similar technique can be used for reflection delays.

In this picture an arbitrary point in phase space within the barrier will evolve along a trajectory determined by the incident momentum. If τ_c/τ_b is a multi-valued function of p, which is generally the case, then trajectories may cross within the barrier and to keep order each trajectory must be tagged with its initial incident momentum. The resulting phase space portrait is found to be very similar to the phase-portrait of the Wigner function.

Detailed comparisons between this approach and exact quantum mechanical calculations indicate that the method is successful at predicting charge build-up in resonant tunnelling quantum wells. It works quite well for situations where T~1, i.e in classical regions or near resonances. The method is also quite accurate at high bias fields where the field acts to destroy some of the interference phenomena which generate charge build-up. This is the intra-collisional field effect in a further guise as noted previously [32]. The description of the carrier distribution within multiple quantum wells is less well-described by the method except near resonances.

If the intra-barrier motion is ballistic and we ignore space charge effects, the details of intra-barrier motion are irrelevant for steady-state calculations. But transients, switching phenomena, hot electron effects within the barrier, dielectric changes, space-charge re-normalisation of the barrier, image-force effects all depend on a detailed picture of the tunnelling.

3.3 Hot Electron Effects

A general review of hot electron effects was given in reference [14]. Here we concentrate on hot electron effects associated with a large transmission coefficient. If T is small and $1/\tau$ is small, we obtain a strong suppression of the tunnelling current. If, on the other hand, T is large and $1/\tau$ is small, the model reduces to the classical equations of transient dominated quasi-ballistic motion typical of staircase systems. Pronounced velocity-overshoot is found from numerical calculations. It should be noted that the injected carriers will cascade down in energy through optical phonon and plasmon emission and the velocity gradually randomises as they progress into the second electrode region. Current continuity is maintained by an equivalent change in f_a. In the absence of a driving field in regions 1 and 2, f_a decays spatially and temporarily over a distance equal to the mean-free path against momentum relaxation and a polarised state is set up. In practise a driving field always exists in these regions so that a current is maintained.

In between the extremes of near-equilibrium tunnelling and ballistic motion we evidently obtain a hot electron regime. This region is easily explored with the previously described Monte Carlo techniques. A space-dependent carrier temperature is feasible provided carrier-carrier collisions are strong.

3.4 Transient Effects

The delay which is introduced into the carrier traversal times for both over-the-barrier and through-the-barrier transmission or reflection will itself lead to an influence on hot electron phenomena. In an earlier Monte Carlo study [14], the spatial velocity distribution in a back-to-back planar doped barrier device is found to be significantly cooled and has reduced overshoot due to delays introduced by reflections at the barriers and by thermalisation of carriers which tunnel through the collector.

It is clear from the foregoing that an estimate of the barrier traversal time τ_b is important. Knowledge of τ_b is also important for the frequency response of tunnelling devices and for a rough guide to space-charge barrier renormalisation. In the semi-classical theory of Sec.4 the simplest version of the traversal time is the classical traversal time for a carrier to cross the potential step formed by the applied potential not including the bare barrier potential. However, the phase time and its generalisations described earlier are more appropriate, particularly for multiple barriers.

3.5 Resonant Tunnelling Devices

Very recently Mudares and Barker have modelled the resonant tunnelling barrier I-V characteristics using a sophisticated space-time-dependent self-consistent ensemble Monte Carlo model (50,000 particles). Here we use scattering rates which allow for degeneracy; the Poisson equation is used to determine the self-consistent potential; the generalised phase-time is used to handle the finite traversal times; the transmission coefficients and traversal times are obtained at each time step from the self-consistent barrier potential.

The expected electron accumulation in the well is clearly seen and has an associated velocity and energy overshoot. The latter are primarily associated with the large potential drop across the barrier. The hot electron spatial inhomogeneity should be directly accessible by the magnetophonon resonant energy relaxation techniques.

Hot electron effects might be expected for large transmission. However this is difficult to achieve in the conventional barrier configuration because the bias field sufficiently distorts the barrier to remove the perfect transmission condition. But this could be achieved in device structures which achieve $T_1=T_2$ at non-zero voltage, for example by using barriers of different widths, or by applying a symmetric bias between well and barriers and sweeping the Fermi level. Figure 8 shows some of the possibilities. An interesting method involves using a resonant barrier in the base of a bipolar transistor (CAPASSO and KIEHL [4]).

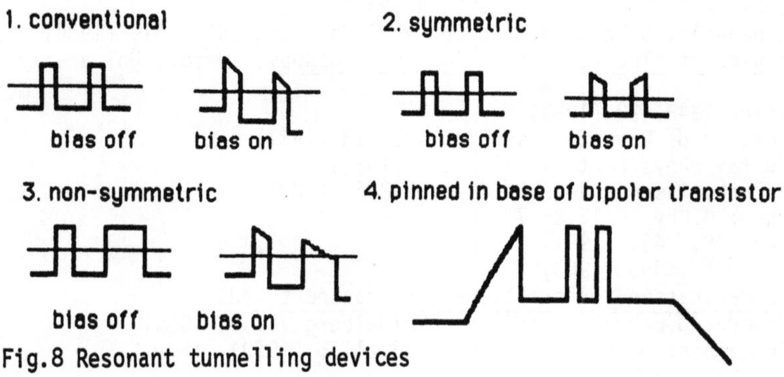

Fig.8 Resonant tunnelling devices

Preliminary Monte Carlo calculations have been made for the I-V characteristics and charge-density profiles, energy and velocity distributions and effective potential for the Capasso-Kiehl device at low bias, near resonance and above resonance. At small bias fields we expect the tunnel injection current to drive the resonance. However, the very strong electron-impurity, electron-electron and electron-hole interactions in the base rapidly thermalise the system and no detectable resonance is obtainable until thermionic emission occurs over the barrier. The strong hot electron effects are not present in the barrier in this case. The charge accumulation is present and does distort the barrier to soften the resonance. The peak-to-valley ratio is disappointingly low. Calculations for a lightly doped base appear more encouraging.

The most promising development in resonant tunnelling is the possibility to use an interdigitated gate electrode in a HEMT configuration. Numerical studies suggest that the self-consistent gate potential digs sufficiently deep to provide a quantum barrier system in GaAs devices. Lateral tunnel and superlattice devices are under active development at Glasgow's nanoelectronics centre.

4. CONCLUSIONS

The semi-classical picture of tunnelling suggests that hot electron effects are to be expected in tunnelling experiments on low geometry devices, particularly near strong transmission. New Monte Carlo techniques permit detailed modelling of such phenomena. Hot electron effects are shown to diminish peak to valley ratios in the I-V characteristics of resonant tunnel devices. Poisson corrections to the barrier potentials soften the resonant tunnelling peaks. The concepts of tunnelling times and non-classical traversal times are central to any treatment of the transient transport which occurs in small structures. Tunnelling and interference phenomena are likely to be of much greater significance than the distortion of scattering processes in the large inhomogeneous fields. A quantal description of scattering in strong inhomogeneous fields is still outstanding.

1. J.R.Barker: in Physics of Non-Linear Transport in Semiconductors, edited by D.K.Ferry, J.R.Barker, C.Jacoboni (Plenum, New York 1980) pp127, 589
2. J.R.Barker: Microelectronics Journal, 17,15(1986)
3. J.R.Barker and K.K.Ferry: Solid State Electronics 23, 519 (1980); 23, 531 (1980)
4. F.Capasso and R.A.Kiehl: J.Appl.Phys.58, 1366 (1985)
5. F.Capasso, K.Mohammed, A Y Cho, R.Hull, A L Hutchinson: Appl.Phys.Lett.47 420 (1985)
6. F.Capasso, K.Mohammed, A Y Cho: J.Appl.Phys.Lett.48, 478 (1986)
7. S.Luryi, F.Capasso: J.Appl.Phys.Lett.47, 1347 (1985)
8. R.F.Kwasnick, M.A.Kastner, J.Melngailis, P.A.Lee: Phys.Rev.Lett.52, 224 (1984)
9. A.C.Warren, D.A.Antoniadis, H.J.Smith, J.Melngailis: IEEE El.Dev.Lett.EDL-6, 294, (1985)
10. A.R.Bonnefoi, T.C.Mcgill, R.D.Burnham: IEEE El.Dev.Lett.EDL-6,636 (1985)
11. L.L.Chang, L.Esaki, R.Tsu: Appl.Phys.Lett.24,593 (1974)
12. T.Sollner, Pllannenwald, D.Peck, W.Goodhue: Appl.Phys.Lett.45, 1319 (1984)
13. E.L.Wolf: Principles of electron tunnelling spectroscopy, (Oxford University London 1985)
14. J.R.Barker: Physica 134B, 22 (1985)
15. J.R.Barker, D.Lowe: J.de Physique, supp 10,42, 293 (1981)
16. J.R.Barker, S.Murray: Phys.Letters, 93A, 271 (1983)
17. P.Carruthers, F.Zachariasen:: Rev.Mod.Phys.55,245 (1983)
18. D.Lowe, J.R.Barker: J.Phys.C 18,2507 (1985)
19. A.Peres: Phys.Rev.24B, 7463 (1981)
20. A.Peres: Phys.Rev.27B, 6493 (1983)
21. R.Landauer: in Localisation, Interactions and Transport, eds B.Kramer, T.Bruynseraeda, G.Bermann (Springer, Heidelberg vol.61 1985)
22. J.P.Vigneron, Ph Lambin: J.Phys A:Math.Gen, 13, 1135 (1980)
23. S.Collins, J.R.Barker: to be published
24. R.E.Borland: Proc.Phys.Soc.78, 926 (1961)
25. A.Peres: J.Math.Phys.24, 1110 (1983)
26. R.Gilmore: Lie groups, Lie algebras and some of their applications, (Wiley, New York, 1974)
27. J.Heading: J.Atmos. Terr.Physics 25, 519 (1963)
28. J.Heading: Proc.Camb.Phil.Soc.74, 161 (1973)
29. M.Vassell, L.Johnson, H.Lockwood: J.Appl.Phys.54, 5206 (1983)
30. J.R.Barker, D.Lowe, S.Murray: in Sub-micron Physics, (Plenum, New York 1983)
31. F.Capasso: Physica, 129B, 92 (1985)
32. J.R.Barker, D.Lowe, S.Collins, S.Murray: Proc.17th Int.Conf.Phys.Semiconductors (Springer, San Francisco 1985); 449
33. C.Mailhiot, D.L.Smith, T.C.McGill: J.Vac.Sci.Technol.B 1, 439 & 637 (1983)
34. S.Collins, D.Lowe, J.R.Barker: J.Phys.C 18, L637 (1985)
35. T.E.Hartmann: J.Appl.Phys.33, 3427 (1962)
36. W.Franz: Phys.Stat.Solidi 22,K139 (1967)
37. E.P.Wigner: Phys.Rev.98,145 (1955)
38. D.Bohm: Quantum Theory, (Prentice Hall, London 1951), 257
39. L.Eisenbud: Dissertation: Princeton (1948)
40. L.A.MacColl: Phys.Rev.40,621 (1932)
41. M.Buttiker, R.Landauer: Phys.Rev.Lett.49, 1739 (1982)
42. K.Stevens: J.Phys.C 16, 3649 (1983)
43. A.A.Sokolow, J.M.Loskutov, I.M.Ternow: Quantenmechanik, (Akademieverlag Berlin, 1964), 113
44. F.T.Smith: Phys.Rev. 118,349 (1960)
45. W.E.Hagston: Phys.Stat.Solidi(b)116,K85, (1983)
46. K.K.Thornber, T.C.McGill, C.A.Mead: J.Appl.Phys.38,2384 (1967)
47. A.Bohm: Quantum Mechanics, (Springer, New York, 1979)

Noise in Microstructures

J.P. Nougier

Centre d'Electronique de Montpellier, (Laboratoire associé au CNRS, UA 391),
Université des Sciences et Techniques du Languedoc,
F-34060 Montpellier Cédex, France

1. INTRODUCTION

Noise in devices is due to local random events, namely scattering mechanisms undergone by the carriers, which produce fluctuations of the current flow, and hence fluctuations of the voltage at the ends of the device. The knowledge of the noise is important mainly for two reasons: From a fundamental point of view, it gives information concerning the transport mechanisms (basic limitations such as the scattering mechanisms with phonons, etc., or technological limitations due to impurities, defects, etc.); from an applied point of view, noise is a fundamental limitation to the sensitivity of the device, so that it is important to study whether it can be lowered or not.

Figure 1 depicts a device, biased between the electrodes M and N, with a d.c. current (water) flow. The angler, sitting in M', is looking at the cork in N', which is trembling because the surface of the water fluctuates, due to stones dropped at point S. This represents a noise source, and the motion of the cork represents the fluctuating potential between electrodes M' and N', superimposed to the d.c. bias. Of course, M' and N' may be identical with M and N. Obviously, the noise between M' and N' depends on the noise source at point S, and also on the way in which the fluctuations, produced at S, propagate towards the electrodes (the fluctuations between M' and N' will not be the same as between M' and N"). This is called the impedance field. As a consequence, for modeling the noise of a device, one needs to know both the local noise sources, and the impedance field. The total noise, between electrodes M' and N', is the sum of the effects of the fluctuations due to all the noise sources inside the device.

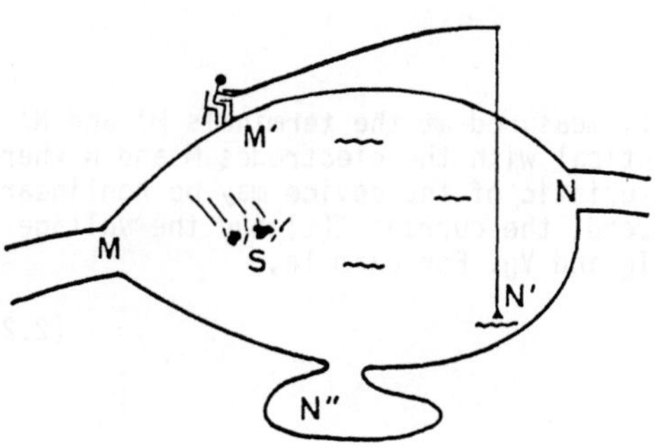

Fig.1. Electronic-like device

Specific problems arise when dealing with short devices, due to hot carrier effects for devices of a few microns length, superimposed to non-stationary effects for devices shorter than one micron. In this paper, the basic definitions for noise quantities are first recalled (Sect.2). The expressions for the noise sources are then given (Sect.3). The impedance field method is explained in Sect.4, together with the practical ways to apply it: using the local field (Sect.5), or the local voltage (Sect.6). The problems encountered when dealing with submicron devices are then reviewed (Sect.7).

2. NOISE: DEFINITIONS AND RELATED QUANTITIES

2.1 Random Functions

A periodic function of time $F_c(t)$ can be expanded into a Fourier series. Let $\overline{F_c(t)}$ be its average value. The component of $F_c(t) - \overline{F_c(t)}$, at frequency f, has a constant amplitude a_f, independent of time. Let us now consider a random function $F_r(t)$, of average value $\overline{F_r(t)}$. $F_r(t) - \overline{F_r(t)}$ cannot be expanded into a Fourier series, but into a Fourier-Stieltjes integral. Its component, at frequency f within the bandwith Δf, is a random function of time $a_f(t)$ (see Fig.2), the amplitude of which suddenly changes at random instants $t_1, t_2, \ldots, t_j, \ldots$. For stationary random functions of second order, two quantities, among others, are of great interest:

- The average amplitude, or its square average

$$S_F(f)\ \Delta f = \overline{|a_f(t)|^2} = \overline{a_f(t).a_f^*(t)} \tag{2.1}$$

where * means the complex conjugate. $S_F(f)$ is the noise spectral density of $F_r(t)$.

- The average time $\tau_c = \overline{t_{j+1} - t_j}$. For durations $t \ll \tau_c$, $a_f(t)$ keeps a constant amplitude. For durations $t \gg \tau_c$, the amplitude of $a_f(t)$ changes, in an unforeseeable way. Thus, τ_c is the correlation time of the random function $F_r(t)$.

This provides a way for noise measurements. Filtering at frequency f within the bandwidth Δf gives $a_f(t)$, then a quadratic detection gives $a_f^2(t)$, and finally an integration gives an output signal proportional to $\overline{a_f^2(t)}$, that is proportional to $S_F(f)$.

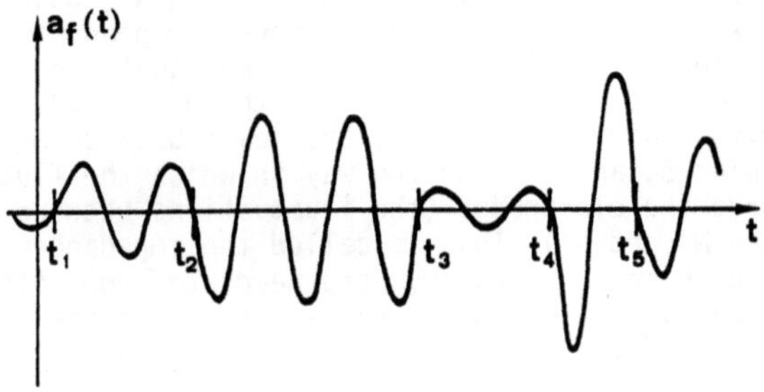

Fig.2. Component, at frequency f, of a random function $F_r(t)$

2.2 Spectral Densities

Let V_0 and I_0 be the d.c. voltage and current, measured at the terminals M' and N' of a device (M' and N' may or may not be identical with the electrodes M and N where the d.c. bias is applied). The $I_0(V_0)$ characteristic of the device may be nonlinear, which generally occurs for short devices. Indeed, the current I(t) and the voltage V(t) are not constant, but fluctuate around I_0 and V_0. For example,

$$V(t) = V_0 + \Delta V(t) \quad . \tag{2.2}$$

The correlation function of V is defined as

$$\Gamma_V(V_0,\tau) = \overline{\Delta V(t).\Delta V^*(t-\tau)}^{\,t}$$

and the spectral density, at the mathematical (algebraic) frequency ν, is

$$\gamma_V(\nu) = \gamma_V(V_0,\nu) = \int_{-\infty}^{+\infty} \Gamma_V(V_0,\tau)\exp(-2j\pi\nu\tau)\,d\tau$$

and the measurable spectral density (noise voltage) at the physical frequency f is

$$S_V(f) = S_V(V_0,f) = \gamma_V(\nu = f) + \gamma_V(\nu = -f) \quad . \tag{2.3}$$

For real quantities

$$S_V(f) = 4 \int_0^{\infty} \Gamma_V(V_0,\tau) \cos(2\pi f\tau)\, d\tau \quad . \tag{2.4}$$

This provides a second method for noise measurements. First the random signal V(t) is sampled, giving $V_j = V(t_j)$, then the correlation is computed, and finally a fast Fourier transform procedure is used to compute $S_V(f_k)$.

2.3 Quantities Related to the Noise

Of course, what has been defined with the voltage fluctuations, leading to the noise voltage $S_V(f)$, can be performed with the current, leading to the noise current $S_I(f)$ around the bias point. Let $\delta V \exp(j\omega t)$ and $\delta I \exp(j\omega t)$ be small a.c. quantities superimposed to the d.c. ones. The a.c. impedance $Z(f) = Z(V_0,f)$ around the d.c. point and the a.c. admittance $Y(f) = Y(V_0,f)$ are defined as

$$Z(f) = \delta V/\delta I \quad \text{and} \quad Y(f) = 1/Z(f) \quad . \tag{2.5}$$

Then, it can be shown that $S_V(f)$ and $S_I(f)$ are related through

$$S_V(f) = |Z(f)|^2 S_I(f) \quad . \tag{2.6}$$

The noise temperature $T_n(f)$, around the d.c. value, is defined by one of the two equivalent relations

$$\begin{cases} S_V(f) = 4\, k_B\, T_n(f)\, \mathrm{Re}\,\{Z(f)\} & (2.7) \\ S_I(f) = 4\, k_B\, T_n(f)\, \mathrm{Re}\,\{Y(f)\} & (2.8) \end{cases}$$

where k_B is the Boltzmann constant, and Re means the real part. It can be easily shown that k_BT_n is the maximum fluctuating power per unit bandwidth, which can be displayed to an output circuit, when impedances are matched. Hence, the noise temperature $T_n(f)$ can be measured.

3. NOISE SOURCES

Due to the relation $j = qnv$, the current through the device fluctuates if at least one of the quantities q, n, or v fluctuates, thus producing Shot noise, Generation-Recombination (GR) noise, or diffusion noise, respectively.

3.1 Shot Noise

Since the electrons are emitted from an electrode at random instants (with a Poissonian distribution), the flow of electron charges q is not constant in time, but fluctuates. This produces a noise current, observable in vacuum valves, p-n junctions, etc., given by [1]

$$S_I(f) = 2\,q\,I \quad . \tag{3.1}$$

The Shot noise is white at usual frequencies.

3.2 Definition of the Local Noise Sources

Let us label $\vec{J}(\vec{r},t)$ the instantaneous local current density of d.c. value $\vec{J}_0(\vec{r})$:

$$\Delta\vec{J}(\vec{r},t) = \vec{J}(\vec{r},t) - \vec{J}_0(\vec{r}) \quad . \tag{3.2}$$

$\Delta\vec{J}(\vec{r},t)$ is randomly distributed. If $\Delta j_{N\alpha}(\vec{r},f)$ denotes the current density noise source along the direction α, i.e., the component of $\Delta j_\alpha(\vec{r},t)$ at frequency f, within the bandwidth Δf, one defines [see (2.1)], the noise source term [2] as

$$S_{j\alpha\beta}(\vec{r},\vec{r}',f)\,\Delta f = \overline{\Delta j_{N\alpha}(\vec{r},f).\Delta j^*_{N\beta}(\vec{r}',f)} \qquad (3.3)$$

where the bar means time average. Quite generally, one assumes that the noise sources at two different points are uncorrelated, which gives

$$S_{j\alpha\beta}(\vec{r},\vec{r}',f) = K_{\alpha\beta}(\vec{r},f)\,\delta\,(\vec{r}' - \vec{r}) \qquad (3.4)$$

where $\delta(\vec{r}' - \vec{r})$ is the Dirac function. Integration of (3.3) and (3.4) over the cross section of a slice of thickness $\Delta x'$ along an a.c. current path (see Fig.3) gives, if ΔI_N denotes the fluctuating current through the slice,

$$\overline{\Delta I_N(x,f).\Delta I^*_N(x',f)} = \Delta A(x)\,K_x\,(\vec{r},f)\,\delta\,(x - x')\,\Delta f \quad . \qquad (3.5)$$

This relation, given in [3], had already been established for some particular types of noise sources [4,5]. Integrating (3.5) over the domain $\Delta x' = \Delta x$, gives the relation between the local noise source $K_X(\vec{r},f)$ and the noise current $S_I(\vec{r},f)$ in the slice at point $\vec{r}$:

$$K_x(\vec{r},f) = S_{Ix}(\vec{r},f)\Delta x/\Delta A(x) \quad . \qquad (3.6)$$

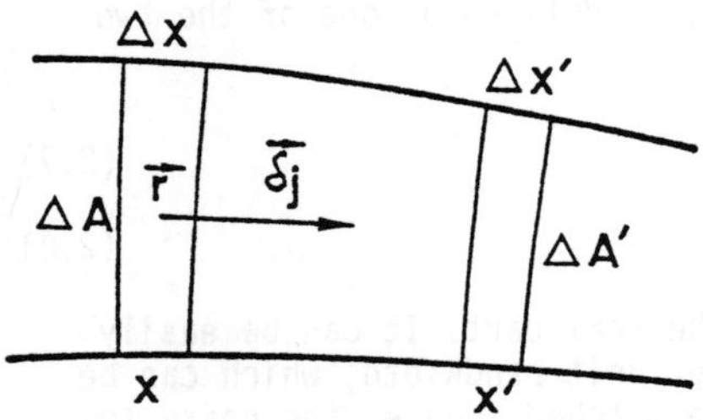

Fig.3. Slice of a current tube; x is the abscissa of the slice Δx along an a.c. current path

3.3 The Local Noise Source and the Local Noise Temperature

The experimental way, for determining the noise source term, obviously follows from (3.6). One determines the noise current of an homogeneous bar in a uniform electric field $\vec{E}$, from which one gets $K_X(E,f)$; then, in an inhomogeneous device with the same impurity concentration, lattice orientation, etc., one takes $K_X(\vec{r},f) = K_X[E(\vec{r}),f]$. However, hot carrier experiments require short pulses in order to avoid thermal heating of the semiconductors, and thus high frequency analysis. At those frequencies, because of matching impedances, the experimental parameter which can be measured is the noise temperature T_n (an extended bibliography will be found in [6,7]). Since T_n is related to S_I [cf. (2.8)], K can be related to T_n [8]. Let ΔY_X be the a.c. of the slice Δx. One has [6,7], from (2.8),

$$S_I(\vec{r},f) = 4\,k_B\,T_{nx}\,(\vec{r},f)\,\mathrm{Re}\{\Delta Y_x\} \qquad (3.7)$$

$$\Delta Y_x = \delta I/\delta V = \Delta A\,\delta j_x\,(\vec{r})/[\Delta x\,\delta E_x\,(\vec{r})] \quad . \qquad (3.8)$$

Equations (3.7) and (3.8), carried into (3.6), give

$$K_\alpha(\vec{r},f) = 4\,k_B\,T_{n\alpha}(\vec{r},f)\,\mathrm{Re}\,\{\sigma_{\alpha\alpha}(\vec{r},f)\} \qquad (3.9)$$

where $\vec{\vec{\sigma}}(\vec{r},f)$ is the a.c. conductivity tensor: $\vec{\delta j} = \vec{\vec{\sigma}}\,\vec{\delta E}$. Equation (3.9) shows that the noise source can be determined experimentally by measuring the noise temperature and the a.c. conductivity, without needing any knowledge on the physical nature of the noise source [8]. Note that the direction α may take different values with respect to the d.c. electric field [8].

3.4 Diffusion Noise

Diffusion noise is due to the fluctuations of the velocities of the carriers. Let $\Delta N(\vec{r})$ be the number of carriers located in the slice of volume $\Delta A \Delta x$. The current through the slice is

$$\Delta I = q\, n(\vec{r})\, v(\vec{r})\, \Delta A \qquad (3.10)$$

where $v(\vec{r})$ is the average velocity of the carriers in the slice, which fluctuates since $\Delta N(\vec{r})$ is finite. From (3.10) one has

$$S_I(x,f) = q^2 n^2(r) \Delta A^2\, S_v(\vec{r},f) \quad . \qquad (3.11)$$

Let $S_{vi}(\vec{r},f)$ describe the fluctuations of the velocity of one single carrier:

$$S_v(\vec{r},f) = S_{vi}(\vec{r},f)/\Delta N(\vec{r}) = S_{vi}(\vec{r},f)/[n(\vec{r})\Delta A \Delta x] \quad . \qquad (3.12)$$

Since, by definition, [6,7], $S_{vi}(\vec{r},f) = 4\, D(\vec{r},f)$, where D is the diffusion coefficient, (3.11) and (3.12), carried into (3.6), give

$$\vec{\vec{K}}_{diff}(\vec{r},f) = 4 q^2 n(\vec{r})\, \vec{\vec{D}}(\vec{r},f) \quad . \qquad (3.13)$$

This result was given in [2,4,5]. Comparing (3.13) and (3.9) gives a relation between the noise temperature and the diffusion coefficient:

$$k_B T_{n\alpha}(\vec{r},f)/q = q\, n(\vec{r})\, D_{\alpha\alpha}(\vec{r},f)/\mathrm{Re}\{\sigma_{\alpha\alpha}(\vec{r},f)\} \quad . \qquad (3.14)$$

This relation is valid for space charge regimes of hot carriers. A particular case of this relation gives the link between the diffusion coefficient and the noise temperature for homogeneous fields [6,7], which, in the ohmic regime, gives back the Einstein relation. It was demonstrated [9] that the diffusion coefficient, defined through the noise as is given here, is identical with the one defined by the spreading of a packet of carriers.

3.5 Generation-Recombination (GR) Noise

GR noise is due to the fluctuations of the density of carriers. Equation (3.10) can be expressed using the number $\Delta N(x)$ of carriers located in the slice $\Delta A\, \Delta x$:

$$\Delta I = q\, v(\vec{r})\, \Delta N(\vec{r})/\Delta x \quad . \qquad (3.15)$$

Assuming that the contribution of the fluctuations of $\vec{v}(\vec{r})$ is negligible, one gets

$$S_I(\vec{r},f) = q^2 v^2(\vec{r})\, S_N(\vec{r},f)/(\Delta x)^2 \qquad (3.16)$$

that is, using (3.15),

$$S_I(\vec{r},f) = j^2(\vec{r}) \Delta A^2\, S_N(\vec{r},f)/\Delta N^2(\vec{r}) \quad .$$

This relation, carried into (3.6), gives

$$K(\vec{r},f) = j^2 \Delta A\, \Delta x\, S_N(\vec{r},f)/\Delta N^2(\vec{r}) \quad . \qquad (3.17)$$

For GR noise involving only two levels [10], $S_N = 4\tau(\mathrm{var}\Delta N)/(1+\omega^2\tau^2)$, and due to the quasi Poissonian character of the fluctuations, $\mathrm{var}\Delta N = \alpha \Delta N$. This gives, for (3.17),

$$K_{GR}(r,f) = \frac{j^2(\vec{r})}{n(\vec{r})}\, \frac{4\alpha(\vec{r})\tau(\vec{r})}{1+\omega^2\tau^2(\vec{r})} \quad . \qquad (3.18)$$

This expression is analogous to that obtained for thermal GR noise [10], except that $j(\vec{r})$ is not necessarily linear with respect to the d.c. electric field, and that $\alpha(\vec{r})$ and $\tau(\vec{r})$ may depend on the local d.c. field. Such a situation occurs, for example, in n-type Si at 77 K. Then, at thermal equilibrium, only a part of the donor impurities is ionized. When a high d.c. field is applied, a fraction of the neutral impurities is ionized, due to Poole Frenkel and/or to impact ionization effect, thus leading to hot carrier GR noise, since this fraction depends on the d.c field, so as α and τ [11] (see also Sect.3.7)

3.6 1/f Noise

In spite of the efforts of a few dozen searchers since a few dozen years (see the proceedings of each conference on noise and on 1/f noise), there does not exist any unique and satisfactory theory for 1/f noise. The oldest, simplest, and perhaps more realistic one, is to represent 1/f as the sum of GR noises produced by subbands of impurities or defects, with a continuous distribution of relaxation times. As a consequence, it may be inferred, from (3.18), that the bulk 1/f noise source can be expressed as

$$K_{\text{bulk }1/f}(\vec{r},f) = \frac{j^2(\vec{r})}{n(\vec{r})} \frac{\alpha_H(\vec{r})}{f^\gamma} \tag{3.19}$$

where α_H is called the Hooge constant [12] (which is in fact a fitting parameter, taking values within a four-order of magnitude domain [13]!). The factor γ is about 1 (generally in the range 0.8 - 1.2).

Very often, the technology of bulk semiconductors is good enough, so that such 1/f noise is absent. However, the technology of surfaces and interfaces is not so well known, and important 1/f noise sources may be present at a given equipotential interface of abscissa x_0. The noise source is then expressed as

$$K_{\text{surf }1/f}(x,f) = \lambda \frac{j^2(x)}{n(x)} \frac{\delta(x-x_0)}{f^\gamma} \quad . \tag{3.20}$$

3.7 Transport Parameters Deduced from Noise Measurements

It has been noticed that measuring the noise temperature and the differential conductivity would get the local noise source in a homogeneous field. A noise analysis in a wide frequency range allows getting detailed information on some transport parameters, such as the diffusion coefficient $D(E)$, free carrier density $n(E)$, mobility $\mu(E)$, lifetime $\tau(E)$, ..., when the experimental conditions are favourable.

As an example, we report, in Fig.4, noise temperature measurements in InP at 300 K. Obviously, two noise sources are present: diffusion noise, and GR noise. The noise current is then, according to (3.13), (3.18) and (3.6),

$$S_I(f) = 4q^2 n(E) D(E) \frac{A}{L} + \frac{j^2}{n(E)} \frac{4\alpha\tau}{(1+\omega^2\tau^2)} \frac{A}{L} \quad . \tag{3.21}$$

If one supposes that the GR noise involves only two levels, the conduction band and one trap, it can be shown [1] that $\alpha = (1-u)/(2-u)$, where $u = n/N_D$. Then, (3.21) depends on only three unknown parameters: $u(E)$, $D(E)$ and $\tau(E)$. At a given field, it is easy to fit the experimental curve $S_I(f)$, using a least square fitting (solid lines of Fig.4) and thus to determine the transport parameters $n(E)$, $D(E)$ and $\tau(E)$, which gives $\mu(E)$ since $\sigma(E)$ has been measured. Figure 5 shows the results obtained for $D(E)$ [14]. Transport parameters were obtained in a similar way in various materials [17,18], see also Chap.4 of [19].

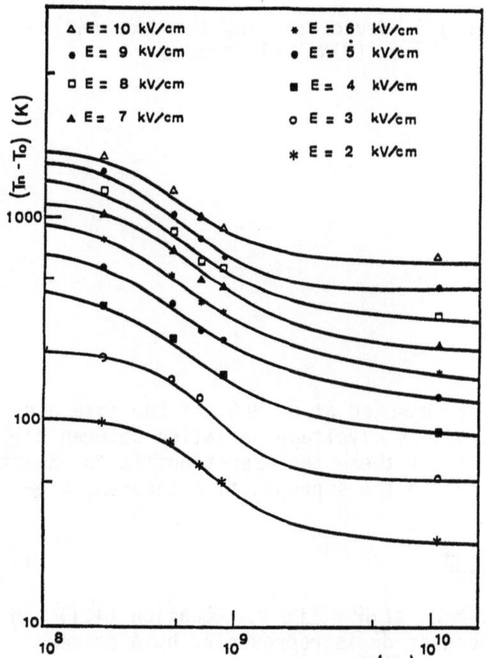

Fig.4. Excess noise temperature versus frequency, at various electric fields, in n-type InP, $N_D = 2.7 \times 10^{15}$ cm^{-3}, T = 300 K

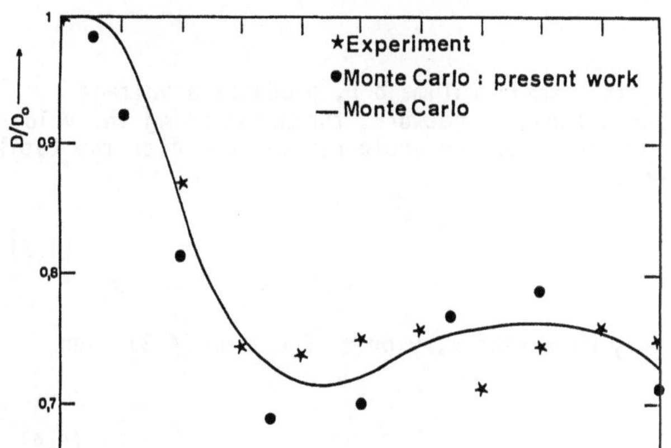

Fig.5. Diffusion coefficient versus electric field in n-type InP, $N_D = 2.7 \times 10^{15}$ cm^{-3}, at 300 K. Comparison between experimental results and Monte Carlo simulation. The ohmic value D_0 is $D_0 = 116$ cm^2/s

4. NOISE OF A DEVICE: THE IMPEDANCE FIELD METHOD (IFM)

4.1 Formulation

The Impedance Field Method (IFM) was first introduced by SHOCKLEY [15] for diffusion noise sources in one-dimensional devices. It was later on extended to GR noise [16], and to three-dimensional devices [2,4] with the transfer impedance formulation. It can be stated as follows [3].

Let us consider a device with a d.c. bias between the electrodes M and N (Fig.6). In order to study the noise between the electrodes M' and N' (which may be identical with or different from M and N), we study the effect of a noise source at point $\vec{r}$. This can be done by injecting a small a.c. current of complex value δI, at point $\vec{r}$, superimposed to the d.c. current. This small a.c. current flows towards N' and produces an a.c. voltage between M' and N': $\Delta V(\vec{r}) = Z(\vec{r},f)\delta I$, where $Z(\vec{r},f)$ is the a.c.

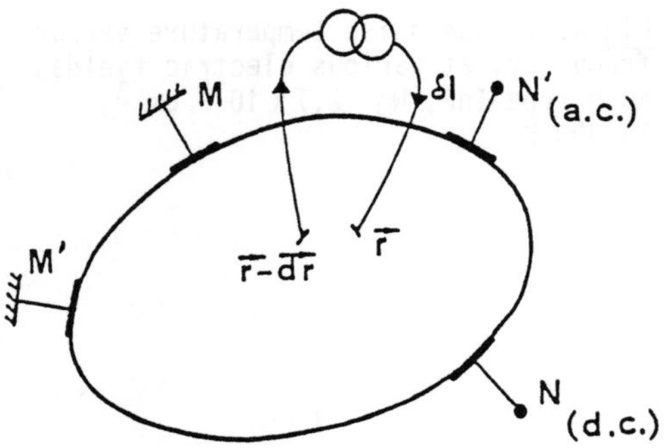

Fig.6. Device showing the d.c. (MN) and a.c. (M'N') electrodes

impedance between N' and M' due to the current injected at $\vec{r}$. Now, if the same a.c. current δI is taken out at point $\vec{r}-d\vec{r}$, it produces a voltage variation between M' and N': $\Delta V(\vec{r}-d\vec{r}) = Z(\vec{r}-d\vec{r},f)\delta I$. The net result of these two operations is to inject δI at $\vec{r}$ and to withdraw it at $\vec{r}-d\vec{r}$, which, due to the superposition theorem, creates an a.c. small voltage between M' and N':

$$\delta V(\vec{r}) = [Z(\vec{r},f) - Z(\vec{r}-d\vec{r},f)]\delta I = \vec{\nabla}Z(\vec{r},f).d\vec{r}\delta I \quad . \tag{4.1}$$

Here, $\vec{\nabla}Z(\vec{r},f)$ is called the impedance field. Now, $\delta I\, d\vec{r} = \vec{\delta j}\, d^3r$. Equation (4.1) can then be written in a matrix form. A vector such as $d\vec{r}$ is represented by a column matrix, a $\tilde{d\vec{r}}$ by a row matrix. Equation (4.1) then writes

$$\delta V(\vec{r}) = \tilde{\vec{\nabla}}Z(\vec{r},f).\vec{\delta j}(\vec{r})\, d^3r \quad . \tag{4.2}$$

Equation (4.2) means that a current, in a small volume d^3r, produces a voltage across the device. The device is then a two-port network, the input being the volume d^3r and the output being M'N'. Integrating over the whole device, one gets the total voltage produced by all these currents:

$$\delta V = \iiint \tilde{\vec{\nabla}}Z(\vec{r},f).\vec{\delta j}(\vec{r})\, d^3r \quad . \tag{4.3}$$

Let us call α the component along every direction x, y or z. Equation (4.3) then writes

$$\delta V = \sum_{\alpha} \iiint d^3r\, \delta j_{\alpha}(r)\, [\partial Z(\vec{r},f)/\partial\alpha] \quad . \tag{4.4}$$

For getting the noise [cf. (2.1)], one multiplies δV by δV^*, which gives

$$\delta V\, \delta V^* = \sum_{\alpha\beta} \iint \frac{\partial Z(\vec{r},f)}{\partial\alpha} \frac{\partial Z^*(\vec{r}',f)}{\partial\beta}\, \delta j_{\alpha}(\vec{r})\, \delta j^*_{\beta}(\vec{r}')\, d^3r\, d^3r' \tag{4.5}$$

where the double integral sign represents indeed an integral of order 6 over $\vec{r}$ and $\vec{r}'$, extended to the whole volume of the device. When dealing with random quantities and not sinusoidal ones, averaging (4.5) [see (2.1)] gives the noise

$$S_V = \sum_{\alpha\beta} \iint \frac{\partial Z(\vec{r},f)}{\partial\alpha} \frac{\partial Z^*(\vec{r}',f)}{\partial\beta} S_{j\alpha\beta}(\vec{r},\vec{r}',f)\, d^3r\, d^3r' \tag{4.6}$$

which can be written in a matrix form:

$$S_V = \iint \tilde{\vec{\nabla}}Z(r,f)\, \vec{\vec{S}}_j\, (\vec{r},\vec{r}',f)\, \vec{\nabla}Z^*\, (\vec{r}',f)\, d^3r\, d^3r' \quad . \tag{4.7}$$

Assuming, as usually, that the noise sources at two different points are uncorrelated [see (3.4)], and integrating over $\vec{r}'$, one gets

$$S_V = \iiint \vec{\nabla} Z(\vec{r},f)\, \vec{\vec{K}}(\vec{r},f)\, \vec{\nabla} Z^*(\vec{r},f)\, d^3r \quad . \tag{4.8}$$

4.2 One-Dimensional Case

One is often dealing with "one-dimensional devices", that is devices where the equipotential surfaces may be considered as parallel planes. This occurs when the cross sections of the equipotential surfaces are low compared with the square of their curvature radius. Then, $\vec{\nabla} Z(\vec{r},f) = \{\partial Z(\vec{r},f)/\partial x, 0, 0\}$, and $\delta\vec{j}(\vec{r}) = \{\delta j(x), 0, 0\}$, so that, once integrated over the y and z coordinates, (4.2) and (4.8) become

$$\begin{cases} \delta V = \int_0^L \nabla Z(x,f)\, \delta I(x)\, dx & (4.9) \\ S_V(f) = \int_0^L A(x)\, K(x,f)\, |\nabla Z(x,f)|^2\, dx & (4.10) \end{cases}$$

where A(x) is the cross section of the equipotential surface at abscissa x. The abscissa refers to the position along a current line, of length L. This formula is the one initially given by SCHOCKLEY [15].

4.3 Direct Numerical Simulation

It is often impossible to compute the impedance field, and hence the noise, in an analytical way (we shall see below how this should be done). Therefore, numerical techniques are necessary. A method was given and applied, by the Lille group, to GaAs MESFETs [20-22], dual gate GaAs FETs [23,24], and TEGFETs [24-27]. It follows from (4.9): If one supposes that $\delta I(x) = 1$ in the slice Δx, and is null elsewhere, (4.9) gives $\delta V = \nabla Z(x,f)\Delta x$. The method is then: i) Write the transport equations. ii) Linearize them to get a relation between the local electric quantity (field or voltage) and the local current $\delta I(x)$. iii) Discretize the length of the device. iv) Set $\delta I(x_j) = 1$ at a point x_j and 0 elsewhere, and compute the voltage δV_j across the device. The impedance field $\nabla Z(x_j)$ is $\nabla Z_j = \delta V_j/\Delta x_j$. v) Compute the noise using (4.10). In the following sections, we shall state more precisely the formalism and give some examples.

Such a method had been used in [28]. The IFM can even be applied without writing the transport equations. This was done in FETs [29], where the transistor was considered as an active line, being made of a chain of elementary FETs. The noise is then considered as the sum of the effects, at the drain and at the gate, of the noises produced by the various slices, which is the basic idea of the IFM, where each slice is an elemental transistor. This uses a circuit representation instead of transport equations.

5. IFM USING THE LOCAL ELECTRIC FIELD

5.1 Formulation

This method was developed by VAN VLIET et al. [2], and applies when the fundamental electrical variable is the local electric field $\vec{E}(\vec{r})$. One writes the transport equations (conduction, continuity, Poisson, etc.) and eliminates the auxiliary variables (such as the free carrier density), so as to get a relation between the local electric field and the local current density, in the form

$$f(\vec{E}(\vec{r})) = \vec{j}(\vec{r}) \quad . \tag{5.1}$$

This relation is usually not linear. Solving (numerically in general) this equation in the stationary case leads to a relation between the d.c. values $\vec{E}_0(\vec{r})$ and $\vec{J}_0(\vec{r})$. Let us apply an a.c. voltage at the terminals M' and N', superimposed to the d.c. voltage applied at M and N. This results in a small local a.c. field $\delta\vec{E}(\vec{r})\exp i\omega t$, and in a local a.c. current density $\delta\vec{J}(\vec{r})\exp i\omega t$, superimposed to the d.c. values $\vec{E}_0(\vec{r})$ and $\vec{J}_0(\vec{r})$. If one sets $\vec{E}(\vec{r}) = \vec{E}_0(\vec{r}) + \delta\vec{E}(\vec{r})\exp i\omega t$, and $\vec{J}(\vec{r}) = \vec{J}_0(\vec{r}) + \delta\vec{J}(\vec{r})\exp i\omega t$ in (5.1) and expands this equation in perturbation, the 0-order terms give the d.c. values, and the first-order terms, in $\delta\vec{E}(\vec{r})$ and $\delta\vec{j}(\vec{r})$, give an equation of the form

$$\hat{L}\,\delta\vec{E}(\vec{r}) = \delta\vec{j}(\vec{r}) \tag{5.2}$$

where $\hat{L}$ is a linear operator. The Green matrix $\vec{\vec{z}}(\vec{r},\vec{r}\,',f)$ of the operator $\hat{L}$ is defined as (here $\hat{I}$ means the unitary matrix and $\delta(\vec{r}-\vec{r}\,')$ the Dirac function)

$$\hat{L}\vec{\vec{z}}(\vec{r},\vec{r}\,',f) = \hat{I}\delta(\vec{r}-\vec{r}\,') \quad . \tag{5.3}$$

Equations (5.2) and (5.3) give, integrating over the whole volume of the device,

$$\delta\vec{E}(\vec{r}) = \iiint d^3r'\,\vec{\vec{z}}(\vec{r},\vec{r}\,',f)\,\delta\vec{J}(\vec{r}\,',f) \quad . \tag{5.4}$$

Integrating on a line (generally a current line) of length L, from M' to N', one gets the a.c. voltage at the terminals:

$$\delta V = -\int_0^L \tilde{\vec{dr}}\,\tilde{\delta\vec{E}}(\vec{r}) = -\int_0^L \iiint d^3r'\,\tilde{\vec{dr}}\,\vec{\vec{z}}(\vec{r},\vec{r}\,',f)\,\delta\vec{J}(\vec{r}\,',f) \quad . \tag{5.5}$$

A comparison of (5.5) and (4.3) shows that

$$\tilde{\vec{\nabla}}Z(\vec{r}\,',f) = -\int_0^L \tilde{\vec{dr}}\,\vec{\vec{z}}(\vec{r},\vec{r}\,',f) \quad . \tag{5.6}$$

5.2 One-Dimensional Devices

When $\vec{\vec{z}}(\vec{r},\vec{r}\,',f) = z(x,x',f)\,\hat{I}$, integrating (5.4) over the y and z coordinates yields

$$\delta E(x,f) = \int_0^L dx'\,z(x,x',f)\,\delta I(x') \tag{5.7}$$

where $\delta I(x')$ is the a.c. current at x'. This shows that, instead of getting a relation between $\vec{E}(\vec{r})$ and $\vec{J}(\vec{r})$, one should, in this case, write the relation between the local field E(x) and the local current I(x), then linearize around the d.c. values so as to get a relation of the form

$$\hat{L}\delta E(x,f) = \delta I(x,f) \quad . \tag{5.8}$$

The Green function is z(x,x',f), solution of

$$\hat{L}\,z(x,x',f) = \delta(x-x') \quad . \tag{5.9}$$

The local field $\delta E(x,f)$ is

$$\delta E(x,f) = \int_0^L dx'z(x,x',f)\,\delta I(x',f) \quad . \tag{5.10}$$

The impedance field writes

$$\nabla Z(x',f) = -\int_0^L z(x,x',f)\,dx \quad . \tag{5.11}$$

It is then easy to get the a.c. impedance Z(f):

$$Z(f) = \int_0^L \nabla Z(x',f)dx' = -\int_0^L \int_0^L z(x,x',f)\,dx\,dx' \quad . \tag{5.12}$$

The noise voltage is given by (4.10).

5.3 Example: Space Charge Limited Current Diode, Single Injection (SCLC)

Let us consider a sclc p^+-p-p^+ diode [2,30-32], where most of the free carriers inside the diode are those injected at one electrode. The conduction equations are

$$I = qpvA + \varepsilon A\partial E/\partial t \quad \text{and} \quad \partial E/\partial x = (q/\varepsilon)(p - N_A) \simeq (q/\varepsilon)p \tag{5.13}$$

where the drift velocity $v = v(E)$ may be not-proportional to E. This gives

$$I = \varepsilon Av(\partial E/\partial x) + \varepsilon A(\partial E/\partial t) \quad . \tag{5.14}$$

The bias point satisfies

$$I_0 = qp_0v_0A \quad \text{and} \quad dE_0/dx = (q/\varepsilon)p_0 \quad , \tag{5.15}$$

that is

$$I_0 = \varepsilon Av_0(dE_0/dx) \quad . \tag{5.16}$$

Once the variation law v(E) is known, integration of (5.16) gives the local field profile $x(E_0)$, hence $E_0(x)$:

$$I_0x = \varepsilon A\int_0^{E_0} v_0(E)\,dE \quad . \tag{5.17}$$

Equation (5.15) gives then p_0 as a function of E_0, and taking dx from (5.16) gives

$$V_0 = -\int_0^x E_0dx = -(\varepsilon A/I_0)\int_0^{E_0} Ev_0(E)\,dE \quad . \tag{5.18}$$

For getting the noise, one sets, in (5.14),

$$I = I_0 + \delta I\exp(j\omega t) \quad \text{and} \quad E = E_0 + \delta E\exp(j\omega t) \quad .$$

These values, carried into (5.14), give for the first-order terms [see 5.8)]

$$\varepsilon Av_0(d\delta E/dx) + [\varepsilon A(dv_0/dx) + j\omega\varepsilon A]\delta E = \delta I \quad . \tag{5.19}$$

The Green function is the solution of

$$\varepsilon Av_0(x)\,dz(x,x',f)/dx + [\varepsilon A(dv_0/dx) + j\omega\varepsilon A]\,z(x,x',f) = \delta(x - x') \quad . \tag{5.20}$$

One first determines the solution of this equation without the right-hand side, using the fact that, according to (5.16), $dx = (\varepsilon Av_0/I_0)dE_0$. One has the integration constant vary, which gives $z(x,x',f)$, where the integration constant is zero, since $z(x=0,x',f)$ should vanish, since $\delta E(x=0) = 0$. Hence,

$$z(x,x',f) = [H(x - x')/(\varepsilon Av_0(x))]\exp\{j\omega\varepsilon A[E_0(x') - E_0(x)]/I_0\} \tag{5.21}$$

where $H(x - x')$ is the Heaviside function, $H = 1$ if $x > x'$, $H = 0$ if $x < x'$. Integration over x, taking once more into account (5.16), gives [cf. (5.11)]

$$\nabla Z(x') = \left[\exp\{j\omega\varepsilon A[E_0(x') - E_0(L)]/I_0\} - 1\right]\Big/ j\omega\varepsilon A \quad . \tag{5.22}$$

In the case of diffusion noise, one has [see (3.13) and (5.16)]

$$K(x,f) = 4q^2 p D(E_0) = 4q\varepsilon D(E_0)(dE_0/dx) \quad . \tag{5.23}$$

Equations (5.23) and (5.22), carried into (4.10), finally give

$$S_V(f) = (16q/\omega^2\varepsilon A)\int_0^{E_0(L)} dE\, D(E)\sin^2\{\omega\varepsilon A[E - E_0(L)]/(2I_0)\} \quad . \tag{5.24}$$

At low frequency, $f \to 0$, one gets

$$S_V(f=0) = (4Aq\varepsilon/I_0^2)\int_0^{E_0(L)} D(E)[E - E_0(L)]^2\, dE \quad . \tag{5.25}$$

This example shows that the knowledge of the electric field dependence of v and D allows getting both the electric field and free carrier distribution inside the device, so as the d.c. $I_0(V_0)$ characteristics, and the noise behaviour, through (5.25) which tells us what is the amount of noise, given a local source (here diffusion), and also what is the contribution of each slice to the total noise.

5.4 Numerical Modeling

It is very unusual to have a model so simple that almost all can be done analytically, as is the case in the previous section. One is usually obliged to perform numerical calculations from the beginning. The formulation is analogous to that explained in Sect.4.3 [33]. One starts from (5.8), the length of the device is discretized, so that (5.8) writes at each x_i, since $\hat{L}$ is linear

$$\sum_j a_{ij}\delta E_j = \delta I_i \quad \text{that is} \quad \vec{\vec{A}}\,\delta\vec{E} = \delta\vec{I} \quad . \tag{5.26}$$

This linear system can be solved, which gives

$$\delta\vec{E} = \vec{\vec{Z}}\,\delta\vec{I} \quad \text{with } \vec{\vec{Z}} = \vec{\vec{A}}^{-1} \quad \text{that is} \quad \delta E_k = \sum_j Z_{kj}\,\delta I_j \quad . \tag{5.27}$$

Now, (5.10), once discretized, writes

$$\delta E_k = \sum_j z(x_k,x_j,f)\,\delta I_j\,\Delta x_j \quad . \tag{5.28}$$

A direct comparison of (5.28) and (5.27) shows that the Green function is

$$z(x_k,x_j) = Z_{kj}/\Delta x_j \quad . \tag{5.29}$$

In the same way, discretization of (5.11) yields

$$\nabla Z_j = -\sum_k z(x_k,x_j,f)\Delta x_k$$

which, by comparison with (5.29), gives

$$\nabla Z_j = -\sum_k Z_{kj}\Delta x_k/\Delta x_j \quad . \tag{5.30}$$

In fact, when the step in x is constant ($\Delta x_k = \Delta x_j = \text{constant}$), the impedance field at point x_j is merely the sum of the elements of the row number j of the matrix $\vec{\vec{Z}}$. This method was verified to work on an sclc diode [33].

6. IFM USING THE LOCAL VOLTAGE

6.1 Formulation

In certain cases, the interesting electrical variable is the local voltage instead of the local electric field. In that case [34,35], as previously, an a.c. voltage is applied to the terminals M' and N', superimposed to the d.c. voltage applied to M and N. Let $\delta V(x)\exp(i\omega t)$ be the a.c. voltage of the equipotential surface, of cross section $A(x)$, crossing a given current line, of length L, at abscissa x, and let $\delta I(x)\exp(i\omega t)$ be the a.c. current through this equipotential surface. As previously, δV and δI are obtained by writing the transport equations, which give, after linearization,

$$\mathscr{L}\delta V(x) = \delta I(x) \tag{6.1}$$

where $\mathscr{L}$ is a linear operator. The Green function $\mathscr{g}(x,x',f)$ of (6.1) is the solution of

$$\mathscr{L}\mathscr{g}(x,x',f) = \delta(x-x') \tag{6.2}$$

and gives

$$\delta V(x) = \int_0^L \mathscr{g}(x,x',f)\,\delta I(x')\,dx' \quad . \tag{6.3}$$

That is, when $x=L$, where L is the length of the current line

$$\delta V(x=L) = \int_0^L \mathscr{g}(L,x',f)\,\delta I(x')\,dx' \quad . \tag{6.4}$$

A comparison between (6.4) and (4.9) readily gives

$$\nabla Z(x',f) = \mathscr{g}(L,x',f) \quad . \tag{6.5}$$

We shall now study an example to see how to apply this method.

6.2 Example: Field Effect Transistor

Let us consider an n-type channel FET (junction gate, MES-FET or MOS-FET). For given values V_G and $V_D = V(L)$ of the gate and drain voltages, with respect to the source, the cross section $A(x)$ of the channel, at distance x from the source, is a function of the local voltage $V(x)$, assuming that the equipotential surfaces are of small curvature, so that a one-dimensional model is valid. One thus may write $A(x) = A[V(x)]$.

Neglecting the diffusion and the displacement current, the transport equations write

$$\begin{cases} I = q\,n\,v\,A & (6.6)\\ d(E\,A)/dx = q\,A(N_D-n)/\varepsilon \quad . & (6.7)\end{cases}$$

This last equation is the Poisson equation, integrated in the volume of the channel between the equipotential surfaces $V(x)$ and $V(x+dx)$. Since the space charge is negligible inside the channel, one has $n \simeq N_D$, and (6.6) and (6.7) become

$$\begin{cases} I \simeq q\,N_D\,v\,A & (6.8)\\ E(X)\,A\,[V(X)] \simeq E(0)\,A(0) \quad . & (6.9)\end{cases}$$

The variation law of the velocity, versus the electric field, will be taken as (6.10), taking $E<0$, $v>0$, and $I>0$,

$$v = -\mu_0\,E/[1-(E/E_c)] \quad . \tag{6.10}$$

This relation, carried into (6.8), gives

$$I[1-(E/E_c)] = -q N_D \mu_0 E A(V) \quad . \tag{6.11}$$

This gives the bias conditions I_0, $E_0(x)$, $V_0(x)$. For getting the noise, one sets

$$I = I_0 + \delta I \exp(i\omega t), \quad E = E_0 + \delta E \exp(i\omega t), \quad V = V_0 + \delta V \exp(i\omega t), \quad \delta E = -d\,\delta V/dx \; .$$

These relations, carried into (6.11), give, for the first-order terms,

$$\begin{cases} a(x)\, d\delta V/dx - b(x)\,\delta V(x) = \delta I(x) & (6.12)\\ a(x) = [q N_D \mu_0 A_0 - (I_0/E_c)]\big/[1-(E_0/E_c)] & (6.13)\\ b(x) = -q N_D \mu_0 E_0 (dA/dV)_0\big/[1-E_0/E_c)] \quad . & (6.14) \end{cases}$$

This is equation (6.1). The Green function $\mathscr{G}(x,x',f)$ [see (6.2)] is the solution of

$$a(x)\, d\mathscr{G}(x,x')/dx + b(x)\,\mathscr{G}(x,x') = \delta(x-x') \quad . \tag{6.15}$$

This solution is obtained by the same method as in Sect.5.3 and writes

$$\mathscr{G}(x,x') = [H(x-x')/a(x')] \exp \int_x^{x'} [b(u)/a(u)]\, du \quad . \tag{6.16}$$

The integral in (6.16) can be calculated by using the variable changes $-E_0 du = dV$, then $(dA/dV)_0 dV = dA_0$, which gives

$$\mathscr{G}(x,x') = H(x-x')\; [1-(E_0(x')/E_c)]\big/[q N_D \mu_0 A_0(x) - (I_0/E_c)] \quad . \tag{6.17}$$

That is, taking into account (6.11),

$$\mathscr{G}(x,x') = -[E_0(x)/I_0][1-(E_0(x')/E_c)]\, H(x-x') \quad . \tag{6.18}$$

According to (6.5), the impedance field is

$$\nabla Z(x') = -[E_0(L)/I_0][1-(E_0(x')/E_c)] \tag{6.19}$$

which gives the a.c. impedance, by integrating over x:

$$Z = -[E_0(L)/I_0][L+(V(L)/E_c)] \quad . \tag{6.20}$$

The noise voltage S_V is given by (4.10), and the noise current is $S_I = S_V/|Z|^2$:

$$S_I = L^{-2}[1+(V(L)/E_c L))]^{-2} \int_0^L K(x')[1-(E_0(x')/E_c)]^2 A(x')\, dx' \quad . \tag{6.21}$$

The diffusion noise is obtained using (3.13) and (3.14), where, according to (6.10), $\sigma = dj_0/dE_0 = -qN_D\mu_0/(1-E/E_c)^2$; (6.21) becomes then

$$S_{I(diff)} = 4 k_B \mu_0 q N_D L^{-2} [1+(V(L)/(E_c L))]^{-2} \int_0^L T_n(x')\, A(x')\, dx' \quad . \tag{6.22}$$

This shows that, roughly speaking, all the slices of the FET bring comparable contributions to the noise, since the slices where E is high, and thus $T_n(x')$ is high, are those where $A(x')$ is small.

The GR noise is obtained by replacing, in (6.21), $K(x')$ by its expression (3.18), and taking into account (6.8), (6.10), and (6.9), which gives

$$S_{I(GR)} = 4q\mu_0 L^{-2}[1+(V(L)/(E_cL))]^{-2}\alpha\tau(1+\omega^2\tau^2)^{-1} I_0 V(L) \quad . \tag{6.23}$$

GR noise of long FETs [$E_cL >> V(L)$] is then proportional to the d.c. power $I_0V(L)$.

According to (3.19), bulk 1/f noise can be deduced from GR noise by substituting α_H/f to $4\alpha\tau(1+\omega^2\tau^2)^{-1}$, which gives

$$S_{I(bulk\ 1/f)} = \alpha_H q\mu_0 L^{-2}[1+(V(L)/(E_cL))]^{-2} f^{-1} I_0 V(L) \quad . \tag{6.24}$$

Bulk 1/f noise is also proportional to the d.c. power.

Contact 1/f noise is obtained by replacing K(x'), in (6.21), by (3.20), which gives

$$S_{I(contact\ 1/f)} = [\lambda/(N_DA(0))] L^{-2}[1+V(L)/(E_cL)]^{-2}(I_0^2/f) \quad . \tag{6.25}$$

This shows that, contrarily to bulk 1/f noise, contact 1/f noise is not proportional to the d.c. power $I_0V(L)$, but to I_0^2, which behaves quite differently in the saturation range of the drain current.

A similar procedure can be applied for calculating the noise current induced in the gate by the fluctuations inside the channel, so as the correlation between the gate and the drain noise currents [35].

6.3 Numerical Modeling

In that case too, it is very unusual to be able to express analytically the impedance field. It is then necessary to perform a numerical approach, which can be done as follows [36]. After discretization, (6.1) writes

$$\sum_j b_{ij}\delta V_j = \delta I_i \quad \text{that is} \quad \overrightarrow{\overrightarrow{B}}\,\overrightarrow{\delta V} = \overrightarrow{\delta I} \quad , \tag{6.26}$$

which gives, after resolution of that system,

$$\delta V_k = \sum_j Z'_{kj}\,\delta I_j \tag{6.27}$$

where $\overrightarrow{\overrightarrow{Z}}' = \overrightarrow{\overrightarrow{B}}^{-1}$. Now, discretizing (6.3), one gets

$$\delta V_k = \sum_j \mathcal{Z}(x_k,x_j,f)\ \delta I_j \Delta x_j \quad . \tag{6.28}$$

A comparison of (6.27) and (6.28) shows that

$$\mathcal{Z}(x_k,x_j,f) = Z'_{kj}/\Delta x_j \tag{6.29}$$

and (6.5) gives $\nabla Z_j = \mathcal{Z}(x_n,x_j,f)$, that is according to (6.29)

$$\nabla Z_j = Z'_{nj}/\Delta x_j \quad . \tag{6.30}$$

One is then able to compute the impedance field, and thus the noise, using the voltage as electrical variable.

7. PROBLEMS RELATED TO SUBMICRON DEVICES

When dealing with submicron devices, the behaviour of the noise is different because some "macroscopic" parameters, which have no effect in "long" devices, are enhanced, but also, and mainly, because new phenomena occur, which are not usually taken into account. Most of these problems are yet unsolved and have just been formulated in the very recent years.

7.1 Unusual Behaviour of "Macroscopic" Parameters

Let us consider the GR noise current of a FET, as given by (6.23). In silicon, $E_C \simeq 10$ kV/cm at 300 K, so that, in the saturation regime, with a drain d.c. voltage $V_D = V(L) = 5$ V, two limiting cases may be examined:

- In long devices, say a junction-gate FET, of length $L \simeq 20$ µm between source and gate, $E_C L \simeq 20$ V, $V(L)/E_C L \ll 1$, and (6.23) shows that $S_{I(GR)}$ is proportional to the d.c. power $I_0 V(L)$, that is increases linearly with the voltage in the saturation range.

- In submicron FETs, with say $L \simeq 0.5$ µm, then $E_C L \simeq 0.5$ V, and $V(L)/E_C L \gg 1$, so that (6.23) shows that $S_{I(GR)}$ is proportional to $I_0/V(L)$, that means it decreases as the drain voltage increases, in the saturation range. Hence, short devices behave quite differently from long devices. However, the behaviour of submicron devices is indeed not characterized by variations due to numerical values, but rather by the appearance of new phenomena, which have no effect at all in the case of longer devices, and should be taken into account for short devices.

7.2 Non-Stationary Effects

The transport equations, used for modeling long devices, are no more valid for submicron devices. The reason is that, when the length is decreased down to a few mean free paths (the mean free path is the average distance travelled by a carrier between two collisions), each carrier undergoes very few collisions when travelling from one electrode to the opposite one, and hence is <u>not</u> in a stationary regime. Its velocity, neglecting fluctuations, cannot be considered as being the drift velocity of the ensemble of carriers. On the contrary, in long devices, each carrier is submitted to a great number of scattering events, and it is then possible to replace its own velocity by the drift velocity v_d, in the sense that the distance travelled by a given carrier within a time interval t is almost tv_d. Thus, in long devices, ergodicity stands, since average values can be computed either by time averaging over a single carrier, or by making an ensemble average, over all the carriers, at a given time. In submicron devices, on the contrary, each carrier not being in its steady state, the system is <u>not</u> ergodic although being in the whole stationary, all average quantities must be obtained by making ensemble averages, which give results which may be quite different from averaging over time for a single carrier.

Then, to the classical transport equations, one should add the dynamical equations, which are approximate equations describing, with a good approximation, non-stationary transport, such as velocity overshoot or undershoot, etc., and give results in reasonable agreement with exact solutions of the Boltzmann equation, for a great variety of situations (temperature, semiconductors, initial conditions, etc.) (see, for example, Chap.8 of [19], and [37,38]. These equations are of the form

$$\begin{cases} d(m\vec{v})/dt = (q\vec{E}/\hbar) - (m\vec{v})/\tau_p(\varepsilon) \\ d\varepsilon/dt = q\vec{E}\,\vec{v} - (\varepsilon - \varepsilon_0)/\tau_\varepsilon(\varepsilon) \end{cases} \quad . \tag{7.1}$$

$\tau_p(\varepsilon)$ and $\tau_\varepsilon(\varepsilon)$ are the momentum and energy relaxation times and are supposed to depend only on energy. As a consequence, they can be determined in a stationary regime and can therefore be deduced from the stationary curves $\varepsilon_s(E_s)$ and $v_s(E_s)$ obtained by solving the Boltzmann equation in a homogeneous field E_s. Namely, by equating to zero the derivatives in (7.1), one gets

$$\tau_p(\varepsilon) = mv_s/(qE_s) \quad \text{and} \quad \tau_\varepsilon(\varepsilon) = (\varepsilon - \varepsilon_0)/(qE_s v_s)$$

which, carried into (7.1), give, along the field

$$\begin{cases} d(mv)/dt = q[E - (v/\mu_s(\varepsilon))] \\ d\varepsilon/dt = q[E\,v - E_s(\varepsilon)\,v_s(\varepsilon)] \end{cases} \quad . \tag{7.2}$$

At a given time t, the drift velocity $v = v(t)$, and the average energy $\varepsilon = \varepsilon(t)$, are supposed to be known. The stationary curves, obtained by a Monte Carlo simulation or by solving the Boltzmann equation, give the field E_S corresponding to the energy $\varepsilon_S = \varepsilon(t)$, as well as the velocity $v_S = v_S(E_S)$. These values, carried into (7.2), give $d(mv)/dt$ and $d\varepsilon/dt$, hence $v(t+\Delta t)$ and $\varepsilon(t+\Delta t)$. A slight modification occurs when studying the steady state along the device. In the above equations, a displacement Δx is associated with a time increment $\Delta t = \Delta x/v$.

As concerning the noise, it is also considered as depending, not on the local field, but on the local energy, so that, for example, the diffusion noise source is given by [21,22]

$$K(x) = 4q^2 n(x) D[\varepsilon(x)] \quad . \tag{7.3}$$

These non-stationary effects have been taken into account for modeling short devices, and particularly FETs (see [20-27]). For example, CARNEZ et al. [21] showed that taking into account non-stationary effects could result in an important diminution of the noise figure of small gate GaAs MES-FETs, as is evidenced in Fig.7 below, taken from [21].

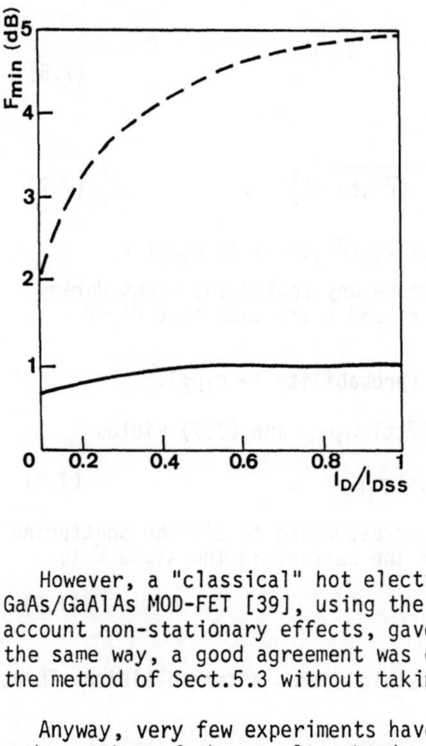

Fig.7. Variation, versus the drain bias, of the minimum noise figure F_{min} of a GaAs MES-FET, 0.4 µm gate length, when taking (full line) and not taking (dashed line) into account non-stationary effects (taken from [21])

However, a "classical" hot electron modeling, performed on a 1.0 µm gate GaAs/GaAlAs MOD-FET [39], using the method described in Sect.6.2, not taking into account non-stationary effects, gave results in good agreement with experiment. In the same way, a good agreement was obtained on a 1.1 µm n^+-n-n^+ GaAs device, using the method of Sect.5.3 without taking into account non-stationary effects [40].

Anyway, very few experiments have been performed on submicron devices [38-42], and a number of the results obtained seems to be due to parasitic effects such as 1/f noise [43] at interfaces, contacts, etc.

7.3 Space Correlations of the Noise Sources

It has been supposed, till now, that the noise sources at two different points were uncorrelated. It will be shown here [44,45] that this is not true for distances of the order of the mean free path, that is for short devices. The reason is that noise is the Fourier transform of the correlation function of the current densities, that is of the velocities of the carriers. Over distances shorter than the distance between two collisions, the velocity of a given carrier is unchanged (if $E = 0$), hence the velocities at these two points are strictly correlated.

Let us use the following hypotheses: i) Two different carriers are uncorrelated. ii) The scattering mechanisms are randomizing. iii) The electric field is null ($E = 0$). The second hypothesis means that the state of a carrier, after a collision, is uncorrelated with its state before the collision. This includes acoustic, non-polar optical, and intervalley phonon scattering. Such a situation is encountered in pure enough silicon and germanium, where impurity scattering is negligible. The third hypothesis ($E = 0$) will be dropped later on.

The current density at point $\vec{r}$ and time t is given by

$$\vec{j}(\vec{r},t) = \vec{j}_0(\vec{r}) + \Delta\vec{j}(\vec{r},t) \tag{7.4}$$

that is, since when $\vec{E} = 0$, $\vec{j}_0(\vec{r}) = \vec{0}$,

$$\vec{j}(\vec{r},t) = (q/d^3r) \sum_i \vec{v}_i(\vec{r},t) \tag{7.5}$$

where the sum extends over all the carriers located in the volume d^3r, around $\vec{r}$, at time t. The noise source $S_{j\alpha\beta}(\vec{r},\vec{r}',\nu)$ is defined [see (2.4) and (3.3)] as the Fourier transform of the correlation function $C_{j\alpha\beta}(\vec{r},\vec{r}',\theta)$, where α and β mean one of the directions x, y and z. Then

$$C_{j\alpha\beta}(\vec{r},\vec{r}',\theta) = \overline{\Delta j_\alpha(\vec{r},t)\Delta j_\beta(\vec{r}',t+\theta)}^{t} \quad . \tag{7.6}$$

That is, according to (7.5) and to hypothesis i):

$$C_{j\alpha\beta}(\vec{r},\vec{r}',\theta) = [(q^2/(d^3r.d^3r'))] \sum_i \overline{v_{i\alpha}(\vec{r},t)\, v_{i\beta}(\vec{r}',t+\theta)}^{t} \quad . \tag{7.7}$$

According to hypotheses ii) and iii), $\Gamma_i = v_{i\alpha}(\vec{r},t)v_{i\beta}(\vec{r}',t+\theta)$ is equal to

* $v_{i\alpha}(\vec{r},t)\, v_{i\beta}(\vec{r},t)$ if the carrier has not undergone any scattering event during the time interval θ (probability p_1), <u>and</u> if $\vec{r}$, $\vec{r}'$ and θ are such that $\vec{r}' - \vec{r} = \theta\vec{v}_i(\vec{r},t)$ (probability p_2).
* 0 if one of these conditions is not fulfilled (probability $1 - p_1p_2$).

The average value of Γ_i is then $v_{i\alpha}(\vec{r},t)\, v_{i\beta}(\vec{r},t)\, p_1p_2$, and (7.7) yields

$$C_{j\alpha\beta}(\vec{r},\vec{r}',\theta) = q^2/(d^3r\, d^3r') \sum_i v_{i\alpha}(\vec{r},t)\, v_{i\beta}(\vec{r},t)\, p_1p_2 \quad . \tag{7.8}$$

Let $P(\vec{k},\vec{k}')$ be the sum of the transition rates corresponding to all the scattering mechanisms involved. The relaxation time $\tau(k)$ of the carrier in the state $\vec{k}$ is

$$1/\tau(k) = \sum_{\vec{k}'} P(\vec{k},\vec{k}') \tag{7.9}$$

where the summation is extended over all the possible states after collision. Then, with $\vec{v}_i = \vec{v}(\vec{k}_i) = (1/\hbar)[\vec{\nabla}_k\varepsilon]_{\vec{k}=\vec{k}_i}$

$$\begin{cases} p_1 = \text{Prob}\,\{0 \text{ collision during } \theta\} = \exp-[\theta/\tau(v_i)] & (7.10) \\ p_2 = \text{Prob}\,\{\theta\vec{v}_i = \vec{r}' - \vec{r}, \text{ within } d^3r'\} = \delta(\vec{r}' - \vec{r} - \theta\vec{v}_i)\, d^3r' & (7.11) \end{cases}$$

so that (7.8) becomes

$$C_{j\alpha\beta}(\vec{r},\vec{r}',\theta) = (q^2/d^3r) \sum_i v_{i\alpha}v_{i\beta}\{\exp[-\theta/\tau(v_i)]\}\delta(\vec{r}' - \vec{r} - \theta\vec{v}_i) \quad . \tag{7.12}$$

Let $f(\vec{k})$ be the distribution function of the carriers, with

$$n = \frac{1}{4\pi^3}\int f(k)\, d^3k \quad . \tag{7.13}$$

The average value $\overline{\varphi}$ of a given function $\varphi(\vec{k})$ is

$$n\overline{\varphi} = \frac{1}{d^3r} \sum_i \varphi(\vec{k}_i) = \frac{1}{4\pi^3} \int \varphi(\vec{k})\, f(\vec{k})\, d^3k \quad . \tag{7.14}$$

Carried into (7.12), this finally gives

$$\begin{cases} C_{j\alpha\beta}(\vec{r},\vec{r}',\theta) = q^2 n\left[\int v_\alpha v_\beta f(k)[\exp(-\theta/\tau(k))]\delta(\vec{r}' - \vec{r} - \theta\vec{v})\, d^3k\right] \Big/ \int f(k)\, d^3k & (7.15) \\ S_{j\alpha\beta}(\vec{r},\vec{r}',\nu) = 4\int_0^\infty C_{j\alpha\beta}(\vec{r},\vec{r}',\theta)\cos(2\pi\nu\theta)\, d\theta \quad . & (7.16) \end{cases}$$

When $\alpha = \beta = x$, $\vec{r}' - \vec{r} = \vec{R} = \{R,0,0\}$, (7.15) and (7.16) give $S_{j//}(R,\nu)$, which can be computed for a given semiconductor. Figure 8 shows $S_{j//}(R,\nu)$ as a function of R, at different frequencies [44]. It is clear, in Fig.8, that the noise is white below 10^{11} Hz and that S_j is null at R = 0 (contrarily to the usual assumption) [cf. (3.4)], is maximum for R ≃ 100 Å ≃ one mean free path, and extends over a few mean free paths.

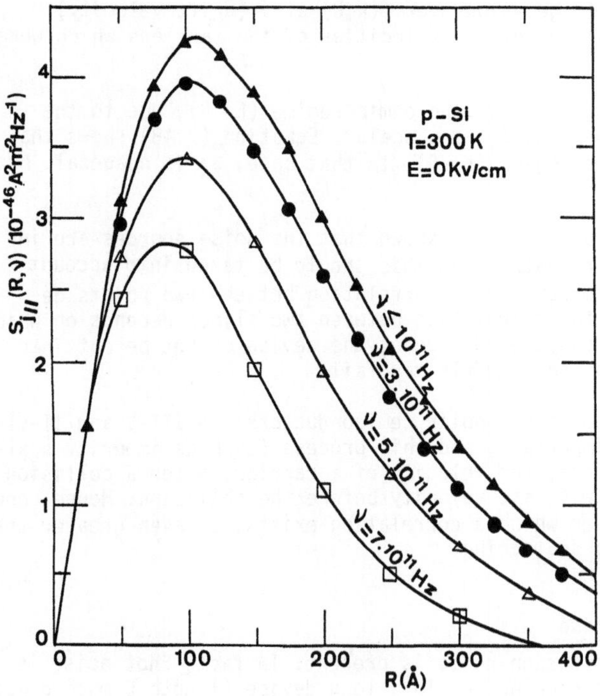

Fig.8. Variation, versus R, and along R, of the noise source $S_{j//}(R,\nu)$, at various frequencies ν, in p-type Si, 300 K, E = 0, showing space correlations

Let k_{th} be the state such that $v_{th} = v(k_{th})$ is the thermal velocity. For $k \gg k_{th}$, one has $f(k) \simeq 0$, and for $\theta \gg \tau$, $\exp -\theta/\tau(k) \simeq 0$. Therefore, for long devices, such that $|\vec{r} - \vec{r}'| \gg v_{th}\tau$, one may set $\delta(\vec{r}' - \vec{r} - \theta\vec{v}) \simeq \delta(\vec{r}' - \vec{r})$, so that (7.15,16) give

$$\begin{cases} S_{j\alpha\beta}(\vec{r},\vec{r}',\nu) \simeq K_{\alpha\beta}(\vec{r},\nu)\delta(\vec{r}' - \vec{r}) \\ K_{\alpha\beta}(\vec{r},\nu) = 4q^2 n D_{\alpha\beta}(\nu) \\ D_{\alpha\beta}(\nu) = \left\langle \dfrac{\tau(k)\, v_\alpha v_\beta}{1 + \omega^2\tau^2(k)} \right\rangle \quad . \end{cases} \tag{7.17}$$

This is the usual expression of the ohmic diffusion coefficient, for randomizing scattering.

Exactly the same theory holds for hot carriers ($E \neq 0$). In that case, (7.15) writes [45]

$$C_{j\alpha\beta}(\vec{r},\vec{r}',\theta) = q^2 n \left\{ \int \Delta v_\alpha \Delta v_\beta f(\vec{k},E) \left(\exp - \int_0^\theta [dt/\tau(\vec{k}(t))] \right) \cdot \delta\left(\vec{r}' - \vec{r} - \int_0^\theta dt\, \vec{v}(\vec{k}(t))\right) \right\} \Big/ \int f(\vec{k},E)\, d^3k \tag{7.18}$$

where

$$\begin{cases} \Delta v_\alpha = v_\alpha(\vec{k}) - v_{0\alpha}(\vec{k}) \\ \Delta v_\beta = v_\beta(\vec{k}) - v_{0\beta}(\vec{k},\theta) \\ \vec{k}(t) = \vec{k} + (q\vec{E}\,t/\hbar) \end{cases} \tag{7.19}$$

In these equations, $f(\vec{k},E)$ is the hot carrier stationary distribution function, $\vec{v}(\vec{k}) = (1/\hbar)\vec{\nabla}_k \varepsilon(\vec{k})$, $\vec{v}_0(\vec{k})$ is the average value over $f(\vec{k},E)$ of $\vec{v}$ (drift velocity) $\vec{v}_0(\vec{k},\theta)$ is the average value at time θ, of the velocities of the carriers which were in the state $\vec{k}$ at time 0.

It follows clearly, from (7.15), that, in the ohmic regime ($E = 0$), due to the fact that $f(\vec{k})$ is symmetric around $\vec{k} = 0$, $\vec{\vec{S}}_j$ is a scalar. Equation (7.18) shows that this is no more true in hot carrier regime ($E \neq 0$). In that case, $\vec{\vec{S}}_j$ is diagonal, but not scalar.

Anyway, in this section it has been clearly shown that the noise sources are indeed correlated over a few mean free paths, and this should be taken into account when modeling submicron devices. Moreover, the correlation between two points depends on their distance, and thus the correlation between two slices depends on which couple of points of the slices is considered. Hence, the device cannot be cut into slices, which means that one-dimensional models may fail.

What has been demonstrated concerns non-polar semiconductors. In III-V and II-VI compounds, optical scattering is important, and this process (such as impurity scattering) is not randomizing. Therefore, the velocity of a carrier, after a collision of this type, is still correlated with its velocity before the collision. Hence, one would expect that the distance, over which a correlation exists, is even greater than in semiconductors with randomizing scattering.

7.4 Change of the Nature of the Noise

Let us consider a device, where diffusion noise is present. In fact, Shot noise is also present, superimposed to diffusion noise. In a long device (length L much greater than the mean free path), each carrier undergoes many scattering events when travelling from one electrode to the opposite one, and diffusion noise prevails $S_{I(diff)} \gg S_{I(Shot\ noise)}$. On the contrary, in devices of lengths smaller than the mean free path, each carrier travels through the device almost without being scattered, therefore the diffusion noise vanishes, and only Shot noise is present: $S_{I(diff)} \ll S_{I(Shot\ noise)}$. This lowering of the diffusion noise, due to short distances responsible for non-stationary effects, is evidenced in Fig.7 by a decrease of the noise figure of the FET, with respect to the value it should have been using "classical" stationary transport. Attempts have been made, in very recent years [46], to evidence such effects in one-port networks. They all failed because in such short devices parasitic effects, such as contact resistance, 1/f noise, etc., canceled a possible variation due to balistic transport in the active region. FETs, or other devices where one electrode acts as a probe, would be probably more suited to show such effects, as was suggested a few years ago [35].

7.5 Discretization Effects

When the dimensions become very short, the system may no longer be considered as a continuum. The states lie then on discrete energy levels. This occurs, for example, in thin epitaxial layers, superlattices, inversion layers, etc. A number of consequences follows.

If the energy levels are to be considered individually, and not statistically, the behaviour then depends on the fine structure of the energy levels. This was evidenced [47] by studying a one-dimensional model of porus membranes. The ions were supposed to jump from one potential well to the adjacent one, and the noise spectrum was strongly dependent on the distribution and on the values of the energies of the wells.

In thin layers with discrete energy levels, the scattering processes differ from bulk properties and the transport coefficients are therefore modified [19]. A recent Monte Carlo simulation, performed on GaAs/GaAlAs quantum wells, showed that the diffusion coefficient D(E) was strongly dependent on the thickness of the layers [48].

In small structures, the density of free carriers must be large, therefore one should take into account carrier-carrier scattering, so as degeneracy. Moreover, the corpuscular model of the carriers may be questioned, since the carrier wavelength $\lambda = 2\pi/k$, with $k \simeq 10^9\ m^{-1}$, is about 60 Å, which is no more negligible compared with epitaxial layer thicknesses.

Finally, the characteristic of a device, as was evidenced in this paper as concerning the noise, is an average value of transport parameters over the whole device. As a consequence, a small defect inside a long device yields a localized perturbation, which, once averaged over the whole device, may give a negligible overall effect. Conversely, the same defect, inside a submicron device, occupies most of the volume of that device, and thus the perturbation it produces affects the behaviour of the entire device. Submicron devices are much more sensitive to defects than long devices. This cannot be modeled or predicted easily!

7.6 Monte Carlo Modeling

It is well known that Monte Carlo simulation is useful for getting parameters in homogeneous semiconductors (bulk or thin layers [19,48]). Its use for modeling devices gives rise to problems related to memory occupancy and time computation, due to the necessity of solving coupled transport and Poisson's equations. As a consequence, the necessity to use a great number of particles for these kinds of Monte Carlo simulations leads to use simplified particle models.

Few attempts have been made to simulate the noise of devices using Monte Carlo techniques. This problem has been solved and applied to a n^+-n-n^+ structure in the following way [49]. The current conservation in a one-dimensional device of uniform cross section writes

$$I(t) = q\,n\,(x,t)\,v\,(x,t)\,A + \varepsilon A\ \partial E(x,t)/\partial t \tag{7.20}$$

where v(x,t) is the drift velocity at point x and time t. After integration over the length L of the device, one gets

$$-\,\partial V\,(t)/\partial t = \varepsilon^{-1}\left[A^{-1} L\, I(t) - q\int_0^L n(x,t)\,v\,(x,t)\,dx\right] \ . \tag{7.21}$$

Now let us consider the $\delta N(t)$ particles located in the slice dx at time t. One has

$$n(x,t) = \delta N(t)/(A\,dx) \quad \text{and} \quad v(x,t) = (\delta N)^{-1}\sum_{i=1}^{N} v_i(x,t)$$

where $v_i(x,t)$ is the velocity of the ith carrier in the slice. These relations, carried into (7.21), give

$$- \partial V(t)/\partial t = (\varepsilon A)^{-1} \left[L\, I(t) - q \sum_{i=1}^{N} v_i(x,t) \right] \tag{7.22}$$

where N is the total number of particles inside the device. Subtracting to (7.22) its average value gives the instantaneous deviation

$$-\partial \delta V/\partial t = (\varepsilon A)^{-1} \left\{ L\, \delta I(t) - q \sum_{i=1}^{N} [v_i(x,t) - v(x)] \right\} \tag{7.23}$$

where v(x) is the drift velocity at point x. Two situations can then be investigated [50]:

a) Noise voltage of an a.c. open circuit: $\delta I(t) = 0$ in (7.23) gives

$$\delta V(t) = q(\varepsilon A)^{-1} \sum_{i=1}^{N} \int_0^t [v_i(x,t') - v(x)]\, dt' \quad . \tag{7.24}$$

Since the Monte Carlo simulations allow one to know the velocity of each particle versus time, $\delta V(t)$ can easily be computed, so as $\delta V(t+\tau)$, and hence the noise voltage

$$S_V(f) = 4 \int_0^\infty \overline{\delta V(t)\, \delta V(t+\tau)}^{\,t} \cos(2\pi f \tau)\, d\tau \quad . \tag{7.25}$$

b) Noise current of an a.c. short-circuited device $\delta V(t) = 0$ in (7.23) gives

$$\delta I(t) = (q/L) \sum_{i=1}^{N} [v_i(x,t) - v(x)] \quad . \tag{7.26}$$

In the same way as previously, this allows determining the noise current $S_I(f)$.

Once the noise $S_V(f)$ or $S_I(f)$ of the particles has been obtained, the noise of the carriers is deduced by dividing the result by the number of carriers represented by each particle.

In spite of its simplicity, this model is questionable. The basic equation (7.20) is valid only for average values, since it is deduced from the Boltzmann equation which involves probabilities, that is the average number of particles in each state. Thus, using (7.20) when describing the noise, is perhaps erroneous, since obviously, due to local fluctuations, the current I should also depend on x, being randomly distributed around its x-dependent average value. In the impedance field method, for example, this is taken into account by the use of x-dependent noise source terms.

Furthermore, in practice, this method is very time consuming, such as any particle modeling of a device. Anyway, this is an interesting way to be explored.

8. CONCLUSION

In the last few years, powerful methods have been developed for modeling the noise of devices, mainly oriented towards the impedance field method. Due to the extension of analytic formulations to numerical methods, it is now possible to model the noise of a great variety of one-dimensional devices where unipolar conduction takes place. These methods could, in principle, be also applied for ambipolar conduction mechanisms, although this has not yet been done till now, up to my knowledge. Extending these models to two- or three-dimensional structures gives rise to problems, at least of technical (computation time), and perhaps even of more fundamental nature.

The models described here can be (and have been) applied to submicron devices, including non-stationary effects such as velocity overshoot. However, in such structures, a lot of problems are yet unsolved, as outlined in Sects.7.3 and 7.5 above. Most of these fundamental problems (mainly linked to discretization effects), together with huge uncertainties related to the absence of precise knowledge of important technical parameters of layers (doping, defects, homogeneities, etc.) do not allow even modeling "first-order" characteristics, such as carrier concentration and electric field profiles, I-V and a.c. characteristics, cut-off frequency, etc. Since it should be kept in mind that *it is completely impossible to imagine to be able to model the noise as long as one is not able to model first-order characteristics*, it can easily be understood that modeling the noise of submicron devices is quite an open field of investigations.

Conversely, studying experimentally the noise of submicron devices is indeed possible, and can bring information both on technical aspects and on physical parameters responsible for the noise observed.

REFERENCES

1. A. Van der Ziel: Noise: Sources, Characterization, Measurement (Prentice Hall, Englewood Cliffs, NJ 1970)
2. K.M. van Vliet, A. Friedmann, R.J.J. Zijlstra, A. Gislof, A. Van der Ziel: J. Appl. Phys. 46, 1804 (1975); J. Appl. Phys. 46, 1814 (1975)
3. J.P. Nougier: In Proceedings of the 6th International Conference on Noise in Physical Systems, Washington 1981, ed. by P.H.E. Meijer, R.D. Mountain, R.J. Soulen (National Bureau of Standards, Special Publication No.614) (1981) p.397
4. A. Van der Ziel, K.M. Van Vliet: Solid State Electron. 11, 508 (1968)
5. K.M. Van Vliet: Solid State Electron. 13, 649 (1970)
6. J.P. Nougier: In Noise in Physical Systems, Springer Ser. Electrophys., Vol.2, ed. by D. Wolf (Springer, Berlin, Heidelberg 1978) p.71
7. J.P. Nougier: In Physics of Linear Transport in Semiconductors, ed. by D.K. Ferry, J.R. Barker, C. Jacoboni (Plenum, New York 1980) p.415
8. J.P. Nougier, J.C. Vaissière, D. Gasquet, A. Moatadid: J. Appl. Phys. 52, 5683 (1981)
9. J.P. Nougier: Appl. Phys. Lett. 33, 89 (1978)
10. K.M. Van Vliet, A. Friedmann, R.J.J. Zijlstra, A. Gisolf, A. Van der Ziel: J. Appl. Phys. 46, 1814 (1975)
11. A. Van der Ziel, R. Jindal, S.K. Kim, H. Park, J.P. Nougier: Solid State Electron. 22, 177 (1979)
12. F.N. Hooge: Phys. Lett. 29A, 139 (1969)
13. L.K.J. Vandamme: In Noise in Physical Systems and 1/f Noise, ed by M. Savelli, G. Lecoy, J.P. Nougier (North-Holland, Amsterdam 1983) p.183
14. D. Gasquet, M. Fadel, J.P. Nougier: In Noise in Physical Systems and 1/f Noise, ed. by M. Savelli, G. Lecoy, J.P. Nogier (North-Holland, Amsterdam 1983) p.15
15. W. Shockley, J.A. Copeland, R.R. James: In Quantum Theory of Atoms, Molecules and the Solid State, ed. by P. Lowdin (Academic, New York 1966) Chap.8
16. A. Rigaud, M.A. Nicolet, M. Savelli: Phys. Status Solidi A18, 531 (1973)
17. D. Gasquet, H. Tijani, J.P. Nougier: In Noise in Physical Systems and 1/f Noise, ed. by M. Savelli, G. Lecoy, J.P. Nougier (North-Holland, Amsterdam 1983) p.165
18. J.C. Vaissière: PhD Thesis, Montpellier (1985) (to be published, available on request)
19. L. Reggiani (ed.): Hot Electron Transport in Semiconductors, Topics in Appl. Phys., Vol. 58 (Springer, Berlin, Heidelberg 1985)
20. B. Carnez, A. Cappy, G. Salmer, E. Constant: Acta Electronica 23, 165 (1980)
21. B. Carnez, A. Cappy, R. Fauquembergue, E. Constant, G. Salmer: IEEE Trans. ED-28, 784 (1981)
22. A. Cappy, R. Fauquembergue, E. Constant, G. Salmer, B. Carnez: Inst. Phys. Conf. Ser. No.56, Chap.3, 175 (1981)
23. E. Allamando, G. Salmer, E. Constant, N.E. Radhy, A. Cappy, B. Carnez: In Noise in Physical Systems and 1/f Noise, ed by M. Savelli, G. Lecoy, J.P. Nougier (North-Holland, Amsterdam 1981) p.177

24. E. Allamando, N.E. Radhy, G. Salmer: Physica B129, 390 (1985)
25. A. Cappy, M. Schortgen, G. Salmer: IEEE Trans. EDL-6, 270 (1985)
26. A. Cappy, A. Vanoverschelde, M. Schortgen, C. Versnaeyeen, G. Salmer: IEEE Trans. ED-32, 2787 (1985)
27. A. Cappy, A. Vanoverschelde, M. Schortgen, C. Versnaeyeen, G. Salmer: Physica B129, 380 (1985)
28. C.R. Brewitt-Taylor, P.N. Robson, J.E. Sitch: Proceedings of the 8th European Microwave Conference (Paris 1978) p.415
29. D. Sodini, D. Rigaud: Proceedings of the 6th International Conference on Noise in Physical Systems, Washington 1981, ed. by P.H.E. Meijer, R.D. Mountain, R.J. Soulen (National Bureau of Standards, Special Publication No.614) (1981)
30. A. Gisolf, R.J.J. Zijlstra: Solid State Electron. 17, 839 (1974)
31. G. Bosmann, R.J.J. Zijlstra: Phys. Lett. A71, 464 (1979)
32. J.P. Nougier, D. Gasquet, J.C. Vaissière, H.R. Bilger: In Noise in Physical Systems, Springer Ser. Electrophys., Vol.2, ed. by D. Wolf (Springer, Berlin, Heidelberg 1978) p.110
33. J.P. Nougier, A. Moatadid, J.C. Vaissière, G. Gasquet: Physica B134, 260 (1985)
34. J.P. Nougier, J.C. Vaissière, D. Gasquet: In Proceedings of the 6th International Conference on Noise in Physical Systems, Washington 1981, ed. by P.H.E. Meijer, R.D. Mountain, R.J. Soulen (National Bureau of Standards, Special Publication No.614) (1981) p.42
35. P. Hesto, J.C. Vaissière, R. Castagné, J.P. Nougier: J. Phys. Paris C7, 235 (1981)
36. J.P. Nougier, A. Moatadid, J.C. Vaissière, D. Gasquet: In Noise in Physical Systems and 1/f Noise 1985, ed. by A. d'Amico, P. Mazzetti (North-Holland, Amsterdam 1986) p.105
37. J.P. Nougier, J.C. Vaissière, D. Gasquet, J. Zimmermann, E. Constant: J. Appl. Phys. 52, 825 (1981)
38. L. Reggiani, J.C. Vaissière, J.P. Nougier, D. Gasquet: J. Phys. Paris C7, 357 (1981)
39. C.F. Whiteside, G. Bosman, A. Van der Ziel, W. Kopp, H. Morkoc: In Noise in Physical Systems and 1/f Noise, ed. by A. d'Amico, P. Mazzetti (North-Holland, Amsterdam 1986) p.167
40. J. Andrian, G. Bosman, A. Van der Ziel, C.M. Van Vliet: In Noise in Physical Systems and 1/f Noise, ed by A. d'Amico, P. Mazzetti (North-Holland, Amsterdam 1986) p.171
41. D. Delagebeaudeuf, N.T. Linh: IEEE Trans. ED-29, 955 (1982)
42. O.K. Mishra, S.C. Palmateer, P.C. Chao, P.M. Smith, J.C.M. Hwang: IEEE Trans. EDL-6, 142 (1985)
43. K.H. Duh, A. Van der Ziel, H. Morkoc: IEEE EDL-4, 10 (1983)
44. J.P. Nougier, C. Gontrand, J.C. Vaissière: In Noise in Physical Systems and 1/f Noise, ed. by M. Savelli, G. Lecoy, J.P. Nougier (North-Holland, Amsterdam 1983) p.15
45. J.P. Nougier, J.C. Vaissière, C. Gontrand: Phys. Rev. Lett. 51, 513 (1983)
46. M. Savelli, G. Lecoy, J.P. Nougier (eds.): Noise in Physical Systems and 1/f Noise (North-Holland, Amsterdam 1983)
47. E. Frehland: Biophys. Struct. Mech. 7, 1 (1980)
48. A.D. Van Rheenen, G. Bosman: In Noise in Physical Systems and 1/f Noise, ed. by A. d'Amico, P. Mazzetti (North-Holland, Amsterdam 1986) p.163
49. J. Zimmermann, E. Constant: Solid State Electron. 23, 915 (1980)
50. J. Zimmermann: PhD Thesis, University of Lille, France (1980) (available on request)

Transport Physics of Multicontact Si MOS Nanostructures

W.J. Skocpol

AT & T Bell Laboratories, Room 4E-330, Holmdel, NJ 07733, USA

1. Introduction

Nanofabrication technology makes possible new types of physics experiments. In this chapter, I describe silicon MOSFET devices with extra voltage-measuring contacts and widths as narrow as 40 nm. These allow exploration of both classical and quantum electron transport phenomena on nanometer length scales.

Because ordinary NMOS transistors contain a quantum-mechanically thin inversion layer of electrons, they have long been understood as an excellent model system for studying the physics of two-dimensional electron transport. More recently, special transistors with narrow inversion layers for studying "one-dimensional" transport have been fabricated by research groups at AT&T Bell Laboratories, Cambridge University, IBM Watson Research Center, Massachusetts Institute of Technology, Philips Research, and Yale University. Our new multicontact devices represent still another step in the evolution of this research. In this chapter, I emphasize transport at relatively high electron densities far above threshold. Other regimes are discussed elsewhere in this volume.

2. Multiterminal Devices

Figures 1-3 show scanning electron micrographs and an interpretive line drawing of a patterned and etched region at the center of one of our narrow, multiterminal MOSFETs. Figure 1 is a top view of the structure, showing a NiCr pattern in the form of a narrow strip with side branches. This has been patterned by liftoff using electron-beam lithography at 120 keV in a Philips EM-400 scanning electron microscope. The minimum feature size here is approximately 50 nm, but features as small as 20 nm can be patterned without too much difficulty. Underneath this NiCr pattern is 50 nm of doped polysilicon, 20 nm of gate oxide, and a lightly-doped silicon substrate. Adjacent material not protected by this NiCr pattern has been removed by reactive ion etching to a total depth of approximately 150 nm, well into the silicon substrate. Figure 2 is a perspective view of the etched structure. The ethereal quality of this micrograph results because the 120 keV SEM beam easily passes through the entire structure, illuminating the interior of the etched grooves on the far side of the narrow strip. Figure 3 is a line drawing of the same structure (without transparency) that depicts the layered structure, including the edge of the electron inversion layer (dashes). Transport through this inversion layer is confined by the etched sidewalls.

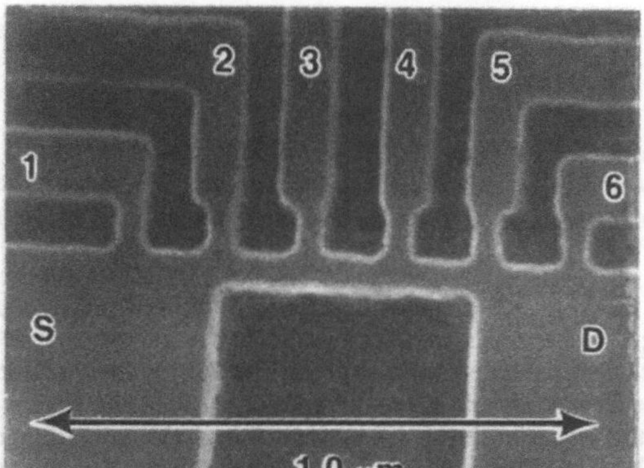

Fig. 1 Top view of submicron region of the gate pattern that defines a narrow (60 nm) MOSFET channel from source S to drain D probed by voltage measuring channels 1-6

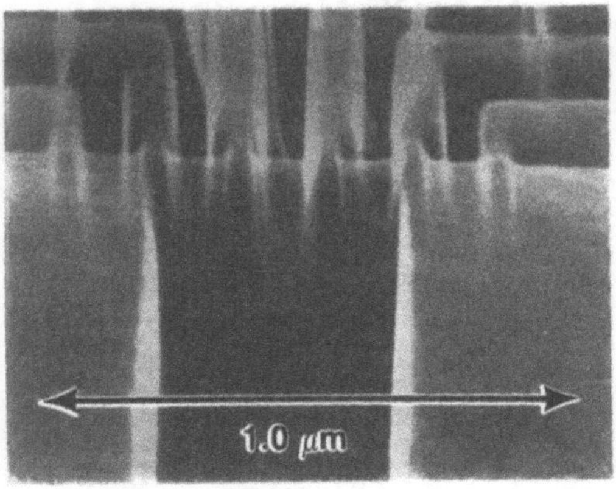

Fig. 2 Perspective view of etched topography of the device structure shown in Fig. 1, so small that it is nearly transparent to the 120 keV scanning electron microscope beam

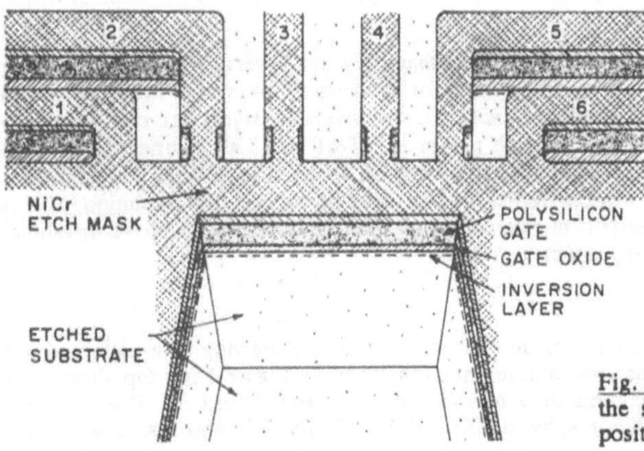

Fig. 3 Schematic nontransparent rendering of the structure in Fig. 2, showing approximate position of the various layers of the MOSFET

The various branches of the inversion layer (labeled S,D, and 1-6 in Fig. 1) continue in a widening series of e-beam patterned rectangles until they each reach one of the eight n+ contacts implanted around the edge of a 30 μm square. Metal leads patterned by optical lithography continue from the contacts and gates of four such devices to 200 μm bonding pads around the edge of a 3 mm by 5 mm chip. The process sequence for fabricating this structure is described in [1] and earlier versions of our technology are described in [2-4].

The basic point of this structure is that current can be passed through the narrow channel from drain (D) to source (S) [electron flow from S to D], and the local potential can be measured along that channel by connecting high-impedance voltmeters to the side branches.

3. Classical Transport Phenomena

3.1 Current-Voltage Characteristics

Despite their narrow width, these nanostructures behave very much as MOSFET models predict, particularly at large voltages and currents. Figure 4 shows various I-V characteristics of a device with the pattern depicted in the bottom diagram (1/8 μm wide, with three 1/4 μm-long segments defined by four narrow voltage probes).

The top set of characteristics corresponding to the entire device has the classic MOSFET shape. At low drain-source voltages the potential drop along the channel is small compared to the gate-to-channel potential. In this regime the characteristics are linear, with a gate-controlled resistance set by the electron density in the channel. The parallel-plate capacitor formed by the gate, the gate oxide, and the channel inversion layer determines the electron density in the inversion layer. A nonzero threshold gate voltage V_T is required for the inversion layer in the channel to first form. As the drain-source voltage V_{SD} begins to approach within V_T of the gate-source voltage V_G, the electron density at the drain end approaches

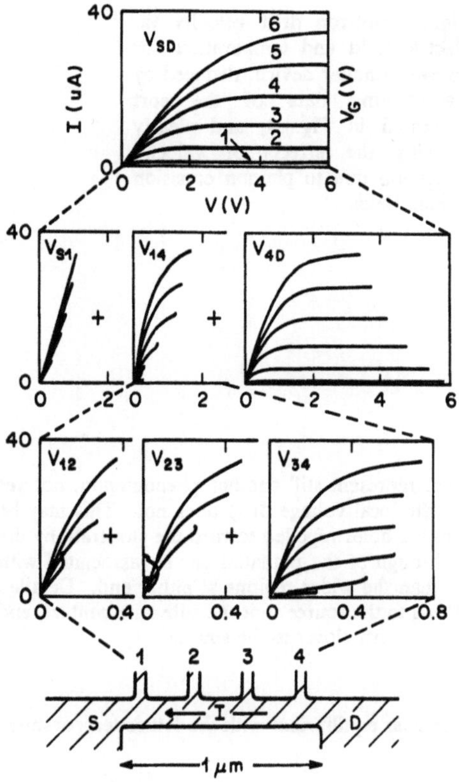

Fig. 4 Current-voltage characteristics of a multiterminal MOSFET with a narrow (125 nm) channel segment and four voltage probes, showing characteristics of the various spatial segments

zero, so that large voltage drops can be absorbed in the drain depletion layer without causing additional current flow. This is called "pinchoff", and accounts for the saturation of the I-V characteristics.

Our extra voltage probes allow us to get inside the device and verify this textbook description. The second set of I-V characteristics in Fig. 4 separates out the voltage drops in the source-side section, the narrow-channel section, and the drain-side section. Saturation and pinchoff obviously occur toward the drain end, while linear characteristics are observed toward the source end, since by comparison the channel potential there never builds up to significant values. The third set of I-V characteristics in Fig. 4 separates out the three 1/4-μm segments of the narrow section, showing that the segment toward the drain is more nonlinear. At low gate voltages, the voltage probes become unreliable because the narrow probe channels have higher thresholds, and begin to turn off.

3.2 Velocity Saturation

In addition to pinchoff, the curvature of the I-V characteristics is a result of saturation of the electron drift velocity at high electric fields. This can be explicitly measured [5], by transforming data such as Fig. 4. The electron velocity is the current per unit width divided by the 2D charge density: $v = I/Wne$. The density in the region between two probes such as V_4 and V_3 can be inferred from the local gate-to-channel capacitance: $ne = (\epsilon/d_{ox})\,[V_G - (V_4+V_3)/2 - V_T]$. The electric field in the region is simply $E = (V_4-V_3)/L_{43}$ where L_{43} is the spacing between the probes. Velocity-field characteristics for the device in Fig. 4 were presented in [5]. Very similar characteristics are shown in Fig. 5, obtained from an even smaller device with 40 nm channel width and 150 nm probe spacing. At low fields, the velocity is a strong function of lattice temperature, because electron scattering by thermal phonons is the dominant scattering mechanism. At high fields, the electrons are accelerated to such high energies that phonon emission dominates, and becomes relatively insensitive to lattice temperature. This phonon emission limits the velocities to approximately 10^7 cm/sec. Previous studies of velocity saturation in silicon by COEN and MULLER [6] and COOPER and NELSON [7] used current and time-of-flight measurements, respectively, in special large MOSFETs with resistive gates having voltage drops compensating for the drain voltage.

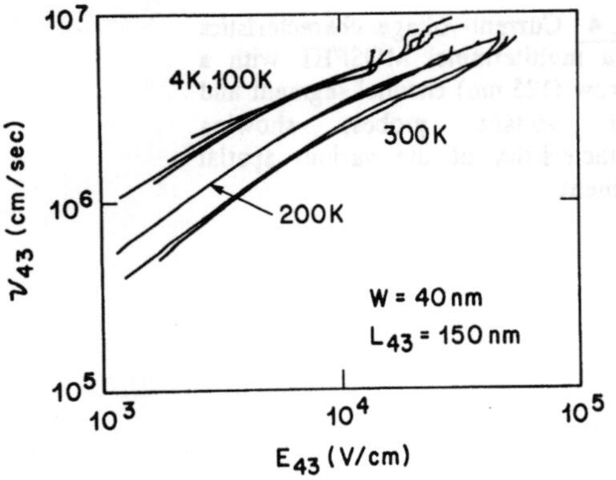

Fig. 5 Electron drift velocity vs. electric field and temperature for an even smaller device, deduced by transforming data of the sort presented in Fig. 4, and clearly showing the effects of velocity saturation due to phonon emission at high fields

3.3 Avalanching or Punchthrough

The squiggles at the high-field end of the characteristics represent still another phenomenon, not yet fully understood. The total current increases rapidly, but the local voltage drop does not. This may be caused by avalanche generation of electrons and holes with the holes repelled toward the substrate by the gate. Alternatively, it may represent a subsurface punchthrough of the potential barrier associated with the narrow channel, which always has a higher threshold than the wider regions at either end. Detailed investigations show that this extra current all eventually flows to the source, not the substrate contact, and indeed appears to be "recollected" even in narrow channel segments closer to the source.

3.4 Device Threshold and Mobility

At small V_{DS} (μV scales), the electron temperature is in near equilibrium with the lattice temperature, and the device is simply a gate-controlled resistor.

As we have discussed elsewhere [8], the electron density in small segments behaves as expected, once an unexplained systematic variation of V_T on width is taken into account. The electron mobility several volts above threshold at liquid helium temperatures approaches 3000 cm^2/V-sec in wide devices. In devices narrower than about 100 nm, it is smaller, corresponding to a mean-free-path of half the width, i.e., boundary limited scattering. It is interesting to note that the particular device shown in Figs. 1 and 2 had a lower mobility than expected from this trend. It had been repeatedly viewed in the scanning electron microscope, in order to investigate the progress of the etching in such narrow gaps. It seems clear that our usual 30-second anneal at 650 C was inadequate to remove all the extra e-beam damage.

3.5 Single-Electron Trapping Phenomena

Even in the reasonably well-understood "classical" domain discussed thus far, small devices begin to reveal interesting new effects. As devices get smaller, it is well known that their low frequency noise increases. Solid state physicists have been trained to look at the average properties of electron transport as the most easily characterizable phenomenon, and frequently to view noise as an impediment to measuring the "correct" result. This philosophy is based on the "law of large numbers" appropriate to bulk size scales. But in a device with only a hundred scatterers, for example, turning just one on and off can have a noticeable effect, and provides a great deal of unique information about the transport that otherwise would be averaged away. One such phenomenon that we have identified and studied extensively is the trapping of single electrons at individual interface traps [9].

Our work shows clearly that there exist electron traps in the gate oxide only a few angstroms from the SiO2-Si interface which can have energies that lie near the Fermi level in the inversion layer. These traps capture and emit electrons with temperature-dependent characteristic times that can be "tuned" through a convenient observation range of fractions of a second to seconds. When the trap is charged, it acts as a scatterer, producing a 1% change if there are 100 scatterers (10^{11} scatterers/cm^2 in a device with 10^{-9} cm^2 area, i.e., 0.1 μm by 1 μm). This phenomenon is most easily studied in devices slightly greater than 0.1 μm wide. In narrower devices the mean-free path is limited by scattering from walls, not interface traps, while in much larger devices there are too many traps with too small an individual effect to easily measure.

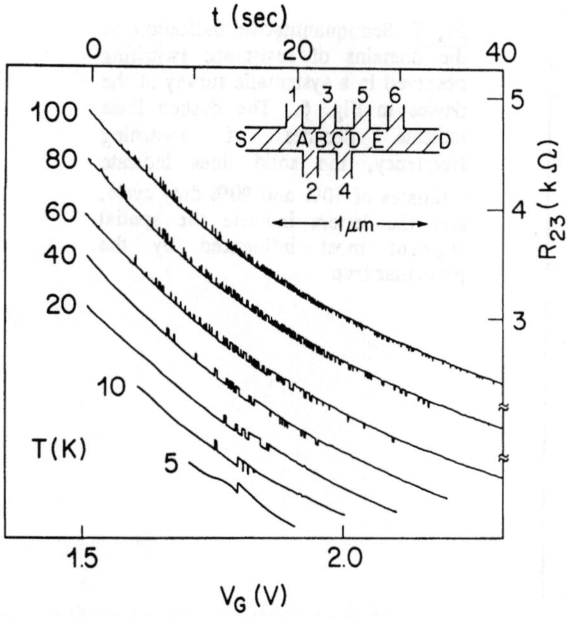

Fig. 6 Domain of resistance switching caused by capture and emission of electrons at a single interface trap located in the gate oxide of spatial segment B. This is equivalent to one scatterer turning off and on. Slow gate voltage sweeps demonstrate the change in duty cycle of the trap occupancy as the trap energy is swept past the electron Fermi level. At low temperatures, k_BT is smaller, as is the activated switching rate

Figure 6 gives a dramatic overview of the domain of influence of a single interface trap. Each trace is a slow sweep of gate voltage that lowers the energy of the trap relative to the Fermi level of the inversion layer electrons. Initially empty a large proportion of the time, the trap begins to be occupied a larger fraction of the time as it approaches within k_BT of the Fermi level. Eventually it is almost always occupied, which corresponds here to the higher resistance state. Each trace sweeps through this full range of "duty cycle". At lower temperatures, k_BT is smaller, corresponding to a narrower range of gate voltage, and the activated switching rates are slower, resulting in less frequent jumps. At the lowest temperature, the trap switches on only once, and stays on. This particular trap is located in segment B between the second and third voltage probes. Other traps in this segment can be observed in other gate voltage ranges, or at other temperatures. Figure 7 indicates the domain of influence in temperature and gate-voltage space of a number of traps observed in various segments of this device. All the easily analyzed, isolated prominent switching sequences have been plotted, although additional switching corresponding to confusing overlaps has been omitted in certain ranges. The areal density of observable traps is of order a few times 10^{10} cm^{-2}, not inconsistent with the observed mean free path and values expected for a clean processing technology.

In detailed studies of the behavior of individual traps, we have used their sensitivity to gate voltage to determine that they are located a few angstroms into the gate oxide [9]. By comparing the effects of lattice heating and electron heating, we have demonstrated that lattice motion is essential to the emission process [10]. We have investigated the statistical properties of the switching, showing that it is truly a "random telegraph signal" [4], and we have verified that it makes a dominant Lorentzian contribution to the noise [11]. The distribution of observed activation energies is consistent with the broad range required to account for 1/f noise in large devices, by superposition of many Lorentzians [9]. By comparing the noise spectra of large, intermediate, and submicron devices, UREN et al.[12]. have made a semiquantitative connection between the magnitude of the noise due to trap switching in small devices and 1/f noise in large devices. Similar phenomena have been studied in small tunnel junctions by ROGERS and BUHRMAN [13-14].

More recently, we have attempted to get information about the spatial range of influence of each trap by searching for correlated switching behavior in adjacent segments of the device. If correlated changes are observed in only two segments, no spatial range is determined, because the trap might lie on the

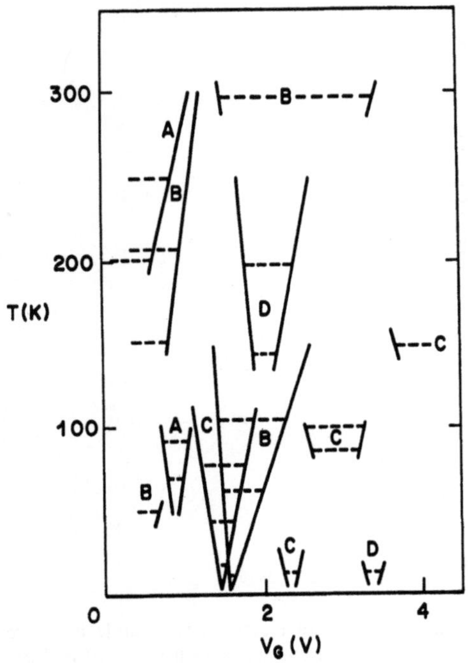

Fig. 7 Semiquantitative indication of the domains of resistance switching observed in a systematic survey of the device in Fig. 6. The dashed lines indicate decades of switching frequency, the solid lines indicate estimates of 10% and 90% duty cycle, and the letters indicate the spatial segment most influenced by the particular trap

boundary, i.e., near the shared voltage probe. We see this occasionally even at room temperature. Correlated effects in three adjacent segments, however, as shown in Fig. 8, are observable only at low temperatures. At first [5] we interpreted this result within the quasi-classical "resistivity dipole" picture proposed many years ago by LANDAUER [15,16], who seems to have been the first to pose and answer the question central to our experiment: "What are the local changes caused by adding one scatterer in the presence of a current?" The phenomena that he considered occur on the length scale of a mean-free-path. In conjunction with compiling Fig. 7, we conducted a systematic search for correlated switching behavior, and found that three-segment correlations were observed only below about 10 K. At 4.2 K such correlations could be detected in three out of five cases. We found that the sign of the effect in adjacent segments was unpredictable, and could change sign as the current was increased; at low currents the changes in all segments were proportional to current (i.e., simple resistance changes) not quadratic as Landauer would expect. Moreover, the mean-free-path length scale studied by Landauer seemed too short. I will return to a possible new quantum interpretation of this experiment at the conclusion of this chapter.

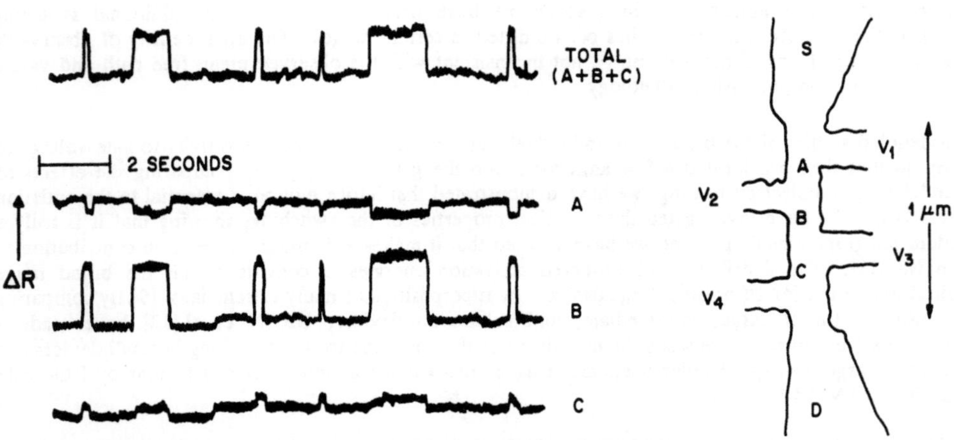

Fig. 8 Observation of correlated switching in three segments of a relatively large device, giving information about the spatial range of influence of a single scatterer (from [5])

4. Quantum Transport

The past decade has seen a major advance in our understanding of quantum transport in disordered systems, particularly of reduced dimensionality. A central role is played by the dimensionless conductance $g = G/(e^2/h)$, where e^2/h is $(25.8\ k\Omega)^{-1}$. A variety of quantum corrections to the classical conductivity have been discovered which are of order $\Delta g \approx 1$. Many of these are systematically reviewed in other chapters, so that my purpose here is to demonstrate their effect in our small Si nanostructures.

4.1 Weak Localization and Interaction Effects

Weak localization is a backscattering phenomenon resulting from time-reversed pairs of closed trajectories that interfere coherently. The relevant length scale for determining dimensionality is the distance diffused before loss of phase information – essentially the inelastic diffusion length $L_{in} = (D\tau_{in})^{1/2}$ where τ_{in} is the inelastic scattering time and D is the diffusion constant. For dimensionality d with respect to the mean free path l, $D = v_F l/d$. The weak localization result for a one dimensional "wire" of length $L >> L_{in}$ and width $W < L_{in}$ is

$$\Delta g = -(L_{in}/L)[1 + L_{in}^2/L_{H1D}^2]^{-1/2}, \tag{1}$$

where $L_{H1D} = 3\Phi_o/\pi HW$. The magnetic field reduces the zero-field effect $-L_{in}/L$ by a factor of two when $3/\pi$ units of flux $\Phi_o = h/2e$ penetrate a length L_{in} of the 1D wire.

Interaction effects are a many-body phenomenon resulting from the self-energy of electron-electron interactions, which are enhanced because of diffusion in a restricted geometry. The relevant length scale is the thermal diffusion length $L_{th} = \pi[D\hbar/k_B T]^{1/2}$ which governs diffusion on time scales corresponding (via the uncertainty principle) to the energy width of partially occupied states at the Fermi surface. Again for the 1D case,

$$\Delta g = -c(L_{th}/L) \tag{2}$$

where the coefficient c depends on screening, and in Si MOSFETs is approximately 0.5. In the relevant range the interaction effect is only weakly dependent on magnetic field.

Figure 9 shows the field dependence of gL at two temperatures and several gate voltages, for a 40 nm-wide channel. (The device actually averaged over 40 such channels in parallel, each 14.3 μm long.) The temperature dependence at high field is primarily due to interaction effects. The dip at low fields is caused by localization effects. The two are comparable in magnitude in these devices. L_{in} can be determined in two ways -- read off directly from the magnitude of the change from zero to high fields, and also computed from the half-width in field based on knowledge of the device width. Figure 10 shows the inelastic scattering length deduced from the half width, versus $G_\square = GL/W$. At high gate voltages where the localization change is a small fractional effect described by (1), the inelastic scattering length determined from the magnitude change also agrees.

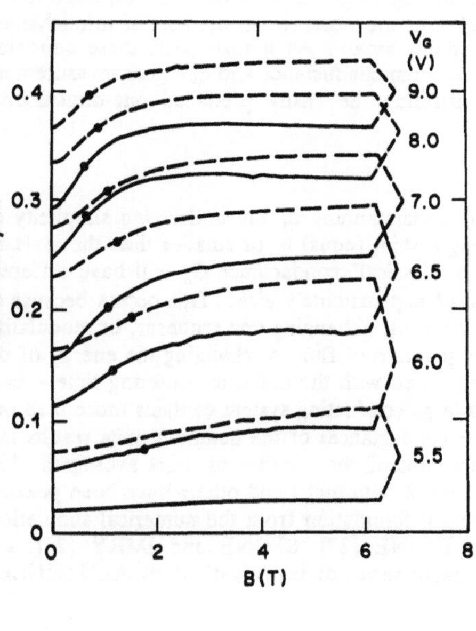

Fig. 9 Normalized conductance per channel as a function of magnetic field and gate voltage at 4.2 K (dashed curves) and 2 K (solid curves), for a device consisting of 40 MOSFET channels 14.3 μm long and 40 nm wide. The variations can be understood in terms of weak localization and interaction effects, as discussed in the text, and used to determine the inelastic diffusion length

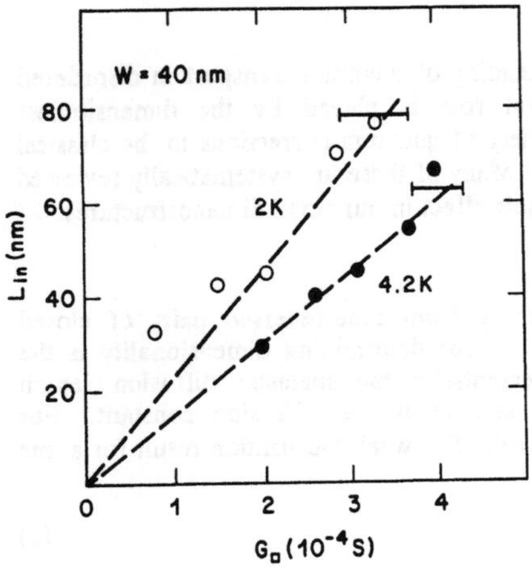

Fig. 10 Inelastic diffusion lengths deduced from the device width of 40 nm and the magnetoconductance halfwidths of Fig. 9. At high gate voltages the length scale deduced from the magnitude change alone (horizontal bar) also agrees. The dashed curves are a fit to the data consistent with a $D/T^{-1/2}$ temperature dependence. The overall magnitude differs at most by a factor of two from previous 2D theories and experiments

The inelastic scattering length of Fig. 10 is quantitatively consistent within a factor of two with the inelastic diffusion length for electron-electron scattering measured by BISHOP et al.[17] in large 2D MOSFETs, and with our previous investigations of the temperature dependence of weak localization and interaction effects in narrow MOSFET channels [18]. The best fit is obtained with the relation

$$\tau_{in}^{2D} = 2.5 \times 10^{-13} \ sec^2K/cm^2 \ D/T \tag{3}$$

that summarizes the data of BISHOP et al.[17], but with the 1D diffusion coefficient $v_F l \approx 2G_{\square}/e^2N_2$ where $N_2 = 2m^*/\pi\hbar^2$ is the usual 2D density of states. There is no evidence here of the 1D correction $\pi W/4\sqrt{2}L_{th}$ discussed by WHEELER et al.[19]. Further experiments are needed, but these experiments already give us a basis to predict the relevant diffusion lengths in our devices. Other systematic experiments on these phenomena in 1D MOSFETs and 1D metal wires include experiments by DEAN and PEPPER [20], SANTHANAM et al.[21], LICINI et al.[22], and BISHOP and DOLAN [23].

4.2 Particle in a Box

In the absence of diffusive scattering, the electron states in an ideal narrow channel will be organized into one-dimensional subbands. With weak scattering, resistance appears, yet evidence of the subband structure may show up in the resistance as the Fermi level is moved through the band structure. Experiments searching for this effect have been reported by SKOCPOL et al.[24] and WHEELER et al.[25], with tantalyzing but ultimately inconclusive results. Our own most recent observations have been reported in [8]. The "problem" with these experiments is that the large conductance structure observed in small devices is frustratingly irregular, while the residual structure observed in the sort of multichannel long devices described in the previous section is frustratingly small. As it turns out, these annoying random fluctuations can tell us more about the nature of quantum conductance and quantum measurement than we would ever expect to learn from simply confirming the easily predicted one-dimensional subbands.

4.3 Universal Conductance Fluctuations

The past year has seen a veritable revolution in our understanding of the underlying simplicity of random scattering in a finite quantum-coherent conducting system (equal to or smaller than the inelastic diffusion length). Nominally similar units with the same "classical" conductance G_o will have different quantum conductances characterized by an rms variation of approximately e^2/h. This occurs because of random quantum interference analogous to laser speckle patterns. Changing one scatterer, or modulating the phases of trajectories by applying approximately one quantum of flux, or changing the energy of the electrons by an amount greater than the energy width associated with the inelastic scattering time – each can result in a totally different interference pattern. If a large conducting system contains more than one such quantum unit, classical series/parallel addition of the conductances of the quantum units results in a total fluctuation that is smaller essentially by the square root of the number of units averaged. This simple picture characterizes in quantitative detail whole files of data that I and others have been puzzling over for several years. It has emerged on a firm theoretical foundation from the numerical simulations and analytical calculations of STONE [26], LEE and STONE [27], STONE and IMRY [28], and FUKUYAMA et al.[29]. It also has emerged from a parallel series of investigations by AL'TSHULER

[30], AL'TSHULER and KHMEL'NITSKII [31], and AL'TSHULER and SPIVAK [32]. Already it has played a crucial role in guiding the search for the h/e quantum interference effects in single small metal rings discussed elsewhere in this volume. Here I will provide a "derivation" of certain key dependences, and show clearcut experimental verification of them.

Figure 11 shows a typical data set at 4.2 K of the resistance as a function of magnetic field and gate voltage (electron energy), for a 300 nm-long double segment of the 60 nm-wide device in Figs. 1-3. In this range the inelastic diffusion length is about 40 nm, less than both the length and the width. A similar data set has been presented by LICINI et al.[33] for preliminary comparison with the Lee and Stone theory. The aperiodic magnetoresistance fluctuations here are characterized by the correlation function shown in Fig. 12. The rms conductance amplitude is 0.06 e^2/h out of the average conductance of 1.1 e^2/h, the half-width in magnetic field is 1.3 T, and the half-width in gate voltage is approximately 0.1 volts (about 0.7 meV in the inversion layer). How is this to be understood?

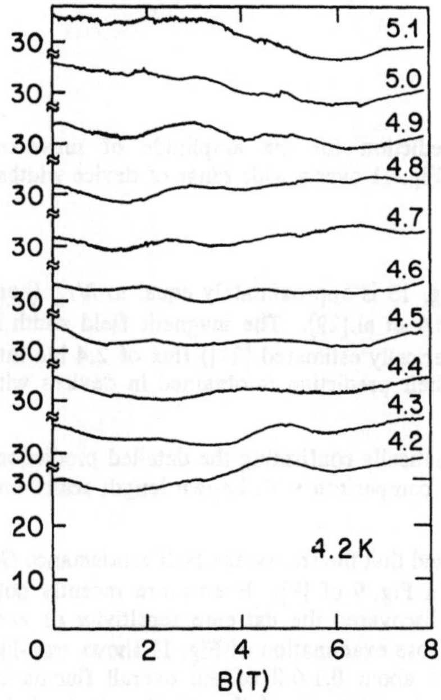

Fig. 11 Aperiodic magnetoresistance oscillations as a function of magnetic field and gate voltage for an 0.3 μm segment of the device in Figs. 1 and 2

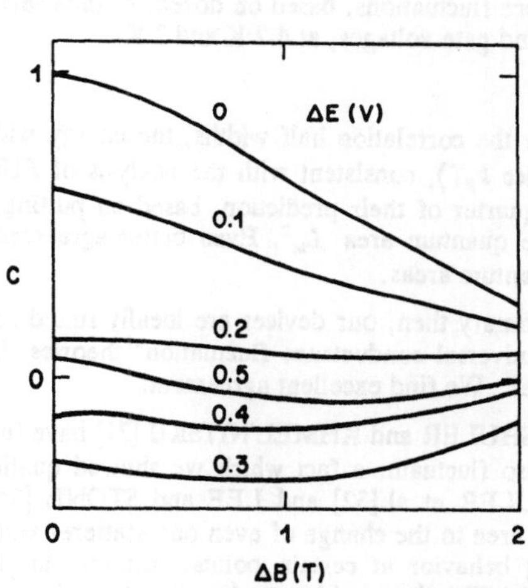

Fig. 12 Correlation of the magnetic field traces in Fig. 11 as a function of displacements in field and gate voltage. The variance at zero displacement is normalized to the square of the observed rms conductance fluctuation 0.06 e^2/h

Consider a 1D conducting system with $W < L_{in} < L$ and an average conductance $<g>$ (averaged over a range of magnetic field and gate voltage, or an ensemble of nominally identical samples, and normalized to e^2/h). It contains $N = L_{in}/L$ quantum subunits of randomly varying conductance $N<g>\pm 1$ connected in series. The resulting combination has random variations about $<g>$ of relative size $N^{-1/2}$ and absolute size $N^{-3/2}$. A 2D system with $W > L_{in}$ can be divided into $M = W/L_{in}$ parallel 1D strips, simply contributing after averaging an extra factor of $M^{-1/2}$ to the relative size and $M^{1/2}$ to the absolute size of the fluctuations. For the conditions in Fig. 11, N $\cong$ 7.5, M $\cong$ 1.5, and thus $N^{-3/2}M^{1/2} \cong 0.04$, in excellent agreement with the observed amplitude, considering the limited degree of averaging achieved.

Figure 13 presents analysis of 39 data sets from four devices of various widths, each with single, double, and triple-length segments, two temperatures (4.2 and 2 K), and several gate voltage ranges. Other points correspond to less systematic data sets from other devices investigated earlier. The observed amplitude agrees well with the simple model presented above for two and a half orders of magnitude of variation, and shows the expected trends with each of the variables that enter the prediction.

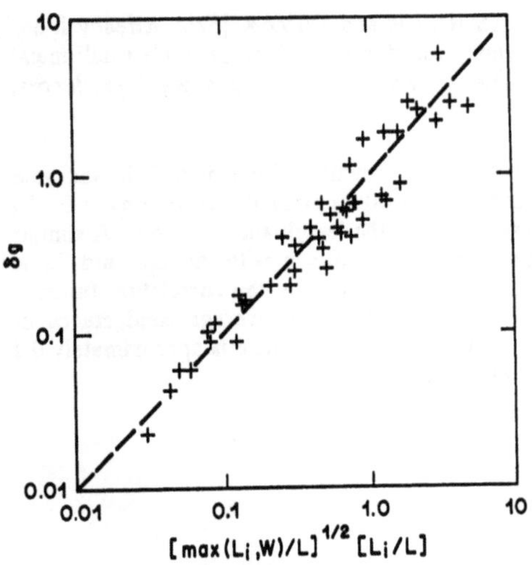

Fig. 13 Excellent agreement between observation and prediction for the amplitude of universal conductance fluctuations, based on dozens of data sets such as Fig. 11 over a wide range of device widths, lengths, and gate voltages, at 4.2 K and 2 K

As for the correlation half widths, the energy width in Fig. 13 is approximately equal to $\hbar/\tau_{in}$ (here about twice k_BT), consistent with the analysis of FUKUYAMA et al.[29]. The magnetic field width is about a quarter of their prediction, based on putting a (numerically estimated [27]) flux of 2.4 h/e into each little quantum area $L_{in}{}^2$. Even better agreement with their prediction is obtained in devices with larger quantum areas.

In summary then, our devices are ideally suited for systematically confirming the detailed predictions of the "universal conductance fluctuation" theories, by direct comparison with known length scales and geometries. We find excellent agreement.

AL'TSHULER and KHMEL'NITSKII [31] have further noted that the transverse Hall conductance G_{xy} should also fluctuate, a fact which we showed qualitatively in Fig. 9 of [8]. Even more recently both AL'TSHULER et al.[32] and LEE and STONE [34] have discovered the extreme sensitivity of each quantum area to the change of even one scatterer within it. Close examination of Fig. 11 shows trap-like switching behavior at certain points, resulting in changes of about 0.1-0.2 of the overall fluctuation amplitude. This is roughly consistent with a given trap affecting just one of the quantum areas being averaged over. Even larger switching is observed in devices consisting of a single quantum area. In the trap-switching correlation experiment described above, the observed range of spatial effect is also roughly consistent with L_{in}. Although these aspects remain to be systematically investigated, we begin to see how all of the various phenomena that we have been exploring can be united in a fundamentally new understanding of quantum transport in small systems.

Acknowledgements

I would like to express my deep appreciation to all of the members of the Holmdel Microscience Collaboration, and particularly to Paul Mankiewich, Rich Howard, and Larry Jackel, without whose vision and persistence there would have been no devices to measure and no lab to measure it in. I also would like to thank Doug Stone for extensive and stimulating discussions as our consultant, and Patrick Lee, Rolf Landauer, Alice White, Bob Dynes, and David Bishop for useful discussions on numerous occasions.

References

1. P.M. Mankiewich, R.E. Howard, L.D. Jackel, W.J. Skocpol, D.M. Tennant: J. Vac. Sci. Technol. B 4, 380 (1986).
2. W.J. Skocpol, A.M. Voshchenkov, R.E. Howard, E.L. Hu, L.D. Jackel, R.W. Epworth, L.A. Fetter, P. Grabbe, D.M. Tennant: Physica 109&110B, 2105 (1982)
3. W.J. Skocpol, L.D. Jackel, R.E. Howard, E.L. Hu, L.A. Fetter: Physica 117B&118B, 667 (1983)

4. R.E. Howard, W.J. Skocpol, L.D. Jackel, P.M. Mankiewich, L.A. Fetter, D.M. Tennant, R.E. Epworth, K.S. Ralls: IEEE Trans. Electron Dev. ED-32, 1669 (1985)
5. R.E. Howard, L.D. Jackel, P.M. Mankiewich, W.J. Skocpol: Science 231, 346 (1986)
6. R.W. Coen, R.S. Muller: Solid-State Electron. 23, 335 (1980)
7. J.A. Cooper, Jr., D.F. Nelson: IEEE Electron Dev. Lett. EDL-2, 171 (1981)
8. W.J. Skocpol, L.D. Jackel, R.E. Howard, P.M. Mankiewich, D.M. Tennant: Surface Sci., to appear
9. K.S. Ralls, W.J. Skocpol, L.D. Jackel, R.E. Howard, L.A. Fetter, R.W. Epworth, and D.M. Tennant: Phys. Rev. Lett. 52, 228 (1984)
10. L.D. Jackel, W.J. Skocpol, R.E. Howard, L.A. Fetter, R.W. Epworth, D.M. Tennant: In Proc. 17th Int. Conf. on Physics of Semiconductors, ed. by J.D. Chadi, W.A. Harrison (Springer, New York 1985) p. 221
11. J. Scofield, private commun.
12. M.J. Uren, D.J. Day, M.J. Kirton: Appl. Phys. Lett. 47, 1195 (1985)
13. C.T. Rogers, R.A. Buhrman: Phys. Rev. Lett. 53, 1272 (1984)
14. C.T. Rogers, R.A. Buhrman: Phys. Rev. Lett. 55, 859 (1985)
15. R. Landauer: IBM J. Research Develop. 1, 223 (1957)
16. R. Landauer: Z. Phys. B 21, 247 (1975)
17. D.J. Bishop, D.C. Tsui, R.C. Dynes: Phys. Rev. B 26, 773 (1982)
18. W.J. Skocpol, L.D. Jackel, E.L. Hu, R.E. Howard, L.A. Fetter: Phys. Rev. Lett. 49, 951 (1982)
19. R.G. Wheeler, K.K. Choi, A. Goel, R. Wisnieff, D.E. Prober: Phys. Rev. Lett. 49, 1674 (1982)
20. C.C. Dean, M. Pepper: J. Phys. C 15, L1287 (1982)
21. P. Santhanam, S. Wind, D.E. Prober: Phys. Rev. Lett. 53, 1179 (1984)
22. J.C. Licini, G.J. Dolan, D.J. Bishop: Phys. Rev. Lett. 54, 1585 (1985)
23. D.J. Bishop, and G.J. Dolan: Phys. Rev. Lett. 55, 2911 (1985)
24. W.J. Skocpol, L.D. Jackel, R.E. Howard, H.G. Craighead, L.A. Fetter, P.M. Mankiewich, P. Grabbe, D.M. Tennant: Surface Science 142, 14 (1984)
25. R.G. Wheeler, K.K. Choi, R. Wisnieff: Surface Science 142, 19 (1984)
26. A.D. Stone: Phys. Rev. Lett. 54, 2692 (1985)
27. P.A. Lee, A.D. Stone: Phys. Rev. Lett. 55, 1622 (1985)
28. A.D. Stone, Y. Imry: Phys. Rev. Lett. 56, 189 (1986)
29. H. Fukuyama, P.A. Lee, A.D. Stone: submitted to Phys. Rev. B
30. B.L. Al'tshuler: JETP Lett. 41, 649 (1985)
31. B.L. Al'tshuler, D.E. Khmel'nitskii: JETP Lett. 42, 359 (1985)
32. B.L. Al'tshuler, B.Z. Spivak: Pis. Zh. Eksp. Teor. Fiz. 42, 363 (1985)
33. J.C. Licini, D.J. Bishop, M.A. Kastner, J. Melngailis: Phys. Rev. Lett. 55, 2987 (1985)
34. P.A. Lee and A.D. Stone, private commun.

1D Structures – Field Confinement Approach

A. Hartstein

IBM Thomas J. Watson Research Center, Yorktown Heights, NY 10598, USA

1. Introduction

The allure of studying 1D structures in physics is very appealing to most people. Textbooks on elementary quantum mechanics are filled with model calculations done on 1D systems because of their relative simplicity. The possibilities of studying 'real' 1D systems is therefore intriguing. The possibility is held out that the simple theories can be tested directly. In addition to this numerous physical properties are dependent on the dimensionality of the system being studied and 1D is the limiting case for transport studies.

The beauty of obtaining a 1D system using MOSFET's lies in the utility of the MOSFET for studying physics. Both inversion layer and accumulation layer MOSFET's have been extremely important in studies of the physics of 2D electron gases [1]. Probably the single most important feature of these systems is the ability to vary the Fermi energy in the system simply by controlling the gate voltage. This can be a very powerful tool indeed. In the MOSFET system an electric field perpendicular to the interface between silicon and silicon dioxide produces an interface charge sufficiently narrow to quantize the electron or hole wave functions perpendicular to the surface. The carriers are still free to move parallel to the surface and thus form a 2D electron gas as long as the temperature is small compared to the energy level spacing for motion perpendicular to the surface.

In recent years it became practical to construct 1D MOSFET samples by confining the 2D electron gas along a second direction in the sample, thus opening up a whole new realm of experimental possibilities. The first such samples were made by a field confinement technique [2]. Subsequently, samples were made using extensions of optical lithography [3], electron beam lithography [4] and shadowing techniques [5]. Each of the techniques has its advantages and disadvantages which will be discussed in the next section. However, each system retains the central advantage of MOSFET systems of being able to readily control the Fermi energy.

The physics which one can study centers on both the 1D nature of the samples and the concurrent fact that the sample area is necessarily extremely small. In some cases the effects of single electron sites can be studied in the transport properties of these systems. Some very general theoretical arguments show that electron states in both 1D and 2D systems are always localized. The electrons do not behave as free electrons although the deviations from free electron behavior can be small if the localization length is much larger than the sample length. The regimes of strong localization and weak localization can be studied in both the 1D and 2D regimes using samples of these narrow types.

The regime of strong localization occurs when the localization length is smaller than the effective sample length. It is characterized by electron states being localized in small regions of space and transport being dominated by electrons hopping between the available states. The dominant conduction mechanism is variable range hopping. In addition to this the effect of individual hopping sites can be extremely important and in fact can dominate the observed characteristics. These can be observed in both 1D and 2D regimes, and examples will be given. Resonant tunneling may also be important in this regime.

Weak localization occurs when the localization length is longer than the effective sample length. The physics in this regime is best described in terms of the interference of electron waves in the sample. Low field negative magnetoresistance is one consequence of these considerations. When samples are small enough, universal conductance fluctuations are additionally observed. This is a consequence of the failure of ensemble averaging for small samples. An example of this behavior will also be discussed in considerable detail.

2. Device Fabrication

There have been essentially two approaches to the fabrication of 1D MOSFET devices. The most obvious technique is to use a narrow enough gate to obtain 1D behavior. Just how narrow this must be depends on the particular physics under study. Optical lithography [6] has been used to produce gates as narrow as 300 nm. Electron beam lithography [7] has been used to produce gates as narrow as 55 nm. These techniques suffer from the drawback that width variations of 5-10% are typically obtained. Shadowing techniques [5] can result in gates as narrow as 10-20 nm, but only 70 nm has been reported.

All of the samples produced with these techniques suffer from a common problem. The conductance curves exhibit a rather large hysteresis as a function of the applied gate voltage and irreproducibility after thermal cycling. This most probably results from charge collecting on the surface of the device adjacent to the narrow gate changing the effective width of the inversion layer channel. The problem is most important for the narrowest devices. An attempt to overcome this problem by defining a mesa structure for the device [4] by reactive ion etching leaves the sides of the device open to similar problems. In the mesa structure, however, the maximum device width is defined.

1D MOSFET's can also be obtained by a technique of electric field confinement of the 2D electron gas in an accumulation layer MOSFET [2, 8]. This technique has both advantages and disadvantages over the narrow gate technique. It requires no special lithography so that samples can be processed by fairly standard and well characterized semiconductor processing techniques. No free surface area or edges are located adjacent to the conducting channel, so that no hysteresis is observed in these devices. The effective width of the 1D channel can be altered by voltages applied to the device and is believed to have been varied between 20 nm and 2 μm in devices of this type.

These electric field confined pinched accumulation layer samples cannot be as easily modelled as the narrow gate samples can. In the narrow gate samples the problem is essentially electrostatics with a well defined gate but somewhat unclear boundary conditions. In the field confined case the modelling is much harder and initial attempts have only very recently been reported [9]. Part of the problem comes from our inability to accurately model the control electrode diffusions on the microscopic scale. In our experiments we have had to infer the effective width of the channel using rather indirect means.

Figure 1 shows idealized top and cross section views of our pinched accumulation layer samples. The substrate is 10 ohm-cm n-type <100> silicon. Two n^+ diffusions are used as the source and drain contacts to the accumulation layer. Two lateral p^+ diffusions are used to control the width of the accumulation layer. The electrons at the surface are in effect pinched into a narrow region because of the electric field applied from these control electrodes and arising from the built-in potential between the p^+ regions and the n-type substrate. The separation between the control electrodes is about 1 μm, and the length of the narrow conducting region is about 10 μm. The gate oxide is 30 nm thick, and the gate is made of aluminum.

In the actual device, shown in Fig. 2, the dimensions are seen to be not nearly as ideal as indicated in Fig. 1. In fact we have no way of ascertaining the effective width variations in this type of device. Actually, we do not even know the width of the conducting channel, except from our

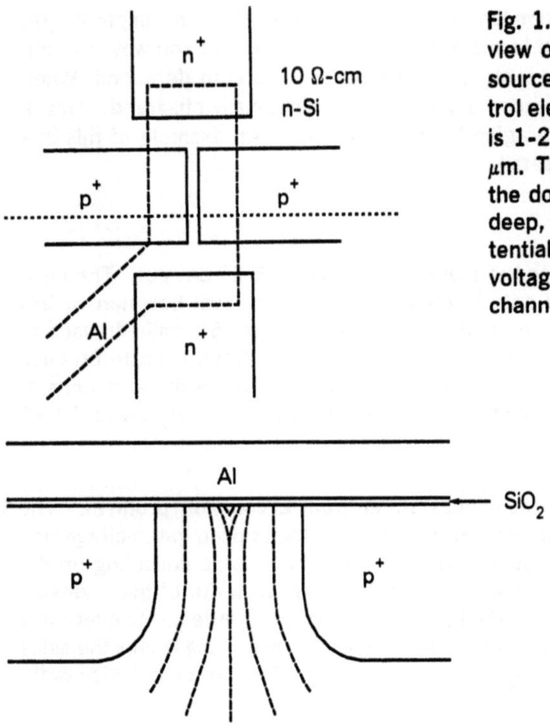

Fig. 1. The upper part shows an idealized top view of a sample. The two n+ regions are the source and drain. The p+ regions are the control electrodes. The width between the controls is 1-2 μm and the length of the controls is 14 μm. The lower part shows a cross section along the dotted line. The diffusions are about 1 μm deep, and the oxide thickness is 30 nm. Potential lines are sketched for a positive gate voltage, indicating the narrow conducting channel

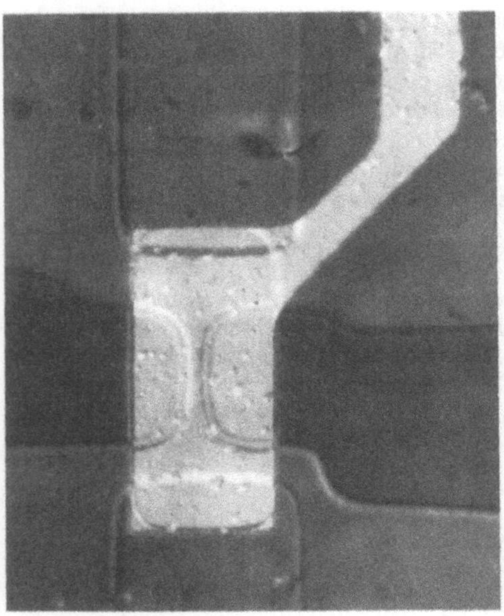

Fig. 2. A micrograph of the device corresponding to the idealized sketch of Fig. 1. The effective channel length is about 10 μm

analysis of the experimental data. The effect of an applied gate voltage is two-fold in samples of this geometry. A positive gate voltage will induce electrons at the surface, thereby changing the carrier density and Fermi energy. Moreover, it will tend to broaden the width of the accumulation layer channel. A negative voltage applied to the control electrodes will pinch the channel to narrower widths, but will also tend to change the threshold voltage and thereby reduce the accumulation layer carrier density.

3. Experimental

These devices can only be measured at low temperatures, in our case T < 10 K. At higher temperatures the substrate is conducting and results in a parallel conduction path between the source and drain contacts. The devices were measured using an AC modulation of the source-drain voltage. The resulting current was synchronously detected using a current sensitive preamplifier feeding into a lock-in amplifier. To avoid hot electron effects the source-drain field was generally kept below 2 mV/cm. Some measurements were made in a dilution refrigerator where good data was obtained at temperatures as low as 36 mK. Other measurements to be described were made in a liquid He_3 cryostat operated at 0.45 K.

Great care was taken to eliminate extraneous noise signals. In our original experiments [2] noise heating made any measurements below about 100 mK questionable. Later refinements [10, 11] of the experiments gave good data as low as 36 mK. For most of the measurements the control voltage was set to 0 V which gave the lowest noise heating. Data were obtained in many different ways to obtain the optimum signal to noise ratios. Some data were taken at constant gate voltage while varying the temperature. This gave the most accurate temperature dependent data. Some data were obtained by fixing the temperature and varying the gate voltage. Our most recent experiments [12] involved fixing both gate voltage and temperature and then varying an applied magnetic field from a superconducting solenoid.

The mass of data needed to fully explore the nature of the observed structure in the conductance curves necessitated the use of elaborate data acquisition systems. In order to obtain these data it was necessary to build a noise free voltage ramp and to carefully interface the equipment to the data acquisition system. The experiment and the computer system were decoupled by use of active filters on all interconnecting lines. These measures were necessary to avoid noise heating of the electrons, particularly at the lowest gate voltages.

Typical data obtained from these samples are shown in Fig. 3. The figure shows the measured conductance as a function of gate voltage for three temperatures. The most notable feature of the data is the large structure observed in the conductance as a function of gate voltage. The structure is reproducible in a given sample, but was sample dependent. This structure was not observed to change over a period of more than six months and numerous temperature cyclings up to 300 K. The relative amplitudes of the peaks can be seen to decrease at higher gate voltages. The widths of the peaks are observed to increase with increasing gate voltage. As the temperature is raised, the peaks broaden and the smaller ones gradually disappear.

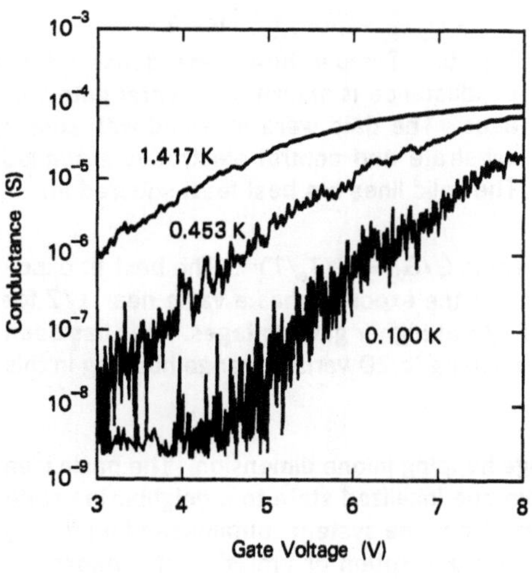

Fig. 3. Conductance is shown as a function of gate voltage for three temperatures. The structure in the figure is real and not noise. In fact the resolution of the figure is insufficient to resolve the sharpest structures

The essential physics of 1D transport is contained in these curves. The low gate voltage, low temperature portion of the figure represents the strong localization regime. The high gate voltage, high temperature regime is best described by weak localization. A smooth transition between these regimes is observed. The curves also contain regions of both 1D and 2D behavior. The details of the temperature dependence of the conductance in the different regimes contains much of this information. The reproducible structure is caused essentially by the extremely small sample size. In the strong localization regime it is caused by the individual statistics of only 1 or 2 hopping events controlling the conductance of the sample. In the weak localization regime it reveals universal conductance fluctuations. In what follows I will describe three of these areas in much more detail.

4. Strong Localization

The data of Fig. 3 look qualitatively like peaks superimposed on a baseline when viewed on a linear scale as shown in Fig. 4. Therefore, it is reasonable to try to understand the overall gate voltage and temperature dependences of the data before concentrating on the conductance structure itself. Figure 5 shows the temperature dependence of representative conductance minima. These data were obtained [2] by fixing the gate voltage and varying the temperature. The temperature dependence at low gate voltages is $\ln G \sim T^{-1/2}$, and for higher gate voltages goes over to $\ln G \sim T^{-1/3}$. At gate voltages higher than those shown in the figure, the temperature dependence becomes $\Delta G/G \sim T^{-1/2}$.

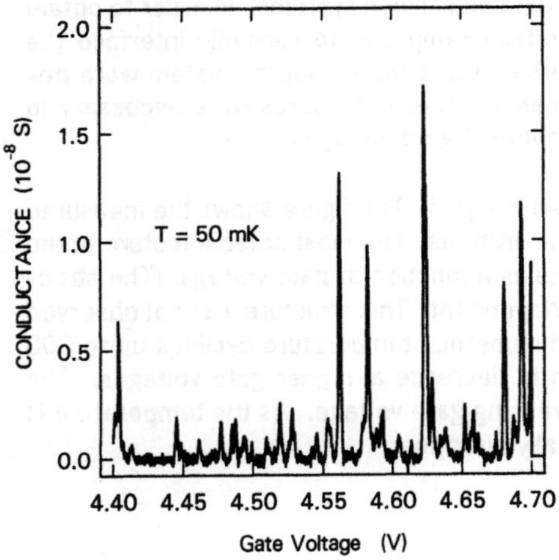

Fig. 4. Low temperature conductance as a function of gate voltage demonstrating the existence of many well isolated individual conductance peaks on a linear scale

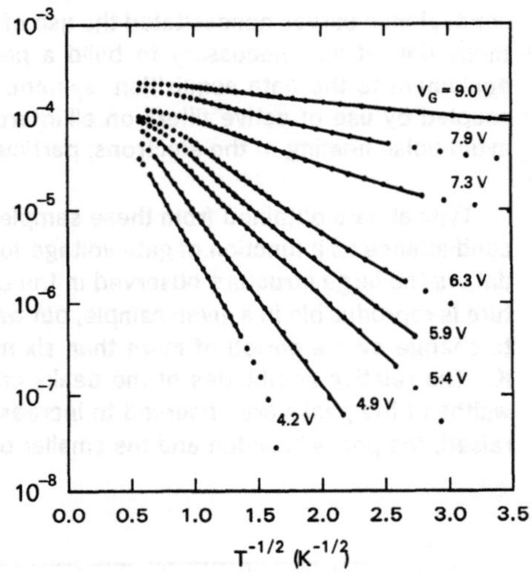

Fig. 5. Temperature dependence of the conductance is shown for several gate voltages. The data were obtained with source, substrate and control electrodes grounded. The solid lines are best least squared fits

A least squared fit to the data was made to the form $\ln G/G_0 = -(T_0/T)^{-n}$. The best fit exponents are shown in Fig. 6. It is clear from this figure that the exponent has a value near 1/2 for low gate voltages and shows a transition to a value of 1/3 at higher gate voltages. This has been interpreted [2] as a transition from 1D variable range hopping to 2D variable range hopping in this system.

It is fairly easy to give a derivation of variable range hopping in one dimension. The basic idea of variable range hopping is that electrons tunnel from one localized state to a neighboring state but do not necessarily favor the nearest neighbor hop. Since the system contains random fluctuations, the energy levels differ from site to site and the absorption or emission of a phonon is

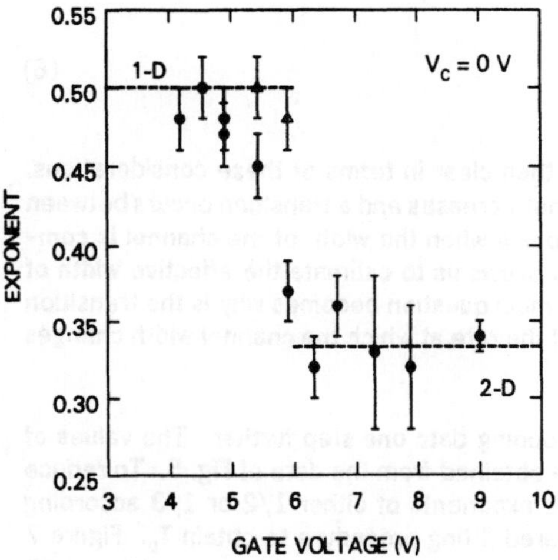

Fig. 6. The exponents that best fit the data in Fig. 5. are shown as a function of gate voltage. The dots are from fits to the entire temperature range, while the open triangles are fits to a restricted low temperature range

necessary to mediate the transfer (phonon assisted tunneling). The probability of electron tunneling from site i to site j is given by

$$P = P_0 \exp(-2\alpha r_{ij} - E_{ij}/k_B T), \tag{1}$$

where α gives the decay of the localized wave functions, r_{ij} is the separation between sites i and j, E_{ij} is the energy difference between the sites, k_B is Boltzmann's constant and T is the temperature. The first term gives the overlap of the wave functions and the second term gives the probability of absorbing a phonon to mediate the transition. There is a competition between these two terms, the former favoring hops to the nearest sites and the latter favoring hops to sites of near energy which on average are not the nearest neighbors.

There is a statistical relationship between the distance an electron must hop to find a state of a given low energy. This is simply given by

$$E_{ij} N_2(E_F) 2 r_{ij} w = 1, \tag{2}$$

where $N_2(E_F)$ is the 2D density of states at the Fermi energy and w is the width of the sample assumed to be smaller than the average hopping distance, r. For simplicity we have assumed that the density of states is uniform across the conducting channel, $N_1 = wN_2$, where N_1 is the 1D density of states. Equation (2) says that there are $1/N_2$ electron states within an area 2rw and energy range E_{ij} from an initial state, on average.

In order to obtain the variable range hopping expressions one has to find the most probable hopping distance, r, which maximizes P. When this is done and cast into the form of a conductance, G, one obtains

$$G = G_0 \exp - (T_0/T)^{1/2} \quad \text{and} \tag{3}$$

$$T_0 = \frac{4\alpha}{N_1 k}. \tag{4}$$

The equivalent expressions derived for the 2D case, obtained by replacing the area 2rw in (2) by πr^2, are

$$G = G_0 \exp - (T_0/T)^{1/3} \quad \text{and} \tag{5}$$

$$T_0 = \frac{27\alpha^2}{\pi N_2 k}. \tag{6}$$

The interpretation of the data in Figs. 5 and 6 is then clear in terms of these considerations. As the gate voltage is increased the width of the channel increases and a transition occurs between 1D variable range hopping and 2D variable range hopping when the width of the channel is comparable to the average hopping distance. The theory allows us to estimate the effective width of the channel at the transition as w = 30 nm [2]. The main question becomes why is the transition as sharp as observed. Perhaps this is an indication of the rate at which the channel width changes as a function of gate voltage.

One can take the analysis of the variable range hopping data one step further. The values of the parameter T_0 as a function of gate voltage can be obtained from the data of Fig. 5. To reduce experimental fluctuations, we assumed values for the exponents of either 1/2 or 1/3 according to the gate voltage range, and then used a least squared fitting procedure to obtain T_0. Figure 7 shows averaged [10] values of T_0 plotted as a function of gate voltage for both the 1D and 2D regimes.

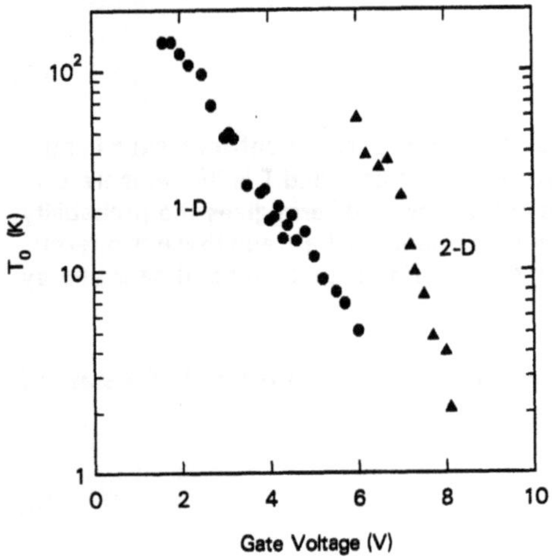

Fig. 7. The variable range hopping parameter, T_0, is plotted as a function of gate voltage for both the 1D and 2D regimes. It is surprising that T_0 varies exponentially with gate voltage, and that the slopes of this dependence vary by a factor of 2 between the 1D and 2D regimes

These results are surprising. Equations (4) and (6) give the expected behavior of T_0 in the two regimes. It was believed that all of the parameters in these expressions were slowly varying functions of the gate voltage. Figure 7 clearly indicates that T_0 is nearly an exponential function of gate voltage. If we assume that N_2 is constant, which has been demonstrated in wide 2D samples at the same carrier densities, we are forced to the conclusion that α must vary exponentially with gate voltage.

We can carry this analysis one step further by determining α from (6) and the data in the 2D regime. If we then apply the criterion that a transition from strong localization to weak localization will occur when the localization length, $1/\alpha$, is equal to the effective sample length, the transition is predicted to occur at a gate voltage of 10 V. We have assumed that the effective sample length is the inelastic diffusion length, 0.3 μm, measured in these samples [12]. This is remarkably close to the observed transition as obtained from the temperature dependent data. Extrapolated data from the strong localization regime correctly predict the onset of weak localization.

5. Origin of the Peaked Structure

With this information we are in a position to gain some insight into just how small these samples are. The width of the channel was estimated to be 30 nm, and the length is about 10 μm. This gives an area of 0.3 μm^2, and assuming a 2D density of states, it gives an estimate of 500 electrons in the channel. If we consider that only those electron states within a few k_BT of the Fermi energy contribute to the conductance, the number drops to less than 50. This led us to suggest [2] that the explanation for these peaks could be found in the statistical fluctuations in the density states due to the extremely small sample size. LEE [13] has expanded on this suggestion and considered the role of statistical fluctuations in small hopping systems in which only one or two hops dominate the conductance because of percolation effects. AZBEL and co-workers [14-16] have considered the possibility that the structure arises from resonant tunneling.

The typical temperature dependence of the conductance for a peak is shown in Fig. 8. It is clear that the conductance shows two separate regions of temperature dependence. The curves can be characterized by three parameters: T_0, the slope for the upper temperature region; T^*, the slope for the lower temperature region, and a temperature, T_t, where the transition occurs. The notation T_0 has been used for the upper temperature region because in this region the peaks have disappeared and the slopes for peaks and valleys are essentially the same. This further suggests that the upper temperature region corresponds to variable range hopping and it is the deviations at lower temperature which need to be explained.

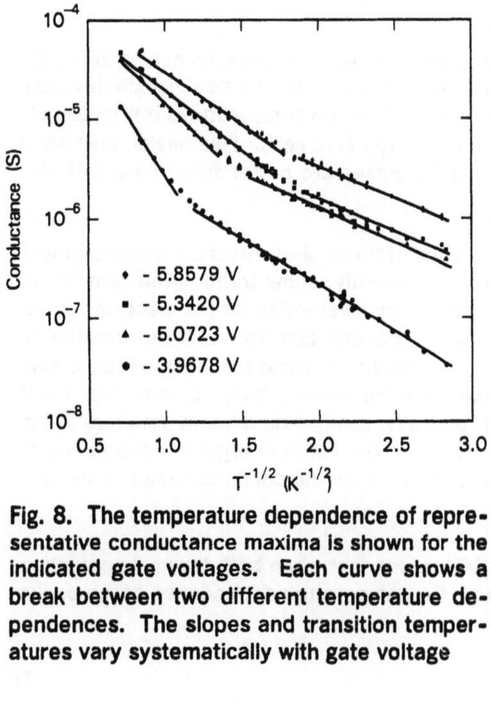

Fig. 8. The temperature dependence of representative conductance maxima is shown for the indicated gate voltages. Each curve shows a break between two different temperature dependences. The slopes and transition temperatures vary systematically with gate voltage

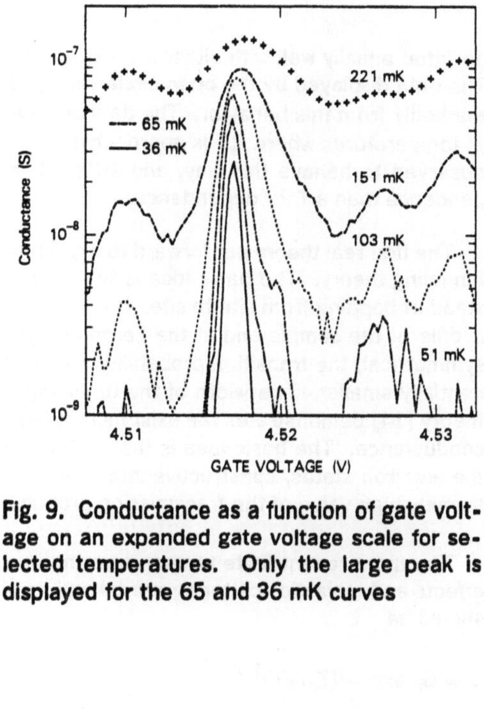

Fig. 9. Conductance as a function of gate voltage on an expanded gate voltage scale for selected temperatures. Only the large peak is displayed for the 65 and 36 mK curves

In order to best examine the nature of the peaks we have looked extensively at a few peaks which show only minimal overlap with other peaks. One such peak is shown in Fig. 9. The largest of the peaks shown in this narrow region of gate voltage is quite well isolated over a reasonable temperature range. Significant overlap sets in above 120 mK. The sides of the peaks are reasonably linear on this semilog plot; and when the gate voltage scale is converted to an energy scale, the magnitudes of the slopes are approximately $1/k_BT$. Note also that the peak positions show a small but systematic shift with temperature.

Figure 10 displays the temperature dependence of the four largest peaks in Fig. 9. Above 200 mK, where the overlap is large, the data are fitted well to the variable range hopping form, (3). For most of the 120 peaks which have been studied, the temperature region below 200 mK can

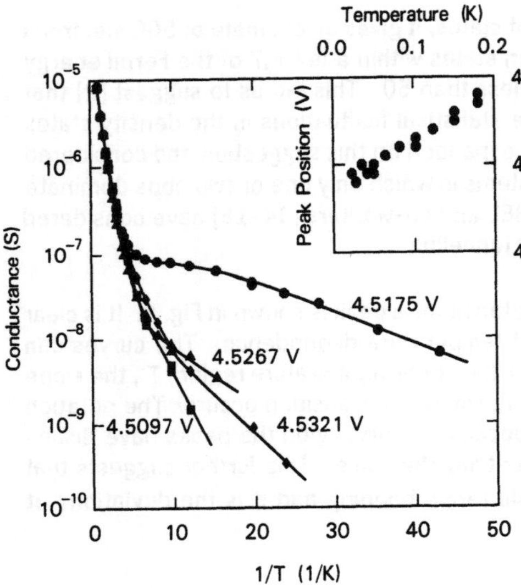

Fig. 10. Temperature dependence of four conductance peaks shown in Fig. 9 at V_G = 4.5097, 4.5175, 4.5267, and 4.5321 V with activation energies 36, 7.2, 13 and 18 μeV, respectively. Solid lines are smooth curves through the data. The inset shows the temperature dependence of the peak position for the peak at V_G = 4.5175 V in the temperature regime where the peak is isolated from its neighboring peaks

be fitted equally well with either a T^{-1} or a $T^{-1/2}$ exponential temperature dependence as in Fig. 8. The data displayed by the open circles in Fig. 10 illustrate one example of a peak which deviates markedly from this behavior. The data are fitted by $\ln G \propto 1/T$ at low temperatures but flatten off at temperatures where k_BT is approaching the activation energy (7.2 μeV). Five peaks have been observed to behave this way, and 10% of the remaining peaks are better fitted by a 1/T dependence than a $T^{-1/2}$ dependence.

The first real theory put forward to explain the existence of the peaked structure is the resonant tunneling theory. The basic idea is that electrons may resonantly tunnel through the sample instead of hopping from site to site. It is known that if an energy level exists for the electrons in the middle of the sample and at the Fermi energy, a resonant transfer can occur. If the situation is symmetrical, the transition probability can be 1, whereas the background tunneling will be exponentially smaller. The width of the tunneling resonance will be exponentially narrow. The T = 0 theory [14] demonstrates the existence of large, reproducible, sample-dependent structure in the conductance. The basic idea is that at the special values of the Fermi energy, corresponding to the electron states, constructive interference of the electron wave functions can lead to the extremely high value of the transmission probability.

The more recent finite temperature theory [15, 16] argues that when both thermal population effects and inelastic scattering are taken into account, the temperature dependence of the peaks should be

$$G = G_0 \exp - (T_0/\eta T)^{1/2}, \tag{7}$$

where $1<\eta<4$ and η is a constant for any one peak, with the 4 corresponding to unity transmission probability. At higher temperatures the peaks merge and the average behavior is governed by the variable range hopping formalism. The theory further predicts that the peak positions should not shift with temperature. Another prediction is that at the lowest temperatures the conductance of a peak should increase with decreasing temperature. One of the basic assumptions inherent in the intermediate temperature portion of this theory has been questioned [17].

The second substantive theory which has been proposed involves a hopping model of the conductance in sparse systems where statistical fluctuations are important. LEE [13] has shown that in a 1D hopping system with parameters comparable to those found in our samples, fluctuations not only in the energy of each state but also in the spatial location of the states can indeed produce structure as large as observed experimentally. The percolation problem reduces to only 1 or 2 critical hops.

When a single critical hop dominates the current, the conductance in the linearized approximation with Boltzmann statistics is

$$G = G_{ij} \exp[-2\alpha_{ij} r_{ij} - (1/2k_B T)(|E_i - \mu| + |E_j - \mu| + |E_i - E_j|)], \tag{8}$$

where μ is the chemical potential. This shows that a peak at low temperature will be simply activated, and that off of the peak the slope of the curves should correspond to $1/k_B T$. If a single peak persists to high enough temperature without overlap from neighboring peaks, the Boltzmann approximation is no longer valid. The simple exponentials in (8) are replaced by Fermi and Bose functions. This can lead to the kind of deviations from a simple activated temperature dependence observed in Fig. 10.

When two hops are taken in series, the situation qualitatively changes. If each hop is described by (8), one can form the series combination and solve for the peak position. One obtains

$$E_p = (E_i + E_l)/2 + k_B T(\alpha_{kl} r_{kl} - \alpha_{ij} r_{ij}) + (k_B T/2) \ln(G_{ij}/G_{kl}), \tag{9}$$

where i, j, k and l describe the four sites involved in the hopping process. This type of peak shifts linearly in temperature with a magnitude and sign given by the asymmetry of the hopping sites involved.

The experimental data as represented by Figs. 8 - 10 are in better agreement with this hopping model than with the resonant tunneling model [11]. The temperature dependence of the main peak shown in Fig. 10 points to this model. The inset to Fig. 10 indicates the temperature dependence of the peak position, which is only consistent with the hopping model. Even though most of the data are consistent with either model, only the hopping model has selected data which point to it alone. In passing we should note that it is possible to have both types of conductance mechanisms acting in parallel so that the experiments do not disprove the resonant tunneling model, they only show that the mechanism has not been directly observed in this system.

6. Universal Conductance Fluctuations

The last example of the physics which is accessible with these types of samples, which I would like to discuss, is the observation of universal conductance fluctuations in the weak localization regime. The initial indication of this phenomenon was the observation that in these narrow 1D samples the reproducible structure continues up to the highest gate voltages, even though a strong localization hopping picture is no longer valid. Reproducible aperiodic structure in the magnetoconductance of small weakly localized metallic wires [18] was reported, and the authors speculated that this structure might be related to the reproducible gate voltage structure observed in narrow MOSFET's. The suggestion was made [19, 20] that the structure might be due to interference between the electron wave functions describing the different conduction paths in the small samples. The sample must be small enough that there cannot be so many paths that the interference effects are washed out by averaging. Different interference paths enclose different amounts of magnetic flux so that as the magnetic field is changed the interference conditions change causing small magnetoconductance structure. In a similar way when the gate voltage is changed, the Fermi energy is changed, changing the interference conditions and giving rise to conductance fluctuations.

The validity of these ideas for the magnetoconductance fluctuations can be tested in a very nice way in the narrow MOSFET samples. All of the different conductance paths which interfere lie in the plane of the surface. When the magnetic field is perpendicular to the surface, the flux penetrates the area surrounded by the paths and the effect should be observable. When the magnetic field is parallel to the surface, no flux penetrates the paths and no magnetoconductance fluctuations should be seen. In order to test these concepts, we measured the angular dependence of the magnetoconductance fluctuations in these samples [12].

Figure 11 shows the magnetoconductance fluctuations as a function of both magnetic field and the perpendicular component of the field. Magnetoconductance fluctuations are clearly evident which depend on the perpendicular component of the applied field. One can track the individual maxima and minima in the structure. Shubnikov-de Haas oscillations are also evident at larger fields and can in fact be used to check the angular calibration of the experiment. These data were obtained in the 1D weak localization regime and at 0.47 K. Figure 12 shows typical conductance fluctuations in the same range, demonstrating that the typical magnitude of the fluctuations (2×10^{-7} S) is the same for both effects.

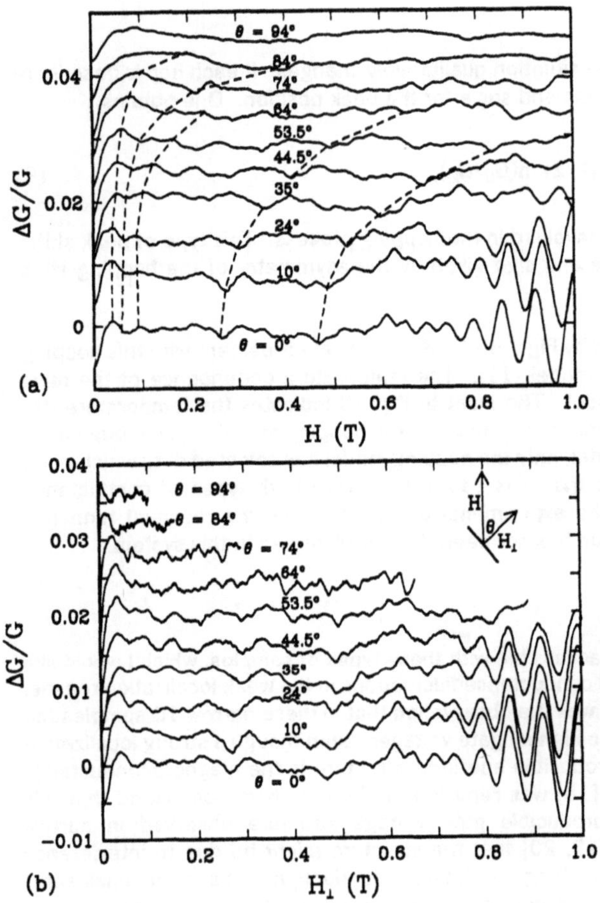

Fig. 11 (a) Fractional change in the magnetoconductance of pinched Si accumulation layer for various values of the angle θ between the magnetic field axis and the normal vector to the sample surface. $V_G = 11$ V, G = 2.3×10^{-4} S and T = 0.47 K $\pm$0.02 K. The curves have been offset for clarity. The dotted lines are used to track several structures from curve to curve as θ is varied. (b) The same data plotted vs the perpendicular component of the magnetic field

AL'TSHULER [19] and LEE and STONE [21] have predicted that at T = 0 and in the limit that the inelastic diffusion length is much larger than the sample length, the magnitude of these universal conductance fluctuations should be of order $e^2/h \sim 3.9 \times 10^{-5}$ S. When the inelastic diffusion length, L_{in}, becomes smaller than the sample length, one can treat the sample as an incoherent average of L/L_{in} samples of effective length L_{in}. In this limit the magnitude of the fluctuations becomes

$$\delta G_{rms} \simeq (L_{in}/L)^{3/2} e^2/h. \qquad (10)$$

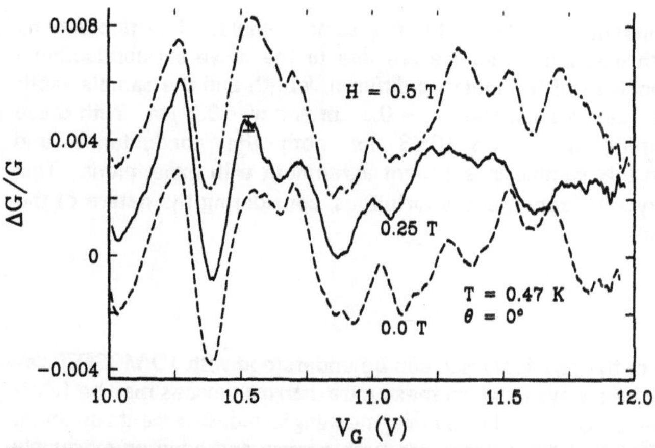

Fig. 12. Fractional change in the conductance of pinched Si accumulation layer as a function of the gate voltage V_G for different values of the magnetic field H. The magnitude of the conductance fluctuations near $V_G = 11$ V is approximately equal to those seen in the magnetoconductance data shown in Fig. 11

The factor of L_{in}/L comes from the conductance change, and the factor of $(L_{in}/L)^{1/2}$ comes from the incoherent averaging of independent segments. It is also possible to have a number of uncorrelated sample segments in parallel which would also reduce the fluctuation amplitudes. For our samples only one parallel segment is present.

In order to quantitatively test this theory we need to know the inelastic diffusion length. This can be obtained from the average low field magnetoconductance [3]. This negative magnetoresistance is due to weak localization effects. At zero field interference between different conductance paths which come back to the origin gives rise to a weak localization effect. The application of a magnetic field destroys the interference effect leading to larger conductance. This effect is similar to the effect which gives rise to the universal conductance fluctuations but involves different interfering paths. The relevant expression for this negative magnetoresistance is

$$\left(\frac{2G_0}{\Delta G}\right)^2 = L^2\{L_{in}^{-2} + \left(\frac{e}{\hbar}\right)^2\frac{w^2H^2}{3}\}. \tag{11}$$

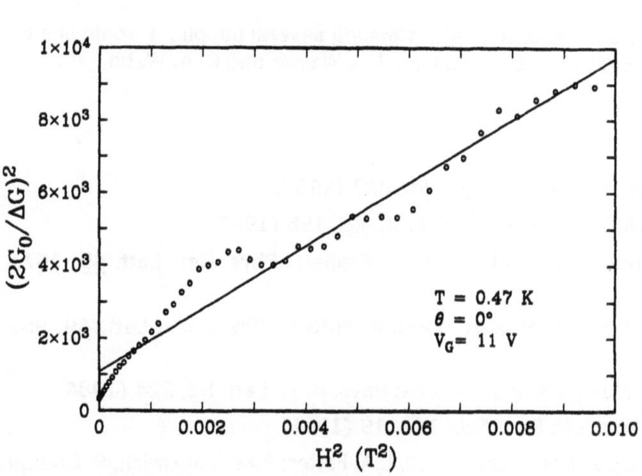

Fig. 13. Inverse square of ΔG, the deviation of the magnetoconductance from it saturation value (at H~0.3 T), plotted against the square of the applied magnetic field. the straight line is the best fit to the data

Figure 13 shows this negative magnetoresistance for the same sample. The fluctuations superimposed on top of the smooth magnetoresistance are due to the universal conductance fluctuations. They complicate the analysis of the inelastic diffusion length and the sample width from this measurement. Using these data we estimate $L_{in} \sim 0.3\ \mu m$ and $w \sim 0.1\ \mu m$. With these numbers (10) predicts a magnitude of $\sim 2 \times 10^{-7}S$ for both the conductance and magnetoconductance fluctuations in this sample in excellent agreement with experiment. This degree of agreement between theory and experiment is fortuitous, considering the nature of the approximations which have been made.

7. Summary

I have tried to give you some flavor of the physics which can be understood with 1D MOSFET devices. We have talked about some of the ways in which these ultra-narrow devices may be fabricated and some of the precautions necessary in order to make meaningful measurements on them. The actual physics centers on the fact that the devices are both narrow and have an extremely small device area.

The fact that they are narrow leads to the 1D effects. In the strong localization hopping regime the device displays 1D behavior as long as the width is smaller than the average hopping distance for the electrons. In the weak localization regime the device behaves as a 1D device as long as the width is smaller than the inelastic diffusion length, an easier criterion to achieve.

The fact that the devices have small area leads to effects due to the failure of ensemble averaging. In the strong localization hopping regime we have seen that the percolation problem leads one to consider only 1 or 2 critical hops in the entire device area in describing the conductance. This leads to large fluctuations in the conductance as a function of gate voltage. In the weak localization regime interference effects that are not completely ensemble averaged lead to universal conductance fluctuations.

This has not been an exhaustive review of the physics which has been explored with 1D samples of the field confinement type. I have only attempted to give some important examples of the richness of the phenomena which have been considered. The field of 2D physics which has been explored using MOSFET's is about 20 years old and still going strong. We have every reason to believe that, similarly, the 1D physics which has been looked at is only the beginning.

8. Acknowledgments

The work which I have discussed was a collaborative effort among several people. I would like to acknowledge the contributions of A. B. Fowler, S. B. Kaplan, J. J. Wainer and R. A. Webb.

9. References

1. T. Ando, A.B. Fowler and F. Stern: Rev. Mod. Phys. 54, 437 (1982)
2. A.B. Fowler, A. Hartstein and R.A. Webb: Phys. Rev. Lett. 48, 196 (1982)
3. R.G. Wheeler, K. K. Choi, A. Goel, R. Wisnieff and D. E. Prober: Phys. Rev. Lett. 49, 1674 (1982)
4. W.J. Skocpol, L.D. Jackel, E.L. Hu, R.E. Howard and L.A. Fetter: Phys. Rev. Lett. 49, 951 (1982)
5. R.R. Kwasnick, M.A. Kastner, J. Melngailis and P.A. Lee: Phys. Rev. Lett. 52, 224 (1984)
6. R.G. Wheeler, K.K. Choi and R. Wisnieff: Surf. Sci. 142, 19 (1984)
7. W.J. Skocpol, L.D. Jackel, R.E. Howard, H.G. Craighead, L.A. Fetter, P.M. Mankiwich, P. Grabbe and D.M. Tennant: Surf. Sci. 142, 14 (1984)
8. C.C. Dean and M. Pepper: J. Phys. C 17, L1287 (1982)

9. A. Ya. Shik: Sov. Phys. Semicond. 19, 915 (1985)
10. A. Hartstein, R. A. Webb, A. B. Fowler and J. J. Wainer: Surf. Sci. 142, 1 (1984)
11. R.A. Webb, A. Hartstein, J.J. Wainer and A.B. Fowler: Phys. Rev. Lett. 54, 1577 (1985)
12. S.B. Kaplan and A. Hartstein: to be published
13. P.A.Lee: Phys. Rev. Lett. 53, 2042 (1984)
14. M. Ya. Azbel and P. Soven: Phys. Rev. B 27, 831 (1983)
15. M. Ya Azbel, A. Hartstein and D.P. DiVicenzo: Phys. Rev. Lett. 52, 1641 (1984)
16. M. Ya Azbel and D.P. Divicenzo: Phys. Rev. B 30, 6877 (1984)
17. A.D. Stone and P.A. Lee: Phys. Rev. Lett. 54, 1196 (1985)
18. C.P. Umbach, S. Washburn, R. B. Laibowitz and R. A. Webb: Phys. Rev. B 30, 4048 (1984)
19. L. Al'tshuler: Pis'ma Zh. Eksp. Teor Fiz. 41, 530 (1985) [JETP Lett. 41, 648 (9185)]
20. A. D. Stone: Phys. Rev. Lett. 54, 2692 (1985)
21. P. A. Lee and A. D. Stone: Phys. Rev. Lett. 55, 1622 (1985)

Excitons in GaAs Quantum Wells: Interface Disorder and Mobility

J. Hegarty

Physics Department, Trinity College, Dublin 2, Ireland

1. Introduction

The interface between GaAs and GaAlAs in two-dimensional quantum wells is characterized by atomic scale roughness which inhomogeneously broadens the lowest energy heavy hole exciton line at low temperatures. Each exciton sees a random potential along the layer which strongly affects its motion. Motion of particles in disordered systems has been a tough problem to solve and has engaged theorists and experimentalists for many years. Motion in two dimensions has been the most uncertain one. In these lectures we will attempt to examine the influence of disorder on two-dimensional excitons in light of classical and quantum approaches to transport and show that there are both localized and delocalized excitons at low temperatures within the timescale of the exciton lifetime. The type of motion will reflect the nature of the interface roughness and in principle be a probe for it. The organization is as follows. The relevant features of Wannier excitons in two-dimensional structures will be discussed to show the origin of the disorder. Next, the essential elements of percolation and localization theories will be reviewed and, finally, results of a variety of laser techniques will be presented from which a general picture of the exciton dynamics will be drawn.

2. Wannier Exciton in Two Dimensions

Wannier excitons in bulk GaAs are observed in optical absorption at low temperatures as sharp lines just below the bandedge. Excitons are bound electron-hole pairs and are hydrogenic in character with a binding energy of $\simeq 4$ meV and Bohr radius of 150 Å. Even though the exciton forms a band, only $k \simeq 0$ excitons can be excited by photons since electrons and holes only at the band edges move at the same velocity and hence can form excitons. The exciton is doubly degenerate due to the degeneracy of the heavy and light hole bands at $k = 0$. In thin layers of GaAs bounded by the higher bandgap material $Al_xGa_{1-x}As$ the excitons are confined to the GaAs layers. When the layer thickness L_z is less than the exciton diameter, the energies of the excitons are quantized and increase as the layer thickness and the mass of the exciton decrease [1]. Instead of a single exciton line, the absorption spectrum is dominated by several sharp lines corresponding to excitation of the quantized excitons. A typical excitation spectrum [3] for a sample grown by Molecular Beam Epitaxy (MBE) is shown in Fig.1 where two closely spaced series of lines are observed corresponding to the heavy and light hole excitons, respectively, whose degeneracy is now lifted by the mass dependence in two dimensions.

The lowest energy heavy hole exciton is well isolated from the rest of the spectrum and is metastable with a lifetime of $\simeq 1$ nsec. All other excitons rapidly relax to this metastable state. Its strong nonlinear optical properties have potential for optical switching applications [2] and the increased binding energy (8 - 10 meV) over three dimensions makes room temperature operation a possibility. Atomic scale fluctuations in the GaAs layer thickness result in the exciton having a slightly different energy from point to point in the layer [4]. Consequently, the profile seen in absorption is an envelope of homogeneously broadened packets of excitons as shown schematically in Fig.2. The homogeneous width Γ_h is determined by the phase memory

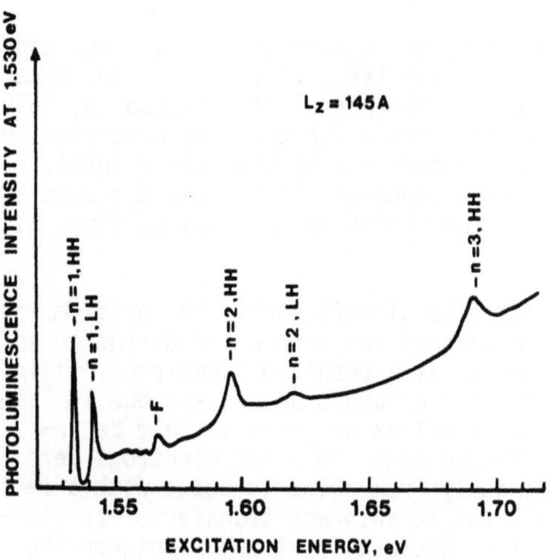

Fig.1. Excitation spectrum of MBE sample with L_z = 145 Å at low T. The excitons are labelled by the quantum number of the participating holes and electrons [3]

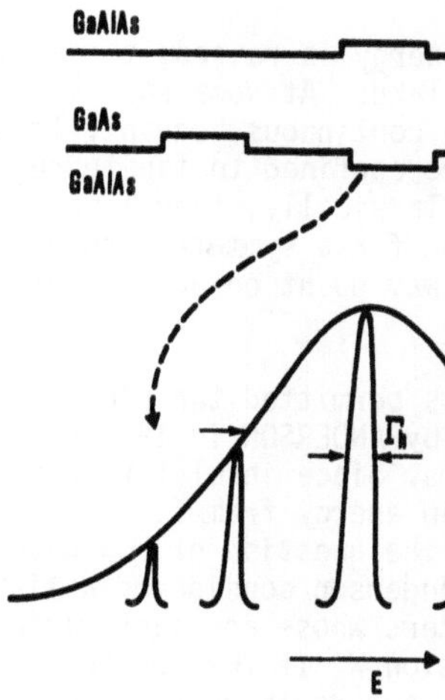

Fig.8. Schematic of the experimental set-up to measure dephasing times

(scattering) time of the exciton. The shape of the envelope will depend critically on the nature of the interface disorder and on the dynamics of the excitons as we will see. We define Γ_i as the full width of the envelope. Γ_i varies from less than 1 meV for L_z = 200 Å to ≈5 meV for L_z = 50 Å.

Interface roughness has two main consequences. Firstly, since each exciton in a layer sees a random potential such as that outlined schematically in Fig.3, the question arises whether the strength of the potential fluctuations is enough to localize the excitons at low temperatures. The localization properties may affect other properties such as nonlinearities and effectiveness of trapping centers. Since it is possible in principle to determine the localization properties from the exciton dynamics it is useful both from a fundamental point of view and from a device point of view to understand the dynamics. Secondly, the shape of the exciton envelope is determined by the surface roughness and any tails in the envelope will affect strongly the usefulness of excitons in switching applications. It is necessary to understand in detail the connection between surfaces and exciton properties in order to control or modify either in a useful way.

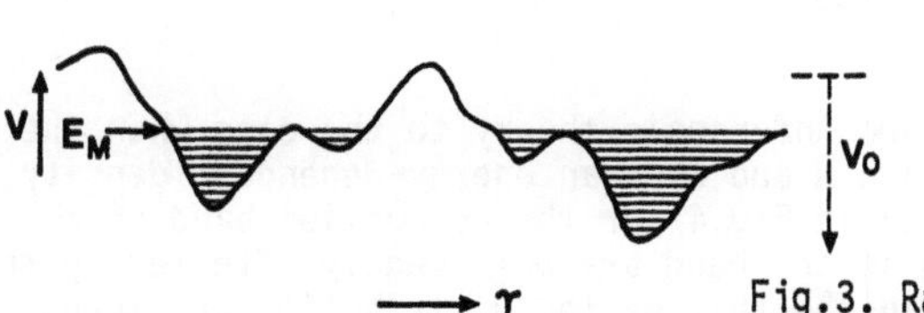

Fig.3. Random potentials of the exciton where $\vec{r}$ is the position in the layer. The horizontal line indicates the classical percolation threshold

3. Percolation and Localization

It is instructive to approach the problem of localization from a percolation point of view. Classical percolation [5] was developed to describe hopping on a discrete regular lattice in which either sites were randomly empty or bonds were randomly missing. Since excitons extend over many lattice sites, however, we adopt a continuum approach. The lowest energy excitons from the distribution of Fig.2 are located in the deepest valley of the two-dimensional topology in Fig.3 and form isolated "lakes". As the energy is raised, the excitons can sample more of the topology and will form larger lakes. At some energy or "level", the lakes will rapidly coalesce and form a single continuous "ocean". This level is the percolation threshold, below which excitons are confined to finite regions and above which they can move over the whole topology. Classically, tunneling through the barriers of the random topology is not allowed and for a symmetric two-dimensional topology the threshold occurs at exactly the half way point on equivalently near the center of the exciton line, as shown in Fig.3.

If tunneling is permitted the picture is modified significantly. The problem was first considered by ANDERSON in 1958 [6] to understand the absence of diffusion in disordered systems. Since insulators were regarded as a result of completely filled bands separated in energy from filled bands and since bandstructure was due to lattice periodicity the question of how disordered lattices or glasses could be insulators arose. Anderson considered a tight-binding model in which electrons were localized at centers whose energies varied randomly from point to point with a rectangular distribution W. If the centers are allowed to interact transfer of an electron between two centers is possible only if the interaction strength between the two centers is greater than the energy mismatch. The greater W, the less likely that an electron on one center can find a nearby state satisfying this condition. Above some value of W transfer between centers is not possible and all states are localized. The dimensionality is crucial to the localization. In one dimension any disorder localizes all states. In three dimensions, two conditions must be satisfied. If the interaction falls off with distance faster than r^{-3} and if W is roughly comparable to the nearest neighbor interaction strength M then all states are localized. The situation in two dimensions was more difficult to assess in this model.

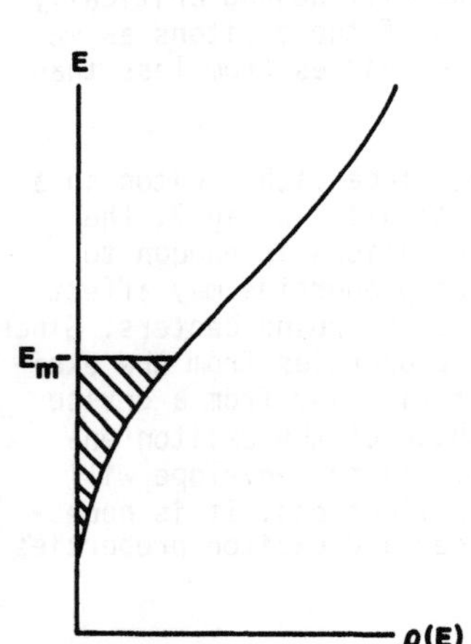

Fig.4. Density of states for the conduction band of a disordered semiconductor showing the position of the mobility edge

MOTT [7] and MOTT and DAVIS [8] extended Anderson's theory to the case when the disorder is much smaller than the bandwidth M and when an energy-dependent density of states is considered such as that shown in Fig.4 for the conduction band of a disordered semiconductor. States high up in the band are only weakly affected by the disorder and may be delocalized. States in the tail of the distribution are more susceptible to the disorder. This can be seen by applying the percolation argument as before. The lower the energy in the tail the fewer and hence further apart are the resonant states. Below a certain energy the states may be localized. MOTT [7] argued that localized and delocalized states could not coexist at the same energy since a localized state would eventually leak into a delocalized state and become delocalized. Consequently, a mobility edge must separate localized from delocalized states. In the case of excitons the picture of Fig.4 can be applied except that the electron and hole must be considered together. If a mobility edge exists in the distribution of Fig.4 and exciton above the edge is more bandlike and acquires more definite values of k. The higher the energy the less the $k=0$ component. Since photons couple only to the $k\simeq 0$ component, the absorption profile below the mobility edge will reflect the density of states but above the edge it will represent the $k \simeq 0$ weighting at each energy.

Mott argued on the basis of the Ioffe-Regel criterion [9] that the mobility changes discontinuously at the edge and a minimum value of the diffusion constant D should exist. The argument is as follows. Since

$$D = \frac{\ell^2}{\tau_s} = \frac{\hbar k \ell}{m} \tag{1}$$

where ℓ is the mean free path and τ_s is the scattering time, a spread $\Delta k \simeq 1/\ell$ in k occurs because of the mean free path. For a delocalized exciton, by the Ioffe-Regel criterion, $k\ell_{min} \simeq 1$ so that $D_{min} \simeq \hbar/m \simeq 1\ cm^2\ s^{-1}$ for $m = m_e$, the free electron mass. In two dimensions, (1) implies a minimum metallic conductivity of $\sim e^2/h$.

Mott's argument of a minimum metallic conductivity has been disputed by ABRAHAMS et al. [10] who claimed that the localized-delocalized transition is a critical point to which scaling theory could be applied. They argue that the conductivity at the mobility edge in three dimensions drops continuously to zero whereas in two dimensions all states should be localized, though only weakly, at 0 K. This latter is in sharp contrast to classical percolation theory which gives a "mobility" edge in two dimensions. Conductivity measurements in the three-dimensional system $S_i : P$ [11] extrapolated to $T = 0$ and in the two-dimensional S_i : MOSFET system [12] qualitatively, at least, support the claims of the scaling theory.

Even though a minimum value for D is not predicted by scaling theory, nevertheless a large value is an indication of delocalization. If the mean free path is roughly the exciton diameter then an elastic scattering time of 10^{-12} s is expected for delocalized excitons with $D \simeq 1\ cm^2\ s^{-1}$. The mean free path will of course depend on the correlation length of the interface disorder. For localized excitons $D = 0$, the scattering is determined by inelastic processes such as by phonons or by recombination so that at low temperatures τ_s should be relatively long. We argue that from the point of view of photons it is the shorter of the elastic and inelastic times that determines the homogeneous broadening Γ_h. Intuitively, a delocalized exciton created through its $k \simeq 0$ component quickly scatters elastically off the random potential into higher k states so that it becomes inaccessible to the photon. Consequently, $\Gamma_h = 1/\tau_s \simeq 1$ meV for delocalized excitons and much less than this for localized excitons at 0 K.

It is clear that the quantities D, Γ_h and τ_s depend strongly on the localization properties of the excitons. Techniques to measure these quantities in multiquantum well layers grown by MBE will now be discussed.

4. Experiments

MBE samples with GaAS layers ranging from 50 - 100 Å in thickness alternating with AlGaAs barrier layers 200 Å thick were studied. The Al concentration was $\simeq 30\%$ and the number of layers was 80 - 100.

4.1 Diffusion

The most direct way of measuring diffusion is by a transient grating [13,14] in which a stationary modulated density or grating of excitons is created by the interference fringe pattern of two overlapping beams obtained from the laser source and tuned to the exciton resonance. The nonlinear properties of the excitons result in a spatially modulated optical response of the material at high exciton densities. A third probe beam can diffract off the grating with an efficiency determined by the grating contrast. If excitons diffuse from the high- to the low-density regions the grating contrast decreases. If the three beams are in the form of picosecond pulses from a modelocked dye laser, the grating decay can be measured by delaying the probe in time after the two write pulses. For a sinusoidal modulation the grating decay time τ_g is given by [15]

$$\frac{1}{\tau_g} = \frac{2}{\tau_0} + \frac{8\pi^2 D}{d^2} \qquad (2)$$

where τ_0 is the exciton lifetime and d is the grating spacing. The dependence of $1/\tau_g$ on d gives D directly and in Fig.5 D is plotted against photon energy across the heavy hole exciton line for $L_z = 205$ Å and for T = 5 K. Even though this is a high-density experiment ($\simeq 10^{10}$ excitons/cm^2) a general trend is obvious [14,16]. D increases from a negligibly low value on the low-energy side of the line to a high value on the high-energy side. The high values of D indicate that the higher energy excitons are delocalized, by the Ioffe-Regel criterion discussed in Sect.3. The high exciton densities, however, preclude any more detailed interpretation other than the direct evidence for delocalization.

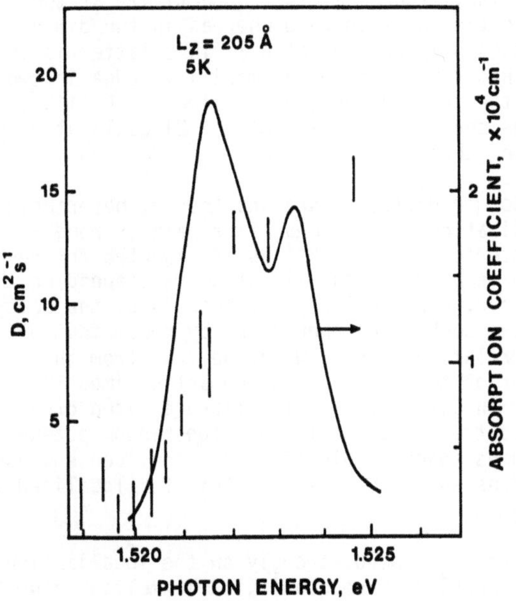

Fig.5. Diffusion D vs photon energy for $L_z = 205$ Å

4.2 Homogeneous Linewidths

The most satisfactory method for measuring Γ_h is Rayleigh scattering [17,18] as it can be performed at extremely low light levels. The scattering arises from the statistical distribution in energy of the exciton resonances from point to point in the quantum well as was shown in Fig.2. A light beam of frequency ω propagating through the layers will encounter strong and sharp resonances at frequency $\Omega(\vec{r})$ where $\Omega(\vec{r})$ is a function of position in the layer. The response $\varepsilon(\omega,\vec{r})$ due to one particular resonance at point $\vec{r}$ is given by

$$\varepsilon(\omega,\vec{r}) = \varepsilon_\infty + \frac{\varepsilon_\infty \omega_{LT}}{\Omega(\vec{r}) - \omega - i\Gamma_h(\Omega)/2} \qquad (3)$$

where ω_{LT} is the longitudinal-transverse splitting and ε_∞ is the background dielectric constant. The fluctuation in $\varepsilon(\omega,\vec{r})$ with $\vec{r}$ gives rise to elastic Rayleigh scattering.

The scattering intensity $s(\omega)$ is determined by

$$s(\omega) \propto \langle |\varepsilon(\omega,\vec{r}) - \langle\varepsilon(\omega,\vec{r})\rangle|^2\rangle_{\vec{r}} \qquad (4)$$

where the < > signifies averaging over all space $\vec{r}$ traversed by the beam. The second term is close to ε_∞ so that

$$s(\omega) \propto \left\langle \frac{\varepsilon_\infty^2 \, \omega_{LT}}{(\Omega-\omega)^2 + \Gamma_h^2(\Omega)/4} \right\rangle_\Omega \tag{5}$$

where averaging over $\vec{r}$ translates to averaging over the inhomogeneous line Ω. Since the absorption coefficient $\alpha(\omega) \propto \mathrm{Im}(\varepsilon)$ then

$$\alpha(\omega) \propto \left\langle \frac{\varepsilon_\infty \, \omega_{LT} \quad \Gamma_h(\Omega)/2}{(\Omega-\omega)^2 + \Gamma_h^2(\Omega)/2} \right\rangle . \tag{6}$$

As a crude approximation, if $\Gamma_h \ll \Gamma_i$ and if $d\Gamma_h/d\Omega \ll 1$ then

$$s(\omega) \simeq K\alpha(\omega)/\Gamma_h(\omega) \tag{7}$$

where K is a constant. A more accurate value for $s(\omega)$ can be obtained [17] by taking account of the attenuation of the light as it penetrates into the layers and by a more accurate estimate of $\langle\varepsilon(\omega,\vec{r})\rangle$ in (4).

In Fig.6 are shown spectra of the light scattered from a sample with $L_z = 51$ Å at different incident laser photon energies. The sharp peak in each spectrum occurs at the laser position. As the laser is tuned to the heavy hole exciton resonance a large signal resonant with the laser grows whose time response is instantaneous. This is the Rayleigh signal. A broader background of luminescence can also be observed. The scattering is the same in all directions so that the correlation length of interface fluctuations $\ll \lambda$, where λ is the wavelength of the light. From the more accurate version of (7) we can calculate the relative magnitude of Γ_h as a function of photon energy and its dependence on energy is shown in Fig.7. At line center a dramatic increase in Γ_h occurs while below this Γ_h is small enough that the scattering is no longer de-

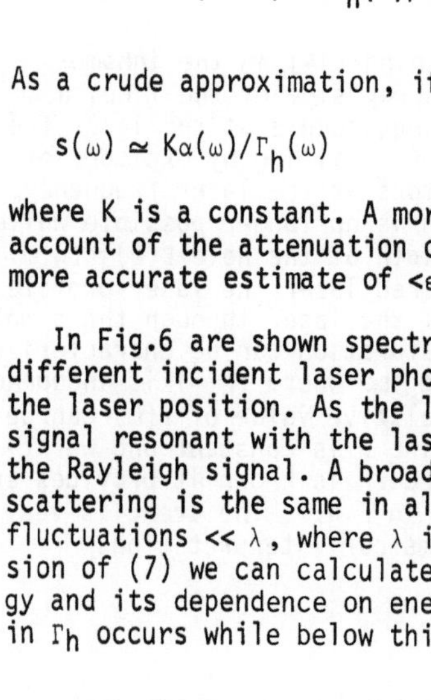

Fig.6. Spectra of scattered light from sample with $L_z = 51$ Å at different incident laser photon energies

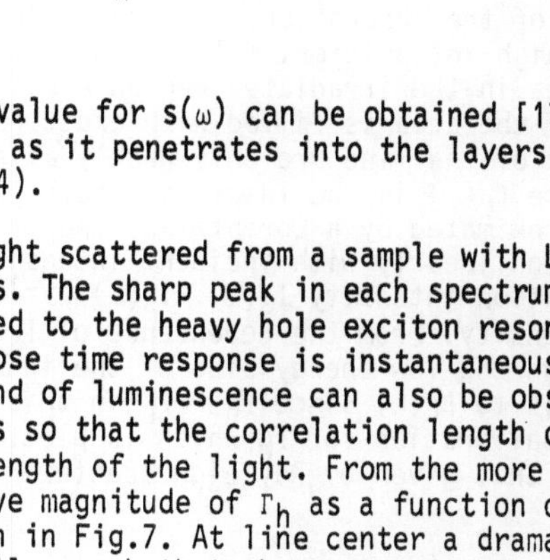

Fig.7. Plots of homogeneous width for $L_z = 51$ Å. The solid line is from Rayleigh scattering and the circles are from saturation. The squares are the activation energies from the temperature-dependent Rayleigh scattering

termined by Γ_h but rather by the shape and sizes of the interface fluctuations [17]. The vertical scale in Fig.7 can be calibrated by holeburning [17] and pulse propagation measurements [19]. Above linecenter Γ_h quickly approaches 1 meV, in agreement with high values of D in Fig.5 and consistent with delocalization of excitons on the high-energy side of the line. As the temperature is raised Γ_h increases strongly on the low-energy side but imperceptibly on the high side [18]. The dependence on temperature between 7 and 20 K follows an activation behavior and the activation energies ΔE are indicated in Fig.7. When the ΔE are added to the corresponding photon energies, they all point to a narrow ($\simeq$1 meV) region near linecenter indicating the existence of a quasi-mobility edge in this region. Below the edge localized excitons are activated to the edge with a consequent strong decrease in τ_s and increase in Γ_h. The fact that the edge is defined only to within 1 meV for $L_z = 51$ Å is an indication of possible macroscopic broadening in the line due to layer-to-layer variations. Similar behavior was observed for $L_z = 102$ Å.

An independent measure of Γ_h can be made from holeburning [14] in the inhomogeneous line. When a narrow laser is tuned to the low-energy side of the line, a saturation of the absorption, and consequently holeburning, occurs at the laser frequency at high intensities. Holeburning occurs because of a filling by excitons of all the area in the irradiated region available to excitons at the laser frequency. When all of the area is filled with excitons, absorption is no longer possible while other parts of the line are only weakly affected. The width of the hole $\Gamma(\omega) \simeq \Gamma_h(\omega) + \Gamma_L(\omega)$ where $\Gamma_L(\omega)$ is the laser linewidth. For a modelocked laser the laser profile can be approximated by a Lorentzian. The transmission of the laser through the sample increases nonlinearly with incident intensity and the saturation can be characterized by a saturation intensity $I_S(\omega)$. $I_S(\omega) \propto \Gamma^{-1}(\omega)$ in the régime where $\Gamma_h(\omega)$ is independent of intensity. From the dependence of $I_S(\omega)$ on ω a relative value of $\Gamma(\omega)$ can be measured. On the low-energy side of the inhomogeneous line Γ is constant and we attribute this to $\Gamma_L(\omega)$ since $\Gamma_h \ll \Gamma_L$ in this region. Since Γ_L is known it provides an automatic calibration for Γ_h which is plotted as points in Fig.7. The trend is very similar to that given by Rayleigh scattering and confirms our interpretation.

4.3 Measurements of Scattering Times

Measurements of Γ_h give accurate results only over a limited but important part of the inhomogeneous line. The approximations used in interpreting the Rayleigh scattering hold only for $\Gamma_h \ll \Gamma_i$ and so break down when Γ_h is more than 1 meV. On the low-energy side, the scattering is determined by factors other than Γ_h [17]. In this régime Γ_h is so small that the corresponding scattering time should be long enough to observe directly in the time domain. One technique capable of measuring dephasing times is transient four-wave mixing [20], a variation of the technique described in Sect.4.1 to measure diffusion. In this technique two beams only are used as shown schematically in Fig.8. The sample is excited by two picosecond pulses of the same frequency in the directions $\vec{k}_1$ and $\vec{k}_2$, respectively. When the pulses overlap in time a grating can be generated from which, in the thin sample limit, second-order diffraction of each beam in the directions $2\vec{k}_1 - \vec{k}_2$ and $2\vec{k}_2 - \vec{k}_1$ occurs. If now pulse 2 is delayed after pulse 1, the signal response in the $2\vec{k}_1 - \vec{k}_2$ directiion decreases as the overlap decreases and the response resembles the pulse autocorrelation profile. In the other direction, however, even when the overlap is zero the response of the medium to the second pulse is determined by the amount of coherence still remaining in the exciton population from pulse 1. The signal decay with increasing delay t_{12} gives a measure of the dephasing time. In these experiments the dephasing time is normally called the transverse relaxation time T_2 to distinguish it from the longitudinal or population relaxation time T_1. In our case we equate τ_s with T_2. In the limit where t_{12} and T_2 are much longer than the pulse length, the response takes the form of a coherent pulse of radiation at time t_{12} after the second pulse. This pulse has been called a photon echo [21] and can be described by an approximation of the general theory. For the pulse lengths used in the present experiment (5 - 10 psec) the dephasing times [22] are such that the photon echo approximation is not strictly applicable and the more general theory must be used. In Fig.9 the decay of the "photon echo" is shown as a function of t_{12} for three dif-

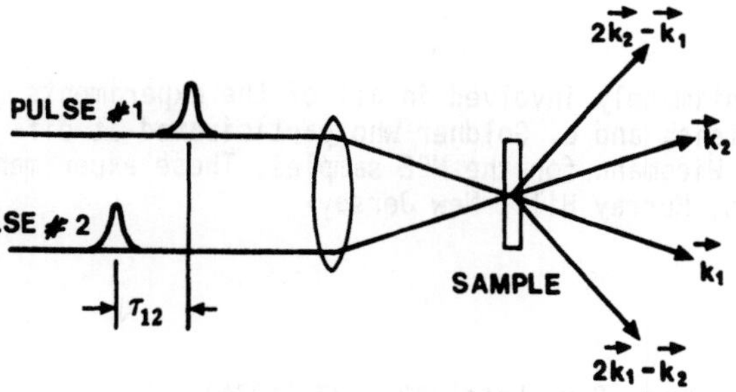

Fig.2. Schematic of the heavy hole absorption line as an envelope of homogeneously broadened exciton packets

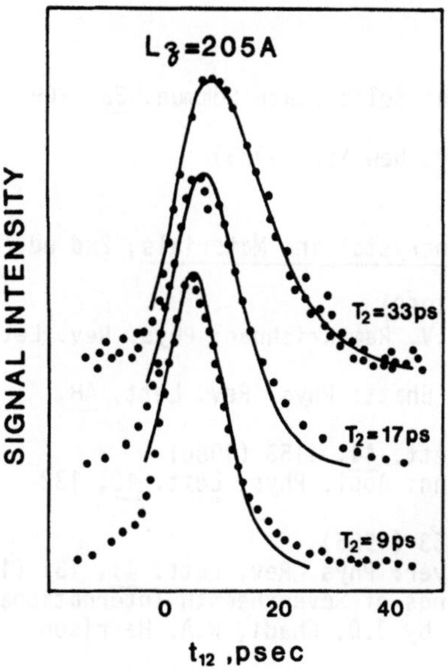

Fig.9. "Photon echo" in a sample with L_z = 205 Å. Response at three photon energies: top, below the peak of the line, middle, at the peak, bottom, above the peak. The solid curves are fits with measured T_2 indicated

ferent laser photon energies within the heavy hole exciton line of the sample with L_z = 205 Å. The curves have been fitted using the treatment of Yajima and Taira [23] where both T_2 and T_1 are adjustable parameters. As can be seen T_2 decreases from about 30 psec on the low-energy side to less than 10 psec on the high side of the line. The shorter values at higher energies are consistent with the large Γ_h discussed in Sect.4.2. The measured values of T_2 on the low-energy side are shorter than expected from the corresponding Γ_h values and have yet to be explained in detail.

5. Conclusions

Several different types of experiments have been performed on the GaAs quantum well system to obtain a handle on the effects of interface disorder on the mobility of excitons within the layer. Diffusion and homogeneous linewidth measurements are in agreement in pointing towards a sharp changeover at linecenter from localized excitons on the low-energy side to delocalized excitons on the high-energy side. The values for D and Γ_h are consistent with those expected for delocalized excitons. Measurements of the scattering times directly support this. The existence of an edge at linecenter is consistent with classical percolation theory. The answer as to whether excitons are delocalized in the classical or quantum sense lies in the nature of the transition region itself. In the present samples the transition region was too broad to clarify this and we must await more controlled samples and lower temperature for the answer. If the excitons are behaving according to classical percolation theory it may be an indication that the correlation length of the interface roughness is too large to allow tunneling. With more detailed results in the transition region a good handle on the nature of the interface should be possible as well as a clearer picture of the localization properties.

Acknowledgement

I wish to thank M.D. Sturge who was intimately involved in all of the experiments discussed here, C. Weisbuch, L. Schulteis, and L. Goldner who participated at different stages and A.C. Gossard and W. Wiegmann for the MBE samples. These experiments were performed while at AT&T Bell Labs, Murray Hill, New Jersey.

References

1 R. Dingle, W. Wiegmann, C.H. Henry: Phys. Rev. Lett. 33, 827 (1974); R. Dingle: Feskörperprobleme 15, 21 (1975)
2 D.S. Chemla: this volume
3 R.C. Miller: private communication
4 C. Weisbuch, R. Dingle, A.C. Gossard, W. Wiegmann: Solid State Commun. 38, 709 (1981)
5 R. Zallen: The Physics of Amorphous Solids (Wiley, New York 1983)
6 P.W. Anderson: Phys. Rev. 109, 1492 (1958)
7 N.F. Mott: Philos. Mag. 13, 989 (1966)
8 N.F. Mott, E.A. Davis: Electronic Processes in Noncrystalline Materials, 2nd ed. (Clarendon, Oxford 1979)
9 A.F. Ioffe, A.R. Regel: Prog. Semicond. 4, 237 (1960)
10 E. Abrahams, P.W. Anderson, D.C. Licciardello, T.V. Ramakrishnan: Phys. Rev. Lett. 42, 673 (1979)
11 M.A. Paalanen, T.F. Rosenbaum, G.A. Thomas, R.N. Bhatt: Phys. Rev. Lett. 48, 1284 (1982)
12 D.J. Bishop, D.C. Tsui, R.C. Dynes: Phys. Rev. Lett. 44, 1153 (1980)
13 J. Hegarty, M.D. Sturge, A.C. Gossard, W. Wiegmann: Appl. Phys. Lett. 40, 132 (1982)
14 J. Hegarty, M.D. Sturge: J. Opt. Soc. Am. B2, 1153 (1985)
15 J.R. Salcedo, A.E. Siegmann, D.D. Dlott, M.D. Fayer: Phys. Rev. Lett. 41, 131 (1978
16 M.D. Sturge, J. Hegarty, L. Goldner: In Proceedings of Seventheenth International Conference on the Physics of Semiconductors, ed. by J.D. Chadi, W.A. Harrison (Springer, Berlin, Heidelberg 1985)
17 J. Hegarty, M.D. Sturge, C. Weisbuch, A.C. Gossard, W. Wiegmann: Phys. Rev. Lett. 49, 930 (1982)
18 J. Hegary, L. Goldner, M.D. Sturge: Phys. Rev. B30, 7346 (1984)
19 J. Hegarty: Phys. Rev. B25, 4324 (1982)
20 P. Ye, Y.R. Shen: Phys. Rev. A25, 2183 (1982)
21 N.A. Kurnit, I.D. Abella, S.R. Hartmann: Phys. Rev. Lett. 13, 567 (1964)
22 L. Schulteiss, M.D. Sturge, J. Hegarty: Appl. Phys. Lett. 47, 995 (1985)
23 T. Yajima, Y. Taira: J. Phys. Soc. Jpn. 47, 1620 (1979)

Nonlinear Optics and Electro-Optics of Quantum Wells

D.S. Chemla

AT & T Bell Laboratories Holmdel, NJ 07733, USA

1. INTRODUCTION

Of particular interest for applications to optical signal processing are the mechanisms by which electrostatic or optical fields are used to change the refractive index or the absorption coefficient of a medium.

The resonances near band-to-band transitions [1,2] and near the narrow excitonic or excitonic-complexes transitions [3] seen in semiconductors have been used to demonstrate enhanced nonlinear optical processes. However the enhancements are clearly seen only at low temperature where the band edge is abrupt or the exciton resonances are still well resolved and the magnitude of the observed nonlinearities are still too small to be really applied to devices. The electric field dependence of the band-to-band absorption in semiconductors is known as the Franz-Keldysh effect [4,5], when excitonic resonances are present a "Stark effect" of the exciton ground state [6-8] is seen by which the exciton peaks are slightly shifted and are broadened owing to field ionization. Applications of these effects have also been limited in practice by the small magnitude of the absorption change observed for reasonable applied fields [9].

The development of materials able to meet the requirements for device applications is thus conditioned by the exploitation of novel physical processes that can enhance the optical nonlinearities well beyond those of "natural" compounds. A promising approach that has attracted much interest in recent years is to fabricate "artificial" materials whose electronic and optical properties can be tailored for specific applications [10].

Within the Effective Mass Approximation (EMA) the dynamics of carriers in a semiconductor mimics that of free particles providing the electon mass m_0 is replaced by the effective masses $m_{e,h}$ and the electrostatic interaction is screened by the dielectric constant of the compound. Consequently, characteristic lengths governing the quantum mechanics of carriers, such as the carrier De Broglie wavelengths and the Bohr radii, are increased by several orders of magnitude over those of the free particles and correspondingly the energies of the electronic levels are reduced. Using the modern techniques of growth that can control the composition of heterostructures on the scale of one monoatomic layer [11,12] and of preparation which can define nanostructures in the 100 Å range [13], it has become possible to change locally the electronic energy structure and to produce quantum size effects that drastically modify the electronic and optical properties [10].

The simplest example of systems where size effects produces fundamental modifications of the optical and electronic properties are semiconductor Quantum Well Structures (QWS). They consist of ultrathin layers of semiconductors (e.g. 1 Å-500 Å thick), with different compositions grown alternately one on the other, Fig. 1. Because the band gaps depend on the composition these structures exhibit a strong modulation of the potential seen by the carriers in the direction normal to the layers. If the depth of the potential step between the low-gap compound and the high-gap one is sufficiently high and if the separation between low-gap one is sufficiently high and if the separation between low-gap layers is large enough, the carriers are confined in the Quantum Well (QW). This reduced dimensionality introduces new selection rules in the optical transitions and changes significantly the excitonic behavior.

In these lectures we present our investigations of the novel nonlinear optical properties and electro-absorption effects seen close to quasi two-dimensional exciton resonances in III-V QWS and we describe their applications to prototype devices.

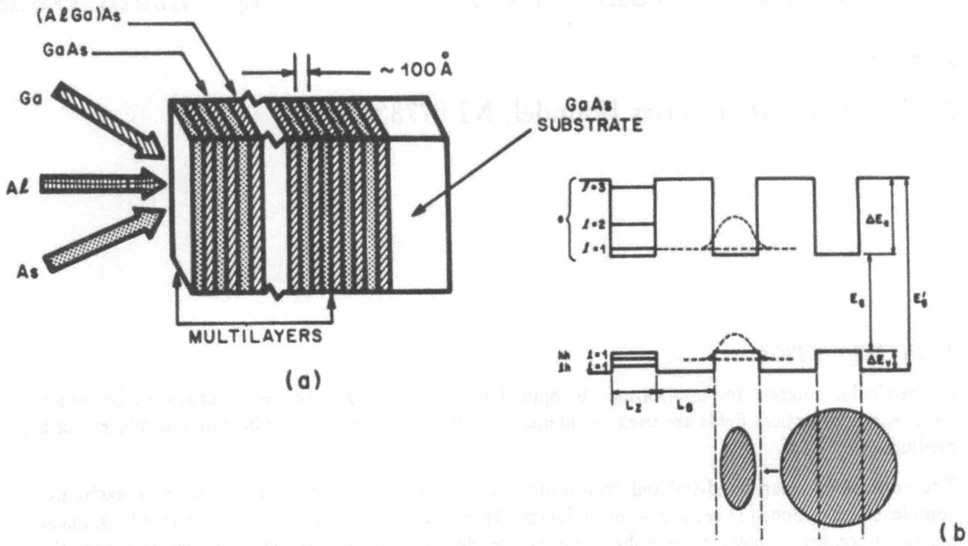

Fig. 1. **Schematic of a semiconductor quantum well structure and of the corresponding disposition of the energy bands along the normal to the layers**

2. *EXCITONIC RESONANCES IN QWS*

The optical linear properties of III-V QWS have been reviewed recently [14,15]. In QWS the envelope wavefunctions of the carriers in the direction perpendicular to the layers are strongly modified by the confinement potential. In the QW they have a sinusoidal dependence on the coordinate along the normal layers and present exponential decay tails outside the QW. In the plane where the carriers move freely the wavefunctions are plane waves. The bulk energy bands split into a series of two-dimensional subbands with steplike density of states. For the conduction band the dispersion in the plane is controlled by the same effective mass as in the bulk. The situation of the valence bands is more complex, the degeneracy of the uppermost valence band is lifted producing the separate heavy-hole (hh) and light-hole (lh) subbands which have highly non-parabolic dispersion in the plane. The optical transitions obey novel selection rules due to parity and angular momentum conservation and the bulk oscillator strengths are modified by the overlap of the EMA wave functions. Close to the fundamental edge the oscillator strengths of the transitions from the hh and lh subbands to the conduction subbands are in the ratio 3/4 and 1/4 for optical fields polarized parallel to the layers and 0 and 1 for polarization normal to the layers.

The correlation between the photo-excited electron and hole produces excitonic resonances at the onset of each inter-subband transition edge and a strong absorption enhancement in the continua. Because the dielectric constants of the QW material and of the barrier material are very close in III-V compounds, the Coulomb interaction has the usual $1/r$ dependence, only the motion and the density of state of the carriers are sensitive to the confinement. If the layer thickness L_Z is smaller or of the order of the bulk exciton Bohr diameter a_0 (e.g. ~300 Å for GaAs) the exciton structure and hence profile of the absorption resonances are strongly modified. For the lowest-energy resonances, the extension of the wave function in the direction normal to the layer is mostly governed by the QW potential and in the plane the electron-hole (e-h) Coulomb interaction plays the major role. This results in an increase of the exciton binding energy that can reach a value 4 times that of the bulk exciton for infinitely narrow QW and is 2.5 to 3 times larger in finite QW thickness. Because the QW thickness has unavoidable fluctuation of the order of one atomic layer, even in the best growth conditions, the exciton resonances are inhomogeneously broadened at low temperature. Typical magnitudes of these parameters in the case of GaAs are: R_Y=4.2 meV for the bulk Rydberg, blinding energies in QWS of the order of E_{1S}~10 meV and Half Width at Half Maximum (HWHM) of the order of 1-2 meV at low temperature [16,18].

The low-energy excitons are well confined in the QW and therefore they mostly interact with the phonons of the QW medium which are not too much modified in the QWS [19,20]. Extended exciton states have also been observed above the band gap discontinuity [21], they correspond to continuum resonances [22] and interact with the phonons of both the QW and the barrier materials [21].

The most spectacular result of these changes in the exciton structure which has important consequences in optoelectronics is the observation of well resolved excitonic resonances at room temperature in III-V QWS [23,25]. As an example of room temperature the absorption spectrum of an L_z~96 Å GaAs/$Al_{0.3}Ga_{0.7}As$ QWS is presented in Fig. 2. The three plateaus of the transitions between the n=1,2,3 subbands are clearly seen and clear exciton resonances preceding each one are resolved. At the first transition the resonances are sharp enough that the hh-exciton and the lh-exciton peaks are separated. In a waveguide where the polarization can be parallel or perpendicular to the layers, the strong dichroism of QW has been verified [26].

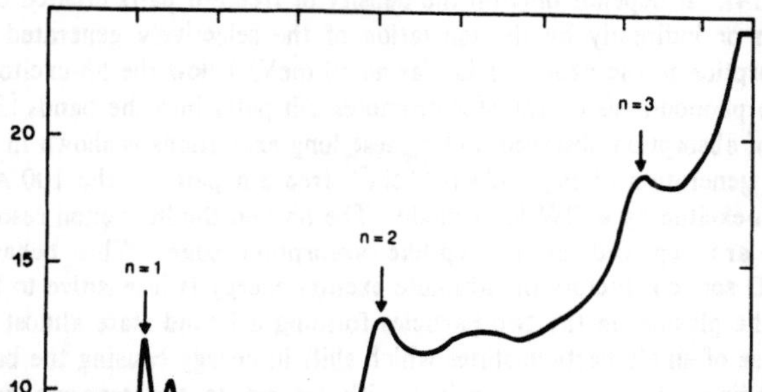

Fig. 2. Room temperature absorption spectrum of 65 periods of 96Å GaAs and 98Å $Al_{0.3}Ga_{0.7}As$ layers QWS

3. *NONLINEAR OPTICAL EFFECTS*

At Room Temperature (RT) quasi-2D excitons have dynamic properties that are qualitatively different from those of low temperature. Once generated by a photon absorption, excitons at RT are in contact with a thermal reservoir whose temperature is 2.5 to 3 times larger than their binding energy. They are therefore very unstable and are ionized in a very short time. The thermal energy is mediated by the thermal phonons, scattering by low-energy acoustic phonons can cause some broadening but the dominant interaction of excitons with the vibrations of polar semiconductors is due to the Longitudinal Optic (LO)-phonons. A single collision with these highly energetic phonons (36 meV for GaAs LO-phonon) will ionize an exciton and release an e-h pair with substantial excess energy (~27 meV). The exciton lifetime is simply correlated to the mean time for collision with thermal LO-phonons [27] that can be deduced from the dependence of the exciton linewidth as a function of temperature. From such measurements it is estimated that at RT excitons live only half a picosecond in GaAs QW [27,28].

The generation of e-h pairs (bound or unbound) modify the optical transitions through several mechanisms, whose effectiveness is different if the pairs are free (plasma) or bounded (exciton gas) [29]. Since the excitons selectively generated by resonant excitation transform very quickly into a plasma, at RT the nonlinear optical response of QW will exhibit strong transients on a time scale of the exciton ionization time.

Because of the exclusion principle transitions originating from empty states or ending in filled states are forbidden. This Phase-Space Filling (PSF) is a blocking mechanism which does not change the nature of the single particle or of the pair states. For free electrons and holes the PSF produces the band filling at the origin of Brustein-Moss shift. In the case of excitons, which behave like Bosons only in the dilute limit, the PSF is also effective and the generation of excitons will prevent further optical transitions to near-band-edge states. Furthermore photoexcitation of e-h pairs produces a screening of the Coulomb interaction that changes the single particle and pair states. For particles obeying the Fermi statistic it is customary to distinguish the classical long-range Direct Screening (DS) and the short-range Exchange Interaction (EI). In three dimensions (3D) the dominant effect of e-h pair photogeneration is the DS of free e-h pairs that causes the bandgap renormalization and the Mott transition: EI is only important in very dense plasmas [29]. Screening by the neutral excitons is much weaker [30,31].

When excitations with long duration compared to the ionization time are used to probe the nonlinear optical response of QW at RT it is found that the latter S is independent of the wavelength and of the duration of the excitation [14]. It depends only on the density of free e-h pairs created directly by excitation in the continuum or indirectly by the ionization of the selectively generated bound pairs. Even the very weak absorption in the band tail (as far as 50 meV) below the hh-exciton peak produces the same effects, once phonon-assisted transfer promotes e-h pairs into the bands [32]. A typical example of the change of absorption observed under these long excitations is shown in Fig. 3, which shows the effects of the generation of $N_{eh} \sim 2X10^{12} cm^{-2}$ free e-h pairs in the 100 Å thick GaAs QW 32meV above the hh-exciton by a CW laser diode. The hh and the lh exciton resonances are completely bleached and are replaced by a step-like absorption edge. This behavior is interpreted as follows. As in 3D semiconductors the absolute exciton energy is insensitive to the e-h density because the effects of the plasma on the two particles forming a bound state almost cancel out. This is not true for the case of single particle states which shift in energy causing the bandgap renormalization. Thus the binding energy of the exciton with respect to the renormalized gap decreases, the absorption weakens both because the e-h correlation diminishes and because the single particle states out of which excitons are built up are more and more occupied. If the plasma density is further increased the bound states eventually disappear leaving only the 2D renormalized bands. Consistently with this model measurements of the saturation density show that about half of the absorption saturates very easily ($I_s \sim 500$ W/cm^2) and the remaining is harder to saturate ($I_s \sim 5$KW/cm^2) [28]. The plasma induced changes in the refractive index have also been determined by four-wave mixing [33,27]. The optical nonlinearity is huge ($\chi^{(3)} \sim 0.06$ esu) and corresponds to changes in the refractive index $n_{eh} \sim 3.7x10^{-19} cm^3$ and changes in absorption coefficient $\sigma_{eh} \sim 7X10^{-14} cm^2$ per e-h pair per cm^3. The effects of the e-h plasma remain as long as

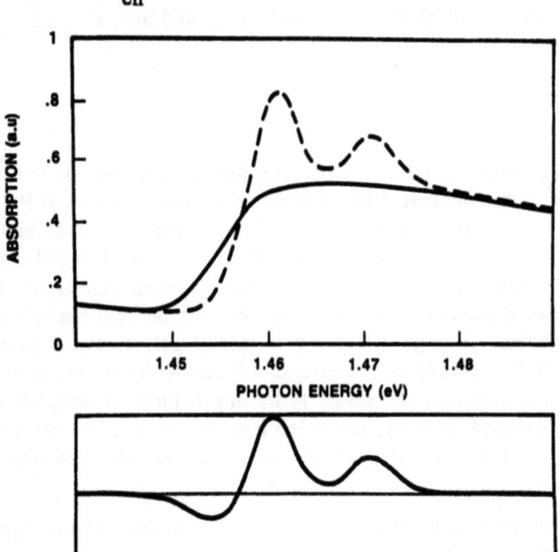

Fig. 3. Absorption spectra of a GaAs QWS with (solid line) and without (dashed line) excitation 32 meV above the gap. The differential spectrum at the bottom of the figure shows the difference between the unexcited and the excited spectra

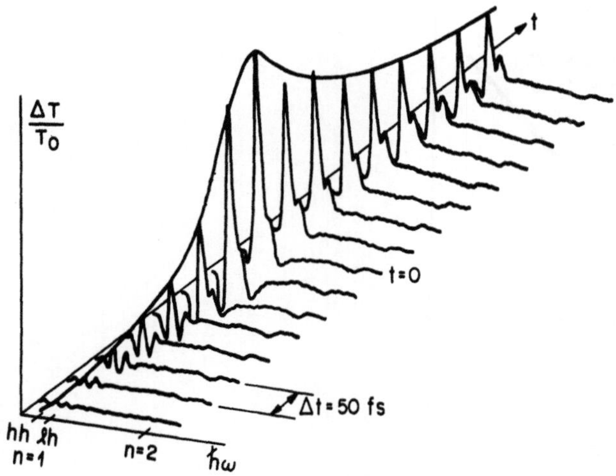

Fig. 4. Absorption differential spectra measured with a broad band (1.45-1.6 eV) 80 fs continuum at various delays after the excitation with a 80 fs pump pulse resonant with the hh-exciton. First the absorption at the peak of the exciton is very efficiently reduced, then after the excitons are ionized by collision with thermal phonons the spectrum evolves toward that measured with long duration excitations

the plasma is present or for about 20 ns in as-grown samples. The recombination time can be reduced (~150 ps) without too much modification of the other optical properties by introducing effective recombination centers, for example by ion bombardment [34]. A simple hard-core real-space filling model [27], that has recently been justified by more elaborate many-body theory [35], has explained quantitatively these experimental results. Numerical calculations based on two-dimensional plasma screening of the excitons are also in good agreement with the experiments [36].

When excitons are selectively generated by resonant excitation a certain type of change is induced in the optical spectrum, and then as the excitons transform into free pairs these changes evolve toward those seen using long duration excitations. Using ultrashort optical pulses to excite and probe the sample in times which are short compared to the exciton ionization time makes it possible to observe this process directly [37,38]. This is seen in Fig. 4 where the differential absorption spectra (absorption of the unexcited sample minus absorption of the excited sample) measured with a wide band (1.45 eV 1.65 eV) 80 fs continuum at various delays after the resonant excitation of the hh-exciton by a 80 fs pump pulse are presented [38]. The density of excitons created is approximately $N_X \sim 2X10^{10}$ cm^{-2}, the interval between each spectrum is 50 fs and each curve corresponds to the integration of 10^6 spectra giving a signal noise ratio > 10^4. A fast and extremely efficient transient beaching of hh-exciton resonance is seen for the first 300 fs. Then surprisingly the resonance recovers and after about 800 fs the changes in the absorption spectra stabilize to the magnitude and shape observed with the long duration excitations and remain constant for tenths of picoseconds. The ionization time of the hh-excitons at RT deduced from these measurements is $t_i \sim 300 \pm 100$fs and thus in good agreement with the estimates from temperature broadening. The most surprising results of these experiments is that the selective generation of excitons produce a much larger reduction of the excitonic absorption than the generation of free carriers. This behavior is in contradiction with theory [29] and experimental results [30,31] in bulk semiconductors. The explanation resides in the very special conditions in which excitons are in QW and at RT. First the DS is strongly reduced in two dimensions, in fact the DS saturates and the screening parameter becomes independent of the carrier concentration at large densities [30]. This decreases the effectiveness of the dominant mechanism by which free e-h pairs reduce the exciton absorption in 3D. In addition the excitons generated with ultrashort optical pulses and observed before they can interact with the thermal reservoir are not in thermodynamic quilibrium with the lattice. They have not yet a "temperature" and occupy the lowest-energy state of the crystal. Then their first interaction with the thermal LO-phonons destroys them and releases a warm plasma, which then thermalizes in less than 1 ps. The PSF and EI are governed by the occupancy of the states that

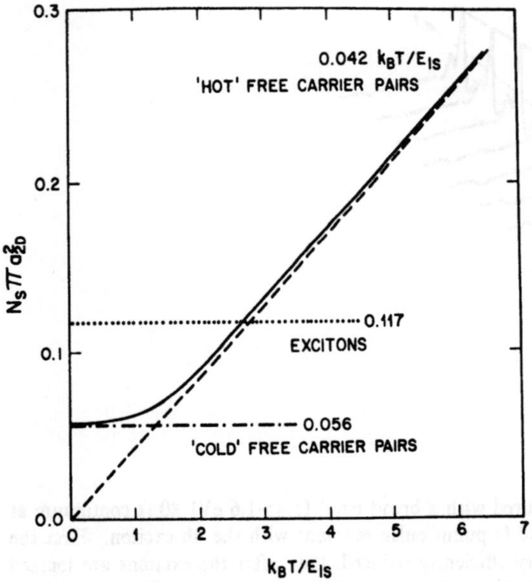

Fig. 5. **Plot of the theoretical saturation density (normalized to the exciton area πa_X^2) for an exciton gas (dotted line) and for a plasma (solid line) versus the temperature (normalized to the exciton binding energy)**

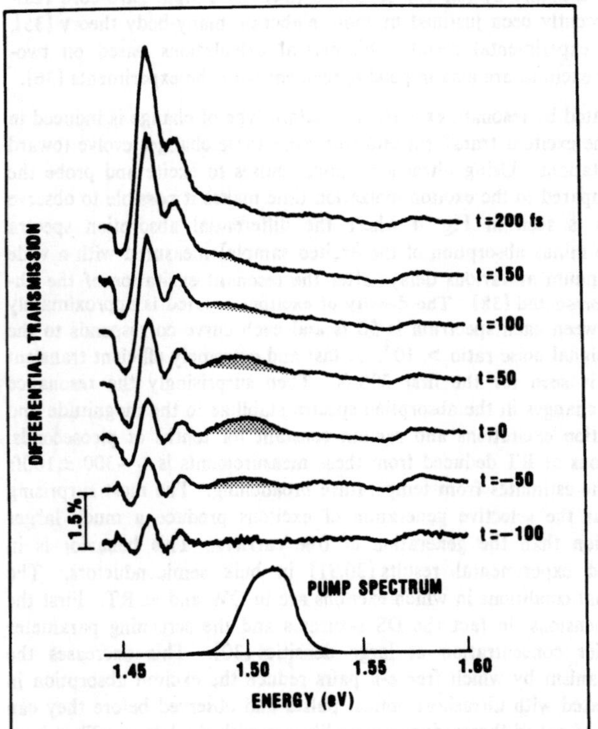

Fig. 6. **Differential absorption spectra for a non-thermal carrier distribution generated by an 80 fs pump centered at 1.509 eV and observed at a 50 fs interval with a broad 80 fs continuum**

enter the exciton wavefunction. For the warm plasma Boltzmann distributions this occupancy is small. As for excitons, it can be shown the N_X bound pairs are equivalent to special distributions of electrons and holes equal to half of the exciton occupation probability in k-space:

$$f_e(k) = f_h(k) = (N_x/2)|U(k)|^2$$

where U(k) is the exciton relative motion wavefunction in the phase space [29]. Of course this distribution overlaps precisely with the portion of the Brillouin zone that enters the exciton wavefunction and therefore produces very efficient PSF and EI. A theory of two-dimensional exciton saturation by e-h plasma and exciton, neglecting the DS, has been developed using the many-body formalism. When only processes of first order in the pair (bound or unbound) density are retained the results can be expressed in a close form [35]. They are shown in Fig. 5. They explain the most salient features experimentally seen, i.e. that at low temperature the plasma is more efficient than the excitons, but that as the temperature increases the plasma efficiency decreases as E_{1S}/kT. For $E_{1S} \sim 9$ meV the model gives an exciton bleaching slightly larger than that of a plasma. The model also gives the good absolute value of the nonlinear cross section for exciton generation, $\sigma_X = 2.5X10^{-13}$ cm^2, as compared to the measured value, $\sigma_X = 10^{-13}$ cm^2.

The most important assumption of the theory is that DS is negligible compared to EI and PSF in QWS. This was verified directly in the following manner. Since PSF and EI are effective only when the carrier wavefunction overlaps with that of the excitons, a nonthermal carrier population generated in the continuum will first interact with the n=1 resonance through DS. Then as the carrier thermalizes to the bottom of the bands the effects of the Pauli principle turn on. This switching of the PSF and EI has indeed been observed [38]. In Fig. 6 the differential absorption spectra seen at 50 fs interval during and after the generation of a non-thermal carrier distribution n=1 continuum less than 1 LO-phonon energy above the gap at a density $N_{eh} \sim 2X10^{10}$ cm^{-2} are presented. The turning on of the absorption at the n=1 resonance as the carrier thermalizes is clearly seen. Simultaneously the effects at the n=2 exciton do not vary in time showing that DS remains essentially constant as the carriers thermalize because the n=2 excitons (which are at higher energy and in any case made of single particle states nearly orthogonal to those occupied by the carriers) are only sensitive to the DS. These results put an upper limit on the magnitude of DS which is at least six time smaller than the sum EI+PSF. In addition, they provide direct information on the carrier scattering times in these important heterostructures. The thermalization-time is measured to be 200 fs and the time it takes the carriers to leave the states in which they are created is less than (but in the order of) 80 fs.

The very large optical nonlinearities seen in QWS at RT have found application to third-order nonlinear optical processes with a laser diode as sole light source [40]. They have also been used as saturable absorbers for modelocking of laser diodes [34], in experiments where stable 1.6 ps pulses at about 1 GHz have been produced.

4. ELECTROABSORPTION IN QWS

Electroabsorption in QWS has been extensively studied and is now well documented; an exhaustive bibliography can be found in [41].

When an electrostatic field is applied parallel to the layer where the confinement potential is ineffective, the only force binding the electron and the hole is the Coulomb force and the effects are similar to those seen in bulk material [42]. The only difference from the case of massive samples is that the exciton binding energy and the in-plane wave function are slightly modified [42]. The dominant process is a field ionization and theoretically a second-order energy shift [43]. As shown in Fig. 7, when a field is applied to a QWS parallel to the layers, the resonances broaden easily and disappear at moderate fields ($\sim$ 2-5X10^4 V/cm). No significant shift is observed even when the sensitive differential spectra are directly measured. The evolution of the spectra are in qualitative agreement with the theory [42]. In this geometry the capacitance is very small so that it can be used to test how fast a change in the excitonic absorption change can be induced in a semiconductor. In recent experiments [44] a potential was applied across the electrodes of a sample and the transmission was probed with a 120 fs-broad continuum light pulse. Then the field in the QWS was shortened by a high-density plasma across the inter-electrode gap produced by a strong 50 fs red pulse. The differential absorption spectra observed at various delays are shown in Fig. 8. The l/e recovery time of the absorption was measured to be 330 fs and seems to be still limited by electric-

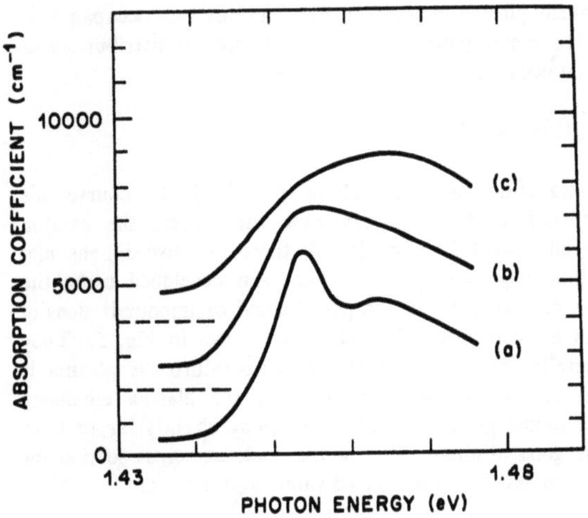

Fig. 7. Absorption spectra for electric fields parallel to the layers. The three spectra have been shifted vertically for clarity, they correspond to (a) E~0: (b) E~1.6X10^4 V/cm; (c) E~4.8X10^4 V/cm

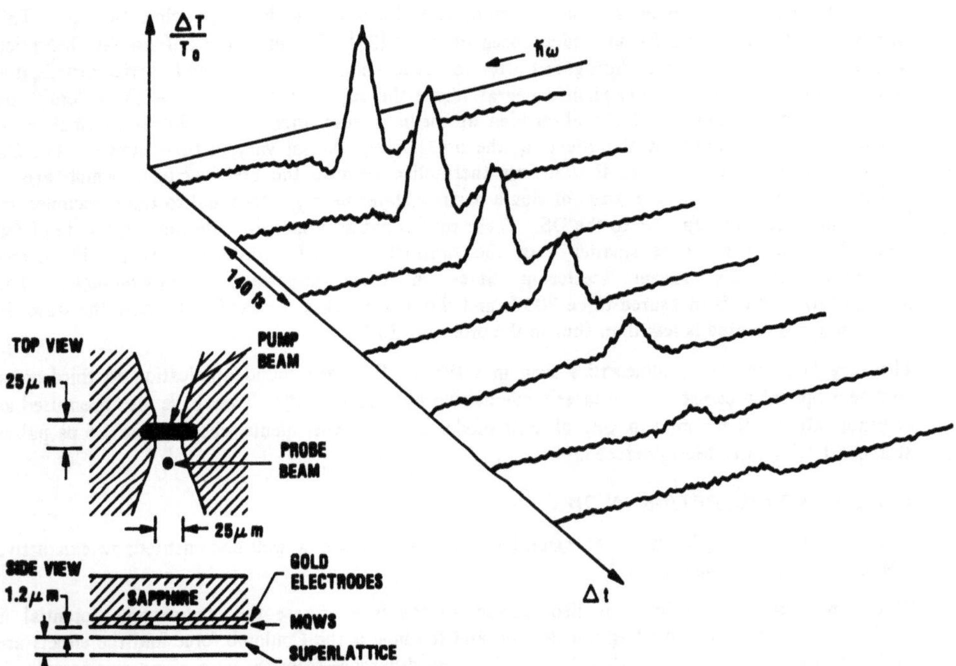

Fig. 8. Time dependence of the electro-absorption for a field parallel to the plane of the layer. The absorption differential spectra are shown as a function of the delay between the 120 fs probe pulse and the 50 fs pump pulse that shorten the electrostatic field

field reflections in the electrodes. The absorption recovery time has to be compared to some characteristic times of the exciton dynamics at RT. For example, a limit on how fast a resonance can recover is set by the mean time for exciton scattering by thermal LO-phonons (300 fs at RT). The change in excitonic absorption should follow the sudden application of a field in a time limited by how long it takes the e-h pair to "feel" its Coulomb correlation (i.e. the exciton classical orbit-time ~150 fs) or by how long it takes the two particles to tunnel away one from another through the Coulomb well (i.e. the mean time-to-field ionization ~380 fs for an 8 kV/cm field [42]). All these values lie in the same range showing that the response measured in these experiments approaches the fundamental limits of the exciton dynamics. The intrinsic limits of the speed of electro-absorption are of course also of great technological interest for high-speed sampling and modulation.

The electro-absorption for fields perpendicular to the layers presents features that are specific for QWS [42]. The electron and the hole charge distributions are pulled apart by the field and pushed against opposite well walls. If the thickness of the QW is large compared to the 3D Bohr radius the pair correlation is lost, but if the QW thickness is less than the 3D Bohr radius the particles are stopped from flying apart by a potential barrier and continue to interact with each other. Therefore, as the QW thickness decreases the electro-absorption should show a continuous transition from the 3D behavior to a strongly quantized one. This has been analytically demonstrated recently for the Franz-Keldysh effect in infinitely deep wells [45]. The transition from the quantum-confined to 3D behavior occurs over a narrow range of thickness for a given field and is accompanied by the growth of symmetry-forbidden optical transitions from which the Franz-Keldysh oscillations originate. In the case of finite well depth, one can expect strong tunneling only for fields large enough to provide the carriers with an energy comparable to the potential barrier height by acceleration of the particles over the QW thickness. Indeed calculations of tunneling resonances give very narrow width for the resonances for moderate fields yet large enough to produce strong shifts of the absorption edge [42].

This behavior is actually observed in GaAs QW [42,46]. A characteristic example of the change in the absorption spectra induced by application of a particular field to a QWS is presented in Fig. 9

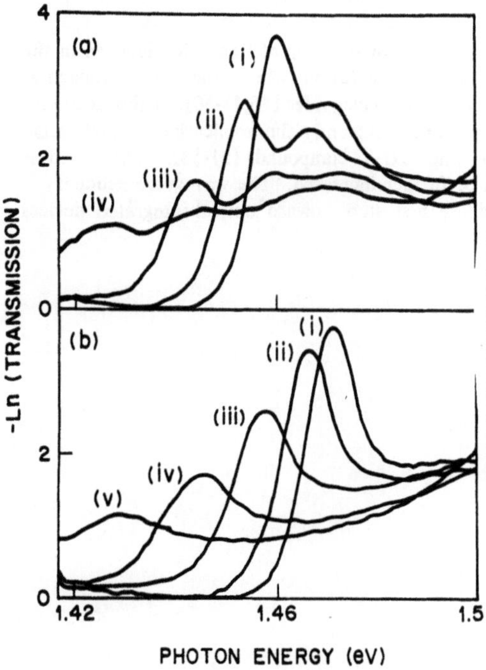

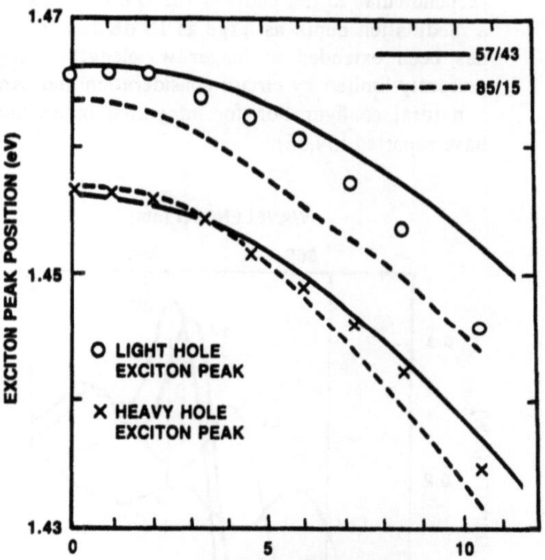

Fig. 9. Absorption spectra of a waveguide containing 2 94Å GaAs quantum wells, as a function of the electric field applied perpendicular to the layers. (a) optical field polarization parallel to the layers; (i) 1.6×10^4 V/cm, (ii) 10^5 V/cm, (iii) 1.3×10^5 V/cm, (iv) 1.8×10^5 V/cm. (b) optical field polarization perpendicular to the layers; (i) 1.6×10^4 V/cm, (ii) 10^5 V/cm, (iii) 1.4×10^5 V/cm, (iv) 1.8×10^5 V/cm, (v) 2.2×10^5 V/cm

Fig. 10. Energy of the excitonic peaks with applied fields perpendicular to the layers. The open circles and crosses are experimental. The lines result from the model discussed in the text which involves no adjustable parameter. The solid lines correspond to a 0.57:0.43 split of the band gap discontinuity between the conduction and the valence bands, the dashed lines are for 0.85:0.15

for the case of a waveguide structure containing two 94 Å QWs. The strong dichroism is clearly apparent, the two excitons are active for polarization parallel to the layers whereas in the perpendicular polarization only the 1 h exciton is present and acquires a larger oscillator strength. When the static field is applied the whole absorption profile shifts toward lower energies, the height of the resonance somehow decreases and the peaks broaden. The magnitude of the effect is surprising, for example the maximum shift seen with yet resolved exciton peaks is about 40 mev at a field of $2.2X10^5$ V/cm. This is 10 times the bulk exciton binding energy and corresponds to 100 times the exciton classical ionization field. This effect can be described as an extreme quantization of the Stark effect, hence its name, the "Quantum Confined Stark Effect" (QCSE) [47]. The overall behavior is now understood in details. The shifts of the resonances are well accounted for by adding the shifts of the single particle states to the change of the exciton binding energy due to the compression of the wavefunctions against the walls of the well [42]. The comparison of the experimental shifts with the prediction of the theory involving no adjustable parameters is shown in Fig. 10. The two sets of curves correspond to the two values of the offsets in the conduction and in the valence bands that have been proposed for the GaAs/AlGaAs system. The changes in height of the resonances and of the continua are well explained by several sum rules that have been demonstrated recently [41].

In order to apply large fields and yet avoid the flow of large currents, the samples are grown as p-i-n diodes with the QWs in the intrinsic region. The field across the QWs is varied by reverse-biasing of the diodes. Because of this geometry the sample also behaves as a photodetector, with an internal quantum efficiency close to unity [42] as shown by the responsivity curves at various bias voltages presented in Fig. 11. This characteristic is important because the photocurrent gives an electrical measure of the absorption that can be utilized in an external circuit to react on the bias voltage, thus providing a mechanism for feedback.

The most simple application of the QCSE is to high-speed modulation of light. Modulation in the 100 ps range has been demonstrated both in the configuration where the light propagates perpendicular to the plane of the QWs and parallel to it in waveguides [46,48-50]; in this geometry a modulation depth as large as 10 dB has been measured. Recently the application to modulation has been extended to longer wavelengths using other III-V compounds [51-53]. The speed is presently limited by circuit consideration and can be further increased. The waveguide geometry is a natural configuration for integrated optics and the first steps toward optical-integrated devices have reported [54,55].

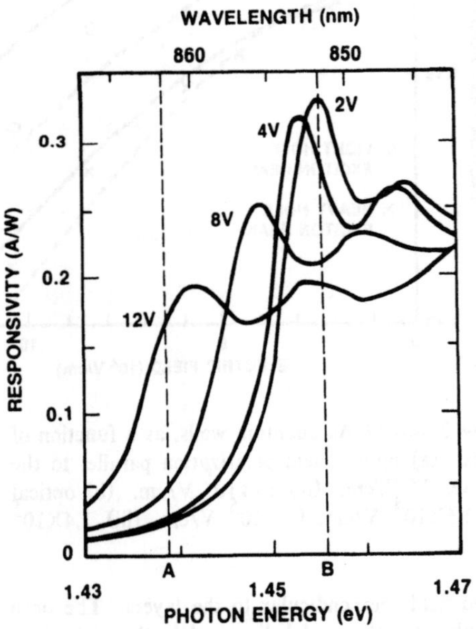

Fig. 11. Responsivity of a p-i-n QW diode as a function of the reverse bias voltage

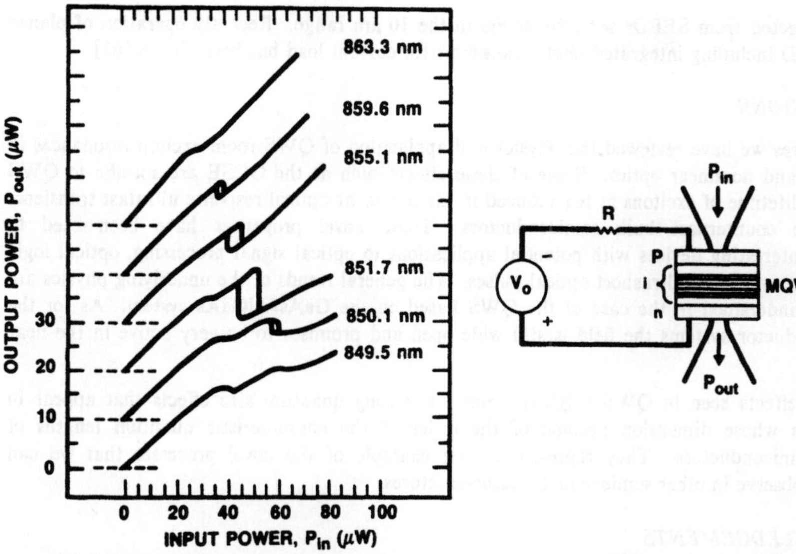

Fig. 12. Transmission characteristics of SEED operating as an optical gate for six wave lengths in the vicinity of the hh exciton peak. The positive feedback is effective enough to produce an optical hysteresis for the four intermediate wavelengths

As mentioned before the possibility to measure the absorption of p-i-n QWS by measuring the photo current leads to new devices. A simple example of such a device is a wavelength-selective voltage-tunable photodetector which operates on the following principle. Light, with photon energy less than that of the hh exciton peak at zero field, incident on the p-i-n QWS produces a photocurrent that is maximized for an applied field just large enough to shift the exciton to the photon energy. If the wavelength is changed then the photodetector response can be maximized again by changing the applied voltage, thus giving means to follow electrically the frequency of the light-source [56,57]. This has potential applications in coherent light-wave communications where it is often necessary to determine simultaneously the intensity and the wavelength of an optical beam.

The Self Electro-optic Devices (SEED) form a novel class of devices that utilize the p-i-n QWS capacity to operate as close-coupled photodetector/modulator. To operate the p-i-n QWS is connected to an external circuit which imposes a relation between the applied voltage and the current passing through the device. These two quantities must also satisfy the response function imposed by the relation between the absorption coefficient and the field across the QWS, hence the possibility of electronic feedback on the optical behavior. The feedback can be positive or negative depending on the wavelength; both have been exploited in optical bistable gates, self-linearized modulators and optical-level shifters and devices have been demonstrated in the propagation mode normal to the QWS as well as in waveguides [46,58-60].

The most interesting SEED operation may be that of low switching energy optical bistable gates whose operation is shown in Fig. 12. In this figure the transmission characteristics for six wavelengths close to the absorption edge for the simple external circuit shown on the right of the figure are plotted. Between 850 nm and 860 nm, the feedback is strong enough to switch the device, hence the optical hysteresis loops seen at the hh-exciton or at the lh-exciton. Similar operations can be observed for different external circuits [37]. The switching as well as the oscillations can be followed optically or electronically. The contrast between the two states is large because the change in the absorption coefficient is typically of the order of $0.5X10^4 cm^{-1}$; 20/1 contrast ratios have been measured in waveguide SEED [46]. The importance of the SEED for optical switching resides in the extremely low operating energy of the device despite the absence of optical cavity, in the tolerance of the SEED to fluctuation in the optical and electrical operating parameters and in the natural structure for large planar arrays. Switching energy as low as 20 fJ (15 fJ electrical and 5 fJ optical) and 20 ns switching time have been demonstrated in large devices (100 μm diameter). The limits for switching energy are expected to be in the few fJ range and a switching time of a few ns can be expected from SEEDs with diameters in the 10 μm range. Recently operation of planar arrays of SEED including integrated photo-transistors for current load has been shown [61].

5. CONCLUSIONS

In these lectures we have reviewed the physics and application of QWS room exciton resonances in electro-optics and nonlinear optics. Some of these effects such as the QCSE are specific to QWS and the short lifetime of excitons at RT induced in the nonlinear optical response ultrafast transients that have no counterpart bulk semiconductors. These novel properties have been used to demonstrate interesting devices with potential applications to optical signal processing, optical logic and convenient sources of ultrashort optical pulses. The general trends of the underlying physics are more or less understood in the case of the QWS based on the GaAs/AlGaAs system. As for the other semiconductor systems the field is still wide open and promises to be very active in the near future.

The unusual effects seen in QWS originate from the strong quantum size effects that appear in nanostructures whose dimension become of the order of the characteristic quantum lengths of carriers in semiconductors. They represent a first example of the novel processes that we can anticipate to observe in other semiconductor nanostructures.

6. ACKNOWLEDGEMENTS

Most of the work reviewed in these lectures has been performed in AT&T Bell Laboratories, in collaboration with D. A. B. Miller, A. C. Gossard, J. S. Weiner, W. H. Knox, T. C. Damen, J. E. Zucker, A. Pinczuk, S. Schmitt-Rink, C. V. Shank, R. L. Fork, J. Shah, T. H. Wood, and C. A. Burrus.

REFERENCES

[1] A. Miller, D. A. B. Miller, S. D. Smith: Adv. Phys., *30*, 697 (1981)

[2] D. S. Chemla, A. Maruani: Prog. Quantum Electron., *8*, 1, (1982)

[3] D. S. Chemla, J. Jerphagnon: "Nonlinear Optical Properties of Semiconductors," in *Handbook of Semiconductors,* Vol. 2, Optical Properties of Solids (North-Holland, Amsterdam 1980)

[4] W. Franz: Z. Naturforsch. *13a*, 484 (1958)

[5] L. V. Keldysh: Zh. Eksp. Teor. Fiz. *34*, 1138 (1958)

[6] J. D. Dow, D. Redfield: Phys. Rev. *B1*, 3358 (1970)

[7] D. F. Blossey: Phys. Rev. *B2*, 3976 (1970)

[8] Q.H.F. Vrehen: J. Phys. Chem. Solids *29*, 129 (1968)

[9] See, for example, G. E. Stillman, C. M. Wolfe, in *Semiconductors and Semimetals,* ed. by R. K. Willardson and A. C. Beer (Academic New York (1977) p.380

[10] For recent reviews see, for example, F. Capasso and B. F. Levine: J. Lumin. *30*, 144 (1985) and D. S. Chemla: J. Lumin. *30*, 502 (1985)

[11] See, for example, L. L. Chang, K. Ploog (eds.): *Molecular Beam Epitaxy and Heterostructures,* NATO Advanced Sciences Institute Series (Nijhoff, Dordrecht (1985)

[12] See, for example, J. B. Mullin, S. J. C. Irvine, R. H. Moss, P. N. Robson, D. R. Wight (eds.): *Metal Organic Vapor Phase Epitaxy* (North Holland, Amsterdam (1984)

[13] See, for example, R. E. Howard, L. D. Jackel, P. M. Mankiewich, W. J. Skocpol: Science *231*, 346 (1986): R. E. Howard, D. E. Prober: *in VLSI Electronics-Microstructures,* Vol V, ed. by N. G. Einspruch ed (1982) and ref. therein.

[14] D. S. Chemla, D. A. B. Miller: J. Opt. Soc. Am. *B2*, 1155 (1985)

[15] R. C. Miller, D. A. Kleinman: J. Lum. *30*, 144 (1985)

[16] C. Weishbuch, R. C. Miller, R. Dingle, A. C. Gossard: Solid State Commun. *37*, 219 (1981)

[17] R. C. Miller, D. A. Kleinman, W. A. Norland, A. C. Gossard: Phys. Rev. *B22*, 863, (1980)

[18] J. Hegarty, M. D. Sturge: J. Opt. Soc. Am. *B2*, 1143 (1985)

[19] J. E. Zucker, A. Pinczuk, D. S. Chemla, A. C. Gossard, W. Wiegmann: Phys. Rev. Lett. *51*, 1293 (1983)

[20] J. E. Zucker, A. Pinczuk, D. S. Chemla, A. C. Gossard, W. Wiegmann: Phys. Rev. Lett. 53, 1280 (1984)

[21] J. E. Zucker, A. Pinczuk, D. S. Chemla, A. C. Gossard, W. Wiegmann: Phys. Rev *B29*, 7065 (1984)

[22] G. Bastard: Phys. Rev. *B30*, 3547 (1984)

[23] D. A. B. Miller, D. S. Chemla, P. W. Smith, A. C. Gossard, W. Wiegmann: Appl. Phys. *B28*, 96 (1982)

[24] J. S. Weiner, D. S. Chemla, D. A. B. Miller, T. H. Wood, D. Sivco, A. Y. Cho: Appl. Phys. Lett. *46*, 619 (1985)

[25] H. Temkin, M. B. Panish, P. M. Petroff, R. A. Hamm, J. M. Vandenberg, S. Sunski: Appl. Phys. Lett. *47*, 394 (1985)

[26] J. S. Weiner, D. S. Chemla, D. A. B. Miller, H. Haus, A.C. Gossard, W. Weigmann, C. A. Burrus: Appl. Phys. Lett. *47*, 664 (1985)

[27] D. S. Chemla, D. A. B. Miller, P. W. Smith: IEEE J. Quantum Electron. *QE-20*, 265 (1984)

[28] D. A. B. Miller, D. S. Chemla, P. W. Smith, A. C. Gossard, W. T. Tsang: Appl. Phys. Lett. *41*, 679 (1982)

[29] For a recent review of semiconductor optical nonlinearities close to the band gap see: H. Haug, S. Schmitt-Rink: Prog. Quant. Electron. *9*, 3 (1984)

[30] G. W. Fehrenbach, W. Schafer, J. Treusch, R. G. Ulbrich: Phys. Rev. Lett. *49*, 1281 (1982)

[31] G. W. Fehrenback, W. Schafer, R. G. Ulbrich: J. Lumin. *30*, 154 (1985)

[32] A. Von Lehnen, J. E. Zucker, J. P. Heritage, D. S. Chemla: to be published.

[33] D. A. B. Miller, D. S. Chemla, D. J. Eilenberger, P. W. Smith, A. C. Gossard, W. Wiegmann: Appl. Phys. Lett. *42*, 925 (1982)

[34] P. W. Smith, Y. Silberberg, D. A. B. Miller: J. Opt. Soc. Am. *B2* 1228 (1985) and references therein

[35] S. Schmitt-Rink, D. S. Chemla, D. A. B. Miller: Phys. Rev. *B32*, 6601 (1985)

[36] S. Schmitt-Rink, C. Ell: J. Lumin. *30*, 585 (1985)

[37] W. H. Knox, R. L. Fork, M. C. Downer, D. A. B. Miller, D. S. Chemla, C. V. Shank, A. C. Gossard, W. Wiegmann: Phys. Rev. Lett. *54*, 1306 (1985)

[38] W. H. Knox, C. Hirlimann, D. A. B. Miller, J. Shah, D. S. Chemla, C. V. Shank: Phys. Rev. Lett. *56*, 1191 (1986)

[39] T. Ando, A. B. Fowler, F. Stern: Rev. Mod. Phys. *54*, 437 (1982)

[40] D. A. B. Miller, D. S. Chemla, P. W. Smith, A. C. Gossard, W. Wiegmann: Opt. Lett. *8*, 477 (1983)

[41] D. A. B. Miller, J. S. Weiner, D. S. Chemla: to be published in IEEE J. Quant. Electr.

[42] D. A. B. Miller, D. S. Chemla, T. C. Damen, A. C. Gossard, W. Weigmann, T. H. Wood, C. A. Burrus: Phys. Rev. *B32*, 1043 (1985)

[43] F. L. Lederman, J. D. Dow: Phys. Rev. *13*, 1633 (1976)

[44] W. H. Knox, D. A. B. Miller, T. C. Damen, D. S. Chemla, C. V. Shank: to be published in Appl. Phys. Lett.

[45] D. A. B. Miller, D.S. Chemla, S. Schmitt-Rink: to be published in Phys. Rev. B (1986)

[46] J. S. Weiner, D. A. B. Miller, D. S. Chemla, T. C. Damen, C. A. Burrus, T. H. Wood, A. C. Gossard, W. Weigmann: Appl. Phys. Lett. *47*, 1148 (1985)

[47] D. A. B. Miller, D. S. Chemla, T. C. Damen, A. C. Gossard, W. Weigmann, T. H. Wood, C. A. Burrus: Phys. Rev. Lett. *53*, 217 (1984)

[48] T. H. Wood, C. A. Burrus, D. A. B. Miller, D. S. Chemla, T. C. Damen, A. C. Gossard, W. Weigmann: Appl. Phys. Lett. *44*, 16 (1984)

[49] T. H. Wood, C. A. Burrus, R. S. Tucker, J. S. Weiner, D. A. B. Miller, D. S. Chemla, T. C. Damen, A. C. Gossard, W. Weigmann: IEEE J. Quantum Electron. *QE-21*, 117 (1985)

[50] T. H. Wood, C. A. Burrus, R. S. Tucker, J. S. Weiner, D. A. B. Miller, D. S. Chemla, T. C. Damen, A. C. Gossard, W. Weigmann: Electron. Lett. *21*, 693 (1985)

[51] K. Wakita, Y. Kawamura, Y. Yoshikuni, A. Asahi: Electon. Lett. *21*, 338 (1985)

[52] K. Wakita, Y. Kawamura, Y. Yoshikuni, A. Asahi: Electon. Lett. *21*, 574 (1985)

[53] T. Miyazawa, S. Tarucha, Y. Ohmori, H. Okamoto: Post deadline paper P1, Second International Conference on Modulated Semiconductor Structures (Kyoto, Japan, Sept. 1985)

[54] S. Tarucha, H. Okamoto: Appl. Phys. Lett. *48*, 1 (1986)

[55] Y. Arakawa, A. Larsson, J. Paslaski, A. Yariv: Appl. Phys. Lett. *48*, 561 (1986)

[56] T. H. Wood, C. A. Burrus, A. H. Gnauck, J. M. Wiesenfeld, D. A. B. Miller, D. S. Chemla, T. C. Damen: Appl. Phys. Lett. *47*, 190 (1985)

[57] A. Larsson, A. Yariv, R. Tell, J. Masaerjian, S. T. Eng: Appl. Phys Lett. *47*, 866 (1985)

[58] D. A. B. Miller, D. S. Chemla, T. C. Damen, A. C. Gossard, W. Weigmann, T. H. Wood, C. A. Burrus: Appl. Phys. Lett. *45*, 13 (1984)

[59] D. A. B. Miller, D. S. Chemla, T. C. Damen, A. C. Gossard, W. Weigmann, T. H. Wood, C. A. Burrus: Opt. Lett. *9*, 567 (1984)

[60] D. A. B. Miller, D. S. Chemla, T. C. Damen, A. C. Gossard, W. Weigmann, T. H. Wood, C. A. Burrus: IEEE J. Quantum Electron. *QE-21*, 1462 (1985)

[61] D. A. B. Miller, J. Henry, A. C. Gossard, J. H. English: to be published.

Quantum Interference Effects in Small Systems: Normal and Superconducting Networks

R. Rammal

Centre de Recherches sur les Très Basses Températures, CNRS, BP 166 X, F-38042 Grenoble Cédex, France

We review some of the recent advances in the study of normal and superconducting networks. Emphasis is placed on the possible use of networks, made of micronic or submicronic units, in order to study some basic physical phenomena: frustration, weak-localization and enhanced backscattering in disordered materials.

1. FINE TUNING OF THE FRUSTRATION IN SUPERCONDUCTING NETWORKS

Frustration [1] means competing interactions and therefore unsatisfied states. Degeneracy, metastability and sensitivity to topology or external parameters are just a few characteristic features of frustrated systems. The three basic concepts are those of gauge invariance, frustration function and curvature. During the last years the manifestation of these concepts in various fields has been the object of intensive studies: magnetic systems (spin glasses), amorphous packing and random networks, optimization problems (traveling salesman problem), biological problems (neural networks), etc. However, in most of these examples frustration and disorder occur simultaneously. As will be shown in the following sections, superconducting networks made of micronic (or submicronic) units, allow a fine tuning of the frustration which enters alone, that is not easy to achieve otherwise. Besides the proper superconducting properties, the studied networks provide a simple physical realization of an almost periodic system, which is relevant for various physical situations. Contact with other physical problems leaves us with a large program for further investigations.

Recent technical and theoretical advances have led to a revival of interest in the magnetic properties of superconducting networks. One line of thought sees these structures as an appropriate tool for studying some specific frustration effects as recalled before. Another one sees the diamagnetic superconducting currents as a sensitive tool for studying the topology of regular or irregular network structures [2,3]. Up to now, attention has been focused on the following physical properties [2,3,4]: i) the critical field for the onset of superconductivity H_c, ii) the zero-field susceptibility, iii) the magnetization [5] and iv) critical current and current distributions [6]. The networks considered fall into three categories: simple finite geometries (ring, lasso, etc.), infinite regular arrays (square lattice, honeycomb lattice, Sierpinski gasket), disordered structures (percolation clusters).

For the sake of simplicity, we shall limit our exposition to the upper critical field studies. In the first section, the theoretical formulation of the problem is presented. The critical temperature variation for various simple geometries, as a function of field, is illustrated in Sect.2. Infinite regular networks (honeycomb, square and Sierpinski gasket) are discussed in Sect.3, on the light of available experiments. In Sect.4., some other physical quantities will be briefly discussed.

1.1 Formulation

The network structures considered in the following are assumed to be made of wires, so thin that they can sustain diamagnetic currents only by loop formation. Fluctuation effects are ignored, the emphasis being on geometrical properties.

For the calculation of the critical field, we assume the validity of the linearized Ginzburg-Landau equation on a strand of wire:

$$\left[i\frac{\partial}{\partial \ell} + \frac{2\pi}{\phi_0}A(\ell)\right]^2 \psi(\ell) = \xi^{-2}\psi(\ell) \tag{1}$$

where ψ is the order parameter, ℓ a curvilinear coordinate, $A(\ell)$ the projection of the vector potential along the strand, ϕ_0 the flux quantum hc/2e and ξ the temperature-dependent correlation length [$\xi(T) = \xi(0)\ T_0/(T_0 - T)$, T_0 = zero field transition temperature].

The boundary condition at the ends or nodes is taken as

$$\sum_p \left[i\frac{\partial}{\partial \ell} + \frac{2\pi}{\phi_0}A(\ell)\right]\psi_p(\ell)\Big|_{\ell=0} = 0 \tag{2}$$

where the sum is over all strands connected to a given node.

The real and imaginary parts of (2) yield two relations at each node. One is simply the Kirchoff law expressing current conservation; the other one is special to superconductivity. This formulation, which has been discussed by various contributors, leads to an eigenvalue problem. Such a problem is best expressed in terms of the order parameter values at the nodes. If node α is linked to n nodes via strands of length $\ell_{\alpha\beta}$ (β = 1 to n), condition (2) at node α may be written

$$\sum_{\beta=1}^{n} \left[-\psi_\alpha \operatorname{cotg}\frac{\ell_{\alpha\beta}}{\xi} + \psi_\beta \frac{e^{-i\gamma_{\alpha\beta}}}{\sin\frac{\ell_{\alpha\beta}}{\xi}}\right] = 0 \tag{3}$$

where ψ_β is the value of the order paramter at node β, and

$$\gamma_{\alpha\beta} = \frac{2\pi}{\phi_0}\int_\alpha^\beta \vec{A}\cdot d\vec{\ell}$$

is the circulation of the vector potential along the strand linking α and β. Illustrative examples of the practical use of (3) will be discussed below. The highest eigenvalue of (3) determines the transition temperature, in the spirit of mean field theory. In fact, the whole eigenvalue spectrum is of some physical interest (see Sect.4), as it is directly related to the Landau level structure of a free electron gas in the same geometry.

1.2 Critical Fields of Some Simple Planar Geometries

a) Ring With or Without Arms

Figure 1 shows the variation of T_c, as a function of applied perpendicular field H, for a ring and for the same ring with an arm (lasso). Periodicity, as a function of magnetic flux through the ring, is conserved in presence of an arm but the piecewise structure of parabolas, corresponding to multiple determinations of the phase winding number around the ring, disappears.

b) Two Connected Loops

Consider now the case of two inequal loops connected in one point. The eignevalue equation is easily derived as

$$\sum_{p=1,2} \frac{\cos\gamma_p - \cos(L_p/\xi)}{\sin(L_p/\xi)} = 0$$

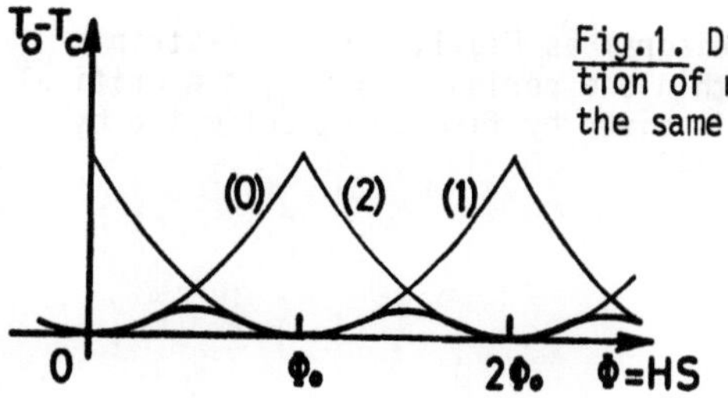

Fig.1. Diagram giving the depression of T_c as a function of magnetic flux, for a ring (thin curve) and for the same ring with an arm (thick curve) [4]

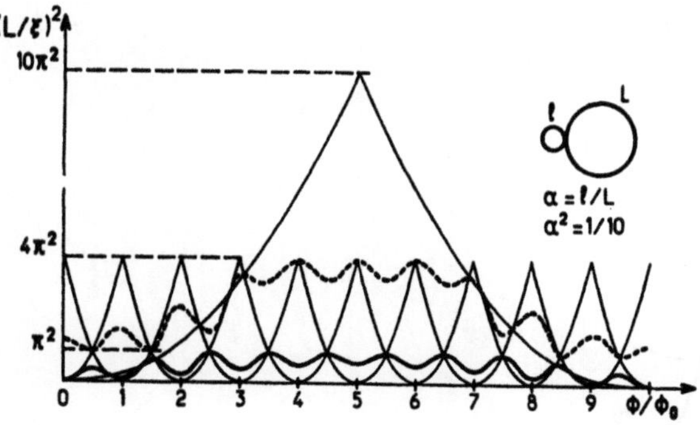

Fig.2. Same diagram as Fig.1 for a double-loop system [4]

where $\gamma_p = 2\pi\phi_p/\phi_0$, ϕ_p being the magnetic flux through loop p, of perimeter L_p. Figure 2 shows the variation of T_c for a double-loop system (areas have been chosen commensurate with a ratio of 10). It is clear that the small field behavior and the maximum depression of T_c are governed by the largest loop, whereas the dominant periodicity as a function of H is controlled by the smaller loop.

c) Strips and the Approach to the Infinite Lattice Limit

Let us recall that the initial variation of T_c, in small perpendicular field, is linear (as in bulk systems) for infinite square lattice, the maximum depression of T_c is determined by $\xi(T) \sim a$, where a is the lattice parameter. The linear variation $T_c - T_{c0} \sim H$ contrasts sharply with the quadratic departure in finite systems. One way to understand this transition between finite and infinite systems is provided by the study of n-strips (Fig.3). Let us consider the case of the two strips (n = 2). Such a 2-strip system is composed of three infinite wires labelled -1, 0 and 1 parallel to the y axis. They are located at x = -a, 0 and +a, with y = ±p. Choosing the gauge with $A_y = Hx$ and introducing the Fourier transform of $\psi_{p,\alpha}$ (α = -1,0,1) as $\psi_{p,\alpha} = \psi_\alpha(k)e^{ikp}$, $-\pi < k < \pi$, we obtain

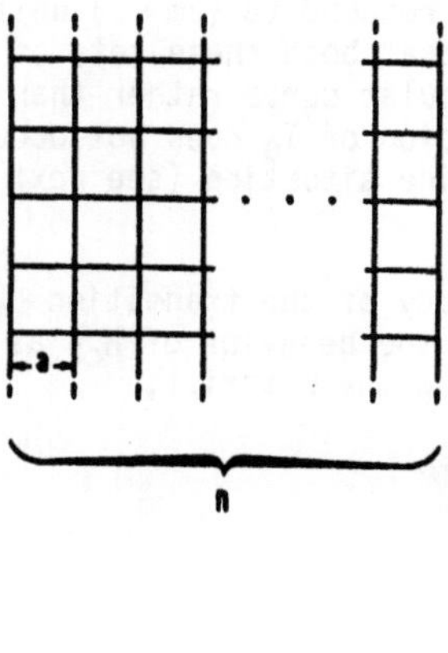

Fig.3. Definition of the n-strip geometry. The vertical extension of the wires is infinite [4]

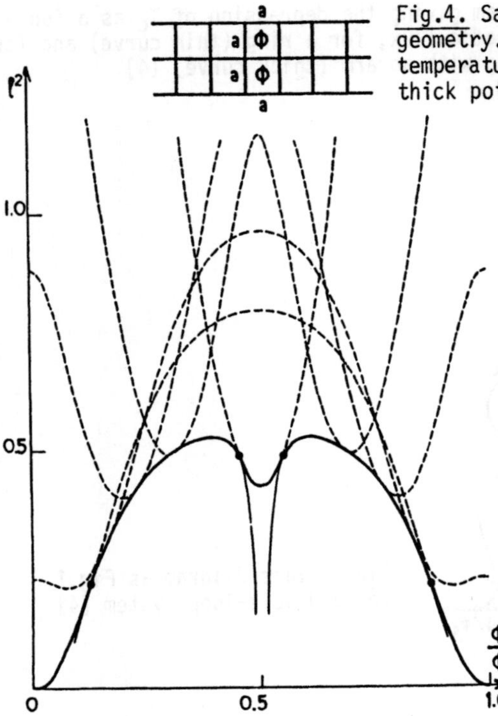

Fig.4. Same diagram as Fig.1, for the 2-strip geometry. Within one period $0<\phi<\phi_0$, the critical temperature is given by five arcs, delimited by thick points [4]

$$[-3\cos\ell+2\cos(k+\gamma)]\,\psi_{-1}(k)+\psi_0(k)=0$$

$$\psi_{-1}(k)+[-4\cos\ell+2\cos k]\,\psi_0(k)+\psi_1(k)=0$$

$$\psi_0(k)+[-3\cos\ell+2\cos(k-\gamma)]\,\psi_1(k)=0$$

where $\ell=a/\xi$ and $\gamma=2\pi(Ha^2)/\phi_0$.

These equations lead to a cubic eigenvalue equation for $\cos\ell$ which can be solved numerically. In Fig.4, $\ell^2(k,\gamma)$ is shown as a function of γ for selected values of k. It is clear from this figure that the nature of solution is more complex than for simple networks and that the band edge $\ell_c^2(\gamma)$ in particular has more structure. In the interval $0<\gamma<\overline{\gamma}_1$ and $\overline{\gamma}_2<\gamma<\pi$, it is identical respectively to the lowest branch of ℓ^2 $(k=0,\gamma)$ and of γ^2 $(k=\pi,\gamma)$, and for $\overline{\gamma}_1<\gamma<\overline{\gamma}_2$ it is the envelope of a family of curves. The positions of the junction points (thick points in Fig.4) are: $\overline{\gamma}_1/\pi\simeq0.25$, $\ell^2=0.216$ and $\overline{\gamma}_2/\pi\simeq0.90$, $\ell^2=0.488$.

The spectrum for $\pi<\gamma<2\pi$ can be obtained from that for $0<\gamma<\pi$ by reflexion about $\gamma=\pi$ and, of course, the whole spectrum is periodic in γ with period 2π. The minima in $\ell_c^2(\gamma)$ at $\gamma=2\pi m$ correspond to situations with m quanta of flux in each fundamental plaquette whereas the minima at $\gamma=(2m+1)\pi$ correspond to $(2m+1)$ units of flux in each pair of plaquettes across the strip. Note that both these sets of minima occur in regions where $\ell_c^2(\gamma)$ corresponds to a particular curve rather than to an envelope. Note also that the point of maximum depression of T_c does not occur at $\gamma=\pi$. This behavior is a precursor of the infinite lattice situation (see next section).

The n-strips provide a convenient laboratory for the study of the transition between finite and infinite systems. Let us consider first the behavior of H_{c2} as a function of ξ (for large ξ). For finite systems, including the n-strips,

$$H_c^2 \sim \phi_0^2/(\pi\xi^2 S) \quad ,$$

where S denotes the effective area. However, for infinite systems, $H_{c2} \sim \phi_0/\pi\xi^2$. The explicit calculation of S gives $S \sim n^2a^2$ (n>>1) for the n-strips. Thus, one can expect a crossover from finite to infinite behavior for $\xi \sim na$. Therefore, the calculation of H_c must be carried out through the full line $\ell_c^2(\gamma)$. In the limit $n=\infty$, $\ell_c^2(\gamma)$ is given by the band edge of the infinite lattice spectrum. The approach to this limit takes place in the following way: The curve $\ell_c^2(\ell)$ will always consist of sections corresponding to single solutions (pure section) of the network equations and sections corresponding to envelopes of many curves (envelope sections). There will be n envelope sections and n pure sections for $0<\gamma<2\pi$ (counting the pure sections around $\gamma=0$ and $\gamma=2\pi$ each as half a section). The section of $\ell_c^2(\gamma)$ for $\gamma=2\pi q/n$ for $q=0,1,\ldots,$ n will always be pure. Thus, in the limit $n=\infty$, there will be pure sections for each rational value of $\gamma/2\pi$ and envelope sections for each irrational value of $\gamma/2\pi$. This picture has been confirmed through a detailed numerical study of the spectrum for $2<n<13$.

1.3 Infinite Regular Networks

For a regular lattice, where $\ell_{\alpha\beta}=a$ is the same for all strands, (3) reduces to

$$(z\cos a/\xi)\psi_\alpha = \sum_\beta \psi_\beta\, e^{-i\gamma_{\alpha\beta}}$$

where z denotes the coordination of the lattice. Thus, the problem reduces to that of Landau levels of a tight-binding model in the same geometry. The energy of Bloch electrons is given by $\varepsilon = z\cos(a/\xi)$. Note that for superconducting networks one is only interested in the lowest eigenvalue.

a) Square Lattice (z = 4)

For a square lattice, there are no eigenvalues outside the range $-4<\varepsilon<4$ and the spectrum shows invariance properties in ε and $\gamma=2\pi\phi/\phi_0$, where $\phi=Ha^2$ is the magnetic flux through the elementary square loop of side a. The Landau level spectrum associated with this model is shown in Fig.5. The edge of the spectrum is a fractal continuous curve exhibiting self-similarity [7]. The genesis of this curve has been discussed in Sect.2. The variation of the lowest eigenvalue as a function of ϕ determines the effect of frustration induced by the magnetic field on the superconducting transition temperature. The frustration parameter is simply the reduced flux

Fig.5. Spectrum of a tight-binding model on a square lattice in a perpendicular magnetic field. Only rational values p/q of the field variable ϕ/ϕ_0 are represented with $q \leq 50$ [7]

ϕ/ϕ_0. Each rational $\phi/\phi_0 = p/q$ corresponds to an amount of frustration in the network giving rise to a particular periodic organization of vortices and currents. Physically, this corresponds to flux quantization phenomena through super unit cells, resulting in a commensurate structure at rational ϕ/ϕ_0. This flux quantization is at the origin of the reduction of T_c without any normal core. Vortices have no normal core and the state of the system is described by a suitable distribution of vortex cores in the holes of the network.

The above theoretical predictions have been checked experimentally on Al networks [8]. The samples were prepared by electron-beam evaporation of pure Al on a silicon substrate. The investigated sample was a square network of Al wires 2 μm wide of periodicity 6 μm, containing about 3 million identical square loops of medium area 36 μm^2. The overall size of the network is 1×1 cm^2. The network pattern was made by reactive-ion etching by optical lithography.

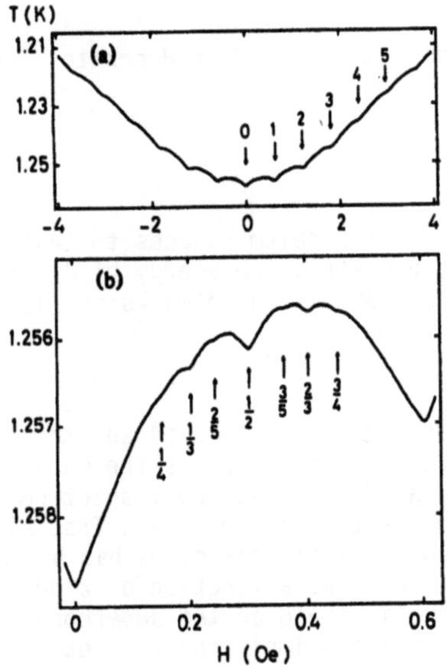

Fig.6a,b. a) Critical temperature vs. magnetic vield for a square network made of superconducting Al. The arrows indicate the magnetic field values corresponding to the integral number of flux quanta ϕ/ϕ_0 per unit cell of the network. b) Magnified view of the first period ($0 < \phi < \phi_0$) of the critical line. In addition to the fundamental dip at $\phi/\phi_0 = 1$, the secondary dips at $\phi/\phi_0 = 1/4$, 1/3, 2/5, 1/2, ..., are indicated by the corresponding arrows [8]

The field-dependent critical temperature $T_c(H)$, as obtained from very precise resistive measurements, is shown in Fig.6. We observe in Fig.6a well-defined oscillations superposed on a parabolic background. This parabolic envelope simply reflects the contribution of the bulk of the Al strands to the critical field, varying as $H_{c0}\ (\lambda/d) \sim (T_{c0} - T)^{1/2}$. Here H_{c0} denotes the thermodynamic critical field, λ is the penetration depth, d the width of the strands, and $T_{c0} = T_c\ (H = 0)$ the bulk critical temperature of Al at zero field. The modulations observed in Fig.6a are the results of a flux-quantization effect. We observed more than six periods for each polarity of the field. These oscillations are reminiscent of Little-Parks oscillations resulting from one-loop quantization phenomena. The magnetic period 0.606 Oe corresponds to one-flux quantum $\phi_0 = hc/2e = 2.07 \times 10^{-7}$ Gcm^2 per unit cell ($a = 6.0 \pm 0.1$ μm deduced from microscope observation).

Figure 6b is a magnification of the critical line discussed above. In addition to the single-loop effect, this plot reflects several features which are characteristic of the network behavior: finite slope dT_c/dH at $H = 0$, cusp at the half-integer values of the reduced flux ϕ/ϕ_0, and down concavity of the modulation. All these features reflect the collective behavior which distinguishes the network from the single loop from the point of view of flux quantization. In addition to these characteristic features, new structures of the critical line are shown in Fig.6.b. As can be seen, well defined dips are observed at $\phi/\phi_0 = 1/4$, 1/3, 2/5, 1/2, ..., in addition to the fundamental dip at $\phi = \phi_0$. These salient features, which are absent in the case of one loop correspond to new flux-quantization phenomena clearly observed with great details.

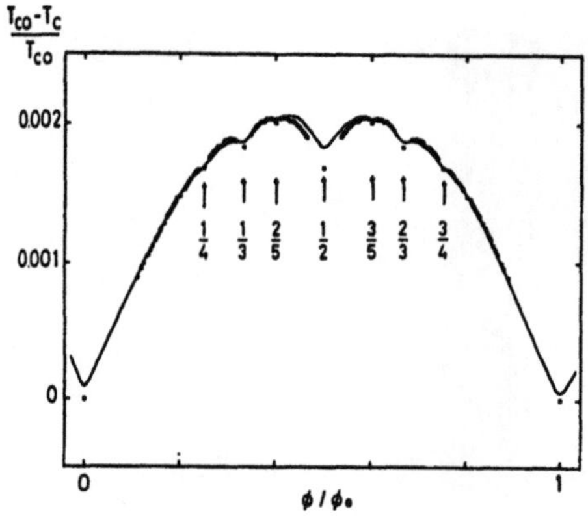

Fig.7. Solid curve, same experimental curve as in Fig.6, with reduced units, $\Delta T_c/T_{c0}$ vs ϕ/ϕ_0, where the parabolic background has been substrated, dotted curve, theoretical values calculated for rational $\phi/\phi_0 = p/q$ ($q < 30$). The best-fit parameters are $T_{c0} = 1.259$ K and $\xi(0) = 3050$ Å, respectively [8]

A close comparison between theory and experiment is illustrated in Fig.7. The continuous curve is the critical line (same data as in Fig.6b) plotted in reduced units $\Delta T_c/T_c$ vs. ϕ/ϕ_0, once the parabolic envelope ($T_{c0} - T_c = AH^2$) has been numerically subtracted. The theoretical values

$$\frac{T_{c0} - T_c}{T_{c0}} \equiv \frac{\Delta T_c}{T_{c0}} = \frac{\xi^2(0)}{a^2} \left(\text{Arcos} \frac{\varepsilon}{4}\right)^2$$

are shown in the same figure. The best fit is obtained from the values of the parameteres $T_{c0} = 1.259$ K, $\xi(0) = 3050$ Å and $A = 0.498$ K/Oe^2. The value of $\xi(0)$ so obtained is in good agreement with that expected for dirty Al, with $\xi(0) = 0.855(\xi_0 \ell)^{1/2}$, if one takes the mean free path $\ell = 800$ Å and the accepted value of the intrinsic coherence length $\xi_0 = 1.6$ µm for Al. Note that deviation from perfect regularity of the network is the main source of discrepancy between theory and experiment. This appears clearly for the fully frustrated network, i.e. at $\phi/\phi_0 = 1/2$.

b) Honeycomb Networks ($z = b$)

The first experimental work on superconducting networks has been done on honeycomb lattices [9]. In this case some additional features have been observed and this in comparison with the square network. The main differences come from very subtle features of the frustration phenomena in this geometry. We direct the reader to the relevant reference for a detailed exposition [10] where a quantitative comparison between square, triangular and honeycomb lattices will be found.

c) Sierpinski Gasket

Very recently, the superconducting to normal phase boundary for a Sierpinski gasket wire network has been measured [11]. The fractal phase boundary curve $T_c(H)$ so obtained was in excellent quantitative agreement with theoretical predictions [12]. In particular, the self-similarity of the gasket has been shown to manifest itself in up to four orders of dilational invariance in the experimentally determined phase boundary. All these features agree with the previous calculations performed in this fractal geometry.

The Al samples (Fig.8) are of tenth order and have linewidths of 0.3 - 0.4 µm. The elementary triangles which generate the network have an area of 1.38 ± 0.01 µm^2. In Fig.9, a dramatic series of oscillations of T_c with field is observed on top of the quadratic background. This complex behavior is reminiscent of the single-loop effect and the phase boundary seen in the square network.

A detailed inspection of the data at the smallest fields, however, reveals a great deal of structure over an exceptionally wide range of fields: $1/500 < \phi/\phi_0 < 5$. From one order n of the gasket (simple triangle at $n = 0$) to the next order ($n + 1$), there is a decrease by a factor of four in the enclosed area, corresponding to a dilation

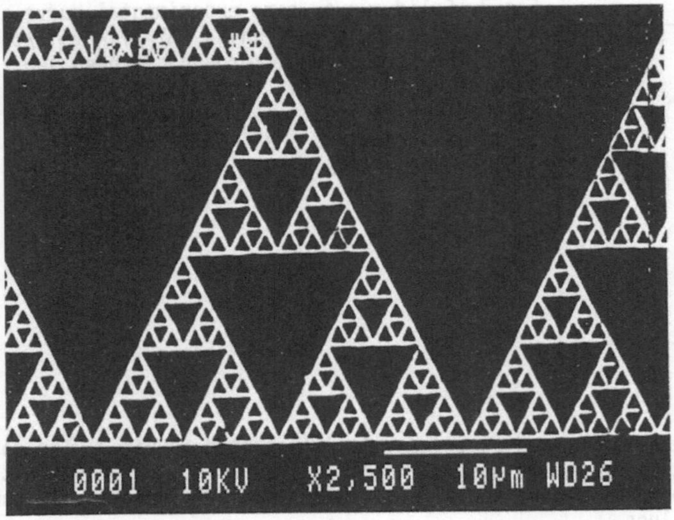

Fig.8. An electron micrograph of a fourth-order section of the Sierpinski gasket. The size bar represents a length of 10 µm [11]

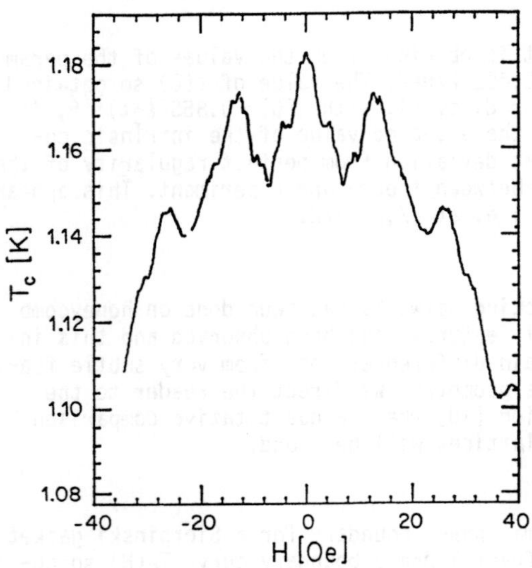

Fig.9. The superconducting normal phase boundary $T_c(H)$ of the Sierpinski gasket [11]

by a factor of four in the flux. In Fig.10b, the field axis has been expanded by a factor of four in order to highlight the structure in T_c at small fields. Two more orders of this self-similarity are demonstrated in Figs.10c and d.

Another new feature in fractal networks is provided by the non-linear dependence of T_c as a function of H. Indeed, in the limit of small fields, the envelope of the phase boundary should obey a power law

$$T_{c0} - T_c \sim H^{\bar{d}/\tilde{d}}$$

where $\bar{d}$ = fractal dimension, $\tilde{d}$ = spectral dimension. This is a simple generalization of the Euclidean network result, where $\bar{d} = \tilde{d} = d$. In the present case, $\bar{d} = \ln 3/\ln 2$ and $\tilde{d} = 2 \ln 3/\ln 5$, this leads to the value $\bar{d}/\tilde{d} = 1.161$ and must be compared with the result of the least squares analysis of the data: 1.17 ± 0.02.

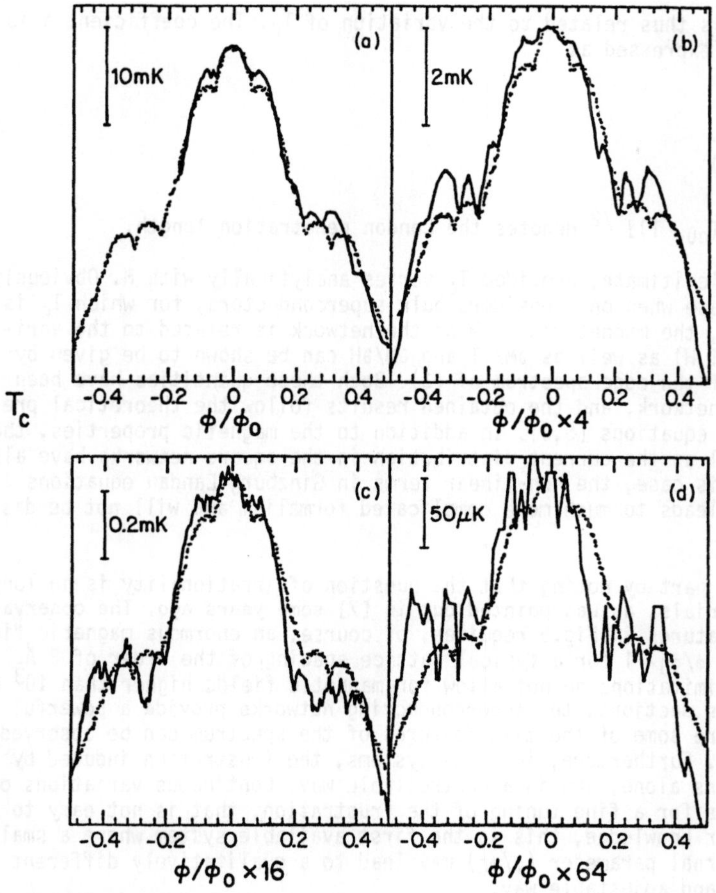

Fig.10a-d. a) $T_c(H)$ vs. ϕ/ϕ_0. The solid line shows experimental data and the points represent the theory for a second-order gasket scaled to fit the date at $\phi/\phi_0 = 1/2$. In b-d) the field axes have been expanded 4, 16 and 64 times, respectively, in order to highlight the small field structure in $T_c(H)$. The self-similar nature of the curve is made apparent by superimposing the second-order theory on each plot [11]

1.4 Other Physical Quantities

For finite systems, and for a small magnetic field H, the critical temperature T_c varies quadratically in H:

$$T_c = T_{c0} - \nu H^2$$

where ν is a constant.

Such a variation is fairly general and can be easily derived by recognizing that, in the vicinity of T_c, the thermodynamic potential G (actually the difference between the superconducting and normal potentials) has a dominant quadratic contribution (this is a second-order transition):

$$\Delta G \sim -\frac{a}{2}[T - T_c(H)]^2 \quad .$$

The magnetization M is obtained from

$$M = -\frac{\partial(\Delta G)}{\partial H}$$

and the susceptibility is thus related to the variation of T_c. The coefficient a is a constant which can be expressed as

$$a = \frac{\phi_0^2}{32\pi^2 T_0^2 \xi^2(0)\lambda^2(0)}$$

where $\lambda(T) = \lambda(0)\ [T_{c0}/(T_{c0} - T)]^{1/2}$ denotes the London penetration length.

This calculation is legitimate, provided T_c varies analytically with H. Obviously, this is no longer the case when one considers bulk superconductors, for which T_c is linear in in H. Besides, the magnetization M of the network is related to the variation of T_c. Indeed, M(T,H) as well as $\partial M/\partial T$ and $\partial M/\partial H$ can be shown to be given by different derivatives of the edge spectrum of (3). Both these quantities have been measured on the square network, and the obtained results follow the theoretical predictions of the network equations [5,6]. In addition to the magnetic properties, the critical current as well as the current distribution in the square networks have also been studied [6]. In this case, the non-linear terms in Ginzburg-Landau equations must be included. This leads to much more complicated formalism and will not be discussed here.

Let us conclude this part by noting that the question of irrationality is no longer artificial in real materials, as was pointed out in [7] some years ago. The observation of the prediced features in Fig.5 requires, of course, an enormous magnetic field of about 10^6 kOe to let $\phi/\phi_0 \sim 1$ for a typical lattice spacing of the order of 2 Å. Nowadays experimental limitations do not allow for magnetic fields higher than 10^3 kOe. As shown in the previous sections, the superconducting networks provide a powerful and unique example, where some of the main features of the spectrum can be observed in realistic conditions. Furthermore, in these systems, the frustration induced by the magnetic field enters alone, and in a controllable way. Continuous variations of the applied field allows for a fine tuning of the frustration, that is not easy to achive otherwise. To our knowledge, this is the first available system where a small modification of an external parameter (ϕ/ϕ_0) may lead to a qualitatively different state in a predictable and adjustable way.

2. TRANSPORT AND INTERFERENCE EFFECTS IN NORMAL-METAL NETWORKS

Recent advances in our understanding of electronic conduction have pointed out deficiencies in traditional thinking. For a metal at a sufficiently low temperature, it is known that the conventional picture in terms of the Boltzmann theory breaks down. Improved understanding of both electron localization and other related effects in a disordered medium has led to experimentally verifiable predictions. As a by-product of Anderson localization studies, it was shown by ALTSHULER, ARONOV and SPIVAK [13] (referred to as AAS hereafter) that the conductivity of a thin-walled cylinder made of dirty metal must be an oscillatory function of the magnetic flux ϕ which penetrates the cylinder. According to AAS, such oscillations of period $\phi_0 = hc/2e$ (the superconducting flux quantum), associated with twice (2e) the electronic charge, can occur in a system of non-interacting electrons that undergo multiple random elastic collisions, as they move along the trajectories within the cylindrical cavity.

In the following sections, we shall describe these puzzling phenomena in some detail. Section 2 will be devoted to a simple picture of weak localization and the physical origin of the AAS effect. In Sect.3, a general formalism is outlined for the calculation of the transport coefficients of a normal-metal network in the weak-localization regime. Simple circuits such as loops and ladders are used to illustrate this approach. Infinite regular networks are discussed in Sect.4, where a close comparison with experimental results is outlined.

2.1 Coherent Backscattering and the AAS Effect

The magnetoresistance (MR) oscillations predicted by AAS and observed later, are actually the manifestation of a specific and new phenomenon in disordered materials.

The physical explanation in terms of coherent backscattering has been advanced in the case of electronic transport [14]. However, the interference effect, obtained originally through an explicit diagram calculation, is really a very general phenomenon in systems with quenched disorder. Indeed, the basic origin must be traced to the amplification of the backscattering during the propagation of waves in randomly inhomogeneous media, where multiple scattering dominates. As long as $\lambda \ll \ell$ (λ is the wave length and ℓ is the mean free path), the first interference corrections to the wave-field energy-transport equation are controlled by the double-passage diagrams (also called fan diagrams). The presence of a magnetic field, which couples to the phase of a wave function, is therefore the most direct method to reveal the interference effects.

In order to be more specific, let us consider the electronic case. At low temperature one must distinguish between the two lifetimes, elastic (τ_0) and inelastic (τ_φ). In other words, τ_0 is the lifetime of the electron in a momentum eigenstate whereas τ_φ is that of an electron in energy eigenstate. At $T \to 0$, $\tau_\varphi \gg \tau_0$. Therefore, an electron in momentum state k can be scattered elastically several times before it suffers an inelastic scattering that destroys its phase coherence. So long as the phase coherence is maintained, elastically scattered waves interfere, giving rise to a backscattering wave. This latter effect is neglected in the usual Boltzmann equation. The interference of most of the amplitudes is not important since the lengths of the trajectories, and hence their phases, are very different, leading to a vanishing interference term. The exception are the trajectories which interesect themselves. Each such broken trajectory can be associated with two electronic states, corresponding to passages around the closed loop in opposite directions (double-passage effect). These two amplitudes are mutually coherent, so that their interference cannot be neglected. Therefore, the correction to conductance is given essentially by the probability of self-intersection for classical diffusion and then a small but finite correction is expected. Large corrections are expected at $d = 1$ and $d = 2$, because of the known behavior $P_0(t) \sim t^{-d/2}$ of the probability of return to the origin, during the time t, for a random walk in d-dimensions.

The above picture is the most important feature of the weak-localization theory of waves in disordered media. The correction to conductance in two dimensions can indeed be obtained as follows. Let ℓ be the elastic mean free path ($\ell = \tau_0 v_F$). This corresponds to a smearing of the Fermi sphere by π/ℓ. The corresponding available area in k-space is $2\pi k_F \cdot (\pi/\ell) = 2\pi^2 k_F/\ell$. We know that during the time t, the electron propagates a distance $\sim (Dt)^{1/2}$. Therefore, not only the state $-\vec{k}$ but all the neighboring states with momentum $-\vec{k}+\vec{q}$, such that $(Dt)^{1/2}.q < 1$, remain coherent. This corresponds to an area π/Dt in momentum space, i.e., the area of the backscattering "spot" shrinks with time as t^{-1} (more generally as $t^{-d/2}$ in d-dimensional spaces). Therefore, the portion of coherent backscattering is given by $(\pi/Dt)(2\pi^2 k_F/\ell) = \tau_0/\pi k_F t$.

The integration over the time interval from τ_0 to τ_φ leads to the conductance

$$G(L) = (ne^2\tau_0/m)\left(1 - \frac{1}{\pi k_F}\ell n \frac{\tau_\varphi}{\tau_0}\right)$$

in two dimensions. The calculation of the correction to conductance can be carried out similarly for other space dimensionalities. Note that at finite frequency ω or in the presence of a magnetic field, the same kind of arguments hold. In such cases, only the cutoff τ_φ must be modified.

The above argument gives a simple physical explanation for the AAS effect. According to quantum mechanics, the wave function and the energy spectrum of charged particles depend on the vector potential $\vec{A}$ of the applied magnetic field. In a thin-walled cylinder, the phase length of the trajectory (around the cylinder) is different for different alternative paths. However, the phase difference for all trajectories in a thin-walled cylinder is the same. This produces through the same arguments presented above oscillations of the sample resistance with the period $\phi_0 = hc/2e$.

This interference effect is expected to occur even when the electrons collide often with defects, i.e., when $\ell \ll L$ (L = perimeter of the cylinder). The low sensitivity to the degree of disorder and the period ϕ_0 of the oscillations distinguishes this phenomenon from Dingle oscillations, which could occur only if $\ell \gg L$ and with period $2\phi_0$. The condition for observing the AAS effect is $L_\varphi = (D\tau_\varphi)^{1/2} \gg L$, where L_φ denotes the diffusion length. The first experimental observation of the AAS effect was done by SHARVIN and SHARVIN [15] (Fig.11), using long hollow cylinders (length ~ 1 cm, diameter ~ 1 µm, thickness ~ 10^3 Å). Oscillations of MR were observed under longitudinal magnetic field at liquid helium temperatures. The original experiment of Sharvin and Sharvin has been repeated since by various groups and for different metals [16]. Today there are no serious doubts about the reality of this phenomenon, which appears as one of the most important concepts in disordered materials. Both theory and experiments have been extended to new geometries such as ladders, regular networks, etc.

In Fig.12 some of the obtained results are shown [17] for an Mg honeycomb network containing about 3×10^6 hexagonal elementary cells (of side $a \simeq 1.5$ µm) and this at various temperatures. In the low-field region, the magnetoresistance R(H) exhibits

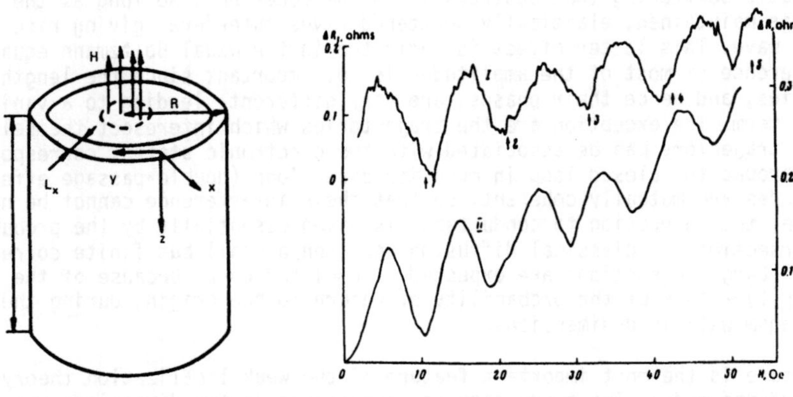

Fig.11. SHARVIN and SHARVIN's experiment showing the predicted AAS effect [15]

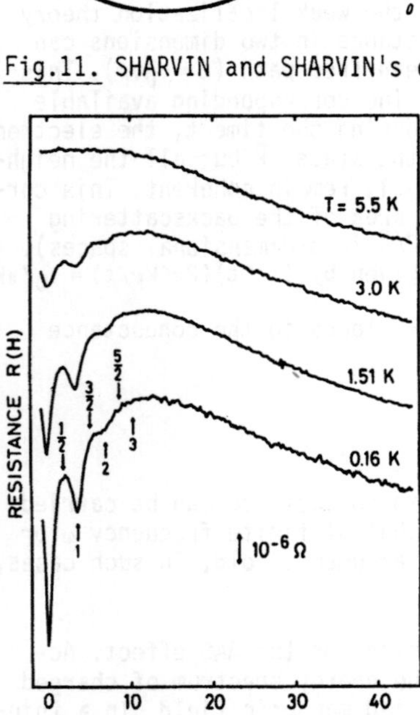

Fig.12. Magnetoresistance R(H) of a Mg honeycomb network. R(H = 0) at 0.16, 1.51, 3.0 and 5.5 K are 0.791369 Ω, 0.791363 Ω, 0.791370 Ω and 0.791434 Ω, respectively [17]

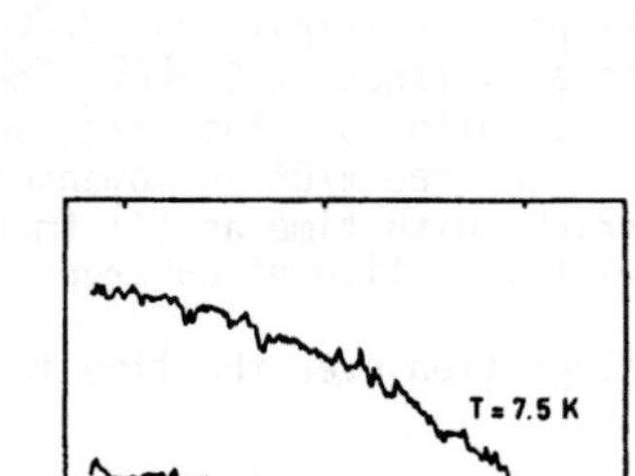

Fig.13. The same plot as in Fig.12, in the region T ⩾ 5 K [17]

oscillations with a period corresponding exactly to one flux quantum ϕ_0 per unit cell. In Fig.13, the same data are shown at high temperatures. It is important to notice that the scale of the change in the conductance is $\Delta R/R^2 \sim 1.3 \cdot 10^{-5}\,\Omega^{-1}$. This value must be compared to the conductance scale $e^2/h = 3.9 \cdot 10^{-5}\,\Omega^{-1}$ of the quantum correction associated with weak localization.

2.2 Network Formalism [18]

The localization correction to the conductivity in the weak-localization regime ($k_F\ell \gg 1$) is given in general by the following expression

$$\Delta\sigma(\vec{r}) = (-2/\pi\nu)\,\sigma_0 C(\vec{r},\vec{r}')$$

where σ_0 is the bulk conductivity of the sample, given by Drude's formula, and ν is the density of states at Fermi level. The equation for the Cooperon, $C(\vec{r},\vec{r}')$ in the presence of a magnetic field (vector potential $\vec{A}$) is

$$\left\{\left[-i\vec{\nabla}_r - \frac{2\pi}{\phi_0}\vec{A}(\vec{r})\right]^2 + L_\varphi^{-2}\right\} C(\vec{r},\vec{r}') = (1/\hbar D)\,\delta(\vec{r}-\vec{r}') \quad .$$

Here D denotes the electron diffusion coefficient and $L_\varphi = (D\tau_\varphi)^{1/2}$ is the length over which dephasing of the electron wave function results from inelastic processes.

As for superconducting networks, the calculation of the Cooperon at different nodes of a given network is provided by the following network equations:

$$\tilde{C}_\alpha \cdot \sum_\beta \coth(\ell_{\alpha\beta}/L_\varphi) - \sum_\beta \tilde{C}_\beta \cdot e^{-i\gamma_{\alpha\beta}}/\mathrm{sh}(\ell_{\alpha\beta}/L_\varphi) = (L_\varphi/\hbar DS)\,\delta_{\alpha,r'} \quad .$$

Here $\gamma_{\alpha\beta} = 2\pi/\phi_0 \int_\alpha^\beta \vec{A}\cdot d\vec{\ell}$, r' is the point where $\Delta\sigma$ is calculated and $\tilde{C}(\vec{r}) = C(\vec{r},\vec{r}')$ (S = cross-section of wires). The above equation was obtained using the wire approximation: $\ell \ll$ wire width $\ll L_\varphi$. Several differences with superconducting networks must be noticed at this level:

i) The source term at $\vec{r}'$, where $\Delta\sigma$ is calculated, acts as an additional node.
ii) $\Delta\sigma$ depends on the position r', and is very sensitive to the whole network structure.
iii) We have a diffusion-like problem instead of a propagation-like one.
iv) iL_φ plays the same role as ξ: the coherence length is an imaginary number.

These remarks, as well as other observations, show the basic differences between normal and superconducting networks. In principle, the above equations provide a program for the calculation of $\Delta\sigma(\vec{r}')$ at any point $\vec{r}'$ and for an arbitrary geometry. In what follows, we shall illustrate such calculations for simple geometries.

a) Single Loop

In this geometry, the source position $\vec{r}'$ is arbitrary and

$$\Delta R/R = \frac{\kappa}{2}\,\mathrm{sh}(L/L_\varphi)/[\mathrm{ch}(L/L_\varphi) - \cos(2\pi\phi/\phi_0)]$$

where $\kappa \equiv 2e^2 L_\varphi/\pi\hbar\sigma_0 S$.

The amplitude of oscillation is exponentially damped at $L \gg L_\varphi$. In the limit $\eta \equiv L/L_\varphi \ll 1$, one obtains $\Delta R/R = \kappa/\eta$ at zero field and the width of the maxima at integer ϕ/ϕ_0 is $\sim \eta$ (sharp maxima).

The presence of arms (Fig.14), usually used to measure the resistance of the ring, alters the amplitude of oscillations in a dramatic way. This effect becomes more and more interesting in the limit $L \ll L_\varphi < b$, where b is the arm length. In such a case the oscillations are still present because of $L \ll L_\varphi$, but the MR oscillations are damped $[\sim\exp(-b/L_\varphi)]$, when compared with the $b = 0$ case. Furthermore, the sharp maxima at integer values of ϕ/ϕ_0 are broadened by the presence of the arm.

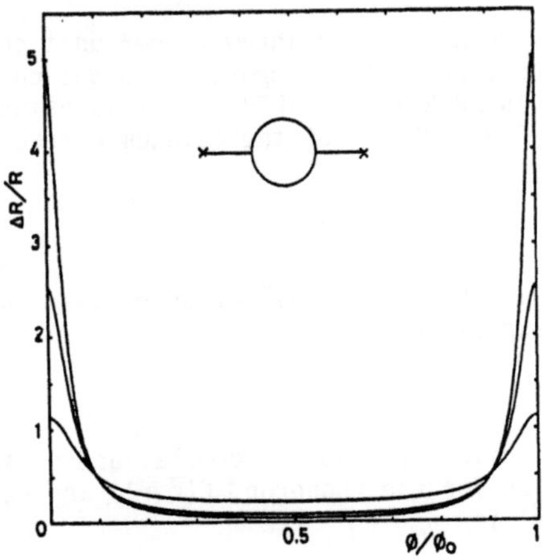

Fig.14. ΔR/R in unit of κ as a function of the reduced flux ϕ/ϕ_0. The length of the loop is kept fixed at $L/L_\varphi = 0.2$. The length of the arms is gradually increased: $b/L_\varphi = 0$, 0.5, 2.0 and this leads to a net damping of the resistance oscillations [18]

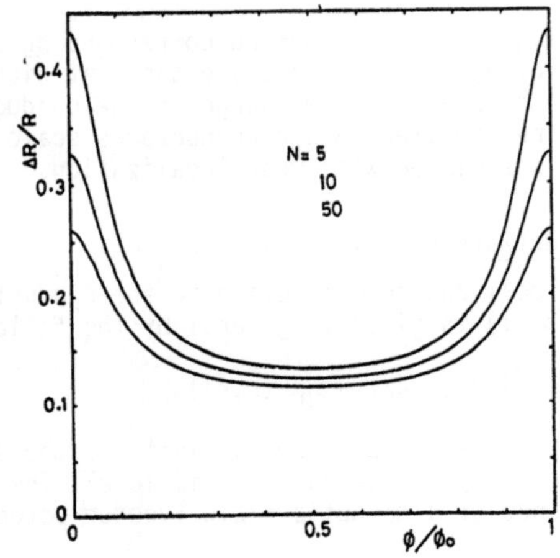

Fig.15. ΔR/R for a simple open ladder, with N elementary loops of side a each ($\eta = a/L_\varphi$). Here, $\eta = 0.2$ and N is set equal to 5, 10 and 50, respectively. Note that ΔR/R decreases as N increases as well as the absence of fine structure [18]

b) Ladders and Infinite Network

In Fig.15, ΔR/R for an open ladder is shown for different sizes N. Note in this geometry the absence of new features of the MR (no fine structure). In Fig.16 three network geometries are compared: single loop, simple ladder and infinite square lattice. Two comments relative to Fig.16 are in order. Firstly, the absence of new interference effects (e.g. secondary maxima at rational ϕ/ϕ_0) in the ladder, as well as in the infinite network, comes from the expression of ΔR/R itself. It is actually a whole integral information over the spectrum (Fig.5) and the period of oscillations is the same. Secondly, the amplitude of the oscillations is strongly influenced by the geometry of the considered network. This is clear at $\eta = a/L_\varphi \ll 1$, where for non integer ϕ/ϕ_0, $\Delta R/R \sim \eta$ in both cases. However, for integer ϕ/ϕ_0, ΔR/R is of the order η^{-1}, η^0 and $\eta \ln(1/\eta)$ respectively, in the single loop, ladder and infinite network.

It appears that the MR does not exhibit any fine structure in an infinite network geometry. ΔR/R is actually related to the Green's functions associated with the tight-binding Hamiltonian, on the underlying network, under an applied magnetic field. This

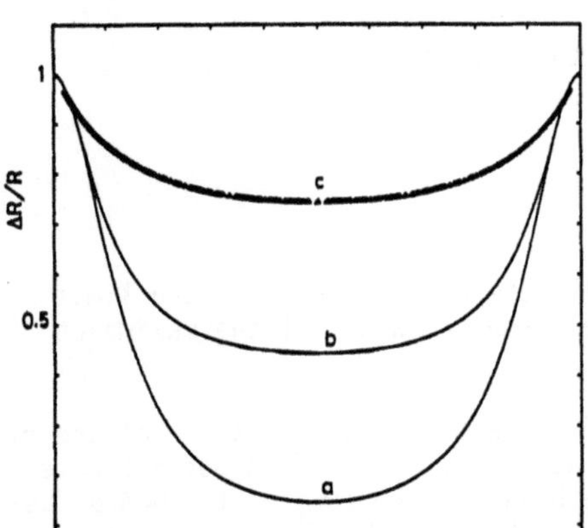

Fig.16. ΔR/R as a function of ϕ/ϕ_0 for a) simple square loop (L = 4a), b) simple ladder c) infinite square network [18]

makes a profound difference with superconducting networks, where $T_c(H)$ was given by the edge of the associated spectrum.

2.3 Infinite Honeycomb Lattices: Comparison with Experimental Data

Presently, a certain number of experiments has been performed on multiconnected networks. In order to make a close comparison with the weak-localization theory predictions, one should take into account three effects, which have been neglected in the previous section.

The first one is the finite width of wires, which is responsible for the damping of the MR oscillations at large field: the backscattering loops are no longer around a very well-defined flux.

The second one is the spin-orbit and spin-flip scatterings which are responsible for a change in both the sign and the magnitude of the MR oscillations. The third effect is given by electron-electron interactions. Indeed, the mutual influence between impurity scattering and screened Coulomb interactions of electrons is known to give important corrections to the conductivity at low temperatures. In the multiconnected networks considered here, the interaction effects on the Cooper channel have to be taken into account, at least for a high enough perpendicular magnetic field: $eH > k_BT/D$. However, in the available experiments, $L_T \equiv (\hbar D/k_BT)^{1/2}$ is very small when compared to L_φ. This allows a good distinction between weak localization effects at small fields and interaction effects at large fields. Therefore, one can assume that $L_T \ll L_\varphi$ and then neglect the interaction effects.

a) Finite Width of Wires

The finite width of wires can be taken into account, through a simple renormalization of the phase coherence length, which becomes a quadratic function of the applied magnetic field:

$$L_\varphi(H) = L_\varphi(H=0)\left[1+\frac{\pi^2}{3}\left(\frac{HwL_\varphi}{\phi_0}\right)^2\right]^{-1/2} \quad .$$

Here w denotes the finite width of wires. This expression is no longer valid when the magnetic flux through the area of a network node ($\sim w^2$) becomes of order ϕ_0. There the MR becomes sensitive to the detailed shape of the nodes and the above approach must be modified. But, this can occur at very high fields and the above expression breaks down before such fields are reached. Actually, this expression is a good approximation as long as $HwL_\varphi/\phi_0 \lesssim 1$.

b) Spin-Orbit and Spin-Flip Scattering

The presence of these processes can be taken into account by separating the singlet and triplet contributions to the Cooperon. The spin-orbit interaction affects only the triplet one. Formally, L_φ has to be replaced by a combinations of L_φ, L_s and L_{so}:

$$L_{triplet} = \left[L_\varphi^{-2} + \frac{4}{3}L_{so}^{-2} + \frac{2}{3}L_s^{-2}\right]^{-1/2}$$

$$L_{singlet} = \left[L_\varphi^{-2} + 2L_s^{-2}\right]^{-1/2}$$

The resulting Cooperon is then given by

$$C = -\frac{1}{2}C_{singlet} + \frac{3}{2}C_{triplet}$$

where $C_{singlet}$ and $C_{triplet}$ are obtained as before by replacing L_φ by $L_{singlet}$ and $L_{triplet}$, respectively.

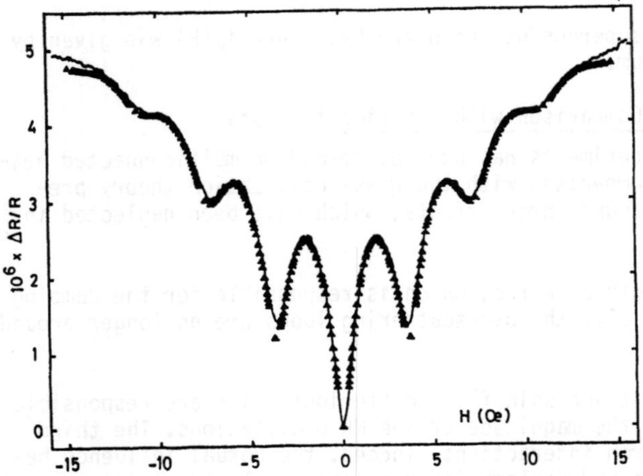

Fig.17. Quantitative comparison between the theoretical results (triangles) and experimental data (solid line), for Cu at T = 133 mK. The hexagonal elementary cells (side a = 1.5 μm) are made of wires of width w = 0.42 μm. In this fit, we have L_φ = 5.36 μm and L_{so} = 3.12 μm, respectively (L_{so} is the spin-orbit length) [18]

A quantitative comparison between the theoretical results and experimental data is shown in Fig.17. The solid line corresponds to Cu, at T = 133 mK, in a honeycomb network geometry. Here the spin-orbit coupling has been taken into account. In this situation the field renormalization of $\eta = a/L_\varphi$ gives

$$\eta^2(H) = \eta^2(H=0) + \frac{4}{81}\pi^2 (\phi/\phi_0)^2 (w/a)^2 \quad .$$

This low-field approximation breaks down at $Haw \gtrsim \phi_0$. The renormalization of η becomes important at large H and is actually responsible for the damping of oscillations.

Other experimental data obtained for different metals, on square and honeycomb networks, have also been analyzed following the same scheme [19]. This allowed in particular the calculation of the temperature variation of L_φ as well as the influence of Kondo-like impurities. The observed deviations, appearing at large H, have also been analyzed in detail.

2.4 Discussion and Other Interference Phenomena

As can be seen, the approach used here gives a set of reliable results for the network geometries in general. In this approach, all traces of randomness are summarized in the phase coherence length L_φ, and the theory is an averaged one. In this respect, fluctuations from sample to sample are neglected and $\Delta\sigma(\vec{r})$ refers actually to an averaged correction to the conductivity. However, recent theoretical and experimental works [20-24], done on single rings and 1D ladders, suggest the existence of universal fluctuations of the magnetoresistance. Some interesting results have been obtained for the resistance probability distribution, such as the second moment, most probable value, etc. Because of space limitations, this topic will not be discussed here in sufficient detail (see next part).

In addition to the self-averaging question, the possibility of the coexistence of two harmonics (ϕ_0 and $2\phi_0$) has been raised recently in very small systems. For instance, MR oscillations with the expected period ϕ_0 have been observed very recently on thin-film rings (1 - 2 μm in diameter). This was the first observation in single rings of oscillations with flux period ϕ_0 at low fields, and this after various attempts in different laboratories. Small periodic oscillations (period $2\phi_0$) have also been observed at higher fields. These new oscillations are the signature of the standard Aharonov-Bohm effect [24] and average out to zero (by central limit theorem), when $\Delta R/R$ is measured on a large number of rings arranged in series. In such a configuration, only the component ϕ_0 survives and this is in perfect agreement with the weak-localization theory.

3. FLUCTUATIONS OF TRANSPORT COEFFICIENTS IN SMALL SYSTEMS

Interference effects involving the flux $2\phi_0 = hc/e$ (Aharonov-Bohm effect) have been observed a long time ago [25] in a two-slit interference experiment involving coherent beams of electrons. MR oscillations in single-crystal whiskers of bismuth periodic in $2\phi_0$ have also been reported at low magnetic fields [26].

Working on very small systems (rings of 3000 Å diameter and 500 Å width) the IBM group [23] observed a sample-specific, aperiodic, reproducible structure in the MR, which was two orders of magnitude larger than the AAS effect and which persisted without decrease in amplitude out to 8 Tesla. The same structure was also seen in wires, showing that, unlike either the AAS effect or the usual Aharonov-Bohm effect, it had nothing to do with the system topology. Furthermore, an asymmetry in the structure was observed when the magnetic field direction was reversed. Numerical simulations [27] of the MR of very small wires revealed the same features observed in the IBM group experiments. This was an explanation of the earlier failures of this group to observe the AAS effect in very small rings. Furthermore, the numerical simulations revealed the non self-averaging nature of quasi-one-dimensional conductors in this size range. Some time later, the AAS effect has been observed on rings of the same size by another group [22]. The issue of this puzzling situation was found very recently. The keywords are: "universal conductance fluctuations" and "non-self averaging" of the conductance of samples of linear size below L_φ (= phase coherence length). Note that $L_\varphi \sim \mu m$ at $T < 1$ K.

3.1 Universal Conductance Fluctuations

Theoretically, if one considers an ensemble of small wires in, say the $(0.1 \times 0.1 \times 1)\ \mu m^3$ size range, the conductance will fluctuate among members of the ensemble. Measuring the conductance g, in units of e^2/h, the following extremely interesting result [20] on the basis of the weak localization theory (weak disorder) has been shown:

$$\langle g^2 \rangle - \langle g \rangle^2 = \text{const}$$

where the angular brackets signify ensemble averaging and the constant is a number of order unity, universal in the sense that it is independent of both disorder and sample size, but it depends very weakly on sample shape, assuming somewhat different (but well-defined) values for squares, cubes and narrow thin wires.

This result is very different from the typical inverse square-of-the-size dependence of the relative fluctuation in classical inhomogeneous conduction. The physical reason for the quantum fluctuations is that below L_φ interference effects are important and that the interference pattern is sensitive to the specific arrangement of the disorder in a given sample. As will be seen below, below a typical temperature (or length scale), the theory becomes non-self-averaging and fluctuations must be taken into account.

This implies that g should be sensitive to a magnetic field (intensity and orientation). This is well illustrated in Fig.18, where the MR aperiodic structure of narrow Si accumulation layers is shown as a function of the magnetic field orientation [28]. The scale of changes of B giving rise to these dependences is very easy to understand in terms of the changes needed to modify the relevant phase differences across the sample by a factor of $\sim 2\pi$. This corresponds to a flux change through the sample of the order of $2\phi_0 = hc/e$.

Likewise, the scale of changes of the Fermi energy, E_F, is given in the electronic diffusion regime by $\sim \hbar D/L^2$, L being the length of the sample.

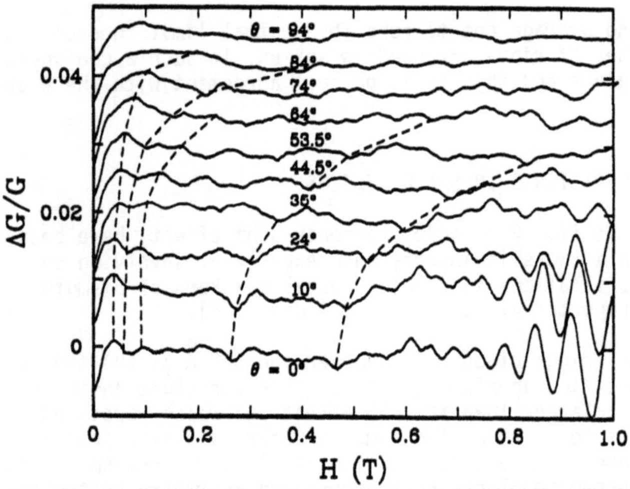

Fig.18. Change in the MR of Si accumulation layer for various values of the angle θ between the magnetic field axis and the normal vector to the sample surface ($T = 0.47$ K ± 0.02 K)

Therefore, a measurement for a specific sample will produce sample-specific, reproducible variations with a universal amplitude of the conductance as a function of H and/or E_F, on the above-mentioned scales. Such measurements should be performed, of course, at temperatures low enough so that both $L_\varphi \gg L$ and $k_BT \ll \hbar DL^2$, i.e., $L \lesssim L_T = (\hbar D/k_BT)^{1/2}$.

The above picture shows that the conductance measurements provide a "spectroscopy" of the sample, which at this length scale looks like a big molecule. The "spectroscopy" of such objects leads naturally to the consideration of random matrices, in conjunction with the multichannel Landauer formula. An interesting analysis of this scattering problem by big objects will be found in [21].

3.2 Periodicity in ϕ : ϕ_0 or $2\phi_0$?

Following [21], let us consider a metal ring in a normal magnetic field. For a given sample, the conductance g can be written as

$$g(E,H) = g_b(E) + g_0(E,H) + \sum_{n=1}^{\infty} g_n(E,H)\cos(2\pi n\phi/2\phi_0 + \gamma_n(E,H))$$

where E denotes the energy of electrons, $\phi = R_0^2H$ is the flux through the hole (R_0 = inner radius of the ring), $\phi_0 = hc/2e$ and g_b is the background Ohmic conductance. In the AAS effect, the odd terms in the Fourier sum ensemble average to zero, whereas the even terms do not. However, numerical simulations suggest that for a ring with a good aspect ratio, g_0 and g_n will vary aperiodically with H, on a field scale, $H_1 \sim \phi_0/S_a$, where S_a is the area of the annulus. This scale is very large when compared to the hc/e oscillation, so that a separation of rapid variation can be done.

Supported by numerical simulations, the following hypothesis seems to be valid. The phases $\gamma_n(E)$ and amplitudes $g_n(E)$ depend on E in a stochastic manner and become uncorrelated when separated by more than an energy correlation range, E_c. This implies that the conductacne g(T) as a function of temerpature is a sum over energies within kT of the Fermi level. Thus, above a crossover temperature $k_BT_c \sim E_c$, g(T) is given by the average of k_BT/E_c uncorrelated $T = 0$ conductance patterns.

Then, if energy averaging is equivalent to an averaging over impurity configurations at $T = 0$, raising the temperature in a single ring will cause a slow transition from $2\phi_0$ periodicity to the AAS ϕ_0 effect.

How to determine E_C? By analogy to optics, E_C may be defined as the energy change necessary so that two electrons which follow the same path of length $\mathcal{L}$ across the sample acquire a phase difference $\sim 2\pi$, i.e.,

$$\frac{\mathcal{L}}{\lambda(E+E_C)} - \frac{\mathcal{L}}{\lambda(E)} = \mathcal{L}\cdot E_C \frac{d(1/\lambda)}{dE} = \frac{\mathcal{L}}{2\pi}\frac{dk}{dE} \sim 1$$

where λ is the electron wavelength. For a metallic diffusion constant D, the time to diffuse the sample length is L^2/D, so $\mathcal{L} \simeq v_F L^2/D$ and thus

$$E_C \simeq \hbar v_F/\mathcal{L} \simeq \hbar D/L^2 \quad .$$

The condition $T_C \sim E_C/k_B$ is equivalent to $L \simeq L_T = (\hbar D/k_B T)^{1/2}$.

From the above analysis, the following picture arises. Because of the different sensitivities of the $2\phi_0$ oscillations and the AAS ϕ_0 effect to T and H, a single normal metal ring exhibits a variety of flux sensitive effects. At $T < T_C$, the dominant periodicity is $2\phi_0$. At $T \gg T_C$ but $L_\varphi \gg 4\pi R_0$, the AAS effect will dominate at low fields, but will give way to weaker $2\phi_0$ oscillations at $H \gg H_1$. Finally, dynamic phase incoherence (e.g. thermal fluctuations) will destroy both ϕ_0 and $2\phi_0$ oscillations and this occurs at high temperatures, because they both require phase coherence around the ring.

References

1. G. Toulouse: In Heidelberg Colloquium on Spin Glasses, Lecture Notes Phys., Vol.192, ed. by J.L. van Hemmen, I. Morgenstern (Springer, Berlin, Heidelberg 1983) p.1
2. S. Alexander: Phys. Rev. B27, 1541 (1983); P.G. de Gennes: C.R. Hebd. Sean Acad. Sci. Ser. B292, 9, 279 (1981)
3. R. Rammal, T.C. Lubensky, G. Toulouse: J. Phys. Paris Lett. 44, L65 (1983)
4. R. Rammal T.C. Lubensky, G. Toulouse: Phys. Rev. B27, 2820 (1983)
5. P. Gandit, B. Pannetier, Y.Y. Wang: to be published
6. Y.Y. Wang, B. Pannetier, R. Rammal: European Phys. Soc., Stockholm, March 1986
7. D. Hofstadter: Phys. Rev. B14, 2239 (1976);
. F.H. Claro, G.H. Wannier: Phys. Rev. B19, 6068 (1979)
8. B. Pannetier, J. Chaussy, R. Rammal, J.C. Villegier: Phys. Rev. Lett. 53, 1845 (1984)
9. B. Pannetier, J. Chaussy, R. Rammal: J. Phys. Paris Lett. 44, L853 (1983)
10. R. Rammal: J. Phys. Paris 46, 1345 (1985)
11. J.M. Gordon, A.M. Goldman, J. Maps, D. Costello, R. Tiberio, B. Whitehed: Phys. Rev. Lett. (1986) (in press)
12. R. Rammal, G. Toulouse: Phys. Rev. Lett. 49, 1194 (1982)
13. B.L. Altshuler, R.G. Aronov, B.Z. Spivak: JETP Lett. 33, 94 (1981)
14. G. Bergmann: Phys. Rep. 107, 1 (1984)
15. D.Y. Sharvin, Y.V. Sharvin: JETP Lett. 34, 272 (1981)
16. F.R. Ladan, J. Maurer: C.R. Acad. Sci. Paris 297, 227 (1983);
Y.V. Sharvin: Physica 126B, 288 (1984);
M. Gijs, C. van Maesendonck, Y. Bruynseraede: Phys. Rev. Lett. 52, 2069 (1984)
17. B. Pannetier, J. Chaussy, R. Rammal, P. Gandit: Phys. Rev. Lett. 53, 718 (1984);
Phys. Rev. B31, 3209 (1985);
see also D.J. Bishop, J.C. Licini, G.J. Dolan: Appl. Phys. Lett. 46, 1000 (1985)
18. B. Doucot, R. Rammal: Phys. Rev. Lett. 55, 1148 (1985); J. Phys. Paris (1986) (in press)
19. B. Doucot, B. Pannetier: in preparation
20. B.L. Altshuler: JETP Lett. 41, 648 (1985);
P.A. Lee, A.D. Stone: Phys. Rev. Lett. 55, 1622 (1985)
21. A.D. Stone, Y. Imry: Phys. Rev. Lett. 56, 189 (1986);
Y. Imry: Europhys. Lett. 1, 249 (1986) and references therein;
B. Doucot, R. Rammal: to be published (1986)

22. V. Chandrasekhar, M. Rooks, S. Wind, D. Prober. Phys. Rev. Lett. 55, 1610 (1985)
23. C.P. Umbach, C. van Haesendonk, R.B. Laibowitz, S. Washburn, R.A. Webb: Phys. Rev. Lett. 56, 386 (1986)
24. S. Datta, M.R. Melloch, S. Bandyopadhyay, R. Noren, M. Vaziri, M. Miller, R. Reifenberger: Phys. Rev. Lett. 55, 2344 (1985)
25. R.B. Chambers: Phys. Rev. Lett. 5, 3 (1960);
A. Tonomura et al.: Phys. Rev. Lett. 48, 1443 (1982)
26. N.B. Brandt et al.: Sov. J. Low Temp. Phys. 8, 358 (1982) and references cited therein
27. A.D. Stone: Phys. Rev. Lett. 54, 2692 (1985)
28. S.B. Kaplan, A. Hartstein: to be published

2D Localisation and Interaction Effects in Semiconductor Structures

M. Pepper

Cavendish Laboratory, Cambridge University, Cambridge, United Kingdom and GEC Research Limited, Hirst Research Centre, Wembley, Middlesex, HA9 7PP, United Kingdom

Semiconductor structures allow a comprehensive range of experiments into the role of disorder on electronic transport /1,2/. Structures containing a 2D electron gas possess the advantage of having a variable carrier concentration, whereas in 3D systems the role of disorder is changed in situ by the application of compression or shrinkage of wavefunction by a magnetic field. The principal structures in this type of work are the inversion layer of the silicon MOSFET, heterojunctions in III-V semiconductors (principally GaAs-AlGaAs) and the GaAs MESFET. The reader is referred to the Review article by ANDO et al /1/ for further details.

We can divide the effects of disorder into two regimes, strong localisation and the so called 'weak' localisation or quantum interference. We will now consider the conduction processes in these regimes.

STRONG LOCALISATION

Here the wavefunction decays with distance r as $\exp-\alpha r$ in the manner first predicted by ANDERSON /3,4/. At low temperatures, when the Fermi energy, E_F, is in the localised states, conduction proceeds by variable range hopping (often termed Mott hopping) /5/. This process combines both tunnelling and thermal activation and the temperature dependence of the conductivity, σ, is determined by the value of $\exp-[2\alpha R+W/kT]$, where R is the hopping distance, $1/\alpha$ is the decay constant of the wavefunction and W is the hopping energy required. When $W \ll kT$ the electron tunnels an increasing distance, R, in order to minimise W. The result is

$$\sigma = \text{constant.} \exp - (C/T)^{1/3}$$

where $C = (3\alpha^2/N(E_F)k)^{1/3}$, here $N(E_F)$ is the density of states at the Fermi energy and k is Boltzmann's constant; this law has been observed in many 2D systems /4/ and an example obtained with a Silicon inversion layer is shown in Fig. 1.

In 3D the $T^{-1/3}$ is replaced by $T^{-1/4}$, reflecting the greater degree of freedom for hopping. Increasing the temperature results in an increased number of carriers being excited to extended states. These carriers dominate the conductivity and $\sigma = \sigma_0 \exp-(W/kT)$, where $W = (E_F-E_C)$, E_C being the energy at which the extended states commence, i.e. the mobility edge.

Theoretical work has established that true extended states do not exist in 2D /6,7,8,9/. It is suggested that states either decay exponentially with distance (as in the strong localisation regime) or as a power of distance. However, at finite temperatures the inelastic scattering length, L_{IN}, cuts off localisation as a localisation length L_0 greater than L_{IN} cannot be observed. This results in an energy which separates localised from extended states, although as the temperature decreases the 'mobility edge' will increase in energy. On the other hand the model of power law localised states, /9,10/, predicts a fixed localisation edge separating exponentially localised from power law localised states. The production of quasi extended states by inelastic scattering results in σ_0 having the value of the 2D minimum metallic conductivity, although at absolute zero the quantity will not exist.

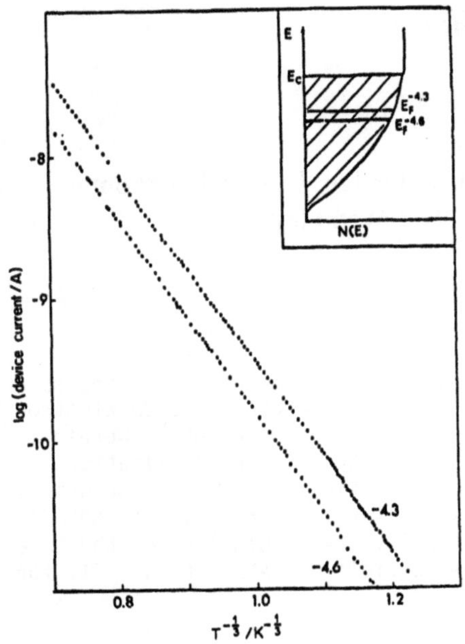

Fig. 1. Measurements of low temperature conductance of a silicon inversion layer showing variable range hopping at low temperatures in the strong localisation regime /4/. The gate voltages are indicated

WEAK LOCALISATION

At the temperatures of experiments $>0.05K$ the localisation of all states in 2D has not been observed, however it is possible to observe the quantum interference which is the precursor to the localisation /11,12/. As in the case of metal rings (discussed elsewhere in this volume) quantum interference proceeds between wavepackets scattered around the same loop but in different directions. However, unlike the ring there is no defined loop size, all values being possible between the elastic scattering length l and L_{IN}. Quantum interference reduces the conductivity which is given by

$$\sigma = \sigma_B - \frac{\alpha e^2}{\pi^2 \hbar} \ \mathrm{Ln}\ (L_{IN}/l) \tag{1}$$

σ_B is the Boltzmann conductivity given by $ne^2\tau/m^*$, n,τ and m^* being respectively the carrier concentration, scattering time and effective mass; α is a constant near unity in value. This quantum correction to the conductivity produces a negative magneto-resistance, /13,14/ as, in a similar way to the ring situation, the magnetic field introduces a phase difference between the different directions around the loop so reducing the interference and enhancing the conductivity.

In equation (1) as L_{IN} varies at T^{-x}, the correction becomes logarithmic in temperature, an example is shown in Fig. 2, the value of x depending on the nature of the inelastic scattering. In semiconductor structures at low temperature this is electron-electron scattering /15,16/. In the absence of disorder, the scattering time τ_{ee} is given by the Landau-Baber formula

$$\tau_{ee}^{-1} = \frac{Ck^2}{\hbar E_F} \tag{2}$$

C is a numerical constant in the range 1-10 and k is Boltzmann's constant. The presence of disorder, i.e. scattering, leads to a disc of momentum indeterminacy around each state in k space of radius $q = (\hbar D/kT)^{1/2}$, where D is the diffusion coefficient of an electron. Consequently a rapid rate of scattering can occur if the momentum change δk is less than q. This results in a scattering rate

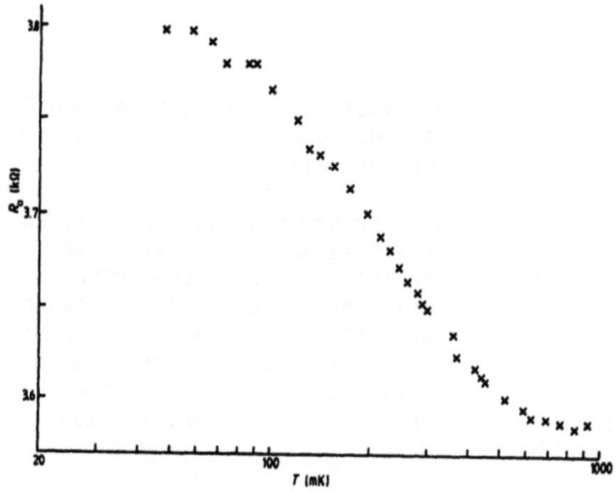

Fig. 2. Sheet resistance versus temperature obtained with a silicon inversion layer at low temperature, /23/, showing the logarithmic correction

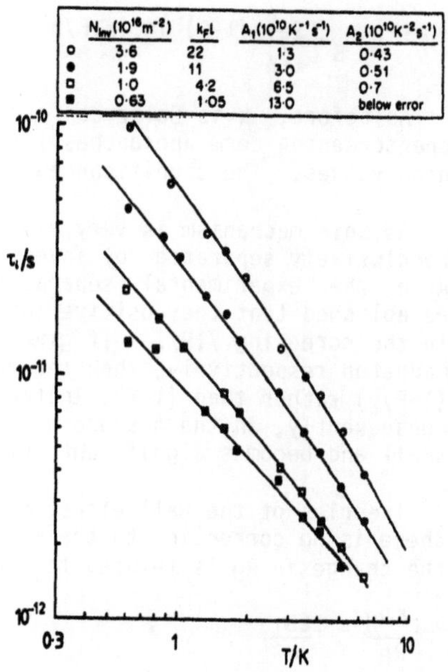

Fig. 3. The inelastic electron-electron scattering time is plotted against temperature, the lines are fitted to $\tau_{ee}^{-1} = A_1T + A_2T^2$ for the stated values of parameters /17/

$$\tau_{ee}^{-1} = 2kT/\hbar k_F l \tag{3}$$

k_F is the Fermi wave vector.

The net rate of scattering is given by

$$\tau_{ee}^{-1} = \frac{2kT}{\hbar k_F l} + \frac{Ck^2}{\hbar E_F} \tag{4}$$

Recent theories /16/, have suggested that small correction terms exist giving a slightly modified rate of electron-electron scattering to that predicted by equation /4/. This might account for small discrepancies between experiment and theory /17/.

The inelastic scattering time can be obtained from the negative magneto-resistance using the formula of HIKAMI et al /13/. The change in the conductivity in a field B, ($\Delta\sigma(B)$), is given by

$$\Delta\sigma(B) = \frac{\alpha e^2}{2\pi^2\hbar}\left[\psi\left(\tfrac{1}{2} + 1/a\,\tau_{e-e}\right) - \psi\left(\tfrac{1}{2} + 1/a\tau\right)\right] \tag{5}$$

ψ is the Digamma function, τ is the elastic scattering time and $a = 4eBD/\hbar$. At low values of B, $\Delta\sigma(B)$ varies as B^2 and as LnB at higher values when the cyclotron length $L_C \ll L_{IN}$. The obtained times are fitted to the relation $\tau_{ee}^{-1} = A_1T + A_2T^2$ and the coefficients A_1 and A_2 extracted. Fig. 3 shows the good agreement between experiment and theory for a silicon inversion layer, obtained from reference 17.

INTERACTION EFFECTS

Corrections to the 2D density of states due to the electron-electron interaction /18/ lead to a logarithmic dependence of σ on temperature

$$\sigma = \sigma_B - \frac{2e^2}{\pi^2\hbar}(1-F)\ln(kT\tau/\hbar) \quad (6)$$

As before, k is Boltzmann's constant, τ is the elastic scattering time and F the screening term approaches 1 for low values of carrier concentration and 0 for high values. The significance of F is discussed elsewhere /12/.

As this mechanism is very similar to quantum interference the two can only be conclusively separated by investigating the magneto-resistance and Hall effect. After the experimental separation of these mechanisms it was theoretically established that the positive interaction magneto-resistance results from a change in the screening /19/. If $g\beta B \gg kT$, where g and β are the spin g value and Bohr magneton respectively, then the temperature dependence of σ becomes determined by $(1-F/2)$ rather than $(1-F)$, In the silicon inversion layer the value of F is ≈ 0.9, consequently, in the absence of a magnetic field the interaction contribution is small and becomes significant only when $g\beta B \gg kT$.

Theories of the Hall effect suggest that in the regime of quantum interference there is no correction to the Hall constant, R_H whereas in the interaction regime the change in R_H is related to the change in conductivity /20/ by

$$\frac{\delta R_H}{R_H} = -\frac{2\delta\sigma}{\sigma} \quad (7)$$

here the conductivity is σ_{xx} and this relation arises because there is no interaction correction to the Hall conductivity σ_{xy}. The interaction Hall effect can be most conveniently compared to the conductivity when the quantum interference is suppressed by a magnetic field /21,22,23/. Results in excellent agreement with theory are shown in Fig. 4. The ratio of 2 is found for $k_F l \gg 1$ and when k_F approaches unity, and the transition to strong localisation, the value of 2 is no longer predicted or found.

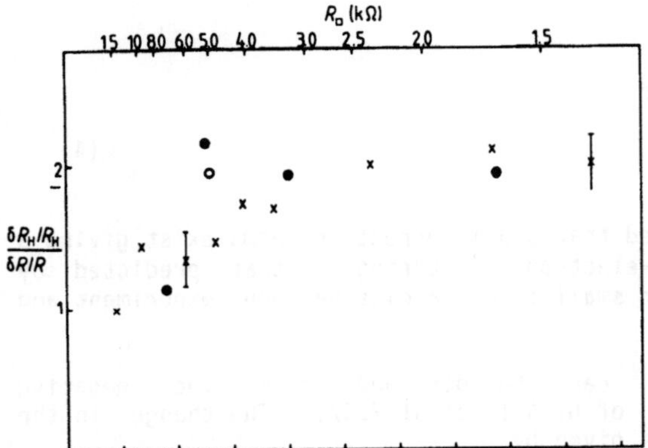

Fig. 4. The ratio of the correction to the Hall constant, R_H, to that in resistance, R, is plotted as a function of $k_F l$, o indicates as a function of temperature, • and x as a function of field for different devices /21/

An experimentally convenient method of observing logarithmic corrections is by increasing the electric field, E, and heating the electron gas to a temperature T_E /24/. It is found that T_E varies as E^x where x is between 1/2 and 1/3. In this way, by initially cooling the sample to a very low temperature the logarithmic corrections can be investigated as a function of increasing electric field as the electron gas heats up.

An example of the use of a magnetic field and electron heating to separate quantum interference and interaction mechanisms is shown in Fig. 5. The results /22/ were taken with a silicon inversion layer which was initially cooled to 0.07K, the resistance was measured as a function of electric field, for different values of magnetic field, B. At zero B the resistance is initially

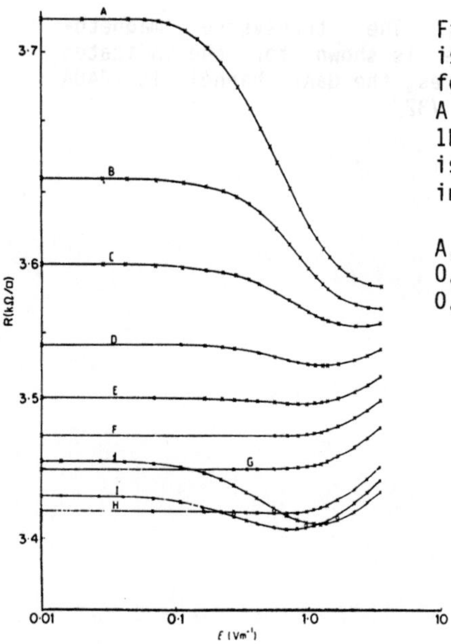

Fig. 5. Resistance in ohms per square is plotted against log electric field for different values of magnetic field. A field of 1.0 volt/metre corresponds to 1K and the ambient lattice temperature is 0.07K. The values of magnetic field in Tesla are as follows:

A, 0; B, 0.008; C, 0.0021; D, 0.47; E, 0.074; G, 0.15; H, 0.26; I, 0.32 and J, 0.52 /22/

constant then decreases with increasing E, saturation occurs when E ≈ 1 volt/metre which corresponds to about 1K. Increasing B gives a negative magneto-resistance and a reduced logarithmic change, until eventually there is a negative magneto-resistance but not a dependence of resistance on T_E. This situation arises as the suppression of quantum interference is determined by the magnitude of the cyclotron length, L_C. When $L_C \ll L_{IN}$, the logarithmic correction is determined exclusively by B /25/, and hence the temperature dependence is lost. Increasing B still further results in the introduction of a logarithmic interaciton correction when $g\beta B > kT_E$, with a positive magneto-resistance. As T_E increases the positive magneto-resistance disappears and is replaced by negative as the quantum interference is still present. The figure shows that as the enhancement of the interaction mechanism is a spin effect and the suppression of interference is orbital in origin so the mechanisms co-exist. However, if there is greater disorder reducing L_{IN} then it is possible for L_{IN} to be less than L_C when the magnetic field starts to enhance the interaction correction. Under these circumstances the corrections are additive, a similar situation exists when a parallel magnetic field is applied which gives a greatly reduced suppression of interference but an enhanced interaction correction.

VARIABLE DIMENSIONALITY

Comprehensive studies of quantum interference, interaction corrections and the metal-insulator transition have been carried out in 3D systems /26,27/. Doped silicon has been a model system here and the lighter mass semiconductors GaAs, InP and InSb have been used in experiments where a magnetic field shrinks the electronic wavefunction and reduces the Bohr orbit of an electron around a donor. The GaAs MESFET, also known as the Schottky gate FET, has been used to study transitions between three and two dimensions /29,30/ as the thickness of a conducting channel, d, is reduced. In this system F ≈ 0.6 and interaction effects are always important in both 3D and 2D. Analysis of the 3D conductivity shows that both interaction and interference corrections are present and that as the temperature decreases the corrections become 2D as their respective length scales exceed the sample thickness. The change of dimensionality can be most clearly seen in the negative magneto-resistance. Fig. 6 shows results obtained at 21.6K and 5.2K when the channel is 440A thick (corresponding to one occupied sub-band). At the highest temperature the inelastic length is shorter than the thickness of the channel and the magneto-resistance is 3D, varying as B^2 at low fields and $B^{1/2}$

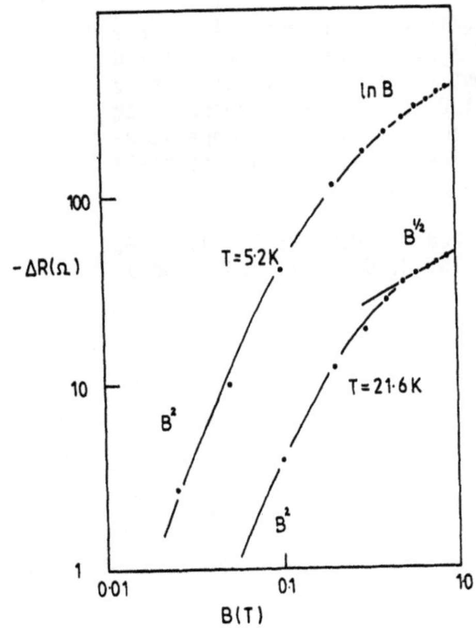

Fig. 6. The transverse magneto-resistance is shown for the indicated temperatures, the GaAs channel is ~440A thickness /32/

at high (a characteristic of 3D). However, at the lowest temperatures the inelastic length exceeds d and behaviour is 2D; varying as B^2, passing into log B.

A parallel magnetic field has no effect on a pure 2D system as no flux is enclosed. However, real systems have a finite thickness and so a parallel field results in a suppression of quantum interference /31/. For fields sufficiently small that $L_C^2 > dL_{IN}$ the magneto-conductance is given by

$$\Delta\sigma(B) = \frac{e^2}{2\pi^2\hbar} \; Ln \left(\frac{d^2 L_{IN} B^2 e^2}{3\hbar^2} + 1 \right) \tag{7}$$

This enables the inelastic length to be extracted; Fig. 7 shows a comparison of the inelastic lenghts extracted from parallel and perpendicular magneto-resistance /32/. As seen, agreement is good; the system is 2D (d ≈ 440A, which is less than L_{IN} for T< 10K), and L_{IN} varies as $T^{-1/2}$ ($\tau_{IN} \propto T^{-1}$) indicating that disorder is determining the rate of electron-electron scattering. In general, as the dimensionality is reduced from three to two the inelastic time increases, reflecting the decreased degree of freedom for inelastic scattering. Increasing the parallel magnetic field eventually results in 3D behaviour when $L_C \ll d$, an

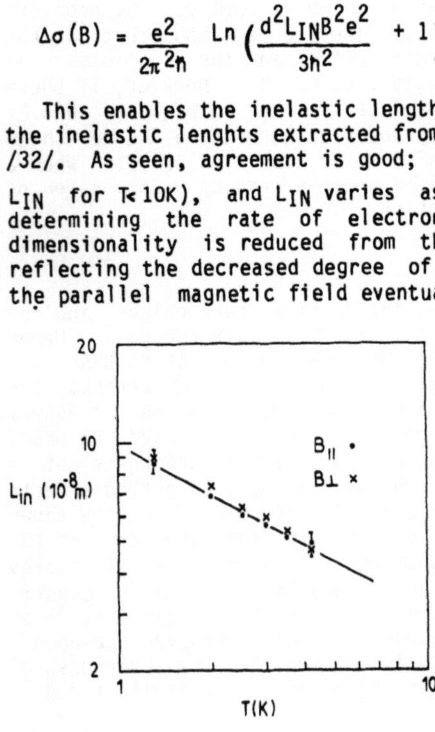

Fig. 7. The inelastic scattering length is plotted as a function of temperature for a channel thickness of ~440A. The close agreement between parallel and transverse methods is clear /32/

intermediate regime between 2D and 3D has the 3D dependence on magnetic field but the 2D temperature dependence /33,34/.

Similar experiments have been performed with 2D to 1D transitions. Here the 2D electron gas in a GaAs/AlGaAs heterojunction has been 'electrostatically squeezed', eventually the length scales exceed the width and the corrections become 1D. In a similar way to 2D behaviour a magnetic field separates interference and interaction corrections. Good agreement with theory has been found and it has been shown that a new mechanism of inelastic scattering dominates in 1D /36/. Here $L_{IN} \alpha T^{-1/3}$ and the scattering of electrons is by fluctuations in the electric field due to fluctuations in charge density /36/. Further details are given elsewhere; at the time of writing it is clear that 1D phenomena contain a great deal of new physics.

REFERENCES

1 T. Ando, A.B. Fowler and F. Stern: Rev. Mod. Phys. 54, 437 (1982)

2 M. Pepper: Contemp. Phys. 26, 257 (1985)

3 P.W. Anderson: Phys. Rev. 109, 1492 (1958)

4 N.R. Mott, M. Pepper, S. Pollitt, R.H. Wallis and C.J. Adkins: Proc. Roy. Soc. A 345, 169 (1975). Also, N.F. Mott and E.A. Davis: 'Electronic Processes in Non-Crystalline Materials', 2nd Edition, Clarendon Press, (1979)

5 N.F. Mott: Metal-Insulator Transitions, Taylor and Francis Ltd., (1974)

6 E. Abrahams, P.W. Anderson, D.C. Licciardello and T.V. Ramakrishnan: Phys. Rev. Lett. 42, 673 (1979)

7 L.P. Gorkov, A.I. Larkin and D.E. Khmelnitzkii, JETP. Lett. 30, 228 (1979)

8 R.V. Haydock: Phil. Mag. B. 43, 203 (1981)

9 M. Kaveh and N.F. Mott: J. Phys. C. 14, L177 (1981)

10 N.F. Mott and M. Kaveh: Adv. Phys. 34, 329 (1985)

11 G. Bergmann: Phys. Rev. B. 28, 2914 (1983)

12 B.L. Altshuler and A.G. Aronov in 'Electron-Electron Interactions in Disordered Systems', Volume 10 of Modern Problems in Condensed Matter Physics, North-Holland, (1985)

13 S. Hikami, A.I. Larkin and Y. Nagaoka: Prog. Theor. Phys. 67, 707 (1980)

14 A. Kawabata: J. Phys. Soc. Japan 49, 628 (1980), also Solid State Comm. 38, 823 (1981)

15 M.J. Uren, R.A. Davies, M. Kaveh and M. Pepper, J. Phys. C. 14, L395 (1981)

16 H. Fukuyama, J. Phys. Soc. Japan 53, 3299 (1984)

17 R.A. Davies and M. Pepper: J. Phys. C. 16, L353 (1983)

18 B.L. Altshuler, D.E. Khmelnitzkii, A.I. Larkin and P.A. Lee: Phys. Rev. B. 22, 5142 (1980)

19 P.A. Lee and T.V. Ramakrishnan: Phys. Rev. B. 26, 4009 (1982)

20 B.L. Altshuler, A.G. Aronov and P.A. Lee: Phys. Rev. Lett. 44, 1288 (1980)

21 M.J. Uren, R.A. Davies and M. Pepper: J. Phys. C. 13, L985 (1980)

22 R.A. Davies, M.J. Uren and M Pepper: J. Phys. C. 14, L531 (1981)

23 M.J. Uren, R.A. Davies, M. Kaveh and M. Pepper, J. Phys. C. 14, 5737 (1981)

24 P.W. Anderson, E. Abrahams and T.V. Ramakrishnan: Phys. Rev. Lett. 43, 718 (1974)

25 M. Kaveh, M.J. Uren, R.A. Davies and M. Pepper, J. Phys. C. 14, L413 (1981)

26 G.A. Thomas: Physica B. 117, 81 (1983)

27 A.P. Long and M. Pepper, J. Phys. C. 17, 3391 (1984); Ibid 17, L425 (1984)

28 G. Biskupski, H. Dubois, J.L. Wojkievicz, H. Briggs and G. Remenyi: J. Phys. C. 17, L411 (1978)

29 M. Pepper: Phil. Mag. B. 38, 515 (1978)

30 D.A. Poole, M. Pepper and H.W. Myron, Physica B. 117-118, 697 (1983)

31 B.L. Altshuler and A.G. Aronov, JETP Lett. 33, 499 (1981)

32 D.J. Newson, C. McFadden and M. Pepper: Phil. Mag. 52, 437 (1985)

33 C. McFadden, D.J. Newson, M. Pepper and N.J. Mason: J. Phys. C. 15, L21 (1982)

34 D.J. Newson and M. Pepper: J. Phys. C. 18, L1049 (1985)

35 T.J. Thornton, M. Pepper, H. Ahmed, D. Andrews and G.J. Davies: Phys. Rev. Lett. 56, 1198 (1986)

36 E.L. Altshuler, B.L. Altshuler and A.G. Aronov: Solid State Comm. 54, 617 (1985)

Optical Nonlinearities in Small Particles and Composite Materials

C. Flytzanis, F. Hache, D. Ricard, and Ph. Roussignol

Laboratoire d'Optique Quantique due C.N.R.S., Ecole Polytechnique, F-91128 Palaiseau, Cedex, France

1 Introduction

The breakthrough of very sophisticated applications of nonlinear optics in modern technology essentially depends on the availability of appropriate materials [1]. To the extent that high values of the nonlinear optical susceptibilities critically depend on the valence electron distribution the line of thought that prevailed up to now was to improve the chemistry of homogeneous bulk materials, inorganic or organic ones. In some cases this endeavor has been crowned [2,3] with success. Although some further progress is expected along this line quite general quantummechanical sum rules indicate that this cannot be stretched too far without abandoning other equally desirable properties of the bulk material. Thus the values of the third order susceptibility in the transparency region of germanium or polydiacetylene [2] the most nonlinear crystals presently known, are of the order of 10^{-10} esu in crystalline phase but drop well below these values in the noncrystalline phase ; higher values can be achieved by approaching resonances, ex. InSb, but then at the expense of high losses and narrow band operation [1].

Recently a different approach has been proposed namely to artificially [4,5] enhance the nonlinear optical nonlineatities by using inhomogeneous materials and in particular the composite ones obtained by dispersing small metal or semiconductor particles in a transparent dielectric for instance a glass or a liquid. The interplay of high optical nonlinearity, appropriate form, size and response time of the nonlinear optical material together with the appropriate mechanical and thermal properties which are very important in long term applications, in many instances, can be more ingenuously optimized with the composite materials than with the homogeneous bulk ones. Furthermore new nonlinear effects are expected in such materials that are not possible in homogeneous ones [6,7]

The study of the nonlinear optical properties of composite materials is interesting in a number of other aspects as it is related to that of small aggregates or clusters which are of paramount importance in many areas in chemistry and metallurgy and in preparing the next generation of miniaturized devices. From a more fundamental point of view the physics of such low dimensionality systems throw a new light on some basic physical principles which are suppressed when we deal with infinitely large systems. Because of the small size of these aggregates, 10-10^3 Å, quite often the electron motion does not find the same space available to develop itself as in the bulk metal or semiconductor and "zero" dimensionality effects become quite pronounced. This lowering of dimensionality and the additional boundary conditions imposed on these particles when embedded in a dielectric confer the composites with properties that are quite different from those of the bulk materials they are issued from.

Because of the predominantly isotropic character of these materials third order nonlinear optical effects will be the most relevant to investigate ; they are also the most sensitive to the dynamics and the dimensionality of the electron motion. In this respect the optical phase conjugation [8] is the most indicated technique to study these materials and furthermore it is closely related to some important applications. In the following we present [4,5,9-12] the main aspects of the cubic optical nonlinearities in two important classes of composite materials mentionned above, namely glasses doped with small metallic or semiconductor crystallites and the corresponding colloids ; the key question that will be addressed here is how these nonlinearities are related to those of the small particles. In parallel we also review some new effects in the nonlinear propagation in inhomogeneous media like intensity autolimitation [6] and self transparency as well as pulse reshaping and sharpening [7] which result from a subtle balance between nonlinear refractive index changes and loss through light scattering.

At this level it is important to state more precisely the composite materials we have in mind. These will be formed by interdispersing small particles of components A and B. Depending on their respective volume fractions p_A and p_B, topologically we may distinguish two classes [see fig. 1]. In the one, when the volume fractions of the two components are comparable, each particle A or B is surrounded by particles of both types ; in the other, when for instance $p_A << p_B$, each particle A is only surrounded by particles B. Their optical properties, when the dimensions of the particles are much smaller than the optical wavelength ($\lambda \sim 5000$ Å) and other characteristic lengths (skin depth etc..), are described in both cases by introducing an effective dielectric constant $\tilde{\varepsilon}$ which however is related differently for the two cases to the dielectric constants of the bulk components ε_A and ε_B.

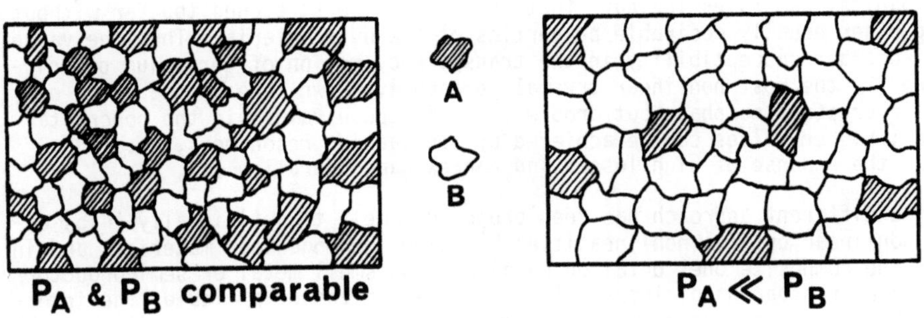

Fig. 1

In the first case, the two components are treated equivalently by assuming both types of particles embedded in a medium with dielectric constant $\tilde{\varepsilon}$ and one obtains Bruggemann's relation [13]

$$p_A \frac{\varepsilon_A - \tilde{\varepsilon}}{\varepsilon_A + 2\tilde{\varepsilon}} + (1 - p_A) \frac{\varepsilon_B - \tilde{\varepsilon}}{\varepsilon_B + 2\tilde{\varepsilon}} = 0 \qquad (1)$$

for spherical inclusions. In the second case, inclusions A are embedded in a medium of dielectric constant ε_B and one obtains the MAXWELL-GARNETT relation [14]

$$\frac{\tilde{\varepsilon} - \varepsilon_B}{\tilde{\varepsilon} + 2\varepsilon_B} = p_A \frac{\varepsilon_A - \varepsilon_B}{\varepsilon_A + 2\varepsilon_B} \qquad (2)$$

valid for spherical inclusions A and $p_A << 1$. This relation is a straight forward consequence of the familiar CLAUSIUS-MOSSOTTI approximation for the local field corrections [15].

Below we shall only consider the second case and will admit that the MAXWELL-GARNETT description is valid. The latter is also intimately related to the MIE theory [16] of light scattering from a spherical particle ; indeed the result (2) can also be derived [17] from the MIE theory when all higher order terms than the dipolar are neglected there and furthermore $p_A << 1$. The complete derivation of MIE'theory [16] with all multipoles included is quite tedious but the dipolar contribution can be easily derived and we outline it below as it illustrates several points we need for the subsequent discussion.

The dipole induced by an applied field E_o on a spherical particle of dielectric constant ε_A and diameter d embedded in a medium of dielectric constant ε_B is [18]

$$p = \left(\frac{d}{2}\right)^3 \frac{\varepsilon_A - \varepsilon_B}{\varepsilon_A + 2\varepsilon_B} E_o \tag{3}$$

and to the extent that $d << \lambda$, ℓ_s (skin depth) the field inside the particle is constant and given by [18]

$$E_1 = \frac{3\varepsilon_B}{\varepsilon_A + 2\varepsilon_B} E_o = f_1 E_o \tag{4}$$

Outside the particle the electric field is the superposition of E_o and the field due to the dipole (3) or explicitly [18]

$$E = E_o + \frac{1}{4}\left(\frac{d}{R}\right)^3 \frac{\varepsilon_A - \varepsilon_B}{\varepsilon_A + 2\varepsilon_B} E_o \tag{5}$$

where R is the distance from the center of the spherical particle (see Fig. 2).

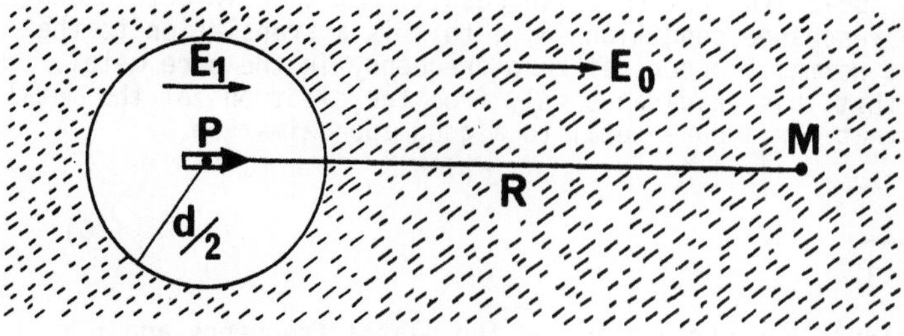

Fig. 2. Spherical inclusion embedded in a dielectric

We notice that p becomes large for frequencies ω_R such that

$$\varepsilon_A' (\omega_R) + 2\varepsilon_B = 0 \tag{6}$$

which is the condition for surface plasmon resonance where ε_A' is the real part of ε_A in $\varepsilon_A = \varepsilon_A' + i\varepsilon_A''$. The presence of such polarizable particles results in additional polarization

$$4\pi\Delta P = 3p_A \frac{\varepsilon_A - \varepsilon_B}{\varepsilon_A + 2\varepsilon_B} E_1 \equiv (\bar{\varepsilon} - \varepsilon_B) E_o \tag{7}$$

where $\bar{\varepsilon}$ is the effective dielectric constant and E_1 is the field acting on the particle. The latter can be obtained by the usual LORENTZ construction and after substitution in (7) one obtains the MAXWELL-GARNETT result (2) which for $p_A << 1$ can also be rewritten

$$\bar{\varepsilon} = \varepsilon_B + 3p_A\varepsilon_B \frac{\varepsilon_A - \varepsilon_B}{\varepsilon_A + 2\varepsilon_B} \tag{8}$$

which exhibits the resonance when (6) is satisfied. The width of this resonance is determined by ε''_A the imaginary part of ε_A and from (8) one can easily extract the extinction coefficient

$$\alpha = 9p_A \frac{\omega}{c} \varepsilon_B^{3/2} \frac{\varepsilon''_A}{(\varepsilon'_A + 2\varepsilon_B)^2 + \varepsilon''^2_A} \tag{9}$$

which also determines the color of the composite material. The above derivation, valid for spherical particles, with slight modifications can be extended to ellipsoidal ones as well provided their dimensions are much smaller than the optical wavelength. It is important to be stressed that the dielectric constant ε_A for a small particle, because of quantum size effects [19-20], is not necessarily identical to that of the homogeneous bulk material.

We proceed now to discuss the nonlinear optical properties of these materials under the assumption that the MAXWELL-GARNETT theory is a good description for their linear optical properties and extend it to the nonlinear case ; we shall discuss separately the metallic and the semiconductor particles.

2 Optical nonlinearities of small metallic particles and metal colloids

Small spherical metallic particles can be obtained now by well established methods [22] and when evenly dispersed in a liquid or glass they give rise to well characterized composite materials [17]. The best known are the ones formed with gold and silver particles ; their linear optical properties have been extensively studied both theoretically and experimentally [17] ; in this section we replace the labels A and B by m (metal) and d (dielectric) respectively. Gold and silver colloids are known to show beautiful colors which result from the dielectric anomaly related to the plasma resonance condition (6) ; this is a translation to the visible of the anomaly that normally occurs at zero frequency in the pure metal. It is essential to stress that this originates only from the electrons of the metallic particles ; we remind that for a pure metal to a good approximation

$$\varepsilon_m(\omega) = \varepsilon_b(\omega) - \frac{\omega_p^2}{\omega(\omega + i/\tau)} \tag{10}$$

where $\varepsilon_b(\omega)$ is the bound electron contribution, ω_p the plasma frequency and τ a time that reflects the scattering processes in the particle and can be written [17]

$$1/\tau = 1/\tau_o + 2v_F/d \tag{11}$$

τ_o is the scattering time in the bulk and v_F is the electron velocity at the Fermi level.

We fix our attention on the component of the induced cubic nonlinear polarization $P^{(3)}_{NLS}(\omega)$ that is relevant to the optical Kerr effect ; the latter will be measured through the optical phase conjugation technique [8]. This component is characterized by a change of the effective dielectric constant $\delta\bar{\varepsilon}$. In line with what was stated above, it is plausible to accept that this change $\delta\bar{\varepsilon}$ results from a change $\delta\varepsilon_m$ of the dielectric constant of the metallic particles [5]; the relation between the two is simply obtained from (8), with A=m and B=d,

$$\delta\bar{\varepsilon} = f_2 p \delta\varepsilon_m \quad \text{where} \tag{12}$$

$$f_2 = (3\varepsilon_d/(\varepsilon_m + 2\varepsilon_d))^2 = f_1^2 \tag{13}$$

With the assumption that only the electrons of the particles contribute which experience the local field (4) one has

$$\delta\varepsilon_m = 12\pi\chi_m^{(3)}E_1^2 = 12\pi\chi_m^{(3)}f_1^2E_o^2 \qquad (14)$$

where $\chi_m^{(3)}$ $(\omega, -\omega,\omega)$ is the third order susceptibility of the metal. One finally has [6]

$$\delta\tilde{\varepsilon} = 12\pi p f_1^4\chi_m^{(3)} E_o^2 \qquad (15)$$

and the relevant nonlinear polarization is

$$P_{NLS}^{(3)}(\omega) = 3pf_1^4\chi_m^{(3)} E_o^3 = 3\,\tilde{\chi}^{(3)}E_o^3 \qquad (16)$$

which can also be obtained differently by using local field considerations ; $\tilde{\chi}^{(3)}$ is the effective third order susceptibility of the composite.

The essential result is that the factor f_1 occurs in the fourth power in (15) and consequently on the resonance condition (6) one has a strong enhancement. Since the field outside the spheres rapidly decreases with the distance (compare (5)) the contribution of dielectric around the metallic particle is not substantial and can be neglected to a first approximation. The enhancement is particularly spectacular when we compare the signal, which is $\sim|P_{NLS}^{(3)}|^2$, on and off resonance : since on resonance $|f_1^R|> 1$ while off resonance $|f_1^{NR}|< 1$ the relative enhancement factor $F= |f_1^R/f_1^{NR}|^2$ may lead to enhancements of several orders of magnitude.

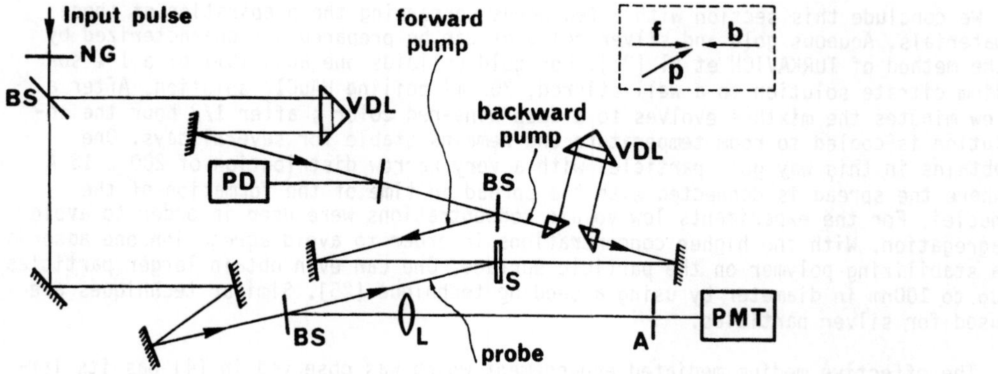

Fig. 3. Typical experimental setup used in optical phase conjugation studies. NG : neutral density filters, BS : beam splitters, VDL : variable delay lines, S : sample, PD : reference photodiode, PMT : photomultiplier tube

This enhancement was strikingly demonstrated in the measurements on silver and gold colloids as well as in the glasses doped with small silver and gold particles [5]. The optical phase conjugation technique was used (Fig. 3) where a weak probe beam together with two equally intense counterpropagating pump beams is incident on the sample to produce a phase conjugated wave whose intensity was measured ; all beams are at the same frequency ω and are issued from the same laser source, a mode locked YAG or Nd/glass laser delivering 27ps and 5ps pulses respectively. The volume fraction p was 5.10^{-6} and $1.5\cdot10^{-6}$ for gold and silver colloids and the diameter of the particles 10nm and 7 nm respectively determined by transmission electron microscopy. Their absorption was well described by (9) using the optical coefficients of the bulk gold and silver with the appropriate corrections related to the size of the particles. For both colloids, the phase conjugated reflectivity near resonance for parallel probe and pump polarizations is by a factor of 5 less than CS_2 but drops by four orders of magnitude when measured off resonance [5]. By varying the delay of either pump, it was found that the main response is instantaneous as expected if the whole contribution in (15) comes from the electrons of the particles. For crossed probe and pump polarizations the signal drops by a factor 64 for the gold colloid which indicates a strong anisotropy in $\chi_m^{(3)}$.

Taking the value $\chi^{(3)}(\omega,-\omega,\omega) = 2\ 10^{-12}$esu in CS_2 at $\lambda = 0.532\mu m$ we obtain from (16)

$\chi_m^{(3)}(\omega,-\omega,\omega) = 1.5 \cdot 10^{-8}$ esu for gold spheres at $0.53\mu m$

$\chi_m^{(3)}(\omega,-\omega,\omega) = 2.4 \cdot 10^{-9}$ esu for silver spheres at $0.40\mu m$

This is the first reported experimental determination of optical Kerr coefficients for a metal. Although the physical effects are not directly comparable, these values are higher by two orders of magnitude than those extracted by third harmonic [23] reflectivity measurements. The finite size of the particles must certainly play a crucial role for this difference ; this was already indicated in connection with the electron scattering mechanism (compare (11)) and the problem has attracted considerable theoretical attention in the case of the linear optical properties of these particles [19-20]. HACHE et al [24] extended the quantummechanical model of a free electron in a sphere to the case of the third order optical coefficients and keeping only the relevant resonant terms were able to account for both the magnitude and the anisotropy of $\chi_m^{(3)}$ for the gold and silver particles. This calculation also allows one to derive the dependence of $\chi_m^{(3)}$ on the size of the particles and experimental work is in progress to determine this dependence.

We conclude this section with a few words concerning the preparation of these materials. Aqueous gold and silver colloids can be prepared and characterized by the method of TURKAVICH et al [21]. For gold colloids one adds 15ml of a 1 % sodium citrate solution to a well stirred, 300 ml boiling $HAuCl_3$ solution. After a few minutes the mixture evolves to a deep wine-red color ; after 1/2 hour the solution is cooled to room temperature and remains stable for several days. One obtains in this way gold particles with a very narrow distribution of 200 ± 15 Å where the spread is connected with the spread in time of the formation of the nuclei. For the experiments low volume concentrations were used in order to avoid agregation. With the higher concentrations, in order to avoid agregation, one adsords a stabilizing polymer on the particle surface. One can even obtain larger particles up to 100nm in diameter by using a seeding technique [25]. Similar techniques are used for silver particles.

The effective medium mediated enhancement which was observed in [4] has its largest impact in the case of the optical Kerr coefficient and opens new ways of obtaining materials with large $\chi^{(3)}(\omega,-\omega,\omega)$. This enhancement is closely related to the surface mediated enhancement[5,26,27] which has been extensively studied in the last decade with one important distinction however : in the composite materials the enhancement resides in the electrons inside the metallic particles that experience the field (4) while in the surface enhancement one considers the electrons on the molecules close to but outside the surface which experience the field (5).

3 Semiconductor microcrystals and semiconductor doped glasses

Composite materials formed by dispersing small semiconductor crystallites in a transparent dielectric or a glass are well known and have found many applications. For instance some color filter glasses such as the sharp Schott and Corning cut off filters are formed from such materials. In these examples the very sharp absorption edge is due to the semiconductor crystallites CdS_xSe_{1-x} that are formed [28,29] inside the glass when Cd, S and Se are introduced into the glass melt in sufficient amounts ; other chemical elements are present as well but these are believed to remain in the glass outside the crystallites. By varying the concentration x the cut-off wavelength can be adjusted over a wide spectral range from that corresponding to pure CdS with band gap energy E_g = 2.55 eV down to that of CdSe with

E_g = 1.82 eV. Other semiconductors can be used as well to obtain composite materials. The sizes of the crystallites in these commercial filters are not well known and certainly show a wide spread as they are usually formed by letting the chemical constituants (Cd, S and Se in the case of CdS_xSe_{1-x}) to diffuse for an extended period t_D in the glass melt that is kept at a constant temperature T ; their sizes then mainly depend on the temperature T and the duration t_D and to a lesser degree on some other factors. Simple models based on the assumption of a brownian type motion for this diffusion allow one to estimate an average size and the spread around it but the uncertainty is considerable ; the particle sizes can range from a few tens up to several hundreds Ångstroms and even larger (50 - 1000 Å). Recently Ekimov et al [30] reported the formation of CuCl and CdS microcrystals of well controlled size in silicate glasses and the observation [31] of quantum size effects on the exciton line of CuCl as the size of these crystallites was swept across the exciton radius. Similar effects were claimed in commercial filters doped with CdS_xSe_{1-x} microcrystals [32]. Other groups [33,35] were able to obtain semiconductor colloids as well, in particular with CdS crystallites, whose size can be well controlled in the range 10 - 300 Å and this opens new possibilities to study some fundamental problems related to exciton dynamics in an isotropically restricted space and will certainly trigger work on other physical aspects and in particular on the nonlinear optical properties of these systems.

Intensity dependent optical properties in these materials, like absorption saturation and optical Kerr effect were reported [36] sporadically in relation with their use in laser technology but no systematic study of their physical origin was attempted then. More recently attention was refocused on these materials in connection with optical phase conjugation measurements [9] and some understanding as to the prevailing mechanisms has been gained since then ; attempts to use these materials in Fabry-Perot cavities for optical bistable devices have also been reported [11] stimulated from the large $\chi^{(3)}$ values (10^{-8} 10^{-9} esu) obtained by the optical phase conjugation technique. Such large values of $\chi^{(3)}$ can be accounted for [4] by using the composite medium approach outlined previously. In addition more recent measurements[10] allowed to clarify the time behavior of these nonlinearities and distinguish different intensity regimes ; both aspects are of central importance in assessing the potential applications of these materials.

The measurements were done with the optical phase conjugation technique as depicted in Fig. 3. For the subsequent discussion we remind that for isotropic media the nonlinear polarisation at frequency ω responsible for the phase conjugated signal can be cast in the form [8].

$$P_{NLS} = A(\theta)(E_f.E_p)E_b + B(\pi-\theta)(E_b.E_p)E_f + C(E_f.E_b)E_p \quad (17)$$

where θ is the angle between the forward pump and probe wavevectors ; apart from some trivial geometrical factors, the coefficients A, B and C are directly related to $\chi^{(3)}$. The two first terms in (17) correspond to spatial gratings of small (λ_s) and large (Λ_s) spacing respectively that are induced in the medium (see Fig. 4)

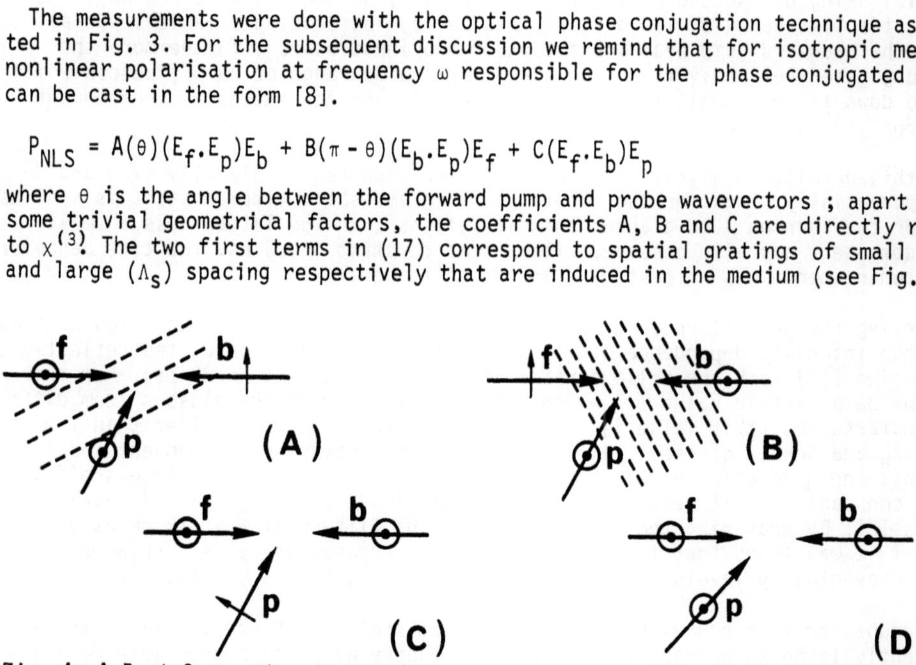

Fig. 4. A,B et C are the three polarisation configurations that allow to isolate the three terms in (17) ; D is the "copolarized" configuration

and the third term has no spatial analogue and is sometimes termed self-diffraction ; by cross polarizing one of the input beams with respect to the other two the three terms can be isolated and related to the dynamics of the induced polarization. It should be pointed out that with sufficiently large pump intensities the coefficients in (17) may become intensity dependent.

In the first measurements [9] of these coefficients by the optical phase conjugation technique Q-switched lasers were used delivering nanosecond pulses with fluences larger than 1mJ/cm^2 ; the frequencies were near but below the band gap of the crystallites in the commercial filters (ex. Corning 3.68 glass). The optical phase conjugated reflectivity was found to be of the order of 10 % which leads to values of $\chi^{(3)} \sim 10^{-8} - 10^{-9}$ esu but the latter were found to be intensity dependent. In the same measurements [9] by cross polarizing one input beam with respect to the other two one was able to assess the relative importance of the three terms in (17). Thus it was found that the large-period and small period gratings result in comparable contributions in these filters whereas in pure CdS only the large-period grating gives the dominant contribution ; the contribution of the third term in (17) was found negligible in both materials. Quite suprisingly, in the case of the filters, the sum of the signals originating from the two spatial gratings significantly exceeded the one measured with all three beams "copolarized". The latter observation confirmed by other measurements [10] is quite puzzling but the former finds a plausible explanation [9] in terms of the diffusion of the photocarriers that set up the gratings : this is sufficiently fast to wash out the small period gratings in pure CdS but not in the small crystallites where the electrons have a restricted space to move. Based on indirect evidence [9] the conclusion was drawn, from these measurements with nanosecond pulses, that the response time was very short, shorter than nanosecond and apparently this was later confirmed [12] using picosecond pulses but this point needed some further clarification as did the observed intensity dependence of the optical phase conjugated reflectivity.

The determination of the time behavior and intensity dependence of the optical nonlinearities along with that of their magnitude is of paramount importance for the potential applications of these materials and to understand the physical origin of their nonlinear polarisation. Such a detailed analysis was initiated by ROUSSIGNOL et al [10] using picosecond pulses from a frequency doubled mode locked Nd : YAG laser ; the time behavior was directly studied by introducing a variable delay between the backward pump beam and the other two input beams ; furthermore the phase conjugated reflectivity was measured for a wide range of pump fluences which extended down to very small fluences ($\sim$ 15 μJ/cm^2) and the intensity independent regime for $\chi^{(3)}$ was clearly observed.

From this detailed analysis it results that the response is globally slow and decays exponentially. Thus for the Schott OG 530 glasses the time constant is $\sim$ 40 ns while for the Corning 3.68 glasses this is $\sim$ 11 ns ; in the latter case a small faster component was also observed with a time constant $\sim$ 0.5 ns. The time behavior of the luminescence was also studied and found to be slow.

By varying the pump fluences over a range extending from 15μJ up to a few hundreds mJ/cm^2 the intensity dependence of $\chi^{(3)}$ and that of the phase conjugated reflectivity were studied [10] with the pulse duration being kept constant ($\sim$ 28 ps). For intensities below a certain value I_s, which somewhat varies with the glasses from different manufacturers (this value is somewhat larger respectively smaller than 1 mJ/cm^2 in Corning and Schott glasses), $\chi^{(3)}$ was found to be intensity independent and $R_c \sim |\chi^{(3)}|$ to exhibit the quadratic regime. For intensities higher than I_s, R_c levels off and remains constant over at least two orders of magnitude implying that $\chi^{(3)}$ varies there as 1/I. By measuring the transmission of the filters at $\lambda \simeq 532$ nm as a function of I it was found that the absorption also decreases as I_s is approached from below and eventually levels off at a low but not negligible value [10,37].

To account for this behavior we may admit that the size of the crystallites is sufficiently large to warrant the use of band theory with the appropriate modifications resulting from the inclusion of many body effects [38] (screening, gap renormalization, excitons etc..) ; in a parabolic band model the conduction band may be assumed initially empty and the valence band filled. Let us designate by α_o and $\chi_o^{(3)}$ the linear absorption and the third order susceptibility respectively calculated with a model based on the above picture and eventually incorporating size and effective (composite) medium effects ; these coefficients will be of the form :

$$\alpha_o = \int d^3k\, A_o(k) \tag{18}$$

$$\chi_o^{(3)} = \int d^3k X(k) \tag{19}$$

where k labels the electron states inside the Brillouin zone (B.Z.) and the integration is made over the first B.Z. The coefficient $\chi_o^{(3)}$ can be estimated using the called blocking or band filling model [39,40,41].

Extending the latter we make the conjecture that the observed behavior as I increases results from the finite modification that occur in the electron and hole distributions in the presence of the radiation. If we call ΔN the number of photocarriers created within a crystallite of volume V we may set

$$\Delta N = \frac{I\alpha\tau}{\hbar\omega} V \tag{20}$$

where α is the absorption coefficient at intensity I and frequency ω and τ is either the recombination time τ_R or the pulse duration τ_p, whichever is shorter. Assuming Boltzmann distribution and neglecting the hole contribution because of the heavier hole mass we may write for the absorption coefficient :

$$\alpha = \int A(k)(1 - f_e(k))d^3k \tag{21}$$

where

$$f_e(k) = 4\left(\frac{\pi\hbar^2}{2m_e k_B T}\right)^{3/2} \frac{\Delta N}{V} e^{-\hbar^2k^2/2m_e k_B T} \tag{22}$$

and the effective nonlinear susceptibility $\chi^{(3)}$ can be derived from

$$\chi^{(3)} \sim \Delta n/I \tag{23}$$

where the refractive index change Δn is related to the absorption change $\Delta\alpha$ by a Kramers-Kronig relation. Inserting (20) and (22) in (21) after integrating over the wavevector k and solving for α and $\chi^{(3)}$ one gets

$$\alpha = \alpha_o/(1 + I/I_s) \tag{24}$$

$$\chi^{(3)} = \chi_o^{(3)}/(1 + I/I_s) \tag{25}$$

with

$$I_s = (2mk_BT/\pi\hbar^2)^{3/2}\ \hbar\omega/4\alpha_o\tau F \tag{26}$$

where F is a form factor which for most purposes can be set equal to unity. From (25) one also obtains a phase conjugated reflectivity $R_c \sim |\chi^{(3)} I|^2$ which beyond I_s levels off with the intensity I in agreement with the experimental findings [10]; from (26) we expect that I_s will increase with temperature.

Clearly with the creation of a finite density distribution of photocarriers an additional absorption sets in that is not accounted for by (24) which only takes into account interband contributions ; this additional contribution due to the intraband motion of the created free carriers whose density and plasma frequency can be simply derived from (20), together with other loss mechanisms that may set in at high intensities (ex. multiphoton absorption), will result in a residual absorption which must be added to (24). This makes an essential difference from the classic absorption saturation that levels off to zero in localized two-level systems. To make the distinction between the two clear we term the mechanism related to (24) and (26) "frustrated saturation".

The above mechanism based on the modification of the electron density distribution by the radiation has additional implications which may be tested experimentally. These are in particular related to the many body effects [38] which depend on the electron density ; thus the band gap and relaxation times will be modified and become intensity dependent as will the exciton characteristics.

As the frequency ω is tuned well above the gap E_g the above model has to be accordingly modified. Recently the light induced changes of the absorption in glasses when the photocarriers are created with an excess energy above the band gap were studied [42] using an excite and probe technique with femtosecond pulses ; an extremely fast relaxation was observed when the excess energy is well above the gap but slows down as the gap is approached. In the latter regime the impurities, presumably surface states in these crystallites [10], play an important role.

The above conclusions were all obtained by working on commercially available semiconductor doped glasses. Access to well controled crystallites either in glasses or colloids will certainly show new effects intrinsically related to the size of these particles and will allow more systematic and careful studies where different parameters like temperature, size, concentration, band gap and others are varied in well defined manner ; this together with the previous considerations on the intensity dependence of $\chi^{(3)}$ may well open the way to a new class of nonlinear optical materials with "tunable nonlinearity".

4 Nonlinear propagation in composite media

The light pulse propagation in composite and other heterogeneous media consisting of two components show some novel features quite distinct from those studied in homogeneous media ; these will occur when the optical nonlinearities are comparable to the fluctuations of the linear refractive index. Thus in the stationary regime one has predicted [6,44] intensity autolimitation and laser induced self transparency due to the suppression of scattering which may lead to several applications in laser technology and in optoelectronics ; when the finite time response of the nonlinearities is also taken into account pulse reshaping or shortening may occur [7].

If the light frequency is in the transparency region of the medium the main loss mechanism in such media is the light scattering which irrespective of the size of the inclusions is proportional to the effective refractive index difference $\Delta\tilde{n}$ between the two components of the composite medium [43]. The key point is that $\Delta\tilde{n}$ can be suppressed for certain critical light intensities when the optical nonlinearities have appropriate sign and magnitude.

A rigorous theoretical analysis of light propagation that takes explicitly into account the diffraction and interference effects along with the optical nonlinearities is very complicated but, at a qualitative level, the effects we mentionned above can be described using a semi-phenomenological single flux approximation [43,44] which leads to

$$\frac{\partial I}{\partial z} + \frac{1}{v}\frac{\partial I}{\partial t} = -\alpha_o I - g(\Delta\tilde{n})I \qquad (27)$$

where v is the light group velocity α_o the linear absorption (or gain) of the medium and g a form factor that depends on the size and concentration of the inclusions and can be computed by using conventional light scattering theory from a spherical particle [16,43]. Now $\Delta\tilde{n}$ can be separated into a linear Δn_L and a nonlinear part Δn_{NL}

$$\Delta\tilde{n} = \Delta n_L + \Delta n_{NL} = \Delta n_L + \Delta n_2 I \quad (28)$$

where Δn_2 is the difference of the optical Kerr effect coefficients between the two components of the medium. The time evolution of Δn_{NL} can be described with a Debye type relaxation equation. The numerical solution of (27) then shows the following main features [6,44,7]

- with positive sign for the product $\Delta n_L.\Delta n_{NL}$, the light attenuates to a certain relative intensity at any input intensity
- with negative sign for $\Delta n_L.\Delta n_{NL}$ and $I > I_o$, the light attenuates only to I_o and when $I = I_o$ a selfsustained transmission takes place (this is the intensity autolimitation and self-transparency effects referred to previously)
- with negative sign for $\Delta n_L.\Delta n_{NL}$, the transmitted intensity shows a step-like dependence on the incident beam intensity at some characteristic optical thickness of the medium which among other things leads to drastic changes of the spatial beam profile.
- when the finite time evolution of the nonlinearity is also included the above effects also lead to drastic changes in the temporal beam profile (a pulse may acquire a rectangular temporal profile or a substantial part will be compressed in time to a much shorter pulse).
- a bistable operation will be obtained when the beam is reflected from a single back mirror.

The above effects and others which are in the process of investigation (light trapping or wave localization in random heterogeneous media) may find some important applications and certainly present an intrinsic fundamental interest.

5 Conclusion

The investigation of the nonlinear optical properties of composite materials is still in its infancy and the previous discussion was only aimed at summarizing the existing results and at extracting some salient features and guidelines for future work. The enhancement of the optical nonlinearities through certain resonances and the importance of the light induced modification of the electron density distribution were particularly emphasized; furthermore the relation between the optical nonlinearities of the composite and those of the metal or semiconductor inclusions was pointed out and exploited to extract information about the latter. The limitations of the above discussion should clearly be borne in mind as the access to well characterized materials is presently a major obstacle in this area and the difficulty in obtaining these few results should not be underestimated. The use of the powerful fabrication techniques for microstructures will certainly remedy this situation and open the way to systematic studies on these materials. From the outset the composite materials present a lot of advantages when compared to the homogeneous ones since one can artificially change a number of characteristics, size of the inclusions, their resonances or band gap, their concentration and even the optical anisotropy by using nonspherical inclusions and thus manufacture materials with "tunable nonlinearity" which are of great importance in many applications of nonlinear optics.

Acknowledgements

The authors acknowledge stimulating discussions with Drs N. Kothari, J. Lukasik and K. Rustagi on different aspects of the material presented above.

References

1 See for instance Nonlinear Optics : Materials and Devices, édit. C. Flytzanis and J.L. Oudar [Springer Verlag, Heidelberg, 1986]
2 C. Sauteret, J.P. Hermann, R. Frey, F. Pradère, J. Ducuing, R.R. Chance and R.H. Baughman, Phys. Rev. Lett. 36, 956 (1976)
3 J.M. Halbout, J.M. Blit and C.L. Tang, IEEE, J. Quant. El. QE-17, 513 (1981)

4 K.C. Rustagi and C. Flytzanis, Opt. Lett. 9, 344 (1984)
5 D. Ricard, Ph. Roussignol and C. Flytzanis, Opt. Lett. 10, 511 (1985)
6 G.B. Altshuler, V.S. Ermolaev, K.I. Krylov, A.A. Manenkov, J. Opt. Soc. Am. B to appear and references given there
7 N. Kothari and C. Flytzanis, to be published
8 See for instance, Optical Phase Conjugation, Edt: R.A. Fisher (Academic Press, 1983)
9 R.K. Jain and R.C. Lind, J. Opt. Soc. Am. 73, 647 (1983)
10 P. Roussignol, D. Ricard, K.C. Rustagi and C. Flytzanis, Opt. Comm. 55, 143 (1985)
11 B. Danielzik, K. Nattermann and D. von der Linde, Appl. Phys. B 28, 31 (1985)
12 S.S. Yao, C. Karaguleff, A. Gabel, R. Fortenberry, C.T. Seaton and G.I. Stegeman, Appl. Phys. Lett. 46, 801 (1985)
13 D.A.G. Bruggeman, Ann. Phys. (Leipzig) 24, 636 (1935)
14 J.C. Maxwell-Garnett, Philos. Trans. R. Soc. (London) 203, 385 (1904) ; 205, 237 (1906)
15 See for instance R. Landauer in Electrical Transport and Optical Properties of Inhomogeneous Media, Eds J.C. Barland and D.B. Tanner (American Institute of Physics, New York, n° 40)
16 G. Mie, Ann. Phys. (Leipzig) 25, 377 (1908)
17 J.A.A.J. Perenboom, P. Wyder and F. Meier, Physics Reports, 78, 173-292 (1981)
18 See for instance, C.J. Böttcher, Theory of Electric Polarization (Elsevier, Amsterdam, 1973) p. 78
19 H. Fröhlich, Physica 4, 406 (1937)
20 A. Kawabata and R. Kubo, J. Phys. Soc. Jap. 21, 1765 (1966)
21 L. Genzel, T.P. Martin and V. Kreibig , Z. Phys. B 21, 339 (1975)
22 J. Turkavich, P.C. Stevenson and J. Hillier, Discuss. Faraday Soc. 11, 55 (1951)
23 W.K. Burns and N.Bloembergen, Phys. Rev. 84, 3437 (1971)
24 F. Hache, D. Ricard and C. Flytzanis, to be published
25 R. Zsigmondy, Z. Phys. Chem. (Leipzig) 56, 65 (1906)
26 See for instance, Surface Enhancement Raman Scattering, eds R.K. Chang and T.E. Furtak (Plenum, New York, 1982)
27 See also the article by D. Ricard in Ref. 1.
28 H.P. Rooksby, J. Soc. Glass Techn. 16, 171 (1932)
29 G. Schmidt, Silikattechn. 14, 12 (1963)
30 V.V. Golubkov, A.I. Ekimov, A.A. Onushchenko & A.V. Tsekhomskii, Fiz. Khim Stekhla 6, 511 (1980)
31 A.I Ekimov, A.A. Onushchenko, A.G. Plyukhin & A.L. Efros, Sov. Phys. JETP 61, 891 (1985) (Zh. E.T.F 88, 1490 (1980)).
32 J. Warnock & D.D. Awschalom, Phys. Rev. 33, 5529 (1985) and Appl. Phys. Lett. 48, 425 (1985)
33 A. Henglein, Ber. Bunsenges. Phys. Chem. 86 241 (1982)
34 R. Rossetti, J.L. Ellison, J.M. Gibson & L.E. Brus, J. Chem. Phys; 80, 4464 (1984)
35 L. Brus in Proceedings of the NATO Workshop on "Homogeneous and Heterogeneous Catalysis", Sept 1985
36 G. Bret & F. Gires, Appl. Phys. Lett. 4, 175 (1964)
37 F. de Rougemont and R. Frey, unpublished results
38 H. Haug & S. Schmitt-Rink, J. Opt. Soc. Amer, B 2, 1135 (1985)
39 see for instance D.A.B. Miller, C.T. Seaton, M.E. Price & S.D. Smith, Phys Rev. Lett. 47, 197 (1981)
40 see for instance B.S.Wherett & N.A. Higgins, Proc. R. Soc. (London) A379, 67 (1982)
41 see also article by B.S. Wherett in Ref. 1
42 M.C. Nuss, W. Zinth & W. Kaiser, to be published
43 A. Ishimaru "Wave propagation and scattering of waves in randomly inhomogeneous media"
44 G.B. Altshuler, V.S. Ermolaev, K.I. Krylov & A.A. Manenkov, Izvestya (Physics) 48, 1534 (1984)

Tunnel Currents and Electron Tunnelling Times in Semiconductor Heterostructure Barriers in the Presence of an Applied Magnetic Field

L. Eaves, K.W.H. Stevens, and F.W. Sheard

Department of Physics, University of Nottingham, Nottingham, NG7 2RD, United Kingdom

The effect of transverse magnetic fields up to 11.5 T on the tunnel current in GaAs/(AlGa)As/GaAs heterostructure barriers is reported. The decrease of the tunnel current with increasing field is shown to arise from an increase in effective barrier height due to the diamagnetic energy. The results are used to test the effective mass and WKB approximations which are generally used to describe tunnelling effects in semiconductors. Recent calculations of barrier traversal times are briefly reviewed. Two of these theoretical models are used to calculate the effect of applied magnetic and electric fields on the barrier traversal time.

1. Introduction

Quantum mechanical tunnelling of electrons through heterostructure barriers is a key feature of many modern electronic devices [1-4]. This paper describes how measurements of the effect of a transverse magnetic field on the tunnel current can be used to test the accuracy of the effective mass and WKB approximations when applied to heterostructure barriers. The paper also discusses a topic that has attracted a great deal of theoretical interest recently, that of the time taken by a tunnelling particle to traverse a barrier. In particular, the procedures used by Stevens and by Buttiker and Landauer for calculating the barrier traversal time will be applied to the case of an electron traversing a heterostructure barrier in the presence of a transverse magnetic field and parallel electric field.

2. Dopings and Compositions of the Multilayers

The structures used in our study were grown by Professor K.E. Singer (UMIST) using MBE. Some of their electrical properties (particularly those associated with hot electron effects) are described elsewhere [5-8]. Briefly, they consist of the following layers: 200 μm thick n^+GaAs substrate doped to 2×10^{18} cm^{-3}; 1 μm n^+GaAs buffer (collector), 2×10^{18} cm^{-3}; 1 μm n^-GaAs layer, 2×10^{15} cm^{-3}; 168 ± 10 Å of undoped (AlGa)As; 1 μm n^+GaAs top (emitter) layer, 2×10^{18} cm^{-3}. The Al concentration varies from 37% (collector side) to 32 ± 2% (emitter). C(V) measurements indicate the presence of a negative space charge density of around 10^{16} cm^{-3} in the (AlGa)As barrier. The structures were processed into the form of mesas with diameters of 200 or 800 μm. The variation of the conduction band edge through the device is shown schematically in Figure 1 under reverse bias conditions. The experiments described here were carried out at low temperatures, usually at liquid helium temperatures (2 to 4.2 K) so that thermionic emission of electrons over the (AlGa)As barrier is negligible and the measured current J(V) is due only to quantum tunnelling of electrons. The large electric field in the depleted n-GaAs region beyond the barrier ensures that effectively all of the tunnelling electrons are collected in the right-hand n^+GaAs contact.

3. The Effect of a Transverse Magnetic Field on the Tunnel Current

For a simple rectangular barrier of potential height V_0 and width b, it is well known that the tunnelling probability of a particle of kinetic energy ε and mass m is given to a good approximation by

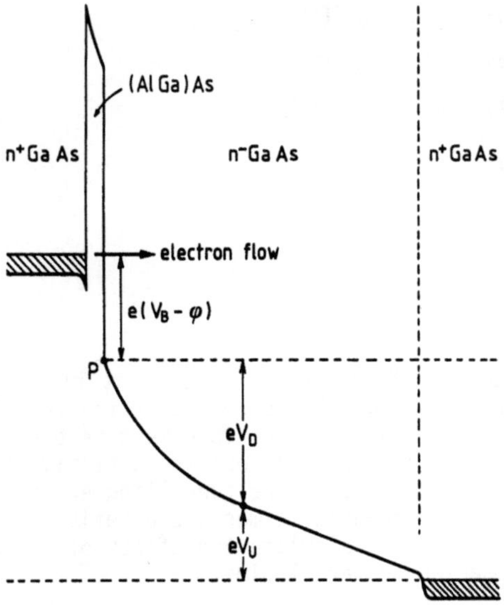

Fig. 1. Schematic diagram (not to scale) showing the variation of the conduction band edge through the device under reverse bias conditions

$$T \approx \frac{16\varepsilon}{V_0}\left(1 - \frac{\varepsilon}{V_0}\right)\exp(-2\kappa b) \ , \tag{1}$$

where $\kappa = [2m(V_0 - \varepsilon)]^{\frac{1}{2}}/\hbar$, for $\kappa b >> 1$. In the effective mass theory of electron tunnelling through a heterostructure barrier, V_0 is set equal to the conduction band discontinuity. The tunnelling effective mass m^* is taken to be close to that of the conduction band of the material forming the barrier. A fuller discussion of the validity of the effective mass approximation and how it can be probed by measurement of the pressure-induced changes of the band structure on the tunnel current can be found elsewhere [9,10].

Any analysis of the tunnelling current in a semiconductor heterostructure barrier system, for example of the type described in section 2 and shown schematically in Figure 1, must take into account the presence of electric fields in the barrier region. These can arise from built-in space charge or from the externally applied bias voltage. Thus in real problems the potential barrier is seldom rectangular. The WKB approximation is a convenient method for estimating the tunnelling probability for a spatially varying barrier potential, $V(x)$, which includes both the band-edge discontinuity and the electrostatic potential energy terms. Neglecting the energy-dependent prefactor, it gives the following expression for the tunnelling probability:

$$T(\varepsilon) \sim \exp[-2\int_0^b \kappa(x)\, dx] \ , \tag{2}$$

where $\kappa(x) = [2m^*(V(x) - \varepsilon)]^{\frac{1}{2}}/\hbar$.

The tunnel current at a particular applied voltage can then be calculated by an integration over the occupied electron states on the side of the barrier which is negatively biased.

The WKB approximation can be extended to include the effect of a transverse magnetic field $\underline{B}$ ($\underline{B} \parallel \underline{z}$) applied perpendicular to the direction of the tunnelling current ($\underline{J} \parallel \underline{x}$). In this case and neglecting spin-effects, the electron Hamilton is of the form

$$H = \frac{(\underline{p} + e\underline{A})^2}{2m^*} + V_0(x) - eEx \ . \tag{3}$$

In this equation $\underline{A} = B(0, x, 0)$ is the magnetic vector potential, $V_0(x)$ represents the step-like heterostructure barrier potential due to the band-edge discontinuity and $\underline{E}$ is the electric field in the barrier region. The equation can be written in the form

$$H = \frac{p^2}{2m^*} + V_0(x) + \frac{e^2B^2x^2}{2m^*} + \frac{p_y eBx}{m^*} - eEx\ . \tag{4}$$

The first two terms give the Hamiltonian in the absence of external fields. The last three terms can be thought of as modifying the form of the barrier potential. Note that p_y commutes with H and is therefore a constant of motion. It is convenient to choose the origin of co-ordinates x = 0 to be the interface between the left-hand n^+-emitter layer and the (AlGa)As barrier. By setting $\underline{A} = 0$ for $x \leq 0$ and $A_y = Bx$ for $0 < x < b$, the dynamic effects of the magnetic field can be ignored in the heavily doped n^+-region. This approximation is justified by the smearing out of Landau level effects due to ionised impurity scattering in this region. In addition, this choice of origin (i.e. of gauge) provides a range of p_y-values for the electrons incident on the barrier which is symmetric about $p_y = 0$. Both $\underline{A}$ and the diamagnetic term are, of course, assumed to vary continuously at the (AlGa)As/n^-GaAs interface and into the n^-region.

In evaluating the tunnelling probability T in the presence of a transverse magnetic field, one would again expect that the principal effect is in the exponential term, rather than in the pre-factor. As can be seen from Figure 2, the effect of the magnetic field is to reduce the tunnelling current for all applied voltages. This can be understood qualitatively in terms of an increase in the effective height of the potential barrier due to the diamagnetic term in (4). There is an analogy here with the ballistic motion of a classical electron in a transverse magnetic field. Both classically and quantum mechanically, the combined effect of the two magnetic terms in (4) is to reduce the kinetic energy of motion along the x-axis and to increase it in the direction parallel to the y-axis. This reduction of the energy of the tunnelling electron along the x-axis is equivalent to an increase in the height of the barrier as given by the third term in (4). By performing a Taylor expansion of the exponential term containing p_y and then integrating over p_y, it can be shown that the tunnelling probability varies as

$$T = T_0 \exp\left\{ - \frac{2\kappa b}{4V_\varepsilon} \left(-eEb + \frac{e^2B^2b^2}{3m^*} \right) \right\}\ , \text{ where } V_\varepsilon = V_0 - \varepsilon\ . \tag{5}$$

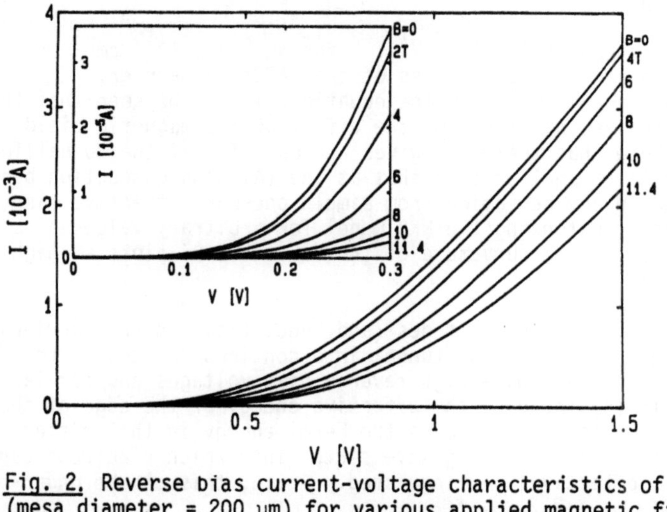

Fig. 2. Reverse bias current-voltage characteristics of the single barrier structure (mesa diameter = 200 µm) for various applied magnetic fields between 0 and 11.4T (T = 4K)

Hence for a fixed voltage on the barrier the tunnel current should decrease with increasing B according to the relation

$$J \propto \exp(-\beta B^2) \ , \tag{6}$$

where $\beta = \dfrac{e^2 b^3 \kappa}{6m^* V_F}$, with $V_F = V_0 - \varepsilon_F$.

Similarly, the additional voltage $\Delta V_B = -b\ E$ that must be applied to the barrier in order to offset the effect of B and maintain a constant tunnelling current is given by

$$e\Delta V_B = e^2B^2b^2/3m^* \quad .$$

The analysis of the magnetic field dependence of tunnelling in our structures is complicated by the fact that the experiment measures the total voltage V applied to the device rather than the barrier voltage V_B. The total voltage is the sum of the voltages dropped across the barrier, the depletion layer and the undepleted region as shown in Figure 1. Assuming that the n^- layer is not fully depleted ($V \leq 1V$) and that the voltage drop V_u across the depletion layer is small, we can write, to a good approximation,

$$V_B = b \left[\frac{2N_D e}{\varepsilon\varepsilon_0} \right]^{\frac{1}{2}} (V - V_f)^{\frac{1}{2}} \quad ,$$

where N_D is the ionised donor concentration in the n-layer and V_f is the flat band bias (V_f is the small positive flat-band voltage due to the negative space charge in the barrier). The following expression for the additional bias voltage ΔV required to offset a magnetic field B and maintain a constant current is then obtained:

$$\Delta V^{\frac{1}{2}} \approx b \left[\frac{\varepsilon\varepsilon_0}{2N_D e} \right]^{\frac{1}{2}} \frac{eB^2}{3m^*} \quad . \tag{7}$$

This relation is tested in Figure 3 which plots the square root of the applied reverse bias voltage against B^2. Note that the initial slope ($B \to 0$) of the curves does not vary greatly over the wide range of tunnelling currents which are shown in the figure: between 0.6 and 100 µA the slope decreases from 2.6 to 1.6 x 10^{-3} $V^{\frac{1}{2}}\ T^{-2}$; at 1 mA it is 1.0 x $10^{-3}\ V^{\frac{1}{2}}\ T^{-2}$.

In comparison, (7) gives a slope of 4 x $10^{-3}\ V^{\frac{1}{2}}\ T^{-2}$ for N_D = 2 x 10^{15} cm^{-3} and m^* = 0.09 m_e, the conduction band effective mass of the (AlGa)As barrier. Considering the approximation made in deriving the equation, it can be seen that the model gives at least a qualitative indication of the effect of the magnetic field on the tunnelling current. Note that the model gives a closer fit if the tunnelling effective mass m^* is assumed to be smaller than that at the (AlGa)As conduction band minimum, as would be expected in any deviation from simple one-band effective mass theory [9-10]. We are presently extending the WKB model for arbitrary values of E and B to see if it can account for the curvature of the $V^{\frac{1}{2}}$ versus B^2 plots at high magnetic fields (see Figure 3).

Note that in Figures 2 and 3, the bias voltages (and hence field in the barrier) are large enough to ensure that the electrons tunnel into continuum states in the conduction band in the n^- region. At low enough reverse bias voltages and for large enough B the magnetic terms in (4) increase the effective energy of the edge of the conduction band in the n^- region to a level above the Fermi energy in the emitter contact. At this point there is no travelling wave state into which electrons can tunnel so that the current is effectively extinguished. Again, this can be understood by a classical analogue. If $e^2B^2b^2/2m^* > \varepsilon$, the energy of the electron as it enters the barrier, then the magnetic field deflects the electron motion so that all

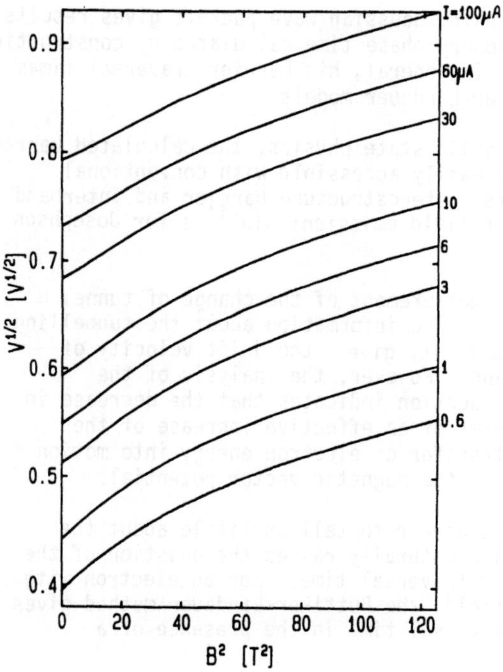

Fig. 3. Plot of $V^{\frac{1}{2}}$ versus B^2 for a range of constant currents between 0.6 and 100 μA (reverse bias) through the device. The slope of the curves are linear in the low magnetic field limit, as predicted by the model discussed in section 4, T = 4 K

of its kinetic energy is parallel to the plane of the barrier. In low bias we see behaviour in qualitative agreement with this prediction (see inset of Figure 2) and we are currently investigating it in more detail.

4. The Stevens-Buttiker-Landauer Theory for the Barrier Traversal Time in the Presence of Magnetic and Electric Fields

One of the motivations of the experimental work described in the previous section was the question of how one describes the motion of a tunnelling electron. The time taken for a tunnelling electron to traverse a quantum mechanical barrier has been the subject of considerable theoretical interest recently. STEVENS [11,12] has considered the problem using Brillouin's concept of a signal velocity corresponding to the propagation of a wave-train with a sharp leading edge. BUTTIKER and LANDAUER [13,14] used a different approach, either by considering propagation through a barrier with a small time-dependent potential superimposed upon it or else a time-independent barrier and a modulated incident wave. BARKER [15] has analysed the problem both semiclassically and with Wigner functions. Earlier work related to the problem can be found in [11-15].

The approaches of Stevens and of Buttiker and Landauer give essentially the same results for a particle moving through a rectangular barrier, the velocity being $= [2(V_0 - \varepsilon)/m]^{\frac{1}{2}}$, where V_0 is the barrier height and ε the kinetic energy of the incident particle. Buttiker and Landauer give a more general form for the traversal time,

$$\tau = \int_0^b \frac{dx}{v(x)} , \tag{8}$$

where $v(x) = [2(V(x)-\varepsilon)/m^*]^{\frac{1}{2}}$ and $V(x)$ is a spatially varying potential. The analysis uses a WKB approximation. The theory can be interpreted as showing that the instantaneous velocity is the same as that of a free classical particle with energy $|\varepsilon - V|$ above the top of the barrier. This implies that, aside from the attenuation, the dynamical properties of a tunnelling particle are very similar to those of a free classical particle moving at an equivalent energy above the barrier.

Barker finds that a numerical solution using Gaussian wave packets gives results which are in good agreement with the stationary phase time calculated by constructing a superposition of stationary waves [15]. In general, his barrier traversal times are shorter than in the Stevens and Buttiker-Landauer models.

For most of the tunnelling problems in solid state physics, the calculated barrier traversal times are much faster than those easily accessible with conventional electronic measuring equipment: $\sim 10^{-14}$ s for heterostructure barrier and interband tunnelling in semiconductors; $\sim 10^{-13}$ s for field emission; $\sim 10^{-11}$ s for Josephson junctions [14,15].

At first sight, one might expect that a measurement of the change of tunnel current in a transverse magnetic field would give information about the tunnelling velocity just as the Hall effect and conductivity give the drift velocity of electrons moving freely in a conduction band. However, the analysis of the experimental results given in the previous section indicates that the decrease in the tunnel current can be understood in terms of an effective increase of the potential height of the barrier due to a transfer of electron energy into motion parallel to the y-axis through the action of the magnetic vector potential.

Although the magnetic field measurements appear to tell us little about the tunnelling velocity, the previous discussion naturally raises the question of the effect of the magnetic field on the barrier traversal time. For an electron with $p_y = 0$ and in the absence of an electric field, the Buttiker-Landauer method gives the following simple expression for the traversal time in the presence of a transverse field:

$$\tau(B) = \int_0^b \frac{dx}{\left(\omega_c^2 x^2 + v_0^2\right)} = \frac{1}{\omega_c} \sinh^{-1} \frac{b\omega_c}{v_0} . \tag{9}$$

In this equation, $\omega_c = eB/m^*$ and $v_0 = [2(V_0 - \varepsilon)/m^*]^{\frac{1}{2}}$ is the tunnelling velocity in the absence of the magnetic field. The use of the WKB approximation implies that $V_0 >> \hbar\omega_c$. In the limit of low B and/or narrow barrier, $\tau(B)$ is given by

$$\tau(B) = \frac{b}{v_0}\left(1 - \frac{6b^2\omega_c^2}{v_0^2}\right). \tag{10}$$

We have also extended the Stevens method to this case, using a WKB approximation to the wave functions, and find that

$$x^2 = \frac{v_0^2 t}{\omega_c} \tan \omega_c t$$

for the position, x, of the front at time t. Setting $x = b$ and $t = \tau$ and for small B, this formula can be rearranged to agree with (10). Hence application of the magnetic field <u>decreases</u> the barrier traversal time whilst at the same time <u>increasing</u> the <u>attenuation</u>. Note also that both analyses indicate that $\tau \sim \omega_c^{-1}$ for <u>sufficiently</u> thick barriers.

Both the Stevens and the Buttiker-Landauer analyses can be extended to include electrons with finite p_y and also the presence of an electric field in the barrier. Taking the direction of electric field which corresponds to a lowering of the mean barrier height and hence an increase in transmission, the traversal time is given by the integral

$$\tau = \int_{x=0}^{b} \left[\frac{2(V_\varepsilon + \frac{1}{2} m^* \omega_c^2 x_c^2 + p_y\omega_c x - eEx)}{m^*}\right]^{-\frac{1}{2}} dx \tag{11}$$

$$= \omega_c^{-1}\left[\sinh^{-1}\left(\frac{\omega_c(b + f)}{v_1}\right) - \sinh^{-1}\left(\frac{\omega_c f}{v_1}\right)\right],$$

where $f = \frac{Bp_y - Em^*}{eB^2}$ and $v_1^2 = v_0^2 - (E/B - p_y/m^*)^2$.

The effect of the electric field alone can be seen more clearly by setting B = 0. In this case (11) gives, for $eEB/V_\varepsilon << 1$,

$$\tau(E) \approx \frac{b}{v_0}\left(1 + \frac{eEb}{2V_\varepsilon}\right) \quad . \tag{12}$$

Hence the Buttiker-Landauer theory predicts that an electric field which decreases attenuation by lowering the mean height of the barrier has the effect of increasing the barrier traversal time. The opposite conclusion was originally reached by STEVENS [12, pp.3658-9], but this has now been traced to an algebraic error in the use of the method of steepest descents. Therefore, to first order in the electric field and second order in the magnetic field, the Stevens and the Buttiker-Landauer theories agree.

5. Conclusion

We have presented data showing the effect of a transverse magnetic field on the tunnel current through single barrier GaAs/(AlGa)As heterostructures. The results are shown to be in at least qualitative agreement with an interpretation based upon effective mass theory and the WKB approximation and upon the assumption that the electrons are not scattered during tunnelling process. The differences between this simple theory and experiment may arise from the effect that the applied magnetic and electric fields have on the (non-exponential) pre-factors of the electron wavefunction and on the density of states involved in determing the tunnel current.

The second half of the paper considers the effect of a transverse magnetic field and parallel electric field on the barrier traversal time. The Stevens and the Buttiker-Landauer models show that the increased attenuation associated with the application of the magnetic field is accompanied with a decrease in the barrier traversal time. Similarly, it is shown that a longitudinal electric field which lowers the mean potential height of the barrier increases the barrier traversal time.

Acknowledgements

This work is supported in part by SERC. We are grateful to Dr. G.A. Toombs and Dr. J.R. Fletcher for many fruitful discussions of this work, to Mr. D.C. Taylor who carried out much of the experimental work described in Section 3, to Professor K.E. Singer for providing the GaAs/(AlGa)As layers and to Dr. G. Hill for processing them.

References

1. L. Esaki: In Proceedings of the 17th International Conference on the Physics of Semiconductors, ed. by J.D. Chadi, W.A. Harrison (Springer, New York 1985) pp.473-483
2. T.C.L.G. Sollner, P.E. Tannenwald, D.C. Peck, W.D. Goodhue: Appl. Phys. Lett. 45, 1319 (1984)
3. M. Heilblum, M.I. Nathan, D.C. Thomas, C.M. Knoedler: Phys. Rev. Lett. 55, 2200 (1985)
4. A.R. Bonnefoi, D.H. Chow, T.C. McGill: Appl. Phys. Lett. 47, 888 (1985)
5. D.C. Taylor, P.S.S. Guimaraes, B.R. Snell, F.W. Sheard, L. Eaves, G.A. Toombs, J.C. Portal, L. Dmowski, K.E. Singer, G. Hill, M.A. Pate: In Proceedings of the International Conference on Modulated Semiconductor Structures, to appear in Surface Sci. (Kyoto 1986)
6. P.S.S. Guimaraes, D.C. Taylor, B.R. Snell, L. Eaves, K.E. Singer, G. Hill, M.A. Pate, G.A. Toombs, F.W. Sheard: J. Phys. C. Solid State Phys. 18, L605-9 (1985)

7. L. Eaves, P.S.S. Guimaraes, F.W. Sheard, B.R. Snell, D.C. Taylor, G.A. Toombs, K.E. Singer: J. Phys. C: Solid State Phys. 18, L885-9 (1985)
8. L. Eaves, P.S.S. Guimaraes, B.R. Snell, D.C. Taylor, K.E. Singer: Phys. Rev. Lett. 53, 262 (1985)
9. L. Eaves, D.C. Taylor, J.C. Portal, L. Dmowski: In Proceedings of the 4th International Winter School on Two-Dimensional Systems, Mauterndorf, Austria 1986 (to be published by Springer, Berlin, Heidelberg)
10. L. Eaves, P.S.S. Guimaraes, B.R. Snell, F.W. Sheard, D.C. Taylor, G.A. Toombs, J.C. Portal, L. Dmowski, K.E. Singer, G. Hill, M.A. Pate: to be published in Microstructures and Superlattices (1986)
11. K.W.H. Stevens: Eur. J. Phys. 1, 98 (1980)
12. K.W.H. Stevens: J. Phys. C: Solid State Phys. C16, 3649 (1983)
13. M. Buttiker, R. Landauer: Phys. Rev. Lett. 49, 1739 (1982)
14. M. Buttiker, R. Landauer: Proceedings of the Europhysics Study Conference on Tunnelling at Low Temperatures (Leuven 1985) pp.5-7
15. J.R. Barker: Physica B134, 22 (1985)

See also these Proceedings for a comprehensive list of references on tunnelling times

Part III

Perspectives in Microfabrication Applications

Scaling Limits of Silicon VLSI Technology

R.H. Dennard

IBM Thomas J. Watson Research Center, P.O. Box 218,
Yorktown Heights, NY 10598, USA

The rapid progress in integration of digital circuits over the last fifteen years has been made possible by continuous reduction of the dimensions of devices and interconnections. The greatly improved density of integrated circuits, combined with growth in size of the silicon chips, has led to very high levels of complexity as measured in logic gates or memory bits per chip. The MOSFET technology is used for the present highest levels of integration known as VLSI, with minimum dimensions of about 1 μm. The progress in miniaturization of devices and interconnections into the submicron regime will lead us inevitably toward dimensions where fundamental limits are encountered.

The purpose of this chapter is to review the limits of miniaturization, as they are presently understood, for MOSFET integrated circuits, and to indicate technology and design directions likely to be necessary when these limits are approached. First, the concept of scaling of MOSFET devices and circuits is reviewed, and an illustration is shown where it has been applied to achieve exploratory circuits with dimensions of about 0.5 μm. A generalized form of scaling, in which the electric field intensity is allowed to be a variable parameter, is also discussed and related to CMOS technology. The subject of limits is introduced by showing how 0.25 μm CMOS devices would be derived from strict adherence to scaling, and how some of the idealizations or assumptions underlying scaling begin to fail at that level of miniaturization. Problems with interconnections at such dimensions are also discussed. Next, the particular constraints of dynamic RAMs at small dimensions are reviewed and technical directions are discussed. Then low temperature operation of MOSFETs is discussed both with respect to improved performance and with respect to extending the limits of miniaturization. Finally, directions in miniaturizing bipolar transistors are reviewed and the projected performance is compared to that for MOSFETs.

1. SCALING OF MOSFET INTEGRATED CIRCUITS

1.1 Basic Scaling with Constant Electric Field

The concept of scaling of MOSFETs is illustrated in Fig. 1, which shows an n-channel MOSFET of relatively large dimensions on the left and a scaled-down version of the same basic device on the right. The basic principle in scaling is to reduce all the device dimensions by a common factor, represented by α, while lowering the applied voltages by the same factor in order to maintain the same electric fields within the scaled device [1]. An important part of the problem is to reduce the depletion regions which surround the n^+ source and drain regions and extend into the substrate. For an optimized device in a given technology, the depletion regions extend over a large portion of the channel length, L, separating the source and drain. These depletion regions merge with the depletion region under the gate produced by a positive voltage applied to turn on the device. The basic function of the gate is to lower the potential barrier between the n-doped source and the p-doped substrate just under the gate to allow electrons to flow out of the source to the drain. This must occur when the gate voltage reaches a certain well-controlled threshold value, V_t, which is typically about 20% of the power-supply voltage for MOSFET circuits. The depletion regions around the source and drain

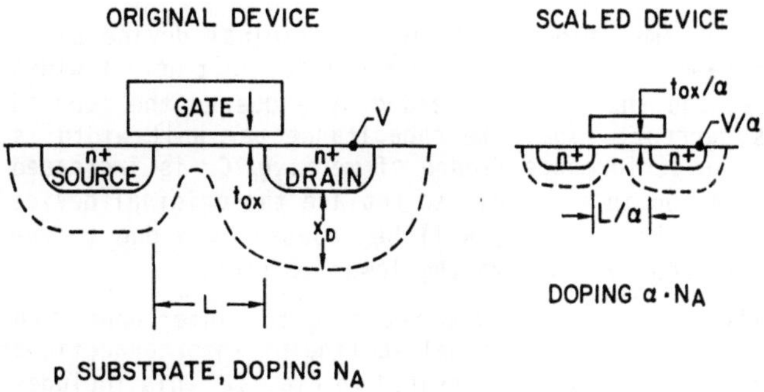

Fig. 1. Principles of MOSFET constant-electric-field scaling

interfere with the gate and affect the magnitude of the potential barrier. Therefore, the channel length, L, of the device on the left cannot be reduced substantially (beyond certain tolerances allowed for manufacturing variations) without causing a severe decrease in the threshold voltage V_t.

In scaling, the depletion region depth is scaled down along with the other device dimensions by increasing the substrate doping concentration, N_A, by the scaling factor α, in addition to reducing the applied voltages, as previously described, by the same factor. For illustration, for an abrupt junction in a uniformly-doped substrate, the depletion depth, X_D, is given by

$$X_D = \sqrt{\frac{2\varepsilon}{q} \, \frac{(V_{D/S} + \psi_B - V_{SUB})}{N_A}} \quad , \tag{1}$$

where $V_{D/S}$ is the applied drain or source voltage, V_{SUB} is the applied substrate voltage (normally negative), and ψ_B is the built-in junction potential. It is seen that X_D reduces approximately by α in the scaled device as long as the applied voltage is large compared to ψ_B. Also, the value of V_{SUB} can be adjusted so that the term $\psi_B - V_{SUB}$ is reduced exactly by α (down to the point that $V_{SUB} = 0$). It is clear that the scaling of depletion regions works in general for non-uniformly doped profiles, when these profiles are scaled in doping concentration and linear dimensions [2].

It is also important that the threshold voltage, V_t, of the scaled device be reduced in about the same proportion as the power supply voltage, V_{DD}, applied to its circuits so that the circuits will operate properly at scaled-down voltage levels. The threshold is given by the well-known equation

$$V_t = (\phi_{ms} + 2\phi_b) - Q_b/C_{ox} \quad . \tag{2}$$

For the commonly-used polysilicon gate with n^+ doping, the term in parentheses (which is the work function of the gate relative to the substrate, ϕ_{ms} , plus a band-bending term, $2\phi_b$) is a small negative value. Since the depletion region under the gate is scaled in very much the same way as the depletion region under the source, the bulk charge, Q_b , within that depletion region is a constant (negative) value. Thus, the threshold scales down because of the increased C_{ox} associated with the scaled-down gate-insulator thickness, t_{ox} . Although these considerations are approximate, it is relatively simple to adjust the ion-implanted doping profile at the silicon surface under the gate to achieve the desired threshold voltage in the scaled device. In very small devices, the electric field may, however, differ appreciably at the silicon surface as will be discussed in Sect. 3.

The conduction of current in the scaled device is very analogous to that in the original device. The scaled gate voltage gives the same electric field in the scaled gate insulator, and therefore the charge density in the inversion layer is the same.

This charge moves laterally at the same velocity as in the original device since the lateral electric field is the same. Therefore, the current per unit device width is the same in the scaled device, and the g_m is increased by α due to the reduced gate voltage needed to get this current. The gate capacitance per unit width is readily seen to be the same, so the performance figure of merit, g_m/C, is increased by α. If the scaled device is inserted in a circuit to replace the original device (reducing L but not the width), the circuit delay will be reduced by α due to the larger g_m and the power will be reduced by α due to the lower voltage.

Normally, however, when we talk of scaling we mean reducing the interconnection dimensions as well as the device dimensions. The ideal scaling of interconnections involves reducing all three dimensions by α, as illustrated in Fig. 2. This includes basically the line widths and spaces, insulator thicknesses, and contact hole dimensions. The increase in resistance effects and in current density were pointed out in our original paper [1]. They become most troublesome only at very small dimensions and will be considered later in Sect. 4 along with other problems in complete integrated-circuit chip design. The most basic thing about scaled interconnections is that the characteristic impedance of an interconnection line is maintained, since all the form factors are the same. Therefore, the capacitance per unit length remains the same as well as all coupling capacitances per unit length to adjacent lines. (This, in fact, is the reason why the thicknesses of the lines should be reduced, besides the fabrication difficulties with thicker lines.)

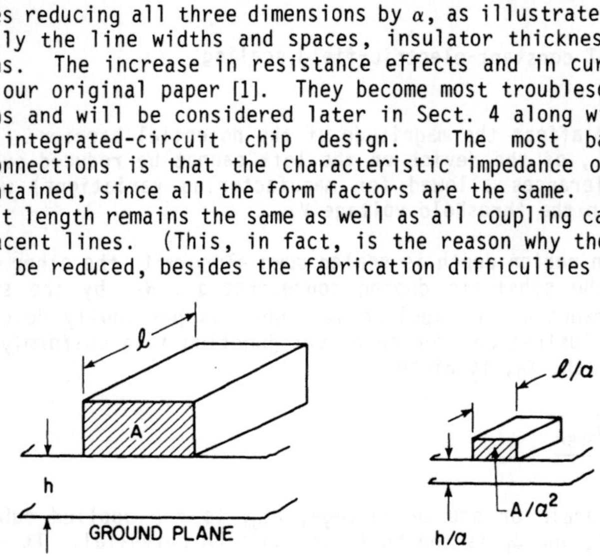

CAPACITANCE	C	C/α
RESISTANCE	R	αR
TIME CONSTANT	RC	RC
CURRENT DENSITY	J	αJ

Fig. 2. Scaling of interconnection lines and the results

When all the integrated-circuit layout dimensions are reduced by α, including the MOSFET widths as well as lengths in this case, the layout density obviously increases by α^2. A given scaled circuit has the total length of its interconnection lines decreased by α, and therefore the wiring capacitances are reduced by the same factor. Since the scaled MOSFETs also have reduced gate capacitance by a factor α, the total capacitance load on a given circuit is less by α. Relating to the previous discussion of g_m, these devices with reduced widths will have the same g_m as the original unscaled devices. Therefore, the g_m/C of fully-scaled integrated circuits increases by α and the circuit delays will be correspondingly decreased. In this case the power consumption per circuit is reduced by α^2 due to reduced current as well as reduced voltage, which keeps the power density constant in the scaled-down circuits.

1.2 An Illustration of Scaling: 0.5 μm NMOS Circuits

In this section, as an illustration of scaling, a brief review is given of the densest experimental integrated circuits which have been fabricated up to this time. These circuits were patterned with a vector-scan electron-beam exposure system with

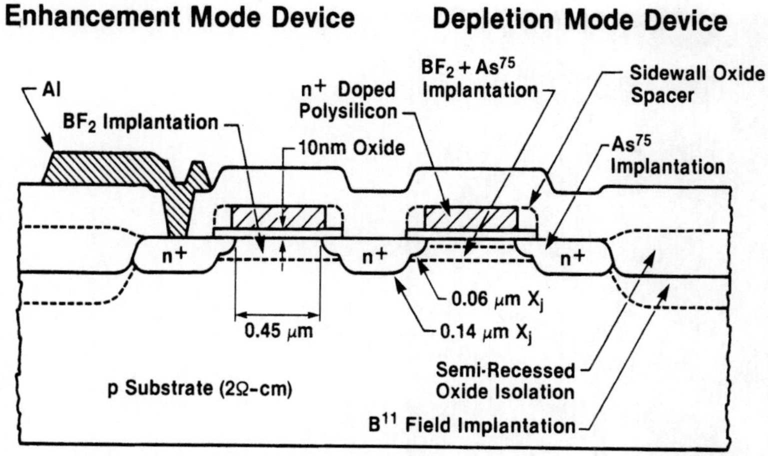

Fig. 3. Cross-section of a 0.5 μm NMOS technology

a capability of producing 0.5 μm features with a standard deviation of $\pm$0.05 μm in both feature size and level-to-level overlay. A scaled enhancement/depletion NMOS technology was developed, as shown in Fig. 3, with a nominal channel length of 0.45 μm [3].

The scaled vertical dimensions include a very thin, 10 nm, gate insulator; in addition, shallow n^+ source/drain regions are illustrated, in which a moderate arsenic ion-implantation dose is used to achieve very shallow junctions near the gate edges, while a heavy dose is used to obtain somewhat deeper and more conductive n^+ regions which are separated from the gate edges by oxide spacers. It should be noted than even better conductivity can be achieved by growing a thin metal silicide layer at the surface of the source/drain and gate regions [4]. In general, all the doping depths and insulator thicknesses in this 0.5 μm NMOS technology are reduced by employing relatively-low processing temperatures and shorter processing times. The doping concentration of the p-substrate is suitably increased.

Although this 0.5 μm NMOS technology is not scaled exactly from any given previous technology, it represents approximately a 4x scaling from typical 2 μm NMOS structures, which operate at 5 volts, and a 2x scaling from experimental 1 μm, 2.5 volt NMOS circuits [5]. Accordingly, the 0.5 μm circuits were designed to operate for a V_{DD} of 1.2-1.4 volts [3]. Figure 4 shows experimental current characteristics for both enhancement and depletion devices with W/L ratios appropriate for inverter or NOR circuits. The scaling of threshold voltage magnitudes for both devices is evident, and suitable characteristics for logic circuits with excellent transfer characteristics and well-defined noise margins are seen.

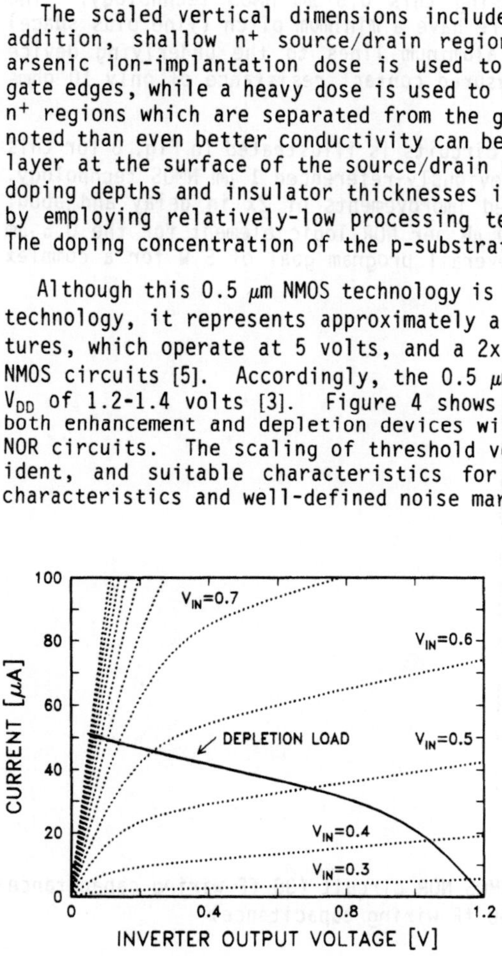

Fig. 4. Current-voltage characteristics of devices in a 0.5 μm NMOS inverter

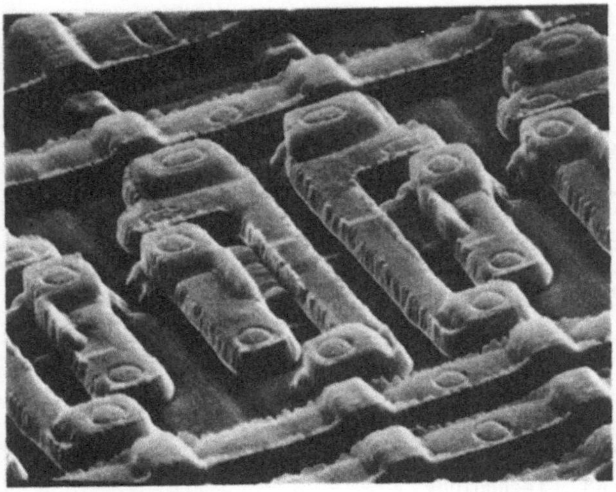

Fig. 5. Scanning-electron micrograph of a portion of PLA decoding circuits built in fully-scaled 0.5 μm NMOS technology

An SEM micrograph in Fig. 5 shows a portion of input circuits for a PLA implementation of an arithmetic logic unit using this 0.5 μm NMOS technology. The lift-off patterned aluminum lines shown here have a minimum pitch (line plus space) of 1.4 μm. The contact holes from the aluminum lines to the underlying device terminals are 0.5 μm x 0.75 μm, and a measured contact resistance of only 10 ohms was achieved.

The performance of typically-loaded NOR circuits is illustrated in Fig. 6 for this 0.5 μm technology in comparison to the previously-referenced 1 μm NMOS technology. These power-delay curves show the expected improvements of 2x in delay and about 4x in power. The power level of about 50 μW per NOR logic element for the 0.5 μm technology is very low, and matches the overall program goal of 5 W for a complex chip with 100 K logic elements.

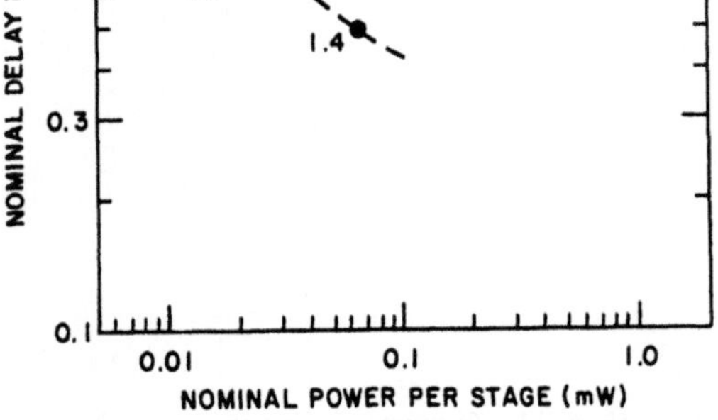

Fig. 6. Measured performance of 0.5 μm NMOS NOR circuit (30 fF wiring capacitance) compared with previous 1 μm technology (60 fF wiring capacitance)

1.3 Generalized Scaling with Electric Field as a Parameter

Although constant-electric-field scaling is applicable to many situations, it is often useful to have a different electric-field intensity within a scaled device. A common problem is trying to maintain a given power-supply voltage from one generation to the next, which must lead to higher electric field within the newer devices having smaller dimensions. Many intuitive approaches to scaling have been proposed, which have ignored the effects on the electric-field patterns within the scaled devices (for example, allowing the magnitude of the depletion regions to increase relative to the other device dimensions, thus causing deleterious changes of V_t with small variations of channel length). Fortunately, there is a general approach to scaling [2], shown in Table 1, which preserves the electric field patterns within the scaled device while allowing a different electric-field intensity by a scale factor ε.

Table 1. Generalized scaling relationships

Physical parameters	Scaling factor
Linear dimensions	$1/\alpha$
Electric field intensity	$\varepsilon \cdot 1$
Voltage (potential)	$\varepsilon \cdot 1/\alpha$
Impurity concentration	$\varepsilon \cdot \alpha$
Wiring current density	$\varepsilon^2 \cdot \alpha$
Gate delay	$1/\varepsilon \cdot 1/\alpha$
Power/gate	$\varepsilon^3 \cdot 1/\alpha^2$

In this very simple concept all the physical device dimensions of a miniaturized device are reduced by a factor α, while the applied voltage and the threshold voltage are multiplied by ε/α. The depletion regions are scaled down, along with the other dimensions, by multiplying the doping concentration within the scaled device by $\varepsilon\alpha$. For ε=1, these scaling relationships clearly reduce to the constant-electric-field case, given in the right hand part of the column.

Obviously, there are practical limits to generalized scaling. Increased electric field (ε>1) is limited by reliability effects such as gate-insulator failure or device degradation due to hot-carrier mechanisms during long-term use. The current density in the devices and interconnection lines also increases, which can lead to electromigration failures.

The current in a scaled device or circuit increases with ε^2 up to the point where it is limited by saturation velocity effects. As shown by the idealized relationships in Table 1, this tends to cause the circuits to operate faster by a factor of ε. However, the power consumption goes up by ε^3 under the same assumptions, and the current density by ε^2. These are severe practical limits. The performance increase with ε is also limited by the extrinsic device and circuit resistances.

To put this subject in perspective, the main commercial products in the past using NMOS technology have kept the electric-field intensity in the devices fairly constant due to constraints on power consumption. At the level of 1 μm technology, the desire to retain a 5 V power-supply standard has led to the widespread development and adoption of CMOS technology because of its lower power-consumption characteristics. Along with the development of graded junctions to alleviate hot-electron problems [6], this has allowed the electric-field intensity to increase by about a factor of two over previous generations of MOSFETs. In future generations of submicron CMOS there will be continued pressure to increase electric field due to proposed power-supply standards, but it probably will increase only gradually with improved understanding of how to deal with the reliability and power problems.

2. PROBLEMS IN SCALING CMOS

The scaling principles were developed originally with NMOS technology in mind. CMOS offers some additional complications. For discussion, consider the n-well CMOS technology as illustrated in Fig. 7. The p-channel devices use the same n-doped polysilicon gate, since the boron in p-doped polysilicon tends to diffuse through the gate oxide layer. Such p-channel devices are more difficult to scale. For a truly complementary device with a p-doped polysilicon gate the small term $(\phi_{ms} + 2\phi_b)$ in the threshold equation, (2), is reversed in sign. However, with an n-doped gate this term is offset by about 1.1 V, giving a threshold voltage of about -1.2 to -1.5 V depending on the doping concentration for this device at the surface of the n-type well. To obtain a threshold voltage of about -0.8 V, suitable for a 5 volt power-supply level, it is necessary to counterdope the n-well with a p-type ion-implanted surface layer as shown in Fig. 7. This surface layer must be very shallow to allow good turn-off behavior, since it is basically a depletion-type device which tends to conduct in a buried channel at the bottom of that layer. To scale this device down, decreasing the magnitude of the threshold voltage requires a heavier implant dose, which must be made even thinner than scaling would dictate in order to turn it off properly. Although this approach can be made to work down to 0.5 μm channel length [7], it appears that a more suitable gate material is needed at smaller dimensions.

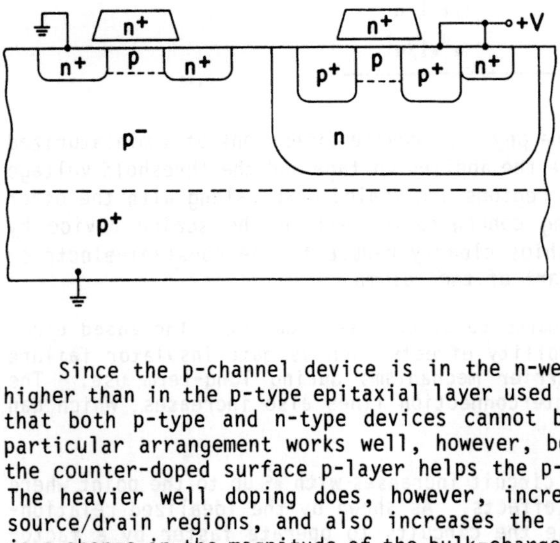

Fig. 7. Illustration of typical n-well CMOS devices

Since the p-channel device is in the n-well where the doping concentration is higher than in the p-type epitaxial layer used for the n-channel device, it is clear that both p-type and n-type devices cannot be optimized in the same way. This particular arrangement works well, however, because the heavier well doping under the counter-doped surface p-layer helps the p-channel device turn off properly [8]. The heavier well doping does, however, increase the junction capacitance of the source/drain regions, and also increases the "substrate-sensitivity" effect, which is a change in the magnitude of the bulk-charge term in (2) when the source potential with respect to the n-well changes, as it does in some circuit applications.

In general, it is possible to have truly complementary structures for both device types in CMOS. One avenue for this is the development of a workable p-doped polysilicon gate. Another, which is being widely pursued, is the development of a gate material with a work function midway between n-doped and p-doped silicon which can be used with both device types. Materials such as tungsten and tantalum silicide fall in this category [9, 10]. The substrate doping concentration can also be made the same by using the "twin-tub" approach [11], which makes more sense when the design approach is the same for both devices with complementary gate electrodes.

The other striking problem indigenous to scaling CMOS is the latchup phenomenon. It has been widely perceived that reducing layout dimensions, and thus bringing p- and n-type devices closer together, will greatly increase the susceptibility to latchup due to increased lateral gain and also due to reduced diffusion time for injected minority carriers. Actually, however, many scaling features are beneficial: reduced voltages, heavier background doping, shallow junctions and wells, and thinner epitaxial layers all favor improvement of latchup behavior. Simulation studies have demonstrated that these factors do indeed control latchup at the level of 0.5 μm dimensions [12]. Since the minimum holding voltage for latchup is about equal to the bandgap potential (in order to forward bias both p and n junctions), it is clear that when the scaled supply voltage approaches that value latchup will not be a problem.

3. SCALING LIMITS OF MOSFETs

At very small dimensions some of the idealizations or assumptions of scaling begin to break down. These limits of scaling pose fundamental limits to further miniaturization. At the beginning of this discussion it should be noted that there is no single factor limiting scaling and no well-defined basic limit; instead, the design measures that are necessary to cope with a number of problems can lead to deterioration of performance.

It is instructive to consider device parameters for 0.25 μm channel-length CMOS derived from scaling and examine the resultant problems. The values of the parameters given in Table 2 are derived from scaling with constant electric-field intensity (ε=1) from typical parameters used in 1 μm CMOS technology operating at 5 V. To make truly complementary devices, a twin-tub structure with n-doped and p-doped polysilicon gates for the n-channel and p-channel devices is assumed. The values given in Table 2 are also roughly consistent with generalized scaling from the 0.5 μm NMOS device of Fig. 3 with values of ε=2 and α=2.

Table 2. Parameters for scaled 0.25 μm CMOS example

Channel length, L	0.25 μm
Gate oxide thickness	5-6 nm
Gate materials	n^+ and p^+ polysilicon
Source/drain junction depth	0.05-0.1 μm
Background impurity concentration	$3\times10^{16}/cm^3$
Power-supply voltage	1.2 V
Threshold voltage	$\pm$0.25 V

The physical problems of fabricating devices with the parameter values shown in Table 2 are not insurmountable in any fundamental way. Improvements in processing techniques are continually being made. The indicated gate insulator thickness of 5-6 nm of course requires great care in terms of defect control to achieve good yield and reliability. The shallow junctions can be achieved with the same method shown in Fig. 3, with deeper junctions spaced away from the edges of the device. New processing ideas for junctions, for example diffusing the dopants out of silicide layers on the surface [13], appear to be promising.

There are several electrical problems which start to become significant in devices of these dimensions, however. One problem relating to the thin gate insulator and the low operating voltage is a reduction of the expected charge in the inversion layer, as illustrated in the experimental curves of Fig. 8, for an n-channel device. This shows the normalized gate-to-channel capacitance, $(dQ_n/dV_g)/C_{ox}$, as a function of gate voltage for a fairly thin gate insulator [2]. This behavior comes about

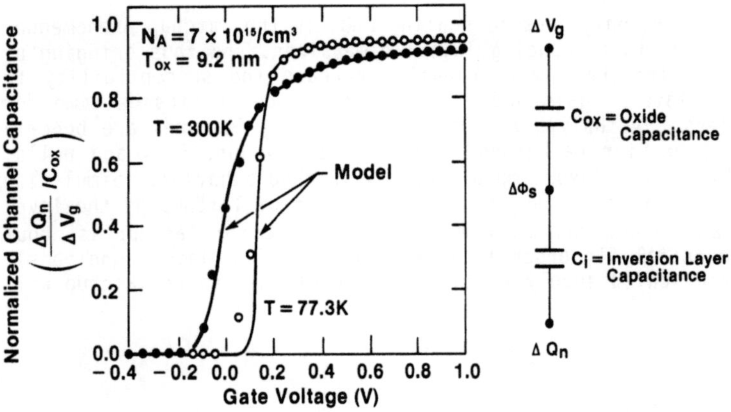

Fig. 8. Normalized capacitance from gate to inversion layer, for a thin gate oxide, versus gate voltage at room and liquid-nitrogen temperatures

because the inversion layer is distributed over a distance below the silicon surface, so that there is a capacitance called the inversion capacitance in series with the gate insulator. The effective depth of the inversion layer decreases at larger gate voltage due to the stronger normal electric field at the silicon surface [14], and the gate-to-channel capacitance approaches C_{ox} . The transition is much sharper at low temperature due to the reduced depth of the inversion layer. For this particular device with a relatively low threshold implant, the threshold voltage corresponds to a channel capacitance of about 0.5 C_{ox}.

A model has been developed for the relationship of Fig. 8, which matches the behavior very well, as shown [2]. Although these curves have been normalized to C_{ox}, they are a function of oxide thickness and also of parameters which affect the normal field. The charge in the inversion layer is given by the integral of this curve up to the value of the applied gate voltage. The model has been used to calculate the inversion-layer charge for device parameters very similar to those shown in Table 2. The resultant curve of charge versus gate voltage is shown in Fig. 9, in comparison with the idealized charge $(V_g - V_t)C_{ox}$, which is valid in large devices with thick oxides. It is seen that the inversion-layer charge is significantly reduced, particularly at low gate drives.

It is important to note that these relationships shown in Figs. 8 and 9 have been derived for the case of little or no lateral field. When the devices have a large lateral field, due to an applied drain-to-source voltage, the inversion layer spreads

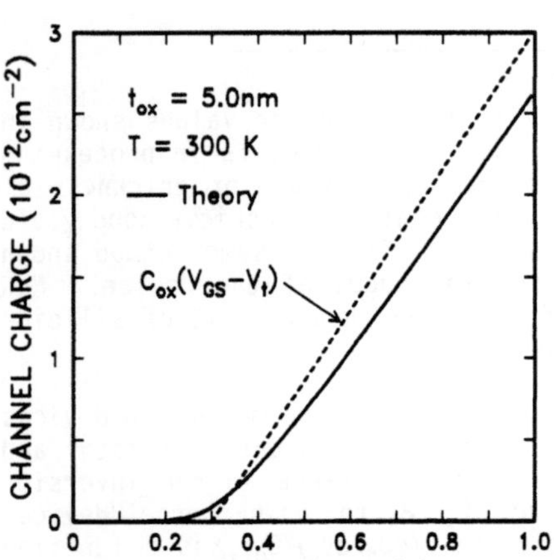

Fig. 9. Modeled inversion-layer charge versus gate voltage in a scaled NMOS device with a 5nm gate insulator

out due to the heating of the electrons [15]. The consequent reduction of charge in the inversion layer tends to decrease the device current, but the mobility is likely to increase as the inversion layer spreads out. The net impact on device current requires better models to evaluate. Also, the reduction in charge is partially compensated in circuit performance by the corresponding decrease of the gate capacitance.

Another important limitation to miniaturization of MOSFETs is illustrated in Fig. 10, which shows a curve of inversion-layer mobility versus average vertical electric field within the inversion layer [16]. This type of relationship has been well established experimentally [17]. The average electric field in the inversion layer can be approximated by

$$E_{in} = \varepsilon_{ox}[\,-(\phi_{ms} + 2\phi_b) + V_t + (V_g - V_t)/2]/(t_{ox}\cdot\varepsilon_{si}) \quad . \qquad (3)$$

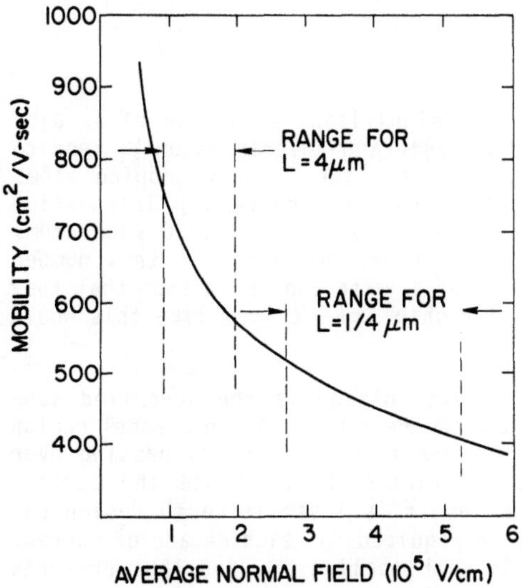

Fig. 10. Measured typical inversion-layer mobility versus average normal electric field, with ranges for different device sizes, using n^+ doped polysilicon gates

Scaling down of n-channel MOSFETs using n^+ doped polysilicon as a gate material leads to higher vertical electric field, because the small non-scaling term in the threshold equation, due to the gate work function, must be offset by higher bulk charge in the substrate (i.e., by higher channel-implant dose). The net result, as shown in Fig. 10, is that the mobility in a scaled 1/4 μm device is considerably reduced in comparison to a larger device. This is also worse because of the higher electric field due to the scaling factor ε=2, which increases V_g and V_t.

Experimentally, the mobility is found to be proportional to $(E_{in})^{-1/3}$ over the range of interest. This model for mobility combined with the model for inversion charge is used to compute the current versus gate-voltage characteristic shown in Fig. 11 for a 0.25 μm MOSFET with similar parameters to the ones given in Table 2. It is seen that the changes of charge and mobility with gate voltage combine in such a way as to give a very linear current versus voltage curve. However, the slope of that curve, which represents the effective mobility which could be used in a simple device model, is only about 390 compared to a value of 550 in a 1 μm n-channel device [5].

There are several problems relating to the value of the threshold voltage, V_t, which must be about 0.25 V from circuit-design considerations. The temperature coefficient of V_t, due to changes in the Fermi level and the band-bending at threshold, reduces only slowly with dimensions in submicron MOSFETs and is about -

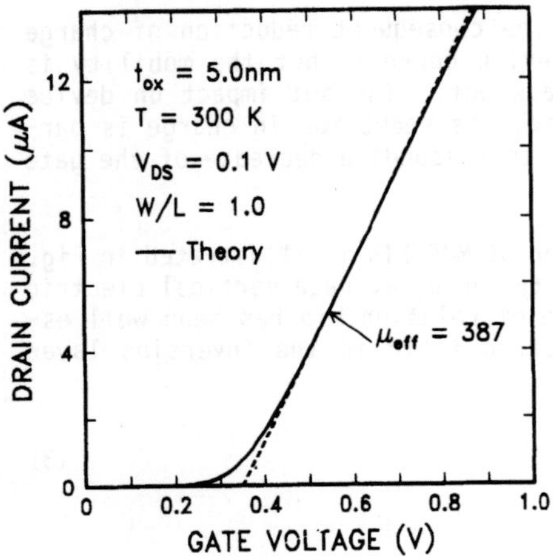

Fig. 11. Calculated current-voltage characteristics for 0.25 μm device including inversion-layer charge and mobility models

0.9 mV/°C for the 0.25 μm devices. This gives a significant variation of V_t over the normal range of operating temperatures. Manufacturing tolerances on V_t tend to scale down with the gate insulator thickness; however, the impurity doping atoms implanted into the surface to set the value of V_t (by way of the term Q_b in equation 2) are subject to statistical fluctuations. There are only about 1000 such atoms in a 0.25 μm-long by 0.25 μm-wide device. The standard deviation in this number is about 3%, which would cause about 5% variation of V_t with the assumption that they are distributed uniformly. Within a very complex chip some devices like this would have 4-5 times the standard deviation.

Still another problem relating to the threshold voltage is the so-called subthreshold current level. As the barrier height from the source to the channel region is raised, by reducing V_g below V_t, the thermally-activated carriers passing over that barrier follow the well-known exponential relationship. Since the surface potential is heavily coupled to the gate, the result is insensitive to design parameters, with about 80 mV change of gate voltage required for each decade of current change. This relationship also changes little with scaling [1], and thus presents a problem as V_t is scaled down, as illustrated in Fig. 12 for the 0.25 μm devices. The leakage current through the devices at $V_g = 0$ is substantial enough for a worst-case room temperature ambient to impact very low-power (battery-operated) applications. Retention time of dynamic RAMs would also be affected (see Sect. 5). Low-temperature operation gives much sharper turn-off behavior [18], with the slope as shown in Fig. 12 increasing inversely with T.

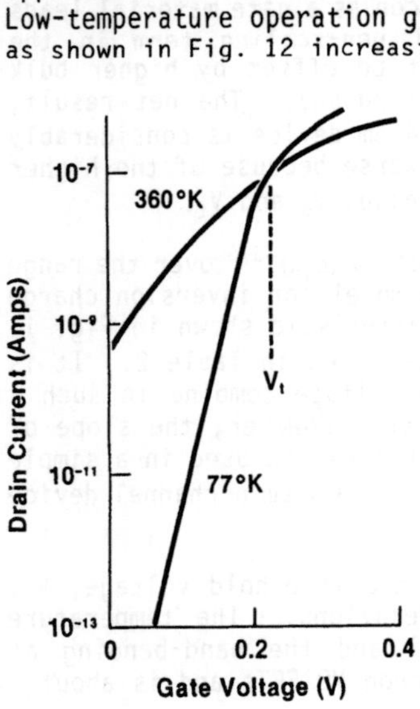

Fig. 12. Illustration of the subthreshold conduction characteristics of a scaled 0.25 μm MOSFET

From the discussion thus far in this section, it can be seen that several problems become significant in 0.25 μm MOSFETs which tend to limit further scaling. For room-temperature operation, the threshold voltage cannot be reduced because of temperature and other variations. It is unproductive to scale the supply voltage below about 1 volt because of this threshold limit. Also, the built-in junction potential would then prevent the scaling of depletion regions in the correct way. Moreover, the inversion-layer capacitance effect is more dominant at lower voltages. Therefore, further scaling is possible only in the framework of constant voltage and increasing electric field. Even then, the insulator thickness can be scaled down only less than a factor of 2 due to the onset of direct tunneling.

There are structural and material changes which can be made in order to minimize some of the problems discussed here. Figure 13 illustrates an improved structure, which is very similar to that used in the 0.5 μm technology. This structure addresses the resistance problems in the very shallow source/drain junctions by adding deeper junctions at a small distance from the gate edge, which can be reinforced by a thin silicide layer [4] to improve conductivity.

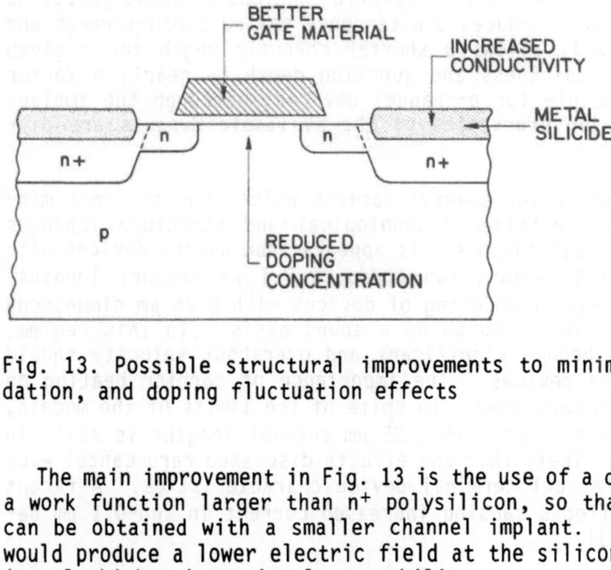

Fig. 13. Possible structural improvements to minimize resistance, mobility degradation, and doping fluctuation effects

The main improvement in Fig. 13 is the use of a different gate material which has a work function larger than n^+ polysilicon, so that the desired threshold voltage can be obtained with a smaller channel implant. As illustrated in Fig. 14, this would produce a lower electric field at the silicon surface which can give significantly higher inversion-layer mobility.

Another advantage of the reduced channel doping is that the threshold voltage is less susceptible to statistical fluctuations. Although less doping leads to a higher

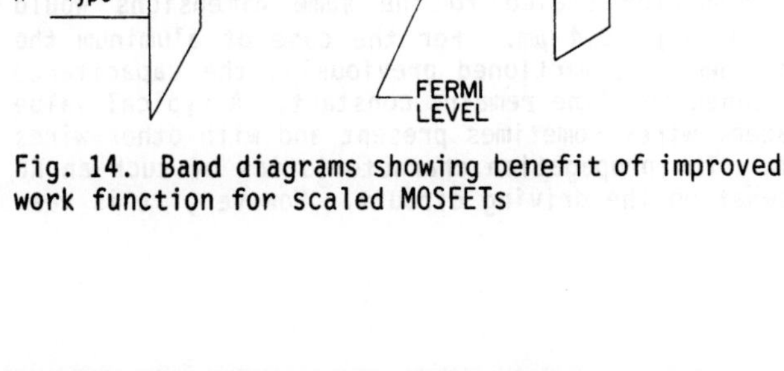

Fig. 14. Band diagrams showing benefit of improved gate material with an appropriate work function for scaled MOSFETs

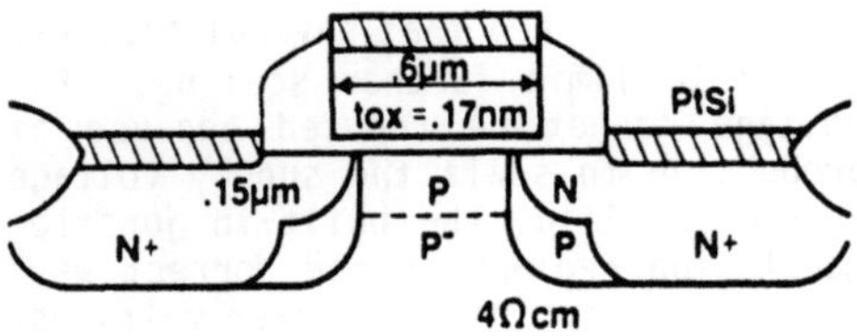

Fig. 15. Improved n-channel MOSFET with enhanced boron doping at source and drain edges (after OGURA et al.[19])

percentage variation, the actual deviation is lower. Unfortunately, materials with optimum work functions are not now readily available for either n-channel or p-channel devices. The mid-gap gate materials mentioned earlier give threshold magnitudes which are too large.

Another kind of structural improvement is illustrated in Fig. 15. The new feature here is the addition of a boron implant at the time the lightly-doped drain implant is performed [19]. The presence of the accentuated boron doping at the edges of the source and drain reduces the amount by which the depletion regions at those junctions extend into the channel region. This reduces the tendency toward punch-through and short-channel effects, and basically allows a shorter channel length for a given set of parameters like insulator thickness and junction depth by nearly a factor of 2. This scheme works in principle for p-channel devices, although the implantation properties and the diffusion properties of the available dopants are different.

In summary, this section has discussed several factors which tend to limit miniaturization of MOSFETs and has identified technological and structural changes which can help cope with some of these factors. It appears that useful devices with good characteristics for some applications can approach 0.1 μm channel lengths. It should be noted that the accuracy of modeling of devices with 0.25 μm dimensions and below requires improvement in order to be on a sound basis. In this regime, non-equilibrium transport should become significant and overshoot velocity should occur at least in portions of the devices. The importance of carrier heating on the inversion-layer depth has been mentioned. In spite of the limits of the models, the measured data on experimental devices with 0.25 μm channel lengths is still in fairly good agreement [20]. It is likely that the effects discussed here cancel each other out to some degree in the actual terminal device characteristics. A recent paper has identified ballistic effects causing increased current in sub-0.1 μm devices at very low temperatures [21].

4. ON-CHIP AND SYSTEM INTERCONNECTION LIMITS

The basic scaling relationships for interconnections on the silicon chip were discussed in an earlier section, and problem areas with respect to resistance and current density were identified. In this section it is desired to put these problems in perspective and understand how they may affect the limits of miniaturization. Interconnection of multiple chips into a system is also of interest.

It is instructive to consider the effects of scaling interconnection lines by an amount comparable to the 0.25 μm CMOS devices discussed in the previous section. The typical thickness of an interconnection line is about the same as the minimum lithographic or device dimension, while the width may be about 1.5 times greater. For 0.25 μm devices, a typical conductor scaled to the same dimensions would therefore have a cross-section of 0.25 μm×0.4 μm. For the case of aluminum the resistance would be about 300 ohms/mm. As mentioned previously, the capacitance per unit length of a scaled interconnection line remains constant. A typical value in an environment with other adjacent wires sometimes present and with other wires crossing above is about 0.2 pF/mm. The propagation characteristics of such an RC interconnection line depends somewhat on the driving circuit. However, it is reasonable to compare the RC time constant to the typical delay time of 0.25 μm CMOS circuits with ideal lossless interconnection lines, which is about 0.3 ns. The length of the scaled interconnection line would have to be a little more than 2 mm to have the same RC product of 0.3 ns and to therefore significantly affect the transmitted signal. Most circuits in this environment would have line lengths of only a fraction of a millimeter and thus would be unaffected.

It can be seen that scaling to 0.25 μm dimensions does not cause any fundamental problems in signal propagation. Circuits on a complex chip would be partitioned so that given functional blocks would occupy a small area with short interconnection lines. An area of 1 mm^2 can contain about 5000 logic gates at these dimensions. These logic gates could be interconnected with lines on two levels for the basic x and y directions on the chips. The outputs of these functional blocks, largely data and signal busses, could be interconnected with perhaps two additional wiring levels. These additional levels would carry the longer lines, which would cross the entire chip in some cases. They could use somewhat larger dimensions to minimize the resistance effects. An alternative is to use repeaters to reshape the signals after a certain transmission length in a very similar fashion to what is commonly done in telephone communications. Of course the rise time and delay requirements for long lines would not be as stringent, since these signals on these lines would be at a higher level of the system hierarchy previously wired up by interconnecting chips on a card or board.

The increased current density in scaled interconnection lines mainly affects the distribution of power supply and ground lines. Considerable work is being carried on to develop interconnection lines with higher current density capability, typically using laminated structures with electromigration-resistant metals or compounds interleaved with lower-resistance aluminum layers [22]. Wide lines on the upper wiring layers would also be used to distribute power to the various functional areas. Again, these lines can be kept somewhat thicker. Finally, multiple contacts to the next packaging level would have to be used to bring power into the chip. Preferably these contacts will be distributed throughout the area of the chip, rather than at the edges.

To take advantage of the higher performance levels of very complex integrated circuits with scaled dimensions it is necessary to feed signals on and off the chip at high rates. The increasing number of such signal lines can be handled by spreading the connections over the area of the chip, for example using solder reflow techniques and "flip-chip" bonding to the next packaging level. Improved packaging will be needed to reduce the inductance of these connections. This is a problem not only of scaled MOSFETs, but is being faced already in the present generation of high-performance bipolar circuits.

5. SCALING OF DYNAMIC RAMS

When studying the limits of miniaturization, special attention must be given to dynamic RAMs because of their particular design problems. One unique problem is the stringent requirements on leakage of charge from the storage capacitor. Because memory arrays on a given chip are getting larger with each successive generation, the storage time between refresh operations is also becoming larger to allow time to sequence through all the word lines (without using a larger portion of the operating cycles for refresh). The subthreshold leakage problem shown in Fig. 12 is not tolerable for dynamic RAM arrays because the charge would leak away from the capacitor through the supposedly turned-off array devices. To hold the charge a sufficient time requires the current in the array devices to be reduced nearly six orders of magnitude below the value at threshold. This prevents scaling of the threshold voltage magnitude of memory-array MOSFETs below about 0.7 volts. This tends to lead toward structures where the array device threshold is different from that of devices used in support circuits. CMOS fits into that picture well if p-channel devices are used in the memory cells, while the n-channel devices are optimized for support circuits. (In present-day technology with n^+ doped poly gates,

Table 3. Scenario for dynamic RAM with quarter-micron dimensions

	Present tech.	Scaled design	Non-uniformly scaled design
Channel length, L [μm]	1.0	0.25	0.25
Power supply [V]	5.0	1.25	2.0
Insulator thickness (device/capacitor) [nm]	25 /15	6.2/3.7	10/5
MOSFET threshold (support/array) [V]	0.8/1.0	0.2/0.25	0.4/0.7
Electric field [MV/cm] [(1)]	3.3	3.3	4.0
Stored charge [fC] [(2)]	200	12.5	15.0

Notes: (1) In capacitor, with grounded plate

(2) With capacitor area scaling with L^2 from 40 fF base at 1 μm

this would make the p-channel device easier to design with the larger threshold magnitude.)

The scaling of dynamic RAMs to 1/4 μm dimensions is examined in Table 3. The values listed in the first column are believed to be typical of designs presently being developed with 1 μm dimensions for 1 Mbit dynamic RAM chips. The second column shows the results of constant electric field scaling down to 1/4 μm dimensions, with reduced voltages. This approach clearly does not work with the scaled array-device threshold because of subthreshold leakage, and merely increasing that threshold magnitude would not work because the word-line voltage would not be adequate to turn it on properly under all conditions. The third column shows an approach in which the power supply voltage is kept larger, relative to the scaled case. Also, the insulators are somewhat thicker to limit the increase in electric field. In this case the array-device threshold magnitude can be set at the required 0.7 volts, and the power supply voltage is large enough to write full voltage levels into the capacitor if word-line boot-strapping is used.

One problem with the non-scaled design in Table 3 is containing short-channel effects and punch-through. The device in Fig. 15 is promising in this regard, as has been described. Assuming the array size (words x bits) remains constant, the stored voltage levels and signal levels should scale down roughly with the supply voltage. Array noises would scale down as well, and the sense amplifiers built in such a 1/4 μm technology would clearly be very responsive to small signals. The signals are well above the thermal noise values. The principal noise source at this level has been shown to still be ionizing radiation [23] either from α-particle emission within the structure or from cosmic radiation. The soft-error problem has been shown to increase somewhat with scaling to smaller dimensions, since the stored charge tends to decrease more than the collected ionized charge from radiation events [24].

The design shown in Table 3 is seen to work well under the assumption of a constant number of words and bits in an array. Since such arrays would be physically much smaller, numerous arrays would be placed on each chip and extra decoding circuits (formerly on the card) would be added. However, most designers appear to want to continue the trend toward larger arrays with more memory cells on each bit line. This requires more stored charge in each memory cell in order to obtain a workable signal on the longer bit lines. This consideration has led to the development of truly 3-dimensional structures where each storage capacitor is formed on all sides of a trench etched into the silicon substrate, and where the other electrode is a polysilicon conductor deposited in the trench [25].

Scaling much below the levels shown in Table 3 is difficult because of the tunneling limit on the thin insulators and the voltage constraint due to the large array-device threshold voltage. The only apparent way to reduce the voltage is to use lower operating temperature to obtain good turn-off behavior with a low threshold voltage as previously shown in Fig. 12.

6. OPERATION OF SCALED MOSFETs AT LIQUID-NITROGEN TEMPERATURE

For some time it has been apparent that there are significant advantages to operating MOSFETs at liquid nitrogen temperature, including increased carrier mobility, decreased leakage, sharper subthreshold turn-off transition, lower interconnection line resistance, and enhanced reliability [26]. There has been renewed interest in low-temperature operation for CMOS technology, possibly because the low power consumption reduces some of the refrigeration requirements. Also, latch-up effects are greatly reduced at low temperature due to lower bipolar gains. The reported performance improvement of about 1.5-2 times has been limited by the lack of optimization of the structures for low temperature to compensate for threshold changes.

On the other hand, the improved device mobilities are offset partially by saturation velocity effects in small devices at relatively high lateral fields and by the effects of parasitic series resistances. Hot-electron degradation has been shown to be more severe at low temperature, which requires lower operating voltage. The cost of developing a satisfactory cooling system has always been a key concern.

In relation to the limits of miniaturization, the subject of low-temperature operation has additional significance since it helps with many of the basic problems which have been described. Lower operating voltages are possible at low temperature because of the sharper turn-off characteristic, because of reduced threshold variation with temperature in a controlled ambient, and because of the sharper transition of the gate-to-channel capacitance (Fig. 8) due to reduced inversion-layer depth. Low-temperature operation also is very effective in minimizing the problems of scaled interconnection lines. The resistivity is greatly reduced in pure-metal conductors by 6-8 times at 77K, and electromigration effects disappear. Even though the built-in potential increases at low temperature due to the shift in Fermi level, it can actually be offset by forward biasing the junctions with a positive substrate bias [2]. In net, it appears that low-temperature operation becomes even more attractive for small-dimension devices and circuits. Liquid-nitrogen temperature specifically is a good choice since it realizes most of the advantages, does not freeze out the substrate completely, and has reasonable refrigeration efficiency.

7. SCALING OF BIPOLAR INTEGRATED CIRCUITS

The subject of scaling of bipolar devices and integrated circuits is considered here to put the previous sections on MOSFETs into perspective. For high-performance applications bipolars are the technology of choice, but they have been limited in integration level by complexity of processing and power dissipation.

The scaling of bipolars is much more complicated than MOSFETs. For a given circuit type, such as high-performance ECL, there is no possibility of reducing the voltages since the base-emitter turn-on voltage is relatively fixed. Thus, the scaling down of bipolars is similar to generalized scaling of MOSFETs, requiring increased electric field, increased doping to control depletion regions, and in-

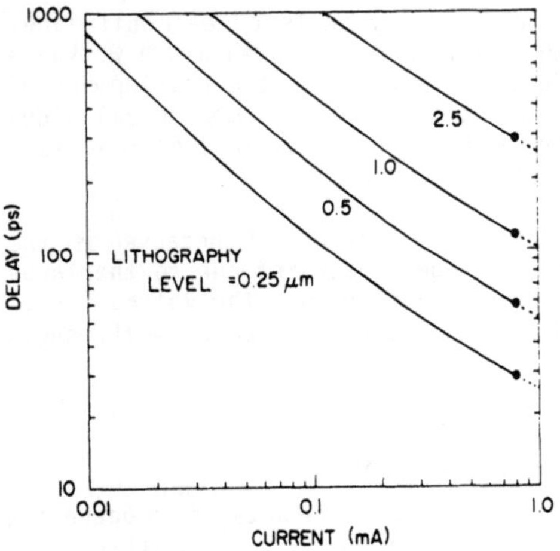

Fig. 16. Performance projections for ECL circuits using scaled bipolar transistors (after SOLOMON and TANG [26])

creased current density. Carried out in a range where the mobility varies greatly with the base doping [26], the scaling is really a design procedure to optimize the device to some degree at each dimension given by the emitter stripe width, a. The solution favored by SOLOMON and TANG [26] is to reduce the base width as $a^{0.8}$, to increase the base doping as $a^{-1.6}$, and to increase the current density (controlled by the circuit) and the collector doping both by a^{-2}. It is clear that bipolar scaling puts a premium on improved metallurgy. To the degree that these goals can be achieved, the principal result, which is shown graphically in Fig. 16, is that the delay decreases linearly with dimension at a fixed current and power level/gate. The power can be reduced as shown by reducing the current levels, but the delay increases accordingly.

Thus it is seen that scaled bipolars have better speed capability than MOSFETs but at much higher power levels. This speed advantage is at the basic circuit level. It seems obvious that the integration level must grow to allow some reduction in propagation delays at the system level in order to match improved basic circuit speed. This will require rapid growth in cooling capability to dissipate the large chip powers unless some other approach is found.

8. CONCLUSIONS

Much as the progress in integrated electronics over the recent past can be understood in terms of scaling, the pathways in the near future are also discernible. However, the limits of miniaturization are rapidly being approached. These limits make scaling much beyond the level of 0.25 μm dimensions difficult at room temperature, but low-temperature operation can extend the limits to about 0.1 μm. Materials and structural changes were discussed which are both beneficial and necessary in order to obtain continued performance improvement with scaling. Although it may sound discouraging to contemplate the idea of limits, it actually should be considered a challenge to step up to the practical materials and processing questions which must be overcome to approach such limiting dimensions in a mass-production manufacturing environment. The chase is by no means over, and the rewards can still be very significant.

ACKNOWLEDGEMENT

Many of the ideas reviewed here come from working with my colleagues as indicated by the references, and I thank them for sharing and discussing these ideas with me. In particular, I acknowledge the contributions of Matthew Wordeman in supplying illustrations and carefully reviewing this paper.

REFERENCES

1. R.H. Dennard, F.H. Gaensslen, H.N. Yu, V.L. Rideout, E. Bassous, A.R. LeBlanc: IEEE J. Solid-State Circuits SC-9 , 256 (1974)
2. G. Baccarani, M.R. Wordeman, R.H. Dennard: IEEE Trans. Electron Devices ED-31 , 452 (1984)
3. M.R. Wordeman, A.M. Schweighart, R.H. Dennard, G.A. Sai-Halasz, W.W. Molzen: IEEE Trans. Electron Devices ED-32 , 2214 (1985)
4. C.M. Osburn, M.Y. Tsai, S. Roberts, C.J. Lucchese, C.Y. Ting: Electrochem. Soc. 1st Int. Symp. on VLSI (Detroit, MI 1982) p.213
5. R.H. Dennard, F.H. Gaensslen, E.J. Walker, P.W. Cook: IEEE J. Solid-State Circuits SC-14 , 247 (1979)
6. S. Ogura, P.J. Tsang, W.W. Walker, D.L. Critchlow, J.F. Shepard: IEEE Trans. Electron Devices ED-27 , 1359 (1980)
7. S.Y. Chiang, K.M. Cham, R.D. Rung: IEEE Int. Electron Device Meeting Tech. Dig. (1983) p. 534
8. G.J. Hu, C.Y. Ting, Y. Taur, R.H. Dennard: IEEE Int. Electron Device Meeting Tech. Dig. (1982) p. 710
9. S. Iwata, N. Yamamoto, N. Kobayashi, T. Terada, T. Mizutani: IEEE Trans. Electron Devices ED-31 , 1174 (1984)
10. V. Schwabe, F. Neppl, E.-P. Jacobs: IEEE Trans. Electron Devices ED-31 , 988 (1984)
11. L.C. Parillo, R.S. Payne, R.E. Davis, G.W. Reutlinger, R.L. Field: IEEE Int. Electron Device Meeting Tech. Dig. (1980) p. 752
12. Y. Taur, W.H. Chang, R.H. Dennard: IEEE Int. Electron Device Meeting Tech. Dig. (1984) p. 398
13. M. Horiuchi, K. Yamaguchi: IEEE Trans. Electron Devices, ED-33 , 260 (1986)
14. F. Stern, W.E. Howard: Phys. Rev. 163 , 816 (1967)
15. D. Schmitt-Landsiedel, G. Dorda: Electron. Lett. 18 , 1041 (1982)
16. G. Baccarani, M.R. Wordeman: IEEE Trans. Electron Devices ED-30 , 1295 (1983)
17. A.G. Sabnis, J.T. Clemons: IEEE Int. Electron Device Meeting Tech. Dig. (1979) p. 18
18. F.H. Gaensslen, V.L. Rideout, E.J. Walker, J.J. Walker: IEEE Trans. Electron Devices ED-24 , 218 (1977)
19. S. Ogura, C.F. Codella, N. Rovedo, J.F. Shepard, J. Riseman: IEEE Int. Electron Device Meeting Tech. Dig. (1982) p. 718
20. W. Fichtner, E.N. Fuls, R.L. Johnston, R.K. Watts, W.W. Wick: IEEE Int. Electron Device Meeting Tech. Dig. (1983) p. 384
21. S.Y. Chow, D.A. Antoniadis, H.I. Smith: IEEE Electron Device Lett. EDL-6 , 665 (1985)
22. J.K. Howard, J.F. White, P.S. Ho: J. Appl. Phys. 49 , 4083 (1978)
23. L.L. Lewin, J.D. Meindl: IEEE Int. Solid-State Circuits Conf. Dig. (1984) p. 160
24. G.A. Sai-Halasz, M.R. Wordeman, R.H. Dennard: IEEE Trans. Electron Devices ED-29 , 725 (1982)
25. H. Sunami, T. Kure, N. Hashimoto, K. Itoh, T. Toyabe, S. Asai: IEEE Trans. Electron Devices ED-24 , 746 (1984)
26. P.M. Solomon, D.D. Tang: IEEE Int. Solid-State Circuits Conf. Dig. (1979) p. 86

Three Part Series on Heterojunction Transistors

P.M. Solomon

IBM Thomas J. Watson Research Center, Yorktown Heights, NY 10598, USA

Introduction

This series of lectures is to be about the exciting and very rapidly developing field of heterojunction devices. The possibilities inherent in heterojunctions i.e. the joining of semiconductor materials of differing band gaps, has been apparent since the invention of the transistor, where a wide bandgap emmitter was proposed by SCHOCKLEY in 1949 [1]. Heterojunctions with well understood electrical properties have been available only since the advent of molecular beam epitaxy (MBE) in the 1970's, and more recently by means of metal-organic vapor phase epitaxy (MOVPE). This series of lectures will deal with high speed transistors using heterojunctions. Unfortunately, there are no good up to date text books on heterojunction devices. SZE [2] deals with principles of semiconductor devices, which can be applied to heterojunction devices, MILNES and FEUCHT [3] discussed heterojunctions, but this book is rather outdated, CASEY and PANISH [4] provide information on heterojunction material parameters, and HARRISON [5] discusses semiconductor band structure, including heterojunctions. Review articles include ANDO, FOWLER and STERN [6], on two-dimensional systems, KROEMER [7] on heterojunction bipolar transistors, SOLOMON [8] on high-speed devices, MORKOC and SOLOMON [9], DRUMMOND, MASSELINK and MORKOC [10] and Linh [11] on the modulation doped FET, HEIBLUM [12] on the hot electron transistor and POATE and DYNES, EASTMAN, and NATHAN and HEIBLUM, in [13], on ballistic transport devices. In Part 1 of this series we will discuss fundamental heterojunction concepts necessary to understand heterojunction devices. Part 2 will deal with heterojunction field effect transistors (HFETs) and Part 3, the heterojunction bipolar transistor (HBT) and the hot electron transistor. Finally, the scaling properties and extendibility of these devices will be addressed. Our examples will focus mainly on the GaAs/(Al,Ga)As system, but the principles apply to many other material systems at present under investigation.

Part 1. Electronic Properties of Heterojunctions

1.1 III-V Bandgap vs. Lattice Constant

Various possibilities of forming heterojunctions of semiconductors of the same lattice constant yet having different bandgaps may be inferred by tracing vertical lines in Fig. 1.1 [14]. The lines joining the binary III-V semiconductor points represent intermediate ternary semiconductor alloys, solid lines correspond to direct, and dotted lines to indirect bandgap. It is seen that the (Ga,Al)As system remains lattice matched over the range of cation admixture; similarly for the Ga and Al with P or Sb. Other systems are only lattice matched at specific alloy compositions, such as the $In_{0.53}Ga_{0.47}As/In_{0.52}Al_{0.48}As$ system which is lattice matched to InP, and the $In_{0.49}Ga_{0.51}P$ on GaAs system.

Strained-Layer Heterojunctions: All of the heterojunction devices we will be discussing occupy only a thin layer at the surface of the semiconductor, mostly less than 1 micrometer in thickness and in some cases, for the channel of a field-effect transistor, or a base of a bipolar

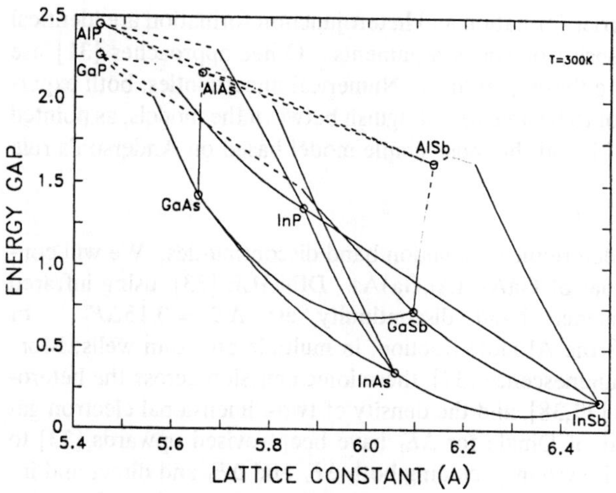

Fig. 1.1 Energy Gap of III-V semiconductors vs. lattice constant, at 300 K, from FRANK [14]. Solid lines indicate a direct gap, chain-dots an X, small dots an L conduction band

transistor, less than 100 nm. For these thin layers the requirements for precise lattice matching may be relaxed. The epitaxial layer is forced to take on the lattice constant of the substrate, and as a result, is under a large strain, equal to the lattice mismatch. The stress builds up as the layer thickness is increased until, in the form of misfit dislocations, the strain is relieved at the interface. The critical thickness of such a layer has been calculated by VAN DER MERWE [15], in terms of the relative hardness and bonding strength of the layers, and also by MATTHEWS and BLAKESLEE [16]. This manner of growth has also been termed pseudomorphic. Pseudomorphic growth offers a new degree of freedom to heterojunction device designers. For instance (In,Ga)As has a smaller bandgap than GaAs and has been used in FET structures with GaAs [17] and AlGaAs [18], and with GaAs [19] in a bipolar transistor. GeSi alloys can also be grown on silicon [20,21] . An important factor to be considered, for these highly strained layers, is the effect of the strain itself on the band structure of the semiconductor. For biaxial compression the bandgap may either decrease [22] or increase [23] depending on the symmetry of the bands. For thick, mismatched layers, the strain is accommodated by misfit dislocations. For instance, MESFET circuits [24] heterojunction FETs [25] and bipolar transistors [26] having 'reasonable' characteristics have been fabricated on GaAs grown on Si. Apparently the misfit dislocations get 'bent out' over the range of a few hundred nm, resulting in material of good crystalline quality.

1.2 The Heterojunction Discontinuity

For predicting what the resulting band diagram will be when two semiconductors are joined it is assumed that the band structure of the component materials is undisturbed right up to the interface. In addition it is assumed that there are fixed discontinuities in the conduction (ΔE_C) and valence (ΔE_V) bands adding up to the bandgap discontinuity (ΔE_G). These discontinuities give rise to the all-important heterojunction barriers in electronic devices. KROEMER [27] has reviewed this subject. There is no general consensus as to the cause of the discontinuities. The first calculations were in terms of Anderson's electron affinity model [28]. Another model based on the LCAO method [5] predicts that most of the discontinuity will be in the conduction band, for heterojunctions sharing a common anion such as the (Al,Ga)As system. Another model [30] assumes that the recipe for Schottky barrier formation and heterojunction formation are identical and can be deduced from effective density of states arguments. Other approaches [31] use pseudo-potential calculations to calculate these quantities. Numerical uncertainties, both experimental and theoretical are so large as not to be able to distinguish between the models, as pointed out by FREEOUF and WOODALL [32], and the most simple model based on Anderson's rule may well be correct

Several methods have been used to determine conduction band discontinuities. We will concentrate on the most-studied system, that of GaAs/(Al,Ga)As. DINGLE [33], using infrared absorption, first determined the valence band discontinuity as: $\Delta E_V = 0.15\Delta E_G$, or $0.18x_{Al}(0 < x_{Al} < 0.4)$ meV where x_{Al} is the Al mole fraction, in multiple quantum wells; More recent measurements have used photoluminescence [34], thermionic emission across the hetero-barrier [35, 36], capacitance vs voltage [37,38], and the density of two-dimensional electron gas [39] (see sect. 1.4). The earlier results of Dingle for ΔE_V have been revised upwards [33] to $\Delta E_V = 0.55x_{Al}$ meV. A summary for this system, including both ΔE_C and ΔE_V and direct and indirect band edges, is given in Fig. 1.2. We will refer to this figure later when discussing heterojunction FETs.

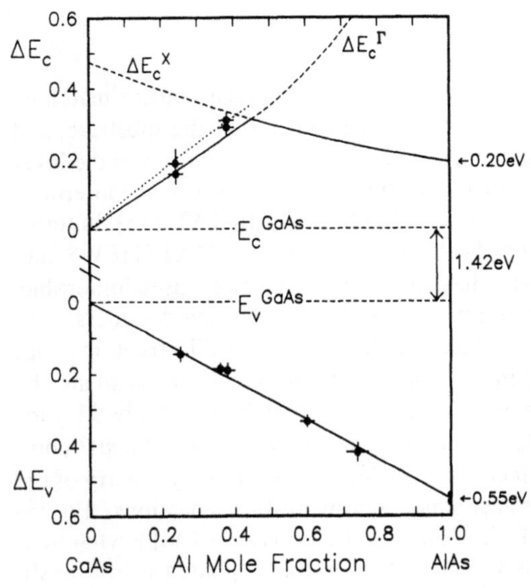

Fig. 1.2 Energy band alignment for the GaAs/(Al,Ga)As system from BATEY and WRIGHT [36]. The solid line for the valence band discontinuity ΔE_C is a straight line fit to their data, and those for the conduction band discontinuity was deduced from $\Delta E_C = \Delta E_G - \Delta E_V$ where ΔE_G is from [4]

1.3 Heterojunction Band Diagram

Given the heterojunction discontinuities, the relevant bandgaps, the doping on either side and the applied bias, we can derive the band diagram across the heterojunction, for semi-infinite semiconductors. We will use the example in Fig. 1.3, of an n-n+ heterojunction, to illustrate the method:

1. Draw the heterojunction in its flat-band condition, with the Fermi levels drawn according to the bulk doping on either side. The built in voltage is the difference between the Fermi levels.
2. Separate the layers and, far from the original interface, draw the bulk bands with the Fermi levels shifted by the applied potential. On the appropriate side construct a line (chain dot in Fig. 1.3b, at an energy of ΔE_C above the conduction band).

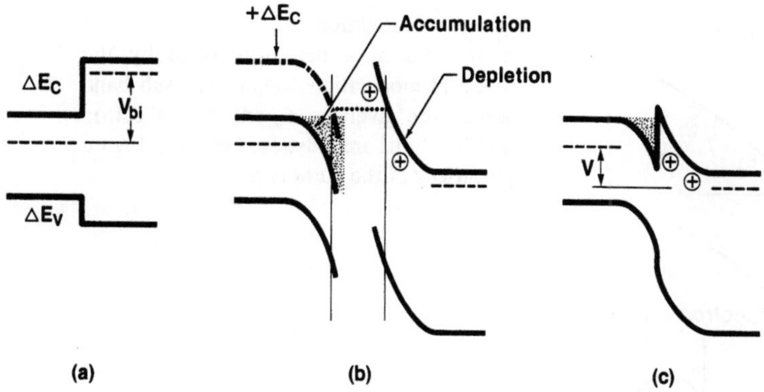

Fig. 1.3 Construction of a heterojunction band diagram (as explained in text) starting from the flat-band condition (a), separating the semiconductors and bending the bands of each independent of the other; then matching fields and energies to determine the interface position (b); then joining the semiconductors (c). Dashed lines indicate Fermi levels. Vertical lines in (b) indicate the projected interface position

3. Bend the bands, including the constructed line, in the direction that would join this line with the conduction band on the other side. The rate of bending is given by (one dimensional, in this example) Poisson's equation

$$\frac{d^2\phi}{dx^2} = \frac{e}{\varepsilon}(N_D - N_A + p - n) \tag{1.1}$$

where ε, N_D, N_A, p, n refer to the permittivity of the layer, donor and acceptor doping densities and free hole and electron concentrations respectively for the layer. Far from the junction the semiconductor is neutral. The net space charge is positive when the bands are bent upwards, negative if downwards. Charge density approaches $e(N_D - N_A)$ when the bands are bent away from the Fermi level. When the band is bent toward the Fermi level a large free carrier concentration of electrons (conduction band) or holes (valence band) is induced.

4. Match the slopes at the same energy as shown by the dotted line in Fig. 1.3b. This is equivalent to matching the electric fields across the interface. Strictly, the displacements should be matched, but the dielectric constants are quite similar for most III-V materials, ranging from 11-14. Draw vertical lines, which will eventually correspond to the interface, at the matching points, and join the semiconductors by horizontal displacement (Fig. 1.3c).

For abrupt heterojunctions (constant composition on each side) the band edges are always parallel to each other. For graded heterojunctions, the assumption of transitivity for heterojunctions leads to a quasi-electric field being generated [40] where the change in the accelerating potential for electrons or holes is given by the change in electrostatic potential plus the composition-induced changes in conduction and valence band energies. The quasi-fields are therefore different for electrons and holes and indeed, may even be opposite.

1.4 Quantum Sub-Bands

Heterojunction confinement of the electron wave function gives rise to a quantization perpendicular to the hetero-interface and the formation of quantum sub-bands. The simplest illustration of this is just the particle in a box where the quantum energies satisfy the relation $E_n = \hbar^2 k_n^2/2m^*$ and $k_n = n\pi/a$, where a is the well width. The energy levels are proportional to

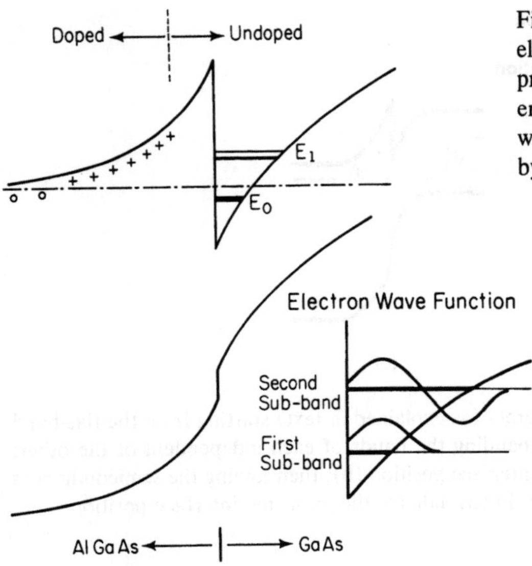

Fig. 1.4 Formation of a 2-dimensional electron gas at a hetero-interface by the process of modulation doping. The sub-band energies are given by E_0 and E_1. Electron wave functions are illustrated in inset. Figure by courtesy of the authors of [9]

n^2. The above simple treatment is approximate in that it assumes idealized boundary conditions and a simple effective mass approximation. The levels correspond to the bottom of sub-bands. In each sub-band the total electron energy is $E = E_n + \hbar^2 k_p^2/2m^*$ where k_p is the wave number in the plane of the layer. The same considerations apply to holes as to electrons, but for holes the situation is rather complicated because of the presence of degenerate light and heavy hole bands. Because this layer is quantized into "two-dimensional" sub-bands, it is commonly called a two-dimensional electron gas (2DEG). The density of states in each sub-band is constant and is equal to $m^*/\pi\hbar sup2$, when both spin states are considered. For wide potential wells (more than a few nm), considerable band bending occurs due to the electronic space charge. This is illustrated in Fig. 1.4. A modulation-doped layer may be formed by doping the larger bandgap material and allowing electrons to spill over into the narrow bandgap material to form quantum channels adjacent to the interfaces [41]. Normally an undoped AlGaAs spacer layer (thickness d_0) separates the channel from the doped AlGaAs. The energy levels in the quantum channels must be derived by self-consistently solving Schroedinger's and Poisson's equations for the quantum well and taking into account exchange and correlation effects for the electrons, which tend to lower the sub-band energies. This was done by ANDO [42], and STERN and DAS SARMA [43]. An approximate formula for the first sub-band energy using the variational method [44] yields

$$E_0 = \left(\frac{3}{2}\right)^{5/3}\left(\frac{e^2\hbar}{m^{*1/2}\varepsilon}\right)\frac{N_d + \frac{55}{96}N_s}{(N_d + \frac{11}{32}N_s)^{1/3}}, \tag{1.2}$$

where N_s is the sheet concentration of the 2DEG and N_d is the sheet concentration of ionized acceptors and traps in the substrate. For high quality samples $N_s >> N_d$ ($\simeq$ 10:1). The sub-band energies are approximately constant with temperature, and the temperature dependence of the Fermi level, at a given sheet concentration, can be found by integrating the Fermi function over all of the sub-bands. The electron density in the channel may be derived by the application of Gauss' law across the hetero-interface [45,46], and accounting for the position of the interface Fermi level. A simple formula, valid at high doping levels, is

$$N_s = \left(\frac{\varepsilon}{ed_0}\right)\Delta E_C - \psi_s - E_D, \tag{1.3}$$

where ψ_s is the surface potential (E_C - E_F) in GaAs ($\simeq$0.1V), E_D is the donor energy in AlGaAs and d_0 is the spacer thickness. The formation of the electron channel is equivalent to that of the inversion layer in the Si MOSFET, except that quantum effects are much more significant in GaAs because of its lower density of states.

1.5 Mobility of the 2 DEG

The most remarkable feature of modulation-doped layers in GaAs-AlGaAs is the very high mobility obtainable, especially at low temperatures. This is because of the physical separation of the ionized impurities from the 2 DEG by the heterojunction barrier and the undoped spacer. The first results of DINGLE *et al.*[41] stimulated a tremendous interest in heterojunction devices, and since then mobilities of $>1\times10^6$ cm ^{2}/V-s [47] have been obtained at low temperatures. Several models [48-51] have been used to determine the mobility of the 2 DEG; yet the precise determination of the magnitude of all scattering mechanisms in the 2 DEG is not settled. The two-dimensionality of the gas has to be invoked to describe screening and acoustic phonon scattering [50]. Nevertheless, in the temperature range of interest to most devices (> 77K), the major feature is that for modulation doped layers with large d_0 (>20 nm) the mobility in the 2 DEG is much the same as in undoped GaAs. The mobility at room temperature is primarily limited by optical phonon scattering; however, this mechanism quickly vanishes as the temperature is reduced. At low temperatures the mobility is limited mainly by remote donor scattering, either in the barrier or the substrate. For low concentrations, and d_0 large, the greater part of the scattering is that due to impurities in the substrate. For higher concentrations this mechanism is less effective because of the increased Fermi velocity of the electrons in the degenerate 2 DEG. Since d_0 must be smaller to achieve higher concentrations, impurity scattering in the AlGaAs will then dominate. This results in a decreasing mobility vs electron concentration at high electron sheet concentrations ($\simeq1\times10^{12}$ cm^{-2}). Large 2 DEG concentrations are needed for high-speed devices so that parameters for the 2 DEG are chosen which maximize 2 DEG concentration rather than mobility. At large electron concentrations ($N_s > 7\times10^{11}$ cm^{-2}) more than one sub-band may be occupied. When this occurs the electron mobility in the 2 DEG decreases because of inter-sub band scattering, as has been clearly observed [52].

1.6 Non-Ohmic Transport

Even at remarkably low electric fields of the order of 10 V/cm, at low temperatures, the transport in a high-mobility 2 DEG layer becomes non-ohmic. The cause is a heating of the electron distribution. SHAH *et al.*[53] reported an equivalence between effects of electron temperature and lattice temperature, at moderate electric fields, where the electron temperature was ascertained by measurement of luminescence. For bulk GaAs RUCH and FAWCETT [54] modelled the electron velocity as a function of electric field, using a Monte-Carlo technique, with results shown in Fig. 1.5a and b. Even though some parameters of GaAs have been revised since then, their results still represent experimental data fairly well. The 2 DEG velocities correspond roughly to those of undoped GaAs. So far the experimental situation has not been very satisfactory, with measured velocities, ranging all the way from 1.4×10^7 [55] to 3.8×10^7 cm/s [56], albeit at temperatures of 77 K vs. 4 K. Most high-speed devices operate with internal fields in the KV/cm range where, as seen from Fig. 1.5b, extremely high values for the low field mobility have little significance. The fall-off in mobility at fields of $\simeq$100V/cm is due to strong interactions with optical phonons in the GaAs. At even higher fields, $\simeq$2KV/cm, k-space transfer to the upper valleys of GaAs can occur (Gunn effect), giving rise to a negative differential mobility, and limiting the peak velocity of the carriers. Other materials such as $In_{0.53}Ga_{0.47}As$ and InP have larger Γ to L band separations (0.55 and 0.6eV respectively, as compared to 0.36eV for GaAs) permitting higher drift velocities. For a modulation doped layer the additional possibility of transfer

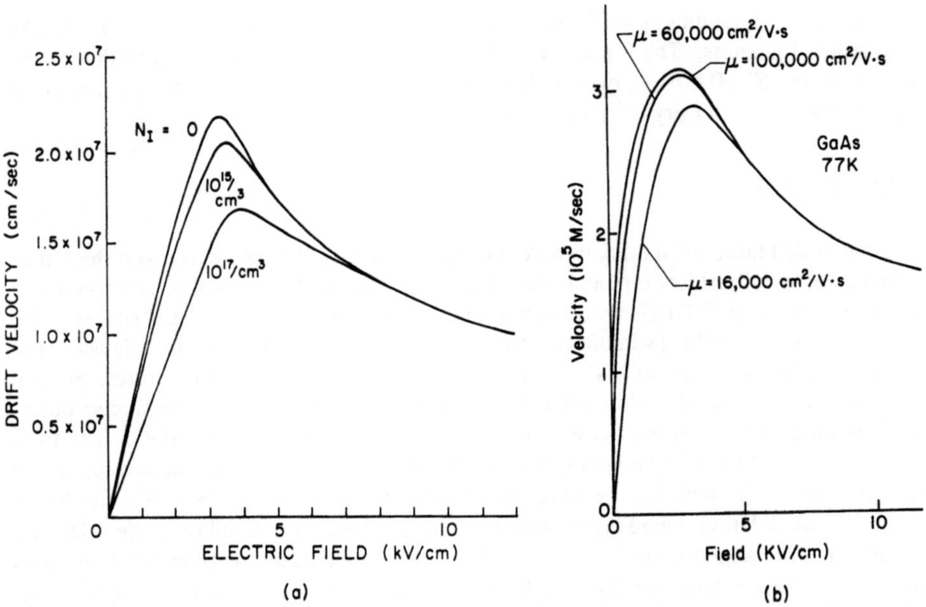

Fig. 1.5 Velocity vs. electric field for GaAs, based on Monte-Carlo simulations by RUCH and FAWCETT [52], (a) at 300 K and (b) at 77 K. Note that the doping parameter, used by RUCH and FAWCETT, was converted to the more relevant low field mobility parameter in (b). Figures by courtesy of L. Eastman.

from the GaAs to the AlGaAs layer may occur. This is known as 'real space transfer' [57]. Ever since the predictions of RUCH [58], velocity overshoot has been anticipated in small devices. When electrons are accellerated suddenly it takes time for the electron distribution to heat up. For a while scattering rates are low, characteristic of the cool, non-equilibrium electron distribution, and drift velocities are higher than steady state values. This effect has recently been modelled in modulation doped FETs by WIDIGER *et al.* using Monte Carlo. A related subject is ballistic (collisionless) transport. This has been modelled extensively by AWANO and TOMIZAWA *et al.*, also using Monte Carlo [59]. Electron velocities may approach the maximum group velocity in GaAs, which is $\simeq 1 \times 10^8$ cm/s. Recently ballistic transport has been observed in the hot electron transistor (see Sect. 3.3).

1.7 Perpendicular Transport Across Heterojunction Barriers

Perpendicular transport is an important aspect in the operation of all heterojunction devices In the following lectures we will deal with it in particular in terms of ballistic transport across the hot-electron device, and gate leakage current in a GaAs gate FET. Another mode of transport, which we will not cover, is resonant tunneling through multiple barriers. Transport across heterojunctions can be described, in most cases, by thermionic emission, or tunneling. The theory of thermionic emission follows the standard formalism in SZE [2],

$$J = A^* T^2 \exp\left[-\frac{e}{kT}(\phi - \frac{E_F}{e}) \right], \tag{1.4}$$

where A^* is the modified Richardson's constant, T the absolute temperature, ϕ the barrier height, measured from the Fermi level. For tunneling, since application of electric fields to heterojunction barriers gives trapezoidal barrier shapes, the Fowler-Nordheim formula applies:

$$J = AF^2 \exp\left[\frac{4}{3}\frac{(2m^*e)^{1/2}\phi^{3/2}}{\hbar F}\right], \tag{1.5}$$

where m^* is the effective mass in the barrier. Formulae 1.4 and 1.5 describe the current flow across the barrier remarkably well, and heterojunctions are much better behaved, in this respect, than Schottky barriers. Using the GaAs/AlGaAs/GaAs system, thermionic emission has been used to find heterojunction discontinuities [35,36], and Fowler-Nordheim tunneling has been verified [60,61]. For high Al mole fractions, tunneling in multiple bands is involved and the situation is considerably more complex than that given by (1.5). In discussing scaling of semiconductor FETs we will see that the minimum thickness for the barrier is an important scaling limit. This is related to the tunneling current where both current and thickness may be minimized by looking for a materials system with a larger barrier and larger tunneling effective mass. Another aspect of perpendicular transport is ballistic injection across a hetero-interface, where the electron is imparted a kinetic energy equal to the conduction band discontinuity, as we will discuss later in Part 3.

Part 2. Heterojunction FETs (HFETs)

2.1 Introduction

There has been much activity over the past few years in the area of heterojunction FETs (HFETs). This has been spurred on by the very high speed results obtained with these devices, with recent ring oscillator delays approaching 5 ps [62,63]. In this lecture we will discuss two heterojunction FET type devices, the MODFET and the GaAs gate FET (SISFET). While the MODFET is by far the better known of the two FETs, and thus far has a proved performance record, the SISFET type devices have properties which make them excellent candidates for future VLSI systems, and the GaAs gate type SISFET is one in which this lecturer is personally involved.

2.2 The FET Principle

Much theoretical modelling done on HFETs recently should not disguise their inherent simple operational principle. All FETs work by modulating a free carrier space charge across a dielectric. The simple formulae for a parallel plate capacitor apply $Q = CV$ and the capacitance by $C = \varepsilon A/w$ where A is the area, ε the permittivity and w the distance between the capacitor plates. The condition for the parallel plate capacitor model to apply is that the minimum lateral dimension (gate length) of the capacitor be much greater than its thickness, and this is also the condition for 'long channel' behavior of an FET (except that the electrical thickness, multiplying by the ratio of the dielectric constants of the semiconducting substrate to the gate insulator, must be used). The insulator thickness may be defined as the total thickness of insulator or depleted semiconductor (again, weighted by its dielectric constant), between the gate and the channel. This is true even when the boundaries of the depletion layers move with application of gate voltage, as long as charge is exchanged only at the layer boundaries. For a 2DEG layer, the electric field lines from the gate are terminated somewhere within the layer. The effect of the layer thickness on capacitance may be expressed as a surface capacitance given by

$$C_s = e\frac{dN_s}{d\psi_s}, \tag{2.1}$$

where ψ_s is the surface potential. The equivalent thickness ε/C_s corresponds to about 5 nm for a GaAs channel at $N_s = 1\times10^{12}$ cm^{-2}, as contrasted to about 2 nm for silicon [64]. By assuming that all the carriers in the GaAs channel are moving with the same velocity, one may derive a simple expression for the transconductance as

$$g_m = \frac{\varepsilon v}{w} \tag{2.2}$$

This formula has also been inverted (with minor refinements) to measure the saturation velocity v. Another relation which is useful, especially for HFETs where there is no doping in the channel, is

$$eN_s = \varepsilon(F_A - F_B), \tag{2.3}$$

where F_A and F_B are the electric fields on the front and backside respectively of the channel.

The formulae for capacitance and transconductance can be used to obtain the familiar relationships for intrinsic time constant $\tau = C/g_m$, intrinsic cutoff frequency $f_T = 1/2\pi\tau$ and transit time $\tau = L/v = \tau_t$ where L is the channel length of the FET and τ_t is the transit time. This last formula connects a circuit property τ to an intrinsic device property τ_t. In a digital circuit these relationships must be modified. The general formula $\tau = CV/I$ or equivalently $\tau = C/G_m$ now relates to large signal quantities. The large signal transconductance, G_m is an average over the logic swing including the region where the device is cut off. The parasitic source resistance R_s affects g_m through $g_{me} = 1/(1 + g_m R_s)$ where g_{me} is the extrinsic transconductance. The capacitance now includes the parasitic fringing and overlap capacitance of the device itself, the capacitance of devices to be driven, and wiring capacitance. Design of device and circuit for highest performance aims to minimize these parasitic effects, by reducing R_s and w, and to increase device width (and power) to match device to wiring capacitance. Under these circumstances, circuit speed is related to τ_t by some multiplying constant.

2.3 The MODFET

The modulation-doped FET (MODFET) is also known as the two-dimensional electron gas FET (TEGFET) [11], selectively doped heterojuction transistor (SDHT) [65] or, referring to its original name, high electron mobility transistor (HEMT) [66], each name revealing some feature of this device. A band diagram of a MODFET is shown in Fig. 2.1.

Threshold Voltage: The similarity between a MODFET and a MOSFET has often been stressed [8-10]. The (Al,Ga)As takes the place of SiO_2. This is true when the AlGaAs is depleted as shown in Fig. 2.1. The ionized donors in the AlGaAs may simply be looked on as a threshold voltage shifting technique. To first order [67] the threshold voltage may be calculated from

$$V_t = \phi_1 - \phi_2 - \frac{eN_D w^2}{2\varepsilon} \tag{2.4}$$

where ϕ_1 and ϕ_2 are the barrier heights at each side of the interface. This assumes that the back field is very small (small level of traps and impurities in the GaAs) and that the Fermi level is near the conduction band at threshold. This latter condition is well fulfilled for low temperature (77 K) FET operation, while at 300 K a 100 mV discrepancy could exist. A more detailed and accurate calculation of V_t is given by TIWARI [68], using the formulation for charge density in the channel of DELAGBEDEUF and LIN [46] who assume occupation of the first and second subbands in the 2DEG. Sensitivities to doping and thickness, from (2.4), are

$$\frac{\Delta V_t}{V_{bi}} = \frac{\Delta N_D}{N_D} = 2\frac{\Delta w}{w} \tag{2.5}$$

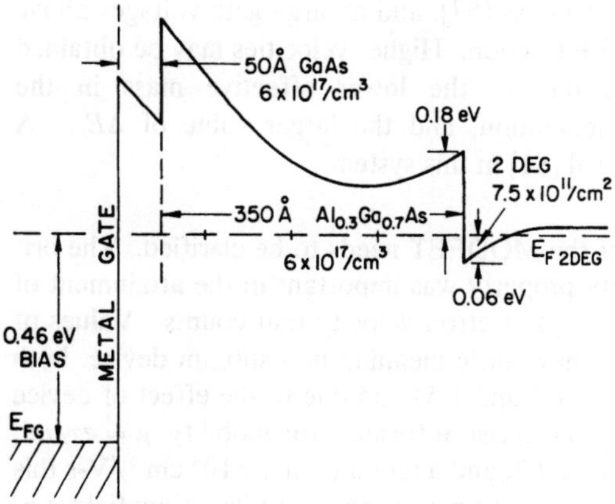

Fig. 2.1 Band diagram of an enhancement mode MODFET under forward gate bias. This MODFET has a 5nm doped GaAs cap on top of the AlGaAs. Figure by courtesy of L. Eastman

where V_{bi}, the built in voltage is given by the last term in (2.4) and is $\simeq$ 0.8 V. For a large scale integrated circuit, a standard deviation in V_t of below $\simeq$20 mV is required; which can be translated into a control requirement of 2.5% for N_d and 1 nm for w. Requirements are even more stringent than this because all causes of variation must be accounted for within the ΔV_t budget. Heavy doping of the AlGaAs is required for the MODFET to satisfy (2.4) for thin layers (as are required for a large transconductance). From 1-2$\times 10^{18}$ cm^{-3} donor concentration is required for doped AlGaAs layers of $\simeq$ 30 nm thick.

Maximum Surface Concentration: The MODFET has the property of a maximum surface concentration. As we see from the band diagram, as soon as the AlGaAs conducts, the Fermi-level is pinned within the AlGaAs at a level depending on the donor binding energy [45]. A further increase in gate voltage will only reduce the gate depletion-layer width and will not change electric fields, or carrier concentrations at the interface. Typical values for maximum carrier concentration are in the 1×10^{12} cm^{-2} range for a doping density of about 1×10^{18} cm^{-3} and a spacer thickness of about 5 nm. The effect of this on a logic device is that the carrier density and current in the device will increase with gate voltage above threshold. As long as the AlGaAs is depleted, the transconductance and capacitance will approximately be a constant. At larger gate biases the AlGaAs will start to conduct, the transconductance will decrease and the capacitance will increase and speed would be reduced.

Various models for the MODFET have been formulated [10,69-73] as well as numerical modelling studies [74-76]. A major difficulty with most models is to be able to model hot electron effects adequately. Some progress has been made in this direction using the Monte Carlo method [75,76].

Performance: The MODFET is the highest speed transistor reported to date and it has shown this property in at least three areas: Ring oscillator delays (6.5 ps) [63,64], cut-off frequency f_T (80GHz) [77], and frequency dividers (13GHz) [78]. In part this property is derived from the superior transport properties of the electrons in the 2DEG channel, since the best results are achieved at a temperature of 77 K where the electrons move at larger velocities than at 300.K, and also because the low resistance of the 2DEG permits the attainment of full device current at small voltages above threshold and lowers the parasitic resistance of the MODFET. High transconductances per unit width have been obtained. This can be estimated from (2.2) using a typical value for w of 35 nm, 12 for the dielectric constant of AlGaAs and 1.5×10^7 for the average velocity for the carriers in the channel. This gives an intrinsic g_m of 400 mS/mm. Some workers [79,80] have obtained even larger numbers by using thinner layers (larger capacitance) and shorter channels (higher velocities). Velocities in the MODFET are limited by real space transfer of hot electrons from the channel to the AlGaAs [57], and at large gate voltages above threshold ($\simeq$0.8V) by the parasitic AlGaAs MESFET action. Higher velocities may be obtained from the $In_{0.53}Ga_{0.47}As/In_{0.52}Al_{0.48}As$ system due to the lower effective mass in the $In_{0.53}Ga_{0.47}As$, the larger Γ to L or X band separation, and the larger value of ΔE_C. A transconductance of 700 mS/mm has been achieved [81] in this system.

The role of mobility in the high speed property of the MODFET needs to be clarified. The original name, HEMT, implied that the high mobility property was important in the attainment of high speed. Others have implied that it is only a high electron velocity that counts. Values of mobility, for large devices, of $> 1\times10^5$ cm ^{2}/V-s have little meaning in a sub μm device, both from the point of view of the field effect (see Figs. 1.4 and 1.5) and due to the effect of device boundaries. This may be appreciated by applying the classical formula for mobility, $\mu = e\tau/m^*$, to a FET. For the effective mass ratio of GaAs of 0.067, and a mobility of 1×10^5 cm ^{2}/V-s this implies a mean free time of 3.8 ps, greater than the transit time in sub μm FETs. Low field mobilities of FETs at 77 K, as inferred from their channel resistance, are of the order of 2×10^4 cm ^{2}/V-s. These mobilities are still much higher than those of a MOSFET ($\simeq$600 cm ^{2}/V-s) or even a GaAs MESFET ($\simeq$ 3000 cm ^{2}/V-s at 300 K). The higher mobility of the MODFET gives higher average electron velocities, when averaged spatially over the channel, and temporally over the logic swing. For a logic circuit a low on-resistance is important to maintain a good noise margin, and this enhances the importance of high mobility. A low mobility will also result in a low peak velocity, as is seen in Figs. 1.4 and 1.5.

Trapping in AlGaAs: The trapping problem in the MODFET is the most serious problem affecting its large-scale logic application. The effect may manifest itself as a bizarre drain current collapse phenomenon at 77 K [82,83]; or less obviously, yet still significantly as shifts in threshold voltages with temperature and time [83,84], and transients both at 77 K and 300 K. The main trapping center is associated with the n type doping in the AlGaAs and has been called the DX center by LANG *et al.* [85]. This has subsequently been extensively studied, and the following set of observations have been made:

- DX centers are associated with all donor species in AlGaAs.
- Trapping effects are increasingly severe as Al mole fraction increases and peaks at 40-50% with a threshold at $\simeq$20%.
- The density of deep traps approaches the donor concentration [84].
- At low temperatures the centers may be ionized by light, and free carriers will remain in the AlGaAs without re-capture (persistent photo-conductivity).
- Thermal emission energies are about 0.4 eV, while photo-ionization energy is $\simeq$1 eV [85].
- Emission energies are constant with Al mole fraction. Capture energies decrease; indeed there is evidence that the filled trap is resonant with the conduction band at a mole fraction $<$ 20% [86].

The trapping properties seem to be associated with the band structure of AlGaAs. Indeed, when a large hydrostatic pressure is applied to GaAs its band structure changes, and becomes similar to that of AlGaAs. Under these circumstances DX-like behavior has been observed [87,88]. The most widely accepted model, due to LANG, LOGAN and JAROS [85], involves a trap which is resonant in the conduction band of AlGaAs. On capturing an electron the lattice relaxes and the trap energy relaxes into the AlGaAs bandgap, for Al mole fraction >20%. Hot electron capture may be seen at even lower mole fractions [86].

Various solutions have been proposed to this problem. Just a careful design of the FET greatly reduces the sensitivity to trapping such as designing so that the AlGaAs is always depleted, and eliminating the possibility of trapping in the parasitic source-drain regions [82,83]. Another strategy is to reduce the Al mole fraction to below 25%. This strategy is effective in reducing trapping, but confinement of the electrons in the 2DEG is also reduced. Another solution to this problem has been to use a superlattice of GaAs and AlAs, rather than an AlGaAs alloy [89]. This way just the GaAs is doped so that there are no DX centers.

Inverted MODFET: Along with the standard MODFET configuration, shown in Fig. 2.1, the inverted MODFET has been experimented with. As seen from (2.3), the charge density in the channel can be modulated by modulating the electric field from either side. This may be visualized in Fig. 2.1 by putting another gate on the 'substrate' side of the interface, with a Schottky barrier to the GaAs. For the inverted MODFET a constraint is placed on the doping and thickness w of the doped part of the buried AlGaAs layer:

$$\left(\frac{eN_D w^2}{2\varepsilon}\right)(1 + 2\frac{d_0}{w}) < \Delta E_C - E_F - E_D \tag{2.6}$$

so that it does not conduct. The constraint on maximum charge in the 2DEG is the same and the sensitivity of V_t to doping and thickness is of the same order of magnitude as in the standard configuration. Because the 2DEG is closer to the surface, a higher transconductance may be achieved in this configuration [90], but at the expense of a large gate leakage current. One problem, as yet not confirmed by experiment, is the susceptibility of the buried AlGaAs layer to trapping of hot electrons.

2.4 The GaAs Gate FET (SISFET)

The GaAs MESFET and MODFET both employ Schottky barriers as a gate insulator, and the GaAs JFET employs a PN junction. Researchers have in vain sought a way to make a MISFET in GaAs in order to realize advantages of large gate voltages and large surface concentrations, as obtainable with the silicon MOSFET. The problem has always been a large, irreducible concentration of surface states at all 'technological' GaAs surfaces, i.e. surfaces with native oxide and deposited insulators, and metallized surfaces. This has prevented the attainment of electron accumulation at the GaAs surface. Other material systems, such as InP [91], have shown more promise in this regard.

Looking at the AlGaAs/GaAs system as an insulator/semiconductor system, it is clear from the high mobilities of MODFET layers that superior transport properties should be obtained from that interface. SOLOMON and HICKMOTT *et al.* [92] measured capacitance vs. voltage for GaAs/AlGaAs/GaAs diodes which revealed curves much like those of MOS capacitors, with little indication of hysteresis. Application of the quasi-static method to look for interface states (effectively comparing a very low-frequency < 1 Hz with a high frequency 1 MHz CV curve) did not reveal any interface states above the noise level [35]. Since then this group [67] and others [93] have made FETs employing n^+ GaAs as a gate and AlGaAs as a gate insulator. FETs have also been made using a metal gate [94,95], a Ge gate [96] and a W_XSi_Y gate [97].

Band Diagram: The structure of the GaAs gate FET, and its band diagram are shown in Fig. 2.2 and 2.3. Comparing the band diagram of the GaAs gate FET with the MODFET, we see that the AlGaAs is undoped, and the metal/AlGaAs Schottky barrier has been replaced by a GaAs/AlGaAs heterojunction barrier. The Al mole fraction is larger, since one wants the barrier

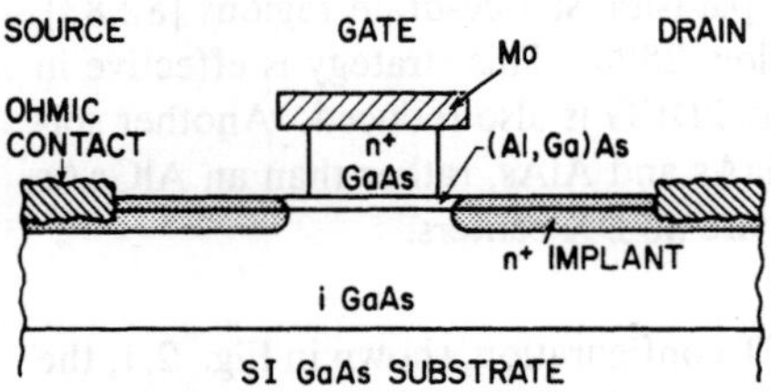

Fig. 2.2 Structure and cross-section of a GaAs gate FET, from SOLOMON, KNOEDLER and WRIGHT [66]. The device was fabricated using a self-aligned ion implantation technique

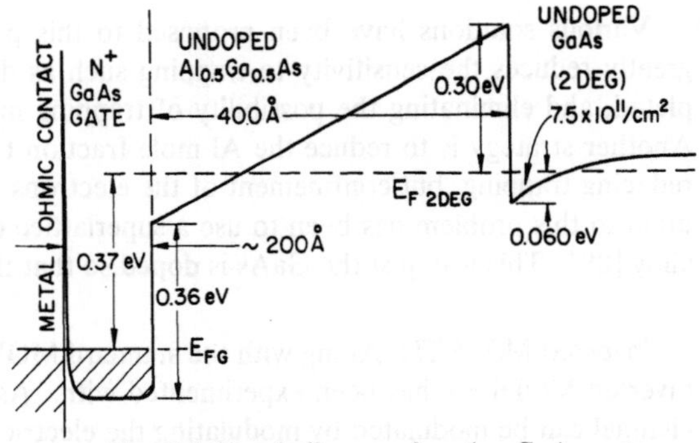

Fig. 2.3 Band diagram for the GaAs gate FET, drawn on the same scale, and similar gate voltage to Fig. 2.1. Figure by courtesy of L. Eastman

height to be as large as possible, and this drives one toward using a mole fraction in the 40-60% range. Being undoped, there is no concern for DX-like traps, or activation of donors, in the AlGaAs.

Zero Threshold Property: According to (2.4), and with no doping in the AlGaAs, the threshold voltage of the GaAs gate FET should be about zero, a property derived just from the symmetry of the structure. This results in an insensitivity of threshold voltage to parameters such as thickness and composition of the AlGaAs, in contrast to the MODFET, as we have discussed.

Performance: As yet most work on the GaAs gate FET has been on the individual device level. The zero threshold property has been verified, and small standard deviations have been obtained. Transconductance has been good, >200 mS/mm at 77 K [67], and the C/g_m ratio indicates good carrier velocities $>2\times10^7$ cm/S at 77 K, which is better than for a MODFET. A reason for the higher velocities could be the better confinement of carriers by the higher Al mole fraction AlGaAs barrier. Channel mobilities measured on long devices are $>1\times10^5$ indicating a good AlGaAs/GaAs interface. P channel devices [98] have been made with good mobilities and transconductance at 77 K. Complementary circuits have been made [99] using Ge for the n-channel and W_XSi_Y for the p-channel device. Another approach to complementary circuits [100] has been to regrow p-gate and n-gate GaAs gate FETs on the same wafer. And yet another, at the expense the n-channel V_t being too high, uses a W_XSi_Y gate for both devices with n and p-type implants [101].

Threshold Shifting: Let us look at the application of this device for logic or memory. Firstly, a zero or slightly negative threshold voltage is not ideal for enhance-deplete logic. A good rule of thumb is that the threshold voltage should be about 1/5 of the logic high level, i.e. about 0.1 V to 0.2V. Means of shifting the threshold voltage may be by doping the AlGaAs, or the substrate, either during epitaxial growth, or afterwards, by means of ion-implantation. An alternative means of shifting the threshold voltage would be to use gate materials of different work functions, such as the Ge gate [96]. When using doping to shift V_t, some sensitivity of V_t to doping and thickness will result, as given by (2.5). As long as only a small shift in V_t is required, V_{bi} is small ($\simeq$ 0.1 V) and sensitivities are small. In contrast to this, for a MODFET V_{bi} is large ($\simeq$ 0.7 V), dictated by the large unbalance in work function.

Gate leakage: Gate leakage is a problem for this device because of the very low barriers $\simeq$0.3 eV vs $\simeq$3 eV for the Si MOSFET. As we have discussed in the first lecture, the mechanism for gate leakage is thermionic emission and Fowler-Nordheim tunneling, both exponentially dependent on barrier height. On considering the band line-up of GaAs/AlGaAs we have concluded that ΔE_C should peak with Al mole fraction in the 40-60% range, and at a value of somewhat under 0.4 V. The barrier height is reduced by about 0.1 V, under forward gate bias, because of the formation of the quantum channel. The low barrier will seriously limit the maximum temperature of operation of the device in logic and memory. To investigate this one can make an assumption of drain/gate current ratios needed for successful operation of logic and memory circuits. For a logic circuit with a large fan-in and fan-out, gate-source leakage current, multiplied by fan-out, loads down the previous stage when the input is high, and drain-gate leakage loads down the drain node when the output is high. The former is probably the dominant effect because of the depletion layer which develops in the substrate under reverse gate bias. With a Fan-Out of 5 for logic, and a load to maximum gate current ratio of 4, the maximum gate/drain current allowable for a tolerable degradation in noise margin is about 1:100, and for a memory cell, about 1:1000. An expression for this ratio, assuming thermionic emission, with the position of the Fermi level being determined by the density of states of the first sub-band, for the gate current, and integrating (2.2) to predict the drain current is

$$\frac{I_G}{I_D} = \left(\frac{L}{\varepsilon v_s V_G}\right)\left(\frac{\pi\hbar^2 N_s}{m^* kT}\right) A^* T^2 \exp\left[-\frac{e}{kT}(\phi - E_0)\right]. \quad (2.7)$$

Application of this to the GaAs gate FET leads to a conclusion that the GaAs/AlGaAs system has a large enough barrier for 350 K (room temperature + chip temperature rise) operation but without much margin for theoretical and fabrication related uncertainties. Of course, operation at 77 K, which may be a prime application for these devices, is unaffected. Other material systems may give larger barrier heights, such as the $In_{0.53}Ga_{0.47}As/In_{0.52}Al_{0.48}As$ system.

Hot Electron Effect: The treatment of gate leakage as independent of acceleration of electrons by the longitudinal electric field is not justified. Hot electrons in the channel may transfer to the gate (real space transfer) and this effect has been exploited by KASTASLKY *et al.* [102] as a negative-resistance electron device. The effect is seen strongly in the GaAs gate FET by FRANK *et al.* [103]. Drain and gate characteristics of the FET are shown in Fig. 2.4 a and b. The gate current is seen to depend strongly on source-drain voltage. The effect is believed to be caused by acceleration of electrons in the channel due to energies greater than the barrier height. A requirement for this condition is that both gate and drain voltage be larger than the barrier height, $\simeq$ 0.3 V. This is to be compared with the data in Fig. 2.4. The hot electron effect is a deleterious one, regenerative in nature and capable of causing circuit related oscillations. Evidence of this is in the 'negative resistance' regions of Fig. 2.4 a. Analysis of the hot electron current reveals that it is unimportant for circuit voltages $<$ 0.5 V, and probably $<$ 1 V for devices with 60 nm AlGaAs thickness. Scaling, with shorter channel lengths and smaller AlGaAs thicknesses will increase this effect, but we need accurate hot electron models, or more device experiments, to estimate its magnitude.

Part 3. Vertical Transport Devices and Device Scaling

3.1 Introduction

In this lecture we will concentrate on two vertical heterojunction devices, the heterojunction bipolar transistor and the hot electron transistor. At the end, we will discuss the issues of scaling semiconductor devices, and in particular, how this applies to all of the devices which we have discussed.

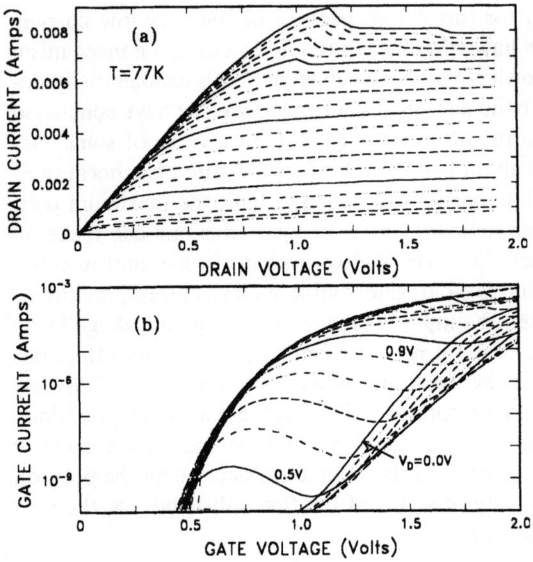

Fig. 2.4 (a) Drain current vs. drain voltage and (b) gate current vs. gate voltage for the GaAs gate FET at 77 K from FRANK *et al.* [103]. The solid lines are separated by 0.5 V increments, and the lowest gate voltage in (a) is 0 V. The gate length and width was 1.5 μm × 25 μm (mask). The gate current increases with drain bias and then saturates. The 'negative resistance' region of the curves is caused by the effect of the drain current on the parasitic source resistance

3.2 The Heterojunction Bipolar Transistor (HBT)

The concept of using a heterojuction in a bipolar transistor is almost as old as the bipolar transistor itself [1]. KROEMER was one of the early pioneers of this device and has written a review article on this subject [7]. The first idea was to use a wide bandgap emitter material in order to improve the emitter efficiency, hence current gain of the transistor. A band diagram is shown in Fig. 3.1a for a graded bandgap emitter (more about this later). Emitter efficiency γ is defined as $\gamma = 1 - I_{hE}/I_{eB}$ where I_{eB} is the electron current injected into the base and I_{hE} is the hole current injected back into the emitter. For a homojunction transistor, non-degenerately doped and with short emitter and base, 1- γ is given by:

$$1-\gamma=\frac{I_{hE}}{I_{eB}}=\frac{N_B w_B}{N_E w_E}e^{\frac{\Delta E_G}{kT}} \tag{3.1}$$

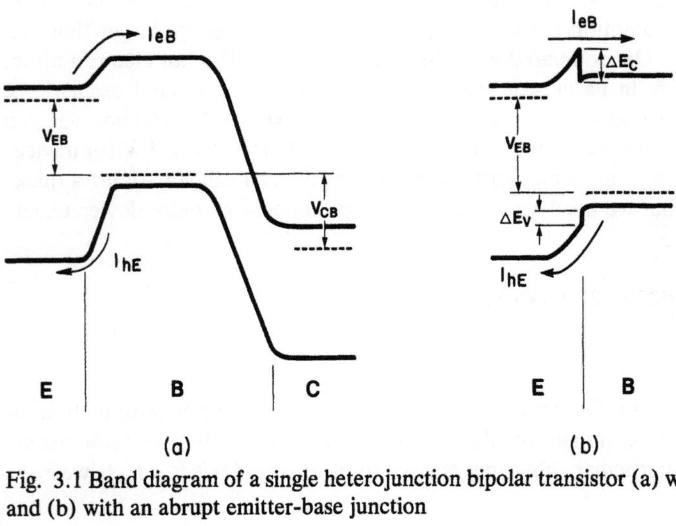

Fig. 3.1 Band diagram of a single heterojunction bipolar transistor (a) with a graded emitter-base junction and (b) with an abrupt emitter-base junction

where N_E, N_B, w_E, w_B are the doping and thickness of the emitter and base regions respectively. ΔE_G is the bandgap difference between base and emitter where, in silicon, the emitter has a smaller bandgap than the base because of heavy doping effects in the emitter. This formula is presented above in a simplified and approximate form partly to illustrate clearly the interplay between base and emitter doping, and bandgap shrinkage in determining the current gain (inverse of 1-γ plus other factors).

A good current gain may only be obtained when the emitter doping is much larger than the base doping. For the HBT, this constraint may be relaxed, since the larger emitter bandgap prevents back-emission of holes into the emitter. This imparts the following advantages to the HBT:

- High current gain.
- High base doping for low base resistance and good resistance to punch-through.
- Lower emitter doping for lower emitter-base capacitance.

Commonly the HBT has been made with a GaAs base and an AlGaAs emitter. According to (3.1), the Al mole fraction need be only about 20% to reduce hole back-emission by three orders of magnitude at room temperature. The advantage of these materials is a high mobility or diffusivity of electrons through base. The higher velocity of carriers in these materials not only reduces the diffusion capacitance (including charge storage in the emitter and collector regions) but through design, may lead to reduction of emitter and collector depletion capacitances as well. This is because the doping in emitter and collector regions must be large enough to carry the current without base pushout (Kirk effect). The higher the velocity of the carriers the greater the current which can be carried or, conversely, the smaller the doping for a given current. This gives a doping reduction proportional to the velocity and a specific capacitance reduction proportional to the square root of the velocity.

Other III-V materials have been used for HBTs such as $In_{0.52}Al_{0.48}As/In_{0.53}Ga_{0.47}As$ [104]. This material system matches long wavelength laser requirements for laser drivers. The silicon transistor with a poly-silicon emitter [105] may be regarded as an HBT of sorts, although the mechanism is still under dispute. GaP [106], and SiC [107], have been tried as emitters of Si bipolar transistor. An old proposal [40] which has now become practical is the use of a Ge, Si alloy for the base of an Si HBT, using strained layer MBE growth.

Grading: There is some confusion in the literature as to what the importance is of grading the emitter-base bandgap. Abrupt vs. graded bandgap situations are shown in Fig. 3.1. For an abrupt heterojunction there is an additional potential spike at the base-emitter junction which reduces injection into the base, hence, current gain. The spike may or may not limit current, depending on the rate of emission of electrons over the spike, their rate of thermalization in the base, and the rate at which the thermalized electrons are carried away by diffusion. The spike will limit the current when

$$1 < \frac{e\phi}{kT} + \ln\left[\frac{eDN_C}{fA^*T^2w_B}\right], \qquad (3.2)$$

where ϕ is the height of the potential spike, D the electron diffusion coefficient in the base, w_B the base width, A^* Richardson's constant and f the fraction of current emitted over the spike which thermallizes into the base. The diffusion constant in GaAs is so large, and the base width so narrow, that the rate limiting step is emission over the barrier, in which case the ratio between electron and hole currents is determined by the valence band discontinuity ΔE_V (replace ΔE_C in (3.1)) which in the case of GaAs/AlGaAs is about 35% of ΔE_G.

Electron Velocity in the Base: The grading of the emitter-base heterojunction, and the composition and doping grading in the base may be used to improve the electron transit-time there. As we discussed in the first lecture, a quasi electric field can be generated by creating an Al gradient in the base of an AlGaAs/GaAs HBT, decreasing in Al content from emitter to collector. The whole bandgap change is effective, since the valence band is kept flat by the high p-type doping. Even a modest Al concentration will do; a grading from 5% Al to zero over 100 nm will give a quasi field of 7.5 KV/cm. Pico-second laser pulse/probe experiments [108] have verified the effect of a grading induced field in improving electron velocities. Another way to speed up electrons is to launch them ballistically from a potential spike at an abrupt emitter-base interface. The group velocity of electrons in GaAs approaches its limiting value of $\simeq 1\times10^8$ cm/s for a potential step in the .15 to .3 V range. Even a 50mV step will give a boost in velocity of 5×10^7 cm/s.

Double Heterojunction HBT: Experiments have been done with a double heterojunction transistor [109], where both collector and emitter had a wide bandgap. The effect of a potential spike on the collector side was to decrease the current gain, but not significantly reduce the collector current at a given base-emitter voltage. The potential spike probably increased recombination of electrons at the base-collector interface, but did not limit collector current flow at room temperature. The double heterojunction transistor has the advantage over the single HBT in that the structure is symmetrical so that transistors may be constructed in either emitter-up or emitter-down configurations. This simplifies the lay-out of circuits such as the ECL gate. With the transistor reversed, the role of collector-base and emitter base junctions are reversed. On the one hand the collector on top is desirable since this minimizes the base-collector capacitance, a major component in determining circuit delay. On the other hand the emitter on the bottom increases the minority carrier injection through this forward biased junction. A way of minimizing this [7,109] is to ion-implant the extrinsic base with acceptors so that the parasitic junction area is within the wide bandgap collector.

Base Resistance: The wide-bandgap emitter in theory permits an increase in base doping almost without limit, and still maintain a large ratio of electron/hole injection. Radiative lifetime τ_r decreases with base doping. Current gain β is related to lifetime in the base by $\beta < \tau_r/\tau_t$. Assuming a base transit time τ_t of 3 ps (see below) and $\beta = 100$, we get $\tau_r > 300$ps corresponding to a doping of 3×10^{19} cm^{-3} [4]. Doping levels in the base have usually been kept below 1×10^{19} cm^{-3} because of concentration dependent Be diffusion [110]. Base sheet resistance is $\simeq 500\Omega/\square$ [111] for a 0.1 μm base. (There has been a demonstration [112] of extremely high base doping where a base sheet resistance of $150\Omega/\square$ was achieved.) These resistances are $\simeq 10\times$ better than Si bipolar transistors, but base contact resistance and extrinsic base resistance have so far been limiting factors.

Recombination and Offset Voltage: One of the problems of present GaAs/AlGaAs technology is leakage current due to recombination in the depletion regions of the p-n junctions as well as surface leakage. This causes a fall-off of current gain at low current densities. Recombination is of the Shockley-Read-Hall type and is worse at the bottom (substrate side) than the top p-n junction [109]. When used in the emitter up mode, the recombination in the collector-base depletion region junction does not affect the current gain, it affects the emitter/collector offset voltage i.e. the emitter-collector voltage V_{CE} when $I_C << \beta I_B$. Under these conditions both emitter-base and collector-base junctions are forward biassed. The value of V_{CE} is determined by the condition that currents across the two p-n junctions are approximately equal and opposite. A leaky collector junction will result in a reduced V_{CB} and an increased V_{CE}. A large area collector junction increases the offset voltage, and a higher bandgap collector reduces it.

Performance: Results achieved with the HBT have been impressive: current-gain cut-off frequency of 45 GHz [113] corresponding to an emitter-collector transit time of only 3.5 ps. A frequency divider of 9 GHz capability has been fabricated as well as NTL logic circuits with delays of 27 ps [114], and ECL circuits with a delay of 65ps [115]. Another circuit family which has been implemented using double heterojunction HBTs is heterojunction I^2L (HI^2L). HI^2L uses transistors in emitter-down configuration. This configuration lays out very densely, and no isolation is required, as is for the conventional HBT. Gate arrays of up to 4000 gate complexity have been fabricated with this technology [116] having gate delays in the 200 ps range.

In spite of the advantages of the HBT as stated above, high speed circuit performance of the HBT has not been much superior to the best Si bipolar transistors. The reason for this is clear when magnitudes of individual components of the equivalent circuit of the device are compared. Devices made up to now show a better base transit time than silicon, base resistances are about the same, and collector capacitances are considerably larger owing to the crude non self-aligned technology employed. An analysis of the delay of a simple non-saturating bipolar circuit shows that the delay can be analysed in terms of the sum of r.c components. (The role of base resistance will further be analysed when we discuss the hot electron transistor.) At low power the load resistance × collector capacitance term is dominant and little advantage is expected to be obtained in using GaAs over Si. Self aligned schemes are now being developed [111,115] and projections for circuit delays are as low as 10ps [117,118].

3.3 Hot Electron Transistor

The hot electron transistor was a concept proposed by C.A. MEAD and others in the early 1960's. No working transistors were produced, owing mainly to the poor state of materials technology at that time. Predictions were made that even in theory the current gain of such a device would be low because of loss of electrons due to scattering in the base, and quantum mechanical reflections at the base-collector interface. Interest was revived when HEIBLUM [12] proposed that such devices, using heterojunctions instead of insulators, should give useable current gains. The device was called a Tunneling Hot Electron Transfer Amplifier (THETA). Versions of this device have been fabricated using planar doped barriers [118,119] and recently, using GaAs/AlgaAs heterojunctions [120,121], devices have been realized having common emitter current gains greater than unity. A common emitter gain 2.3 was obtained at a temperature of 4 K. The structure and principle of operation of this device can be understood with the aid of the band-diagram in Fig. 3.2. Electrons are injected through a tunnel barrier from emitter into base acquiring the kinetic energy of eV_{BE}. The electrons transit across the base and are collected by the collector barrier ϕ as long as $\phi < V_{BE}$. For a narrow base width of about 30 nm most of the electrons cross the base ballistically, i.e. without collisions. This has been shown to be the case [121] by analysing the energy of the electrons arriving at the collector by applying a retarding field to the collector.

Performance: The advantage of this device is the very short transit-time of the electrons across the base. With a base width of 30 nm and a velocity of 1×10^8 cm/s, the transit time is only 30 fs, corresponding to a frequency response of 5.3 THz. Several factors combine to make the actual frequency response much poorer than this. Transit across the collector barrier region, which is wider than the base, must also be included. Since the collector is AlGaAs, the velocity will be lower than in GaAs, and transfer to the L or X valley more probable. This may increase the transit time to a few tenths of picoseconds. Current densities will have to be increased greatly to prevent time constants from being dominated by the capacitance of the emitter and collector barriers. For instance for an emitter tunneling barrier of width w_E, a collector barrier of width w_C, a collector/emitter area ratio of m and voltage excursions of ΔV and $2\Delta V$ on emitter and col-

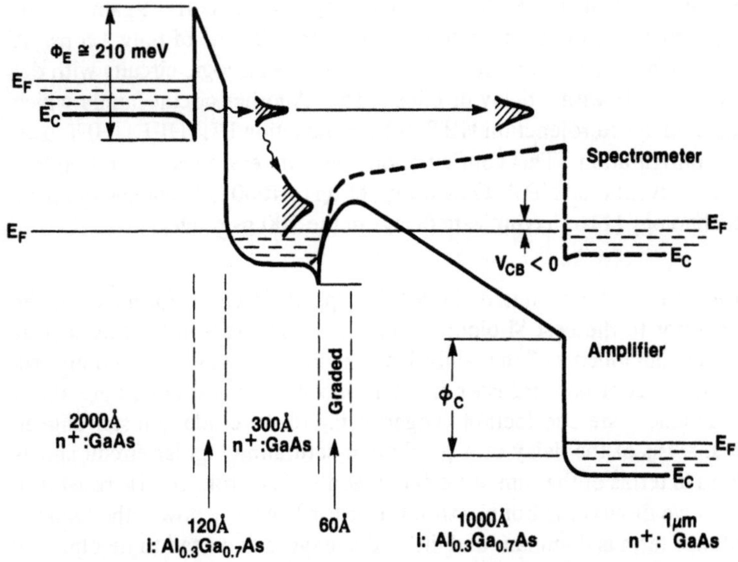

Fig. 3.2 Structure and band diagram of a tunneling hot electron amplifier, from Heiblum *et al.* [118]. The solid lines show the band configuration when used as an amplifier and the dashed lines when used as a spectrometer.

lector respectively, the time to charge up these capacitances is $(1/w_E + 2m/w_C)\varepsilon\Delta V/J$ and is 0.3 ps for an emitter current density of 1×10^6 A-cm^{-2} with $\Delta V = 0.1$ V, $w_E = 10$ nm, $w_C = 50$nm and $m = 5$. A similar, related issue is the base-resistance × capacitance time constants. These time constants may be expressed in terms of base sheet resistance R_s as

$$r_b C_E + 2r_b C_C = \varepsilon a^2 R_s\left(\frac{3}{7}\frac{1}{w_E} + \frac{3}{2}\frac{1}{w_C}\right), \tag{3.3}$$

where a is both the emitter width and base contact-emitter spacing. For R_s of 500 Ω/□ and a = 0.25 μm, a time constant of 0.25 ps is obtained. A 2 DEG base [12,122,123] has been proposed as a method to reduce the base resistance. While this would work in a small-signal mode, the 2 DEG is unable to carry enough current to charge the depletion capacitances rapidly through an appreciable voltage. It appears, therefore, that the key to realizing sub-ps switching times in a THETA-like device would be the attainment of feature size of ≃ 1/4 μm, with concomitant high current densities.

Metal Base Transistor: The original proposals for the hot electron transistor involved the use of a metal base to reduce base resistance. (In those days they were talking of switching times of about 10 ps.) Work is progressing at some laboratories [124,125] on an Si-$CoSi_2$-Si device. Mean free paths in the metal are very short, and so far only a very small electron transmission has been seen.

3.4 Device Scaling and Extendibility

It is important to be able to predict how device design must change when dimensions are reduced, and how far this scaling process may be pushed. Discussion of this subject is found in [8,64]. Devices are scaled both to increase densities and speeds. Since silicon devices become faster on scaling, it is pointless to scale the III-V heterojunction devices without expecting an increase in speed. For the horizontal transport devices (HFETs) the speed increase is straight forward, since transit-times are reduced. For the vertical devices, horizontal scaling reduces the base resistance × depletion capacitance time constants.

The simple guiding principle to device scaling is that horizontal/vertical aspect ratios should be large or constant as one reduces horizontal dimension. This includes layer thicknesses, depletion layer widths, screening lengths, etc. To reduce depletion layer widths, doping concentration has to be increased (for constant voltage scaling, as the square of the dimension). For the FETs the electrical insulator thickness must be much less than the channel length. The electrical channel thickness includes the thickness of the 2 DEG and the electrical screening length in the gate. For the MODFET this amounts to about 5 nm, and for the GaAs gate FET about 10 nm. The thickness of the AlGaAs is limited by tunneling. For the SISFET, where relatively high Al mole fractions are used, AlGaAs thicknesses may eventually be reduced to ≃150nm. It is not as easy to reduce the thickness in the MODFET because of problems associated with heavy doping of the AlGaAs which would be required to adjust the threshold voltage. At present the minimum thickness is ≃ 30 nm. Thickness is related to channel length in that as channel length is reduced, the parallel plate component of the capacitance is reduced, whereas the fringing capacitance component tends to remain constant. The fringing capacitance limits the device speed, and also contributes to an increase in output conductance, since the drain can modulate the carrier concentration in the channel. To obtain good 'long channel' behavior, i.e. dominance of the parallel plate component, the ratio of channel length to thickness should be kept above ≃8. This implies a minimum channel length of ≃ 0.2μm for the SISFET and a bit more than that for the MODFET. Possibly an improvement in the short-channel behavior of HFETs may result by applying a heterojunction confinement to the channel backside, which could be translated into smaller channel lengths. A limitation on the large signal performance of HFETs is the limited sheet concentration in the channel before transfer of electrons into the AlGaAs, for the MODFET, or before gate leakage current becomes excessive, for the SISFET. MODFETs are limited to a little over 1×10^{12} cm^{-2} for a single channel, while considerably higher concentrations, $> 2 \times 10^{12}$ cm^{-2} may be achieved for the SISFET.

Because of the very short base transit-times, compared to other circuit time constants, associated with the HBT and THETA devices, base widths would not have to be reduced until sub-μm horizontal dimensions are achieved. At that point the base resistance, or load resistance × depletion capacitances will become comparable to transit times. Emitter and collector dopings would have to be increased to handle the higher current densities. Contact to the emitter is another critical area of the technology which will have to be addressed. When smaller base widths and larger dopings are eventually needed (probably below 1/2 μm for the HBT) the problem for the HBT will be how to control p-type impurity diffusion. The limits for these devices would probably be related to limits on current density.

The problem of doping fluctuations, often discussed in the context of device limits, is especially severe for the MODFET. This is because the dopant ions in the MODFET are unscreened and can give rise to large fluctuations in charge concentration for the conducting channel, and large fluctuations in potential for the non-conducting channel. For instance, in a 20 nm thick region doped to 3×10^{18} cm^{-3} there are only 24 ions per cube, giving rise to local fluctuations of ≃20%.

Concluding Remarks

In this series of lectures I attempted to cover important properties of heterojunctions as related to transistors, and criteria for designing and estimating the performance of some important heterojunction transistors. In doing so I have not been able to address many aspects of this rich topic, including resonant tunneling structures, opto-electronic devices, complementary FETs etc., but I hope these lectures will stimulate further interest in these topics. I am indebted to my colleagues for furnishing me with much important information. In particular I would like to thank D. Frank, J. Freeouf, M. Heiblum, P. Mooney, F. Stern, T. Theis, S. Tiwari, J. Woodall and S. Wright, of IBM, and L. Eastman, currently on sabbatical at IBM from Cornell University.

References.

1. W. Shockley; U.S. Patent No. 2569347, Filed 26th June, 1948.
2. S.M. Sze; *Physics of Semiconductor Devices*, 2nd ed. (Wiley New York 1981)
3. A.G. Milnes and D.L. Feucht; *Heterojunctions and Metal-Semiconductor Junctions* (Academic, New York 1972)
4. H.C. Casey and M.B. Panish; *Heterostructure Lasers* (Acedemic, New York 1978)
5. W.A. Harrison; *Electronic Structure and Properties of Solids*(Freeman, San Francisco 1980)
6. T. Ando, A.B. Fowler, and F. Stern; Rev. of Mod. Phys. #54 437 (1982)
7. H. Kroemer; Proc. IEEE 70, 13 (1982)
8. P.M. Solomon; Proc. IEEE 70, 489 (1982)
9. H. Morkoc and P.M. Solomon; IEEE Spectrum, 21, 28 (1984). P.M. Solomon and H. Morkoc; IEEE Trans. Electron. Dev. ED-31, 1015 (1984)
10. T.J. Drummond, W.T. Masselink and H. Morkoc; Proc. IEEE, (in print)
11. N.T. Linh; in *Applications of GaAs MESFETs* R. Soares, J. Graffeuil and J. Obrégon (Artech House 1983) Chap. 9
12. M. Heiblum; Solid State Electron. 24, 343 (1981)
13. T.E. Bell; (Sr. Assoc. Ed.) IEEE Spectrum, 23, 36 (Feb. 1986)
14. D. Frank of IBM T.J. Watson Research center (unpublished) Compiled from data in Landolt-Bornstein, *Numerical Data and Functional Relationships in Science and Technology* (Springer Heidelberg, New York 1982) Vol. 17a
15. J.H. van der Merwe; Surf. Sci. 31, 198 (1972)
16. J.W. Matthews and A.E. Blakeslee; J. Crys. Growth, 27, 118 (1974)
17. J.J. Rosenberg, M. Benlamri, P.D. Kirchner, J.M. Woodall, and G.D. Pettit; IEEE Electron Dev. Lett. EDL-6, 491 (1985)
18. A. Ketterson, M. Moloney, W.T. Masselink, C.K. Peng, J. Klem, R. Fischer, W. Kopp, and H. Morkoç; IEEE Electron Dev. Lett. EDL-6 , 628 (1985)
19. W.P. Dumke and J.M. Woodall; US Pat. No. 4518979, 1985, L. Ramberg; Cornell University (unpublished)
20. E. Kaspar; Appl. Phys. A28, 129 (1982)
21. J.C. Bean, L.C. Feldman, A.T. Fiory, S. Nakahara and I.K. Robinson; J. Vac. Sci. Technol. A2, 436 (1984)
22. R. People, Phys. Rev. B32, 1405 (1985).
23. C. Mailhiot and D.L. Smith; (to be published)
24. M. Akiyama, S. Nishi and K. Kaminishi; The Second International Conference on Modulated Semiconductor Structures (Kyoto, Sept 9-11, 1985)
25. R. Fischer, T. Henderson, J. Klem, W.T. Masselink, W. Kopp, H. Morkoc and C.W. Litton; Electron. Lett. 20, 945 (1984)
26. R. Fischer, J. Klem, J.S. Gedymin, T. Henderson, W. Kopp and H. Morkoç; 1985 Intennational Electron Device Meeting (Washington D.C., Dec. 1-4, 1985) Digest p.332.
27. H. Kroemer; J. Vac. Sci. Technol. B2, 433 (1984); Surf. Sci. (1986) (to be published)
28. R.L. Anderson; Solid State Electron. 5, 341 (1962)
29. J.L. Freeouf and J.M. Woodall; Appl. Phys. Lett. 39, 727 (1981)
30. J. Tersoff; Phys. Rev. Lett. B30, 4874 (1977)

31. W. Pickett and M. L. Cohen; J. Vac. Sci. Technol. 15, 1437 (1978)
32. J.L. Freeouf and J.M. Woodall; (to be published in Surf. Sci)
33. R. Dingle; in *Festkörperprobleme / Advances in Solid State Physics*, Vol. 15, edited by H.J. Queisser (Vieweg, Braunschweig 1975) p. 21.
34. R.C. Miller, A.C. Gossard, D.A. Kleinman, and O. Munteanu; Phys. Rev. B29, 3470 (1984)
35. T.W. Hickmott, P.M. Solomon, R. Fischer, and H. Morkoç; J. Appl. Phys. 57, 2844 (1985)
36. J. Batey and S.L. Wright; J. Appl. Phys. 59, 200 (1986)
37. H. Kroemer, Wu-Yi Chien, J.S. Harris, Jr., and D.D. Edwall; Appl. Phys Lett. 36, 295 (1980)
38. H. Okumura, S. Misawa, S. Yoshida, and S. Gonda; Appl. Phys. Lett. 46, 377 (1985)
39. W.I. Wang, E.E. Mendez and F. Stern; Appl. Phys. Lett. 45, 639 (1984)
40. H. Kroemer; Proc. IRE 45, 1535 (1957)
41. R. Dingle, H. Stormer, A.C. Gossard, and W. Wiegman; Appl. Phys. Lett. 31, 665 (1978)
42. T. Ando; J. Phys. Soc. Japan 51, 3900 (1982)
43. F. Stern and S. Das Sarma; Phys. Rev. B30, 840 (1984)
44. F. Stern; Phys. Rev. B5, 4891 (1972)
45. F. Stern; Appl. Phys. Lett. 43, 974 (1983)
46. D. Delagbeaudeuf and N.T. Linh; IEEE Trans. Electron Dev. ED-30, 207 (1983).
47. E.E. Mendez, P.J. Price, and M. Heiblum; Appl. Phys. Lett. 45, 294 (1984)
48. T. Ando; J. Phys. Soc. Jpn 51, 3900 (1982)
49. K. Lee, M.S. Shur, T.J. Drummond and H. Morkoç; J. Appl. Phys. 54, 6432 (1983).
50. P.J. Price, Surf. Sci. 113, 199 (1982).
51. T. Ando, J. Phys. Soc. Jap. 51, 3893 (1982)
52. H.L. Störmer, A.C. Gossard, W. Wiegmann; Solid State Commun. 41 707, 1982
53. J. Shah, A. Pinczuk, H.L. Stormer, A.C. Gossard and W. Wiegmann; Appl. Phys. Lett. 44, 322 (1984)
54. J.G. Ruch and W. Fawcett; J. Appl. Phys. 41, 3848 (1970)
55. W.T. Masselink, W. Kopp, T. Henderson and H. Morkoç; Proc. IEEE/Cornell Conf. (Ithaca, N.Y., Jul. 29-31, (1985), p.136
56. M. Inoue, M. Inayama, S. Hiyamizu and Y. Inuishi; Jpn. J. Appl. Phys. Lett. 22, L213 (1983)
57. K. Hess, H. Morkoç, H. Schichijo and B.G. Streetman; Appl. Phys. Lett. 35, 469 (1979)
58. J.G. Ruch; IEEE Trans. Electron Dev. ED-19, 652 (1972)
59. Y. Awano, K. Tomizawa, N. Hashizume, and M. Kawashima; Electron. Lett. 18, 133 (1982)
60. D. Delagebeaudeuf, P. Delecluse, P. Etienne, J. Massies, M. Laviron, J. Chaplart, and N.T. Linh; Electron. Lett. 18, 85 (1982)
61. T.W. Hickmott, P.M. Solomon, R. Fischer, and H. Morkoç, Appl. Phys. Lett. 44, 90 (1984)
62. Electronics, p.22 (Dec. 16, 1985), on work by S.S Pei *et al.*
63. N.C. Cirillo, Jr. and J.K Abrokwah; 43rd Annual Device Research Conference (U. Colorado, June, 1985) session IIA-7.
64. P.M. Solomon; *The Physics of VLSI*, AIP Conference Proceedings No. 122 ed. by J.C. Knights, Ed. p. 172.
65. J.V. DiLorenzo, R. Dingle, M. Feuer, A.C. Gossard, R. Hendel, J.C.M. Hwang, A. Kastalsky, V.G. Keramidas, R.A. Kiehl and P. O'Conner; IEEE Intern. Electron Dev. Meeting (Dec 1982) Technical Digest p.578
66. T. Mimura, S. Hiyamizu, T. Fujii and K. Nanbu; Jpn. J. Appl. Phys. 19, L225 (1980)
67. P.M. Solomon, C.M. Knoedler, and S.L. Wright; IEEE Electron Dev. Lett. EDL-5, 379 (1984)
68. S. Tiwari; IEEE Trans. Electron Dev. ED-31, 879 (1984)
69. K. Lee, M. Shur, T.J. Drummond and H. Morkoç; IEEE Trans. Electron Dev. ED-31, 30 (1984)

70. K. Lee, M.S. Shur, T.J. Drummond and H. Morkoç; IEEE Trans. Electron Dev. ED-30, 207 (1983)
71. B. Vinter; Appl. Phys. Lett. 44, 307 (1984)
72. M.H. Weiler and Y. Ayasli; IEEE Trans. Electron Dev. ED-31, 1854 (1984)
73. S. Tiwari; IEEE Trans. Electron. Dev., (May, 1986)
74. J.Y. Tang; IEEE Trans Electron Dev. ED-32, 1817 (1985)
75. D. Widiger, K. Hess, and J.J. Coleman; IEEE Electron Dev. Lett. EDL-5, 266 (1984)
76. A. Cappy, A. Vanoverschelde, J. Zimmermann, P. Philippe, C. Versnayen and G. Salmer; Rev. Phys. Appl. 18, 719 (1983)
77. P.C. Chao, P.M. Smith, U.K. Mishra, S.C. Palmateer, K.H.G. Duh, and J.C.M. Hwang; Proc. IEEE/Cornell Conf. (Ithaca, N.Y., Jul. 29-31 (1985), p.163
78. S.S. Pei, N.J. Shah, R.H. Hendel, C.W. Tu and R. Dingle; Proc. 1985 GaAs IC Symp., p. 129
79. T. Mimura, K. Nishiuchi, M. Abe, A. Shibatomi and M. Kobayashi; IEEE Intern. Electron. Dev. Meeting (Dec 1983) Digest p.99
80. L.H. Camnitz, P.J. Tasker, H. Lee, D. van der Merwe and L. Eastman; IEEE Intern. Electron Dev. Meeting (Dec 1984) Technical Digest p.360
81. K. Hirose, K. Ohata, T. Mizutani, T. Itoh and M. Ogawa; 12th Intern. Symposium on GaAs and Related Compounds (Karuizawa, Sept. 23-26, 1985) (proceedings to be published)
82. R. Fischer, T.J. Drummond, J. Klem, W. Kopp, T.S. Henderson, D Perrachione and H. Morkoç; IEEE Trans. Electron Dev. ED-31, 1028 (1985)
83. A. Kastalsky and R.A. Kiehl; IEEE Trans. Electron. Dev. ED-31, 414 (1986).
84. P.M. Mooney, P.M. Solomon and T.N. Theis; Inst. Phys. Conf. Ser. No. 74, 623 (1985)
85. D.V. Lang, R.A. Logan, and M. Jaros; Phys. Rev. B19, 1015 (1979)
86. T.N. Theis, B. Parker, P.M. Solomon and S.L. Wright; APS. March Meeting (Las Vegas, CA., Mar. 31 - Apr. 4, 1986)
87. M. Mizuta, M. Tachikawa, H. Kukimoto and S. Minomura; Jpn. J. Appl. Phys. 24 L143 (1985)
88. M. Tachikawa, T. Fujisawa, H. Kukimoto, A. Shibata, G. Oomi and S. Minomura; Jpn. J. Appl. Phys. 24, L839 (1985)
89. T. Baba, T. Mizutani and M. Ogawa; Jpn. J. Appl. Phys. 22, L627 (1983)
90. Cirillo &eal; IEEE Electron Dev. Lett. EDL, (Feb 1986)
91. H.H. Wieder; 70th Annual Conf. Phys. Compound Semicond. Interfaces (Estes Park, CO., 29-31 Jan. 1980) Proceedings p. 1009
92. P.M. Solomon, T.W. Hickmott, H. Morkoç and R. Fisher; Appl. Phys. Lett. 42, 821 (1983)
93. M.K. Matsumoto, M. Ogura, T. Wada, N. Hashizume and Y. Hayashi; Electron. Lett. 20, 462 (1984)
94. T.J. Drummond, J. Klem, D. Arnold, R. Fisher, R.E. Thorne, W.G. Lyons and H. Morkoç; Appl. Phys. Lett. 42, 615 (1983)
95. Y. Katayama, M. Morioka, Y. Sawada, K. Ueyanagi, T. Mishima, Y. Ono, T. Usagawa and Y. Shiraji; Jpn. J. Appl. Phys. 23, L150 (1984)
96. K. Arai, T. Mizutani and F. Yanagawa; IEEE Electron Dev. Lett. EDL-7, 158 (1986)
97. K. Arai, T. Mizutani and F. Yanagawa; Jpn. J. Appl. Phys. 24, L623 (1985)
98. K. Matsumoto, M. Ogura, T. Wada, T. Yao, Y. Hayashi, N. Hashizume, M. Kato, T. Endo, and H. Inage; Electron. Lett. 21, 580 (1985)
99. K. Arai, T. Mizutani and F. Yanagawa; 12th International Symposium on GaAs and Related Compounds (Karuizawa, Sept. 23-26, 1985) (proceedings to be published)
100. K. Matsumuto, M. Ogura, T. Wada, T. Yao, Y. Hayashi, N. Hashizume, M. Kato, N. Fukohara, H. Hirashima and T. Miyashita; IEEE Electron Dev. Lett., EDL-7, 182 (1986)
101. N.C. Cirillo, M.S. Shur, P.J. Vold, J.K. Abrokwah; IEEE Electron Dev. Lett. EDL-6, 645 (1985). N.C. Cirillo, Jr., M.S. Shur, P.J. Vold, J.K. Abrokwah, R.R. Daniels, and O.N. Tufte; IEEE Intern. Electron Dev. Meeting, (Washington DC, Dec. 1985) Digest p. 317

102. A. Kastalsky, S. Luryi, A.C. Gossard and R. Hendel; IEEE Electron Dev. Lett. EDL-5, 57 (1984)
103. D.J. Frank, P.M. Solomon, D.C. La Tulipe, Jr., C.M. Knoedler and S.L. Wright; unpublished
104 R.J. Malik, J.R. Hayes, F. Capasso, K. Alavi, and A.Y. Cho; IEEE Electron Dev. Lett. EDL-4, 385 (1983)
105. T.H. Ning and R.D. Isaac; IEEE Trans. Electron Dev. ED-27, 22051 (1980)
106. S.L. Wright, H. Kroemer and M. Inada; J. Appl. Phys. 55, 2916 (1984)
107. K. Sasaki, S. Furukawa and M.Rahman; IEEE Intern. Electron Device Meeting (Washington DC, Dec. 1985) Digest p. 294
108. B.F. Levine, W.T. Tsang, C.G. Bethea and F. Capasso; Appl. Phys. Lett. 41, 470 (1982)
109. S. Tiwari, S.L. Wright and A. Kleinsasser; 43rd Annual Device Research Conference (U. Colorado, June, 1985) Session VA-6. (A more complete version of this is being prepared for publication)
110. D.L. Miller and P.M. Asbeck; J. Appl. Phys. 57, 1816 (1985)
111. T. Izawa, T. Ishibashi and T. Sugeta; IEEE Intern. Electron Dev. Meeting (Washington DC, Dec. 1985) Digest p. 328
112. J. Lievin, C. Dubon-Chevalier, G. Leroux, J. Dangla, F. Alexandre and D. Ankri; 12th Intern. Symposium on GaAs and Related Compounds, (Karuizawa, Sept. 23-26, 1985) (proceedings to be published)
113 P.M. Asbeck, A.K. Gupta, F.J. Ryan, D.L. Miller, R.J. Anderson, C.A. Liechti and F.H. Eisen; IEEE Intern. Electron Dev. Meeting (San Francisco CA., Dec. 1984) Digest p.864
114. P.M. Asbeck, D.L. Miller, R.J. Anderson, R.N. Deming, R.T. Chen, C.A. Liechti and F.H. Eisen; 1984 GaAs IC Symposium, Digest 133
115. M.F. Chang, P.M. Asbeck, D.L. Miller and K.C. Wang; IEEE Electron Dev. Lett. EDL-7, 8 (1986)
116. H.T. Yuan, J.B. Delaney, H-D Shin, and L.T. Tran; 1986 ISSCC (San Francisco, CA., Feb. 1986) Digest p. 74
117. S. Tiwari; 1985 GaAs IC Symposium, Digest p. 95
118. R.J. Malik, M.A. Hollis, L.F. Eastman, D.J. Woodard, C.E.C. Wood and T.R. AuCoin; Proc. Conf. Active Microwave Dev. (Cornell Univ., N.Y. 1981)
119. J.R. Hayes, A.F.J. Levi and W. Wiegman; Electron. Lett. 20, 851 (1984)
120. N. Yokoyama, K. Imamura, T. Oshima, H. Nishi, S. Muto, K. Kondo and S. Hiyamizu; IEEE Intern. Electron Dev. Meeting (San Francisco, CA. 1984) Digest. p.532
121. M. Heiblum, M.I. Nathan, D.C. Thomas and C.M. Knoedler; Phys. Rev. Lett. 55, 2200 (1985)
122. P. M. Solomon; European Patent Office Gazette, No. 92645, Filed 11/02/83
123. S. Luryi; Electron. Lett. 6, 178 (1985)
124. J.C. Hensel, A.F.J. Levi, R.T. Tung, and J.M. Gibson; Appl. Phys. Lett. 47, 151 (1985)
125. E. Rosencher, S. Delage, Y. Campidelli, and F.A. D'Avitaya; Electron. Lett. 20, 764 (1984)

CAD: Overview and Perspectives

R. Sénéor

Centre de Physique Théorique, Ecole Polytechnique,
F-91128 Palaiseau Cedex, France

1. INTRODUCTION

Computer Aided Design (CAD) started at the beginning of the sixties. It appeared at the same time as the introduction of multiprogrammed, time-shared graphic terminals. It was applied to different domains of industrial activity such as for example engineering problems, realisations of electrical networks or of integrated circuits. Even if the subject of my talk is only related to CAD for Very Large Scale Integration (VLSI) the general principles governing it apply as well to other industrial purposes. There are 4 phases:

i/ Conceptual project. It starts by the definition of a product. This is the result of commercial, scientific or technical demands.

ii/ Project analysis

iii/ Detailed definition of all the components

iv/ Project documentation

These 4 phases are interrelated. When the project is complex, and this is the case for most of the Large Scale Integration (LSI) or VLSI projects, the conception is hierarchical and follows either an ascending or a descending path. The ascending route starts from elementary components and generates more and more complex blocks up to the final product. The descending route is more logical starting from the functional structure expected from the final product and integrates progressively the technological and environmental constraints. Very often the different steps of this route are modelled (for instance in terms of logical, analog or electrical models).

1.1 What Is CAD?

CAD was originally defined as the use of computers in a complex association with man in order to realize through a graphic terminal the industrial (or architectural or ...) drawings necessary for the fabrication of a product. The present evolution is toward suppression of this intermediate step. The result of CAD is a direct driving of machines ensuring some of the fabrication levels.

1.2 Why Use CAD?

We list some of the reasons [1].

i/ To increase creativity

The use of CAD allows one to investigate in an economical way many possible solutions. It appears as an experimental tool.

ii/ To increase quality

The increase of quality is due to the fact that by CAD the designer has the tools to analyse the characteristics of functioning at different levels of elaboration of his project.

iii/ To reduce delays and costs

iv/ To realize extremely complex structures

These last two points are evidently the main reasons for CAD in LSI or VLSI. We give some examples as a partial justification: Motorola spent 5 M$ and

60 man-years (my) in developing the M6800 family between 1972 and 1974 but it took around 52 my consisting of 20 my (and indeed 2 men) for the definition of the M68000 and 8 my for the design of circuitry (done by about 8 persons), the other important posts being the layout (6 my), the breadboard (5 my) and the simulations and tests (3 + 3 my). The Intel 8086 family took 5 to 10 M$ and 2 years.

On the cost level, the 64K bit dynamic RAM resulted from a 4 M$ project, in 1979 the 68000 project was estimated at 30 M$ but in 1983 the HP 9000 project (32 bit microprocessor + I/O buffer + memory controler + 128 kbit RAM + clock driver) rose up to 100 M$. In the beginning of the eighties it was estimated that a microprocessor chip of 100 000 devices would take 60 my to layout and another 60 my to debug. At this level there is an important parameter: the device ratio which measures the total number of devices relative to the full-custom designed ones. For the I8086 it was 4.4:1. An increase of this ratio will considerably decrease the number of man-years required to realize a project. This ratio is clearly increased by the progress made in CAD. In our present state of knowledge it is related to the use of regular structures [6].

v/ Finally, there is also the idea of proposing a portable tool allowing non-professionnal users (i.e. persons who are neither professionnal designers nor circuit analysts) to realize a product ready for industrial implementation with only the knowledge of general parameters and functions such as cost and functional specifications.

Our purpose is to present a (partial) overview of CAD for VLSI systems. Clearly, as computers have been entering the every-day life of circuit designers for more than 20 years, we cannot treat many deserving topics. We describe in Sect.2 traditional CAD tools as they are applied in the major current technologies (bipolar, nMos or CMOS). They were introduced, independently, for specific purposes and are generally unrelated. In the third section, a corpus of ideas and methods of design more specifically related to very large circuits is presented, computers are no more adequate tools but the essential instruments of the design. The last section emphasizes the importance of new ideas coming from statistical physics and presents some of the main trends in design such as Silicon Compilation (which provides a striking analogy between design and sofware).

2.TRADITIONAL COMPUTER AIDS TO DESIGN

2.1 History of CAD

CAD is a term which appeared in the late 1950's. The field was completely established by the 1960's when new concepts appeared, in particular the idea of a computer being able to design other computers. Important investments were made in the defense, areospace and semiconductor industries.

However the next decade up to the 1970's proved to be a big disappointment in the use of CAD as a consequence of unrealistic hopes.

The hardware revolution of the 1970's completely changed the situation by allowing the introduction of small- and low-cost systems. The increase in storage capacity, the low cost of hardware and the consequent appearance of minicomputers gave rise to the introduction of commercial CAD systems. This revolution corresponds at the same time to an increase in complexity of the Integrated Circuits (IC). CAD was progressively introduced at all levels of elaboration of an IC, the main uses being digitalization, simulation and interactive graphical editing, the main characteristics being that all these systems interact only with difficulty or were realized to deal only with specific and limited aspects of design.

At the beginning of the 1980's another revolution started centered on the idea that design must be analogous to and indeed reduced to programing in some high-level language. It was the result of a remarkable technological revolution (the silicon industry), progresses in software (new languages such as C, LISP...) and at the same time it appeared as a possible answer to the question of designing and testing very large scale devices.

We will describe the usual approach to CAD which was progressively developed at the end of the sixties.

In VLSI design, CAD enters traditionally in three ways:
-Special purpose architectures and algorithms for simulation
-Modeling and process simulation
-Testing and design for testability
In practical situations, they are strongly mixed and appear at each hierarchical level through top-to-bottom and down-to-up paths.

For very complex designs such as microprocessors, the ideal route is the descending one and unless specified we will stuck at this scheme.

Design from top-to-bottom corresponds to different levels each of them being associated with some degree of abstraction or of description of the circuit. The highest level corresponds to a behavioral specification of the product i.e. a definition of the architecture of the device and of the logical functions to be performed. Then functionality is represented electrically and finally as a layout of geometrical patterns. At each step, one has to check that the system works as desired and implements correctly the function desired. For example, an electrical description of some logical function has to be simulated to see if the circuit operation and performance are the expected ones, but one has also to extract from this description an equivalent logical description and compare it with the initial logical function. This way mixing top-to-bottom and down-to-up differs drastically from the ideal approach to design (fig. 1).

Most of CAD were primitively developed as tools of simulation and comparison.

2.2 The Custom Design Tool Box

Regardless of the design method chosen there is a certain number of standard tools that can help the designer. These computer aids for design concern simulation, layout, design verification and testing [5,7,11].

2.2.1 Simulation

Architecture is verified by various levels of system simulations which include, taking account of the progress of the design
i) functional simulation (behavioral level)
ii) register transfer level simulation
iii) logic gate level simulation
iv) timing simulation
v) circuit simulation.

We will only discuss the last three points.

a/ Circuit Simulation

Circuit simulators give very accurate predictions of circuit performance for circuits of relatively small size and are used in VLSI to check some parts of

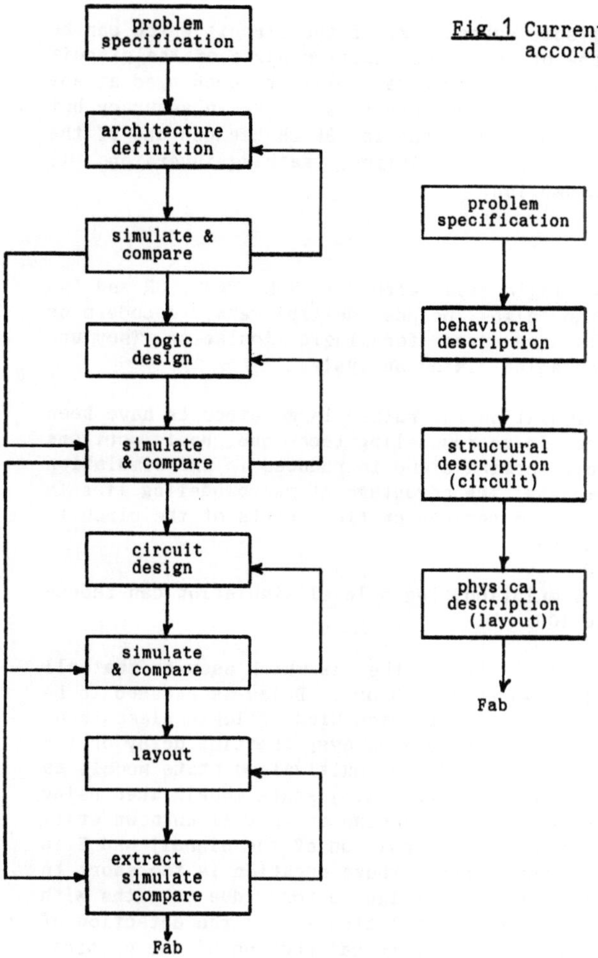

Fig. 1 Current and ideal design approaches according to Weste and Eshraghian [11]

the design (hours of Cray 1 computer time are necessary for the simulation of circuits with a few thousand components).

A typical simulation program is SPICE from NAGEL [9]. The description of the components and the interconnections is entered in this type of program and transformed by the circuit simulators into a set of equations. The electrical behavior of the components is modeled in terms of ideal elements such as resistors, capacitors, sources, ... Constraints between the elements are governed by Kirchoff current and voltage laws. Solutions of the circuit equations, which can be linear or nonlinear, are given through algorithms (the nonlinear case is generally done using the iterative linear approximation of Newton-Raphson) which use intensively the sparsity of the matrix circuit equations [19].

It has been observed that the simulation time for circuits containing n nonlinear devices is proportional to $n^{1.2}$. The memory requirement is also very important (of order n). Thus, this type of program can only be used for simulation of some parts of VLSI chips.

b/ Timing Simulation

This is a simplified circuit simulation. Simplification is obtained in two ways. One is ignoring some feedback paths (for example in a MOS inverter one would ignore the capacitance feedback from the output node to the input node).

The other (latency) is identifying some subnetworks of the circuit which can be processed independently with the idea of bypassing inactive parts of the circuit (typically around 80 percent of the circuit variables are unchanged at any time). A typical simulator is MOTIS [21]. The result is a loss in accuracy but an execution time two orders of magnitude less than for SPICE because one of the most time-consuming parts of circuit simulation, matrix inversion, is drastically reduced by the first assumption.

c/ Logic Level Simulation

This concerns simulation at the logic level with NOT, NOR, NAND, OR and AND gates as primitives but they can also include multiplexers, decoders or registers. Simplified gate models are used for logic simulation. They use parameters coming from a previous detailed timing analysis.

Extension of logic level simulation for rather large circuits have been obtained by introducting macromodels. In macromodeling technique, basic portions of a circuit are modeled at their terminals and introduced in the simulation process instead of the device models. The advantage of macromodeling lies in differentiating the critical parts from the non critical parts of the circuit. The non critical parts are then modeled.

Compared to electrical level simulation, logic level simulation can reduce the computer time by a factor 10 to 100.

The lowest level of logic simulation is the one which assumes that all logical elements can assume only two states: 0 or 1. Delay is assumed to be nominal, i.e. a single delay value is assigned to each kind of logic elements or modeled by introducing, for example, rise and fall delays, inertial delay or the notion of delay ambiguity. Similarly there can be multivalued state models as for example three-value (0,1,X) or four-value (0,1,X,E) state models when using rise and fall delays; here X represents the unknown or initial unknown state (corresponding to model initialization and transition of the signal) and E is the unpredictable state due to spikes (signals whose duration is too short to produce a change of states) or hazards (design errors due to paths with different delay times). The four-valued state simulation allows the detection of critical states of the circuit which may end with malfunction of the physical device. Finally, the whole system works by sequencing the events; there are two ways of driving the system. One, called static levelling, was by sorting the gates in such way that all signals on which a gate depends have been computed before it is processed (loops are treated apart). The other one, called dynamical levelling, labels the order of event processing at a given time by the circuit condition at that time [8,20].

d/ Mixed Mode Simulation

This combines several simulation capabilities. It is in fact a multi-level simulation. Critical paths are simulated through detailed device level circuit simulations. Then they are embedded in larger circuit parts which will be simulated through timing simulation for which crude estimation of timing delay is enough. Finally, the whole circuitry is submitted to logic simulation. Regular parts of the circuit such as PLAs or ROMs can be simulated at the functional level. Communication between the different levels is ensured by using a common data language i.e. a common hierarchical description language. For example, in SPLICE - a mixed-mode simulation program which combines logic, timing and circuit simulation - nodes of the circuit network can share four properties: type, value, fanin and fanout. The type can be logic, timing or circuit; the value specifies the voltage or the logic level and so on [20].

e/ Timing Analysers

A timing analyzer or verifier automatically enumerates all possible timing paths, analyses their timing delays and reports on the worst of them. The direction of signal flows is also given. In the Genesil Silicon Development System, the timing analyzer can analyse 2000 transistors/mn with an accuracy of 10% of SPICE. In this system, timing analysis is modelled on a representation of the circuit at the transistor level (with all their intrinsic parameters) [22].

Table 1 summarizes the expenses in time and the needs in storage capacities required by the different types of simulation [22].

Table 1. Run time and memory requirement for simulation of one instruction cycle in a VLSI chip.

Level of Simulation	CPU Time	Memory in MB
RTL	<< 1 mn	<< 0.5
Logic	10 mn	4
Timing	8 hours	30
Circuit	6 months	250

2.2.2 Layout

At the layout level, CAD can enter in three different manners depending on the degree of integration of CAD in the process. This notion of integration will make more sense to the reader at the end of this section.

a/ Handcrafted Mask Layout

The design is divided amongst designers who are experts in logic, circuit or process details. A sequence of hand drawings is then produced with the repeated use of basic circuit structures, this information being then digitized using a computer. The correct hand-drawn layout is finally obtained by the use of an interactive computer graphic terminal. This gives the very high figures (in man-years) quoted for microprocessor design at the end of the seventies. For example the time for handcrafted design and layout of random logic was estimated to be 10 devices per day and per person. With an average of 30 percent of random logic in a VLSI of 10^5 devices, this gives more then 80 my.

b/ Layout of Regular Structures

This is a semi-automatic technique. Using predesigned chips or libraries of standard cells the final layout is reduced to a problem of placement and routing of the interconnections. In the following, we give three examples of such structures.

While circuit design and layout is roughly 10 times faster than traditional hand-drawing techniques, the trade-off is, for the first two examples, an area penalty of the order of 10 to 30 percent.

Gate arrays.
This is a method to provide many custom codes with the shortest possible design time and at the lowest cost. A gate array chip contains a large number of identical sites. Each site consists of a certain number of transistors forming logic functions. Some of the possible connections are left open. The final product is personalized by completing the connections. Usually arrays of cells are separated by routing channels (for instance metal horizontally and polysilicon vertically) as well as mask programmable I/O pads. Connections are made through one or two metalization layers and contacts. Related to that is a placement and routing problem.

There are CAD systems which perform placement, routing, simulation and testing in an efficient way.

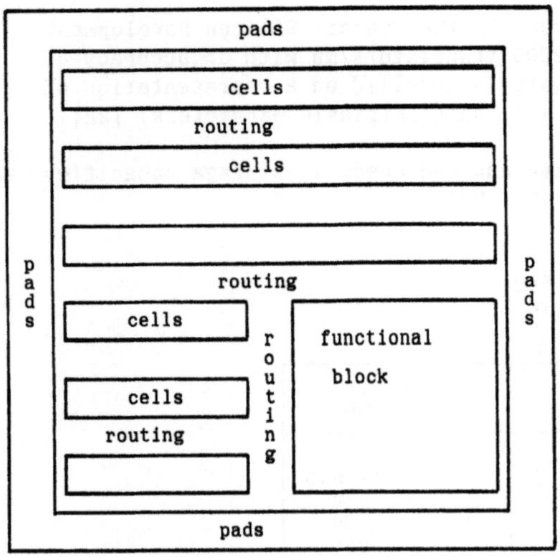

Fig. 2 A standard cell floor plan

Standard cell design. Standard cell chips are made with predefined logic and layout cells which can be elementary SSI cells as well as functional blocks such as RAM, ROM or PLAs (Fig. 2). If a priori, a full mask set is required, it is possible by designing a large custom chip to personalize it through specific PLA or ROM.

As for the gate array, the main problem treated with computer help is placement and routing (fig. 2).

Among other regular structures which belong to VLSI design we may also quote Progammable Logical Arrays (PLA) and Storage/Logic Arrays (SLA).

Storage / Logic Arrays. This last structure is typical of recent CAD use. Unlike PLA in which the AND and OR arrays occupy separate planes, SLA are AND and OR arrays folded together containing flip-flops distributed throughout the array with arbitrary row and column breaks . This results in a segmentation of SLA leading to independent submodular structures such as memory or combinatorial logic and reduced space density (they can be as dense as full-custom ICs.) since the distributed gates can perform both logic and interconnection.

c/ Symbolic Layout

As an input the layout is represented in a symbolic way describing the designers'intent. Compaction is then done using an algorithmic program which includes automatically design rules and minimal geometry constraints. To deal with large structures a hierarchical procedure is introduced which allows treating portions of a circuit already laid out as new entities characterized only by their interconnections.

We will describe the symbolic layout approach in more detail in the next section.

2.2.3 Design Verification

Once the layout of a circuit portion has been achieved a design-rule check is done.

a/ Design Rule Checkers and Electrical Rule Checkers

This checks that the design rules are observed in manually designed layouts. Almost all design rule checkers (DRC) need computer time which behaves as $n^{1.5}$, with n measuring the circuit size complexity. This means that for large circuits checking is done hierarchically. A circuit portion which is design-rule safe is then treated as an entity which is not reinvestigated internally in the next checking.

Electrical rule checkers (ERC) have to check that there are no illegal connections of transistors; for example, MASKAP transforms "rectangles" into schematics i.e. gives a netlist "a la SPICE".

b/ Circuit Extractors

They are an essential part of the DRCs and ERCs. For example, a typical circuit extractor will extract the electrical structure from the layout level. It will check the crossing of mask layers to infer the existence of transistors. Active areas are recognized as geometric shape intersections. Parasitic capacitances and resistances will then be reported as well as transistor connectivity. Circuit extractors imply the use of description languages suitable for a hierarchical analysis. Typical descriptions at the mask level are Caltech Intermediate Form (CIF), GDSII of Calma or Electronic Design Interchange Format (EDIF) formats.

One of the main problems in this way up is the reconstitution of modular structures from low level descriptions. This is clearly related to the content of information and the structure of the format used as data base.

There also exist extractors corresponding to higher levels as for example functional extractors from logic to functional or from switch level to boolean.

Design rule checkers and circuit extractors will be used less and less with the development of hierarchical design which produces "correct by construction" elements.

2.2.4 TESTING

Testing is still one of the main problems in actual design and layout of ICs. Methods of testing circuits need to be incorporated in VLSI design and this task should be done at each level of architectural consideration. Testing of VLSI is rendered particularly difficult for two reasons: low pin-to-gate ratio (the number of pins is roughly the square root of the number of gates) and large number of elements.

This last point is well illustrated by the case of testing combinatorial logic. A logical circuit with n inputs and m storage elements needs 2^{n+m} test vectors to be exhaustively tested. As remarked by WILLIAMS [10] n=25 and m=50 gives, for an application rate of one test vector per microsecond, a test time of one billion years (10^9).

Testing enters in three different ways in VLSI design:
i) test generation
ii) fault simulation
iii) design for test.

Test generation is the task of generating a minimal number of test vectors to verify the behavior of a circuit and to some extent the quality of its nodes. It relies on fault modeling and simulation.

a) Fault Simulation. Networks may contain defects and faults. A fault will yield incorrect results for a given input pattern. A defect can give rise to a fault. For an AND gate there may be as many as 30 potential defects: open paths, shorted paths, missing elements... . Each defect may generate a fault but faults can also be generated by the occurrence of multiple defects. Thus the problems faced by designers is to model the occurrence of faults, then to generate a test for each of these faults and then to simulate them.

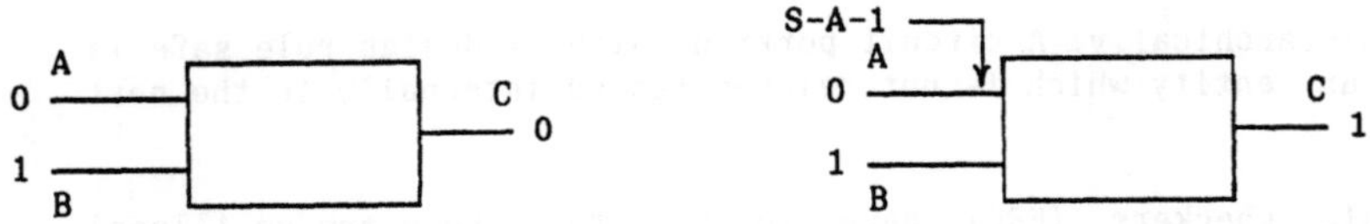

a) Fault free AND gate b) Faulty AND gate (Stuck-At-1)

Fig. 3 Test for input stuck at fault

Usually the fault models retained are the Stuck-At model and models corresponding to faults caused by open- or short-circuited networks. Regarding the Stuck-At model, it is generally assumed that each fault only appears alone.

The Stuck-At model assumes that a logic gate input or output is fixed to either a logic 0 or a logic 1 (Fig. 3).

In fig. 3 b), the A input is perceived as a 1, thus the input A=0 and B=1 is considered as a test for the A input S-A-1.

To give some figures, a 10,000 gate chip has roughly 20 000 stuck-at fault opportunities. There are of the order of 20 000 open circuit opportunities and as many as 20 000 bridging opportunities. This leads to 60 000 possible different faults. There can be 3.6 10^9 double faults.

The problem of test verification is concerned with finding measures of the effectiveness of a given set of tests. Fault simulation leads to run times of the order of N^2 and therefore to high costs. It also has to be remarked that the cost for detecting and diagnosing a fault at different levels goes up at a rate of 10 per level.

b) Design for Testability. An answer to the problem of increasing complexity of chips and of lower values of the pin-to-gate ratio is through design for testability. There are two key words which govern device testability: controllability and observability. Controllability is the ability to set or reset every internal node of the circuit. Observability is the ability to observe either dirctly or indirectly the state of any node of the circuit. The There are two methods:

i) Ad Hoc Testing. A typical example is the partition of long combinatorial parts since testing time grows exponentially with the length.

ii) Structured Design for Testability. The method is to design specific additional features which enhance controllability and observability. These designs are Level Sensitive Scan Designs (LSSD), built-in signature analysers and built-in pattern generators. A general reference is [10]; it is updated for CMOS circuits in [11].

3. VLSI DESIGN

Ideally a design is realized by first developing a complete specification of the system, then a behavioral description followed by a structural description specifying the circuit and finally, before the fabrication steps, by a physical description leading to the layout. The behavioral description ensures the correct implementation of functions. The structural description allows the specification of all parameters that affect performance.

3.1 Device Representations

3.1.1 The behavorial representation

The behavioral representation of a device is the description of the outputs in response to a given set of inputs. This description is technology independent.

The representation can correspond to different levels of abstraction. Devices can be described alternatively by the logical function implemented, by a high-level language description of the function or by a high language description of the input-output correspondence. For example, a NOR gate can be described by

```
      F = (A + B)'                    on the logical level
      out = (A + B)'                  in a high-level language describing
                                      the function
```

For complex structures the information content can vary greatly. Some of the internal or optional specifications of the system can be omitted. The use of a high level language sometimes also allows the representation of the system as an algorithm.

3.1.2 The structural representation

This representation specifies how components are interconnected in order to perform a certain function. It can be a graphical or a high-level language description. Again the information content can vary considerably, the distinction between this level and the behavioral level not being completely clear. For example in the language MODEL (an hardware description language) an NOR gate in CMOS is specified by [23]

```
        Part   nor   (a,b) → out
        Signal i1
           Nfet   out   a   vss
           Nfet   out   b   vss
           Pfet   vdd   a   i1
           Pfet   i1    b   out
        End
```

The part of the device consisting of the NOR gate "nor" has its inputs "a" and "b" specified with the resulting output being "out". The transistors, which correspond to lines 3 to 6, are coded as follows:

transistor type   drain connect   gate connect   source connect.

Therefore the third line corresponds to an n-mosfet with drain = out, gate = a and source = vss. The second line indicates the existence of an internal signal "i1".

3.1.3 The physical representation

It is used to define how particular parts of the circuit have to be constructed in order to yield the specific structure and behavior expected. There can be both a graphical and a high-level language description. The graphical description is done in terms of rectangles representing different layers: diffusion, polysilicon, metal .... These rectangles represent the lithography necessary for fabricating a transistor. The overlap of rectangles leading to transistors is also shown graphically (fig. 4).

A transistor is characterized by a rectangle with three points representing: source, drain and gate. The electrode gate rectangle shares the same gate point.

As an example of a high level language we can give the Intermediate Circuit Description Language (ICDL). It was among one of the first descriptions

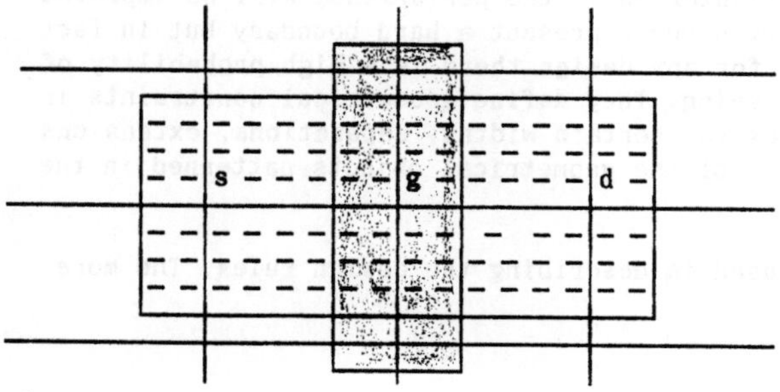

Fig. 4   A transistor

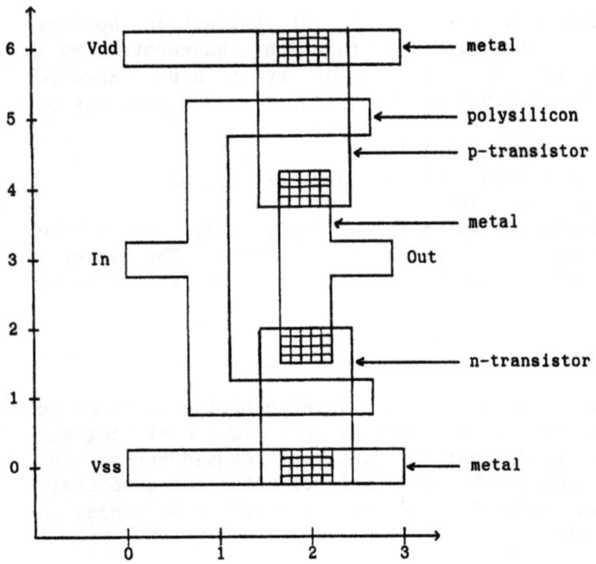

Fig. 5 An inverter

of this type [15]. One uses a virtual grid to specify a relative topology. Connectivity and relative placement are given. The primitive objects are: transistors, contacts, wires, pins, points, instances, capacitors, resistors... We give an example (fig. 5):

This inverter can be coded as

```
begin inv
t1:     device n  (2,1)  or=north
t2:     device p  (2,5)  or=south
        wire alum  (0,0)  (3,0)
        wire alum  (0,6)  (3,6)
        wire poly  (1,1)  (1,2)
        wire poly  (1,1)  (1,5)
        wire poly  (0,3)  (1,3)
        wire poly  (1,5)  (2,5)
        wire alum  (2,2)  (2,4)
        wire alum (2,3)  (3,3)
        contact md   (2,6)
        contact md   (2,4)
        contact md   (2,2)
        contact md   (2,0)
end
```

3.2 Layout Design Rules

Design rules concern the photomasks used in the fabrication of an IC. They are generally chosen in order to optimize yield and geometry without compromising the reliability of the circuit. In fact, they always represent a compromise: with conservative rules the circuit will function but generally with lower-circuit performance; with tighter rules the performance will be improved but with diminution of yield. They never represent a hard boundary but in fact rough tolerances which insure that for any design there is a high probability of success in fabrication and functioning. They define geometrical constraints in the form of minimal allowable values for certain widths, separations, extensions and overlaps in the description of the geometrical objects patterned in the various levels of a system.

Several approaches have been used in describing the design rules. The more common ones are:

i) <u>The micron rules</u>. They correspond to a given state of technology and are given as a list of minimum feature sizes and spacings for the masks required in a given process. A typical manufacturer set of micron rules extends over more than than fifty pages and contains the rules at each mask level as a function of the geometry, the particular cases, the technical specifications of each processes and the precision of each level assignment. For example, in HMOS2 the minimum gate extension is 2 µm and each mask level assignment is known with ± .5 µm.

ii) <u>The α and β rules</u>. The minimum grid size is described by α. The basic feature size is given by β.

iii) <u>The λ rules</u>. They are based on a single parameter λ. All dimensions are given in term of a length unit λ. They were popularized by C. Mead and L. Conway at the end of the seventies by extrapolating the micron rules for current nMos technology of minimum feature size 5 µm. They are well adapted to a change of scale . At least to first order (with respect to scaling) a complete design can be described in terms of λ, the particular choice of λ being made at the end. In 1978, λ was about 3 µm.

We now show some of the λ rules for the case of nMOS and of CMOS (fig. 6).

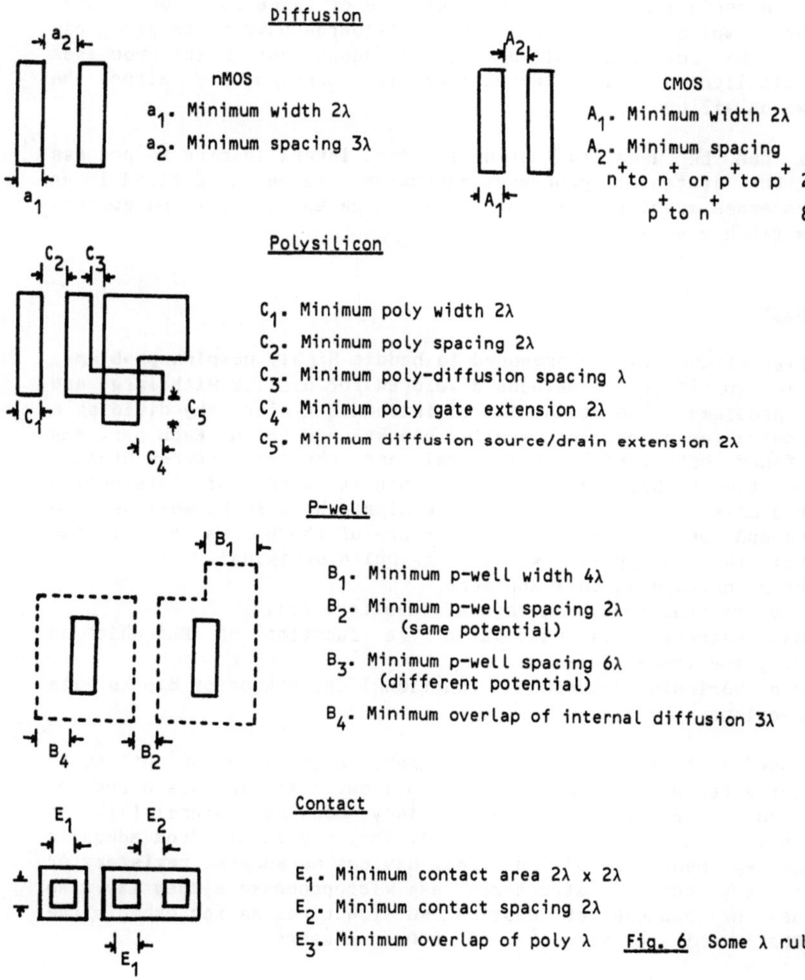

<u>Fig. 6</u> Some λ rules for nMOS and CMOS

3.2.1 Remarks

a) Minimum Spacing Between Diffused Regions (of the same type).
The width 3λ is to avoid overlap of the depletion layers. It is an average. If one or both regions are at ground potential then the width can be smaller since the depletion regions will be smaller. The minimum spacing between polysilicon lines can be taken to be 2λ since there are no depletion layers.

b) Overlap of Polysilicon over Diffused Region to Form a Transistor.
To avoid source-drain shorting the gate extension ought to be 2λ beyond the boundary of the diffused aera. This is a conservative value which can be reduced to 1.5λ.

c) Separation Rules for p-well in CMOS.
To avoid bird-beak effects, field oxide should overlap p-wells at least 3λ. This can be a very conservative rule.

d) Contacts.
Since they may involve different mask layers, the misalignment can add randomly leading to a registration proportional to $\sqrt{n}$ where n is the number of layers. The rules for contact are therefore very conservative.

These remarks are to convince the reader that the λ-rules are very conservative. They present the advantage of being independent of the process. However increase in performance may demand that some of these rules be relaxed or replaced by more complex ones. The fact that safeguards have to be given can be illustrated by the following picture which shows that in the processes uncertainty due misalignments or fluctuations may considerably alter the geometry when drawn on silicon.

The λ rules can be used to define abstract layers leading to process parameterization. For instance, processes and parameters can be defined in an abstract way. Processes specify diffusion, active area and contact; parameters specify spacing, width and extension.

3.3 Structured Design

The structured design strategies are proposed to handle highly complex problems. They are close in spirit to the methods developed for dealing with large and complex software problems. The idea is to introduce a hierarchy dividing a module into submodules (or blocks into smaller blocks) and so on. Each part may have a well defined behaviorial, structural and physical representation depending only on the highest level blocks within it is made of. Interaction between different blocks at a given level of the hierarchy must be well defined and ought not depend on the internal structure of the blocks (this is the concept of locality). The analogy with software is quite striking:

to concatenation corresponds cell abutment.

to iteration corresponds regular arrays of identical cells.

to conditional selection is associated the function of PLA which is determined by placing the transistors.

to reduction of variables corresponds functional definition of blocks with limited outer connections.

One of the hardest tasks for current designers is probably the ability to cut a given complex structure into hierarchical pieces. A key in this direction is the introduction of regular structures. They can be natural (like in memories) where all unit cells are identical. They may result from adequate optimization in the implementation of functional devices as adders, registers or multiplexers. For very complex structures as microprocessors they may also appear in the use of numbers of identical substructures as for example the extensive use of PLA in chips of more than 100 000 components.

3.4 Symbolic Design

Symbolic layout systems consist of designing cells or macrocells (in some of them) in a symbolic manner without having to care about the geometric design rules. Generally a set of algorithms for compaction is incorporated into the systems. The symbolic drawing is done, according to the system, with or without the designer fixing the placement of components.

The first symbolic layout system, STICKS (WILLIAMS [12]), appeared in 1977. There are now many systems (CASL from Applicon, STICKS from Calma, VTI from VLSI Technology) which implements in various manners the goals described above. We now review some of them.

a) Coarse Grid Symbolic Layout

The chip surface is divided into a uniformly spaced grid in both the X and Y directions. The design process consists of placing layout symbols on the coarse grid. The grid size represents the minimium size or spacing of features. Design rules may be violated.

b) Gate Matrix Layout

This is a technique which combines the preceding point of view with a standard cell regular layout. Vertical lines can be polysilicon (S) or diffusion (D). Horizontal rows are transistors or metal tracks. Metal can also run vertically. A character-symbolic description is introduced which characterizes nodes and segments of wires (fig. 7).

```
    S D S D S D S D S D S D
Vss - + + - + - + - + × + -
      : I   I   I   I I I        N  n-channel transistor
      : I   I   × × N I N        +  metal-poly or metal diffusion
      : N   I   I   I   I           crossover
      × N × + - + × N × ×        ×  contact
      : I   I   I   I   I        I  polysilicon
      + + × N × + × N × + -      :  vertical metal
      : I   I   I   I   I        -  horizontal metal
      × N   N × + - ×
```

Fig. 7 A gate matrix layout

c) Sticks Layout

Graphical symbols are placed relative to each other without any reference to a grid. They are interconnected by colored sticks representing interconnection layers on the mask level. Macro-expansion and/or general-purpose algorithms translate sticks to rectangles (on the geometrical level). Correct mask spacings are obtained via compaction processes, sizing, scaling or stretching.

In many of the systems (STICKS, VTI, MULGA ...), the graphical symbols can be elements of interchange formats (such as CIF for VTI) or rectangles and polygons on mask layers, sticks or symbolic characters, placed or not placed on a virtual grid.

We now discuss some of the layout compactors.

d) Layout Compactors

Roughly speaking there are currently two types of compactors: one, the shear-line, which acts on physical layouts and the other one, 1-D linear programming, which modifies the symbolic representation. In both cases, neither the topology nor the relative placement are changed. The goal of all these compactors is to minimize area.

The shear-line compactors work in each direction separately. In a given direction the compaction reduces the white (column locations with no entries) spaces which are perpendicular to the direction. Directional compactions are in general non-commutative.

In the 1-D linear compactor, in each dimension of compaction, components appear as vertices in a graph, the arcs being weighted with the design-rule limits or the constraints of the user. The placement is then found for each component starting from the end points of the graph.

3.5 Interchange Formats

Interchange formats represent the connections between different stages of realisation of a design. For example, schematic capture described as a netlist format must be put into a form readable by silicon foundries' layout systems. They were originally introduced for the transfer of data and should not be confused with symbolic layout languages (such as ABCD an extension of ICDL in the MULGA integrated design system [11]). A typical example is CIF that we now describe briefly.

CIF is a standard machine-readable file which gives an intermediate representation between symbolic layout language and formats suitable for output devices. The advantage of CIF is that it allows a combination of different designs and is general enough to serve as a common denominator for various integrated system projects.

In CIF primitive geometric objects such as polygons, flashes, wires and boxes are described in terms of positions, directions and sizes using a specific syntax. For example, a wire (fig. 8) of width 10 and defined by the path [(1,1) (3,1) (3,2) (5,2)] is described by: W10 1 1 3 1 3 2 5 2 .

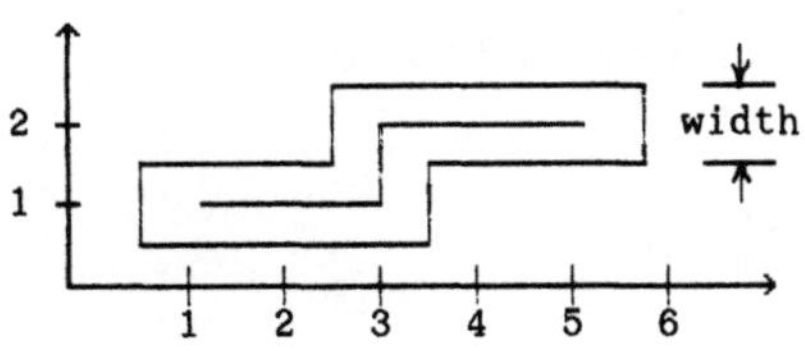

Fig. 8 A wire

Each primitive geometric element is labeled by the name of the fabrication mask (layer) to which it belongs. For instance

ND	represents	nMOS diffusion	
CD	" "	CMOS " "	or thinoxide
PD	" "	pMOS " "	
NP	" "	nMOS polysilicon	
CP	" "	CMOS " "	
NM	" "	nMOS metal1	
CM	" "	CMOS " "	
NN	" "	nMOS metal2	
NI	" "	nMOS depletion mode Implant	

Measurements are given in terms of a unit of length which can be specified by the designer. This unit can be λ. Frequently used items ("symbols") are defined. Symbols can be called at given positions, rotated, translated ... .Algorithmic definitions (for example a decomposition of wires into boxes) can be given.

Remark: there is a widely used color code related to CIF that was proposed by the Jet Propulsion Laboratory and which can also be implemented on a computer graphical description of intermediate forms.

It is interesting to quote EDIF as an attempt to propose a standard interchange format to replace CIF which is relatively limited and GDSII or others which are private. EDIF expresses design data in the form of a hierarchy that is detailed at its lower levels and highly abstract at its upper levels.

Its syntax is similar to LISP, but it is not a programming language. However because of its LISP-syntax and its hierarchical structure EDIF will probably appear as a natural candidate for the transmisssion of design data in a Silicon Compilation environment.

4. TRENDS

4.1 Some Combinatorial Problems

In this section we will discuss some optimization problems of combinatorial structure which appear in the physical design of chips. These are of two types: placement and interconnection. Indeed these problems are already present in the physical design of computer systems where a third type of problem, partitioning, may appear.

If a system specification leads to a physical design too large to fit on a single chip, the design has to be partitioned into groups of circuits small enough to fit on single chips (this is, for example, the case for the HP9000 Series 500 fully-integrated 32-bit processing system which is based on 5 nMOS VLSI chips).

The assignment of specific locations for components on a chip (or a card) is the placement.

Circuits are then connected by wires the location of which have to be precisely defined. All of them, partitioning, placement and routing are NP-complete problems. Problems arise due to the fact that these processes involve generally conflicting goals. For example:

- Placement has generally to be done in order to minimize the wire length, but it may happen that shortest wires leads to congested areas of wires.
- Interconnection also has minimization of wire length as its goal but this may result in multiple crossing and consequently lead to parasitic effects.

These two examples also show that the two problems are not particularly independent if one remembers that the original goal is to minimize the chip area without affecting any of the functional specifications.

We will focus in this section on these two problems and present a new approach borrowed from statistical physics.

4.1.1 The Placement Problem

A general principle used in solving the placement problem is to try to put close together components which have to be connected. This will shorten the length of the connecting wires, thus requiring a small area.

Two typical placement algorithms are frequently used and combined, namely the quadratic assignment algorithm and the min-cut algorithm. They correspond o two different approaches:

i) A constructive approach where one starts with some "seed" cells [17] and adds other ones with the requirement that they have to be as close as possible to cells already placed.

ii) Iterative improvement. Starting from an arbitrary situation one improves it by successive rearrangement of small groups of cells.

In the quadratic assignement problem a cost function is given by the sum of the two dimensional distances (weighted by the number of required connections) between a module and its interconnected neighbors; the cost function is minimized by interchanging the modules. This algorithm is particularly inefficient for widely varying module sizes. The min-cut algorithm is based on successive partitions of the plane by cutting lines. The cost function in the min-cut placement is based on the minimum of interconnections crossing a cut-line. Again it is particularly suited to modules of equal size.

Extensions of these placement algorithms can be achieved for general cells (i.e. of different sizes) and adapted to VLSI design with block hierarchy. A typical example of the problems appearing in VLSI design is the chip floor plan optimization [16].

4.1.2 The Interconnection Problem

This problem can be subdivided into subproblems: wire segment generation, ordering, layering and routing but we only discuss the last one, referring the reader to [17] for the first three.

Routing is the problem of finding in an optimal way paths between points in a signal net. The challenge is to perform 100% routing (it is an essential condition for the existence of the circuit). Optimization should not only be regarded with respect to a cost function, such as minimal lengths of wires, but may also concern computer time. However in routing algorithms the primary criterion is that it has to be 100%, i.e. without the need of manual intervention which is extremely time consuming.

We describe two widely used routing algorithms.

i) The Lee algorithm. If one wants to connect two cells A and B, one first introduces a grid whose mesh characterizes the minimum spacings due to the technological constraints. Cells which already contain wires are withdrawn. Starting from one end point, let us say A, one then labels the cells in such a way that a cell adjacent to cell i has label i+1. Costs related to density of wires and to electrical parameters such as crossover or parallel conductor capacitance or contact resistance can be introduced by assigning weights to the cells. The problem of this wavefront grid expansion router is that it is time consuming and needs substantial storage capacity.

ii) The Hightower line search algorithm. Again one wants to go from a point A to a point B. One first constructs through A two perpendicular lines. The algorithm tries to find an escape point A_1 on one of these lines such that from it there is an escape line which extends beyond the boundaries surrounding A. One then repeats this operation starting from A_1.

A third method, channel routing tries to avoid the sequential nature of the previous ones. A solution is obtained by partitioning the routing in 2 parts: global routing (channel routing) and track assignment. We will not describe it.

Generally these algorithms work with two metal layers and a possible choice is to ask that all vertical segments be in one layer and all horizontal segements in the other one. The connection between horizontal and vertical parts is done through windows.

We now present a recent approach to these problems which has its origin in some problems of condensed-matter physics.

4.1.3 Simulated Annealing

Simulated annealing is a method introduced by KIRKPATRICK et al. [2] in 1983 to treat problems arising in optimal design of computers. It has been applied to partitioning, component placement and wiring of electronic systems.

Up to this period many of the problems of practical interest belonging to the class of the so-called NP-complete problems (i.e. if N is the characteristic number of the problem: number of blocks to be placed or to be wired..., the computing time for getting an exact solution increases faster than N to any power) were treated using heuristic methods generally specific to the problem. There were two basic strategies developed with the heuristic approaches:

i) Divide and Conquer. One divides the problem into solvable subproblems and tries to patch them back together. To be efficient this method needs a natural partitioning of the initial problem into mutually disjoint subproblems of manageable size.

ii) Iterative Improvement. One starts with the system in a given configuration. One then applies a standard rearrangement procedure to all parts of the system until one discovers a configuration which improves the cost function. One then starts with this new configuration and so on and so forth. The disadvantage of this method is that one is always going downhill on the cost function surface, therefore if the starting configuration corresponds to a local minimum one is stuck in this configuration.

The aim of simulated annealing is, roughly speaking, to cure the basic defect of the second method and not to be restricted to specific problems as in case i), nevertheless following a similar strategy. It works analogously to some behavior in the physics of condensed matter.

a) The Statistical Mechanics Approach

Consider a physical system in the limit of low temperature. We may ask what is the nature of the state it will reach: fluid, crystalline, glassy? In practical contexts the low-temperature state is obtained by carefully annealing. Following Kirkpatrick et al, we may describe the procedure to get a low-temperature state by growing a single crystal from a melt as "first melting the substance, then lowering the temperature slowly and spending a long time at temperatures in the vicinity of the freezing point. If this is not done, and the substance is allowed to get out of equilibrium, the resulting crystal will have many defects, or may form a glass with no crystalline order and only metastable locally optimal structures". To find the low-temperature state by calculus is then an optimization problem with the cost function being the internal energy of the system. The interest of the statistical physics analogy is that there is a notion of temperature which allows one to weight the possible configurations of the system by the Boltzman factor exp(-E/kT), E being the energy of the configuration. In this context, iterative improvement would have meant rearrangement from one configuration to another, the new configuration being acceptable only if it lowers the energy. It is like quenching from high temperature to low one. On the otherhand, simulated annealing will mean to possibly allow for configurations not energetically favorable, and therefore the introduction of a temperature in our optimization problem. Finally, and this was the origin of the method, optimization in computer science is done with constraints (which may even seem contradictory) concerning non-identical elements leading to non-regular structures and consequently non-unique ground states, exactly as "frustated" systems (i.e. systems in which interactions favoring different and incompatible kinds of ordering are present) such as spin glasses exhibiting competition between ferromagnetic and antiferromagnetic spin ordering and leading to many nearly degenerate ground states.

The analogy between spin systems and optimization problems is given by Table 2.

The method used is an extension of the Metropolis Monte-Carlo probabilistic algorithm. In each step, one atom is moved or the spin at a site is changed. The new configuration is accepted either if the energy is lowered or if, ΔE (>0) being the unfavorable change in energy, $\exp -\Delta E/kT \geq r$, r being a random number in [0,1]. There are many such steps at a given temperature. The temperature is then lowered and another sequence of changes is performed. The system is said to

Table 2. The analogy between physical systems and optimization problems

Spin or Atomic System	Optimization Problem
internal energy	cost function
positions of atoms or spin values	parameters
physical temperature	temperature as a control parameter
cool into stable low temperature energy state	find a nearly optimal configuration

be frozen if there are no new configurations accepted in the vicinity of some temperature. The probalistic aspect of the procedure allows uphill steps, the temperature playing the role of a control parameter - the lower being the temperature the harder it is the chance to escape from a valley.

Simulated annealing was used for the first time by KIRKPATRICK et al. [2,3] in partitioning, placement and routing. We give an idea in the case of global wiring of the connection with statistical mechanics and discuss the method.

i) <u>Global Wiring</u>. The system to be wired is represented by a model grid. The nodes represent the sinks and the sources to be connected. The wires are routed along the links which connect adjacent nodes. Blockages are introduced by prefilling some of the links. The aim is to do routing in the most uniform possible way.

A cost function is introduced,

$$F = \sum_{\nu} m_{\nu}^2$$

where m_ν is the number of wires (including blockages) on the ν link. For L-shaped paths, i.e. with only one bend, F can be given the form of a spin glass model.

In fact, let $\epsilon_{i\nu} = +1$ if connection i uses the link ν in the positive way, $\epsilon_{i\nu} = -1$ in the opposite orientation and 0 otherwise. Consider then the incidence matrix

$$a_{i\nu} = \begin{cases} 1 & \text{if the connection i goes through the link } \nu \\ 0 & \text{otherwise .} \end{cases}$$

Then $a_{i\nu} = [\epsilon_{i\nu}]^2$, and if $\mu_i = \pm 1$, the number of wires along the νth link is given by

$$m_\nu = \sum_i a_{i\nu}(\epsilon_{i\nu}\mu_i + 1)/2 + m_\nu(0)$$

where $m_\nu(0)$ is the number of blockages and of straight lines. Then F can be rewritten as (up to a constant)

$$F = \sum_{i,j} J_{ij}\, \mu_i\, \mu_j + \sum_i h_i\, \mu_i \quad ,$$

which is exactly the form of a spin glass Hamiltonian. Both J_{ij} and h_i have vanishing average. It is their fluctuations that determine the nature of the states which minimize F. The two possible values of μ_i correspond to the two possible states of an L-shaped path. Each trial in the Monte-Carlo algorithm corresponds to a random choice of a connection and a change of its state.

This method has been successfully compared to other methods such as iterative improvement and applied by IBM for the placement and routing of some middle-size components.

ii) <u>Discussion of the method [18].</u> There are two problem to be understood for an efficient application of simulated annealing (SA). One is how high the initial temperature must be before starting annealing. The other is how low the system must be cooled to achieve a good result. For typical systems it seems that the situation is the one shown in Fig. 9.

There is a minimum energy, E_0, below which the system never goes and a maximum energy E_∞. There is a transition at low temperature $T=T_0$ and a transition at high temperature T_∞. Above this temperature $\langle E \rangle \approx E_\infty$. The low temperature T_0 can be estimated from a study of the first excited states. The high temperature T_∞ can be defined as being the temperature for which the specific heat is ≈ 1. Finally E_∞ can be identified with the average energy, i.e. the energy corresponding to the maximum number of states. The only parameter which is hard to estimate is the minimum energy (An estimate can be nevertheless obtained in some case by using phenomenological inputs). Otherwise, as soon as

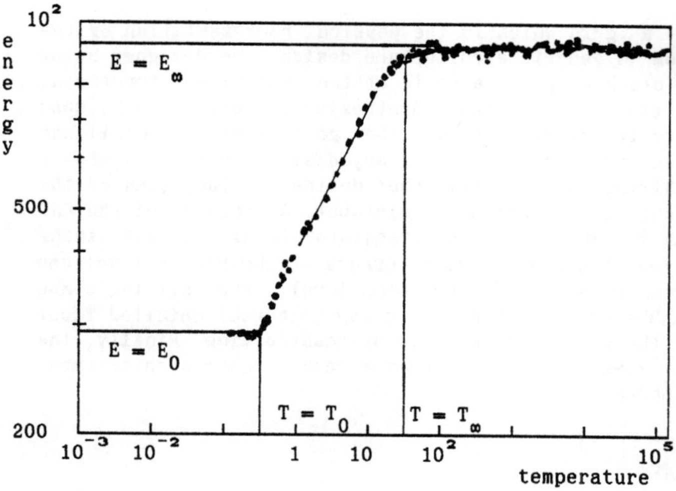

Fig. 9 Energy vs. Temperature for a Typical Problem

these four parameters are known, the energy/temperature curve is also known. There is another parameter which has to be fixed, namely the rate of decrease of temperature. We will not discuss it. There are some problems for which simulated annealing is not efficient in comparison with other algorithms. A numerical experiment on smaller size problems has to be done to fix the values of the parameters for simulated annealing and then to compare this method with other ones. The comparison makes sense because SA gives more highly peaked distributions (as a function of energy) than the traditional methods. Therefore to decide if SA achieves an energy lower than some given one faster than iterative improvement (II), one compares [28] the characteristic time θ_{SA} to n θ_{II}, where n is the average number of times II as to be performed with a characteristic time θ_{II}. The fact that n > 1 is a consequence of the fact that the range of energies covered by SA lies generally in the tail of the II distribution.

4.2 Silicon Compilers

Silicon compilation is an evolutionary technology that manages VLSI complexity by using hierarchical methodology on higher levels of abstraction. More precisely we can view it as the use of an optimizing transformation program that produces mask-level descriptions of manufacturable VLSI circuits from an intelligible input language description.

Depending on the level of specification one starts to compile, one can, following GROSS [13], distinguish two classes of compilers:

i) Limited spectrum compilers
- place and route systems
- compilers producing regular structures: gate arrays systems, PLA generators, SLA generators etc...
- silicon assemblers
- logic synthesizers.

ii) Broad spectrum compilers

4.2.1 Limited Spectrum Compilers

As an example of a new limited spectrum compiler we propose the SLA compiler.

Recently the answer of CAD to the design problem of SLA comes [27] from an interactive two-level Silicon Compiler (from Cyrrus Logic, Inc.). Software for the first level runs on Apollo color graphic workstations while that for the second level is run on a Vax-11/750. The two software tools are a high-level language GLIDE and a symbolic editor STRIDE. Circuit design is performed by placing logic symbols on a grid which is the physical representation of the actual chip. Starting from a generic architecture design, the designer using GLIDE transforms it into a block diagram which is at the same time a functional specification and the initial floor plan. Synthesis of each block, done hierarchically, allows circuit simulation at the gate-level of detail and therefore permits detection and correction of any design errors. A symbolic layout is displayed by STRIDE with labels that define the functions of the modules. These labels belong to a 30-character alphabet. At this level the CAD system synthesizes the SLA design down to the transistor level. Circuit timing and correctness is then simulated and analysed. Errors at the STRIDE level can be corrected by local changes or changes at the GLIDE level. After all logic and timing verifications, the CAD system compiles the array to a real detailed floor plan. The correctness of the circuit is ensured by construction. Finally, the CAD system produces the tapes for generating the masks. Major architectural changes can be done in few hours.

4.2.2 Broad Spectrum Compiler

The key word in this approach is structural design methodology. Functional elements are in one to one correspondance with structural units. Each structural unit corresponds in turn to a single contiguous area of the chip. The functional elements are the "blocks". The behavior of higher level blocks built up by composing lower level ones can be derived in a precise way from the behavior of the more primitive ones.

Silicon Compilation is defined as the generation of a chip from a functional description of the chip. A recursive definition of blocks can be given: a block is either a primitive block or a composition of primitive blocks; in particular, the behavorial description of a compilable chip must be representable as a chip. Composite blocks are made by connecting several primitive or composite blocks.

Indeed Silicon Compilers are different from usual programs generating design and simulation in the sense that they have built in a direct correspondance between the behavorial and the physical structure. Systems are simulated through a universal hierarchical simulator (UHS). A UHS can result from a high level language such as LISP, or C or Pascal through which processes can be conveniently represented, each process containing both functional and data attributes as well as connections.

A Silicon Compiler generates either mask geometry or simulation by passing the functional description to a general compilation mechanism. It is the procedures and the data which are introduced in the successive definition of a block which finally generates the polygons and the rectangles implementing the mask geometry of the block. Similarly simulation is processed through the functional description of the elements of the block. This last part is one of the main advantages of silicon compilation since high level design and low level description such as mask geometry result from the same mechanism.

Since technology is changing quite rapidly, the longevity of a compiler should not depend on the particular description of elements at a given time (for example a cell library) but rather in its ability in adaptating new technologies to work for identical functional blocks. There is a great analogy between Silicon Compilers and programming languages. A set of "codes" has to be defined: the cell set. This choice is crucial since any possible block must be unambiguously generated from it. The correctness of the language has to be checked at all levels, not only on the level of design rules but also on the level of electrical rules or others. Interactive cell design systems are presently being constructed which are trying to implement a more semantic technology independent representation maintaining connectivity, circuit schematic and mask geometry [4,13].

The above point of view is quite abstract in the sense that no current systems are able to implement such an approach. It is also controversial since some authors think that the only general approach will be by starting from a structural specification instead of a behavioral one. The FIRST Silicon Compiler [26] belongs to the first class, the MacPitts Silicon Compiler [24] belongs to the second category. The advantage of the last one is that very little knowledge is required of the designer. However, both belong in fact to the class of limited spectrum compilers.

An intermediate approach is developing quite rapidly: it is the notion of Silicon Compilation Environment. This covers integrated system which may consist of several Silicon Compilers and simulation CAD tools interrelated by the use of a high-level language. Most of the design is automated but still the designer has to make choices among the possibilities offered by the system. A typical example is the Genesil Silicon Development System. In such a system high-level architectural descriptions in terms of structures known by the system: PLAs, ROMs, ALUs, registers and data paths are synthesized on the layout level. At the same time the function and the performance of the specified architecture is given (through a simulation and a timing model). In such a system one minute is needed to compile few thousand transistors and a 50 000 transistor chip can be simulated in ten minutes.

To have an idea of the current status of the Silicon Compilation systems one can quote the result of a design challenge organized in 1985. The challenge was to redesign in one month using a silicon compiler the CMOS 6502 microprocessor from Western Design Center which took a year to be designed. The chip was described by giving a functional description of internal registers and signals as well as signal timing characteristics. The addressing modes and instructions of the microprocessor were given with a summary table of the operational codes, the number of cycles and the memory requirement of each instruction. Some of the results are shown in Table 3 [25].

<u>Table 3</u> Comparison of four implementations

Source	Technology	Area (mil^2)
MetaLogic (MetaSyn)	nominal 3-μm nMOS	296×263
Seattle Silicon Technology (Concorde)	WDC 2.5 μm CMOS SST 3 μm CMOS WDC 3μm CMOS	221×187 256×217 262×222
Silicon Compilers[1] (Genesil)	nominal 3μm CMOS nominal 3μm CMOS	144×242 78×148
Western Design Center	WDC 3μm CMOS	137×168

[1] I/O pads are included in the area.

In this table, Silicon Compilers explored two versions, one cycle-compatible and one function-compatible using microcodes and reducing the data-path to minimum (this gave the low area figure).

The example given above (and many other recent achievements) shows that the design of complex structures can be handled in a short time using integrated CAD systems. Another answer for dealing with the complexity problem can also result from modifying the chip architecture. This is for example the case of the microprocessors with reduced sets of instruction (RISC) in which case the complexity, if any, is postponed to the software level.

Acknowledgements

It is a pleasure to thank S. Solla and J. Feldman for their help during the preparation of this manuscript.

REFERENCES

[1] N. Giambasi, J-C. Rault, J-C. Sabonnadiere: Introduction à la conception assistée par ordinnateur (Hermes, Paris, 1983)
[2] S. Kirkpatrick, C.D. Gelatt, Jr., M.P. Vecchi: Optimization by simulated annealing. Science, 220, 671-680 (1983)
[3] M.P. Vecchi, S. Kirkpatrick: Global wiring by simulated annealing. IEEE Trans. CAD, CAD-2, N°4, 215-222 (1983)
[4] C.A. Mead: "Structural and Behavioral Composition of VLSI," in VLSI Design of Digital Systems, p. 3, (North Holland, Amsterdam, 1983)
[5] G.R. Rabbat: Hardware and Software Concepts in VLSI, (Van Nostrand, New-York, 1983)
[6] S. Muroga: VLSI System Design, (Wiley, New York, 1982)
[7] D.F. Barbe (ed): Very Large Scale Integration VLSI, (Springer, Berlin, Heidelberg 1980)
[8] H. Sakuma, Y. Fujinami, T. Kurobe: "Nelsim: A Hierarchical VLSI Design Verification System", in VLSI 83 - VLSI Design of Digital Systems, ed. by F. Anceau, E. J. Aas, p. 109, (North-Holland, Amsterdam 1983)
[9] L. W. Nagel: "SPICE2: A Computer Program to Simulate Semiconductor Circuits", Memo ERL-M520, University of California, Berkeley CA (1975)
[10] T. W. Williams: "Design for testability" in Computer Design Aids for VLSI Circuits ed. by P. Antognetti, D.O. Pederson, H. De Man , Nato Advances Study Institutes Series (1981)
[11] N. Weste and K. Eshraghian: Principles of CMOS VLSI Design, A Systems Perpestive, (Addison-Wesley, Reading MA 1985)
[12] J. D. Williams: STICKS-A Graphical Compiler for High Level LSI Design, AFIPS Conference (1978)
[13] T. Whitney, C. Mead: "Pooh: A Uniform Representation For Circuit Level Designs" in VLSI 83 - VLSI design of digital systems, ed. by F. Anceau, E.J. Aas (North-Holland, Amsterdam 1983)
[14] Silicon Compilers: Electronic Engng. 57, 9, p.135 (1985)
[15] C.A. Mead, L.A. Conway: Introduction to VLSI Systems (Addison-Wesley, Reading MA 1980)
[16] K. Ueda, H. Kitazawa and I. Harada: "CHAMP: Chip floor plan for hierarchical VLSI layout design. IEEE, CAD-4, NO 1 p. 1985
[17] M.A. Beuer and H.W. Carter: "VLSI Routing" in Hardware and Software Concepts in VLSI, ed. by G.R. Rabbat, (Van Nostrand, New-York, 1983)
[18] S.R. White:"Concepts of Scale in Simulated Annealing" in Proc. IEEE Int. Conf. on Computer Design, Port Chester, New York, p. 652 (1984)
[19] A.L. Sangiovanni-Vincentinelli: "Circuit Simulation" in Computer Design Aids for VLSI Circuits ed. by P. Antognetti, D.O. Pederson, H. De Man , Nato Advances Study Institutes Series (1981)
[20] A.R. Newton: "Timing, Logic and Mixed-mode Simulation for Large MOS Integrated Circuits" in Computer Design Aids for VLSI Circuits ed. by P. Antognetti, D.O. Pederson, H. De Man , Nato Advances Study Institutes Series (1981)
[21] B.R. Chawla, H.K. Gummel, P. Kozak: "MOTIS - An MOS Timing Simulator" IEEE Trans. Circuits and Systems, 22, 12, p. 901 (1975)
[22] E.K. Cheng: Verifying Compiled Silicon, VLSI Design, October, p. 70 (1984)
[23] Lattice Logic Ldt.: "Designing with Gate Arrays", Edition 3.2 LLL, Edinburgh (1982)
[24] J.M. Siskind, J.R. Southand, K.W. Crouch: "Generating Custom High Performance VLSI Design from Succinct Algorithm Description" in Proceedings of Conference on Avanced Research in VLSI, January, p. 28 (1982)
[25] S. Evanczuk: Results of a Silicon Compiler Design Challenge, VLSI Design, July, p.46 (1985)

[26] P. Denyer, D. Renshaw, N. Bergmann: "A Silicon Compiler for VLSI Signal Processors" in Proceedings of the 1982 European Solid State Circuits Conference (1982)
[27] "Storage/logic Arrays Finally get Practical" in Electronics, January, p. 29 (1986)
[28] S.A. Solla: Private Communication

Silicon-on-Insulator Technology Leading Towards Three-Dimensional Integration of Microelectronics

H. Ahmed

Microelectronics Research Group, Cavendish Laboratory, Cambridge University, Madingley Road, Cambridge, CB3 0HE, United Kingdom

1. Introduction

Thin silicon layers of device-worthy quality, grown on electrically insulating layers, provide an attractive material for many VLSI circuit applications. There are several benefits to be gained compared with fabrication in a bulk silicon wafer. The isolation of devices from each other is governed by the properties of the insulating substrate rather than through pn junctions. Capacitances between devices and from devices to the substrate are reduced so that faster devices can be fabricated. It is also observed that latch-up in devices is virtually eliminated. Devices may be fabricated with both higher operating voltage and higher breakdown voltage. In general, lower power dissipation and higher packing density are also achieved in SOI material. The junction areas in devices are very substantially reduced so that tolerance to radiation is much greater in silicon-on-insulator devices compared with devices in bulk silicon. Another important development that has followed from this new technology is that layers of devices may be stacked on top of each other, thus enabling three-dimensional integration of microelectronics based on silicon as the device material.

Several methods have been proposed for the manufacture of silicon films on insulators. The established technology uses silicon-on-sapphire substrates, but there have been many advances recently in techniques for producing layers of single crystal silicon on the native oxide of silicon which is easily grown on relatively inexpensive silicon wafers. A very promising technique is one in which polysilicon layers deposited on silicon oxide are re-crystallized by means of a thermal treatment into single crystal silicon. This technique offers the possibility of forming layers of silicon on top of each other separated by insulating layers. An alternative approach is the SIMOX process in which oxygen is implanted and buried under the silicon to separate a thin single crystal silicon layer from the underlying substrate. This approach does not lend itself very easily to 3-D microelectronics and will not be considered in detail. Some of the more important features that determine the practical viability of re-crystallized layers are the crystalline quality, the types of residual defects, the maximum area of usable re-crystallized material and the mechanical properties of the wafers after heat treatment, such as extent of wafer warpage and build-up of material during the re-crystallization.

2. Techniques of Re-Crystallization

The general feature of most of the techniques of re-crystallization is that a thin polysilicon film is first sandwiched between two or more layers, one of which is the underlying SiO_2 layer. The layer structure is heated isothermally to a high temperature of around 1000°C. A small region of polysilicon is melted with a powerful and localised heat source and this molten zone is moved across the substrate. The polysilicon re-crystallizes into single crystal material as it solidifies behind the molten front. The polysilicon is connected with the underlying silicon base by etching windows in the oxide layer to ensure that the desired crystal orientation is obtained

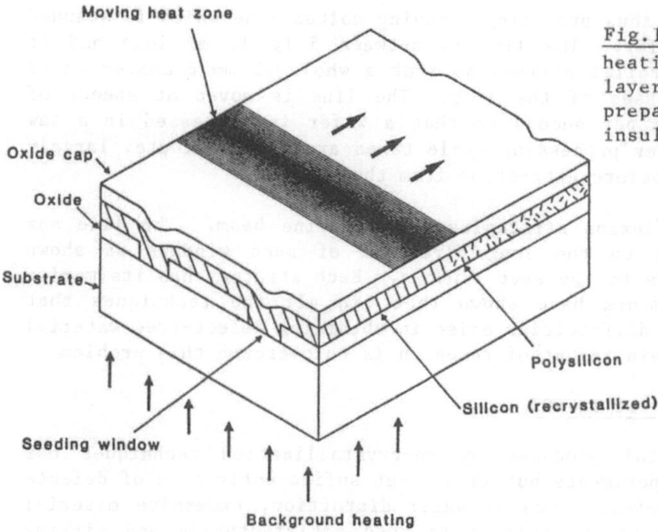

Fig.1 Dual-source heating process and layer structure used to prepare silicon-on-insulator layers

from a seeding region. The wafer is heated by a background source in order to minimise the requirements placed on the moving heat zone. A general schematic view of the technique is shown in Fig.1.

Many heat sources have been developed in combination with background heating to carry out this re-crystallization process. Examples are scanned laser beams [1], moving stages with graphite strip heaters [2], line beams formed with high intensity flash lamps, and electron beams. In particular a dual electron beam technique as shown in Fig.2, has been developed for heating samples rapidly by impact from both the back and front surfaces of the wafer [3]. This has been extensively used to re-crystallize polysilicon films whose properties have been characterized in detail [4]. The wafer is supported in a moveable carrier which is thermally isolated and designed so that both the front and back surfaces are accessible. The lower beam deposits energy uniformly so that the wafer temperature rises to a temperature of about 1000°C throughout the wafer. The upper electron beam is scanned rapidly in one direction to simulate effectively a line source of energy. This line is scanned

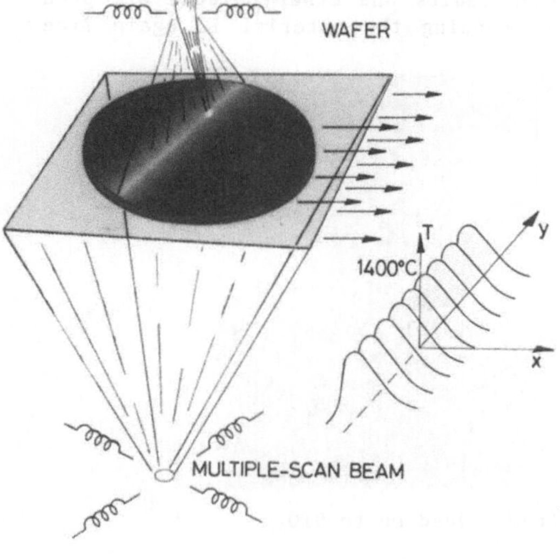

Fig.2 The dual electron beam technique

orthogonally to its length, thus producing a moving molten zone which is scanned in a stripe across the wafer. The line is between 5 to 10 mm long and it is deflected in several parallel stripes so that a whole 75 mm diameter wafer is covered by stitching passes of the line. The line is moved at speeds of several tens of centimetres per second so that a wafer is processed in a few seconds. The whole of a wafer processing cycle takes around one minute, largely to allow the wafer to cool before extraction from the machine.

There are two basic deflexion strategies for the line beam. The line may be scanned either parallel to the long direction of seed windows as shown in Fig.1, or at right angles to the seed windows. Each strategy has its merits and disadvantages. Experiments have shown that, in all the techniques that have been developed so far, difficulties arise in obtaining defect-free material over large areas, and the main thrust of research is to overcome this problem.

3. Material Quality and Devices in SOI

The quality of the material produced by re-crystallization techniques has improved since the early experiments but is not yet sufficiently free of defects over large areas. Gross defects such as wafer distortion, excessive material flow, de-wetting and insufficient melting to seed satisfactorily and sinking of the oxide into the substrate have all been overcome to some degree. Crystalline defects of several types continue to be in evidence in all the techniques developed so far.

The ideal mechanism of growth is that a small region of single crystal in the seed window is melted and seeds the polysilicon in contact with it which also melts. The melt continues over the oxide for a distance of a few tens of microns as defect-free single-crystal silicon. After some tens of microns of lateral growth, low angle grain boundaries form, and the crystal quality deteriorates. The crystal orientation corresponds with the seed window material. Figure 3 shows a cross-section of a crystal grown on an oxide layer from a seed window and Fig. 4 shows a plan view of SOI regions between narrower seed windows. These pictures were taken after etching to decorate the defects and can be clearly seen to be relatively free from defects. The integrity of the layers is maintained and the melting in the seed region has penetrated only a little under the oxide.

With scanning parallel to seed windows two melt fronts emerge from each seed window and solidify towards the centre of the SOI stripe. Where they meet a line of defects is frequently observed. For seed windows spacings of around 40 μm, almost defect-free material (except for the central defect) may be grown. For wider spacings stacking faults and other defects are seen after SECCO etching. With perpendicular scanning the material is again free

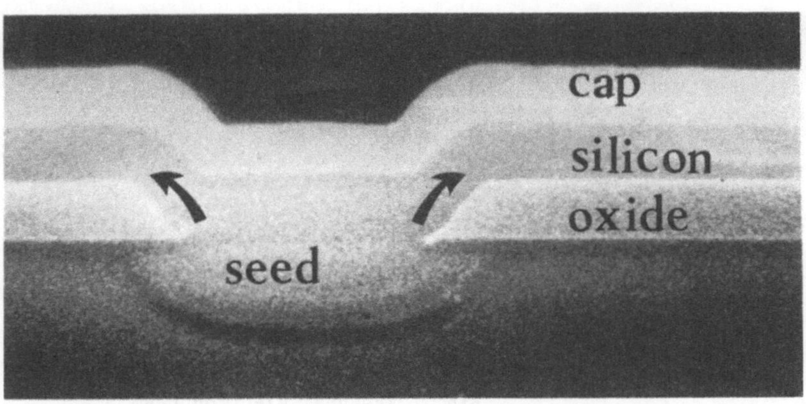

Fig.3 Silicon-on-insulator layer grown from a seed on to SiO_2

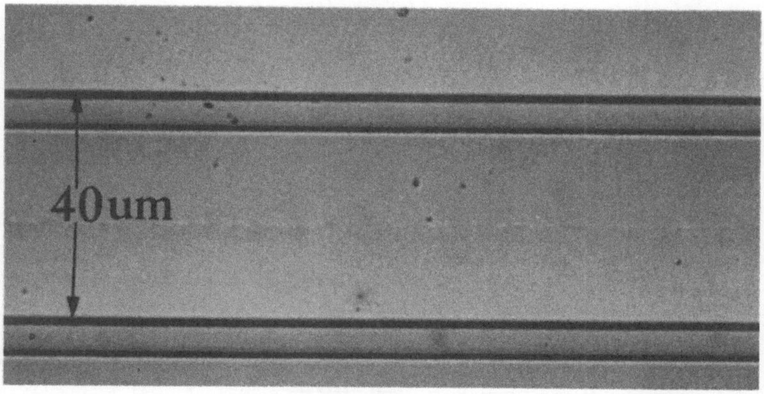

Fig.4 Surface of re-crystallized silicon after etching shows only central defect. Re-crystallized (100) silicon using a background temperature of 1000°C and a 45W/mm line beam swept at 35cm/s

from defects for several tens of microns from the seed window, but there is considerable movement of material causing the surface to be uneven after re-crystallization.

Areas have been produced which are large enough to fabricate not only test devices but also simple circuits such as shift registers and ring oscillators. The main central defect can be kept away from areas in which it would affect device action. There are occasional signs of twinning and stacking faults in the crystal. They can arise from fabrication difficulties and could probably be eliminated altogether and certainly minimised by careful preparation of the material before re-crystallization. In some instances low-angle grain boundaries are formed in the material but they are widely separated. Devices have been made in materials of this type and experiments indicate that working devices can be obtained. Characteristics of devices fabricated in SOI material have been compared with those of devices in bulk silicon and in SOS wafers. Generally n-channel mobility of around 500 to 600 cm^2/V-s and p-channel mobilities around 200 cm^2/V-s have been reported. Results are generally better than for SOS material and also compare favourably with results from devices in bulk silicon. Tests on ring oscillators indicate reduced propagation delay and other improved performance characteristics. Most device results obtained so far show SOI to be a very promising material for VLSI circuits. The future prospects will be greatly improved if wafers of SOI material with few defects between seeding areas can be prepared in large quantities.

4. Three-Dimensional Integration of Devices

The most direct approach to the integration of devices in three dimensions is to fabricate a first layer of devices in the bulk silicon substrate and to make a second layer of similar devices in a re-crystallized silicon layer, which is electrically isolated from the bulk substrate by an oxide layer [5]. Devices in these two layers may then be interconnected in each plane and from layer to layer. In the next level of sophistication the first layer of devices is also made in silicon-on-oxide and the second layer of devices is placed in a second film of re-grown silicon-on-oxide. In each approach one gains advantages of greater packing density and higher speeds through shorter interconnection paths, but probably the most important advantage for the future will be in easing the design of complex circuits by computer automation using true silicon compilers working through a hierarchical approach to circuit design, Fig.5. It is also possible to envisage more than two layers of devices but the fabrication difficulties become formidable and the returns from three-

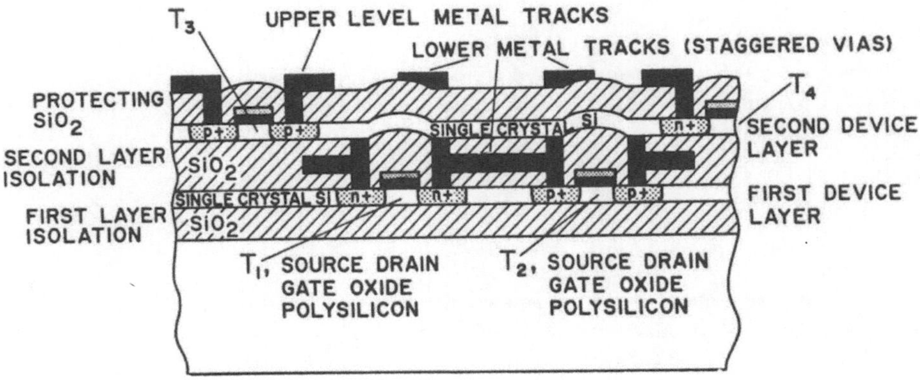

Fig.5 Three-dimensional microelectronics: multiple interconnection of devices in two layers

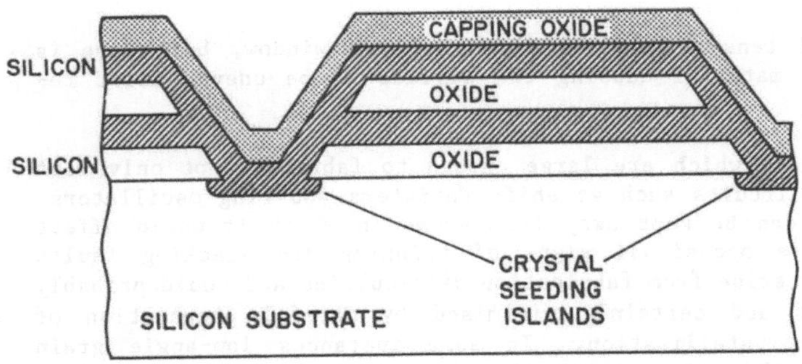

Fig.6 Layer structure used for re-crystallizing two polysilicon layers simultaneously from a substrate seed

dimensional integration of standard devices in conventional circuits diminish. Figures 6 and 7 show approaches to material processing required to achieve two SOI layers. The seed required to re-crystallize the polysilicon may be drawn either from the substrate or the intermediate layer. Very careful control both in the time-temperature cycle and spatially on the substrate is required to achieve these structures. Several successful attempts have been reported on three dimensional integration of relatively simple CMOS circuits using re-crystallized layers [5], but many difficulties still remain to be overcome before there is any prospect of commercial developments.

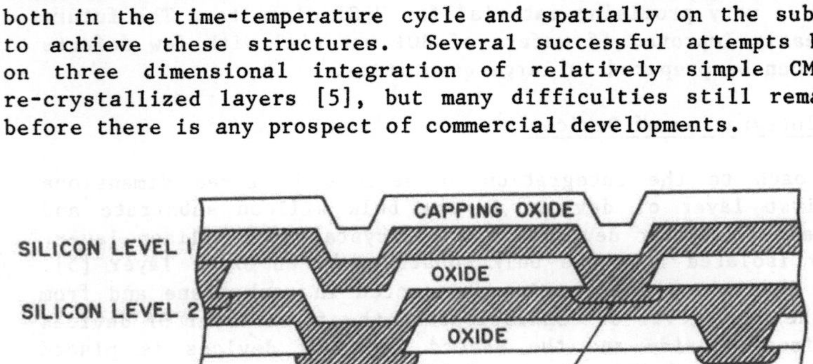

Fig.7 Sequential re-crystallization process for a second layer grown on an initial layer of re-crystallized silicon

More significant advantages are gained from the 3-D approach when one is able to combine several different and hitherto exclusive processing technologies in the same wafer. Examples are bipolar with CMOS, power devices with signal processing and control circuits, optical receivers such as charge-coupled devices with memories, signal processing and interfacing circuitry, and perhaps optical, physical, chemical and biological sensors with suitable signal processing. Some of these possibilities require special structures which create considerable problems in materials science, but the advantages of integrating sensing, signal processing and interfacing in one chip are very significant and worth gaining in the long term. A visualization of a chip design which incorporates several levels of devices is shown in Fig.8.

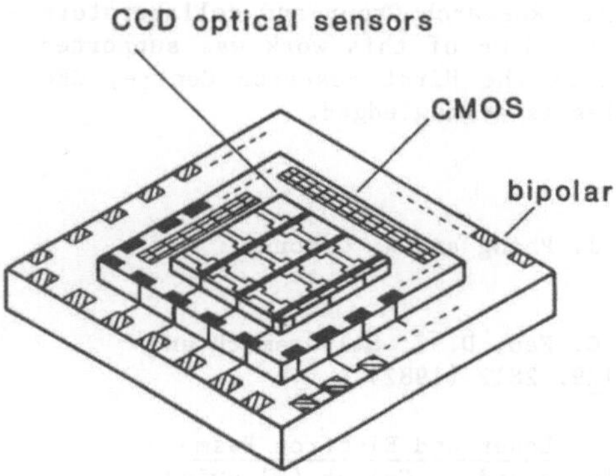

Fig. 8 Three-dimensional microelectronics: concept for a single chip for receiving visual information, signal processing and interfacing

A straightforward example of 3-D integration is one in which an array of power transistors capable of operation at greater than 100 V and carrying currents of several amperes could be incorporated in a layer which is isolated from a CMOS signal processing layer by an oxide layer which provides electrical insulation and an earthed ground plane which provides shielding of CMOS circuits from transient surges during switching cycles of the power devices. Figure 9 shows the type of structure that is required for the so-called 'smart power' devices using 3-D integration. Multiple layers also make it possible to incorporate heatsinks into structures which could overcome one of the main limitations to greater levels of integration.

Finally, the techniques developed for lateral growth of single-crystal silicon layers open another possibility in mixing semiconductors. For example, III-V compounds can be used for high-speed or optical devices and linked to silicon devices in a single chip. The main problem with this concept is that GaAs has to be processed at much lower temperatures than silicon if dis-association of the material through the loss of arsenic is to be avoided.

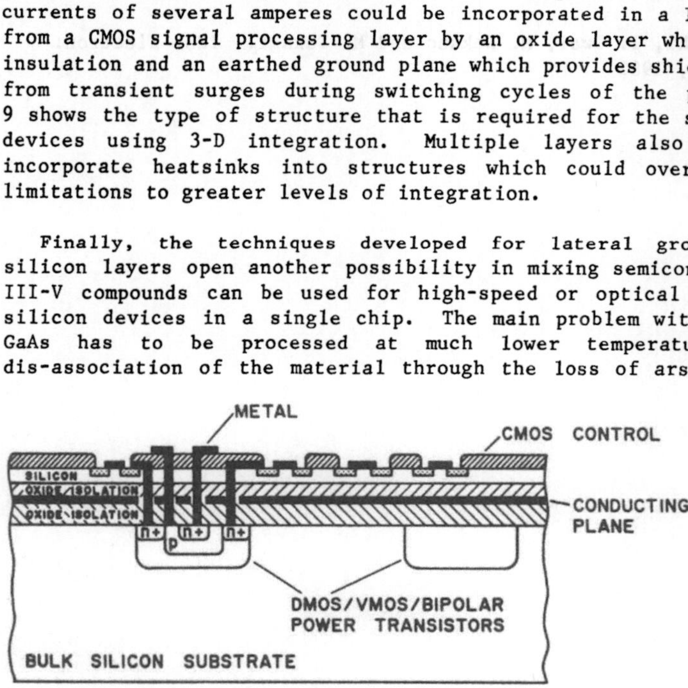

Fig.9 Concept for single chip realization of power devices and signal processing

Other processing incompatibilities also exist but there do not appear to be any fundamental difficulties in combining several different materials in a future semiconductor technology.

5. Conclusions

SOI could become an important material for VLSI in the near future and its development could lead to chips with multiple layers of devices. Applications for these chips arise in cameras, many types of sensors, power systems and in any situation where a very compact design is essential.

ACKNOWLEDGEMENTS

The many contributions of the Microelectronics Research Group and collaborators at GEC and British Telecom are acknowledged. Some of this work was supported by an ESPRIT grant. The financial support of the Hirst Research Centre, GEC (UK) and British Telecom Research Laboratories is acknowledged.

REFERENCES

1. J. F. Gibbons, K. F. Lee, T. J. Magee, J. Phang and R. Ormond: Appl. Phys. Lett. 34, 831 (1979)

2. M. W. Geis, H. I. Smith, B-Y Tsaur, J. C. Fan, D. J. Silversmith and R. W. Mountain: J. Electrochem. Soc. 129, 2812 (1982)

3. R. A. McMahon, J. R. Davis and H. Ahmed: Laser and Electron Beam Interaction with Solids, ed. by B. Appleton and G. Cellar (Elsevier, London 1982), p.783

4. J. R. Davis, R. A. McMahon and H. Ahmed. J. Electrochem. Soc. 132, 1924 (1985)

5. S. Kawamura, N. Sasaki, T. Iwai, M. Nakano and M. Takagi: IEEE Electron. Dev. Lett. EDL-4, 10, 1983

The Metal Base Transistors

E. Rosencher

Centre National d'Etudes des Télécommunications, CNS,
Chemin du Vieux Chêne, B.P. 98, F-38243 Meylan Cedex, France

I. INTRODUCTION

One of the major challenges in modern microelectronics is the creation of microwave devices or very high speed logic switches working in the 100 GHz range. It is now clear that device performances are no more limited by the transit time of the carriers in the active region of the structures. For instance, a p–n–p transistor with a 1 μm base width could, in principle, operate in the 100 GHz range (for a saturation velocity of 10^7 cm/sec as in most common semiconductors) while the best results obtained in Si, using a super self-aligned technology (S.S.T.), are in the 20 GHz range [1]. This limitation is clearly due to the $R_{acc}C_p$ time constant of the device, R_{acc} being the access resistance and C_p the parasitic capacitance of the device.

The improvement in device speed has been mainly obtained through the decrease of C_p, using highly performant self-aligned technology. As far as the base resistance is concerned the shrinking of bipolar device dimensions is impeded by the following effects :

i) The depleted layers on both sides of the narrow base overlap so that the collector depletion region is in direct contact with the emitter depletion region : this is the punch-through effect.

ii) This may be prevented by the use of high doping level in the base, which also decreases the base access resistance. However, these high doping levels are detrimental to the emitter efficiency, and thus to the current gain.

In order to alleviate these difficulties, ROSE [2], ATTALA and KAHNG [3] and GEPPERT [4], proposed in the early 60's to replace the semiconductor base by a metallic one, resulting in a Semiconductor-Metal-Semiconductor (SMS) device also named Metal Base Transistor (MBT). This structure has been investigated in extensive details by SZE et al. [5] : a cantilever beam was used as a micromanipulator to make a contact between a smoothly pointed and freshly cleaned semiconductor needle and a thin metal film evaporated on semiconductor substrates. The substrate served as the collector and the needle as the emitter. There was, of course, no hope of integration for such devices. The question also arose whether functioning of these structures was not controlled by pinholes in the metal base or impeded by the high non-ideality of the emitter diode.

In this lecture, we shall describe how such structures can be obtained as monolithic devices using molecular beam epitaxy and advanced reactive ion etching processes. In Sect. II, we shall address the physical principles of the SMS transistor. The fabrication technology will be described in Sect. III. In Sect. IV, the results of electrical characterization measurements will be discussed, with emphasis on the properties of transport in very thin crystalline metallic films and the influence of pinholes in the metal base.

II. PHYSICAL PRINCIPLES

The SMS transistor consists basically of two back–to–back Schottky diodes. The schematic band diagram is shown in Fig. 1. In such a device, the electrons are injected into the metal base by thermionic emission from the forward–biased emitter Schottky diode. A fraction of these carriers crosses the metal base via ballistic transport, i.e., at a constant energy ΔE above the metal Fermi energy E_F. They are finally collected by the reverse - biased collector Schottky junction. Three physical factors control the device performance :

1) The base transport loss due to electron–electron and electron–phonon scattering,

2) The electron–phonon backscattering in the collector junction,

3) The quantum mechanical reflection at the metal–semiconductor interface.

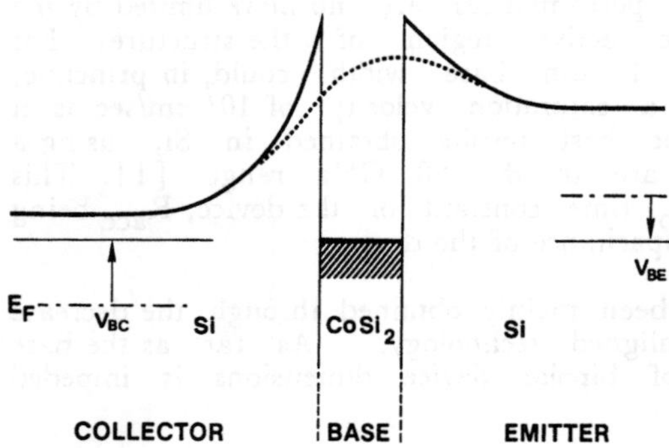

Figure 1 Schematic energy band diagram of an SMS transistor in a Semiconductor–Metal–Semiconductor junction (solid line) and in a semiconductor pinhole channel (dotted line)

Table I Ballistic mean free paths λ_B and ohmic mean free path ℓ_σ determined in Au, Ag and Pt [7].

METAL	ℓ_σ (Å)	ℓ_B ($E - E_F \sim 1$ eV) (Å)
Ag	310	265
Au	220	220
Pd	88	87

II.1. Loss in the Metal Base

Since any scattering in the metal virtually precludes collection of an injected electron [5], the base transfer efficiency α_B may be described by a ballistic mean free path λ_B :

$$\alpha_B = \exp - L_B/\lambda_B, \tag{1}$$

where L_B is the base thickness and λ_B is related to the excited state lifetime of the electron.

The ballistic mean free path λ_B is thus given by

$$\frac{1}{\lambda_B} = \frac{1}{\ell_p} + \frac{1}{\ell_i} + \frac{1}{\ell_e} \equiv \frac{1}{\ell_\sigma} + \frac{1}{\ell_e}, \tag{2}$$

where the various mean free paths are ℓ_p for phonon scattering, ℓ_i for impurity scattering and ℓ_e for electron–electron scattering. Since the two former scattering mechanisms are almost independent of energy and the ℓ_e decreases roughly as the inverse square of ΔE [6], ℓ_σ obtained from ohmic conduction measurements is the practical upper limit to ballistic mean free paths in the metal base. As listed in Table I, the values of ℓ_σ and λ_B are quite close in many metals. It thus seems that the electron–electron scattering is of little importance in the transport mechanism through the metal base.

II.2. Loss in the Collector Junction

The electrons injected into the collector experience over a distance x_m an electric field in the opposite direction : x_m is the position of the potential energy maximum in the semiconductor due to the image force barrier lowering [7] (see Fig. 2). The collection efficiency α_c, due to electron–phonon interaction in the collector, is given as the probability for an electron to reach x_m without being scattered by optical phonons. Then, α_c may be written as

$$\alpha_c = \exp - x_m/\lambda_{op}, \tag{3}$$

where λ_{op} is the optical phonon mean free path in the semiconductor. Typically, x_m varies from 20 Å down to 2 Å when the electric field at the interface increases from 10^5 V/cm up to 10^7 V/cm and λ_{op} is 65 Å in Si.

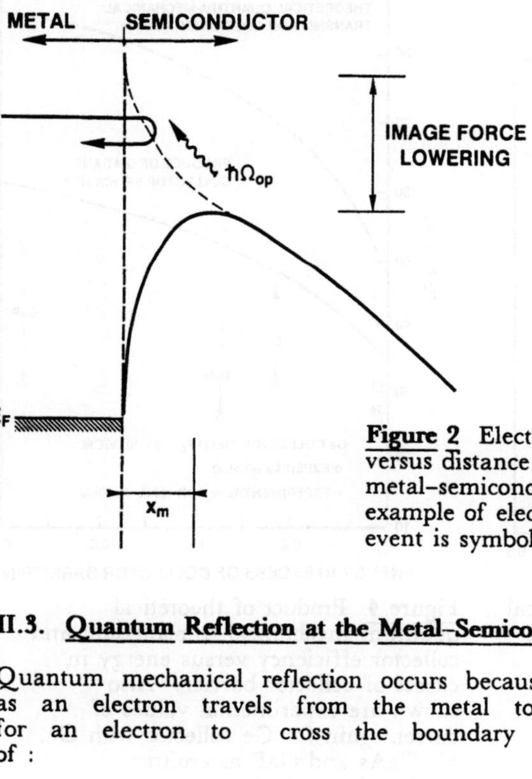

Figure 2 Electron potential energy versus distance from the collector metal–semiconductor interface. An example of electron backscattering event is symbolized

II.3. Quantum Reflection at the Metal–Semiconductor Interface

Quantum mechanical reflection occurs because of the sudden change in potential as an electron travels from the metal to the semiconductor. The probability for an electron to cross the boundary is obtained by writing the continuity of :

i) The electron quantum probability density $\Psi(x)$;

ii) The quantum probability current $J(x)$ at the interface, which is given by ;

$$J(x) = \frac{q\hbar}{2m^* i} \left(\Psi^* \frac{d\Psi}{dx} - \Psi \frac{d\Psi^*}{dx}\right), \tag{4}$$

where q is the electronic charge, $\hbar$ is the Planck constant and m^* is the effective mass in each material.

In a nearly free electron approach, matching $J(x)$ at the boundary condition means that the electrons have two different group velocities in the metal and the semiconductor and that the electron flux must be conserved despite these different velocities. Since group velocity is much larger in metals ($\sim 10^8$ cm/s) than in semiconductors ($\sim 10^7$ cm/s), the effect of Quantum Mechanical Reflection (QMR) is expected to be important.

The exact calculations are however very hazardous, since the results are highly affected by the shape of the metal-semiconductor potential discontinuity, which is poorly known. Figure 3 gives the quantum mechanical transmission α_q at a Si/Au interface as a function of excess energy above the barrier calculated for three different potential barrier models. One notes indeed that the expected influence of QMR is important and that the sensitivity of the numerical results to the model used makes a precise prediction very difficult [8].

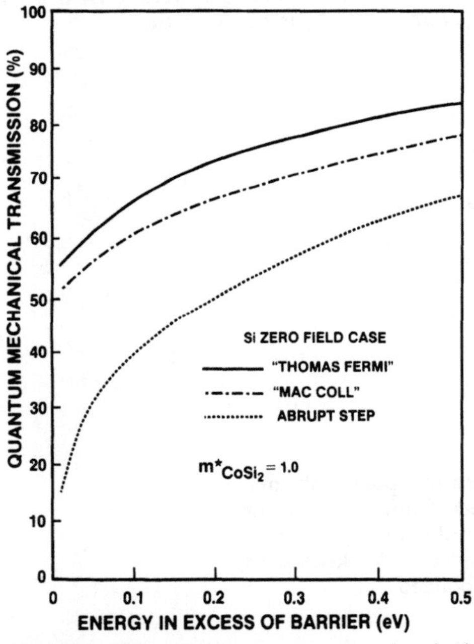

Figure 3 Zero-field quantum-mechanical transmission for an Si/Au barrier height of 0.63 eV and image force dielectric constant of 11.7 as a function of energy in excess of the barrier for an abrupt-step, Mac Coll and Thomas-Fermi barrier model

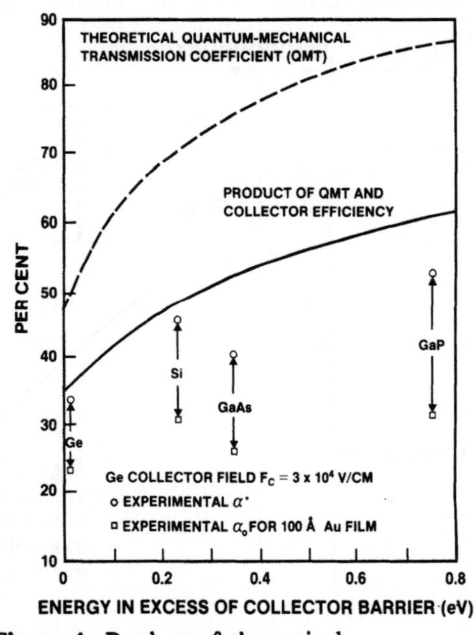

Figure 4 Product of theoretical quantum-mechanical transmission and collector efficiency versus energy in excess of collector barrier. Also shown are experimental values of current gain for Ge collector with Ge, Si, GaAs and GaP as emitter semiconductor [7]

Finally, the common base current gain α, i.e., the ratio I_C/I_E of the collected to the injected current is given by,

$$\alpha = \alpha_B \times \alpha_c \times \alpha_q, \tag{5}$$

also written as

$$\alpha = \alpha_o \exp - L_B/\lambda_B, \tag{6}$$

where α_o is the extrapolated value of α to zero film thickness. Figure 4 shows the calculated values of α_o as a function of energy in excess above the collector barrier together with some experimental values of α for Ge, Si, GaAs and GaP, as determined by SZE et al. [7]. Because of this rather low value of α, the SMS transistor can attain power gain only by means of impedance gain.

III. DEVICE REALISATION

In order to obtain monolithic SMS devices, evaporation or epitaxial growth techniques must be used. Some attempts to use amorphous or polycristalline evaporated films have been made but the device quality of the deposited metal/semiconductor structure was very poor. There has been a long search for a metal which could be epitaxially grown on top of Si, with good stability and electrical properties. Up to now, only some silicides have been found to fulfill those conditions : $NiSi_2$ with a lattice mismatch $\delta a/a$ of 0.4 % with Si and a resistivity of 30 μΩ.cm and $CoSi_2$ with $\delta a/a$ of 1.4 % and a resistivity of 15 μΩ.cm [9, 10].

In order to obtain a reasonable α gain, very thin metal films (in the 100 Å range) must be used. Only ultra-high vacuum evaporation techniques, with the help of surface analysis such as low electron energy diffraction and Auger spectroscopy, can satisfy such criteria.

The first SMS epitaxial transistors have been obtained using an $Si/CoSi_2/Si$ heterostructure [11]. <111> oriented Si wafers are submitted to a cleaning procedure which allows the production of atomically clean surfaces. Co layers are electron gun evaporated under a pressure less than 5×10^{-8} Torr with the sample kept at room temperature and then annealed at 650°C (± 25°C) during 10 mn in order to obtain the $CoSi_2$ layers. These films are crystalline, as checked in situ by LEED observations. It has been shown that continuous and uniform films could thus be obtained, up to a $CoSi_2$ thickness of 200 Å. Above this limit, the elastic energy due to stretching of the $CoSi_2$ lattice becomes overwhelming and the films get torn, yielding a milky aspect of the metal

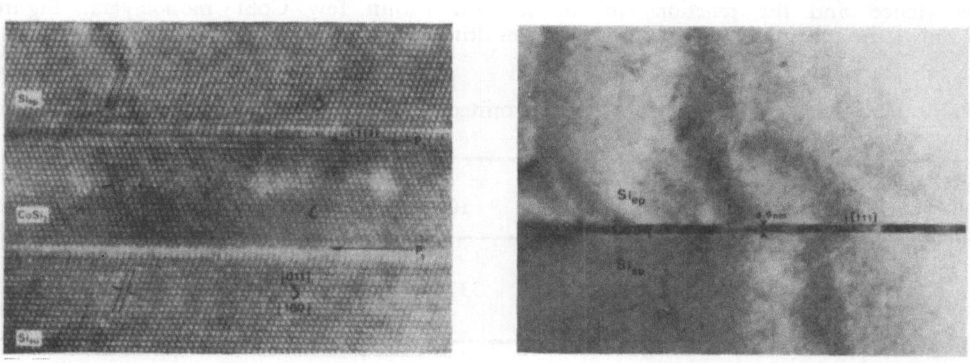

Figure 5 High resolution transmission electron micrographs of an $Si/CoSi_2/$ Si heterostructure at two different magnifications [14]

surface [12]. Silicon layers are then epitaxially deposited and doped by Sb coevaporation under a pressure in the 10^{-9} Torr range.

Two High Resolution Transmission Electron Micrographs (HRTEM) of an $Si/CoSi_2/Si$ heterostructure are presented in Fig. 5a and b. Figure 5a demonstrates that for such small $CoSi_2$ film thicknesses (~ 70 Å), the $CoSi_2$ lattice is indeed strained in order to fit the Si lattice. Figure 5b shows moreover that the transition between $CoSi_2$ and Si materials occurs within one monolayer over long distances. Such HRTEM observations have shown that continuous sandwiches can be obtained for a silicon growth temperature in the 650°C range while pinholes are present when the growth temperature exceeds 750°C.

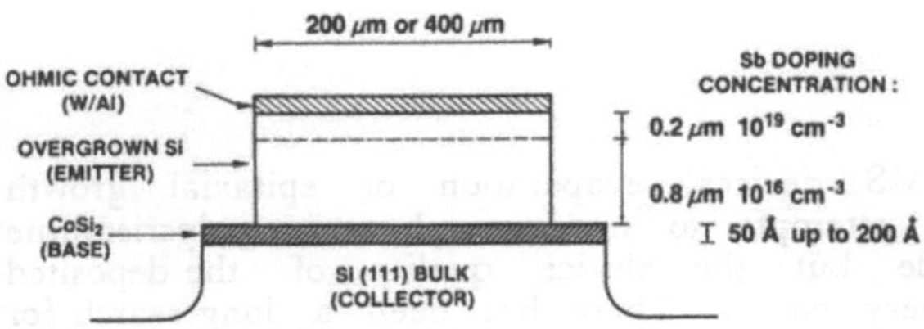

Figure 6 Schematic cross section of a mesa SMS transistor

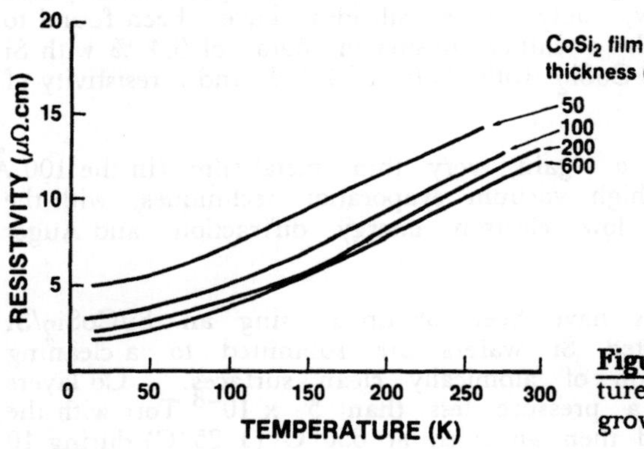

Figure 7 Resistivity versus temperature curves in $CoSi_2$ films epitaxially grown on top of Si substrates [13]

W/Al layers are then ex situ evaporated in order to yield a good ohmic contact on the overgrown Si layer. The structures are processed in a sequence of steps involving photolithographic masking and plasma (SF_6) Reactive Ion Etching (RIE). With the help of reflectivity measurements, 1 µm of Si layer can thus be etched and the reaction can be stopped within few $CoSi_2$ monolayers. Figure 6 shows the schematic cross section of a resulting mesa davice.

Table II Mean free paths ℓ_σ determined at 4 K and 300 K for different $CoSi_2$ film thicknesses, L_B.

$CoSi_2$ film thickness (Å)		50	100	200	600
ℓ_σ (Å)	4 K	205	337	519	559
	300 K	73	86	107	101

IV. ELECTRICAL PROPERTIES

IV.1. Transport Properties in $CoSi_2$ Films

We have seen that the determination of the ohmic mean free path ℓ_σ is of prime importance for the prediction of SMS device properties. Figure 7 shows the resistivity ρ versus temperature T curves in $CoSi_2$ films for different film thicknesses L_B. From Hall voltage measurements, the mean free path ℓ_σ can be deduced and its values are listed in Table II at 4 K and 300 K. One notes that

i) $CoSi_2$ thin films follow MATTHIESSEN's rule, i.e., the resistivity may be written as

$$\rho(T) = \rho_i + \rho_P(T), \tag{7}$$

where ρ_i is the residual resistivity due to impurities at low temperature and ρ_P is the resistivity due to phonon scattering, independent of the film thickness.

ii) The mean free paths ℓ_σ are larger than the metal film thicknesses and slighly dependent on L_B. This means that specular scattering prevails in longitudinal electron transport properties, due to the smoothness of the $Si/CoSi_2$ interface.

iii) The ℓ_σ value is in the 100 Å range at room temperature, which is expected to be a good approximation of the ballistic mean free path λ_B [13].

IV.2. I–V Characteristics

Figure 8 shows the common base characteristics, i.e., the collector-base voltage V_{CB} curves for different emitter currents I_E, in an $Si/CoSi_2/Si$ heterostructure with a 70 Å thick base. These characteristics are similar to those of the bipolar transistor. The intersection point V_{th} with the $I_c = 0$ axis depends on the difference between the two Schottky barrier diodes, which is still uncontrolled. Experimentally, V_{th} is found to vary between a few 10 mV up to 200 mV [14].

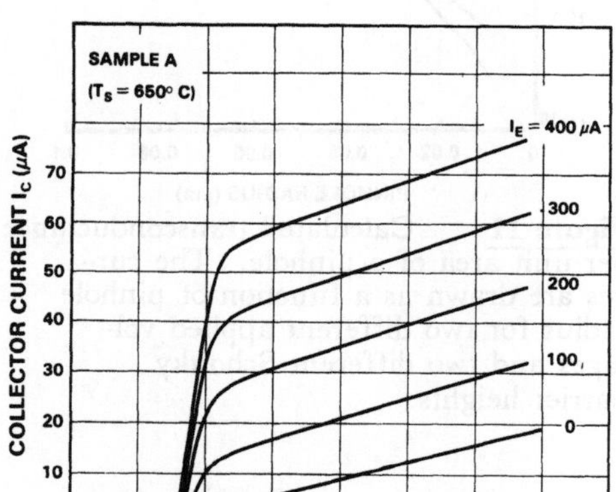

Figure 8 Common base current–voltage characteristics of an $Si/CoSi_2/Si$ transistor with Si layer grown at 650°C

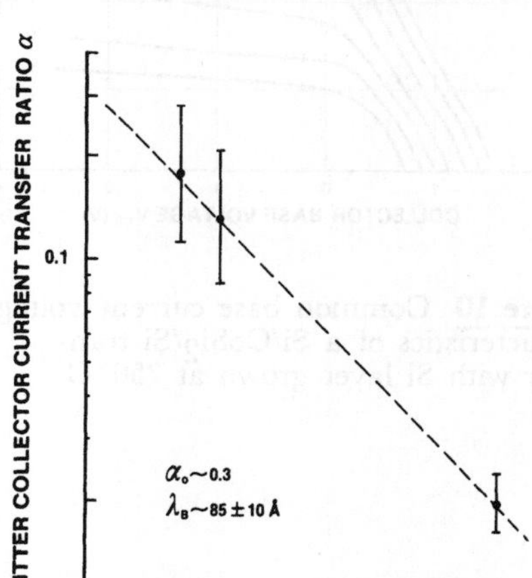

Figure 9 Common base current gain α as a function of $CoSi_2$ film thickness

Figure 9 shows the common base current gain α as a function of base thickness L_B : the dispersion of α over each wafer is indicated by error bars. Using (6), one finds that $\alpha_o \sim 0.3$ and $\lambda_B \sim 85$ Å ± 10 Å, which is indeed in good agreement with the mean free path value ℓ_σ. The value of α_o, due to quantum reflection and collector phonon–backscattering, confirms the results of SZE et al. [5].

IV.3. Influence of Pinholes in the Metal Base

When the overgrown Si layers are deposited above 750 °C, the α gains increase dramatically. Figure 10 shows the common base characteristics of an $Si/CoSi_2/Si$ structure identical to the one of Fig. 8, except for the Si growth temperature (T_{Si} = 750 °C) : a gain of 0.8 is obtained. However, pinholes in the metal base are revealed by HRTEM and scanning electron microscopy in these structures. The increase of α is thus due to the potential barrier lowering in the semiconducting pinholes between the emitter and the collector junctions, as indicated in Fig. 1.

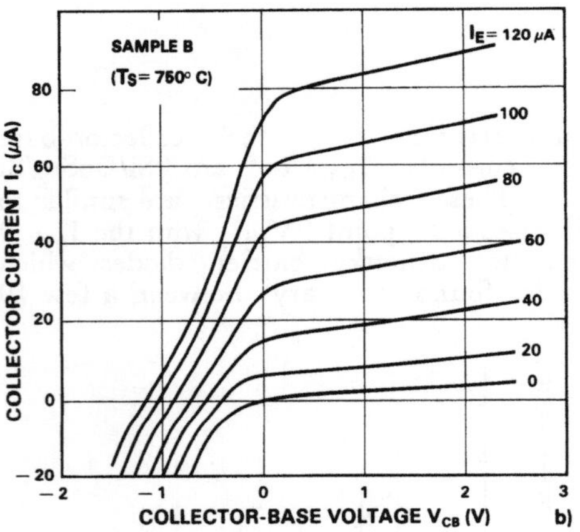

Figure 10 Common base current–voltage characteristics of a $Si/CoSi_2/Si$ transistor with Si layer grown at 750 °C

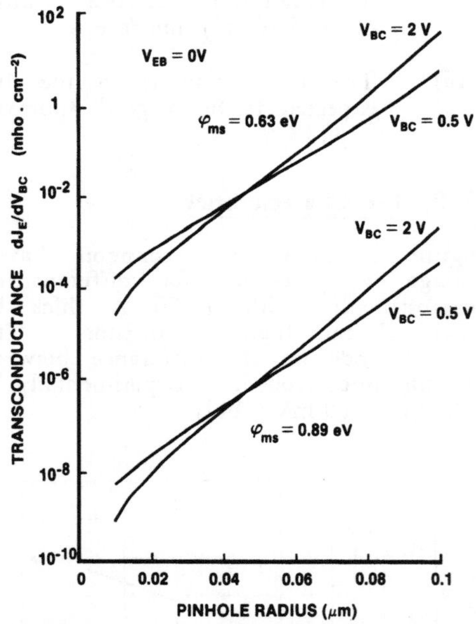

Figure 11 Calculated transconductance per unit area of a pinhole. The curves are drawn as a function of pinhole radius for two different applied voltages and two different Schottky barrier heights.

Such pinholes may be electrically detected with great sensitivity. Indeed, if no discontinuities are present in the metal base, the emitter current should not be influenced by the collector bias, since the collector potential is screened by the metal base within a Thomas Fermi distance (~ 0.5 Å). This is not the case in a semiconducting channel where the barrier height between emitter and collector junction is a function of collector bias. The value of the transconductance ($\partial I_E/\partial V_{BC}$) is thus a direct measurement of the density of pinholes in the metal base. The influence of a pinhole radius on the transconductance of a typical SMS transistor has been calculated. The results, presented in Fig. 11, illustrate the sensitivity of this measurement : for instance, the transconductance values of 10^{-8} mho obtained in the SMS transistor of Fig. 7 (2.5×10^{-3} cm^2) correspond to a single pinhole of 1000 Å radius in the whole structure [15]. The good performances of SMS transistors containing pinholes are auspicious for the realisation of Si/$CoSi_2$/Si permeable base transistors. In such devices, a metallic grid, defined by photolithography techniques, is imbedded in the semiconductor. The current modulation is performed by the pinch-off of space charge layers in the semiconducting channel. Good results have already been obtained using a tungsten grid in GaAs, despite the high lattice mismatch between both materials [16].

V. CONCLUSION

We have described the physical principles as well as the practical realisation of metal base transistors. The Si/$CoSi_2$/Si heterostructures are grown by molecular beam epitaxy, with a $CoSi_2$ film thickness between 50 Å and 200 Å. The device gain has been shown to be limited by basic physical effects, such as quantum reflection at Si/$CoSi_2$ interfaces. Gain enhancement may be expected by designing a proper potential barrier shape at the collector-base interface. Moreover, these structures open the way to epitaxial Si/$CoSi_2$/Si permeable base transistors and allow the physical study of electron transport in very thin metallic films.

REFERENCES

1. T. Sakai, S. Konaka, Y. Kobayashi, M. Sujuki, Y. Kawai : Electron. Lett. 19, 283 (1983)
2. A. Rose : Interim Rept., No. 6A RCA (June 1960)
3. M.M. Atalla, D. Kahng : IRE - AIEE Solid State Device Conference (July 1962)
4. D.V. Geppert : Proc. IRE, 50, 1527 (1962)
5. S.M. Sze, C.R. Crowell, G.P. Carey, E.E. Labate : J. Appl. Phys. 37, 2690 (1966)
6. A.H. Wilson : In Theory of Metals, Cambridge Univ. Press, London, New York (1955) p. 264
7. S.M. Sze : In Physics of Semiconductors Devices, Wiley- Interscience, New York (1969) Chap. 11
8. C.R. Crowell, S.M. Sze : J. Appl. Phys. 37, 2683 (1966)
9. S. Saitoh, H. Ishiwara, S. Furukawa : Appl. Phys. Lett. 37, 203 (1980)
10. R.T. Tung, J.M. Poate, J.C. Bean, J.M. Gilson, D.C. Jacobson : Thin Solid Films, 93, 77 (1982)
11. E. Rosencher, S. Delage, Y. Campidelli, F. Arnaud d'Avitaya : Electron. Lett. 20, 762 (1984)
12. F. Arnaud d'Avitaya, S. Delage, E. Rosencher, J. Derrien : J. Vac. Sci. Technol. B3, 770 (1985)

13. S. Delage : Ph. D Thesis, Université Paris VI (1985)

14. E. Rosencher, S. Delage, F. Arnaud d'Avitaya, C. d'Anterroches, K. Belhaddad, J.C. Pfister : Physica 134B, 106 (1985)

15. J.C. Pfister, E. Rosencher, K. Belhaddad, A. Poncet : Solid State Electron. (to be published)

16. C.O. Bozler, G.D. Alley : Proc. IEEE, 70, 46 (1982)

An Integrated Microfabrication System for Low-Dimensionality Structures and Devices

H. Ahmed

Microelectronics Research Group, Cavendish Laboratory, Cambridge University, Madingley Road, Cambridge CB3 0HE, United Kingdom

1. Introduction

The preparation of structures and devices that explore and exploit phenomena associated with low dimensionality requires a new range of fabrication techniques and equipment. There is a need to look beyond the tools developed for the fabrication of silicon integrated circuits since the requirements now are for thinner layers of materials, of higher crystalline quality and controlled to a higher level in both composition and desired impurity content. Secondly, better resolution is required in lithography and it must be carried

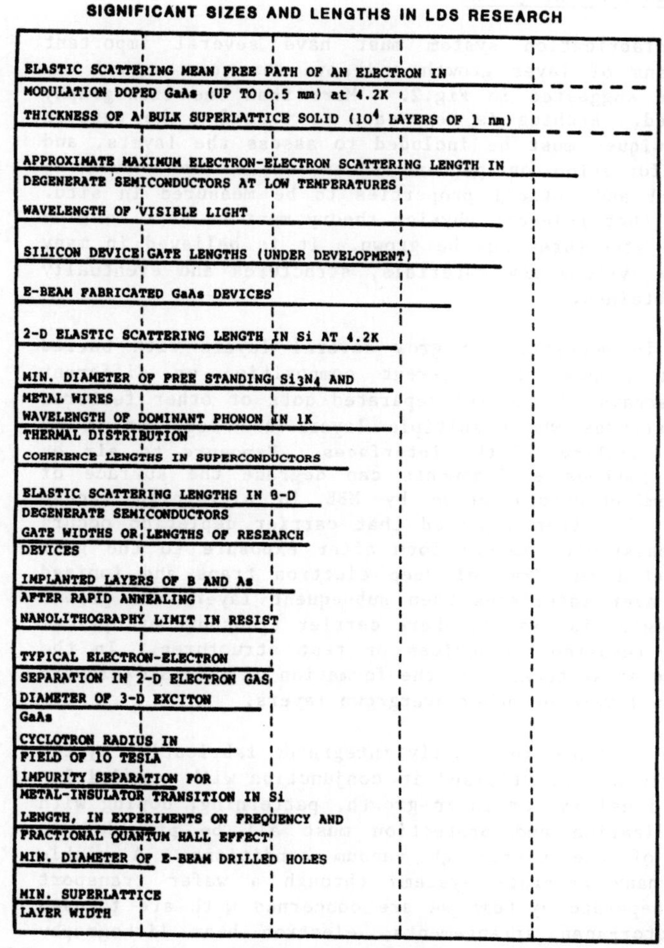

Fig.1 Some significant sizes and lengths in physics and microfabrication

out under more stringent conditions than those practised currently in microelectronics. For example, it is not always permissible to use organic resist layers as transfer media between lithography source and substrate.

The range of layer thicknesses, lateral dimensions and physical phenomena that are relevant is illustrated in Fig.1. Epitaxial layers as thin as 0.5nm have been grown for superlattices using molecular beam epitaxy (MBE) and implanted layers have been confined to shallow regions less than 25nm deep into the silicon substrate for hot-electron devices. Fine lines around 10nm wide have been fabricated and made into gratings or grids. Free standing wires <100 nm in two of their dimensions have been made. Transistor structures with sub-0.1 μm gate lengths have been fabricated. These dimensions are, in some instances, less than some of those in the physical phenomena listed in Fig.1, such as the typical electron-electron separation in a 2-D electron gas [1]. So far they have been obtained in separate experiments and not combined in a single, comprehensive system.

The most direct way to make the structures that are needed for low-dimensionality research activity is to use a fully integrated system for fabrication in all three dimensions, both in layer growth and in patterning the layer.

2. Totally Integrated Microfabrication System

A totally integrated microfabrication system must have several important features. It must have means of layer growth and more than one method may need to be incorporated, as suggested in Fig.2. More than one lithography technique may also be needed. Etching is necessary to transfer the pattern and several diagnostic techniques must be included to assess the layers, and to provide quantitative information on pattern dimensions. Means must be provided to allow electrical and optical properties to be measured in situ. The illustration also shows that relevant physics theory must be kept in mind so that correctly conceived structures may be grown. It is believed in many laboratories that from such systems new materials, structures and eventually devices will inevitably be obtained.

For many structures it is necessary to grow several layers with buried gratings or grids between layers of different composition or different materials. In some cases arrays of closely separated dots or other features may be needed. In all instances where multiple layers are required it is imperative to maintain the quality of the interfaces. Exposure to air or even to carefully selected gaseous environments can degrade the surface of GaAs or other compound semiconductors grown by MBE in ultra-high vacuum conditions. For example, it has been reported that carrier depletion occurs at the surface and high resistivity layers form after exposure to the gas. This has also been interpreted in terms of deep electron traps and ionised accepter levels which form near interfaces when subsequent layers are grown. This surface or interface region is exactly where carrier transport or optical interaction is subsequently required in devices or test structures. In the case of silicon MBE, problems arise because of the formation of thin SiO_2 layers at interfaces between silicon layers or other overgrown layers.

Most of the problems are overcome in a fully integrated fabrication system in which focused beam techniques are utilised in conjunction with a multiple-chamber MBE system. In an ideal system layer growth, patterning, doping with impurities, etching, metallization and protection must all be provided in situ without interruption of the ultra-high vacuum conditions. A viable arrangement is to couple many separate systems through a wafer transport tracking arrangement. The separate systems we are concerned with are focused ion beam columns with vector-scan arrangements, electron beam lithography and microscopy systems, rapid thermal annealing via electron beams and

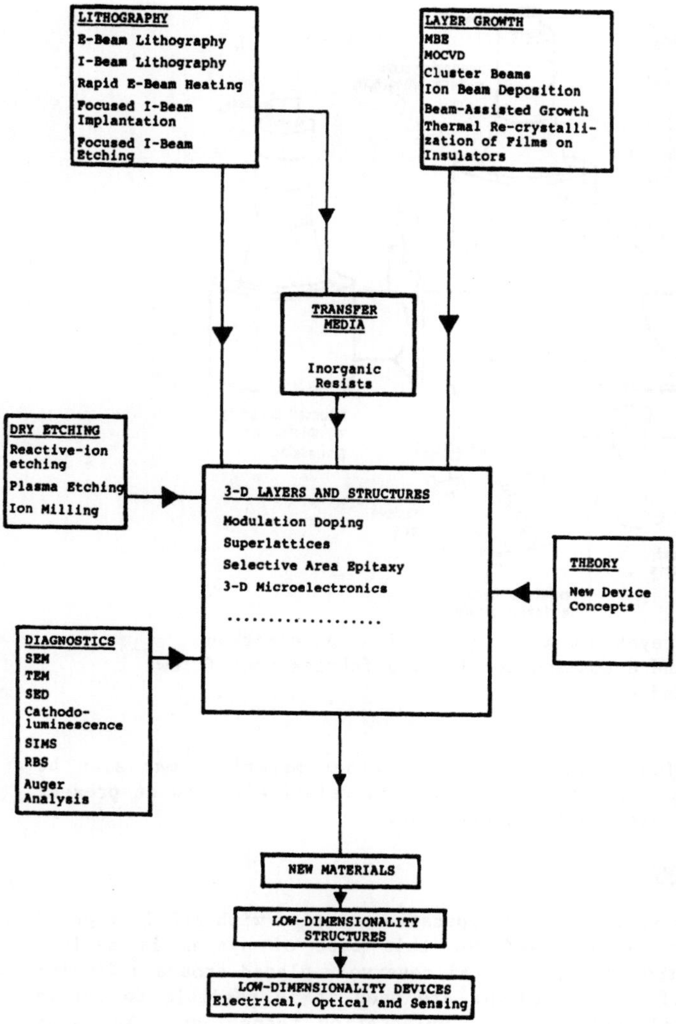

Fig. 2 Techniques of layer growth, microfabrication and and diagnostics required to explore the physics of low dimensionality which will lead to new materials, structures and devices

facilities such as plasma etching, evaporation and ion milling. A conceptual visualization of a facility which is under planning and construction is shown in Fig.3.

The current performance specifications for some of the techniques combined in this system and their limitations are outlined in the next section.

3. Molecular Beam Epitaxy

In this technique an ultra-high vacuum environment enables the growth of layered structures of extremely high purity, high order of crystalline perfection and uniformity in thickness of layers. The most important materials at present are multiple layers of $Al_x Ga_{1-x} As$ and GaAs grown by combining beams of Ga, As and Al landing on a heated substrate which is rotated to obtain uniform layers. The Ga and Al atoms interchange and rapid shuttering of the beams is carried out to change the properties so that the value of x is controlled. At present a thickness of 5Å can be obtained in layers with accurate control of composition between layers. The layers may be doped with impurities from other sources in the chamber. A vacuum level down to 1×10^{-11} Torr ensures that the impingement rate of unwanted atoms on the substrate is negligible and the mean free path of the species from the effusion sources is very long.

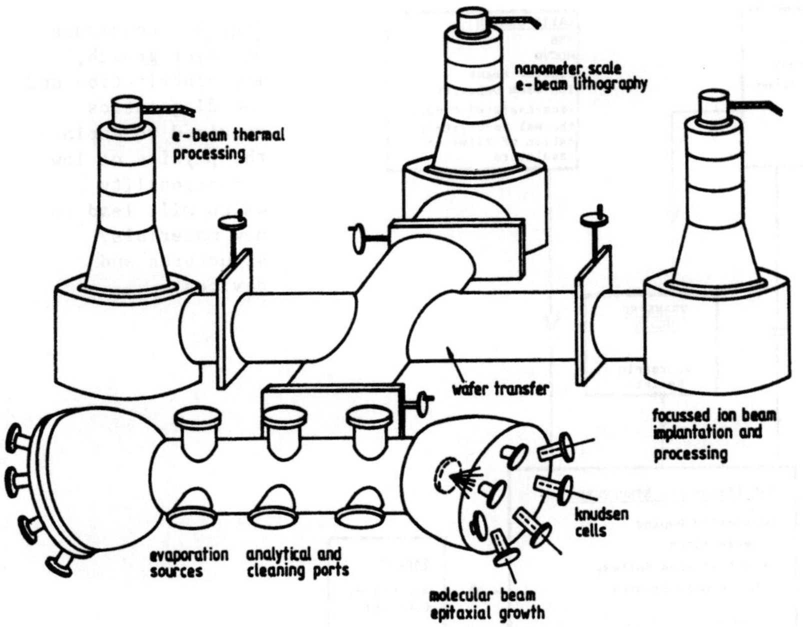

Fig.3 A totally integrated system combining MBE, ion and electron lithography, implantation and annealing for in-situ fabrication of low dimensionality structures and devices

Silicon MBE systems are also available and many other materials may also be grown using molecular beams. The rate of growth is relatively slow at present but high throughput systems are now being designed.

4. Focused Ion Beam Lithography

The most viable lithography system for UHV operation coupled with MBE is focused ion beam lithography. The source used for this type of system is derived by field emission of ions from a liquid metal reservoir placed around a finely-pointed carrier needle. If certain alloys are used it is possible to obtain beams of ions that are relevant in GaAs implantation technology. An alloy of Au-Si has been used extensively to provide Si (n-type) doping in GaAs and an alloy of Au-Si-Be can be used to provide both n-type and p-type dopant beams from the same source. The beams are separated by mass filters incorporated in the ion column. Electrostatic lenses are used to focus the beam to diameters as small as 250Å, and the beam is deflected over a field on the substrate which is typically around 0.25 mm x 0.25 mm, but fields may be joined together to cover a larger area of the substrate. A schematic diagram of an ion beam column is shown in Fig.4 in which focusing, mass-separation and deflexion and blanking are all available [2].

The beam energy may be varied from a few keV to 100 keV, although there is loss of resolution at lower beam voltages. In addition to implantation the focused ion beam may also be used to sputter away parts of the substrate. This enables layers to be cut through right down to the semi-insulating substrates so that mesas of GaAs or AlGaAs may be formed. Care has to be taken to ensure that material is not re-deposited on critical parts of the structure. Figure 5 shows sub-micron lines fabricated in an organic layer using a focused ion beam.

Finally, the focused ion beam may be used to image the structure formed by the beam and the MBE. Here the beam is raster-scanned over the surface and the secondary electrons are collected to form a picture of the surface

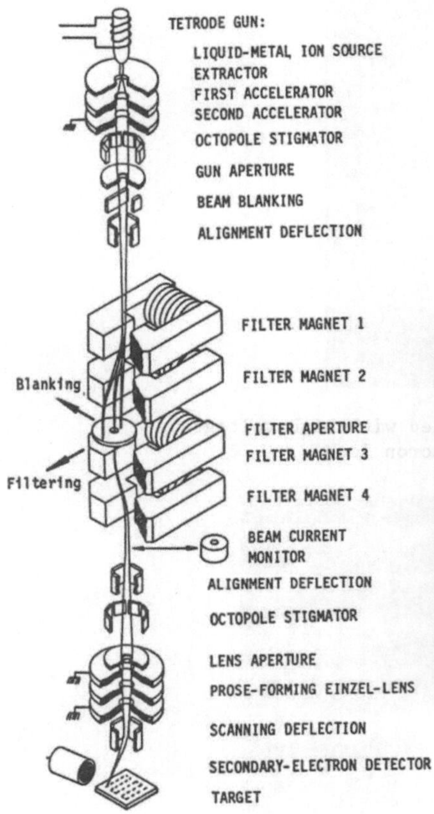

Fig.4 A focused ion beam system with mass selection for implantation, sputtering and inspection of structures

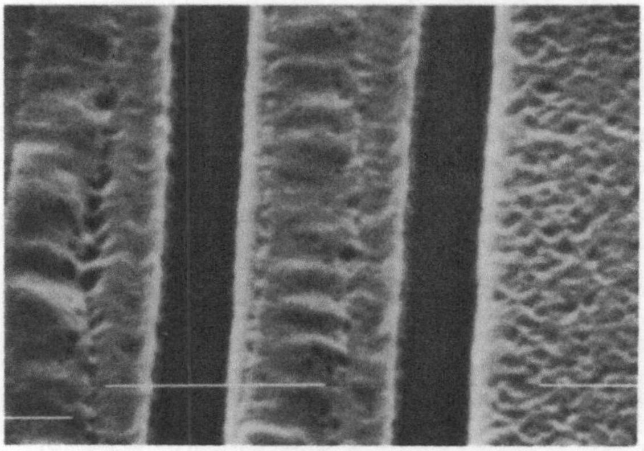

Fig. 5 Sub-micron patterns exposed with the focused ion beam system. The bar is 1 micron long

as in scanning electron microscopes. The main difficulty is that repeated scans can erode the surface but this can be minimised by modern electronic image storage and processing techniques. A few scans are sufficient to give enough information to produce a useful result. Takamori et al. [3] have reported the first successful combination of MBE and focused ion beam for implantation. They also report improvements in layer quality in their system compared with growth followed by exposure to air before lithography. Unfortunately the ultimate resolution obtainable from ion beams is limited by the large energy spread from the sources leading to excessive chromatic aberration and noise effects which arise from having insufficient ions in a very finely focused beam. Another problem with ion beams is that alinement of several layers on top of each other is difficult because registration marks are damaged by the beam during inspection.

5. Nanometre-Scale Electron Beam Lithography

Electron beams are capable of forming structures as small as 10 nm, but require a medium on the substrate and an additional etching or deposition step to transfer the pattern into the substrate. This makes it difficult to carry out electron beam lithography in a fully integrated system. If inorganic resists can be developed which may be deposited on the substrate in a vacuum environment then electron beam exposure becomes practical. Ideally, exposure should lead to changes in the resist including the breakdown of the exposed parts into volatile gases so that development is simultaneous with exposure. In such systems the resolution limit of ion beam systems is easily overcome by using electron beams at 100 kV and beam diameters around 1 nm. Figure 6 shows lines as thin as 20 nm fabricated by 100 kV electron lithography,

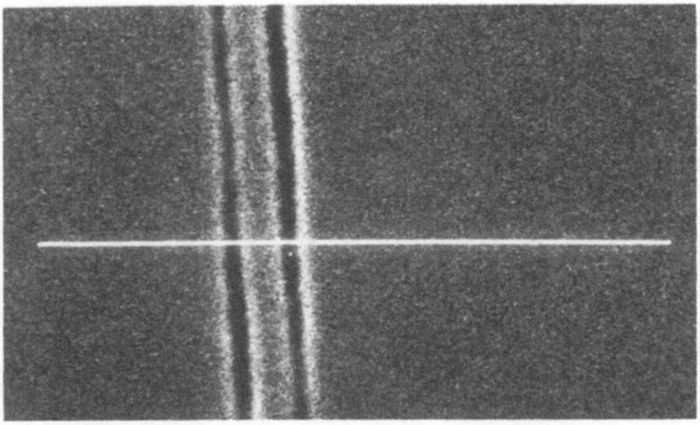

Fig. 6 Closely spaced 20 nm wide lines fabricated with high voltage electron lithography in resist. The bar is 1 micron long

Fig. 7 Grid structure fabricated in resist to demonstrate the feasibility of forming buried grids in MBE grown material. 50 grid squares/(μm)^2

and Fig. 7 shows a grid in resist which is the type of structure that must be buried in MBE grown layers (4).

The second application of a focused electron beam column is high resolution inspection of the parts previously fabricated. This can be achieved by scanning electron microscopy at resolution levels around 20 to 30Å and is an essential feature of an integrated system, since the dimensions and placements of structures cannot be measured by other means after fabrication.

In an alternative scheme the electron and ion beams may be combined in the same system [5] and designed to impinge on the same point on the substrate. This scheme enables fabrication with the ion beam and both inspection and registration of layers by the electron beams.

6. Rapid Electron Beam Annealing

Rapid annealing of ion implanted regions is essential if crystal quality and small dimensions are to be maintained simultaneously. Ion implantation causes both electron excitation and atomic displacements in the substrate lattice and this damage is usually removed by a protracted anneal cycle. Such annealing

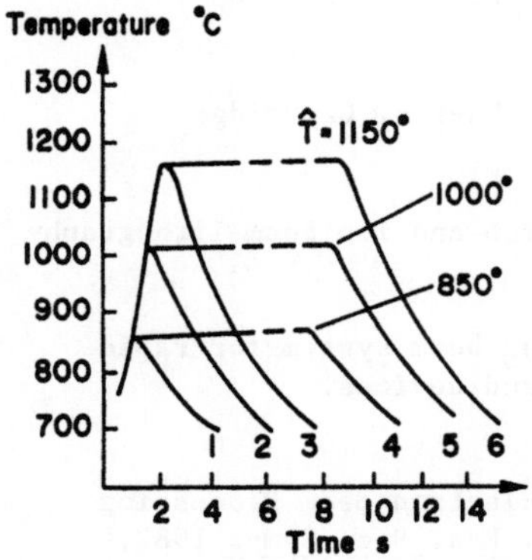

Fig.8 Time temperature cycle obtained from a rapid electron beam annealer. 1-3 are fastest wafer heating cycles and 4-6 demonstrate controlled heating periods

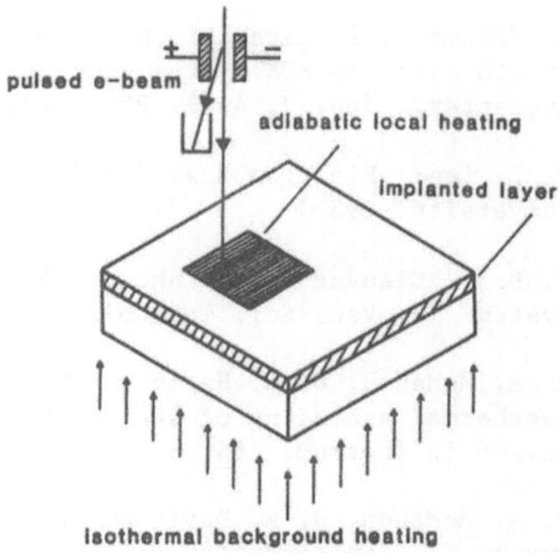

Fig.9 Dual electron beam annealing system enables two-second isothermal heating and 1-100 ms adiabatic heating of the top surface

must be carried out at high temperatures, at which GaAs would normally disassociate, in order to obtain device-worthy crystalline quality. Techniques such as capping of the layers with Si_3N_4 or heating in an over-pressure of arsine are used to obtain satisfactory results. These are not appropriate for an integrated system but the alternative method of very rapid isothermal annealing of the substrate is very promising. This is a short-cycle method in which the heating cycle lasts for times of a few seconds only [6]. Figure 8 shows typical time-temperature cycles. The technique has been extensively applied to annealing implants in silicon and even shorter cycles of only a few milliseconds for re-growth can be obtained while diffusion is restricted to distances less than 0.1 nm. using the two-beam techniques illustrated in Fig.9. The technique has also been combined with very low energy ion implantation to achieve highly doped regions in layers as thin as 25 nm.

7. Conclusions

Molecular beam epitaxy systems for layer growth combined with focused beam systems enables a fully integrated approach to microfabrication. Such systems yield high quality materials without interface problems inherent in growing multiple layers in which the vacuum conditions are interrupted. They also minimise defects from dust particles since all the processing is in vacuum. Clean room space requirements are minimised since only the entry and exit to the system need to be kept permanently under clean room conditions. This type of system is expected to evolve gradually for many applications as processing requirements move further towards smaller dimensions and higher standards.

ACKNOWLEDGEMENTS

Many discussions with G. A. C. Jones, J. R. A. Cleaver and R. A. McMahon are acknowledged. The support of the SERC and IBT-Dubilier is acknowledged.

REFERENCES:

(1) Private communication, M. Pepper.
(2) J. R. A. Cleaver, P. J. Heard and H. Ahmed: "Scanning Ion Beam Lithography with a Magnetic Ion Species Filter", Proc. Microcircuit Engineering '83, ed. by H. Ahmed, J. R. A. Cleaver and G. A. C. Jones (Academic, London 1983)

(3) A. Takamori, E. Miyauchi, H. Arimoto, Y. Bamba and H. Hashimoto: GaAs growth using an MBE system connected with a 100kV UHV maskless ion implanter. Jpn. J. Appl. Phys. 23, 8 (1984)

(4) K. L. Lee: Electron Beam Nanolithography, Ph.D. Thesis, Cambridge University (1982)

(5) J. R. A. Cleaver and H. Ahmed: A combined electron and ion beam lithography system. J. Vac. Sci. Technol. B.3 1 (1985)

(6) R. A. McMahon, D. G. Hasko and H. Ahmed: Electron beam system for rapid isothermal annealing of semiconductor materials and devices. Rev. Sci. Instrum. 56, 6 (1985)

(7) R. A. McMahon, J. R. Davis and H. Ahmed: "Dual Electron Beam Processing System for Semiconductor Materials" in Proc. Mat. Res. Soc. Symp. 1982, ed. by B. R. Appleton and G. K. Celler (Elsevier, New York 1982)

Fabrication of Gate Array Interconnect Structures Using Direct-Write Deposition Processes

F. Mitlitsky, J.C. Whitehead, A.F. Bernhardt, and B.M. McWilliams

Special Studies Program, LP Project, Lawrence Livermore National Laboratory, P.O. Box 5503, L-271, Livermore, CA 94550, USA

General principles of laser direct-write deposition processes are reviewed. Device interconnection of CMOS gate arrays by means of computer-controlled, laser-induced thermochemical surface reactions is described. Maskless, automated, five-minute interconnection of 1000 gate CMOS array dies is described using discretionary laser-induced chemical vapor deposition. Eight 125-stage ring oscillators written on a single CMOS gate array die were shown to have the device-limited performance of similar patterns manufactured by a photolithographically patterned aluminum-silicon alloy. These results suggest the feasibility of using this method, Laser Pantography (LP), for rapid implementation of prototype and limited volume semi-custom VLSI circuits immediately after their design is completed.

I. Introduction

The desirability of using local, laser microchemical techniques for maskless fabrication and alteration of integrated circuits has been recognized. Laser direct-write processes capable of local deposition or etching of a wide variety of materials on semiconductor substrates have been demonstrated [1-7]. They use a focused laser beam to supply energy to a micron-scale reaction zone. The local reaction of the ambient gas may be driven by direct photo-absorption, or by local heating of the substrate. The former class of reactions (photolytic), which have characteristic rates orders of magnitude slower than those of the latter class (pyrolytic), will not be considered here. In one of the earliest reported pyrolytic experiments, a CO_2 laser was used to deposit polysilicon (poly) by pyrolysis of silane [8]. However, the deposited lines were ~50 μm wide.

The use of shorter wavelength lasers is needed to achieve feature sizes small enough for integrated circuits (ICs). In 1981, two groups [9,10] reported using an Ar^+ laser to deposit micron-scale poly lines (from silane) at speeds up to 100 μm/s. Linewidths as narrow as ~0.2 μm have been achieved [11]. In consideration of these reaction rates, the use of local laser microchemical deposition was suggested mainly for minor alterations, such as mask repair and restructuring of previously lithographed ICs.

To date, single-trace circuit alteration [12], photolithography mask repair [13], and device connection [14] on CMOS gate arrays have been demonstrated, as well as the fabrication of single devices by laser deposition, doping, and etching [15]. Our work has emphasized rapid reactions capable of writing rates of at least 1000 μm/s laterally, and which are suitable for fabrication of IC interconnects. We previously reported a poly deposition reaction in this line-writing speed range [14]. This paper reports the use of that reaction to rapidly interconnect complex circuits on 1000 gate CMOS arrays.

---

*This work was performed under the auspices of the U.S. Department of Energy by Lawrence Livermore National Laboratory under Contract No. W-7405-Eng-48.

II. Chemical Reactions for Laser Deposition

A variety of laser-induced interconnection formation processes has been studied [1-10]. These fall into two broad classes: photolytic and pyrolytic processes. The photolytic processes are based on local laser-induced photodecomposition reactions occurring either at the substrate and/or in the gas phase. The pyrolytic processes are based on chemical vapor deposition reactions which occur due to local heating from a focused laser beam incident on the substrate. In applications for IC manufacturing, pyrolytic processes have the advantages of exhibiting substantially higher rates with presently available laser technology and, since they are based on chemical reactions already used in IC manufacturing, they interface more readily with typical IC processes and materials.

There are two basic groups of laser-driven pyrolytic processes which might be applied to interconnection of device structures in integrated circuits. One group is the surface-catalyzed unimolecular thermodecomposition reactions, such as those involving carbonyls and silanes. Examples of particular interest are

$$Ni(CO)_4 \rightarrow Ni(s) + 4CO \tag{1}$$

$$SiH_4 \rightarrow Si(s) + 2H_2 \tag{2}$$

$$Si_2H_6 \rightarrow 2Si(s) + 3H_2 \tag{3}$$

These reactions are based on reactants which at room temperature have vapor pressures of at least a few hundred Torr and deposit materials which have high melting points, thus potentially allowing relatively fast deposition rates (as explained below). The other groups of reactions are "bimolecular" reactions, typically involving reduction of metal halides. Two reactions of this type are

$$WF_6 + 3H_2 \rightarrow W(s) + 6HF \tag{4}$$

$$WF_6 + \frac{3}{2}Si(s) \rightarrow W(s) + \frac{3}{2}SiF_4 \tag{5}$$

Since tungsten and tungsten silicide are attractive interconnect materials for next-generation VLSI/ULSI circuitry, processes which rapidly generate high quality W and WSi_2 interconnections could be very useful.

Tungsten deposition by silicon reduction of tungsten hexafluoride [reaction (5)] offers the interesting possibility of being used as a bulk process following direct-write of a silicon interconnect structure by a laser process based on reactions (2) and (3). This approach could greatly increase the conductivity of the laser-patterned polysilicon structure at the cost of only a small amount of processing.

An inherent advantage of direct-write laser processes is their sequential nature which allows online changes of the interconnect structures for circumventing defects in devices or interconnect structures. In general, such changes can require both etching and deposition processes. For removal of silicon and various metal structures, processes based on halogen containing gases have been studied and reported [20,21,22].

III. Reaction Kinetics for Laser Deposition

Since laser-induced deposition processes proceed in a sequential manner, one of the most important issues to be considered is reaction kinetics. Laser-induced pyrolytic deposition processes can be thought of as involving six consecutive steps:

1. Local heating of the substrate by absorption of laser energy.
2. Transport of the reactant gas to the laser-heated portion of the substrate;
3. Adsorption of the reactant on the substrate surface;
4. Decomposition of the reactant gas on the heated portion of the surface;
5. Desorption of the volatile product from the heat surface; and
6. Transport of the product gas away from the substrate surface.

The rate-limiting factors of such processes are often set by the chemical reaction rate after the nucleation phase of the process which is described by an Arrhenious-type relation of the form:

$$R = R_o e^{-T_A/T} P^n \tag{6}$$

where P is the reactant gas pressure, T is the substrate temperature, n is the reaction order and T_A is the activation temperature.

At low temperatures, the deposition rates are limited by the surface reaction rate of (6). Clearly this process can be accelerated by increasing the incident laser power, and hence the surface temperature. As this is done, the surface reaction rate may eventually outstrip the ability of the gas to deliver reactants and/or to remove reaction products. As a result the deposition rate does not continue to climb with laser power, but eventually saturates.

A general treatment of the transport of reactant and product gases during laser deposition is difficult. The character of the transport depends critically on two parameters: whether the reaction absorbs or liberates gas molecules; and the ratio of the hot spot size, ρ, to the molecular mean-free path, λ.

If the Knudsen number, $Kn = \lambda/\rho$, is very large, then the effects of gas transport are easily calculated. In this molecular flow regime, reactant species are supplied to the surface at a rate of

$$F_i = \frac{1}{4} n_i v_i \tag{7}$$

where F_i is the number of molecules per second per area striking the surface, v_i is the average molecular speed of reactant species i, and n_i is its molecular density.

In this flow regime, the ejected molecules free-stream away from the hot spot, creating a local atmosphere of reaction products. However, because the ejection speeds are greater than those of reactant gases and are more effectively directed, these product densities are low, and should not choke the reaction. Hence, in the molecular flow regime, the deposition rate saturates at a level dictated by the slowest influx rate. For Si deposition by silane, this rate is $\approx$75 μm sec^{-1} $Torr^{-1}$.

While the molecular flow regime is easy to analyze, it can only be assumed for large Knudsen numbers, which for micron-scale spots implies pressures below about 10 Torr. Since we generally operate at higher gas densities in order to increase the deposition rate, the actual molecular flow is strongly influenced by gas collisions.

Accurate calculation of the gas transport in the fluid regime is difficult, involving both diffusion and convection of multiple species. The analysis is even more complicated in the $p \approx 100$ Torr regime, in which transport is in transition between the molecular and fluid regimes. Accordingly, we will only qualitatively discuss transport at such gas densities.

The flow will strongly depend upon the reaction balance of gas molecules. If the reaction absorbs more gas molecules than are liberated, then reactants can be supplied to the hot spot at sonic velocities. The specific influx rate ranges from the molecular limit of (7), to a rate about 70% higher in the fluid limit. If there are no product gases, the gas transport rate can then be monotonically increased by raising the gas density, until the surface rate of (5) is reached. Product gases will have to diffuse away from the hot spot against the reactant influx, and if present can thus choke off the reaction.

Most reactions, including the ones discussed earlier, liberate at least as many molecules as they absorb. In the case where there is no net molecular generation, the gas convection is relatively weak. Here the transport of reactants and products is a diffusion limited process. This case was analyzed for constant gas diffusion constants in [3]. The specific transport rates scale as D/ρ, a considerably lower velocity than the sonic rate. Furthermore, one cannot arbitrarily increase the deposition rate by increasing gas density. Once the reactant density dominates the total density, then the diffusion coefficient will decrease, and the transport limited deposition rate will saturate.

IV. Gate Array Interconnect Process Description and Apparatus

The gate array, made by the *selective oxide silicon gate CMOS process*[16], is designed for single level metallization with underlying poly tunnels for routing.

A ~1 μm thick aluminum-silicon alloy on the gate array wafer is pre-patterned using conventional photolithography to leave I/O pads with interconnected drivers, power distribution, plugs in the contact vias, and a few simple logic gates interconnected at the periphery. After the metal patterning is complete, the surface is coated with a thin film of undoped amorphous silicon (a-Si). For this particular gate array personalization technology no loss of generality is suffered by prepatterning the metal as described in the preceding paragraph because this portion of the metallization is independent of the circuit design. The few interconnected gates are added as a process control monitor to allow electrical pretesting of each die before the laser direct-write personalization step.

The substrates are personalized by discretionary laser-induced pyrolytic deposition of degenerately doped poly. The reactant gas used in this process is a mixture of silane, disilane, and a dopant gas. Phosphine is used when n-type doping is desired. Diborane is substituted for phosphine to obtain p-type doped poly. Total reactant gas pressure of ~1 atmosphere is typically used. The deposited material has a resistivity of $\sim 3\times10^{-3}$ ohm-cm, as reported previously [14]. The poly line segments, deposited at a lateral scan rate of 1000 μm/s in all cases, are generally ~3-4 μm wide and ~1 μm high (Fig. 1).

The apparatus used for laser-writing in the work being reported here is similar to that previously described 17, with a 5145Å Ar^+ laser focused through the reaction chamber window by a 20X microscope objective to a ~1 μm spot on the substrate. One new feature is to move the beam over a 512 micron square field by an orthogonal pair of acousto-optic scan crystals. Direct writing is accomplished in real time using programmable hardware which controls beam position, writing speed, and beam power.

The software required to automatically generate the command sequences necessary for discretionary laser writing of complex patterns has been developed. There are two principal software packages which do this: they are known as the Assembler and the Monitor.

The Assembler translates an industry-standard CIF (Caltech Intermediate Format) file, which contains a geometric description of a circuit design, into a set of instructions called scan code. This scan code specifies the position, velocity, direction, and power level of the laser beam. The Assembler is cognizant of the special processing required for certain device features and alters the scan code process parameters to account for this. The scan code also includes the information to position and align the substrate after each 512 micron square scan field has been written.

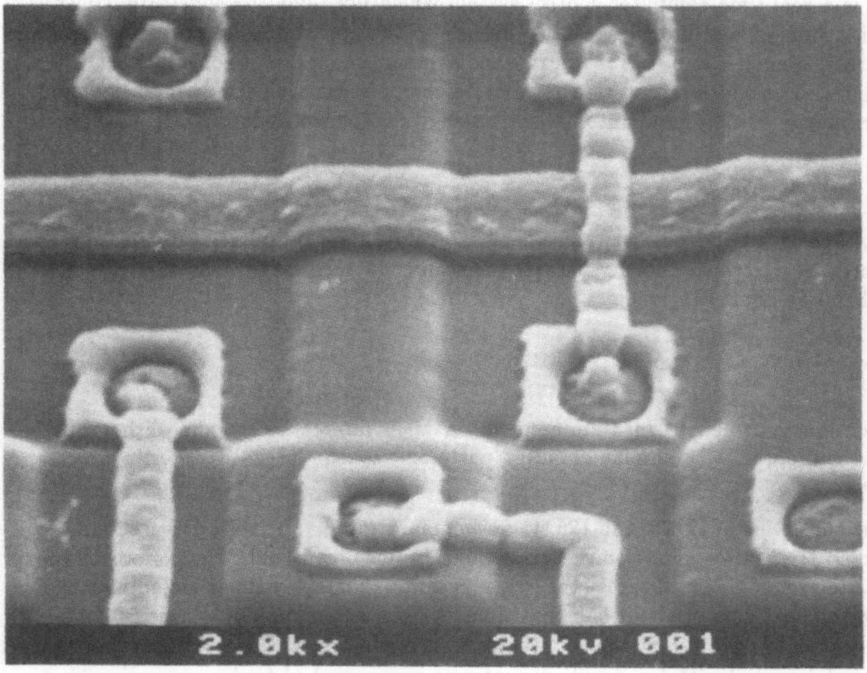

Figure 1. **SEM of laser deposited polysilicon interconnects on a CMOS gate array. Photolithographic preprocessing of Al-Si alloy includes power bus (large horizontal trace) and filled vias.**

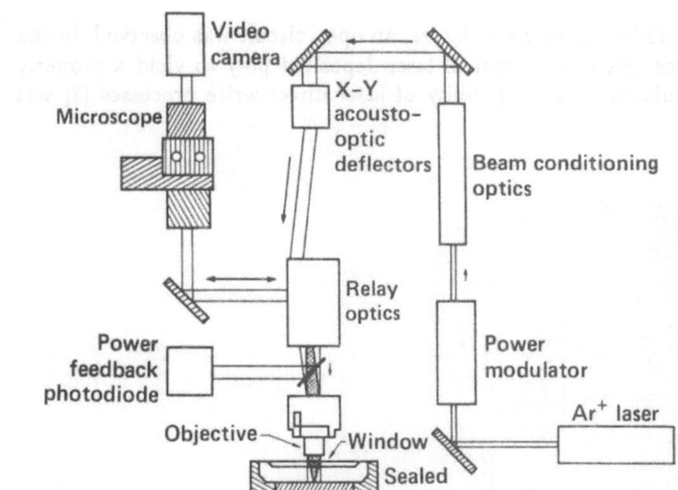

Figure 2. **Layout of apparatus used for polysilicon direct write by solid state laser scanning. (The darker stipling indicates the path of the laser beam. The lesser shaded path is that of white light for imaging.)**

The Monitor is a real-time computer control system which positions the substrate, controls the direct-write process, and provides a user interface to the apparatus. It accepts scan code produced by the Assembler and uses it to drive the scanner, beam power controller, and X-Y translation slides. The slides reposition the reaction chamber for each scan window. One or more gate arrays may be patterned automatically using the Monitor under high-level operator control.

Once a circuit is designed and physical placement and routing are complete, a minicomputer running the software described above generates the scan code necessary to run the LP apparatus in less than one hour.

V. RESULTS

Digital logic circuitry patterned in aluminum-silicon alloy fabricated by photolithography was tested and compared with similar circuitry patterned by the laser direct-write process described above.

Ring oscillators consisting of 17, 35, and 125 stages were fabricated. The periods were found to have the expected 1:2:7 ratios. The nominal cycle time obtained by dividing the periods by the number of stages was ~6 ns in all cases (Fig. 3), suggesting gate delay times including signal propagation to the next gate of about half that time. A light micrograph of a typical 17 stage ring oscillator segment fabricated by the LP process is shown in Fig. 4. Laser induced damage, poor for circuits fabricated with laser writing processes were searched for by comparing identical ring-oscillator circuits fabricated by conventionally-lithographed aluminum-silicon. No significant difference in oscillation frequency between laser- and conventionally-interconnected gate-arrays was observed. Furthermore, both n-type and p-type laser-deposited poly ring-oscillators had the same oscillation frequency.

On one die consisting of eight 125-stage ring oscillators, an open circuit was observed during electrical testing. This gap was closed by a spot of laser-deposited poly to yield a properly functioning circuit. The postulated repair capability of laser direct-write processes [3] was thus verified.

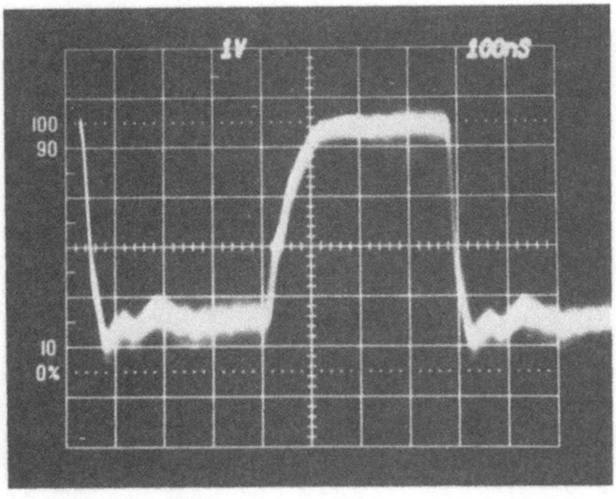

Figure 3. Oscillograph of output of 125-stage ring oscillator interconnected by the laser direct-write process, operating at 5 volts. The period of 750 ns reveals an average inverter cycle time of 6 ns.

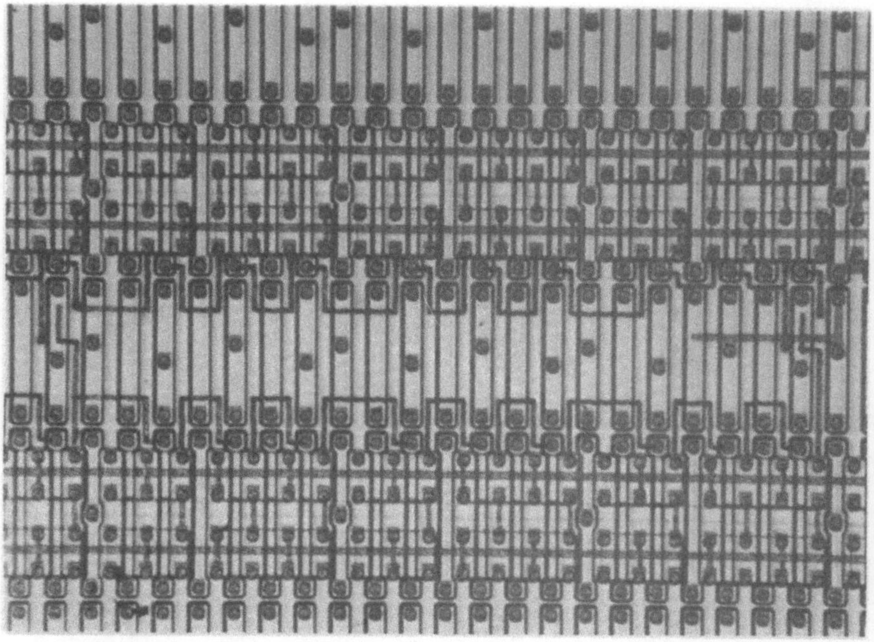

Figure 4. Light micrograph showing a single scan window portion of a 125-stage ring oscillator circuit. Device interconnects were fabricated by laser direct-write of n-doped polysilicon. Power, buses, via plugs, and output trace at left are photolithographed Al-Si alloy.

Interconnection of an entire gate array requires writing 50 to 100 scan windows of roughly 400-500 microns on a side. Each window required 4 - 5 seconds, including stage translation and settling. The time required to personalize a gate array is therefore a few hundred seconds. In particular, writing eight 125-stage ring oscillators was achieved in just under 5 minutes. Circuits with traces running between scan windows were demonstrated. Reliable connection of traces running between windows was achieved by superimposing traces at scan window boundaries.

Circuits with more complex interconnect structures, consisting of shift registers and counters, have been fabricated. In particular, several chips were fabricated having five 16-stage shift registers and one 16-stage counter on them. A light micrograph of one of these circuits fabricated with laser-deposited polysilicon is shown in Fig. 5. The basic building block of these circuits were D flip-flops and transmission gates. The counter and one of the shift registers were wired to output buffer ports at each stage to monitor sequential operation by electrical probing after interconnection. Electrical testing verified that these circuits functioned properly at a clock frequency of 1 MHz.

VI. DISCUSSION

Typical digital circuits often require high density interconnect structures, in contrast to circuits used in our evaluation of the process properties. In particular, the 5 minute-per-die figure cited is minimal for the poly process. Maximum possible density interconnection of these 4 mm square die with a pitch of 11 μm would take $(4000\ \mu m)^2/(11\ \mu m)(1000\ \mu m/s) = 1455$ s or ~24 minutes for line writing, with the worst-case addition of up to ~2 minutes for stage translation and settling. A more realistic time estimate for single level personalization of a complex circuit on gate arrays of these dimensions is 10-15 minutes.

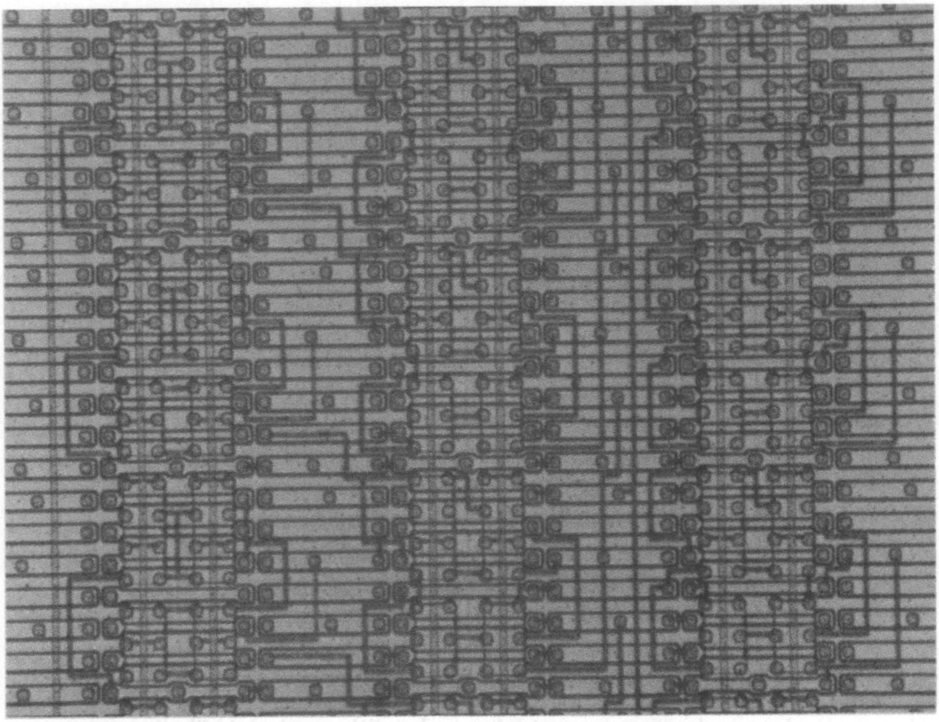

Figure 5. Light micrograph of part of shift register circuit described in text. Portions of two scan windows are shown.

All viable laser pyrolytic deposition processes for use in fabrication of IC interconnect structures need some means of accounting for the dramatic differences in thermal conductivity and optical absorption normally found between device contact, routing channel, and tunnel contact areas. Deposition of a thin film of a-Si as an absorbing layer greatly reduced the effects which underlying structures had on process parameters.

Since aluminum-silicon has very high thermal conductivity compared with a-Si, the a-Si served the dual function of heat retention (over vias and power buses), in addition to absorbing the incident laser power. The reaction still ran cooler over aluminum-silicon regions than over oxide regions, because a-Si does not completely shield the effect of the high thermal conductivity of the aluminum-silicon. The net result is that slightly less material will be deposited over aluminum-silicon regions, for a given dwell time, compared with oxide regions. Although we had the ability to alter process conditions over various structures, highly reliable laser direct-write interconnection of circuits was achieved using a single value for incident laser power everywhere.

The use of degenerately doped poly is limited to short interconnects, due to its low electrical conductivity. This process flow can be modified to achieve higher conductivity material by removing the a-Si in bulk after laser direct-write personalization. Selective tungsten CVD [18] may then be used to grow tungsten only where poly lines have been deposited. In particular, a thick selective tungsten CVD process [19] could yield conductivities sufficient for interconnecting high performance VLSI MOS circuits.

ACKNOWLEDGMENTS

Dave Ashkenas, Steve Farmwald, Rod Hyde, Irving Herman, Jim Kaschmitter, Harlan Lau, Jerrold Rittmann, and David B. Tuckerman made major contributions in helping to understand the engineering processes and equipment used in work reported. Contributions made by Bonnie K. Carnahan, Frederick D. Cook, Donald E. Duerre, Irving P. Herman, Richard Mendonca, A. Elon Ormsby, Wayne E. Tindall, and Jimmy C. Trevino are greatly appreciated. Also, thanks to Deborah Ross for expert assistance in preparing the manuscript.

Lowell L. Wood and A. Carl Haussmann have been instrumental in making this effort possible. The work was sponsored by DoE (through LLNL's IR&D Program) and DoD (through the VHSIC Program Office under Contract No. FY1175-85-N-5695, monitored by J. P. Letellier). John C. Whitehead was supported by the Fannie and John Hertz Foundation. Fred Mitlitsky was aided by Prof. Richard M. Osgood, Jr. and the Columbia University Electrical Engineering Department.

REFERENCES

1. R. M. Osgood, Jr: *Ann. Rev. Phys. Chem.* **34**, 77 (1983)
2. D. J. Ehrlich and J. Y. Tsao: *J. Vac. Sci. Technol.* **B1**, 969 (1983)
3. D. J. Ehrlich and J. Y. Tsao: In *VLSI Electronics: Microstructure Science* **7**, N. Einspruch, ed., (Academic, New York 1984), pp. 129-164.
4. R. J. von Gutfeld: In *Laser Applications* **5**, (Academic Press, New York 1984), pp. 1-67.
5. T. J. Chuang: *Surf. Sci. Rep.* **3**, 1 (1983)
6. W. W. Duley: *Laser Processing and Analysis of Materials* (Plenum, New York 1983)
7. I. P. Herman: *Proceedings of the Laser Processing and Diagnostics Conference*, Linz Austria, July 15-19, 1984.
8. C. P. Christensen and K. M. Lakin, *Appl. Phys. Lett.* **32**(4), (1978), pp. 254-255.
9. D. J. Ehrlich, R. M. Osgood, Jr. and T. F. Deutsch: *Appl. Phys. Lett.* **39**(12), 957 (1981).
10. D. Bauerle, P. Irsigler, G. Leyendecker, H. Noll and D. Wagner: *Appl. Phys. Lett.* **40**(9), 819 (1982).
11. D. J. Ehrlich and J. Y. Tsao, *Appl. Phys. Lett.* **44**(2), 267 - 269 (1984).
12. D. J. Ehrlich, J. Y. Tsao, D. J. Silversmith, J. H. C. Sedlacek, R. W. Mountain, and W.S. Graber, *IEEE Electron Device Letters*,**EDL-5**, 32 (1984).
13. D. J. Ehrlich, R. M. Osgood, Jr., D. J. Silversmith, and T. F. Deutsch, *IEEE Electron Device Letters*,**EDL-1**, 101 (1980).
14. B. M. McWilliams, H. W. Chin, I. P. Herman, R. A. Hyde, F. Mitlitsky, J. C. Whitehead and L. L. Wood, *SPIE Proceedings Vol. 459, "Laser Assisted Deposition, Etching, and Doping,"* January 22-27, 1984.
15. B. M. McWilliams, I. P. Herman, R. A. Hyde, F. Mitlitsky, and L. L. Wood *Technical Digest—Conference on Lasers and Electro-Optics*, Baltimore, Maryland, May 17-20, 1983.
16. Semi Process, Inc. CMOS Masterchips Manual, 1983.
17. B. M. McWilliams, I. P. Herman, F. Mitlitsky, R. A. Hyde and L. L. Wood, *Appl. Phys. Lett.*, **43**(10), 946 (1983).

18. N. E. Miller and I. Beinglass. *Solid State Technology*, **25**(12), 85-90 (1982).

19. R. S. Blewer and V. A. Wells. *Proc. First International IEEE Multilevel Interconnection Conference*, June 21-22, 1984.

20. D. J. Ehrlich, R. M. Osgood Jr., and T. F. Deutsch, *Appl. Phys. Lett.*,**36**, 698 (1980).

21. J. I. Steinfeld, T. G. Anderson, C. Raiser, D. R. Denison, L. D. Hartsough, and J. R. Hollahan, J. *Electrochem. Soc.*, **127**, 514 (1980).

22. D. J. Chuang, J. *Chem. Phys.*,**72**,6303 (1980).

Electronic Neural Computing

L.D. Jackel, J.S. Denker, H.P. Graf, R.E. Howard, W. Hubbard, D. Schwartz, B.L. Straughn, and D.M. Tennant

AT & T Bell Laboratories, Holmdel, NJ 07733, USA

Models for neural function have suggested new ways of designing special-purpose processors. These electronic neural networks mix analog and digital processing. For some applications neural networks may have speed advantages over conventional architectures. This paper provides an introduction to electronic neural computing and describes circuit implementations.

1. INTRODUCTION

System and circuit considerations will probably limit digital computer cycle times to a few tenths of a nanosecond, speeds that are within an order of magnitude of today's fastest machines. The way to further increase computing power is through parallel computation. The first generation of massively parallel machines made from arrays of microprocessors is now commercially available.

Among the tasks for which advanced computers are targeted are problems that animals solve with little apparent effort. We suspect that they do this by massively parallel analog computation. It is appropriate that we ask if there is a niche for analog electronic parallel computing, possibly with architectures based on neuronal models. In this paper we describe some simple theoretical models of neural function, describe how they can be applied to specialized computation tasks, and discuss early work in the fabrication of integrated electronic neural-network circuitry.

2. ELECTRONIC NEURONS

Figure 1 shows a classical drawing of a biological neuron. The neuron consists of a structure that receives inputs (the dendrite), a structure that transmits outputs (the axon), and the cell body. When the neuron is excited by its inputs it sends out pulses through its axon. The pulse rate depends upon the degree of excitation. Connections from the axon of one neuron to the dendrite of the next occur at points called synapses. Some of the synaptic connections are excitory and encourage the receiving neuron to fire, while other synapses are inhibitory and discourage firing. The output pulse rate depends on the time-averaged algebraic sum of the inputs, weighted by the strength of the appropriate synapses. The output is usually assumed to be a sigmoid function of the input sum. This means that for a range of inputs, the output is a linear function of the input, but for large or small input sums, the output saturates, at either the maximum or minimum firing rates, respectively.

This kind of response can be modeled electronically. Rather than using pulse rates, the electronic neuron communicates by producing a voltage level proportional to its activation level. Inputs are fed through resistive "synapses" to a wire "dendrite" where they are algebraically summed. This sum is fed to an amplifier with a sigmoid response. The amplifier's input capacitance provides a time-averaging of the neuron's response. The amplifier output travels through the wire "axon" to resistive synapses with other neurons.

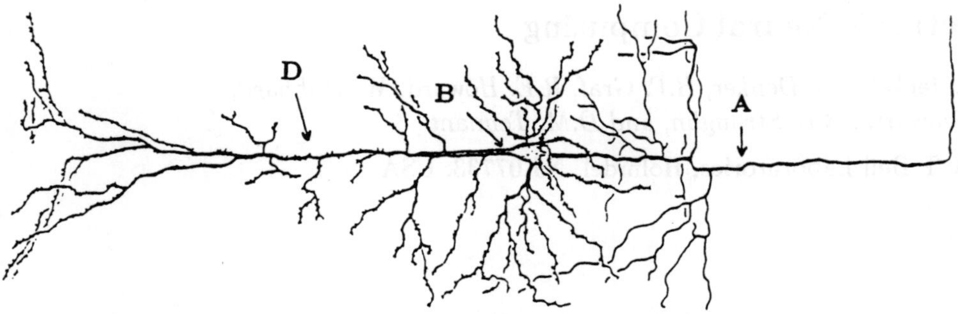

Fig. 1. A real neuron, as drawn by Ramon y Cajal. A is the axon, B is the cell body, and D is the dendrite. The axon extends far beyond the edge of the drawing. The cell body is about 10 microns across

Now we have the basic electronic component of a neural network. To the extent that this neuron mimics real biological neurons, we should be able, at least in principle, to mimic aspects of brain function electronically. The key, of course, is determining how to connect one neuron to another to obtain such features such as processing, memory, and fault tolerance.

3. NEURAL NETWORKS

A rather general network of electronic neurons is shown in Figure 2. Here each neuron can have a connection to any other. In order to provide both excitation and inhibition through simple resistive connections, the neurons are modified to have two axons, one of which is the negative of the other. Synaptic connection to the positive line produces excitation, while connection to the negative line produces inhibition. The configuration space for n neurons is an n-dimensional hypercube: each neuron's output can range continuously from the low saturated value, which we will define as the "0" level, to the high saturated value, defined as the "1" level. Thus at any instant, the state of the system can be represented as an n-dimensional vector within the hypercube.

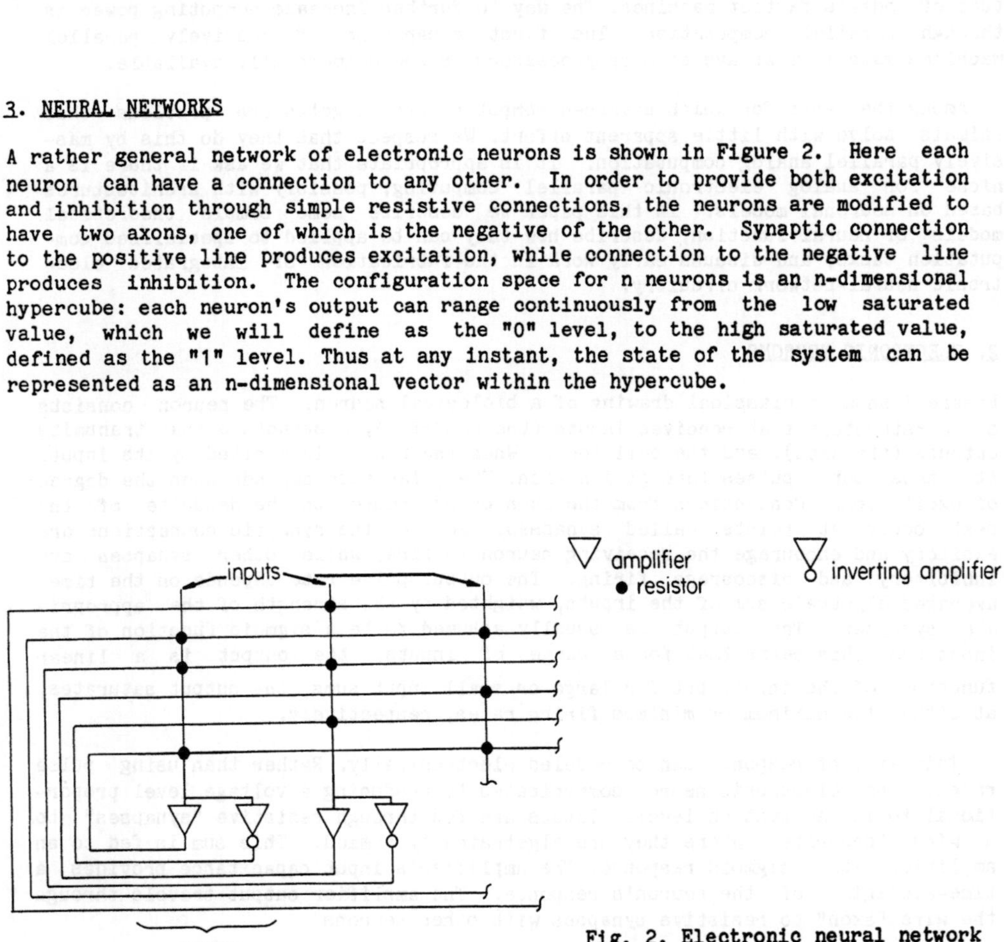

Fig. 2. Electronic neural network

HOPFIELD /1/ has analyzed the circuit of Fig. 2 for the case of symmetric connections (i.e. the resistor connecting the dendrite of neuron i to the axon of neuron j is the same as the one that connects j to i.) He found that the equation of motion for the neural net is identical to that of a massless particle moving through a viscous fluid on a potential defined on an n-dimensional space. If the system is launched at some point in the hypercube it will move "downhill" to a region of lower potential, coming to rest at a local potential minimum. In most applications these local minima are near the hypercube corners, getting closer to the corners as the amplifier sigmoid transfer function approaches a step function. The hypercube thus divides into "basins of attraction": all system trajectories that start within a particular basin of attraction end at the same point, the local minimum of that basin. The shapes of the basins depend upon the synapse resistors, the bias currents fed into the dendrites, and the amplifier transfer function.

4. ASSOCIATIVE MEMORY

By proper choice of synapses, it is possible to have this neural net function as an associative memory. Each neuron represents one bit; a complete memory word, a string of bits, is recalled by presenting the memory with starting state, a key that contains a subset of the bits of the complete word. (This kind of memory differs from usual random-access computer memory where a memory word is recalled by presenting the address for that word. Associative memory is thought to embody some aspects that are similar to human memory.) In the hypercube picture, the memory corresponds to the minimum of the basin of attraction. The entire basin of attraction is the set of all keys for that memory. Notice that it is possible to have keys with wrong bits (bits that differ from those in the sought memory) so long as enough bits are correct to launch the system in the right basin.

The major issue in designing a neural-net associative memory is choosing the synapses so that potential minima correspond to the desired memories. As one tries to increase the number of stored memories, undesired minima crop up in regions between the desired basins of attraction. Worse yet, with heavy memory loading the location of desired minima are shifted and errors in recall become severe. There are a number of algorithms for efficient synapse coding to provide as large as possible memory capacity /2-12/. For N neurons, generally about N/20 to N/4 N-bit words can be stored and still maintain good associative recall. Capacity improves if slight errors in recall can be tolerated /13,14/.

There are others circuits for performing associative memory besides the scheme shown in Fig. 2. A simpler circuit using resistive connections was described some years back by KOHONEN and is discussed in his recent book /2/. It remains to be seen which, if any, of the methods proposed for associative memory become practical alternatives to more conventional architectures.

5. NEURAL-NET OPTIMIZERS

A very promising offshoot from the work on associative memory that seems to depart from the original biological motivation is the use of the neural net as an optimizer. HOPFIELD and TANK /15/ showed that a neural network could be used to find a deep minimum of a function of many variables. For this application the synapses are chosen to create an energy surface that represents the function to be minimized. The network is started at some point in the hypercube and finds the minimum of the basin of attraction associated with that starting point. Convergence is usually achieved within about ten network time constants. For a practical circuit this may take tens of microseconds or less.

HOPFIELD and TANK /15/ found that low gain in the amplifiers was useful for reducing the basins of attraction of shallow local minima. There is no guarantee that the network will find the global minimum of the objective function since the settling point of the network depends on the starting point. However, it is observed empirically that for at least some functions, the deepest minima have the largest basins. Hopfield and Tank demonstrated the neural-net optimizer on the well known NP-complete Traveling Salesman Problem. The network found good solutions quickly, but for tours with a large number of cities the network did not find the optimal tour. Recently, Baum /16/ suggested a method for iterating the neural-net optimizing procedure to obtain more nearly optimal solutions.

TANK and HOPFIELD /17/ extended their work on optimizers to show that linear programming could also be done by a neural net. In this case the network always finds the optimal solution, within the accuracy of the electronic components, but for a large number of variables, convergence may require many network time constants.

There are still many unanswered questions about neural-net optimizers. No one knows how the performance varies as the number of variables, usually the number of electronic neurons, becomes large. Some help may come from the theory of spin glasses since the spin-glass Hamiltonian looks very much like the potential for the neural network. Recent results on spin glasses were discussed in a 1985 Les Houches Winter School /18/. BALDI and BAUM /19/ use concepts of ultrametricity that were recently applied to spin glasses /20/ to argue that there are a relatively small number of large basins of attraction for problems like Traveling Salesman.

6. NETWORK FABRICATION

To demonstrate neural-network operation and to evaluate network fabrication procedures we have begun a program to make high-density integrated neural-net chips.

Conventional memory chips use at least one transistor per bit. In a neural network, a synapse (the elementary unit of storage) can be made using a simple resistor, which can be several times smaller than a transistor, and requires many fewer fabrication steps. This opens the possibility of neural-network chips with very large, dense memories. The synapses are not 100% efficient at storing information, but the neural network is both a processor as well as a memory.

Our first fabrication effort centered on making a resistive synapse network on a single chip with the amplifiers off the chip and connected to it by ordinary wires. Figure 3 shows such a synapse chip that connects to an array of 22 off-chip neuron amplifiers. Most of the chip patterning was done by conventional 2 micron-linewidth optical lithography. Synapses were patterned by e-beam lithography. The horizontal wires are the dendrites and the vertical wires are axons. The wires are made of tungsten films about 0.1 micron thick. The two planes of wires are separated by a layer of polyimide insulator. Using e-beam lithography and plasma etching, a hole is formed in the polyimide where a synapse is desired. This hole is filled with a thin film of amorphous silicon which forms a resistor between the two tungsten wiring planes. For this chip, all synapses that are formed, have the same value, typically 1 megohm. Variations on the synapse value can be achieved by varying the area of the etched hole in the polyimide. Associative recall of four 22-bit words and their complements was performed by this chip.

A conventional chip carrier with 44 leads was used to connect the chip in Fig. 3 to the rest of the circuit. For a circuit with enough neurons to have interesting computational power, most of the circuit must be on a single chip. The inputs and outputs must be multiplexed because even today's largest chip carriers only have about 200 leads. A completely integrated neural network circuit is shown in Fig. 4.

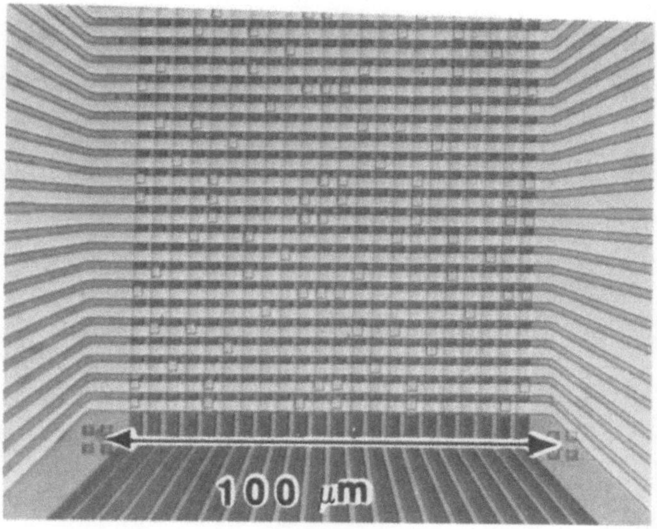

Fig. 3. Micrograph of a 22 x 22 resistive synapse array made with 2 micron lines

Fig. 4. Integrated CMOS neural-network chip

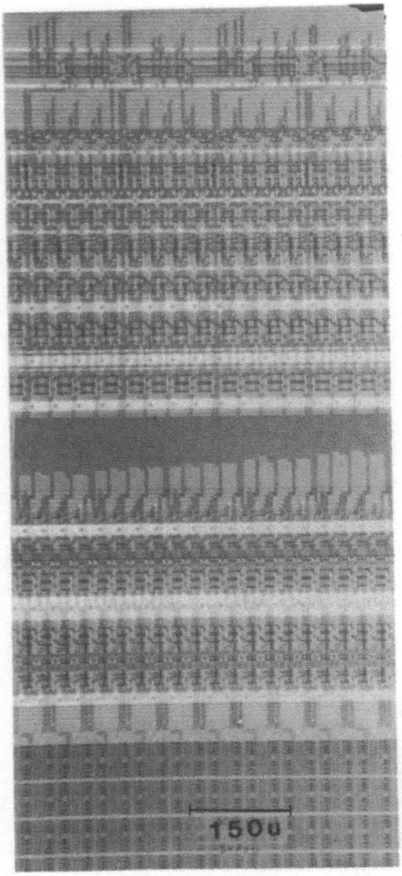

Fig. 5. Magnified view of a section of the chip displayed in Fig. 4 showing a sample of the CMOS circuitry

This chip, fabricated in 2.5 µm silicon CMOS technology, contains 256 electronic neurons plus the input/output control to load and unload the data over a 16 bit bus. Each of the neurons consists of an amplifier with an inverting and a noninverting output. The interconnections between the neurons are made through a matrix of crossing polysilicon and aluminum wires. This matrix fills the big square in the center of the chip shown in Fig. 4. Surrounding the matrix are the amplifiers arranged symmetrically on all four sides. Between the amplifiers and the pads are address decoders, buffers and control logic forming a multiplexer. The whole chip contains about 25000 transistors and 130000 synapse sites. It occupies 7 mm x 7 mm of silicon area. A magnified view of a section of the chip containing active components is shown in Fig. 5.

Information is stored in the circuit by placing amorphous silicon resistors at selected crosspoints in the matrix. We use electron beam lithography in the last fabrication step to pattern the resistors. This allows an easy change of the stored information from one chip to the next.

The multiplexer is used to initialize the system. After the circuit has relaxed to a stable condition, the state of each amplifier is read out with the multiplexer. Simulations indicate that it takes around 1 µm until the circuit reaches a stable state from a set initial state.

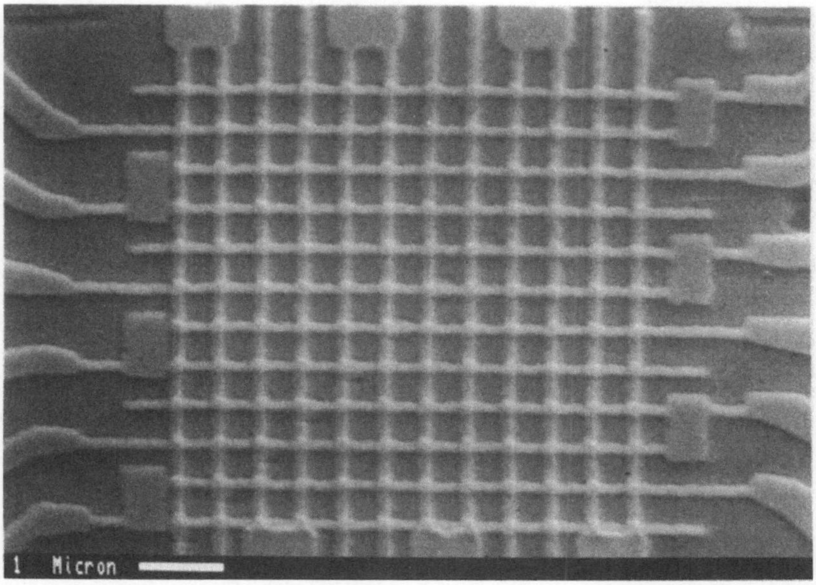

Fig. 6. Micrograph of a 12x12 resistive array patterned by e-beam lithography

The circuit described above is our first attempt at complete integration. About equal areas are required for the synapses and the active CMOS electronics. For this size circuit, the area required by the active elements scales roughly linearly with the number of neurons. (When the number of neurons becomes very large the area of each amplifier must be increased to drive an increasing number of synapses.) However, the area required by the synapses scales as the square of the number of neurons. For a circuit with more neurons, the synapse area will dominate unless finer lithography is used for the synapses. Such a situation is quite feasible because, as mentioned earlier, individual synapses can be shrunk to very small dimensions. As a demonstration of the compactness that can be achieved in a resistive synapse matrix, we fabricated a 12 x 12 synapse array using e-beam lithography. The test chip is shown in Fig. 6. The same materials were used for this circuit as were used in the circuit of Fig. 4. For diagnostic purposes synapses were only formed on the diagonal. The entire matrix, with 144 potential synapses, fits in an area required for one bit of conventional silicon Dynamic Random Access Memory (DRAM). With this kind of technology, a chip with 1/2 billion synapses per square centimeter could be made.

7. CONCLUSION

Neural networks are a very promising unconventional architecture for information processing, but it is still too early to tell where these electronic neural networks will find their best use. In order to evaluate the performance of electronic neural nets, we have begun a circuit fabrication program to build integrated electronic neural-network circuits. Early results show the circuits work as promised.

ACKNOWLEDGEMENT

We thank John Hopfield and David Tank for many helpful discussions.

REFERENCES

1. J.J. Hopfield: Proc. Nat. Acad. Sci. USA 81, 3088 (1984)
2. T. Kohonen: Self-Organization and Associative Memory (Springer, Berlin, Heidelberg 1984) and the prior references cited therein
3. See J.S. Denker: To appear in Physica D (1986) for an overview
4. D.O. Hebb: The Organization of Behavior (Wiley, New York 1949)
5. J.A. Anderson: In Disordered Systems and Biological Organization, ed. by E. Bienenstock, F. Fogelman, G. Weisbuch (Springer, Berlin, Heidelberg 1986)
6. B. Widrow, M.E. Hoff: WESCON Convention Record, Part IV, 99-104 (August 1960); see also B. Widrow: In Self-Organizing Systems 1962, ed. by M.C. Yovits, G.T. Jacobi, G.D. Goldstein (Spartan, Washington D.C. 1962)
7. A.S. Lapedes, R.M. Farbert: To appear in Physica D (1986)
8. S. Grossberg: Studies of Mind and Brain (Reidel, Dordrecht 1982)
9. L. Personnaz, I. Guyon, G. Dreyfus: In Disordered Systems and Biological Organization, ed. by E. Bienenstock, F. Fogelman, G. Weisbuch (Springer, Berlin, Heidelberg 1986
10. F. Crick, G. Mitchison: Nature 304, 111 (1983)
11. J.J. Hopfield, D.I. Feinstein, R.G. Palmer: Nature 304, 158 (1983)
12. D. Kleinfield, D.B. Pendergraft: Private communication
13. D.J. Amit, H. Gutfreund, H. Somplinsky: Phys. Rev. Lett. 55, 1530
14. R.J. McEliece, E.C. Posner, E.R. Rodemich, S.S. Venkatesch: submitted to IEEE Trans. Inform. Theory
15. J.J. Hopfield, D.W. Tank: Biol. Cybernetics 52, 141 (1985)
16. E. Baum: Private communication
17. D.W. Tank, J.J. Hopfield: IEEE Circuits and Systems (in press)
18. E. Bienenstock, F. Fogleman, G. Weisbuch (eds.): Disordered Systems and Biological Organization (Springer, Berlin, Heidelberg 1986)
19. P. Baldi, E. Baum: Phys. Rev. Lett (in press)
20. M. Mezard, G. Parisi, N. Siurlas, G. Toulouse, M. Mirasoro: Phys. Rev. Lett. 52, 1156 (1984)

Part IV

Poster Session, Abstracts

Fabrication of Submicron Structures Combining E-Beam and Deep-UV Exposure

J. Romijn and J.E. Mooij

Department of Applied Physics, Delft University of Technology,
NL-Delft, The Netherlands

In order to obtain metallic structures of large overall dimensions containing details smaller than 0.1 µm in a limited area, we have developed a two-stage process for exposure and development of one double resist layer. The layer consists of Elvacite 2041 on top of Elvacite 2021, both 60 nm thick. The large pattern is obtained by deep UV contact exposure through a chromium mask on quartz. Small details are exposed in a modified Philips PSEM 500. The SEM goniometer stage is used to position with 1 µm accuracy with respect to markers in the large pattern. The finest lines obtained so far have a width of 30 nm in resist and 40 nm in aluminium films 40 nm thick.

Fabrication of Short-Gate GaAs MESFETs by Electron Beam Lithography

G. Bernstein and D.K. Ferry

Center for Solid State Electronics Research, Arizona State University,
Tempe, AZ 85287, USA

We have used an ISI-100B scanning electron microscope to fabricate GaAs MESFETs with gate lengths between 30 and 100 nm. Direct write was achieved with 40 kV electron exposure of 950 000 molecular weight PMMA and a liftoff process. All fabrication levels were performed using EBL with a desktop computer used to control all timing and vector scan settings. Output was via two 12-bit D/As. The gate metal was 40 nm thick gold and fit in 2.3 to 3 micron source-drain spacings. Maximum g_m achieved was 85 ms/mm (with calculated intrinsic g_m of 110 ms/mm) for a 45 nm gate device. This is consistent with modeling performed by W. CURTICE at RCA in Princeton, New Jersey, for parameters similar to ours. It was found that pinch-off was achieved for gate voltages of -3 volts and drain voltages up to 2.5 volts. It is believed that charge injection into the substrate caused lack of pinchoff after that point.

2D Josephson Junction Networks

B.J. van Wees, J.G.F. Kablau, and J.E. Mooij

Department of Applied Physics, Delft University,
NL-Delft, The Netherlands

We have fabricated square 2D arrays of superconducting islands, coupled with Josephson junctions. Array sizes are up to 100 000 elements. Junctions are all-niobium tunnel junctions of size 1 x 0.2 µm. Fabrication is through oblique evaporation from two directions, using a self-aligning technique involving free-hanging bridges 1 x 1 µm at a distance of 0.5 µm from the substrate. Measurements on the arrays give indications of a Kosterlitz-Thouless-like transition. In a perpendicular magnetic field rich structure is observed in the resistance as a function of field. Minima occur when the flux through one elementary cell of the array equals m/n times the flux quantum h/2e, with m and n integers. Arrays with different widths have been compared.

Electric Field Heating of Supported and Free-Standing AuPd Fine Wires

C.G. Smith[1] *and M.N. Wybourne*[2]

[1]Cavendish Laboratory, Madingley Road, Cambridge CB3 0HE, United Kingdom
[2]GEC Research Ltd., East Lane, Wembley, Middx. HA9 7PP, United Kingdom

Using electron-beam lithography, fine AuPd wires of 50 nm and 30 nm thickness have been fabricated over a length of 10 µm on a Si_3N_4 substrate. The resistance rise as a function of temperature fits well to one-dimensional Coulomb interactions for low temperatures. On etching away the substrate the functional form of the resistance rise changes and cannot be fitted to theory. It is found that for a supported wire the energy loss is limited by the electron-phonon scattering rate which is proportional to the square of the electron temperature. For a free-standing wire the heat loss is limited by diffusion of electrons and phonons along the wire. Recently 100 nm width and 5 nm thickness have been made free standing over a length of 15 µm.

High-Frequency Limits of Quantum Noise Detectors Based on Superconducting Tunnel Junction Mixers

D. Winkler, B. Nilsson, W.R. McGrath, T. Claeson, J. Johansson, E. Kollberg, and S. Rudner

Departsments of Physics and Electron Physics,
Chalmers University of Technology, S-41296 Gothenburg, Sweden

Quasiparticle mixers (based on superconductor-insulator-superconductor tunnel junctions) have successfully been used in high-frequency detectors up to a few hundred GHz. Their sensitivities approach the quantum noise limit. This work concerns the high-frequency limit of such devices. The intrinsic limits were probed in a scale-type experiment where aluminium tunnel junctions were investigated at about 0.85 of the superconducting gap frequency f_g. Calculations were based on Tucker's theory of quantum mixing and used parameters extracted from the experiment. The conversion efficiency is affected by the decreased effective nonlinearity of the mixer element at high frequency, as fewer photon points fall in the gap voltage region. Conversion gain may still be possible up to 2 f_g. Josephson interference and instabilities may cause extra noise and a deteriorated conversion, but the Josephson currents could be suppressed by a magnetic field. Non-equilibrium effects are large in high quasiparticle injection rates in Al junctions but they had no detrimental effect on the conversion. They should pose a negligible problem in a high T_c superconductor. The pair breaking by high frequency photons did not cause any substantial differences between experimental conversions and values calculated from the theory that does not take the effect into account. High T_c mixers should be able to operate well into the THz range.

Technological obstacles to a high-frequency operation include coupling problems and the shunting effect of the tunnel capacitance. The latter motivates the use of series coupled, sub-micron size tunnel junctions. A progress report of a 750 GHz mixer is given in the poster.

Photoresponse and Transport Properties of a 'Quasi' Graded Gap Superlattice P-I-N Diode

N.R. Couch

GEC Research Ltd., Hirst Research Centre, Wembley, United Kingdom

A GaAs p-i-n diode, where the intrinsic region consists of a five-period asymmetric GaAs/AlAs superlattice, has been grown by molecular beam epitaxy. The current-voltage characteristics show negative differential resistance (n.d.r) in forward bias up to room temperature. Both the positions and magnitudes of the n.d.r. features are fitted by a simple resonant tunnelling model using the indirect (X) band gap of the AlAs. In reverse bias very low dark currents are observed ($\sim 6 \times 10^{-8}$ A/cm^2 at -3 V and 300 K). The photoresponse to a 400 fs light pulse reveals two response times. The faster is limited by the RC time constant of the large mesa, giving a rise time of 80 ps and a decay rate of 225 ps. At long times a slower decay rate is observed. This long time tail is commensurate with the time taken to tunnel via the direct (Γ) band gap through the thickest AlAs barrier. For future device applications, it is expected that this form of structure will provide the basis for fast, enhanced sensitivity photodetectors.

Disorder and Two-Dimensional Electronic Sub-Bands

D.J. Newson

Cavendish Laboratory, Madingley Road, Cambridge, United Kingdom

In M.E.S.F.E.T. and J.F.E.T. semiconductor devices, electrons are confined by the potentials from the depleted region. This results in the formation of quantised sub-bands. A study has been made of the electrical resistivity of such devices. The effect of disorder has been investigated by using devices of different doping levels. When the doping level is low, life-time broadening reduces the observability of sub-bands.

Magnetic depopulation effects have been studied in very high mobility InAlAs/InGaAs heterojunctions: The experimental results show the importance of life-time broadening even in this high mobility system.

A Model for the Oscillatory Structure in the J(V) Characteristics of GaAs/(AlGa)As Tunneling Structures

D.C. Taylor and L. Eaves

Department of Physics, University of Nottingham, University Park, Nottingham NG7 2RD, United Kingdom

The low-temperature J(V) characteristics of n^+GaAs/(AlGa)As/n^-GaAs/n^+GaAs heterostructures show remarkable oscillatory structure, with period $\Delta V = \hbar\omega_L/e$, where $\hbar\omega_L = 36$ meV is the longitudinal optic phonon energy /1 - 4/. In this poster we show that the structure can be explained using Ohm's law, Poisson's equation and some basic principles of semiconductor physics /5,6/, without involving exotic effects such as magnetopolaronic corrections /7/ or selfenergy corrections to the density of electron states /8/.

In reverse bias, a depletion layer exists in the n^- layer adjacent to the (AlGa)As due to the high electric field in the barrier region and to the relatively low current flowing. The total voltage drop V across the device that divides into three contributions V_B, V_D, V_U across the barrier, depleted and undepleted regions respectively. This division is determined by the conditions of continuity of current density J and electric field E at the interfaces. Electrons are injected through the tunnel barrier into the depletion region with considerable kinetic energy which is lost by longitudinal optic phonon emission. The energy of these phonons is very well defined due to the requirements of momentum conservation. Electrons arriving at the interface between the depleted and undepleted regions have a kinetic energy $e(V_B+V_D) - n\hbar\omega_L$, where n is the number of LO phonons emitted. With increasing V, V_B+V_D passes the threshold for emission of $n+1$ phonons. Thus the energy of the electrons arriving at the undepleted region changes from hot to cold. Energy relaxation will thus change the conductivity of the electron gas in the undepleted region from a hot value σ_h to a significantly lower cold value σ_c. It is this variation of the impedance of the undepleted region that gives rise to the oscillatory structure. We may picture the behaviour in terms of two distinct J(V) curves for the device, one calculated for the case when the undepleted layer has a cold conductivity σ_c and the other for a hot conductivity σ_h. The oscillatory structure in the observed J(V) characteristics corresponds to transitions between these two curves whenever $e(V_B+V_D)/\hbar\omega_L$ reaches an integer value. Since $V_U \ll V$ the observed oscillations are effectively periodic in the applied voltage V.

The poster will explain how the model can account for the overall form of the oscillations and the temperature and magnetic field dependence of their amplitude.

References

1. T.W. Hickmott, P.M. Solomon, F.F. Fang, F. Fischer, H. Morkoc: In Proceedings of the 17th International Conference on the Physics of Semiconductors, ed. by J.D. Chadi, W.A. Harrison (Springer, New York 1985) p.417
2. T.W. Hickmott, P.M. Solomon, F.F. Fang, F. Stern, R. Fischer, H. Morkoc: Phys. Rev. Lett. **52**, 2053 (1984)
3. P.S.S. Guinaraes, D.C. Taylor, B.R. Snell, L. Eaves, K.E. Singer, G. Hill, M.A. Pate, G.A. Toombs, F.W. Sheard: J. Phys. C**18**, L605 (1985)
4. L. Eaves, P.S.S. Guimaraes, B.R. Snell, D.C. Taylor, K.E. Singer: Phys. Rev. Lett. **53**, 262 (1985)
5. D.C. Taylor, P.S.S. Guimaraes, B.R. Snell, F.W. Sheard, L. Eaves, G.A. Toombs, J.C. Portal, L. Omowski, K.E. Singer, G. Hill, M.A. Pate: Proceedings of the International Conference on Modulated Semiconductor Structures, Kyoto (Surface Science (1986)

6. L. Eaves, P.S.S. Guimaraes, F.W. Sheard, B.R. Snell, D.C. Taylor, G.A. Toombs, K.E. Singer: J. Phys. C**18**, L885 (1985)
7. C.B. Hanna, E.S. Hellman, R.B. Laughlin: Bull. Am. Phys. Soc. **31**, 394 (1986)
8. J. Ihm: Bull. Am. Phys. Soc. **31**, 396 (1986)

A Minute Metallic Sphere Close to Flat Metal Surface System: A Scanning Tunneling Microscope Problem

M.T. Michalewicz and J. Mahanty

Department of Theoretical Physics, The Research School of Physical Sciences, The Australian National University, Canberra, A.C.T. 2601, Australia

A general formalism for evaluating the interaction potential of a charged particle with a solid is developed using linear response theory. The detailed form of this potential at small distances from the surface where the image description fails, is essential for interpretation of experimental data on image potential surface states in metals, surface resonance states in metals, in scattering of charged particles and also MVM tunnel effect experiments.

In the present study, the formalism is applied to first obtain the interaction with a metal sphere (20 Å ≤ R ≤ 600 Å) and the interaction with a system of two flat surfaces and finally with a sphere and a flat metal surface.

The latter geometry is particularly relevant in formulating the theory of Scanning Tunneling Microscopy - a new and very promising experimental technique for non-destructive surface structure and local work function measurement.

A Calculation of the Effect of Subband Structure on the Thermopower of a Quasi-1D Wire

M.J. Kearney

University of Warwick, Coventry CV4 7AL, United Kingdom

A Boltzmann treatment is used to obtain the thermopower at low temperatures of a quasi-1D wire. The subband structure is found to give rise to large positive thermopower for Fermi energies near subband minima, and the structure should remain observable for temperatures below about 10 K. Lifetime broadening is investigated, and the 1D density of states is shown to have a marked variation with energy for $(k_F l)_{10} \gtrsim 5$.

The Effect of Electron-Electron Scattering on the Distribution Function in Semiconductors

N.S. Wingreen, C.J. Stanton, and J.W. Wilkins

Clark Hall, Department of Physics, Cornell University, Ithaca, NY 14853, USA

We determine the effect of electron-electron scattering in semiconductors by solving the Boltzmann equation with an electron-electron collision term that relaxes the distribution to a heated, displaced Maxwellian. The Maxwellian has the same number, average velocity, and average energy as the instantaneous distribution. The effective electron-electron relaxation rate is solved for self-consistently assuming a screened Coulomb interaction. The electron distributions we find are strongly affected by electron-electron scattering, but differ significantly from heated, displaced Maxwellians.

This work was supported by the Semiconductor Research Corporation. Two of us (C.J.S and N.S.W) gratefully acknowledge the support of the Fannie and John Hertz Foundation.

Index of Contributors